P9-EEK-251

Phillip Gendreau

The elephant — largest of land mammals.

# THE BOOK OF POPULAR SCIENCE

Volume 3

Distributed in the United States by
THE GROLIER SOCIETY INC.

Distributed in Canada by
THE GROLIER SOCIETY OF CANADA LIMITED

Cover photograph: colorful blossom of the large anthered nightshade (*Solanum machrantherum*). It was photographed at an altitude of approximately 6,000 feet in the southern part of Mexico.

Walter Dawn

No part of THE BOOK OF POPULAR SCIENCE may be reproduced without special permission in writing from the publishers

COPYRIGHT © 1963 BY GROLIER INCORPORATED

Copyright © 1963, 1962, 1961, 1960, 1959, 1958, 1957, 1956, 1955, 1954, 1953, 1952, 1951, 1950, 1947, 1946, 1945, 1943, 1941, 1939, by

GROLIER INCORPORATED

Copyright, 1931, 1930, 1929, 1928, 1926, 1924, by

THE GROLIER SOCIETY

Copyright © in Canada, 1963, 1962, 1961, 1960, 1959, 1958, 1957, 1956, 1955, 1954, 1953, 1952, 1951, 1950, by

THE GROLIER SOCIETY OF CANADA LIMITED

PRINTED IN THE UNITED STATES OF AMERICA

13

# CONTENTS OF VOLUME III



**(Continued on next page)**

# CONTENTS OF VOLUME III (continued)

# SCIENCE GROWS UP (1600-1765)[1]

BY JUSTUS SCHIFFERES

### THE FIRST OF THE MODERNS

THE fires of learning had been rekindled in the 16th century and thirst for a new kind of knowledge — the kind of factual knowledge that we today call scientific — had been awakened. The revolution in astronomy and the rediscovery of anatomy had set the stage for the rapid development of modern scientific thought. On to this stage strode Galileo Galilei, who has been justly described as "the first of the moderns" in science. He lived a long and fruitful life. Born in Pisa in 1564, just two decades after the publication of Copernicus' REVOLUTIONS and Vesalius' ANATOMY (1543), Galileo lived and worked fruitfully until 1642, the year that his greatest successor, Isaac Newton, was born.

We think of Galileo as the first of the moderns because he broke the strangling hold of the ancient authorities — Aristotle, Ptolemy and Galen — upon scientific thought. He boldly showed up the many errors of these ancients and against their authority he appealed to something new: the evidence of experiments — the credibility of what one can see with his own eyes. He thus helped to brush away the cobwebs of speculation that had long kept men's minds in darkness, error and ignorance of the ways of the world of nature about them.

At the very core of his thinking was his insistence on measuring things. He measured whatever he could by whatever means, crude as they might be, that were available. He looked for quantities, not qualities. When he was a young man, for example, he sat in the cathedral at Pisa and observed the motion of a lamp hanging from the roof by a long chain. It swung to and fro. But whether the swing (the arc) was long or short, it appeared to take the same length of time. Galileo timed the swings by counting his own pulse beats and thus hit upon an important scientific truth.

He played a truly outstanding part in the development of modern science. He helped to win men over to the Copernican hypothesis. He was the first man to see the true face of the moon, the first to observe the infinite multitude of stars beyond sight of the naked eye. While he did not actually invent the telescope, he perfected it to a point where it became a scientific instrument for scanning the heavens. Many modern scholars believe that "the most real part of the glory of this great man" was his work in mechanics, the science that deals with the action of forces on bodies. In this field he exposed the errors of Aristotle and laid the foundations of modern theory.

Galileo was born at Pisa on February 15, 1564. He was the son of a poor nobleman, who was interested in mathematics and music but contemptuous of authority in both. When young Galileo entered the University of Pisa, he intended to study medicine; but he found himself fascinated by mathematics and physical science and forsook medicine for these subjects. His intellectual brilliance was quickly recognized even though his "radical ideas" were suspect.

At the age of twenty-five, Galileo began to lecture on mathematics at the University of Pisa. He soon shocked his colleagues beyond measure. For one thing, he lectured in Italian instead of in Latin, the language that was almost universally

Three unequal weights were dropped by Galileo from the top of Pisa's Leaning Tower. They hit the ground at about the same time (that is, discounting air resistance).

employed by learned men. But what was even worse, he openly attacked the "errors of Aristotle." It is said, for example, that by a very simple experiment he disproved the Aristotelian teaching that bodies fall with speeds proportional to their weights. He dropped three unequal weights from the top of the Leaning Tower of Pisa; contrary to Aristotle's theory, they all hit the ground at about the same time (air resistance being discounted). Relations between Galileo and his colleagues became so strained that he was glad to accept an invitation to lecture at the University of Padua. He began to teach in December 1592, and there he stayed for eighteen years, one of the most popular university lecturers in Europe.

Galileo had already become a convert to the doctrines of Copernicus. In a letter to the astronomer Johannes Kepler, written in 1597, he asserted: "I esteem myself fortunate to have found so great an ally [Copernicus] in the search for truth . . . I have been for many years a follower of the Copernican system. It explains to me the reasons of many phenomena which are quite incomprehensible according to the views commonly accepted."

Because they had no telescope with which to view the heavens, the early disciples of Copernicus were hampered in their efforts to prove their master's teachings by direct observation. Naturally Galileo was greatly interested to hear that a Dutch spectacle-maker had devised a lens system that made it possible to see distant objects distinctly. He looked into the matter and confirmed it. He made a thoroughgoing study of the theory of refraction; then he constructed a number of telescopes with his own hands.

As he swept the heavens with these crude instruments, he made a number of breath-taking discoveries. He disclosed them to the world at large in a little book, THE SIDEREAL MESSENGER, which he published in 1610. Its title page proudly announced that it unfolded "great and marvelous sights," which it proposed "to the attention of everyone, but especially philosophers and astronomers."

The infinite multitude of the stars was revealed for the first time through Galileo's telescope. "Beyond the stars of the sixth magnitude," he wrote, "you will behold through the telescope a host of other stars, which escape the unassisted sight, so numerous as to be almost beyond belief." He charted many of the hitherto unseen stars in Orion's belt and among the Pleiades. The Milky Way, he demonstrated, "is nothing else but a mass of innumerable stars planted together in clusters."

He noted for the first time the essential difference between the appearance of the fixed stars and the planets. The stars appeared through the telescope as blazes of light; they shot their beams on every side. On the other hand the planets glowed like perfectly round discs.

What Galileo considered one of his most important discoveries was that the planet Jupiter had four satellites, or moons, "never seen from the very beginning of the world up to our own time." (We know today that Jupiter has twelve satellites.) Further observation showed him that these satellites were constantly revolving about the planet. Galileo almost discovered the rings of Saturn. Viewed through his imperfect telescope, this planet appeared to consist of "three spheres which almost touch each other." He also noted that the planet Venus shines in the reflected light of the sun and has different phases.

Aristotle and those who unquestioningly accepted his scientific doctrines had considered the moon to be a smooth, round ball; Galileo's telescopes revealed that it was a "body like the earth," with a rugged surface crisscrossed with mountains and valleys. He even estimated the heights of the moon's mountains from the lengths of the shadows they cast.

Galileo also trained his telescopes on the sun. He shares with other observers, notably Kepler, Scheiner and Fabricius, the honor of having discovered the sunspots. Some believe that his blindness in the later years of his life was caused by his frequent telescopic observations of the sun.

In 1610 Galileo left Padua for Florence, where he became professor for life at the university and mathematician extraordinary to the Grand Duke of Tuscany. In the years that followed, an important part of his income came from the sale of scientific instruments — particularly telescopes and balances — that he made with his own hands.

In 1613 Galileo published a LETTER ON SUNSPOTS in which he revealed his belief in the Copernican system. His enemies accused him of heresy; but he defended his views tenaciously, even citing Biblical texts to prove that Copernicus was right. The authorities at Rome warned him to stay away from theological debate. A few years later the Inquisition banned all books supporting the theory that the earth moves around the sun; the Pope, Paul V, warned Galileo not to hold, teach or defend the Copernican system.

For a time Galileo held his peace. In 1623, however, he rather cautiously defended Copernicanism in a little treatise on comets called THE ASSAYER. Since his remarks on the Copernican system were tucked away in a few paragraphs, they seem to have attracted no particular attention. But in 1632 Galileo apparently felt that the time had come to bring the issue out in the open; in that year he published his DIALOGUE ON THE TWO CHIEF SYSTEMS OF THE WORLD (that is, the Copernican and Ptolemaic systems).

In this work he set forth the evidence in behalf of the Copernican theory. As the title indicates, the treatise is in the form of a dialogue among three persons: Salviati, who is a Copernican, Simplicio, who blindly follows the teachings of the ancients, and Sagredo, who is neutral. The author professes to give the arguments pro and con; but he has obviously stacked the cards in favor of the Copernican theory. In the debate the supporter of Copernicus

Galileo displaying his newly invented telescope to the Venetian senate.

Alinari

has all the better of it and Simplicio comes off a poor second. The supposedly impartial third party in the discussion is obviously shaken by Salviati's arguments.

It is easy to follow Galileo's own viewpoint in this dialogue. He attacks the idea that the heavens are immutable; he cites the appearance of new stars and of sunspots. Even the heavenly bodies, he claims, follow natural laws. His basic claim, however, is that the Copernican system explains the functioning of the universe more logically than the Ptolemaic system.

For example: Ptolemy had had to introduce a highly complicated system of epicycles in order to explain the apparent and observed halts and backward journeys of the planets in the heavens. The Copernican hypothesis explains these motions very simply; it points out that they are mere *appearances* resulting from the annual revolution of the planets, including the earth, around the sun.

THE DIALOGUE ON THE TWO CHIEF SYSTEMS OF THE WORLD created a furore. The book was denounced as an out-and-out defense of Copernicanism instead of the debate that it was supposed to be; it was banned and its author was summoned to Rome. In February 1633, he arrived in the Holy City and was taken into custody. Realizing his danger, the old philosopher publicly recanted his belief that the earth moves; he promised never again to indulge in any speculations that would bring him under suspicion of heresy. Yet tradition has it that as he left the court room, he shook his head and muttered under his breath, "Still it moves." Galileo was sentenced to repeat the Seven Penitential Psalms and to a term of imprisonment. But the prison sentence was commuted and he was allowed to retire to the little village of Arcetri, near Florence.

**This diagram, which shows the "force of a vacuum," is from the *Dialogues on Two New Sciences*, by Galileo.**

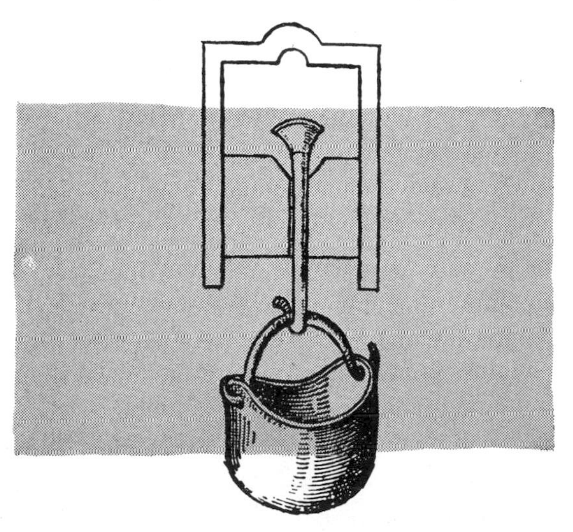

He accomplished some of his finest work in this retreat. He here assembled his lifelong work on mathematics; by 1636 he had completed his DIALOGUES ON TWO NEW SCIENCES, the foundation for much that is basic in the modern study of mechanics. Since his books were banned in Italy, he had the DIALOGUES published at Leiden, Holland, in 1638.

Before Galileo's time it was believed that bodies in motion must come to rest unless some force or other keeps them moving. As a matter of fact, the older philosophers and scientists had felt impelled to invent such a force — "a prime mover" — in order to account for the continued motion of the heavenly bodies. Galileo now advanced the idea that once a body is in motion it must continue to move at the same speed and in the same direction, unless a force from the outside acts upon it. He thus extended the idea of inertia (or resistance to change) to bodies in motion. He maintained also that when a force acts upon a body the effect is the same whether the body is in motion or at rest. Galileo made many experiments with falling bodies, usually by rolling balls down a wooden incline. He was greatly interested, too, in the motion of projectiles. He showed that if a projectile forms an angle of less than 90 degrees with the horizon when it is cast upward, it describes the geometrical figure known as a parabola.

We can mention here only a few of the other accomplishments that have given Galileo so high a rank among scientists. He made improvements in the construction of the compound microscope. He developed a balance scale apparatus to weigh the air; with it he disproved another one of Aristotle's errors: namely, the theory that a large amount of air weighs less than a small amount. Aristotle believed the "essence" of air to be levity, the opposite of weight.

A practical engineer, Galileo measured the strength of materials under tension — both beams of wood and bones. He experimented with the heat-measuring apparatus that we now call the thermometer.

Like other scientists, Galileo had his fair quota of failures. He had correctly assumed that the velocity of light is not infinite, as Aristotle had maintained. To demonstrate his theory he set up some blinking lanterns, which he placed on widely separated hilltops. Naturally the experiment failed; the speed of light (about 186,280 miles a second) is altogether too great to be measured in this crude way.

Galileo also failed to clear up the mystery of the water pump. Various philosophers of the medieval period had sought to explain the rise of water in such a pump as due to the fact that "nature abhors a vacuum." This meant that as the water rose, a vacuum was formed behind it and nature supplied water post-haste in order to fill up the vacuum. Galileo did not see that the rise of water in a water pump was caused by atmospheric pressure; in fact, he set out to measure the force of the vacuum in question.

By 1637 Galileo had become totally blind, but he still doggedly continued his scientific researches with the help of his disciples. He died on January 8, 1642.

The name of Johannes Kepler is linked with that of Galileo as a co-worker in the effort to win men over to the teachings of Copernicus. Kepler set forth these teachings in the EPITOME OF COPERNICAN ASTRONOMY (1618), a remarkably clear and interesting exposition. He and Galileo were on terms of mutual esteem; you will recall that it was in a letter to Kepler that Galileo revealed his devotion to the teachings of Copernicus.

Poor Kepler's life was a constant struggle against ill-health and poverty. He had to cast horoscopes to supplement his meager earnings. "Mother Astronomy," he said, ruefully, "would surely have to suffer hunger if Daughter Astrology did not earn their bread." He was born in 1571 at Weill in Wuerttemberg, Germany. His father was an impoverished army officer. His mother seems to have had a knack for getting herself into trouble; in later years Kepler had to defend her against a charge of witchcraft.

Kepler attended the great Protestant university at Tuebingen, where he was converted to the doctrines of Copernicus by Michael Maestlin, professor of mathematics and astronomy. Kepler received his Master's degree in 1591, and then accepted a post as teacher of mathematics and astronomy at Graz. He was required, as part of his professorial duties, to prepare a yearly almanac complete with weather predictions and astrological data.

#### Kepler served Rudolph II of Bohemia

At the turn of the century he became an assistant of Tycho Brahe (see page 373, Volume 2) at Prague. After Brahe's death in 1601, Kepler was appointed to Brahe's place as imperial mathematician and court astronomer to Rudolph II of Bohemia, who was also the Holy Roman Emperor. Unfortunately, Kepler's salary was rarely paid. To obtain necessary funds he taught at Linz and Ulm and for a time was in the service of the Catholic general Wallenstein. He died of a fever at Ratisbon in 1630.

Utilizing both his own extensive observations and the data collected by Brahe, Kepler calculated a set of astronomical tables that were far more accurate than any then in existence. These RUDOLPHINE TABLES (1627), so named in honor of the patron King Rudolph II, supplanted the medieval ALFONSINE TABLES (see page 309, Volume 2) and the PRUSSIAN TABLES, which had been published in 1551 under the auspices of the Duke of Prussia.

But Kepler was no mere calculator and compiler; he won the proud title of "lawmaker of the heavens." For he discovered three laws which exactly describe the orbits of the planets and which furnished further proof of the authenticity of the Copernican theory. These laws — they are called Kepler's Laws — may be briefly stated as follows:

(1) Each planet describes an ellipse, the sun being at one focus.

(2) The straight line joining the planet to the sun sweeps over equal areas in equal intervals of time.

(3) The squares of the periodic times of any two planets are to each other exactly as the cubes of their median distances. (The periodic time of a planet is the time the planet takes to make a revolution about the sun. Median means average.)

The mathematical proof of Kepler's Laws need not concern us here; they represented a lifetime of labor.

Kepler did some fundamental research in optics, dealing particularly with the properties of lenses and the theory of vision. His treatises on THE OPTICAL PART OF ASTRONOMY (1604) and DIOPTRICS (1611) are considered to be the starting points of modern optical research. He also made important contributions to pure mathematics. His calculations had a vital bearing on the later development of the calculus; he invented his own system of logarithms, which greatly simplified the preparation of the RUDOLPHINE TABLES.

Not all of Kepler's teachings were sound. He clung to the old Pythagorean notion of "the music of the spheres" (see page 134, Volume 1). He said there was a relationship between the proportions of the solar system and various musical scales. His lifelong desire to find harmonious geometric relations in the various movements of the planets occasionally led him into absurdities. Such vagaries were in striking contrast with the core of his work.

## *TWO THEORISTS OF THE NEW SCIENCE*

While Galileo and his disciples labored in the vineyard of scientific experiment, working with their hands and seeing with their eyes, other men tried to spin in words a philosophical explanation of this yet incomplete and misunderstood way of dealing with and looking at the world and universe about them. Philosophy was an old business; science a new one. The philosophers of science who now arose called their subject "natural philosophy" — or an attempt to give a systematic explanation to the natural (as opposed to the supernatural) world about them. Two of the philosophers or theorists of the new science were legalistic Sir Francis Bacon and keen-witted René Descartes.

Bacon was essentially a man of words — a lawyer who eventually rose to be lord chancellor of England; a concise stylist both in Latin and English whose ESSAYS, still read with appreciation, helped to shape the character of the English language as we use it today. He did not set out to be a philosopher of science. He came to it because of his philosophic passion for, as he put it, "the knowledge of causes and secret motions of things." Like Columbus, whom he admired, he aspired to discover some new world. That he should have been interested in "secret motions" is evidenced by the checkered intrigues and motives of his political career. As a lawyer, he was interested in the "rules of evidence"; he threw out "hearsay" evidence, idle speculation and immaterial verbiage. He respected experiment because it gave eye-witness answers more credible than the testimony of authorities — even Aristotle himself. It may be said that Bacon fashioned his philosophy of experimental science out of the rules of evidence.

But it must also be pointed out that this "natural philosopher" recognized from the start the limitations of natural science. "Nature to be commanded must be obeyed," he wrote. "The subtlety of Nature is greater many times over than the subtlety of [man's] senses and understanding . . . or argument."

Bacon, later dubbed Baron Verulam, was born in London in 1561. His father was Sir Nicholas Bacon, Lord Keeper of the Great Seal under Queen Elizabeth. At twelve, young Francis entered Trinity College at Cambridge. He went to France in 1576, but returned to England upon the death of his father in 1579 and took up the

Above: Sir Francis Bacon, first Baron Verulam. This brilliant man was one of the foremost scientific theorists of the seventeenth century. Left: title page of his *New Method* (*Novum Organum*).

study of law. He tells us that his "birth, rearing and education had all pointed him not toward philosophy but toward politics." Politics became his career; philosophy, his avocation.

His political career was marked by incessant intrigue. As a member of Parliament he incurred Queen Elizabeth's displeasure by proposing an unpopular measure. He was befriended (and given an estate) by the Earl of Essex, Elizabeth's favorite; a few years later, to regain the Queen's favor, Bacon took an active part in prosecuting Essex on a treason charge.

Bacon was one of the commissioners who arranged for the accession of James VI of Scotland to the throne of England as James I, and it was in the reign of this monarch that Sir Francis first won substantial recognition. He became in turn solicitor general, attorney general and finally, at the age of fifty-seven, lord chancellor. But his extravagance proved his undoing; continually in need of money, he was not too scrupulous about the ways in which he obtained it. In 1621 he was accused of bribery and corruption in office. He humbly confessed his wrongdoing and was condemned to be imprisoned in the Tower of London and to pay a fine of 40,000 pounds. Bacon stayed in prison only two days, however, and the King remitted his fine. The chastened lord chancellor admitted that "it was the justest judgment that was in Parliament these two hundred years." The last five years of his life were spent in retirement and were devoted to his studies.

The works in which Bacon takes up the cudgels for the philosophy of experiment are AN INTRODUCTION TO THE INTERPRETATION OF NATURE (1603), THE ADVANCEMENT OF LEARNING (1605), THE NEW METHOD (*Novum Organum,* 1620) and THE NEW ATLANTIS (1624). This last-named work describes a mythical island whose governors are more interested in controlling nature than in ruling men. A feature of this scientist's Utopia is a scientific institution called Solomon's House, which carries on an ambitious program of scientific research.

Bacon believed that before men could make scientific progress, they must first discard certain prejudices, which he calls "idols." For him there were four idols: "idols of the tribe" — prejudices that are common to all men; "idols of the cave" — prejudices that are peculiar to individuals; "idols of the market place" — prejudices arising from the influence of words upon men's minds; "idols of the theater" — prejudices that arise from the adoption of definite systems of thought.

Once he has rid himself of these prejudices, the man of science must begin to collect facts by means of scrupulous observation and experiment. He should draw up three tables of data. First, there should be a table of positive instances: a list of all the phenomena in which a given property — say, heat — is present. Secondly, there should be a table of negative instances: a list of phenomena in which the property is not present. Finally, there should be a table of degrees of comparison: cases in which the property is present to a greater or lesser extent.

What is to be done with this data, once it is collected? Bacon criticizes those who are content to experiment endlessly, without trying to seek general laws. He is just as critical of "reasoners," who build up elaborate systems on the basis of a few observations or experiments or who dispense with these entirely. "The men of experiment," he says, "are like the ant; they only collect and use. The reasoners resemble spiders, who make cobwebs out of their own substance. But the bee takes a middle course; it gathers its material from the flowers of the garden and of the field, but transforms and digests it by a power of its own." The bee is the symbol of the true scientist, who experiments and reasons, too.

When a respectable body of scientific truth is acquired, it should not be buried in the archives of the mind. Scientific discoveries should have a practical goal: they should serve to better the lot of mankind. Yet Bacon does not believe that each experiment should be judged on the basis of its immediate practical results. He realizes the value of scientific experimentation for its own sake in order to establish a backlog of scientific data, even though he maintains that, in the long run, science should serve mankind.

Bacon preached far more effectively than he practiced. To be sure, he anticipated modern theory by the statement that "heat itself, its essence and quiddity, is motion and nothing else." (Today we would say that heat is a form of energy.) But on the whole his contributions to scientific discovery were negligible. His few efforts at experimentation were pretty primitive and almost entirely unsuccessful. Nor did he fully appreciate the victories that science had already won. He rejected the theories of Copernicus; he disparaged the teachings of William Gilbert, the "father of electricity," and those of William Harvey, the discoverer of the circulation of the blood.

#### Bacon's role in the advance of science

Bacon's role was to instill in others a thirst for scientific inquiry. Robert Boyle, John Locke and other great spirits of the seventeenth century freely acknowledged their debt to him. The first historian of the Royal Society, Thomas Sprat, attributes to Bacon's writings the scientific ferment that led to the foundation of the Society. In Bacon's own words, he "rang the bell which called the wits together."

Like Bacon, René Descartes was dissatisfied with the philosophical methods of inquiry that were in vogue in his time, and he too set forth his ideas of the methods that the true seeker after truth should employ. He was born at La Haye, in Touraine, France, in 1596. His family, of the lesser nobility, was well-to-do and cultured. He was educated at the Jesuit school at La Flèche; afterward he attended the University of Poitiers, spent some time in the fashionable world of Paris and entered the army.

He was not interested in soldiering. He preferred to associate with mathematicians and to ponder over mathematical problems. Finally he left the army to devote himself to his mathematical studies. In 1628 he left France for Holland, where he lived for twenty years. "In what other country," he asked, "can you enjoy such perfect liberty, sleep with more security?" His fame as a mathematician and philosopher grew apace with each of his publications. In 1649 Queen Christina of Sweden invited him to Stockholm to be her tutor in philosophy. The rigors of the northern winter undermined his health and he died in 1650, five months after his arrival at the Swedish court.

Descartes set forth his ideas on scientific method in his DISCOURSE ON METHOD, which was published in 1637. In this work he has given us an account of the youthful dissatisfactions that set him on the trail of a more satisfactory method of scientific inquiry. "From my childhood," he wrote, "I have been familiar with letters; and as I was given to believe that by their help a clear and certain knowledge of all that is useful in life might be acquired, I was ardently desirous of instruction. But as soon as I had finished the entire course of study . . . I completely changed my opinion . . . Though I was studying at one of the most celebrated schools in Europe, in which I thought there must be learned men, if such were anywhere to be found . . . I was led to take the liberty of judging of all other men by myself and of concluding that there was no science in existence of such a nature as I had previously been given to believe."

Of all the studies that he pursued, only mathematics appeared to give that superior certainty at which he was aiming. He proposed, therefore, to adopt the mathematical method of analysis as the method of science. Analysis has been important in scientific research ever since, but only as *one* method of describing phenomena or arriving at scientific truth. To Descartes analysis seemed particularly desirable because it breaks down the complex and confused patterns found in the world of nature and of ideas into simple elements.

In seeking truth, he felt, an investigator should take nothing for granted; he should doubt everything that he observes. But where will the doubting stop? According to Descartes it stops with the fact of doubting itself. For if a man doubts, he thinks, and to think is to exist. "I think, therefore I am" (*Cogito, ergo sum*) was Descartes' assumption of the ultimate certainty.

Once Descartes arrived at this certainty, he set up four rules of straight thinking:

"The first was never to accept anything for true which I did not clearly know to be such; that is to say, carefully to avoid haste and prejudice and to include nothing more in my judgment than what was presented to my mind so clearly and distinctly as to exclude all ground of doubt.

"The second, to divide each of the difficulties under examination into as many parts as possible and as might be necessary for its adequate solution.

"The third, to conduct my thoughts in

such order that, by commencing with objects the simplest and easiest to know, I might ascend little by little and, as it were, step by step to the knowledge of the more complex . . .

"And the last, in every case, to make enumerations so complete and reviews so general that I might be assured that nothing was omitted."

These rules for straight thinking are as valid today as they ever were; they remain an integral part of scientific method. But they are not the whole method.

Descartes utilized these rules — or so he imagined, at least — in MAN, in THE WORLD and in the PRINCIPLES OF PHILOSOPHY in order to explain the nature of the physical world in which we live. All the universe, he said, is filled with matter, and this matter is caught up in vortices, or whirlpools of motion. The visible, physical shapes of the universe — sun, planets, earth and its inhabitants — are created by the vortices. God created matter and set the vortices going, but their motions are regulated by unchangeable laws.

For Descartes, animals are mere machines; even man himself is largely an "earthly machine" (see page 13). Yet he feels that this mechanistic doctrine cannot represent the whole truth. He maintains that the thinking human being has a mind that is distinct from his body. In other words, he arbitrarily assumes the existence of mind, as well as of matter.

Descartes made some notable contributions to optics, meteorology and geometry in three appendixes to his DISCOURSE ON METHOD. He held that light obeys the same mechanical laws as the body in motion. He described its reflection from a mirror surface with the simile of a bouncing tennis ball. He explained the position of primary and secondary rainbows, but could not account for their colors; that remained for Newton to determine. Descartes was a fine mathematician; he won particular fame for his invention of analytical geometry. We discuss the development of this branch of mathematics in another section of this chapter.

By 1650, the year of Descartes' death, science had developed a philosophy — experimental — and a method — mathematical; it had also set up a goal — the improvement of the lot of mankind. It had won a place in an expanding world; it was prepared to go forward to new triumphs.

## *THE CIRCULATION OF THE BLOOD*

"The blood is constantly passing through the lungs into the aorta [the large blood vessel leading from the heart] . . . There is a passage of blood from the arteries to the veins . . . A perpetual motion of the blood in a circle is brought about by the beat of the heart."

In these words the English physician William Harvey summarized one of the high landmarks in the history of medicine and physiology: the discovery of the circulation of the blood. We take this idea so much for granted today that it is hard to realize why its demonstration was so long delayed. The lingering influence of Galen was chiefly responsible. He had taught that the blood ebbs and flows and that it moves from the right to the left side of the heart by oozing through fine pores. He had also held that the blood produces some mysterious sort of "vital spirit" (*pneuma*). These erroneous doctrines persisted for many centuries.

In the sixteenth century, several men were on the verge of discovering the real nature of the blood's movement. Among them were Realdo Columbus, Michael Servetus, Hieronymus Fabricius and Andrea Cesalpino. (The Italians still credit Cesalpino with the discovery.) However, Harvey alone made the necessary experiments and measurements that proved once and for all that the blood actually circulates through the body and that it does not alternately ebb and flow like the tides.

William Harvey was born at Folkestone, in the south of England, in 1578. After a fine classical education at the grammar school in Canterbury and at Cambridge, he departed for Padua to study

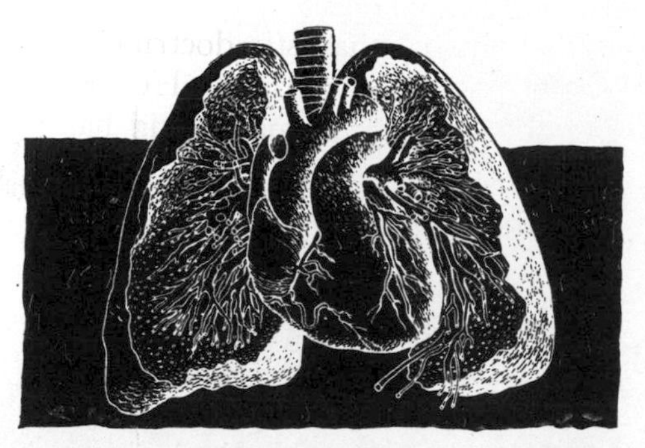

**Harvey studied blood flow through the heart and lungs.**

medicine. His master was the eminent Fabricius, whose teachings on anatomy and embryology were quite advanced for the time. While at Padua, Harvey was elected president of the "English nation" — the fraternity of English lads attending the great Italian university. After receiving his doctor's degree in medicine in 1602, Harvey returned to England to become a physician in St. Bartholomew's Hospital in London and a lecturer at the Royal College of Physicians. He later became physician to the Stuart kings James I and Charles I. Among his official duties as court physician was the examination of women accused of witchcraft; he was required to see if their bodies showed signs of burns — supposedly an indication of congress with the Devil. Harvey never found any such burns!

During the English civil wars he fled with his royal master, Charles I, to Oxford. His London house, containing his research notes, was destroyed; these notes included, among other things, a discussion of the reproduction of insects. At the battle of Edgehill in October 1642, Harvey is said to have taken refuge prudently under a hedge and to have read a book while the fighting was going on; later, he cared for the wounded. He returned to London after the execution of Charles I in 1649, and thereafter he lived in retirement. He was unanimously elected president of the Royal College of Physicians in 1654 but declined the post in view of his advanced age and feeble health. He died in 1657.

Harvey had a most successful private practice; he numbered Francis Bacon and other notables among his patients. His chief claim to fame, however, rests upon his work as an original investigator and scientific writer. His studies on the circulation of the blood he reported in a slender volume, written in Latin and published in 1628: AN ESSAY CONCERNING THE MOTION OF THE HEART AND BLOOD IN ANIMALS. Harvey had presented his findings in his lectures at the Royal College of Physicians long before this time. The delay in publication was due probably to his desire to accumulate further proofs of the circulation of the blood.

Harvey has set forth the difficulties of his task in the opening words of the ESSAY: "When I first gave my mind to vivisections, as a means of discovering the motions and uses of the heart, and sought to discover these from actual inspection and not from the writings of others, I found the task so truly arduous, so full of difficulties, that I was almost tempted to think, with Fracastoro [see Index], that the motion of the heart was to be comprehended only by God."

**Harvey examined the action of the heart**

But Harvey was not discouraged. With the perseverance of the true scientist, he proceeded to examine the action of the heart in all kinds of living creatures — hogs, sheep, dogs, toads, frogs, serpents, eels, small fishes, crabs, shrimps, snails and shellfish. His vivisections led him to the conclusion that the blood circulates continuously through the body — that it flows from the heart to the arteries, from the arteries to the veins and from the veins back to the heart again.

Harvey discovered that the heart is a muscle which serves as a pump, forcing blood into the arteries. Its most important act is contraction (systole), when the forcing of blood into the vessels can be felt as a pulse beat. The opposite opinion, held by Galen and his followers, was that the moment of rest (diastole) was the crucial period, for that was supposed to be the time when the "vital spirits" were being mixed with the blood.

Harvey's description of the actual mechanism of the blood's circulation is as-

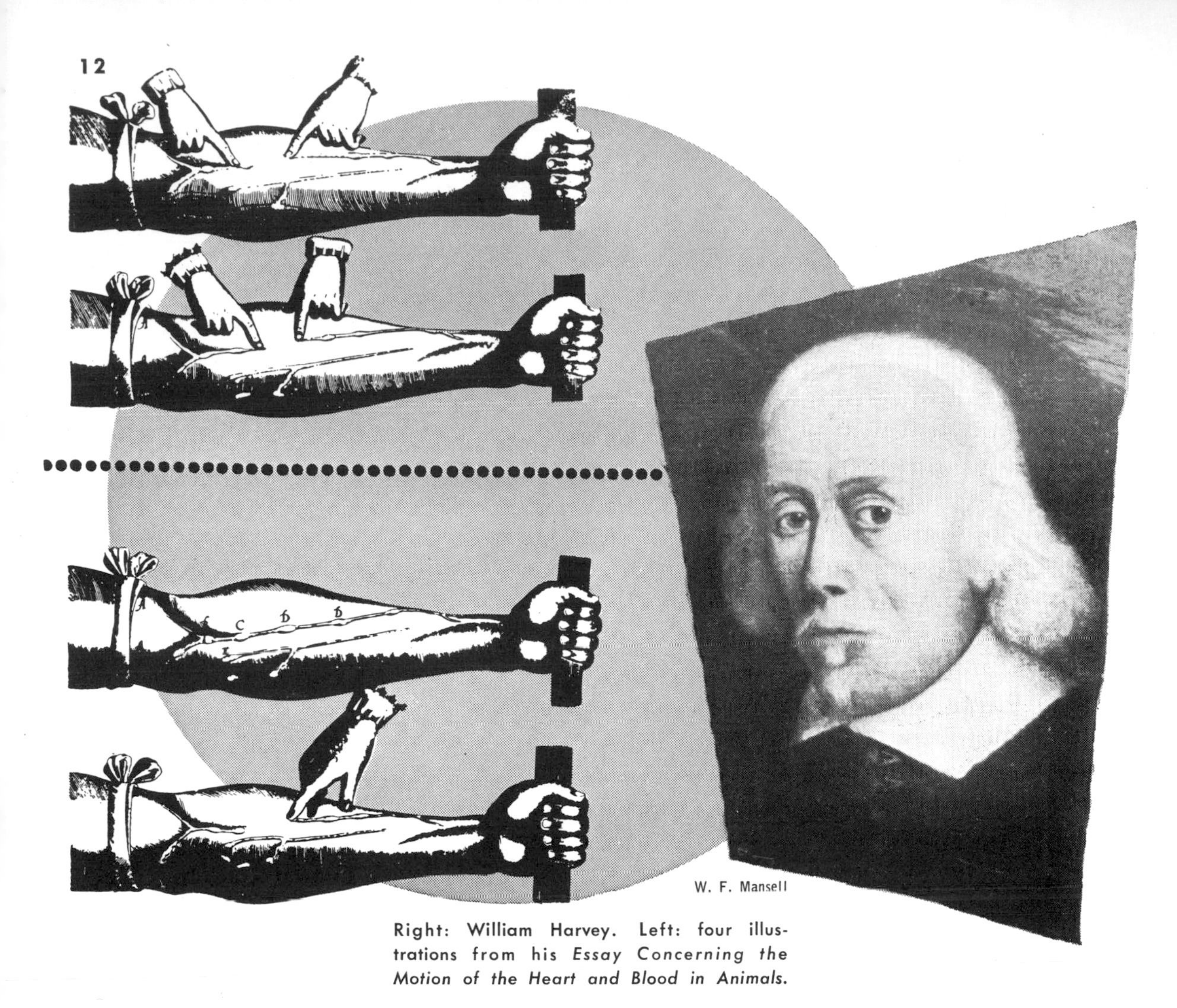


**Right: William Harvey. Left: four illustrations from his *Essay Concerning the Motion of the Heart and Blood in Animals.***

tonishingly accurate. He pointed out that there are four chambers in the heart — two ventricles and two auricles. The left ventricle contracts and the blood it contains passes into the artery called the aorta. The blood then makes its way to smaller arteries and finally enters the veins. It is returned to the right auricle of the heart. The right auricle contracts; the blood it contains passes into the right ventricle and then through the pulmonary artery to the lungs. From there it passes through the pulmonary veins into the left auricle and then into the left ventricle, where the cycle begins again.

The clinching demonstration of the circulation of the blood — once the action of the heart was established — was made by a study of the quantity of blood passing through the heart. Let us assume, says Harvey, that when the heart is contracted, a certain small amount of blood, say half an ounce, is forced out of the heart into the aorta. "This quantity, by reason of the valves at the root of the vessel, cannot return. Now in the course of half an hour the heart will have made more than a thousand beats." Suppose we multiply the amount of blood expelled from the heart at each beat by the number of beats in half an hour. We shall then have "a larger quantity in every case than is contained in the whole body." Where can this blood go unless it circulates?

There remained certain gaps in Harvey's theory. He had no microscope, so that he could not see how the blood passes over from the arteries to the veins through the network of the tiny blood vessels known as capillaries. This was shown a few years after his death by Malpighi and Leeuwenhoek, whose work is discussed in a later chapter. Nor did Harvey understand what happens to the blood in the lungs; like Aristotle, he believed that it was "cooled" there. The true understanding of the physiology of respiration had to wait upon the further development of the science of chemistry.

Harvey made one other important contribution to science: his work on embryology. Like Aristotle and Fabricius before him, he patiently studied the growth of chicks in eggs opened on consecutive days after they were laid. The sum of these researches was included in a book ON THE GENERATION OF ANIMALS, published in 1651. The lack of a microscope hindered Harvey in this work also; but he did lay down one doctrine that became important later. "All animals," he said, "even those that produce their young alive, including man himself, are evolved out of an egg." He believed, however, that certain insects and other lower forms of life sprang spontaneously into life from decaying matter.

## *MECHANICAL EXPLANATIONS OF THE HUMAN BODY*

Harvey's explanation of the circulation of the blood was destined to have important consequences. For by demonstrating the mechanical nature of the circulation with its pumps (auricles and ventricles) and its valves, he opened the way for a mechanical explanation of the working of living bodies as a whole.

There is a striking exposition of the idea that living things are mechanical devices in MAN, by René Descartes. For him animals are pure machines; men are machines for the most part. In fact, he compares the human body with certain ingenious man-made devices. "You may have seen," he says, "in the grottoes and fountains which are in our royal gardens, that the simple force with which the water moves when issuing from its source is enough to set in motion various machines and to make instruments play or utter words, according to the different arrangement of the tubes which convey the water. And so one may well compare the nerves [of man] with the tubes of these fountains; the muscles and tendons with the other engines and springs which move the machines . . . Breathing and similar acts are like the movements of a water clock or a mill, which the ordinary flow of water keeps going continually."

Descartes hastens to assure us that if the body proper is a machine, it is governed by soul, or mind. The mind comes into contact with the body proper within the *conarion,* later identified with the tiny pineal gland, at the base of the brain; from there it is radiated through the rest of the body by means of the nerves and even the blood.

Descartes' mechanistic theory represented philosophical speculation rather than the result of scientific research. In the following generation the Italian Borelli made a far more scientific attempt to apply the principles of mechanics to the workings of the bodies of living organisms.

Giovanni Alfonso Borelli (1608–79) was a native of Naples. He became a professor at Naples and later at Rome; he also worked at the Accademia del Cimento in Florence. He spent the last years of his life in a monastery. His best-known work, the treatise ON THE MOVEMENT OF ANIMALS, was published in 1680–81. In it, he tried to apply the principles of mechanics to explain the movements of man and the lower animals; he looked upon such living organisms as combinations of levers and weights.

He was particularly interested in the movements of human beings — walking, running, jumping, lifting and the like; he also discussed the flight of birds, the swimming of fishes and the crawling of insects. He began in each case with an analysis of the movements of single muscles and wound up with an account of the mobility of the animal as a whole.

To be sure, Borelli realized the limitations of the mechanistic explanations of animal movement. He conceded, for example, that the laws of mechanics could not account for the contraction and swelling of the muscles, on which movement depends. His own explanation was that stimulation by the nerves was involved. Fluid from the nerves, according to his theory, mixes with the blood in the muscles and brings about a sort of fermentation; this fermentation brings about muscular contraction and expansion. The correct explanation is given elsewhere in THE BOOK OF POPULAR SCIENCE.

We realize today that while the principles of mechanics apply to the human body, they are not the whole story by any manner of means. It is true that the heart is a living pump, that its valves serve much the same function as the valves of an air pump, that the bones and the muscles form an extensive system of levers and weights. But the circulation of the blood and the workings of the muscles involve a vast number of phenomena that have nothing to do with the basic laws of mechanics.

Still, Harvey and Borelli and other physiologists who followed their methods accomplished a great deal. They sounded the death knell of those mysterious "vital spirits" that had held back the progress of physiology for so many centuries.

## *"PHILOSOPHICAL INSTRUMENTS"*

Telescopes, microscopes, barometers, thermometers and chronometers have all become indispensable tools of scientific research. The "scopes" (from the Greek *skopos*: watcher) are viewing instruments; the "meters" (from the Greek *metron*: measure) are measuring instruments. To see clearly and to measure accurately are two necessary steps in the scientific process.

The seventeenth and eighteenth centuries saw the invention and development of a good many "philosophical instruments," as scientific viewing and measuring devices were then called. Often, indeed, these instruments had to be devised in order to put specific scientific theories to the test. It was impossible, for example, to give a positive demonstration of the Copernican hypothesis until the telescope was invented; the proof that nature does not abhor a vacuum required the invention of the barometer.

Instrument-making often went hand in hand with original research. Great scientists like Galileo, Huygens, Boyle, Hooke and Newton did not scorn to construct scientific instruments with their own hands. It was only gradually, in fact, that the trade of instrument-making developed apart from the scientific laboratories in which investigations took place.

Among the scientific instruments which have helped most to advance the cause of science is the telescope. Its origin is rather obscure. As far as we know, it was invented in 1608 by Hans Lippershey, a Dutch spectacle-maker of Middelburg. (Another Dutch spectacle-maker of Middelburg — Zacharias Janssen — has also been credited with the discovery.) It is said that Lippershey happened to pick up a pair of spectacle lenses from his workbench. Gazing through the lenses at a distant steeple, he was amazed and delighted to see how near at hand the steeple appeared to be. The first telescope built by Lippershey was something like a modern opera glass. This is a double lens system, with one lens — the eyepiece — near the eye and another lens — the object glass — at the end of a hollow tube pointing to the object to be viewed.

Lippershey's telescope came to the attention of Galileo in 1609. "A report reaching my ears," he wrote in 1610, "that a Dutchman had constructed a telescope, by the aid of which visible objects, although at a great distance from the observer, were seen distinctly as if near . . . I resolved to inquire into the principle of the telescope and consider how I might invent a similar instrument. This I did through a deep study of the theory of refraction. I prepared a tube, at first of lead, in the ends of which I fitted two glass lenses, both plane on one side, but on the other side, one spherically convex and the other concave . . . I obtained an excellent instrument which enabled me to see objects almost a thousand times as large and only one-thirtieth of the distance in comparison with their appearance to the naked eye." Galileo's instrument was based on the same general principles as the Dutch telescope from which it was derived, though he improved it considerably. To this day this type of telescope is called either a Dutch or Galilean telescope.

The astronomer Johannes Kepler designed a more efficient type which he called an astronomical telescope; in this, a second convex lens was to be placed between the object glass and the eyepiece. Kepler never actually built any such instrument; the first to construct a true astronomical telescope based on Kepler's principle was the Jesuit father Christoph Scheiner (1575?–1650). Scheiner also developed what he called a helioscope, or sun viewer, a modification of the Galilean telescope. He put this in a dark room and pointed it in the direction of the sun. He obtained an image of the sun's disc on a white surface behind the telescope; he was thus able to show to others the spots on the sun's surface. Scheiner published an account of his experiments in 1630.

The Galilean and astronomical telescopes were refracting telescopes — that is, they bent, or refracted, the incoming rays of light. Their lenses were not very

It is said that Hans Lippershey happened to look at a church steeple through a pair of ordinary spectacle lenses, and that it was then that the idea of the telescope came to him.

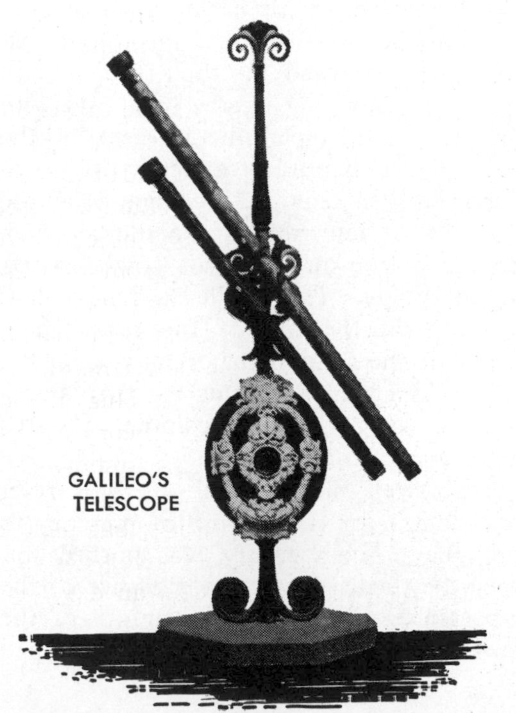

GALILEO'S TELESCOPE

satisfactory; their defects included color fringes and aberration, or scattering of the rays of light. To overcome these defects, very long telescopes were built. The German astronomer Johann Hevelius built an instrument 150 feet long which was supported by a tower. Isaac Newton thought that there was no way of overcoming the defects of refracting telescopes. He therefore designed a reflecting telescope which reflected the rays of the light source from a mirror. He built and polished the first telescope of this kind in 1669.

While the telescope has extended man's vision in space, the microscope has made him aware of a new world of miniature living things and structures. There are two principle types of microscopes: simple and compound. The simple microscope is just a magnifying glass, a single lens which magnifies the object under view. The compound microscope has at least two lenses — an object glass and an eyepiece — arranged at either end of a hollow tube.

The Greeks were familiar with the magnifying glass. They often used their magnifying lenses as "burning glasses," directing and magnifying the rays of the sun. The magnifying glass was also known to the Arabs and to Roger Bacon; the compound microscope, however, was not invented until after 1590. Credit for it is usually given to Zacharias Janssen, the Dutch spectacle-maker, who may also have been the inventor of the telescope. Janssen is said to have chanced upon the compound microscope by the happy accident of holding one lens over another and noting the improved magnifying effect. Galileo was among the first to use this type of microscope for scientific purposes; with it he studied the motion of insects and the compound eyes of insects.

**Robert Hooke's compound microscope**

A far better compound microscope was constructed by crotchety Robert Hooke about the middle of the seventeenth century. The two lenses of the instrument were flat on one side and convex on the other. There was a drawtube arrangement by which the distance between lenses could be increased or decreased, thus sharpening the focus. The microscope was mounted on a stand. The object to be viewed was fixed on a pin set vertically upon the stand; a small lamp provided illumination.

Simple microscopes, however, continued to be constructed in the seventeenth century; they sometimes gave better results than the imperfect compound lenses of that day. Their excellence depended on the painstaking care given to polishing (usually with pumice) the curves on the glass. Among the best of the seventeenth-century simple microscopes were those built by the Dutchman Anton van Leeuwenhoek. The best of his lenses magnified objects about 270 times their original size.

In the chapter on Science Grows Up II, Volume 3, we tell how Leeuwenhoek, Hooke and other men of science used the microscope to win new information about tiny living things and about structures too small to be seen by the naked eye.

The seventeenth century saw the invention of the barometer, the instrument that measures atmospheric pressure. This pressure is really the weight of the mass of air that envelops the earth for miles above its surface. The inventor of this device, Evangelista Torricelli (1608–47), was a gifted young Italian nobleman, who became a pupil of Galileo a short time before the master's death. Torricelli was particularly interested, as Galileo himself had been, by the fact that water cannot be raised by a common suction pump to a height greater than about thirty-four feet. As we saw in the section on The First of the Moderns, the popular explanation of the rise of water in such a pump was that "nature abhors a vacuum." In other words, as "empty space" is created behind the rising column of water, nature's horror of this "empty space," or vacuum, causes other water to fill the gap. Galileo is said to have wondered why nature's horror of a vacuum stopped abruptly at thirty-four feet. Torricelli decided to examine the problem.

He began experimenting with mercury, the only metal that is a liquid under ordinary temperature conditions. Since mercury is more than 13 times as heavy as water, Torricelli reasoned that nature's "horror of a vacuum" would raise a column of mercury in a tube to less than $1/13$ the height of a corresponding column of water. To verify his conclusions, he prepared a glass tube about 2 yards long and sealed it at one end; he then filled it with mercury. Stopping up the open end with his finger, he dipped the tube under the surface of a quantity of mercury in a bowl. Torricelli then removed his finger from the tube. A certain amount of the mercury ran out of the tube into the bowl, leaving an empty space (now called a Torricellian vacuum) at the top of the tube. The height of the column that remained in the tube measured about 30 inches; that is, less than $1/13$ the height of a column of water under equivalent conditions.

Torricelli now worked out the correct explanation for the column of mercury in the tube. The mercury was pushed upward, he theorized, by the pressure of the atmosphere upon the open surface of the bowl of mercury, into which the tube had been inserted. Torricelli knew that we live at the bottom of a sea of air which, as Galileo had proved, has definite weight. The weight of this overhanging mass of air is great enough, Torricelli guessed, to push mercury up 30 inches in a tube, but it is not great enough to push it beyond that point. It can push water up about 34 feet but no higher.

Torricelli noted that the height of the mercury column in his simple apparatus fluctuated from day to day. He correctly attributed this to changes in the pressure of the air. His apparatus, therefore, provided a means of measuring this pressure. In other words, he had constructed a true barometer, though the name was not given to the instrument until later.

Torricelli's untimely death in 1647 prevented his providing definite proof of his theory. This proof was supplied in the following year through a famous experiment directed by the great French philosopher-mathematician-physicist Blaise Pascal (1623–62). The scene of the experiment was the Puy-de-Dôme, a high mountain in the French province of Auvergne; the time was the month of September, 1648. Since Pascal was in Paris at the time, his obliging brother-in-law F. Périer took charge of the proceedings on the mountain. He provided himself with witnesses, including some churchmen, a lawyer and a doctor — "irreproachably honest men, so that the sincerity of their testimony should leave no doubt as to the . . . experiment."

Périer set up a Torricellian barometer at various places along the mountain slope, as he climbed with his witnesses from the foot to the summit of the peak. If Torricelli's theory of atmospheric pressure was correct, the mercury in the tube would drop as greater heights were reached, since there would be less weight of air to press upon the open bowl. Sure enough, that is just what happened. "We were carried away with wonder and delight," wrote Périer, reporting to his illustrious brother-in-law on the results of his experiment.

In the years that followed, many improvements were made in the Torricellian barometer, but the essential feature of the

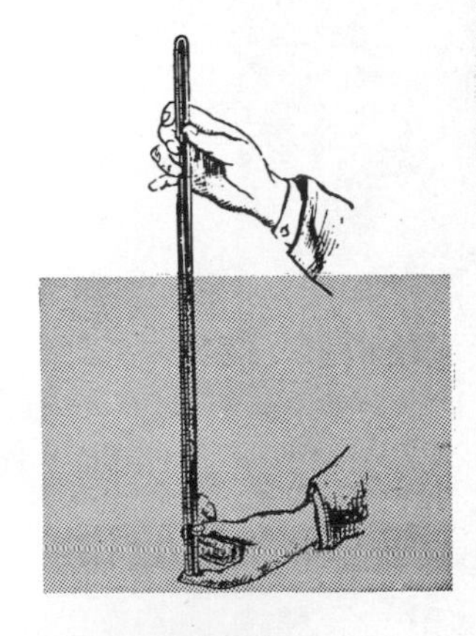

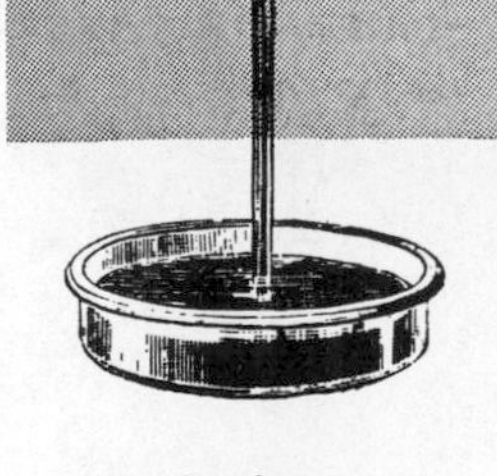

Torricelli's barometer. He sealed one end of a glass tube and filled the tube with mercury. Stopping up the open end with his finger, he dipped the tube under the surface of mercury in a bowl.

instrument — a tube of mercury exposed to atmospheric pressure — has been retained. Another type of barometer, the aneroid barometer, in which atmospheric pressure is registered on a thin diaphragm instead of on a column of mercury, was first suggested by Leibniz about 1700.

Another "philosophical instrument" that proved to be extremely useful in the study of the atmosphere was the air pump, invented by the German Otto von Guericke (1602–86). As a young man he lived amid the horrors of the Thirty Years War. He fled from his native city of Magdeburg when it was sacked in 1631; later he returned to the city and became its burgomaster.

His first air pump consisted of a cask filled with water and caulked so that the outer air could not enter. The water was withdrawn from the cask by a pump containing two leather valves and worked by two strong men alternately pulling and pushing the piston. Guericke reasoned that after the water was withdrawn from the cask, a vacuum would exist within it. The first air pump experiment was a failure because of leakage. Guericke then tried to pump the air out of a copper sphere without first filling it with water. Unfortunately the supposed "sphere" was not perfectly spherical; as a result, it collapsed

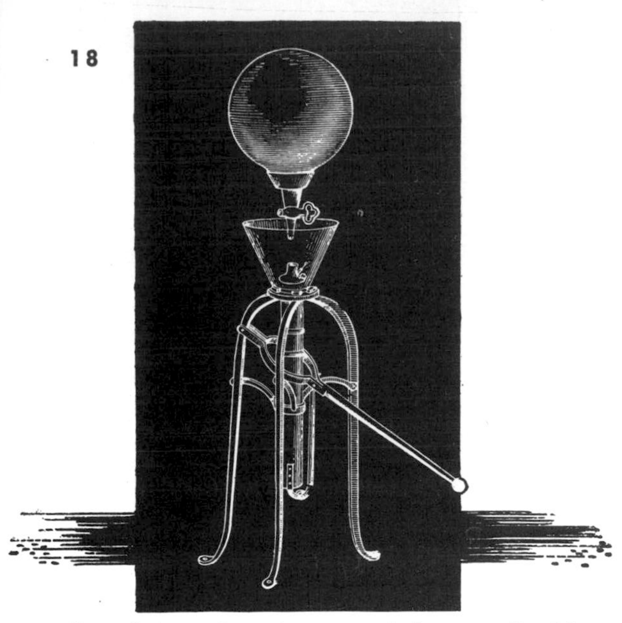

One of the earliest air pumps of Otto von Guericke.

inward because of the pressure of the outer air after a certain amount of air had been withdrawn from it. At last, however, Guericke perfected an air pump that could bring about a fairly high vacuum. (Remember that not even the most modern pump can create a *perfect* vacuum, which would be the *complete* absence of matter from a given space.)

Guericke performed many interesting experiments with his air pumps. The most spectacular was the one performed with the famous "Magdeburg Hemispheres" before the Imperial Diet at Ratisbon in 1654. The "Hemispheres" were two hollow bronze half-spheres, carefully fitted together at the edges. They were exhausted of air through a stopcock which was closed after a high vacuum had been obtained. The pressure of the outer air caused the hemispheres to be pressed tightly together. Two teams of eight horses each (some say sixteen horses each), pulling in opposite directions, could not draw them apart.

Guericke also invented a machine for producing frictional electricity. It consisted of a sulfur sphere mounted on a horizontal axle between two supports so that it could be rotated. It was "excited" or electrified when a hand was placed upon it as it turned. Guericke noted that the rubbed sphere would attract light objects, such as paper and feathers. When he brought his finger near it, there was a crackling sound and occasionally tiny sparks were produced. Certain bodies, he found, could be electrified if they were brought close to the rubbed sphere, without actually touching it. All these phenomena are explained in our article on Electricity, in Volume 6.

A most useful "philosophical instrument" is the thermometer. The invention of this device is generally attributed to Galileo. One of his disciples, Castelli, wrote that he saw the great Pisan using a primitive sort of thermometer, called a thermoscope, in lecturing in the year 1603. This instrument consisted of a bulb-shaped glass vessel, about the size of a hen's egg, fitted to one end of a tube. Galileo would warm the bulb between his hands and he would then insert the open end of the tube in a vessel of water. "As soon as the ball cooled down," says Castelli, "the water rose in the tube to the height of a span above the level in the vessel. This instrument he [Galileo] used to investigate degrees of heat and cold."

In 1632 a French physician, Jean Rey, suggested the use of liquid thermometers, in which temperature would be measured by the expansion or contraction of a liquid and not by the expansion or contraction of the air. A bulb was fitted to a tube, as in Galileo's apparatus, but the position of the bulb was reversed; it was now at the bottom of the apparatus. The bulb was filled with water and the expansion or contraction of the water served as an index of temperature. The top end of Rey's tube was not sealed. The Grand Duke Ferdinand II of Tuscany sealed the tube and substituted wine for water. As this type of thermometer, elaborately designed, was used by the members of the Florentine Accademia del Cimento (see the Index), it came to be known as the Florentine thermometer.

These early thermometers had no fixed scales. Huygens suggested that the freezing point of water or its boiling point might be used as a fixed point from which a definite scale of temperature could be built up. Other experimenters, including H. Fabri and C. Renaldini, suggested using two fixed points for such a scale.

It was in the year 1714 that Gabriel Daniel Fahrenheit (1686–1736), a modest German scientist, introduced the simple mercury thermometer and temperature scale that still bears his name. He used mercury for the thermometric fluid and found that the result "answered his prayer." As the low point on his thermometer scale — "zero" — he took "the most intense cold obtained artificially in a mixture of water, ice and sal ammoniac, or even sea salt." As an upper fixed point, he took "the limit of heat which is found in the blood of a healthy man" — a figure which he placed at 96 degrees. A third and middle fixed point was the temperature of a mixture of ice and pure water. This was labeled "freezing point"; it corresponded to 32 degrees. Later he adopted the boiling point of water (212 degrees at sea level) as another fixed point.

Other temperature scales were soon suggested; some of these, too, still remain in use. The Réaumur scale, introduced by R.-A. F. de Réaumur (1683–1757), a French naturalist, in the year 1730, set 80 degrees between the freezing and boiling points of water. Anders Celsius (1701–44), a Swedish astronomer at Upsala, perfected the first centigrade (100-degree) thermometer in 1742. In his thermometer, the temperature of melting ice was set at 100 degrees and the temperature of boiling water at zero. In the modern centigrade scale, the freezing point corresponds to zero; the boiling point, to 100 degrees.

A "philosophical instrument" that has become essential for everyday living is the clock. The ancient Greeks had water clocks and hourglasses for measuring time; the sundial also is of ancient vintage. But many generations of men were required not merely to develop and make clocks and watches but to discover the mechanical principles upon which they were based. The foundations of the modern clockmaker's art were laid by the Dutch scientist Christian Huygens in a little book, THE CLOCK, published in 1658. In this work Huygens tells how he constructed an accurate timekeeping instrument by applying a pendulum to a clock driven by weights. The weights kept the clock going, but the pendulum regulated the rate of movement. It required unusual mastery of the new science of dynamics for Huygens to discover the shape of the curve in which an isochronous (that is to say, equal-timed) pendulum must beat.

Many other "philosophical instruments" were invented in this pioneer period of science. There was the sounding instrument and the sea-water sampler of Hooke, the hydrometer of Boyle, the marine clock and the micrometer of Huygens and a thousand others. The men of science in those days were doing their utmost to extend the range and the accuracy of their senses and thus to provide a surer foundation, not only for their scientific research but also for their theories.

## *MATHEMATICS BECOMES PRACTICAL*

The scientist often uses the findings and methods of mathematics in order to solve his problems. As Descartes put it: "All the sciences that have for their end investigations concerning order and measure are related to mathematics, it being of small importance whether this measure be sought in numbers, forms, stars, sounds, or any other object." In the seventeenth and eighteenth centuries a number of new mathematical tools were put into the hands of scientists.

For one thing calculations were enormously simplified by the invention of logarithms. In this practical system of calculation, problems in multiplication and division can be worked out by the much simpler processes of addition and subtraction; roots (square roots, cube roots and so on) can be extracted by simple division. In the system of logarithms called common, or briggsian, logarithms, 10 is taken as the base. The number 100 is $10^2$ (ten to the second power); the exponent 2 is the logarithm (also called simply log) of 100. The number 1,000 is $10^3$; the exponent 3 is the logarithm of 1,000. Suppose we wish to multiply 100 by 1,000. We simply add 2 (the logarithm of 100) and 3 (the logarithm of 1,000). That gives as the answer logarithm 5, which is equivalent to $10^5$ or 100,000. By using previously calculated and published tables of logarithms, it is possible to obtain the logarithms of all numbers and to apply them in calculations.

John Napier (1550–1617), Baron of Merchiston, is generally regarded as the inventor of logarithms, although a Swiss mathematician, Joost Buerg (1552–1632), devised another system independently. Napier published an account of his invention in 1614 in THE DESCRIPTION OF THE MARVELOUS CANON OF LOGARITHMS. Napier's friend, Henry Briggs (1561?–1631?), a professor of mathematics in Gresham College, helped Napier perfect the system. In his ARITHMETIC OF LOGARITHMS (1624), Briggs prepared tables giving the common logarithms of 30,000 numbers. Today logarithmic numbers provide indispensable shortcuts for workers in many fields.

Napier provided another method of simplifying calculation: the device known as Napier's bones. These consisted of notched rods like those shown on this page; they were used for both multiplication and division.

Calculation is simplified enormously nowadays by machines. The very simplest of these is the abacus, or counting frame, which was known in antiquity. However, the earliest calculating machines, in the ordinary use of the term, go back to the seventeenth century. In his MATHEMATICAL DISCIPLINES, published in 1640, J. Ciermans claimed to have built a mechanical calculating device, but since he gave no details of its operation, his claim is doubtful. In 1642 Pascal invented an adding machine of which a replica still exists. Neither Pascal's machine nor those that were introduced later in the century by Moreland and Leibniz worked well; it was not until much later that satisfactory calculating devices were developed.

The invention of analytical geometry by René Descartes provided scientists with another valuable kind of mathematical tool. Analytical geometry is based on the fact that the position of a point in space can be fixed by determining its distance from two fixed intersecting lines, or from three fixed intersecting planes. It is said that the idea of this branch of mathematics came to Descartes while he lay idly in bed watching a fly darting hither and thither in a corner of the room. It suddenly came to him that he could define the position of the insect in space at any given moment. All that would be necessary would be to determine its position from each of the three planes formed by two adjacent walls and the ceiling.

Analytical geometry has innumerable applications. One of its most familiar uses is in the charts known as graphs. In these, all sorts of useful data can be set forth by noting the distances of a group of points

This calculating device, known as Napier's bones, was used widely in the seventeenth century. The device consisted of notched rods and served for both multiplication and division.

From A. Wolf, "A History of Science, Technology and Philosophy in the Sixteenth and Seventeenth Centuries," published by Allen and Unwin, Ltd., and The Macmillan Company

from two fixed lines, called axes, usually set at right angles to each other.

The development of differential calculus furnished another important mathematical tool. First proposed by Kepler and by Cavalieri, a disciple of Galileo, it was worked out more usefully by Newton and Leibniz. Calculus represents a mathematical method of dealing with infinitesimal quantities, so small that they cannot be worked with the methods employed in elementary mathematics.

The seventeenth century saw the beginnings of the theory of probability. It was born at the gaming tables. A Parisian gambler of noble birth, the Chevalier de Méré, had sought the help of Blaise Pascal to find out how to establish betting odds. Pascal, intrigued by the problem, decided to discuss it with one of the most brilliant mathematicians of the day, Pierre Fermat (1601–65). In the course of the correspondence between Pascal and Fermat over the Chevalier de Méré's problem, the foundations of the theory of probability were laid down.

Today this theory plays an important part in scientific investigation. In the case of many natural phenomena, such as the interplay of forces in subatomic worlds, the only way we can determine what is likely to happen in a given experiment is to determine the degree of probability.

### The beginnings of the statistical method

The statistical method also originated in the seventeenth century. It was first called political arithmetic; its originator was the learned Sir William Petty (1623–87). Political arithmetic began as an effort to organize information in a way that could be easily grasped by men of affairs, rulers and parliaments. It discussed things in terms of the group rather than of the individuals who made up the group. It was soon discovered that one could describe more accurately the group as a whole than the individuals of which it was made up. For example, it could be said with a fair degree of certainty that a certain number of people of a certain age would die in a year; but it could not be predicted which individuals would die.

One of the outstanding pioneers of the statistical method and vital statistics was Captain John Graunt (1620–74), a citizen of London. In 1662 he published, probably with the aid of Sir William Petty, a remarkable volume entitled NATURAL AND POLITICAL OBSERVATIONS ON BILLS OF MORTALITY. The bills of mortality were weekly records of births (christenings) and, more particularly, deaths and funerals. These records of numbers and causes of death noted by old women, called "searchers," had been kept by the parish clerks of London since the great plague year 1603. They had been put together in tables, under such headings as "plague," "notorious diseases" (apoplexy, epilepsy, stone and so on) and "casualties" (bleeding, grief, drowning, hanging, vomiting and so on).

Graunt was apparently the first man to analyze these lists and figures so that they might serve to improve the welfare of his city and his countrymen. By reducing "several great confused volumes into a few perspicuous tables," Graunt was able to note, among other things, the excess of male over female births, the high mortality rate in the early years of life and the fact that the death rate in the city was greater than in the country. Graunt pointed out the alarming fact that 36 per cent of the children born in London in the seventeenth century died before the age of six.

Statistics have been applied to a great variety of scientific problems since the days of Petty and Graunt. They have provided the scientist with a tool for testing a good many of his experimental conclusions, particularly in physics, chemistry and biology. The statistical method is particularly useful in the field of vital statistics, which has to do with records and changes in the birth rate, the death rate, with population shifts and the like; it is a *must* in the preparation of adequate insurance tables. Statistical method only tests scientific truth; it does not discover it, as do the methods of "experimental science" practiced by Galileo and praised by Sir Francis Bacon.

*Continued on page 265.*

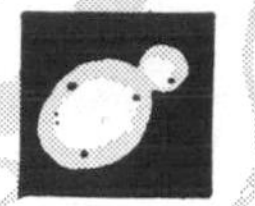

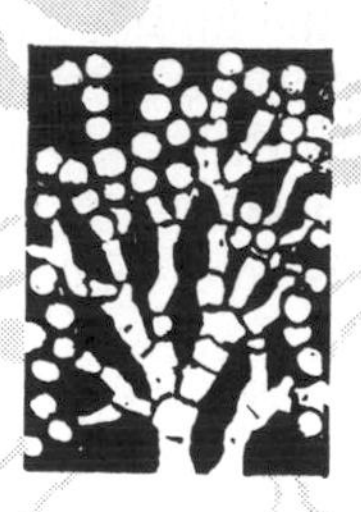

# THE FUNGI

## Molds, Yeasts, Mildews, Mushrooms and Their Kin

EVERYONE is familiar with certain fungi, particularly such forms as mushrooms, yeasts and molds on bread. The spores of these plants are always in the air around us; together with pollen grains and other suspended particles, they are often the cause of allergies.* A great many fungi are beneficial to man; a considerable number are injurious to crop plants, commercial products and even to man himself. Hence the branch of botany that deals with the fungi is of particular interest. It is called mycology, from the Greek *mykes:* "mushroom," and *logos:* "science." Specialists in this field are known as mycologists.

The fungi are among the most primitive forms of the plant kingdom; they are included among the thallophytes, or plants with a thallus. A thallus is a simple plant body that has no roots, stems, flowers and seeds — structures we commonly associate with the higher plants. The thallus of a fungus is usually made up of branching threads, called hyphae. The name "mycelium" is given to the sum total of the hyphae. The fluffy growth of a mold on bread is the part of the mycelium that extends above the surface of the bread. The hyphae that provide nourishment for the mold are within the bread or upon its surface.

Most fungi reproduce in two different ways — sexually and asexually. The mycelium of many species produces both asexual spores and sexual structures. The asexual spores form a sort of dust made up of tiny particles; they are called conidia (singular: conidium), from the Greek word *konis,* meaning dust. When they germinate, they give rise to a mycelium similar to the one from which they themselves are derived. The conidia are commonly scattered far and wide by wind, rain, insects, birds, man and other agents. Many of them are so resistant that they can be blown for hundreds or even thousands of miles without losing their ability to germinate. This is one of the reasons why many species of fungi are found so widely distributed over the earth.

Fungi show a great deal of variation in sexual reproduction. In all cases, however, the essential feature is the bringing together and fusing of two cell nuclei. In

*An allergy represents an exaggerated reaction of the human body to certain foods, to substances that are inhaled, to insect stings, to contact with various plants and so on. See the chapter The Allergies, in Volume 4.

many of the higher fungi, the sexual spores are produced in conspicuous fruiting bodies, of which the familiar mushrooms are a good example. The fruiting body of a fungus is only a part of the complete plant; it arises from hyphae which are underground or in the wood of a decaying log.

The algae and most higher plants have a green pigment, or coloring matter, called chlorophyll. In cells containing this pigment, water (absorbed from the soil) and carbon dioxide (derived from the air) react in the presence of sunlight to form an essential food element — a sugar, called glucose.* This substance serves as a basic material from which the plant manufactures other vital foods. The fungi lack chlorophyll. Hence they cannot make their own food, but are dependent upon other organisms (living things) for their supply. Some exist as parasites, growing upon other living organisms and obtaining nourishment directly from them. Others feed upon dead organic matter, such as decaying leaves or dead wood, on which they grow. Such fungi are called saprophytes ("plants feeding on rotting things").

True fungi are unable to take in solid foods as animals do; they must change such foods before they can use them. They do this by secreting enzymes (substances that speed up chemical transformations) into their surroundings. The enzymes break down foods consisting of complex organic materials into simpler substances that can be dissolved in water. When these substances are in dissolved form, the fungi can absorb them.

Some species of fungi are very much restricted in their choice of foods — that is, in the organic materials upon which their enzymes can act. One small fungus group, for example, is known to occur only on the castoff horns and hooves of animals. Other species produce many different kinds of enzymes, which enable them to grow upon a wide variety of substances.

The enzyme-forming activities of fungi are a vital factor in the natural process upon which the very existence of life upon the earth depends — the process of decay. These activities convert the complex organic matter present in the dead bodies of plants and animals into simpler substances that plants can absorb. In the plants, these substances are used in the manufacture of food elements for both plants and animals.*

Man puts the enzyme-forming activities of certain fungus species to work in brewing, baking and cheese-making. The enzymes of other species bring about the decay of lumber and textiles and cause numerous diseases of animals and plants.

* This process is called photosynthesis ("putting together in the presence of sunlight"). See Index, under Photosynthesis.

* Animals obtain these substances by eating plants or by devouring plant-eating animals.

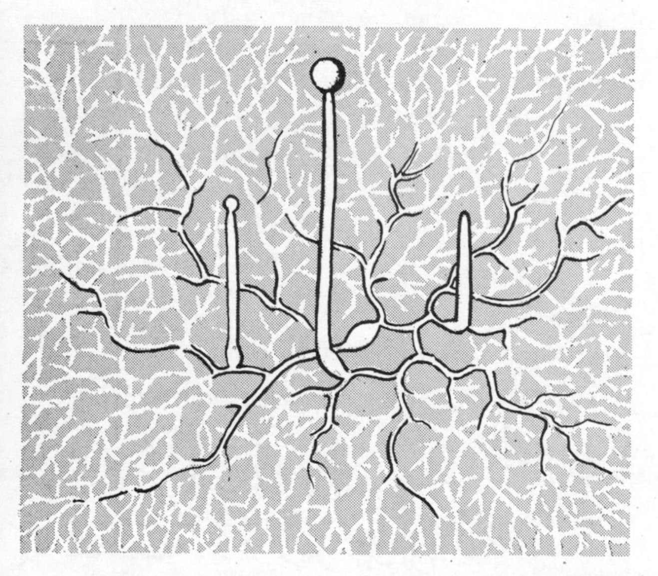

**A black bread mold, showing the branching of the mycelium. Note the three sporophores (spore-bearing organs).**

**Fruiting bodies of the common mushroom *(Agaricus campestris)*. Spores are produced in such bodies.**

A great many seed plants — trees and shrubs — have a remarkable relationship with certain fungi. The mycelium of these fungi invades the roots of the plants, but instead of harming them it helps them. It absorbs water and dissolves minerals from the soil and transports them to the roots. The seed plant, in turn, supplies food to the fungi. Fungi of this kind are called "mycorrhizal" (fungus-root). The relationship between root and fungus in this case is a good example of symbiosis — an association beneficial to both partners. (See Index, under Symbiosis.)

Certain substances, obtained from fungi, have proved invaluable in medicine. The antibiotic penicillin, derived from the small blue-green fungus called *Penicillium,* has saved countless lives and relieved much suffering. The story of penicillin and other antibiotics is told elsewhere in THE BOOK OF POPULAR SCIENCE. (See Progress in Medicine, Volume 10.)

No one is sure how the fungi originated. Since certain lower forms resemble the green algae in some respects, especially in form and mode of reproduction, some mycologists believe that the fungi arose from the algae. They argue that the fungi became parasites or saprophytes as they lost their chlorophyll and the ability to manufacture their own food. Other botanists hold a different view. They point out that certain species of the lower fungi are animallike in many ways; they believe, therefore, that the fungi evolved from the protozoans, the most primitive group of animals. Before we shall be able to have a clearer idea of the origin and evolution of the fungi, we shall have to learn more about the species already known to man. Perhaps the discovery of new species will help fill the gaps that now exist in our knowledge.

Nearly 50,000 species of fungi have been described. Some authorities estimate that this figure represents less than half the total number likely to be in existence. It is entirely likely, therefore, that we shall discover many new and interesting species in the course of time. Large parts of the earth, including Africa, much of South America, the South Pacific and Asia, remain comparatively unexplored as far as fungi are concerned.

In the following discussion of the fungi, we shall begin with the slime molds, which are recognized as the most primitive and animallike. Then we shall take up in turn the algalike fungi, which include the downy mildews and the black-bread-mold group; the sac fungi, containing such forms as yeasts, blue and green molds, powdery mildews, sphere fungi, cup fungi and truffles; the club fungi, including rusts and smuts, gill fungi, puffballs and stinkhorns; and finally the imperfect fungi.

### Slime molds (Myxomycetes) *

The slime molds, or Myxomycetes (*myxa* means "slime" in Greek), are commonly found on rotting logs, old stumps and decaying humus. When the spore of a slime mold germinates, it produces a cell that propels itself through the water by means of a whiplike projection called a flagellum. (Sometimes there are two.) The flagellum then disappears and the cell begins creeping around like the primitive

* You will note that the scientific names for the different classes of fungi, with the exception of the imperfect fungi, all end in "mycetes," a combining form meaning "fungi." The first part of the name indicates the kind of fungus. We explain the derivation of this part of the name in every case.

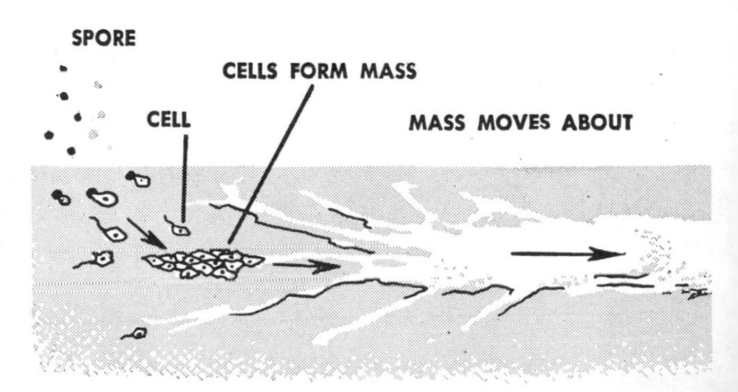


**The development of a slime mold. The spore germinates and produces an amebalike cell, which propels itself through the water by means of one or two flagella (whiplike projections). After the flagella are lost, the cell starts creeping about like an ameba. When cells of this type come in contact with others of the same species, they generally fuse. They come to form a mass, called the plasmodium, which creeps about and feeds. Finally it flows out onto an exposed surface. Upright extensions arise, developing into stalked sporangia, or spore cases.**

one-celled animal called the ameba.* Like the ameba, the slime mold changes its shape constantly and takes in solid particles. When a slime-mold cell comes in contact with others belonging to the same species, they generally fuse. The product of such fusion may then gradually enlarge while creeping about and feeding on bacteria and fungus spores. This feeding stage is called the plasmodium; it may grow to a comparatively large size on decaying logs or humus. Finally, it flows out onto an exposed surface. Here it appears as a soft thin layer or a hemispherical mass; it varies in color — white, yellow, orange, red — according to the species. The plasmodium now produces upright extensions. These develop into small stalked sporangia, or spore cases, containing numerous spores. In some species the entire plasmodial mass may develop into a single fruiting body.

Slime molds have been the subject of a number of basic science studies. The plasmodia, which have no cell walls around them nor cell divisions within, are an excellent source of relatively pure protoplasm for laboratory study.

### Algalike fungi (Phycomycetes)

This is an important group of parasitic fungi; the scientific name Phycomycetes means "seaweed fungi" in Greek. (The seaweeds are marine algae.) The algalike fungi cause diseases of crop plants and ornamentals (plants cultivated for decorative purposes) and are often serious pests.

*Downy mildews.* These fungi are so named because they produce a downy growth on the surface of infected plant parts, usually the leaves. The infection cells that appear on these downy areas are blown about by the wind and cause other plants to be infected.

One of the best known of the downy mildews is the late blight of potatoes. When the white potato was first carried to Europe from South America many years ago, it was as a botanical curiosity. Later, it became a staple crop throughout Europe and the British Isles, especially in Ireland. In 1845, a devastating blight struck down the potato plants; the tubers, whether harvested or left in the field, rotted in large numbers. The famine that resulted left about half a million dead in Ireland and brought about the migration of nearly two million Irish, principally to America. It was with some difficulty that mycologists were able to convince the authorities that the disease was caused by a parasitic fungus — a downy mildew.

About 1865, another downy mildew was accidentally introduced from North America into Europe. Grapevines in France had been attacked by the root louse. It was hoped that this pest might be foiled if the French grape were grafted onto the American grape rootstock, since this was resistant to the insect. Unfortunately, the downy mildew of grape was brought into France together with the rootstock. The French grape proved highly susceptible to this fungus; the disastrous epidemic that resulted nearly ruined the French wine industry.

The discovery of the cure was as much of an accident as the introduction of the disease. One day, the French mycologist Alexis Millardet (1838-1902) was walking in the country near Bordeaux when he

* As a matter of fact, some authorities think that the slime molds should be considered as animals, related to the protozoans, the group to which the ameba belongs. We describe the protozoans elsewhere; see Index.

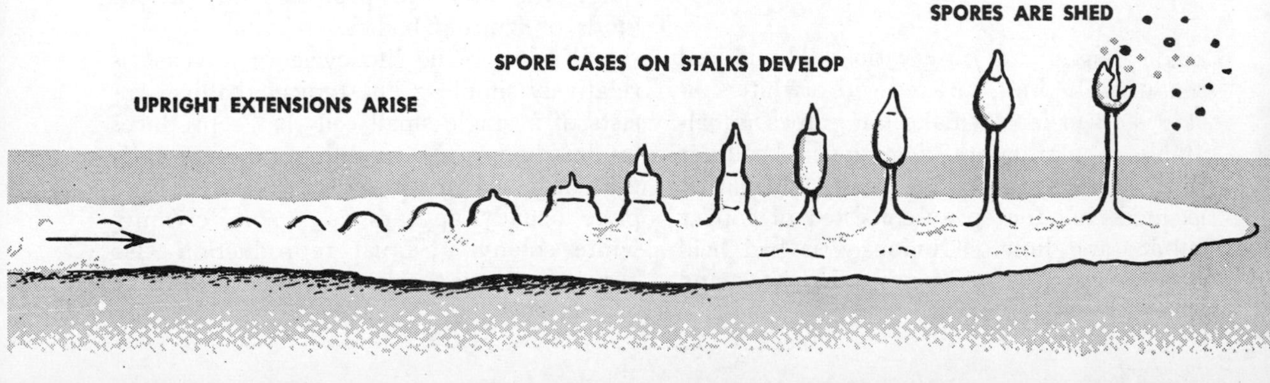

UPRIGHT EXTENSIONS ARISE

SPORE CASES ON STALKS DEVELOP

SPORES ARE SHED

The black specks on the loaf of bread shown above are due to black bread mold. Below: how this familiar mold develops on bread. Note the rounded spore cases (the sporangia) at the tips of erect filaments (the hyphae).

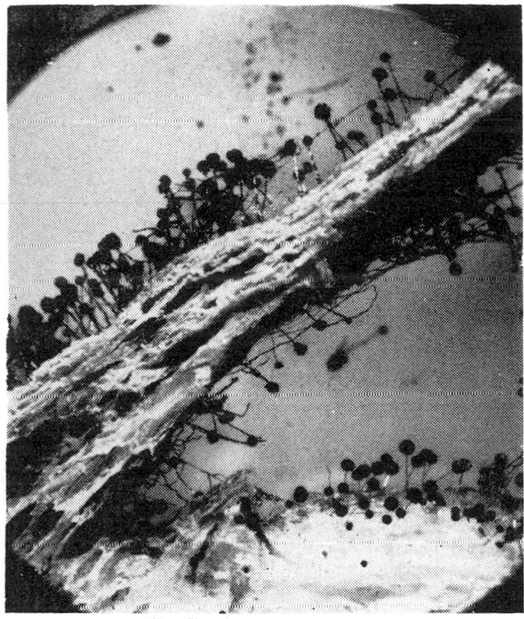

The Fleischmann Laboratories, Standard Brands Inc.

made a significant observation. He noticed that certain vines, sprayed with a white substance in order to make the grapes unpalatable to marauders, were much healthier than unsprayed vines nearby. Millardet learned that the spray consisted of copper sulfate and lime. Obviously, it had held the downy mildew of grape in check and was an effective fungicide (fungus-killer). Under the name of "Bordeaux mixture," it was used extensively in the years that followed to protect plants from various kinds of fungi. It is still considered to be one of the most useful fungicides.

Downy mildews cause various other plant diseases. They attack tobacco, cabbage, cucumbers, squash and various other plants.

*Black-bread-mold group.* The molds of this group are generally called mucors. There are many species; most of them occur as fluffy growths on bread, fruits, vegetables and preserved foods. They are also found in the soil and in decaying vegetation. They are white at first, but they soon become dark as their spores mature in large numbers.

Black bread mold is probably the best known example of these fungi. It develops on stale bread — not so frequently as in the past, however, because of the use of preservatives in bread. Great quantities of tiny black-bread-mold spores are produced in spherical cases called sporangia. Any one of these spores, when carried by wind or air currents to a suitable substance, can germinate. It then produces a fluffy mycelium similar to the one from which it came. This same fungus is destructive to market vegetables and fruits, and causes a rot of sweet potatoes.

### Sac fungi (Ascomycetes)

The "Asco" part of the scientific name of the sac fungi comes from the Greek *askos,* meaning "sac" or "bladder." All the plants belonging to this group have a number of asci (singular: ascus), or spore sacs. These are a product of sexual reproduction. In the higher sac fungi, such as the cup fungi, the asci are produced in various kinds of fruiting bodies.

*Yeasts.* The life cycle of a yeast is relatively simple. The typical thallus consists of a single small cell that reproduces by budding. The budding mass of cells that develops on the surface of a culture plate usually appears as a soft, creamy-white colony. Sexual reproduction also occurs in many species. It begins with the fusion of two similar cells; the end result is the production of several spores (usually four) within the characteristic sac, or ascus. These spores are very resistant.

There are many kinds of yeasts; they commonly occur in nature wherever simple sugars are available. They are especially to be found on fruits, in the sap from trees and in the soil, particularly in orchards and vineyards. The best-known are baker's and brewer's yeasts, which both belong to the genus *Saccharomyces*. Brewer's yeast has been bred and selected for its high alcoholic yield. When the yeast is placed in a sugar solution and is deprived of free oxygen, it carries on a special type of respiration called alcoholic fermentation. During this process, carbon-dioxide gas is given off and alcohol is produced. Brewer's yeast is the principal organism used in making beer, liquors and commercial alcohol. A different species is used in wine making.

When baker's yeast is placed in bread dough, it causes the dough to "rise" or expand by producing carbon-dioxide gas. Upon being heated, the gas pockets in the dough expand still more, making the final baked product light and porous. The alcohol produced is driven off by the heat.

Not all yeasts are beneficial; a few species cause serious diseases of man. *Candida albicans* is responsible for thrush, a disease of the mouth and throat. Other yeasts produce infections of the respiratory system. Fortunately, such diseases are not very common.

*Blue and green molds.* These are among the most common and widespread of all fungi. They occur on many different kinds of organic substances, including foods and textiles. In humid areas, they are frequently responsible for the mildewing of clothing. Although some blue and green molds reproduce sexually, the majority lack the sexual stage. They reproduce only by means of asexual spores (conidia). These are produced in chains on upright branches of the mycelium. The large numbers of mature spores give these fungi their characteristic colors — greenish, blue-green or blue, and sometimes yellow, tan or black. The blue and green molds can be readily distinguished from the black-bread-mold group because the mycelium is not fluffy and does not spread so rapidly.

A well-known genus is *Aspergillus;* another is *Penicillium,* from which the antibiotic penicillin is derived. *Penicillium* has been put to various other uses by man. Roquefort cheese is made with one species of *Penicillium;* Camembert cheese, with another. The blue color of Roquefort cheese is due to the presence of great numbers of spores. A species of *Aspergillus* is the primary source of citric acid, widely used in flavoring candies and fruits. Various other species of *Aspergillus* and certain *Penicillium* species also produce this acid.

**The mold known as *Penicillium notatum,* from which the antibiotic penicillin is derived, is shown below. At the right we see a close-up of the mycelium of this fungus. The chains of conidia, or spores, are borne at the tips of hyphae.**

*Powdery mildews.* As the name of these fungi indicates, they give a powdery white appearance to the infected parts — usually the leaves of various plants. This is caused by the presence of large numbers of asexual spores (conidia), produced in chains and distributed mainly by wind. These spores develop throughout the summer and cause the host plants to be reinfected during this period. In late summer and fall, the sexual stage appears in the form of little brown dots over the infected areas. These structures contain the ascospores, or sexual spores, which survive during the winter and cause new infections in the spring. The powery mildews are plant parasites. They cause diseases of crop plants and ornamentals, including the grape, apple, rose and lilac.

*Sphere fungi.* This is the largest group of sac fungi. The fruiting bodies that result from sexual reproduction are small and spherical; they have necklike extensions through which the spores escape. Asexual spores (conidia) are also commonly produced in this group. Some of the species are parasitic, while others live on dead organic matter. Among the parasitic varieties is the chestnut-blight organism. Introduced into New York from Asia about 1900, it has completely destroyed the American chestnut. Others species of the sphere fungi cause Dutch elm disease, oak wilt, apple scab and ergot of cereals.

One of the most useful species from the standpoint of basic science has been a pink bread mold known as *Neurospora*. A great deal of research has been done on the heredity of this organism. These studies have contributed greatly to our present understanding of the genes, the hereditary units of all organisms. It has been shown that the genes control chemical reactions within the cell and thus regulate the growth and development of the organism.

*Cup fungi.* This is another large group of sac fungi; its members are called cup fungi because the fruiting bodies are generally cup-shaped. They vary greatly in size from tiny, barely visible structures to conspicuous and sometimes brightly colored fleshy cups or saucer-shaped bodies. The larger ones are often edible. Some related forms, such as the morels, are not cup-shaped, but have stalks and conical, pitted caps. Most mushroom-gatherers consider them a delicacy.

The majority of cup fungi live on dead wood, humus and soil. Some of them cause destructive diseases. The brown-rot fungi, for example, may do millions of dollars' worth of damage annually to peaches and other stone fruits, as well as to apples and pears.

An odd group of fungi frequently classified with the cup fungi are the truffles. Their fruiting bodies are usually spherical rather than cup-shaped, and — what is particularly notable — they occur underground. They are mycorrhizal; as we pointed out, this means they are associated with the roots of higher plants, in such a way that the relationship is beneficial to both kinds of organisms. In the United States, truffles appear to be uncommon except in the Pacific-coast area. They occur more frequently in Europe.

Truffles are considered a prime delicacy. In Europe, they are sought out by means of specially trained pigs and dogs, which can locate them by smell. Most truffles that are gathered for the market are about one to two inches in diameter and show a dark, roughened exterior. When cut open, they have a distinctive aromatic odor. Their flavor is imparted to foods that are seasoned with them.

**Below is shown the fruiting body of a truffle. The mycelium of this plant is associated with the roots of various trees and shrubs in a mutually beneficial relationship. We discuss on a previous page this relationship — symbiosis.**

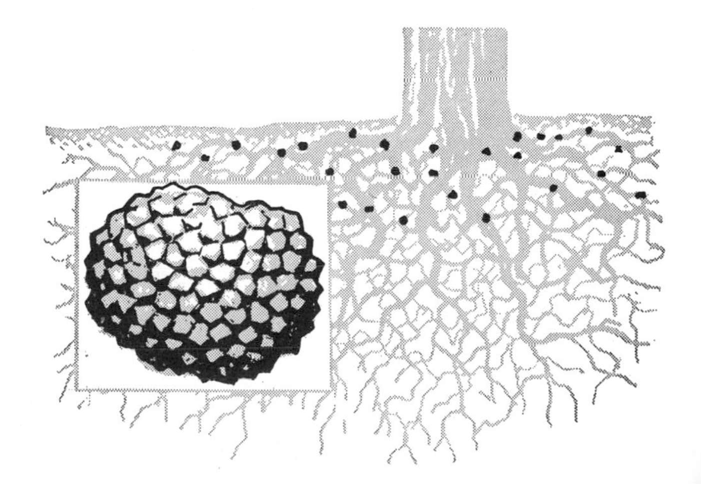

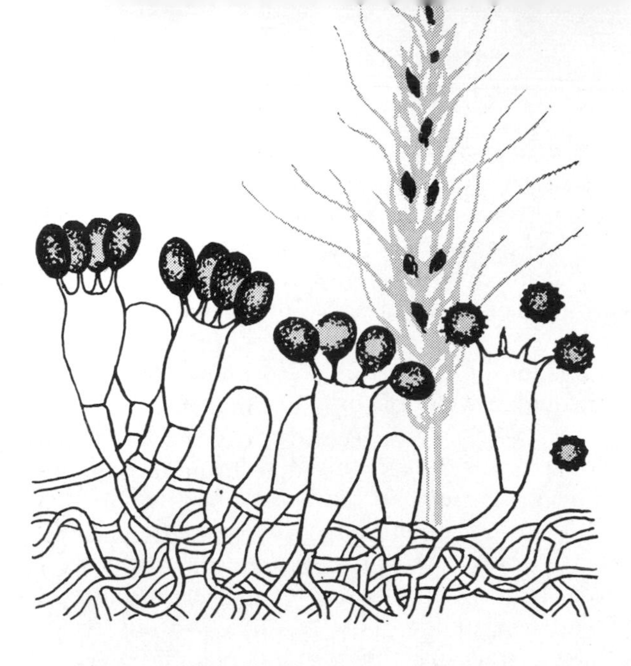

**How spores are produced in wheat smut. The large cells are the club-shaped basidia, or spore-producing organs. Each basidium bears four spores, each supported on a stalk. The basidium at the right is shedding its spores.**

## Club fungi (Basidiomycetes)

The club fungi are the most highly developed group in the evolution of the fungi. The first part of the scientific name "Basidiomycetes" is derived from the characteristic spore-producing organ, which is called the basidium and is club-shaped.

*Rusts and smuts.* These are parasitic groups belonging to the lower orders of club fungi. The rusts are so called because one of their spore stages has a bright rust-brown color. There are a large number of species and they attack a wide variety of higher plants, many of which are valuable to man. One of the most destructive of the rusts, from our viewpoint, is the wheat-barberry rust. In order to complete its life cycle, this fungus must alternate between two host plants. It produces certain spore forms on wheat and different forms on barberry. Many rusts have this complicated life cycle. They cause much damage to cereals, vegetables, fruit trees and some forest trees. Careful preventive measures help keep them in check.

The smuts have received their name because of the black masses of spores they produce on infected plant parts. Their life cycle is much simpler than that of the rusts; they are able to complete their development on the same host plant. Smut spores are dry and easily blown about from one plant to another, or from one field to the next. They are quite resistant; those of some species pass the winter on the ground. The best-known smut fungus in the United States is probably corn smut, which turns corn kernels into tumorlike structures full of black spores. Other important smuts attack such plants as wheat, oats, barley, rye and sorghum.

*Gill fungi.* As we shall see, these highly evolved forms are often referred to as mushrooms. They are among the commonest and most conspicuous of all fungi, occurring abundantly in woods and meadows. The vegetative portion of a gill fungus — the mycelium — occurs in the ground and lives on organic wastes. At certain places in the mycelium swellings are produced. These develop into small sporophores (spore cases). In some of the more common species, these are enclosed in wrappers called volvae. Under proper conditions of warmth and moisture, a sporophore bursts through the earth, rupturing the volva, and it expands in umbrella fashion. The fully developed fruiting body consists of a stalk and cap. Under the cap are attached a number of thin gill-

**Longitudinal section of *Amanita caesaria*, a favorite edible mushroom, showing some of its important structures.**

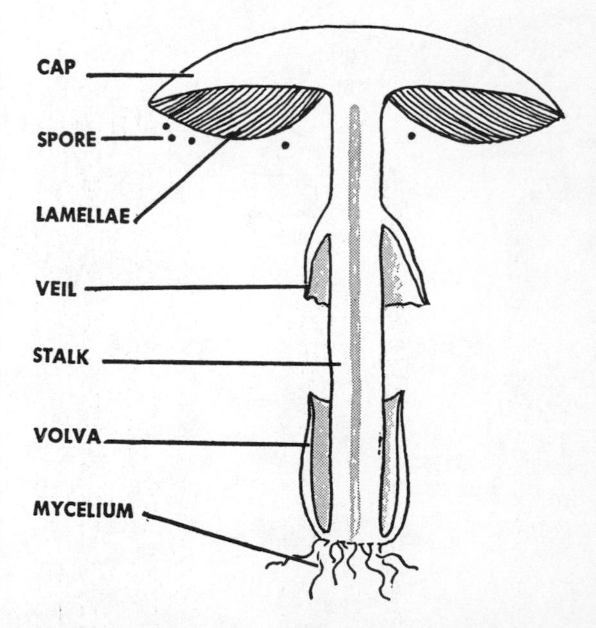

like plates, called lamellae, which radiate from the center. It is on these lamellae, or "gills," that the spores are produced. Gill fungi are extremely varied in size and color; many of them are among the brightest and most attractive of the fungi.

*Coral fungi.* The fruiting bodies of these plants are much branched and sometimes strikingly resemble coral in appearance. They are creamy, tan, yellow, brown or even purplish in color and fleshy in texture. Coral fungi can be found most often in temperate areas in woods made up of deciduous trees — those that lose their leaves every autumn.

*Bracket or leather fungi.* Their fruiting bodies often have a leathery texture. They are often found on the sides of dead trees and logs; they grow in shelflike fashion, one shelf rising above another. Bracket fungi are an important factor in wood decay.

*Pore fungi.* These plants are also important as wood-decay organisms. The spores are produced in pores on the underside of the fruiting bodies (sporophores). These bodies vary in shape. Some are thin layers; others resemble gill fungi or bracket fungi. Pore fungi may cause great damage to lumber and to wooden structures; certain species attack trees.

*Puffballs.* Most of these forms occur as saprophytes on the ground, on humus or on rotting wood. The fruiting body of a typical puffball, such as *Lycoperdon,* ranges in shape from spherical to pearlike; it is whitish at first and turns brown at maturity. Inside the mature fruiting body are millions of tiny spores. These are expelled through a small opening at the top by pressure from wind, raindrops or animals. Most puffballs are only a few inches high. The giant puffball, however, may reach a diameter of four and a half feet.

*Stinkhorns.* These fungi, which are related to the puffballs, are well named for their carrionlike odor. Most stinkhorns have a stalk, usually with a cup at the base, and a cap covered with a dark, slimy mass of spores. It is this spore mass that causes the unpleasant odor.

*Bird's-nest fungi.* They are also related to the puffballs. They produce small cups, containing several tiny bodies in which the spores are borne. The entire cup structure resembles a miniature bird's nest with several tiny eggs. These fungi may be found on humus, dead wood and the dung of herbivorous animals.

*Mushrooms and toadstools.* The name "mushroom" refers to various fungi that have fleshy fruiting bodies, growing on

The drawings on these two pages show a variety of mushrooms. Those at the extreme right — *Amanita phalloides* and *Amanita muscaria* — are poisonous.

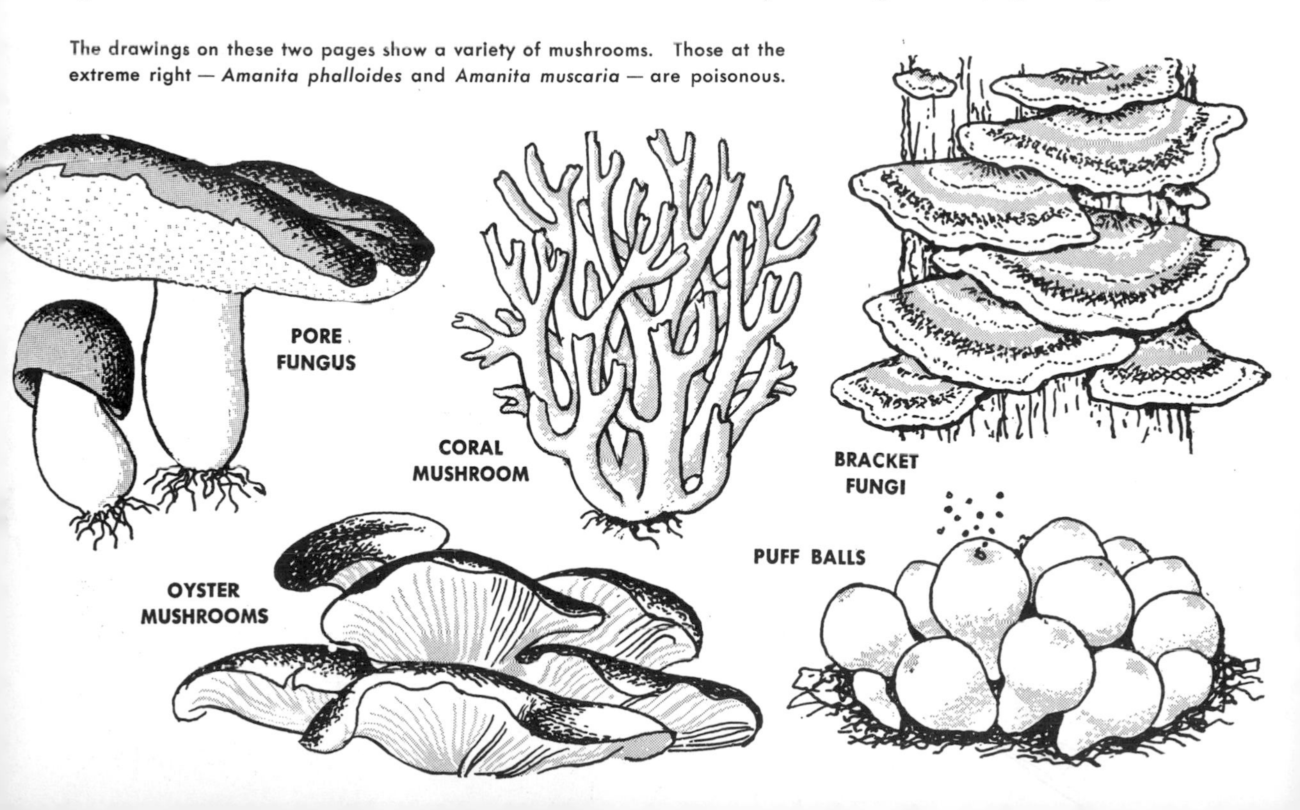

the ground, or on dead wood or on trees. The name is used most often in connection with the gill fungi described above. However, it is also applied to various other club fungi and even to certain sac fungi, such as the morels. Some persons reserve the name "mushrooms" for the edible species of these fungi; they call the poisonous species toadstools. Most students of the fungi, however, speak of poisonous and edible mushrooms.

Contrary to the belief of those unfamiliar with mushrooms, the great majority of species are edible and relatively few are very dangerous to eat. However, since some of the poisonous species are widespread, the amateur mushroom hunter should be careful in choosing specimens for the table. He should purchase a mushroom guide* and learn to identify a few common edible species; he should pick only these mushrooms and shun all others. Only when he can positively identify other edible varieties should he add them to his list. Unfortunately, there are no easily applied tests for distinguishing between edible and poisonous mushrooms, if one is not sure of their identity.

*A useful guide is the *Fieldbook of Common Mushrooms,* by William S. Thomas, G. P. Putnam's Sons, 1948, New York.

The safest and certainly one of the most palatable is the common edible mushroom *(Agaricus campestris),* which may be bought at the grocery store. Its wild variety is common in nature, especially in meadows and lawns. Other wild edible mushrooms that are easy to identify are the morels, the coral fungi and the oyster mushroom, which forms whitish fruiting bodies with stubby stalks on dead logs. The puffballs should also be included. One of the most delicious of all mushrooms is the bright orange-yellow Caesar's amanita *(Amanita caesaria),* which was a favorite of the Roman emperor Nero.

Unfortunately other species of *Amanita* are among the most poisonous varieties. The fly amanita *(Amanita muscaria)* is quite similar in color to *Amanita caesaria.* It has a bright orange-red cap, covered over with whitish flecks. Like most species of *Amanita* it has a veil around the stem, just below the cap. The base of the stalk is enlarged and bulbous. The deadliest of all mushrooms is the destroying angel *(Amanita phalloides),* which ranges in color from white to grayish; it has a veil and also a prominent cup at the base of the stalk. Although it is easy to recognize, it is probably the commonest cause of serious mushroom poisoning in North America. A

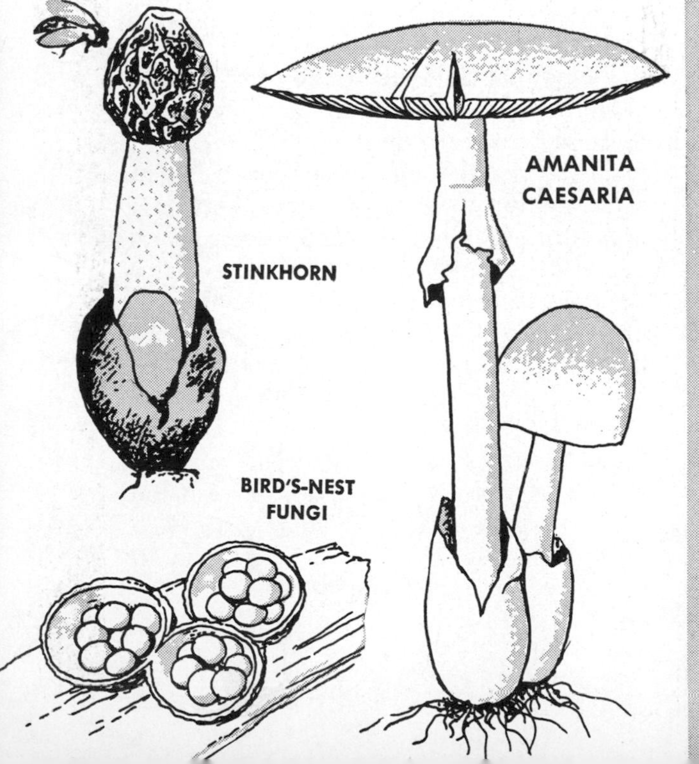


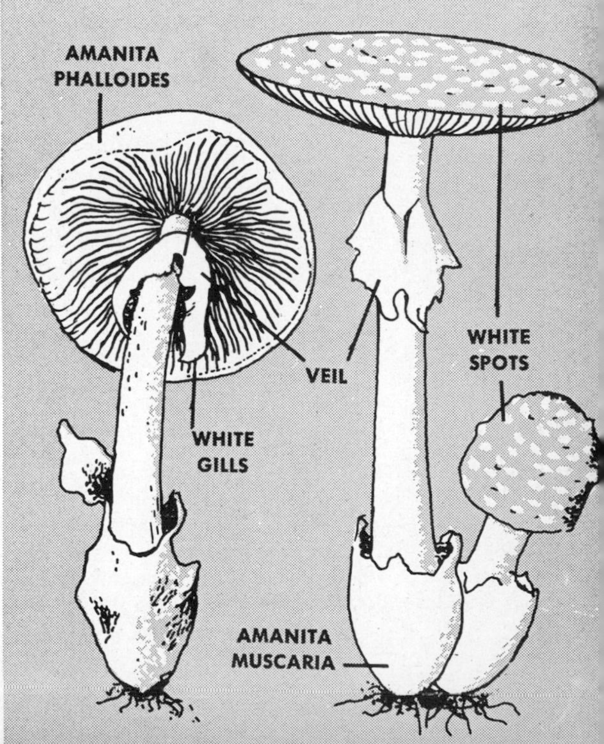

single forkful is sometimes sufficient to cause an agonizing death.

Symptoms of poisoning by mushrooms are often delayed for hours after the fungi are eaten — perhaps for an entire day. Whenever mushroom poisoning is suspected, a doctor should be called at once. In the meantime, any effective means of emptying the stomach of its contents is recommended.

It is interesting to note that certain species of mushrooms may cause nausea in some persons, usually without serious results, while they may be eaten by others with no ill effects whatsoever. It is possible that a food allergy is involved here.

Mushrooms may be grown for the market in caves, abandoned mines, dark cellars and similar places. In the United States, most commercial growing is done in specially constructed houses in which the temperature and humidity can be controlled. The mushroom beds are generally made by spreading a layer of well-rotted plant manure and loam over a deep layer of animal manure. After fermentation has taken place, the mushroom "spawn" is planted. This consists of the mycelium of the fungus, encased in bricks made of animal manure and loam. The mushrooms are harvested while still firm and before the spore dust has started falling from the gills.

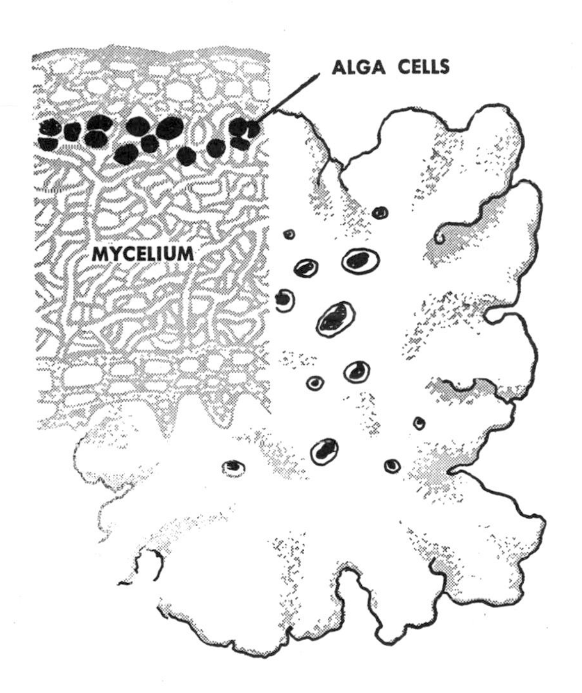


**The large drawing, above, shows the thallus, or plant body, of a lichen. In the insert, we see a cross section of the thallus, with the filaments of the mycelium making up the framework. Note the alga cells in the mycelium.**

### Imperfect fungi (Fungi imperfecti)

This group receives its name from the fact that the sexual stage is lacking and reproduction is by means of asexual spores (conidia) only. Most imperfect fungi seem to have evolved from the sac fungi through the loss of the ability to reproduce sexually.

The imperfect fungi form a very large and important group. They occur widely in nature either as saprophytes on many different kinds of organic matter or as parasites of higher plants. Some cause mildewing of fabrics; others, molds of foods. The parasitic species are often very destructive to crop plants and ornamentals. Most fungi that cause disease in man belong in the group of the imperfect fungi. They are responsible for all of the ringworm diseases, including athlete's foot.

### Lichens

A lichen is a combination of a fungus and an alga. Most often, the fungus is a member of the sac fungi, while the alga is a green or blue-green species. The thallus, or plant body, that results from the combination of the two is quite different from anything that could be produced by the fungus or the alga growing alone. The hyphae of the fungus make up the framework of the plant body; the algal cells or filaments occur within this framework. The two organisms — fungus and alga — live together in a mutually beneficial relationship. By the process of photosynthesis, the alga manufactures all the organic food that they both require; the fungus brings in water and minerals and offers protection to the alga. Lichens are most commonly found on trees; they also occur on rocks and on barren ground. They furnish fodder for reindeer and, in some places, for cattle. Certain kinds of lichens yield litmus, used as a chemical indicator.

*See also Vol. 10, p. 272:* "General Works."

# IMPORTANT METALS

## Valuable Resources in the Earth's Crust

AMONG the most useful and important substances in the earth's crust are the metals. Man has used them to make his tools and weapons ever since, thousands of years ago, he learned the superiority of metal to stone. So valuable are metals to man that some have served as milestones of human progress. Thus, we give the name of Bronze Age to the period when men first discovered this alloy of copper and used it on a wide scale; we speak of the epoch when iron implements came into general use as the Iron Age.

In the language of chemistry, a metal is an element that has certain "metallic" properties. (A mixture of two or more metals is called an alloy, to distingush it from a pure metal.) All the metals except mercury are solid at ordinary temperatures. Most of them have low melting points, as compared with nonmetals. Metals are usually lustrous, have a definite crystal structure and are much better conductors of heat and electricity than nonmetals.

This last property of metals is due to the way their atoms are made up. The electrons in the outer shells of metallic atoms are always few in number and loosely bound to the nucleus; thus, they are quite easily given up by the atoms. (We discuss electronic shells in the article Inside the Atom, in Volume 1.) In fact, when a small amount of energy is absorbed by a group of metallic atoms, such as those inside a copper wire, a number of electrons are liberated. These free electrons form a kind of electrical gas inside the metal, similar to the molecules of an ordinary gas. When an electric field is applied to the metal, the free electrons drift in the direction of the field, thus making up an electric current. The metallic property of giving up outer electrons is known as electropositiveness, since the loss of negatively charged electrons leaves the atoms with a net positive charge.

Metals conduct heat so well because of the same free-electron gas. The "hotness" of a body is merely an expression of how fast the particles inside it are moving or vibrating — the greater the average speed, the "hotter" the body. When a metal object is heated, the increased speed is absorbed by the electron gas and quickly carried to all parts of the body.

Most metals withstand great pressure before they are deformed, and resume their original shapes almost perfectly when the pressure is removed — that is, they are hard and elastic. However, heat weakens a metal and makes it more plastic; it can then be molded into new shapes much more easily. Even when they are cold, most metals have some degree of plasticity; this is shown by the fact that many metals are malleable, that is, can be hammered out into thin sheets without crumbling. Another plastic quality that many metals show is ductility — a piece of the metal may be drawn, without breaking, into a fine wire by stretching its ends apart.

Of all the known elements, the great majority are considered metals. Elements, such as chlorine and oxygen, that have no metallic properties are called nonmetals. Certain elements, such as arsenic and germanium, have only a few metallic properties and show them rather weakly; these are sometimes called metalloids.

There are two special groups of metals, known as the alkali and alkaline-earth groups, that differ greatly from the more typical metals. They are taken up in another article of THE BOOK OF POPULAR SCIENCE. These metals are amazing for their softness and their extreme electropositiveness; most of them are violently active when pure. Lithium, sodium, potassium, calcium and magnesium are the leading representatives of these groups.

The remaining metals include some, such as iron and aluminum, that are as abundant as common clay and familiar to almost everyone. Others, such as osmium and iridium, are rarities in the earth's crust and were virtually unknown until recently.

**Gold — the most famous of metals**

Gold was probably discovered by men long before any of the other metals, since large nuggets of pure, uncombined gold are sometimes found close to the surface of the earth. The scarcity of this beautiful yellow metal, its brilliance and unchanging nature have made it a standard of value since the earliest times.

Before the rarer elements were known, gold was considered the most "noble" of metals, for it can be dissolved only by aqua regia, selenic acid and a few other substances. It is among the heaviest of the metals (osmium is the heaviest), with a specific gravity of 19.32; this means that a volume of gold weighs 19.32 times as much as an equal volume of water. As a conductor of heat and electricity, gold is surpassed only by silver and copper. All three metals belong to a single chemical family, Group IB of the periodic table, known as the copper group.

There is no more malleable and ductile metal than gold. A single grain ($1/438$ of an ounce) can be drawn into 500 feet of wire. A piece of gold can also be beaten to less than $1/100{,}000$ inch in thickness. Such fabulously thin pieces of metal, known as gold leaf, are used to decorate leather, wood, metal and ceramics in the finest kind of gilding. So thin is gold leaf that it transmits light.

Gold is too soft to be used pure, even in "solid gold" objects; for extra hardness, it is usually alloyed with its sister metals, copper and silver, and often with platinum. The best gold jewelry is 600 to 800 fine, that is, contains 600 to 800 parts of gold per thousand; United States coinage gold was 900 fine. Gold purity is also expressed in carats, or parts per 24. Thus, an ounce of 18-carat gold alloy contains $18/24$ of an ounce of gold. Alloying metals produce a variety of beautiful colors in gold. Platinum and palladium give white; aluminum gives a purple color; copper, a deep yellow or reddish color; and silver, a pale yellow or green.

Like most of the pure metals, gold has a rather low melting point (1,063° C.) and boiling point (2,600° C.). Hot gold vapor has a deep purple color.

The costliness and structural weakness of gold limits its uses almost entirely to the arts. In addition to solid gold and gold leaf, coatings of lustrous, corrosion-resistant gold are given in many ways: plating with gold sheets, electroplating, painting or dipping in molten gold and spraying with gold vapor. Radioactive colloidal gold is sometimes used in treating cancerous growths or tumors. Gold alloys are preferred by many people for filling large dental cavities. We tell about the mining and refining of gold elsewhere in THE BOOK OF POPULAR SCIENCE.

**Silver — the pale sister of gold**

Silver has been known to man almost as long as gold itself. Silver is only slightly more active chemically; the metal is found free in nature almost as abundantly as gold, and the two often occur in a natural pale yellow or white alloy, called electrum. The symbol of silver, Ag, comes from its Latin name, *argentum*. Silver also occurs in the widely distributed mineral argentite, or silver sulfide.

Pure silver is not so inert as gold, but neither air nor water will corrode it. (Silver tarnishes because traces of hydrogen sulfide gas in the air react with it to produce a thin film of black silver sulfide.) The luster of silver is unsurpassed and it is the whitest of all metals. For all these reasons, it was considered the next most noble metal after gold by the

Surface plant of a nickel mine in Ontario. About 80 per cent of the world's nickel comes from this area.

Internat. Nickel Co.

## METALS THAT ARE FOUND IN CANADA

George Hunter

George Hunter

View of mining operations at the cliffs of the Steep Rock Iron Mines, west of Port Arthur, Ontario. High-grade iron ore is extracted from these great mines.

Much of the world's uranium is processed from the pitchblende deposits of Canada. Left: surveying for pitchblende at the Ace Mine, in Saskatchewan.

medieval alchemists and still serves as a standard of value in many countries. Some of the oldest coins in the world are made of this metal. Perhaps even more widely than gold, silver has caught the fancy of artists and craftsmen; finely wrought filigree, massive bracelets and medallions, delicate vessels and tableware and every other imaginable ornament are included in the silversmith's art.

Though it is harder and stronger than gold, silver is soft when compared to most other minerals and metals — its hardness is rated 2.5 on the Mohs scale (see Index). Silver sheets have been beaten to less than $1/10{,}000$ inch in thickness, and a single gram of silver (about $1/28$ of an ounce) can be drawn into more than a mile of wire. Thus, silver is second only to gold in malleability and ductility. As a conductor of heat and electricity, silver knows no equal; for this reason, it is used in the finest electrical wires and contacts, when the high price of the metal is not important. Silver melts at 960.5° C. and boils at 1,950° C., giving a blue vapor.

Too costly for any wide-scale industrial use, unalloyed silver is also too soft for most artistic purposes. The purest sterling silver is 925 fine (See Index, under Sterling silver). Silver coins of the United States are 900 fine. Copper is the alloying element in these cases, greatly increasing silver's strength; as much as 45 per cent copper can be added without affecting the whiteness of the alloy. Silver coatings are made by most of the processes used for gold.

Since silver is more actively electropositive than gold, it forms a number of chemical compounds. Silver nitrate, $AgNO_3$, is perhaps the most important, as a source for all the others. It is called lunar caustic (silver was known to the alchemists as Luna, the moon) and is used in medicine for cauterizing warts and other growths, as an antiseptic and for treating hemorrhages and ulcers. Compounds of silver with chlorine, bromine and iodine are very sensitive to light and are used in manufacturing photographic emulsions for coating plates and films.

**Copper — for saucepans, solenoids and statues**

Red-colored copper is the first and most common member of the copper group. The name copper and its chemical symbol, Cu, are taken from the Latin word *cuprum*. Like its sister metals, copper was discovered by man in the late Stone Age. Bronze, an alloy of copper and tin, is much harder and stronger than pure copper. This most valuable alloy had been developed by 3,000 B.C. As we have pointed out, it was used widely (for weapons and for various kinds of tools) in the cultural stage called the Bronze Age.

At present, more copper is used by the world than any other metal except iron. All of this must be smelted from copper ores, consisting mainly of sulfides and oxides, which are plentiful in the soil and igneous rocks throughout the world. Pure copper, which melts at 1,083° C. and boils at 2,336° C., is more active than silver, but less so than all the other common metals; free copper is very rare in nature.

The extreme malleability of copper (third among the metals) is rendered even greater when a little zinc is added; this alloy, known as Dutch metal, can be beaten into a foil resembling gold leaf. Copper is one of the best conductors of heat and electricity, second only to silver. Since copper is also more abundant, cheaper and stronger than silver, it is the choice metal for electrical wire, coils, contacts and armatures. When copper is exposed to air or water, a thin film of basic copper carbonate, or patina, forms on its surface; this is hard and insoluble and protects the metal from further corrosion. Patina is responsible for the beautiful greenish color of copper roofing and bronze statues. In statuary it is treasured for its artistic value.

Relatively pure copper is strong enough for pots and pans, but for more rugged duty it must be alloyed. Mixtures of copper and zinc are known as brass; they are hard, strong, machinable and hold a bright finish. Different brasses are used for cartridges, steam condensers, musical instruments and fittings exposed to the corrosive action of sea water.

Alloys of copper with other elements are known as bronze. A common bronze, containing 10 per cent tin, has been cast into statues since ancient Roman times. In recent years, alloys of copper and new metals, such as beryllium and aluminum, have been developed. These vie with steel in strength and endurance; they are being used for gears, propellers and other heavy-duty parts. An alloy of copper, manganese and nickel — Manganin — has the same electrical resistivity at almost any temperature, making it ideal for resistors in delicate electrical instruments.

Almost all copper salts are poisonous to animal life, in high enough concentrations; copper-arsenic compounds are used as powerful insecticides. However, small amounts of copper are vital to the metabolism of some organisms. Copper is part of the compound hemocyanin, found in the blood of certain crustaceans and mollusks; it is also found, in small quantities, in vertebrates, including man.

**The basic metal of the modern world — iron**

If we could choose a single metal on which the modern world depends most, it would certainly be iron. Pure iron (symbol Fe, from the Latin *ferrum*: "iron") is little more than a chemical curiosity and has almost no uses; yet, when we add carbon and other elements to iron, we create the marvelous alloys known as steel, whose special qualities and uses have filled many books. Iron is not a modern discovery; it has been known and used by men all over the world since the beginning of the Iron Age.

Since iron is much more active than the copper-group metals, it almost never occurs free in nature. It is extracted from iron oxide ores, such as hematite and magnetite; these form a good proportion of our earth, for iron is the fourth most abundant element in the earth's crust.

Metallurgists find it impossible to produce absolutely pure iron, because the attraction between iron and certain elements is so strong. But a very high-purity iron can be achieved by electrolytic refining or by a method known as the carbonyl process. This iron appears silvery, melts at 1,535° C. and boils at 3,000° C. It is highly lustrous, a fair conductor of heat and electricity, moderately malleable and exceedingly ductile.

The native strength and hardness of iron, which is more than twice that of gold, can be made twenty times greater if the proper alloying elements are added. Unlike the copper-group metals, pure iron exists in four different crystal arrangements, depending on the temperature. The various crystal forms in a piece of steel influence its structural properties; and the effects are multiplied when various elements are added. Like many other metals, iron can be hardened by cold working; but unlike many others, it can also be hardened by heat treatment, since this produces changes in its crystal structure.

**Solving the problem of rust**

Perhaps the only fault of iron as a structural metal is its chemical activity; pure iron will be dissolved by almost any strong acid, and is attacked by oxygen. The layer of rust, or iron oxide ($Fe_2O_3$), that forms is easily penetrated by more oxygen, and the rusting may continue until the iron is entirely eaten away. However, iron may be protected by coatings of corrosion-resistant materials such as zinc, tin, terne or enamel, or by the addition of elements that make the iron itself resistant, such as chromium or nickel. The varieties of steel, their uses and the effects of different alloying metals are discussed in the article The Story of Steel, in Volume 1.

Among its other virtues, iron is the most paramagnetic substance known; that is, it increases the strength of a magnetic field passing through it, to a greater degree than any other substance. When properly treated, iron may become permanently magnetic itself. Electromagnets, transformers and solenoids of all kinds are supplied with magnetic cores of iron.

Iron compounds are valuable mostly as a source of the metal, but some have

## METALS ARE FOUND IN

British Information Services

A native worker panning in a north Nigerian stream for tin concentrate, which contains the valuable mineral columbite.

Caterpillar Tractor Co.

A bulldozer working in the pit of a bauxite mine, near Sweet Home, Arkansas. Bauxite, an ore of aluminum, is one of the more important metals.

St. Joseph Lead Co.

Scene in a Missouri lead mine. Lead is used in pipes and in storage-battery plates; its compounds are made into paints.

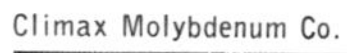

Climax Molybdenum Co.

## EVERY CORNER OF THE GLOBE

Denver and Rio Grande Western Railroad

Part of the workings of a zinc mine at Gilman, Colorado. The ore is dug out of the flanks of Battle Mountain.

National Lead Co.

Magnetic drum separating titanium and iron-ore particles from rock at a titanium mine at Tahawus, New York. Rock particles drop from the drum into the trough beneath.

Anaconda Mining Co.

Miner drilling dynamite holes preparatory to blasting at Butte, Montana. Copper has been used since antiquity.

Separating molybdenum concentrates from rock particles in flotation cells. The molybdenum concentrates float in a layer of oil; the rock particles sink to the bottom.

special uses of their own. Ferric oxide, $Fe_2O_3$, is used in making the cheaper red pigments and the most expensive rouges. A number of iron compounds, such as ferrous sulfate (green vitriol) and ferric ferrocyanide (Prussian blue), are used to color inks, ceramics and glass. It is because of the iron in hemoglobin that this pigment of red blood cells combines with oxygen and carries it to the body tissues.

### The modern wonder metal — aluminum

Leaving the renowned ancient metals, whose uses are as old as history itself, we come now to a truly modern one — aluminum, which was first isolated in 1825. It was not until the late nineteenth century that methods for refining aluminum on a large scale were developed. Aluminum is a highly active element with a powerful affinity for oxygen. Hot aluminum burns brilliantly in the air, like a magnesium flare. Combustion also occurs at lower temperatures, when the aluminum is finely powdered, and yields tremendous heat. This powder is used in Thermit mixtures, for welding steel and in incendiary bombs.

Aluminum is the third most abundant element in the earth's crust; alumina (aluminum oxide, $Al_2O_3$) accounts for about 15 per cent of the crust's weight. Of course, since aluminum is so active, it is never found free. Some of its natural compounds are very common minerals, such as the feldspars and micas, kaolin (clay) and cryolite. Rubies and sapphires are almost pure alumina, and corundum is totally pure. Corundum is even harder than tungsten carbide, and can be scratched only by diamond or Carborundum; pure corundum, or the impure form known as emery, makes an excellent abrasive. Bauxite, or hydrated alumina, is the chief ore from which aluminum is refined.

Aluminum is too active to be liberated from its oxide ore by the common chemical methods. Instead, the Hall-Héroult process is used. Pure alumina is dissolved in molten cryolite; a heavy current is passed through the solution and reduces the aluminum, which then settles to the bottom and is tapped off. The pure metal is silvery white, slighty harder and stronger than copper (2 to 3 on the Mohs scale) and a little less malleable and ductile. It melts at 659.7° C. and boils at 2,057° C.

As a conductor of heat and electricity, aluminum is about 60 per cent as efficient as copper. But the truly outstanding property of this metal is its lightness; the specific gravity of aluminum is only 2.7, as compared with 8.9 for copper and 19.3 for gold. Adding the virtues of lightness and strength to all its other qualities, aluminum is fast replacing copper in long high-tension electrical wires.

Although aluminum itself is quite active, objects made of aluminum are not easily corroded. A super-thin film of aluminum oxide is formed on any surface exposed to water or air; this film is hard and insoluble in most acids and alkalies, and thus protects the object from further corrosion. These qualities make aluminum an ideal metal for pots and pans, cooking utensils and decorative metal trim. Aluminum, fabricated into rolls of thin foil, is used for packaging and preserving foods and other perishable items.

Strong alloys of aluminum have been developed, such as Magnalium, Duralumin and Alcoa-75S, that are comparable to soft steel but retain their lightness and corrosion resistance; they are used in airplane and automobile parts. Aluminum alloys are also acquiring added importance in the building industry, as they are used for frameworks, facings and all kinds of prefabricated sections.

### Some softer metals with many uses

Together with copper and iron, aluminum completes the list of really outstanding metals, considering the staggering quantities of each that the world uses. Zinc, lead and tin head the list of remaining metals. All three are very much alike, though zinc is a little more active than iron; tin and lead are somewhat less so.

All of these metals lack tensile strength and are soft and malleable, except that zinc tends to be brittle when cold.

Lead is so soft that a man can weld two pieces together merely by pressing them against each other in his hands. Zinc is the best electrical conductor of the three, but all are rather poor, compared to other metals. Zinc melts at 419.5°, lead at 327.4° and tin at 231.9° C. Because lead and tin melt so easily, they are ideal for cheap, rapid castings. Printing type metal is a lead-tin-antimony alloy; the most common kind of solder is half lead and half tin.

#### Zinc, tin and lead act as defenders

All three of these elements are used as protective coatings on other metals, particularly iron. Zinc and tin can be applied by electroplating or dipping in the molten metal; these are the most common methods of making tin plate and galvanized iron (zinc-coated iron). (Zinc boils at so low a temperature that objects can be coated in zinc vapor, a process known as sherardizing.) Lead is usually cast or mechanically bonded in sheets. Terneplate, metal coated with 80 per cent lead and 20 per cent tin, is applied by the hot-dip method and is used for gasoline tanks.

Most of the world's zinc is employed in galvanizing iron and making brass. However, zinc also serves us in dry-cell batteries and in refining gold. The oxides and carbonates of zinc, and those of lead, have long been used in making white paints and pigments and in coloring ceramics.

During Roman times, lead became the foremost metal for water-carrying pipes and has remained a leader to this day. In fact, the word "plumbing" is taken from the Latin word for lead, *plumbum*. Today, copper pipes are used for drinking water, and lead pipes only for wastes and chemicals, as lead compounds are very poisonous. Flexible lead tubing is also employed to protect underground cables. Lead has an extraordinary ability to absorb atomic radiations; it serves as shielding for atomic piles, cyclotrons and X-ray machines. Much lead is also used in the manufacture of storage-battery plates, bullets and tetraethyl lead, the compound that is added to gasoline to prevent knocking.

Unlike lead and zinc, tin is a bright metal; tin plate and tin foil are highly reflective. Mixed with mercury, tin is used for silvering inexpensive mirrors. The many useful alloys of lead and tin vary from soft antifriction metals, such as babbitt, to sturdy serviceable pewter.

#### Nickel, a tough and versatile metal

No coin of the United States is less easily tarnished than the common "nickel" (which contains only 25 per cent nickel — the balance is copper); this fact emphasizes the truly stainless qualities of nickel and nickel-plated objects. Nickel-copper alloys were used for coins thousands of years ago, but pure nickel was unknown till modern times. The lustrous, silvery white metal is very resistant to chemical attack, extremely tough and strongly magnetic. Many new alloys have been developed to take advantage of these properties. Nickel steels are among the strongest and toughest known. Another nickel alloy, known as Alnico, makes excellent permanent magnets. Alloys with chromium and iron, such as Chromel and Nichrome, are used for electrical heating elements in irons, toasters and the like. Monel, a bright, stainless alloy with copper, serves in many modern kitchens, restaurants and chemical plants.

#### The strangest metal of all — mercury

The Greeks called mercury "water silver" (*hydrargyrum,* from which the chemical symbol, Hg, is derived); the equivalent in English is "quicksilver." These names describe accurately the only metal that is liquid at room temperature (it melts at —38.9° C.); mercury is 13.5 times denser than water, and thus the heaviest liquid known. It is sometimes considered noble due to its inertness; it will not wet glass, and is only slightly volatile. (Even so, the vapor is extremely poisonous and should be handled with great caution.) All these qualities give mercury a unique importance in many scientific devices. Chief among these are the instruments for measuring fluid pressure, such as barometers, hydrometers and vacuum gauges. (To measure atmospheric pressures by water requires a column about 36 feet tall, while 32 inches of mercury suffices, because of its greater density.) Since it expands like a typical metal when heated, mercury is used in the best liquid thermometers.

Mercury has the highest electrical resistivity of all metals, and, in fact, is used as the international standard of resistance. Ionized mercury vapor is a much better conductor than the liquid; both are used in the high-power electronic rectifier tubes known as ignitrons (see Index). Mercury-vapor discharge tubes are a good source of ultraviolet light; when coated with special chemicals, they serve as fluorescent lamps. Mercury dissolves many metals at room temperature, yielding alloys known as amalgams. This process is employed in the extraction of silver and gold, in the preparation of certain kinds of tooth fillings and in the silvering of mirrors.

### The noble metals of the platinum group

Platinum was first introduced into Europe from South America during the eighteenth century; it takes its name from the Spanish word *platina,* or "little silver." It is the most abundant of six rare, noble, heavy metals known as the platinum group; the other five are ruthenium, rhodium, palladium, osmium and iridium. Osmium is the densest substance known, with a specific gravity of 22.5.

The inertness of the platinum metals is famous; they are attacked only by hot aqua regia and the strongest alkalies. Platinum is stronger, harder and melts at a higher temperature (1,773.5° C.) than gold, but is quite malleable and the third most ductile metal. Very expensive and enduring jewelry is fashioned with this metal. Crucibles and wires of platinum are used for making the most sensitive chemical tests. Platinum is a good electrical conductor; it serves for contacts and wires in electron tubes and heating elements. In surgery, platinum and its alloys with osmium and iridium are used for artificial bones; because of their inertness they will not poison the body and are not dissolved by the body fluids. The permanent standards for weights and measures are made of similar alloys. Osmium-iridium is replacing gold for the material of fountain-pen points. Tantalum is a rare metal often substituted for platinum alloys in constructive surgery. Tantalum carbide is so hard that it serves for the tips of steel drills.

Tungsten is the hardest, strongest and most difficult to melt of all metals, melting at 3,370° C. and boiling at the fantastic temperature of 5,900° C. Like its twin metal, molybdenum, it is alloyed in the hardest, heat-resisting steel for cutting tools. Tungsten is a fine electrical conductor and, due to its negligible evaporation even at white heat, serves for the filaments of incandescent lamps and electron tubes. Tungsten is also used in the anticathode targets of X-ray tubes.

### New uses for metals in the Atomic Age

In this era of nuclear science and electronics, some metals are finding hitherto undreamed-of applications. The metalloid germanium has helped open whole new vistas in the complex field of electronics; its semiconducting properties are used in the tiny new electronic devices called transistors (see Index). Cadmium, a more familiar metal, serves to absorb atomic radiations, particularly neutrons. Cadmium rods are employed to control the rate of fission in atomic piles.

The most spectacular rise to fame has of course been that of uranium. This little-known metal, which looks somewhat like steel but is more brittle, was long familiar only to potters, who used its oxides to produce colored glazes of yellow, brown and green. Chemists also knew that many uranium compounds become fluorescent. Then it was discovered that uranium is naturally radioactive, disintegrating into radium. Recently, a new era was begun with the discovery that uranium could undergo nuclear fission and release vast stores of atomic energy for use in war and peace. (See Index, under Atomic Energy.)

*See also Vol. 10, p. 279:* "Elements."

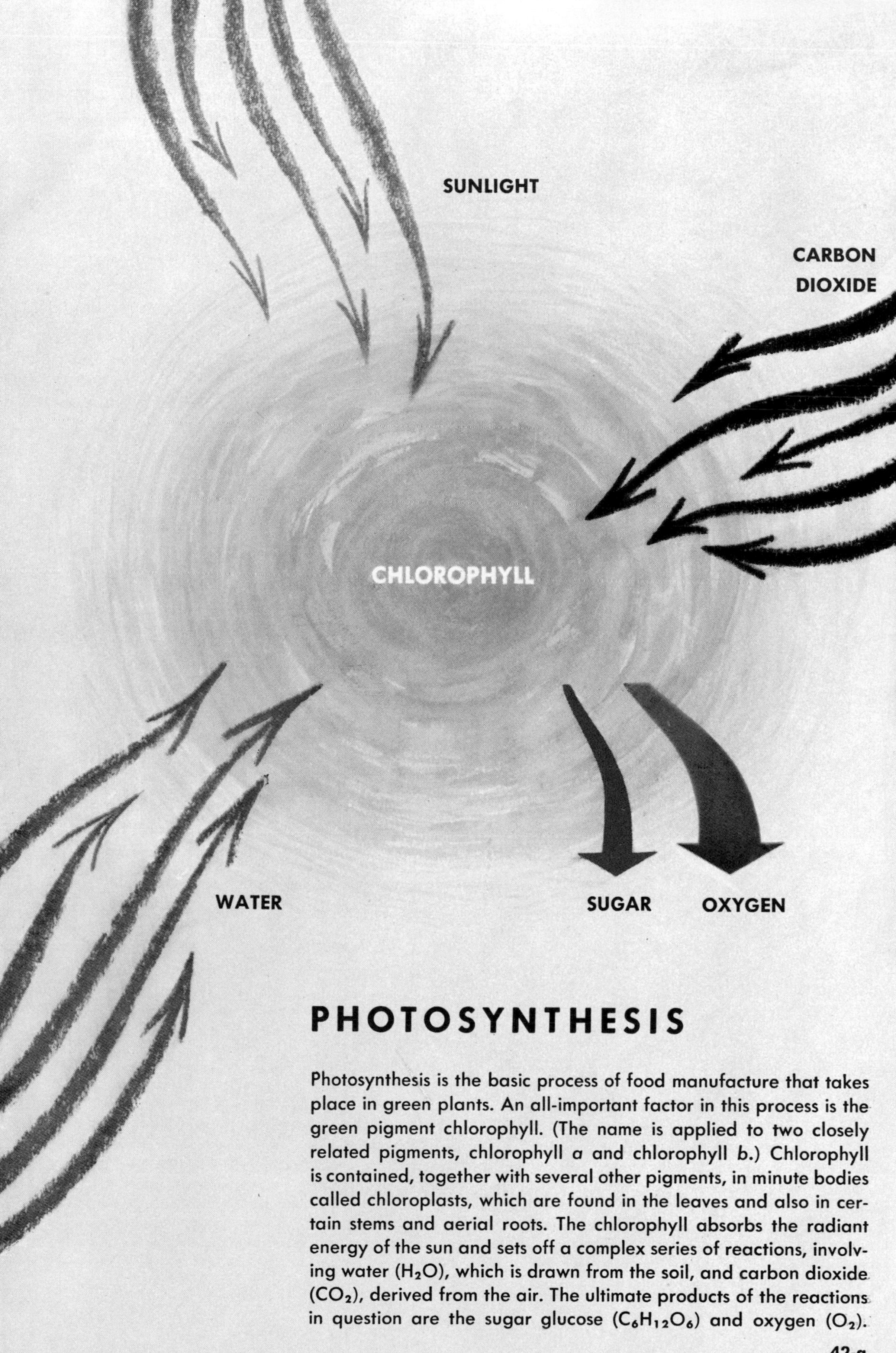


# PHOTOSYNTHESIS

Photosynthesis is the basic process of food manufacture that takes place in green plants. An all-important factor in this process is the green pigment chlorophyll. (The name is applied to two closely related pigments, chlorophyll *a* and chlorophyll *b*.) Chlorophyll is contained, together with several other pigments, in minute bodies called chloroplasts, which are found in the leaves and also in certain stems and aerial roots. The chlorophyll absorbs the radiant energy of the sun and sets off a complex series of reactions, involving water ($H_2O$), which is drawn from the soil, and carbon dioxide ($CO_2$), derived from the air. The ultimate products of the reactions in question are the sugar glucose ($C_6H_{12}O_6$) and oxygen ($O_2$).

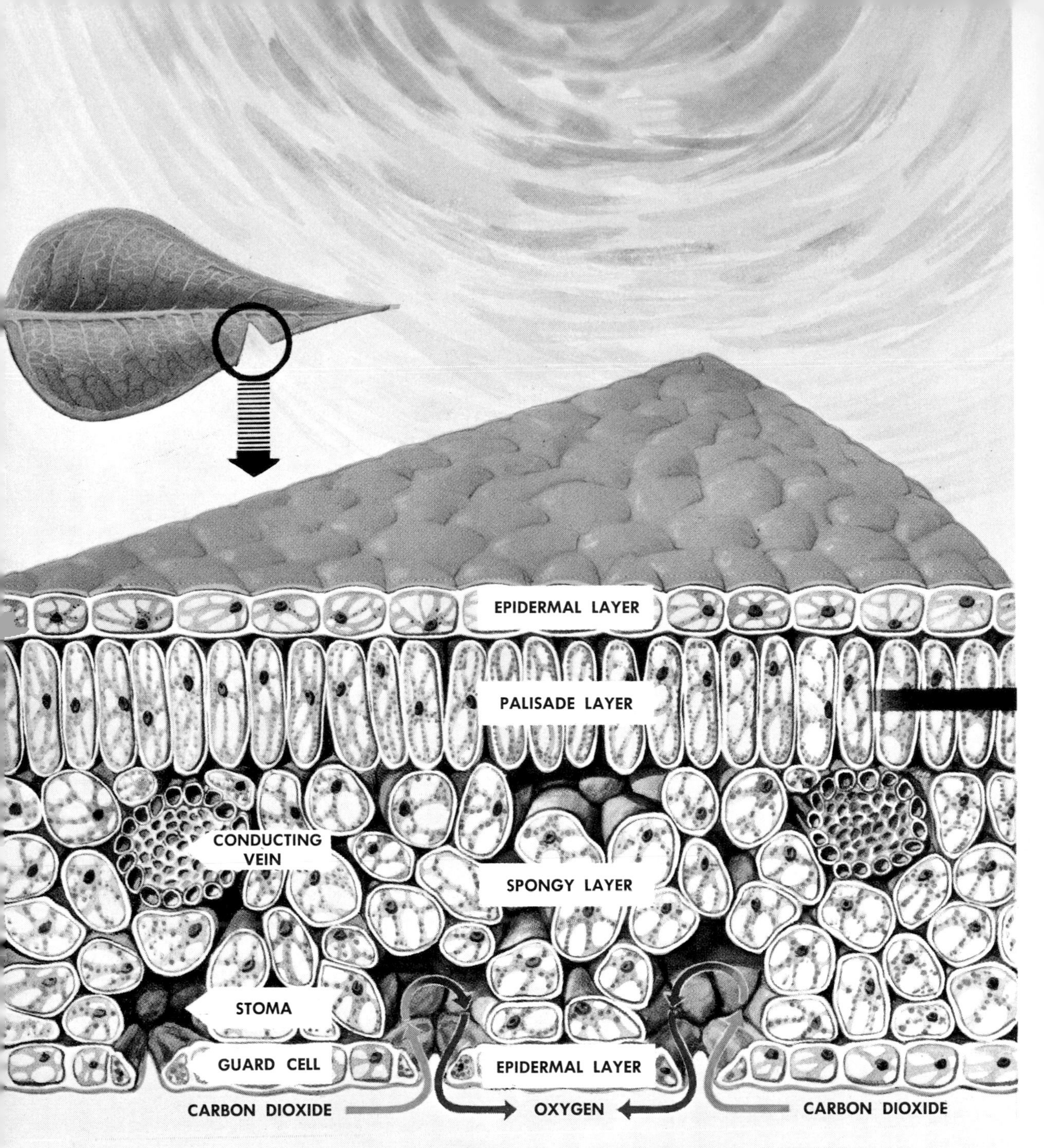


In the higher plants, it is particularly in the leaves that photosynthesis occurs. The above diagram shows a cross section of a typical leaf. Both the upper and lower surfaces of the leaf contain an epidermal, or surface, layer of cells. The epidermal layer has a number of openings, which are called stomata (singular: stoma); each of them is flanked by a pair of guard cells. The stomata can be opened or closed through changes in the shape of the guard cells. Gaseous exchange between the inside of the leaf and the outer air takes place through the stomata; carbon dioxide can be drawn in and oxygen released, or vice versa. The cell tissue between the upper and lower surfaces of the leaf is called mesophyll. It is made up of two layers. The first of these—the palisade layer—consists of long and quite narrow cells. The second layer is known as the spongy layer; it is made up of loosely packed cells. The guard cells and the cells of the palisade and spongy layers all contain chlorophyll; hence they all take part in the process of photosynthesis. The leaf veins, containing the conducting tissues known as xylem and phloem, run through the mesophyll. Xylem conducts water and dissolved substances throughout the leaf; the food that has been manufactured in the leaf through photosynthesis is carried in the phloem to the other parts of the plant.

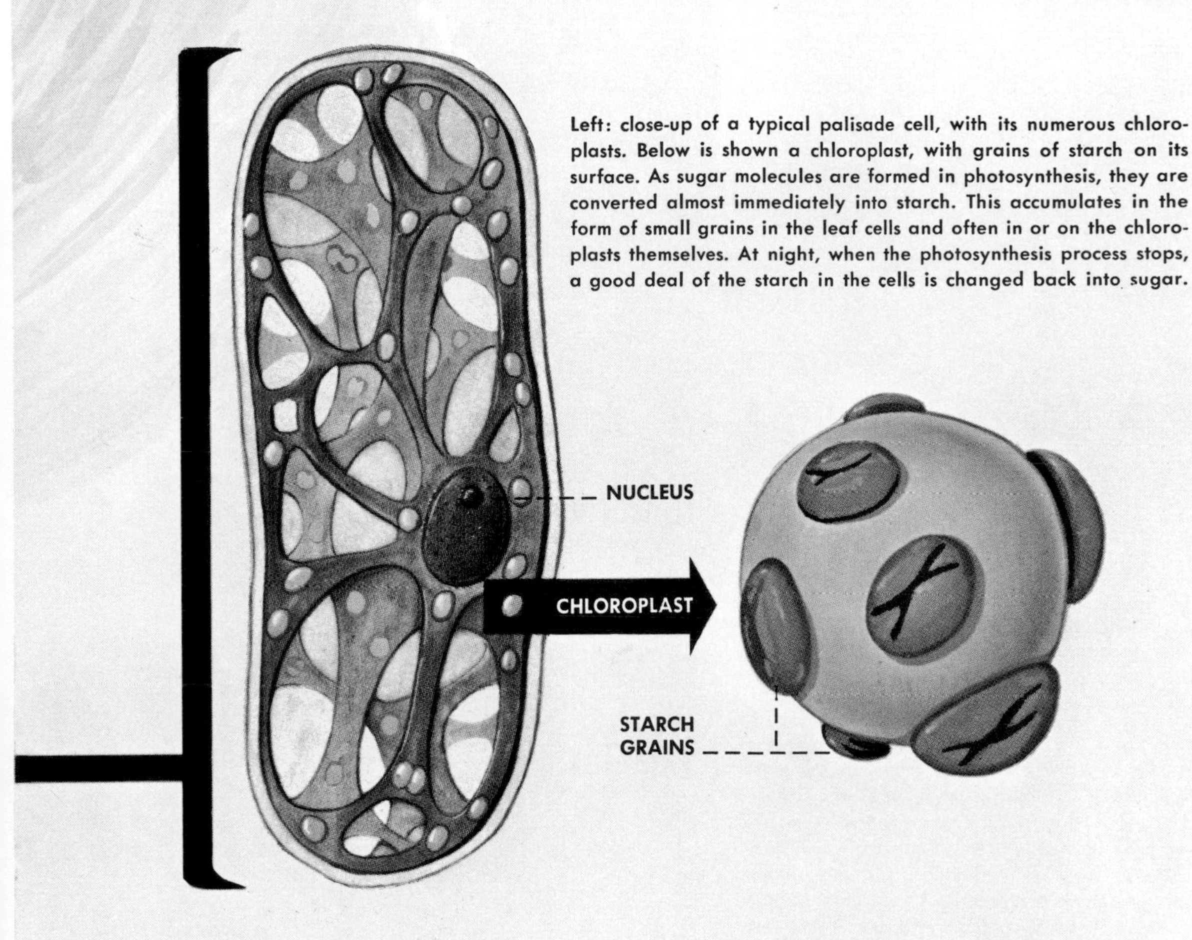


Left: close-up of a typical palisade cell, with its numerous chloroplasts. Below is shown a chloroplast, with grains of starch on its surface. As sugar molecules are formed in photosynthesis, they are converted almost immediately into starch. This accumulates in the form of small grains in the leaf cells and often in or on the chloroplasts themselves. At night, when the photosynthesis process stops, a good deal of the starch in the cells is changed back into sugar.

In the daytime, the gas carbon dioxide ($CO_2$) enters the leaf through the stomata, and it is used in the photosynthesis reaction. Part of the oxygen ($O_2$) that is formed in the reaction is used within the leaf itself and in other parts of the plant in the continuing process of respiration. The rest of the oxygen passes to the outer air through the stomata. During the hours of daylight, consequently, green plants give off oxygen and take in carbon dioxide.

At night, since there is no sunlight, photosynthesis ceases. Respiration goes on, however. Since the oxygen necessary for respiration is not supplied now by photosynthesis, it must be drawn in from the outer air. The carbon dioxide produced in respiration is no longer used in photosynthesis. It collects in the leaf and passes out of it through the stomata. As a result of all this, green plants draw in oxygen during the nighttime and give off carbon dioxide.

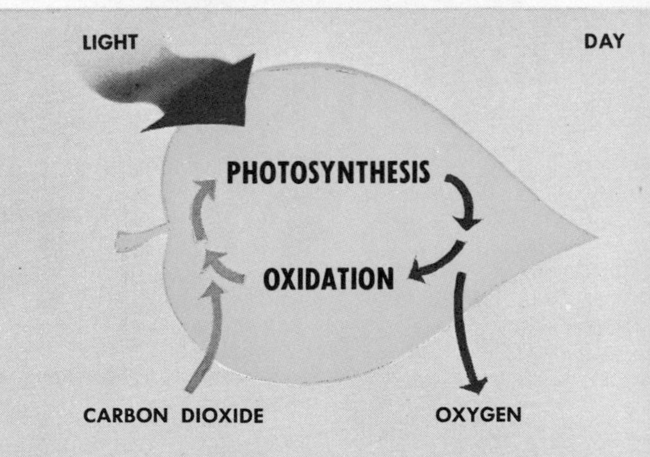


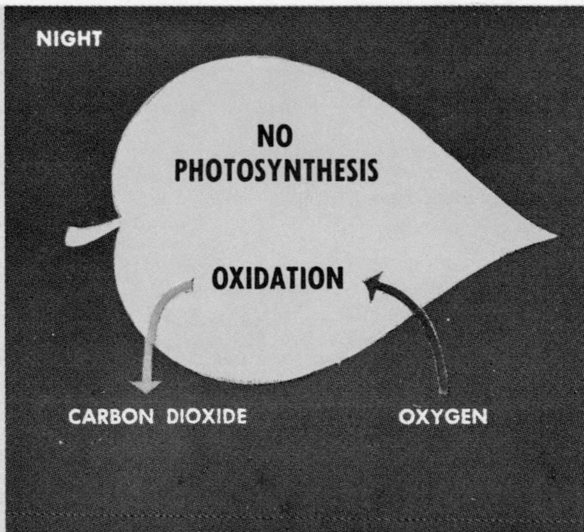

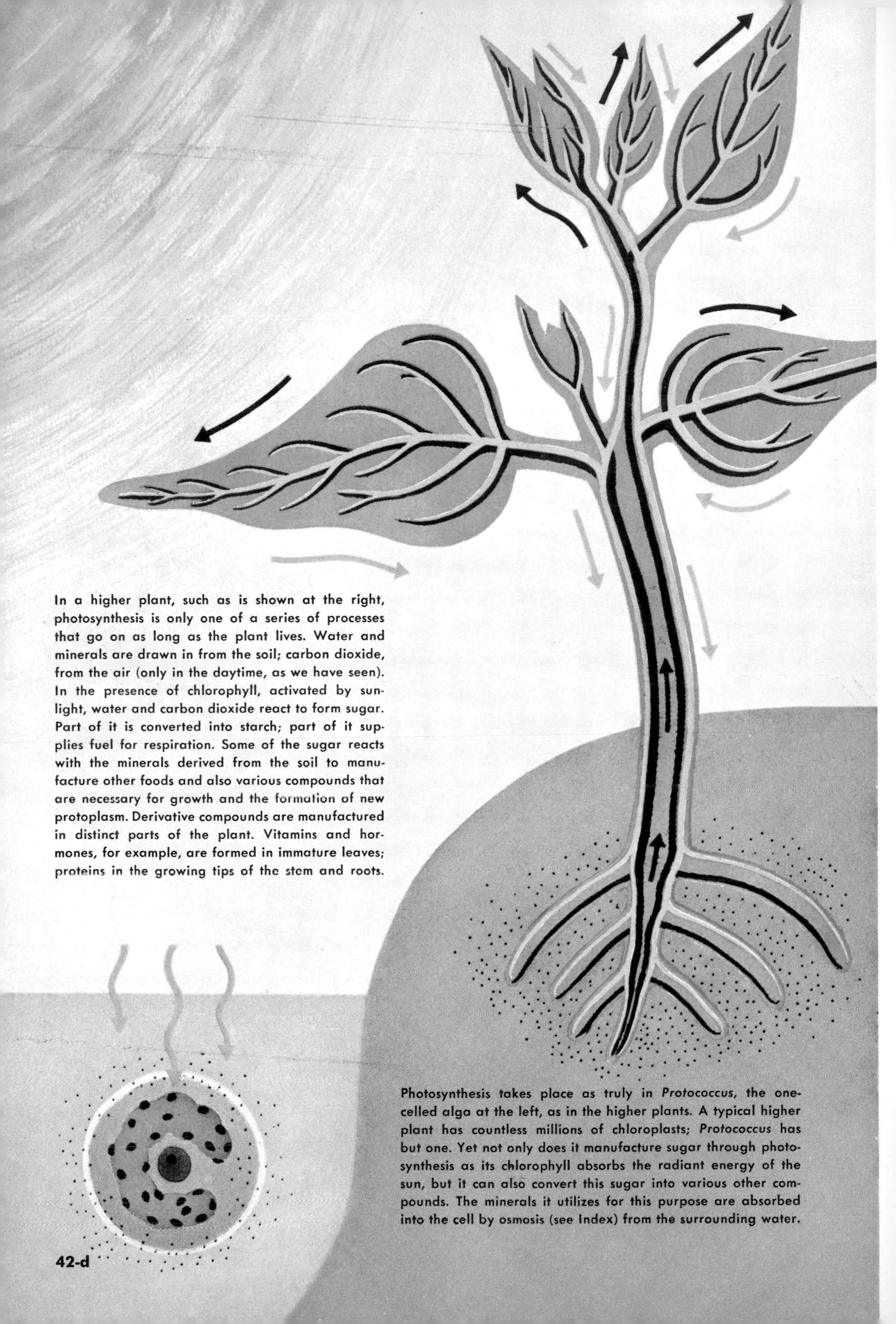

In a higher plant, such as is shown at the right, photosynthesis is only one of a series of processes that go on as long as the plant lives. Water and minerals are drawn in from the soil; carbon dioxide, from the air (only in the daytime, as we have seen). In the presence of chlorophyll, activated by sunlight, water and carbon dioxide react to form sugar. Part of it is converted into starch; part of it supplies fuel for respiration. Some of the sugar reacts with the minerals derived from the soil to manufacture other foods and also various compounds that are necessary for growth and the formation of new protoplasm. Derivative compounds are manufactured in distinct parts of the plant. Vitamins and hormones, for example, are formed in immature leaves; proteins in the growing tips of the stem and roots.

Photosynthesis takes place as truly in *Protococcus*, the one-celled alga at the left, as in the higher plants. A typical higher plant has countless millions of chloroplasts; *Protococcus* has but one. Yet not only does it manufacture sugar through photosynthesis as its chlorophyll absorbs the radiant energy of the sun, but it can also convert this sugar into various other compounds. The minerals it utilizes for this purpose are absorbed into the cell by osmosis (see Index) from the surrounding water.

# The Vital Processes of Plants

## Food Manufacture, Digestion and Other Phenomena

by HARRY J. FULLER

THERE are certain important differences between animals and plants. First and foremost, most plants are able to manufacture their own food supply by a series of chemical reactions known as photosynthesis and by various other chemical reactions that depend on photosynthesis. No animal can do so, unless certain lowly forms of life called phytoflagellates, which can manufacture food, are animals, as many zoologists believe. Animals generally have a maximum size and a rather definite form, which do not change appreciably after the animals become mature; in plants, both size and form are quite variable and depend upon the nature of the environment. The cell walls of most plants are made up of a carbohydrate compound called cellulose. Practically all animals lack cellulose; the only exceptions are the lowly tunicates, whose coat, called tunicin, is almost identical with cellulose. Finally, most plants (not all) are firmly fixed in the soil in which they grow, while most animals (not all) can move from one place to another.

The differences between plants and animals, then, are quite striking. Yet the similarities between them, particularly when we consider their vital processes, are perhaps even more remarkable. Both plants and animals utilize foods as sources of energy and as materials for the building up of their living substance — protoplasm. Both are able to digest complex foods. Both produce regulatory substances, known as hormones, which influence various processes of development. Both grow as new cells are formed from old cells, or as cells are enlarged or as they become differentiated. Both take part in reproduction processes that lead to the development of offspring. Both require water, in varying quantities, for the normal processes of life. Both transport materials within their bodies. Both possess the property of irritability — that is, sensitivity to the conditions of the environment; both are able to react to these conditions.

When we deal with the vital processes of plants, therefore, we must not think of these processes as characteristic of plants alone. We must bear in mind that most of them are found, to a greater or lesser extent, in all living things. A rose plant assimilates and digests food, respires, grows, reproduces its kind and responds to the varying conditions of its environment; so do earthworms, chimpanzees and men.

### The processes of plant metabolism

The word "metabolism," as applied to plants and indeed to all living things, has a wide range of meaning. It includes all the chemical reactions that have to do with the vital processes. It has two aspects: anabolism and catabolism. Anabolism, or constructive metabolism, is the sum total of the processes involved in the manufacture of food, growth and reproduction and the replacement of worn-out tissues. The counterpart of anabolism is catabolism, or destructive metabolism. It refers to the chemical reactions that cause the breakdown of organic substances, with the release of the energy that had been stored up within them.

All phases of metabolism are derived ultimately from the boundless energy sup-

Grapevine on a red-maple sapling. Photosynthesis is taking place in the leaves in the presence of sunlight; oxygen is released to the atmosphere.

U. S. Forest Service

plied by the sun. This solar energy is transformed through the process of photosynthesis into the stored energy of foods. The stored energy is later released and is used for growth, movement, reproduction and the repair of bodily tissues.

An important metabolic process in plants is the accumulation in their substance of materials obtained from the outer world. These materials are water, oxygen, carbon dioxide and soil nutrients.

Water is absorbed by the surface cells of roots and is transported into the central portions of the roots. From there it passes through a conducting tissue, called xylem, which extends upward from the roots into the stems and then branches from the stems into the buds, leaves, flowers and fruits.

Carbon dioxide, which usually makes up about 0.03 per cent of the atmosphere, enters plant bodies chiefly through pores in the surfaces of leaves. The opening and closing of these pores, which are called stomata, are regulated by the expanding and contracting movements of the special cells that surround them. The stomata are usually open in the presence of sunlight and are closed during the hours of darkness.

The soil nutrients that seem to be necessary for the normal metabolic activities of all green plants include nitrogen, phosphorus, sulfur, potassium, iron, manganese, zinc, boron, calcium, magnesium and molybdenum.

These elements are absorbed as electrically charged particles (ions), which result from the splitting up of the molecules of various chemical compounds found in the soil. Like water, the nutrient ions are absorbed by the roots, move into root xylem and then rise with water through the xylem of stems into leaves and other aerial parts of plants.

We know something about these essential elements, though much remains to be learned. Nitrogen, for example, is found in the molecules of all proteins, as well as of chlorophyll, a green pigment that plays a vital part in photosynthesis. Many proteins contain phosphorus, which is directly involved when food is converted into living protoplasm. Magnesium is present in chlorophyll molecules. Zinc is essential in the formation of growth hormones. Though iron is not found in

chlorophyll molecules, it plays an important part in their manufacture. It is also directly involved in the processes that lead to the release of energy from foods.

The substances drawn from the air and the soil are the raw materials that the plant uses in the manufacture of food. The first step is photosynthesis (putting together in light). In this process, water, absorbed by roots from the soil, and carbon dioxide, obtained from the atmosphere, react in the green tissues of plants in the presence of sunlight to form a simple sugar, glucose. This substance serves as a basic material from which the plant manufactures other organic compounds, as we shall see.

The process of photosynthesis takes place only in cells that contain the green pigment chlorophyll. In many plants, including most trees and shrubs, chlorophyll is present exclusively in the leaves; it is therefore only in the leaves of such plants that photosynthesis can take place. In other plant species, such as corn, zinnia and tobacco, not only the leaves but the stem tissues as well contain chlorophyll and are therefore capable of photosynthesis. In still other varieties of plants, including tomatoes and grapes, there is chlorophyll in young fruits, which take part in the food-manufacturing process.

The chlorophyll of all higher plants is present in the protoplasm of cells in minute bodies, called chloroplasts, which are spherical or ovoid in shape. The chloroplasts absorb the radiant energy of sunlight, the only form of energy that promotes photosynthesis. As a result of chemical and physical reactions that are incompletely known, they bring about the manufacture of the simple sugar glucose from water and carbon dioxide, as we have seen.

Photosynthesis may be represented by a simplified chemical equation as follows:

$$\text{Energy} + 6H_2O + 6CO_2 \rightarrow C_6H_{12}O_6 + 6O_2 \uparrow$$

Energy + six molecules of water + six molecules of carbon dioxide yield one molecule of glucose + six molecules of oxygen, which escape

The gaseous oxygen that results from photosynthesis makes its way to the outer air through the stomata of the leaves. The above equation indicates the raw materials and the products of photosynthesis and the proportions of each. It does not give an idea of the intermediate chemical reactions, which are numerous and only imperfectly understood.

The glucose that is manufactured in photosynthesis does not accumulate to any marked degree in green cells as glucose. It is used as a source of chemical energy or is transformed into other chemical compounds. Some of the glucose is converted into starch, which is stored in the form of microscopic grains in plant cells; some is changed into cellulose, which is used in the construction of plant cell walls. A certain amount, as a result of a series of chemical processes, is transformed into fats and oils. Glucose reacts with nitrogen and often with phosphorus and sulfur as well, to form amino acids, from which proteins are synthesized. It is transformed into still other organic compounds.

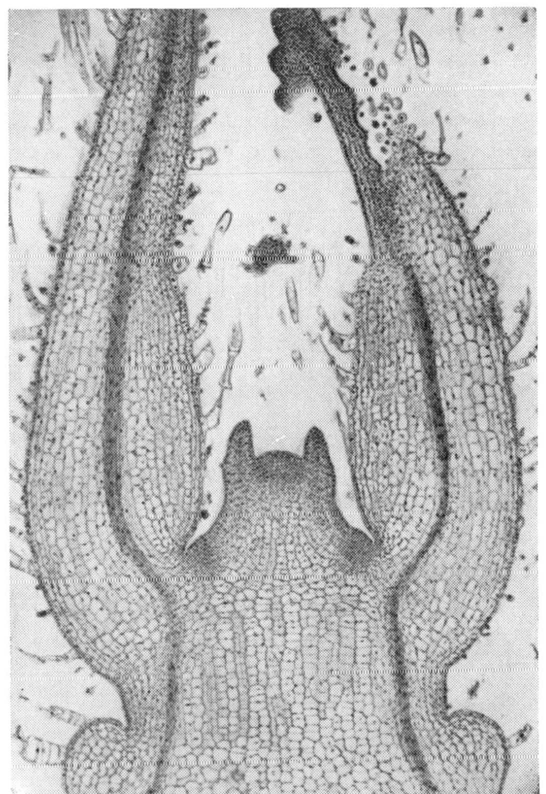

General Biological Supply House Inc.

Stem tip of a coleus plant, showing meristematic tissue. Meristematic cells have unusually thin walls.

The process of photosynthesis is just as vital to animals as it is to plants, since they too are dependent upon the basic substances derived from the process. They must obtain these substances from plants, since they cannot manufacture them in their own bodies. Herbivorous animals meet their needs by eating grasses, leaves, grains, fruits and other parts of plants. Carniverous animals devour animals that have incorporated the essential food elements in their bodies by eating plant tissues.

Not only is photosynthesis the ultimate source of food for animals, but it is also the major source of the oxygen that they require for respiration. If we keep in mind the myriads of plants, large and small, in which photosynthesis takes place while the sun is shining, we shall have some idea of the vast quantities of oxygen released to the atmosphere as a result of the process.

### All plants do not carry on photosynthesis

Certain plant species lack chlorophyll and so cannot carry on photosynthesis. These species, which are mostly fungi and bacteria, cannot make organic compounds from simple inorganic substances of air and soil, as the green plants can. Therefore, like animals, they must obtain from green plants, directly or indirectly, the food substances that these green plants manufacture.

Many of the foods stored in plant tissues as the products (direct or indirect) of photosynthesis have a complex chemical structure or are insoluble in water. Therefore they cannot be quickly or easily utilized to provide energy or to build new tissues. They must first be converted into simpler foods or into water-soluble foods by a series of metabolic processess called digestion.

These processess are controlled by special substances called enzymes, which are manufactured by living cells. There

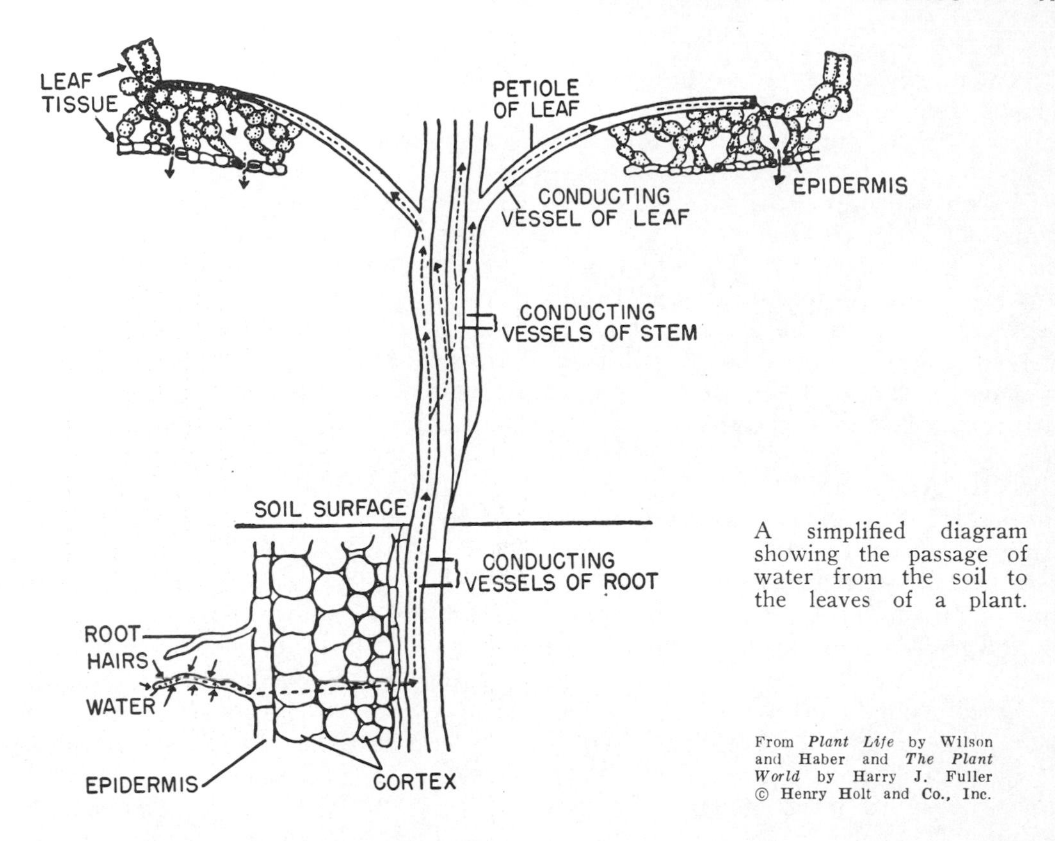


A simplified diagram showing the passage of water from the soil to the leaves of a plant.

From *Plant Life* by Wilson and Haber and *The Plant World* by Harry J. Fuller © Henry Holt and Co., Inc.

are many different kinds of enzymes, but all have certain features in common. They are highly efficient; a small amount of enzyme can digest a large amount of food very quickly. They are highly specialized; each type of enzyme ordinarily digests only one type of food or several types of chemically related foods. None of them are used up in the digestive processess that they promote.

There are three major groups of digestive enzymes in plants: carbohydrate-digesting, fat-digesting and protein-digesting. Among the most widely occurring plant enzymes are amylase (or diastase), which digests starch to malt sugar, and invertase (or sucrase), which digests cane sugar (sucrose) to glucose (grape sugar) and fructose (fruit sugar).

Digestion is especially active in plants when seeds sprout. The reason is that the stored food of seeds is chiefly in the form of starch, fats and complex proteins, which must be converted into simpler substances in order to provide energy and build protoplasm. Digestion also takes place when many fruits, such as bananas, ripen, and when potatoes and other tubers sprout. In these cases starch is converted into sugars.

Living plants require large amounts of chemical energy for growth, reproduction, reactions to external stimuli and other vital processess. This energy is made available through the process of respiration, in which foods (chiefly glucose) are chemically broken down and their stored energy released. A part of the energy liberated by respiration is heat energy, which is given off into the air from the body surfaces of plants; most of it, however, is chemical energy. The reacting substances and products of respiration, as it occurs in most plants, may be represented by this simplified chemical equation:

$$C_6H_{12}O_6 + 6O_2 \rightarrow 6H_2O + 6CO_2 + \text{energy}$$

One molecule of glucose + six molecules of oxygen yield six molecules of water + six molecules of carbon dioxide + energy

As in the case of the equation we used to represent photosynthesis, this equation shows none of the intricate and incompletely known intermediate chemical reactions of the entire process.

You will note that the respiration equation is the reverse of the one representing photosynthesis. The major differences between photosynthesis and respiration are shown in the table on this page.

This sort of respiration, which occurs in most plants, is called aerobic; it occurs in the presence of the free oxygen gas of the atmosphere. Some of the lower plants, such as yeasts, and certain bacteria that normally live in the absence of free oxygen carry on anaerobic, or incomplete, respiration, which is also called fermentation. In this, sugars are broken down in the absence of oxygen. The products that are formed are carbon dioxide and alcohol or some other compound, such as lactic acid. Anaerobic respiration is not limited to the lower plants. It may also occur in higher forms, if these are deprived of free oxygen.

It was formerly believed that the aerobic and anaerobic types of respiration had very little in common. Actually, they are closely related. They both begin with the chemical conversion of glucose into an intermediate organic compound, pyruvic acid. In the presence of oxygen, pyruvic acid is converted into carbon dioxide and water; this happens, of course, in the case of the higher plants, which carry on respiration in the presence of oxygen. If oxygen is absent, as is generally the case with yeasts and bacteria, the pyruvic acid is changed into carbon dioxide and alcohol, or lactic acid or some other compound.

The yeasts, which are primitive one-celled fungi, are responsible for alcoholic fermentation. They derive the energy for their growth and reproduction from their anaerobic breakdown of sugar. The brewer is primarily concerned with the alcohol that they form in this process; but from the viewpoint of the yeasts, the alcohol is only a waste product, since it is the energy released in respiration that is vital to them. The carbon dioxide that is produced in alcoholic fermentation is given off in bubbles, which escape from the liquid or other medium in which the yeasts live. Carbon dioxide bubbles liberated by yeasts in dough cause the dough to rise.

The various processes that make up respiration are controlled by a group of enzymes called respiratory enzymes; these operate as controlling mechanisms through a series of very complex chemical reactions. Some the lower plants, which normally live under anaerobic conditions, that is, in the absence of oxygen, are unable to carry on aerobic respiration when free oxygen is available. The reason is that they lack certain enzymes required for the completion of aerobic respiration.

Photosynthesis and respiration account for various interesting phenomena involving leaves and other plant parts; among other things, they explain why green plants give off oxygen during the day and carbon dioxide at night. Leaves and other green parts of plants carry on both photosynthesis and respiration during the day, when they receive light. Under these conditions photosynthesis proceeds more rapidly than respiration. The plant draws in much more carbon dioxide than it releases, and much less oxygen than it releases. The result, therefore, is that the plant gives off oxygen during the daytime.

At night, with the disappearance of light, photosynthesis comes to an end; but

| PHOTOSYNTHESIS | RESPIRATION |
|---|---|
| 1. Occurs only in light. | 1. Occurs in light and in darkness. |
| 2. Stores energy in food. | 2. Releases energy from foods. |
| 3. Occurs only in green cells. | 3. Occurs in all living cells. |
| 4. Reacting substances: water and carbon dioxide. | 4. Reacting substances: glucose and oxygen. |
| 5. Products: glucose and oxygen. | 5. Products: carbon dioxide and water. |

Standard Oil Co. (N. J.)

Amer. Mus. of Nat. Hist.

Above: a lotus plant, a member of the water lily family, thriving in a cypress swamp.

The fruit cactus (right) flourishes in soils in which the water supply is extremely limited.

respiration continues, since it is independent of light. The result is that the plant draws oxygen from the atmosphere; it releases carbon dioxide (an end product of respiration) to the atmosphere but does not release oxygen. At night, therefore, plants deplete the supply of oxygen and add to the supply of carbon dioxide.

The word "assimilation" is applied to a group of metabolic processes in which nonliving organic compounds (that is, nonliving considered individually) are converted into living protoplasm. Among these nonliving compounds are amino acids, various fatty compounds, chlorophyll and other pigments, and vitamins. The processes of assimilation are very complex and are not clearly understood. Since they involve the release of energy, they are closely linked with respiration.

#### The growth of plants

In plants, as in animals, growth is a cellular process. As we have already noted, it involves the production of new cells from pre-existing ones, the enlargement of the newly formed cells and, finally, the differentiation of enlarging cells into the mature tissues of the body. Obviously increase in size is only one phase of the total growth process in living organisms.

In the higher plants, growth tissues, known as meristematic tissues, are located

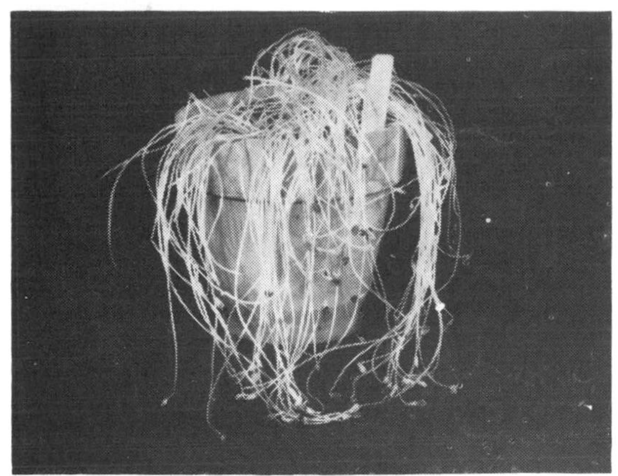

Photos, William G. Smith, Jr. — Boyce-Thompson Institute for Plant Research

Buckwheat seedling that has grown under conditions of total darkness.

in the buds of stems, in young flower parts and young fruits and immediately above the tips of young roots. Meristematic tissue occurs in the stems and roots of many plants as a growth layer (cambium) just outside the tissue called xylem. In the outer portion of tree bark a meristematic tissue called cork cambium produces cork cells, which make up the rough, hard surface tissue of woody stems and roots. A meristematic tissue (pericycle) lying in the central mass of root tissues produces new cells that develop into root branches.

Growth requires abundant chemical energy, which is provided by respiration. The growing plant must also have proteins and other organic compounds for the building up of protoplasm, as well as cellulose and various related compounds for the construction of cell walls. Conditions that interfere with photosynthesis or that cause stored foods to be depleted are quickly reflected in diminished growth.

Meristematic cells have thin, elastic walls, abundant protoplasm and a high level of metabolic activity. The creation of new cells begins with the formation of two nuclei from the original nucleus of the meristematic cell. As these two nuclei, both like the parent nucleus, are produced, a new wall is formed across the cell, and two cells are produced, each with its own nucleus. Millions of new cells are created by division during the growth of a plant. We describe the processes involved in greater detail in another chapter of THE BOOK OF POPULAR SCIENCE.

The newly formed cells increase rapidly in size, partly through the assimilation of food into their protoplasm, partly through the accumulation of water in the cells. The cell walls are stretched in the process.

In the third phase of growth — cell differentiation — division of labor occurs among the newly formed, enlarging and maturing cells. As a result, the various permanent tissues are formed; these include conducting, strengthening, food-making and storage tissues. Ordinarily, mature tissue cells remain undivided, retaining their characteristics and their functions throughout their lives. Under certain conditions, however, some of them may also undergo cell division, thus contributing to further growth. This often occurs in the formation of roots on stem cuttings and leaf cuttings.

Growth in plants is influenced to a large extent by their genes — the hereditary factors that determine how a plant is to grow and to what extent it is to develop. A corn plant, for example, is always a corn plant, with characteristic growth, behavior and structure. It is true that shifting conditions of the environment may influence the rate of growth, the total amount of growth and the yield of the plant. Yet these factors can never overcome or alter its basic qualities — its behavior as an annual plant, the form and structure of its leaves, stems, flowers, grains and other parts, or, in other words, its "cornness."

The growth of plants is under the general control of auxins, or growth hormones, which are synthesized by plants and which are present in all their growing tissues. Only one growth hormone, indole-3-acetic acid (IAA), has thus far been chemically isolated from plants; but there are possibly others. IAA directly influences cell enlargement, leaf fall and fruit fall, the growth of fruits from the ovaries of flowers, the mutual interactions of buds and various other growth phenomena.

Certain synthetic organic compounds that are chemically related to IAA are now available in commercial preparations.

They are used to hasten the formation of roots on stem cuttings, to reduce or prevent premature fruit fall from trees, to produce seedless fruit and to overcome the dormancy of tubers, bulbs and other underground stems. They serve to kill weeds by accelerating their respiration and thus depleting their food reserves; they promote the healing of wounds left by pruning; they hasten tissue unions in grafts; they retard the growth of trees in nursery storage.

The growth of plants is influenced to a certain extent by the conditions of the environment. Among the most important of these factors are the availability of water, oxygen, carbon dioxide and soil nutrients. Soil and air temperatures are also vital factors; so are light intensity and light duration (the number of hours of light received every day), poisons, mechanical factors (such as pressure and contact) and the attacks of parasites.

Different species of plants are affected in various ways by these external factors. Some plants, such as water lilies, are able to grow only in water; others, including most cacti, thrive in soils in which the water supply is extremely limited. The majority of plant species grow best in soils of moderate to abundant moisture content. Different species of plants also react differently to temperature conditions. Plants that are natives of far northerly latitudes are able to withstand low winter temperatures without injury, but cannot endure high summer temperatures; plants of tropical origin must have warmth throughout the year.

Plants also have different soil nutrient requirements. For example, some species require high concentrations of nitrogen, while others thrive only if relatively low concentrations are present. Since plants react in different ways to the food elements in the soil, they require different kinds of fertilizers for maximum production.

Plant growth is influenced to a marked degree by the gases present in the atmosphere. The concentrations of carbon dioxide and oxygen in the atmosphere fluctuate but little; therefore these gases, which are essential in plant metabolism, are usually available to plants in adequate quantities. Sometimes, however, unwanted gases accumulate. In heavily industrialized areas, where oil and coal are burned on a large scale, and in regions of hot springs and active volcanoes, gases that are not ordinarily found in the atmosphere may be present in harmful concentrations. They may hold back plant growth and may even kill unusually sensitive plants. These harmful gases include carbon monoxide, sulfur dioxide, sulfur trioxide, hydrogen sulfide, chlorine and ethylene.

The action of geotropism. The flowerpot at the left has been kept in a horizontal position for some time. The main stem of the plant has bent upward so that it is now parallel to the stem of the other plant, which has remained in the normal position throughout.

Since light is essential in photosynthesis, it is an all-important factor in plant growth. Most plants cannot live long in the absence of light, for under such conditions they manufacture no food and therefore starve. Plants that are kept for a time in darkness or in very dim light develop long and weak stems. Their leaves remain stunted and undeveloped, and they fail to develop chlorophyll in their tissues. The intensity of light is also important. For example, most range grasses, sunflowers, corn, wheat and milkweeds are sun-loving plants, which grow best in bright sunlight and cannot survive in partial or general shade. On the other hand, mosses, ferns, cocoa trees and many woodland wild flowers cannot endure full sunlight for a prolonged period of time. They can live and grow normally only when they are shaded during all or most of each day.

#### Plant reactions resulting in movements

Both plants and animals, as we have seen, possess the property of irritability: that is, they are sensitive to various factors of the environment, and they can react to these factors by a series of movements. In some ways, such reactions are quite similar in plants and animals. In both these forms of life the reactions are brought about by stimuli, such as temperature, light, contact, chemical agents, electricity, gravity, water and gases; in fact, the only important stimulus that affects animals and not plants is sound. In both animals and plants, the stimuli are frequently transmitted from one part of the body (a receptive zone) to another part (a reactive zone).

In some important respects, however, plant reactions resulting in movements differ from the corresponding animal reactions. For one thing, the reactions of all higher plants involve only the movements of individual parts and never the movement of the entire body of the plant from one place to another. In other words, most plants do not have the power of locomotion, which most animals possess. Again, the reactions of most animals result from the contraction and extension of muscles; those of plants, from shifts in growth rates or changes in internal water pressure. In the third place, the higher animals have specialized sense organs, which receive external stimuli. Plants (with a few exceptions) have no such specialized structures. They receive stimuli from the outer world rather generally through their more newly formed tissues. Finally, in most animals, the effects of stimuli are transmitted by specialized nerve tissue; in plants, by diffusing chemical substances.

Plant movements are brought about either by growth reactions or by turgor reactions. Growth reactions are caused by changes in growth rates in different parts of a plant, such as a stem or a root. Turgor reactions usually result from sudden and extensive changes in water pressure within certain localized tissues.

These two kinds of reactions differ in various respects. Growth reactions occur in all higher plants, wherever there are actively growing tissues; turgor reactions are found only in certain families of flowering plants, including the legumes, wood sorrels and several other groups. Growth reactions are the chief means by which the higher plants adjust to their environment and are definitely beneficial to the plants; most turgor movements (not all, however) do not appear to benefit the plants in which they occur. Growth reactions require hours, days or even weeks for their completion. Turgor movements are relatively rapid; they may be completed within a second or two or, at most, within thirty minutes.

The most important growth reactions are known as tropisms (from the Greek *trope,* meaning "a turning"). Tropisms of plants are bending movements of more or less cylindrical organs, such as stems, roots, leaf stalks and flower stalks; they result from different rates of growth on the opposite sides of the reacting organs. There are different kinds of tropisms. These are indicated by an approximate Greek word placed before "tropism"; the introductory word represents the type of stimulus. For example sunlight brings about phototropism (*photos* means "light"); the earth's gravitational force, geotropism (*ge* means "earth"); water, hydrotropism (*hydor* means "water"); contact, thigmotropism (*thigma* means "touch"). The normal growth hormone, indole-3-acetic acid, is involved in all tropisms.

Phototropism and geotropism have been more thoroughly investigated than the rest. Phototropism takes place when a plant grows in such a way that one side receives more illumination than the other, as in the case of a plant set upon a window sill. Under such conditions, the plant will turn its stems and leaves toward the sun. If such a plant is to grow symmetrically, it should be turned regularly (if it is in a flowerpot), so that the different sides will face the sun in turn.

You can observe the action of geotropism, brought about by the stimulus of the earth's gravitational force, when you place a plant in a horizontal position. Within a few hours or days, the main stem of the plant will begin to bend upward, assuming the normal upward direction of growth; the main roots will bend vertically downward.

In such cases there can be no question of purposeful activity. A plant growing on a window sill does not turn its leaves and stems toward the light because it is consciously seeking the light; a plant set horizontally does not struggle to attain an erect position. In either case, the movements result from the unequal distribution of growth hormone in the reacting organs.

In a plant growing vertically the hormone is formed in the tip of the stem and ordinarily diffuses downward into rapidly enlarging cells; it diffuses equally on all sides. If, however, light intensity is greater on one side than on the other, an unequal distribution of growth hormone occurs; the less brightly illuminated side will receive more of the growth hormone than the other side. The darker side will grow more rapidly than the opposite one, and, as a result, the stem will bend.

A shift in hormone distribution also occurs if a plant is placed in a horizontal position. The lower side of the stem will receive more hormone than the upper one; hence growth will occur more rapidly on the lower side, and the stem will slowly bend upward. Similar inequalities in the distribution of growth hormone in plant tissues, corresponding to shifting external

Left: mimosa plant in its normal condition. Right: the same plant just three seconds after it was touched.

General Biological Supply House Inc.

stimuli, are responsible for other types of tropisms. Botanists still do not understand how the external stimuli bring about movement of the growth hormone.

Turgor movements occur in the flowers of some species of plants, such as moonvine, morning-glory and four-o'clock, and in the leaves of other species, including clovers, beans, locusts and various plants that trap and digest insects. Apparently hormones are involved in turgor movements, just as they are in the case of tropisms. Little is known of these hormones. They seem to be chemically different from indole-3-acetic acid, the hormone that is involved in tropisms.

The turgor reactions of the flower petals and leaves of such plants as clovers and locusts are usually called sleep movements, because they are stimulated by the daily changes of light intensity as night alternates with day. The leaflets of white clover, for example, occupy a horizontal position between sunrise and sunset. At nightfall they point upward from the stem and maintain that position during the night; at dawn they return to their daytime horizontal position. In garden beans, the leaflets point downward from the stem at night and have a horizontal position during the day.

Leaf reactions such as these result from changes in the internal water pressure of small masses of tissue, located at the bases of the leaflets, in response to changing light conditions. Ordinarily these reactions are rather rapid; sleep movements are usually completed within thirty minutes after a pronounced change in light intensity. The exact significance of these movements is unknown; most authorities assume that they do not benefit the plant.

The plant called the Venus's-flytrap ensnares insects in its leaves as a result of rapid turgor movements in the leaf halves. Each leaf has a hinged portion, whose inner surface contains hairs sensitive to contact. When an insect alights on these hairs, the contact stimulus causes the leaf halves to close rapidly. The reaction is usually completed within two seconds and effectively traps the insect; the leaf then digests the softer parts of the victim's body. The Venus's-flytrap is one of the few plants with specialized sensory devices — the hairs on the inner surface of the leaf. Obviously in this case the turgor movement is beneficial to the plant.

The leaves of the sensitive plant (*Mimosa*) fold in a striking fashion when they are touched or when they are subjected to electric shock or to changes in light intensity. These reactions, like those of the Venus's-flytrap leaves, are almost instantaneous; they result from rapid turgor changes in cushions of cells at the bases of the leaflets. It is hard to see how these reactions benefit the plant.

### Reproduction in flowering plants

There is another group of vital processes that are of the utmost importance in the development and survival of plant species — those that have to do with reproduction. We have dealt in detail with this topic in other chapters of THE BOOK OF POPULAR SCIENCE. (See Index, under Reproduction and Flowers.)

Recent investigations have thrown much light on the formation of flowers and on other phases of reproduction in flowering plants. Flowers are really clusters of organs that have to do with the processes of reproduction. The formation of flowers is influenced by a number of factors: the availability of stored food in the plant, the temperature conditions and the light conditions — particularly the periods of light duration, or photoperiods.

Large amounts of food are required for the formation of flowers, for the reproductive processes occurring within flowers and for the development of fruits and seeds from flowers. This food is used partly to provide the necessary energy, partly to build up new protoplasm and cell walls. Every flowering plant, therefore, must pass through a vegetative period (varying in length in different species), during which it absorbs raw materials, manufactures and stores food and grows, before it can form flowers.

The plant does not necessarily begin to flower after a certain quantity of food reserves has accumulated. Recent research has shown that many plants require exposure to certain definite temperatures or photoperiods, or both, before they can form flowers. For example, winter-wheat plants and apple trees must have at least a brief exposure to low winter temperatures before they can flower normally; that is why they cannot be successfully grown in the tropics. Some plants require short photoperiods; others, moderate ones; still others, long ones.

The tendrils of the passionflower provide support for the plant and help it to cope with its environment.

Roche

We do not know just what these critical temperatures and photoperiods have to do with the formation of flowers. However, it appears likely, on the basis of numerous experiments, that a specific flowering hormone is produced within plants under the proper conditions of temperature and light. This flower-inducing hormone has received the tentative name of "florigen"; its chemical composition has not yet been determined.

After flowers have appeared and have opened, their reproductive processes begin. The first of these processes is pollination — the transfer of pollen grains (which produce sperms, or male sex cells) by insects, birds or wind from the stamens (male parts) to the pistils (female parts) of flowers. Within the pistil, pollen grains form tubes; these reach the ovules, or immature seeds, located in the ovary, at the base of the pistil. Fertilization then occurs. The ovary begins to grow under the stimulus of the growth hormone contained in the fertilizing pollen. During this period, large amounts of food and often of water move into the ovaries and ovules. At last the ovary becomes a mature fruit, and the ovules it contains become mature seeds.

In many plants, the development of fruits from ovaries and of seeds from ovules exerts such a drain upon the foods stored in roots, stems and leaves that these vegetative parts die. This commonly happens in annual plants, in biennials during the second year of their life and in some perennials. For instance, century plants have a vegetative period as long as forty years; after that, they flower once, produce seeds and then die. There is sketchy evidence that, in some species at least, the repressive effects of fruiting upon vegetative growth may be due to the action of some kind of hormone.

Many phases of plant reproduction are still mysterious to us. It is becoming increasingly clear, however, that hormones play important roles in reproduction.

*See also Vol. 10, p. 273:* "Physiology of Plants."

# SIMPLE MACHINES

## Basic Tools That Make Us More Powerful

by ALEXANDER JOSEPH

MOST people, perhaps, think of a machine as a complicated device with hundreds of moving parts and powered by some sort of heat engine or by electricity. As a matter of fact, even such simple tools as a hammer and a screw driver are also machines. To a scientist, a machine is any object that allows us to do work with less effort; less, that is, than if we did the job with nothing but our own bodies. You may have to strain yourself in lifting a heavy weight by hand; yet you can often push the weight up an inclined plane, such as a ramp, very easily. The amount of work you do in both cases is equal, but the inclined plane permits you to use less effort; therefore, it is a machine.

If we were to take apart a complicated machine, such as a typewriter, we would find it made up of a number of simpler elements. There are actually just six of these basic elements, which we call simple machines; these are the lever, the inclined plane, the wedge, the screw, the wheel and axle and the pulley.

What do we mean by saying that a machine allows us to use less effort in doing a given amount of work? First of all, we must see what a scientist means when he speaks of "effort" and of "work." No object moves without being pushed or pulled in some way. A door opens when we push against it; a window shade comes down when we pull it by a string; a bomb drops from an airplane because it is pulled down by gravity. In the language of physics such a push or pull is called a force; and "effort" means the same thing as "force."

If you braced yourself against the Empire State Building and pushed with all your might, you would not succeed in moving it. This is because your force is

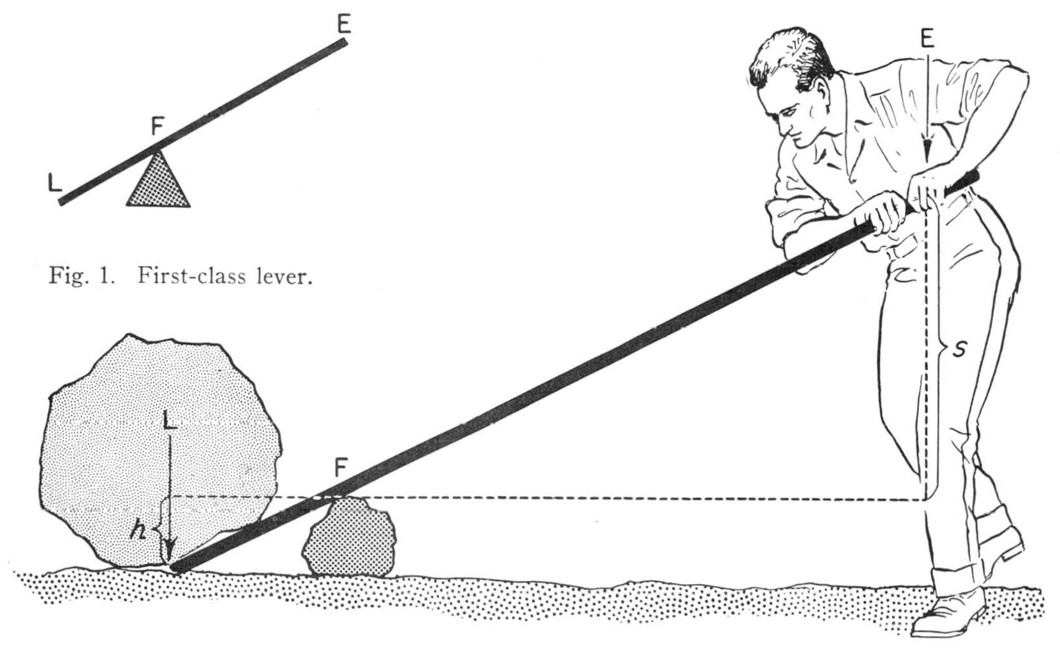


Fig. 1. First-class lever.

not great enough to overcome the resisting force of the building. You are certainly exerting effort, but a physicist would say that you are not doing work. For him, work is performed only when a force results in *moving* a body. To move a body, and thus to do work, we must exert a steady *force* that is greater than the resisting force of the body — the force, such as gravity, friction or inertia, that prevents the body from being moved. We must also exert this force through a definite *distance*. Merely holding a bar bell over your head is not work because your force is not moving anything; however, *lifting* the bar bell (moving it against the force of gravity) is work in the physical sense.

To find the amount of work we do in performing any task, we multiply our force by the distance through which it acts. Thus if a man pushes against a piano with 50 pounds of force and continues to push while he moves the piano 10 feet, he performs 500 foot-pounds of work — 50 pounds of force times 10 feet.

Mechanical work is a form of energy; as such, it cannot be destroyed nor can it be created out of nothing. A simple machine has no energy of its own and cannot do work by itself. It is only when we perform work upon it that it will perform work, in turn, upon some other object. The machine never adds to the sum total of work that is accomplished, since that would mean creating energy out of nothing. The law of machines states this fact in more scientific language: it tells us that the work output of a machine is always equal to the work input. (In reality, every machine has a certain amount of friction. Since some of the work input is wasted in overcoming this friction, the work output is a little less than the work input.)

Once we understand this law, we can see why machines multiply our strength, or force. We know that the work input and output of a machine must be the same, and that each of these is equal to a force multiplied by a distance. Let us suppose that a force of 100 pounds acts through a distance of 1 foot, and another force, of 1 pound, acts through a distance of 100 feet; each force is doing the same amount of work, but one is 100 times larger than the other.

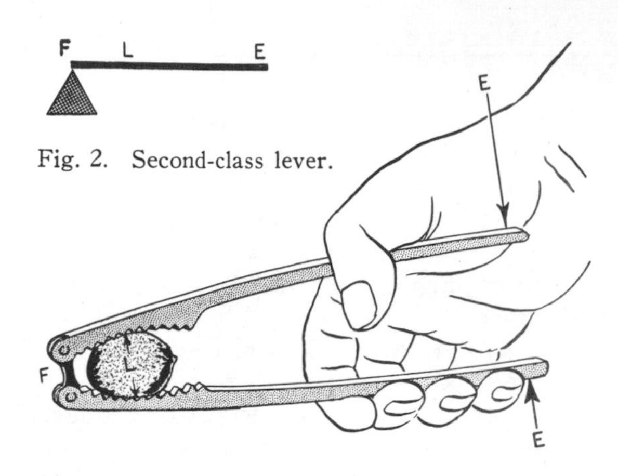

Fig. 2. Second-class lever.

We can use a small force at one end of a simple machine to move an object against a much greater force (such as that of gravity) at the other end of the machine, if the object does not move *as far* as we do. The ratio between the resisting force on the object that is moved and the force that is applied to this particular object is called the mechanical advantage.

#### The three classes of levers

The first of the simple machines is the lever, which is merely a rigid bar, capable of turning about one point, or axis. One of the most common levers is the crowbar, a long iron or steel bar that is used to move heavy objects or to pry things apart. Figure 1 shows a man lifting a boulder with a 6-foot crowbar. One end is placed under the load L and the bar is braced, at a point near the load end, against a small rock that serves as a pivot. The point F, resting against the pivot, is called the fulcrum. The man applies his effort E to the upper end of the crowbar. This kind of arrangement, with the fulcrum, or pivot point, between the load and the effort is called a lever of the first class.

We can divide the lever into two parts: the effort arm, extending from the fulcrum to the effort end, and the load arm, from the fulcrum to the load end. Let us suppose, in this example, that the effort arm is 5 feet long and the load arm is 1 foot, and that the weight, or downward force, of the boulder is 200 pounds.

How much force must the man use in order to lift this weight?

If he pushes his end of the crowbar down through a certain distance $s$ (the effort distance), the crowbar will raise the boulder through the distance $h$ (the load distance). The man will be doing $E \times s$ foot-pounds of work, which will be equal to the $L \times h$ foot-pounds of work done by the machine. From the length of the lever arms we can see that $s$ is 5 times longer than $h$; therefore, the effort will be 5 times smaller than the load. The man will have

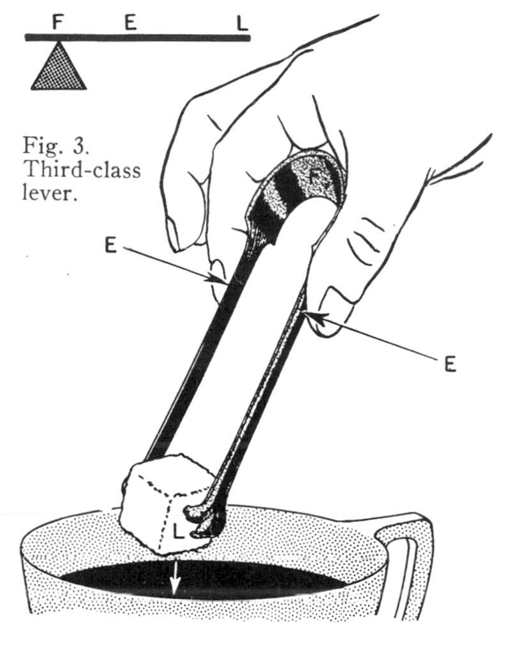

Fig. 3. Third-class lever.

to exert a force of only 40 pounds to raise the 200-pound boulder. The mechanical advantage of this crowbar is 200 divided by 40, or 5; applying effort to the long arm, you could always move a load 5 times greater than the effort.

The mechanical advantage of every simple machine is found in the same way: we just divide the load by the effort required to move it, or we divide the effort distance by the load distance, which amounts to the same thing. In the case of levers, we can also divide the effort arm by the load arm. (In the above example, 5 ft. $\div$ 1 ft. $=$ 5.)

A pair of scissors is a combination of two first-class levers, attached to a common fulcrum. No doubt you have noticed that it is much easier to cut something with the scissors if you place the material close to the pivot point. When you do this, you are shortening the load arm while the effort arm remains the same; thus you increase the mechanical advantage and the load requires less effort.

In some levers, the load is placed between the fulcrum and the effort; these are known as levers of the second class. The ordinary nutcracker shown in Figure 2 is a common example. Actually, it is made up of two second-class levers (the upper and lower parts of the nutcracker in Figure 2), whose fulcrums are joined together. The load is the nut, which resists being crushed. We apply the effort by squeezing the handles together. In this case, and for all second-class levers, the effort arm consists of the entire length of the lever. The load arm of each lever is the distance from the nut to the fulcrum; since the effort arm is longer than the load arm, the nutcracker has a mechanical advantage of more than one.

A familiar second-class lever is the rowboat oar. At first glance it might seem that an oar should be a first-class lever. See if you can find out why it really belongs to the second class. As a clue, remember that the purpose of the oar is to move the boat, and not the river!

Grasping a lump of sugar with a pair of tongs, as shown in Figure 3, involves the use of the lever. Tongs and tweezers belong to the third class of levers — here the effort is applied between the load and the fulcrum. The fulcrum (F) is located where the two branches of the tongs join, and the load is placed at the opposite end, between the jaws. (Each branch is itself a third-class lever.) In third-class levers, it is the load arm that takes up the entire length.

Since the effort is always applied somewhere in the middle, the effort arm is always shorter than the load arm. This means that the mechanical advantage of a third-class lever is always less than one; in effect, we must use more pressure on the tongs to keep the cube from slipping than we would have to apply if we grasped

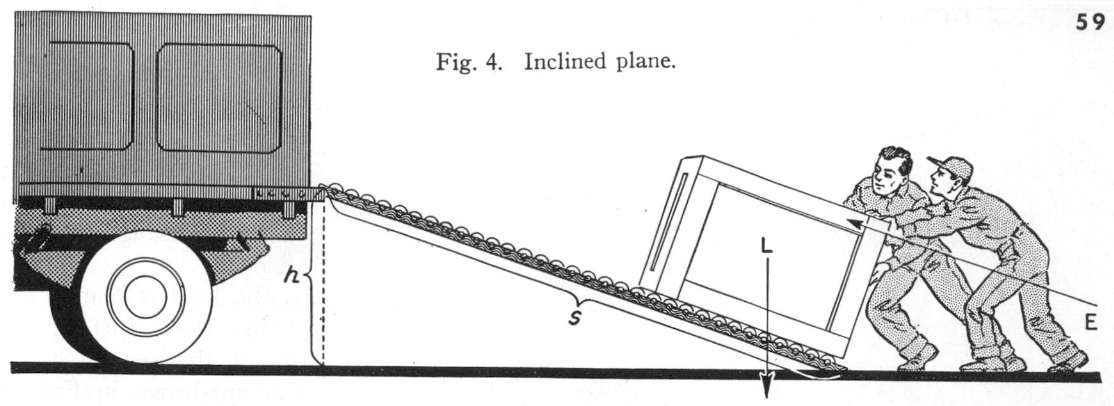


Fig. 4. Inclined plane.

the cube directly with our fingers. This can sometimes be very useful; tweezers are ideal for handling stamps, insects and other delicate objects because they lessen the crushing force of our fingers.

#### The inclined plane has many uses

In Figure 4 the workmen are pushing a 400-pound crate up a loading conveyor into a truck. This conveyor is nothing more than an inclined plane with roller wheels on it for reducing friction. An inclined plane lessens the effort required to lift a load. If the men tried to lift the crate to the height $h$ by themselves, they would have to apply 400 pounds of force — an almost impossible feat. If $h$ were 4 feet, they would be doing 1,600 foot-pounds of work.

Using the inclined plane they do the same amount of work; but their effort is applied over the longer distance $s$, the length of the plane. If this length were 16 feet, they would need only 100 pounds of force between them to do the work. Dividing the load of 400 pounds by the effort of 100 pounds, we find a mechanical advantage of 4. For any inclined plane, the mechanical advantage is the length of the incline divided by its vertical height at the top.

Railroad terminals often use inclined planes, in the form of gradually sloping ramps, instead of stairs. In mountainous country, roads are constructed in long, gradual slopes so that automobiles may ascend the mountains without imposing too great a strain on their motors.

#### The wedge is really an inclined plane

The simple machine called the wedge is really a small inclined plane. Instead of pushing a load up the incline, the entire wedge is driven under, or into, the load, as shown in Figure 5. The wedge is ideal for separating two objects that are held together by a great force. For the boulder and the earth this force is gravity; for the two halves of the log the force is the rigid strength of the wood itself.

Fig. 5. Wedges.

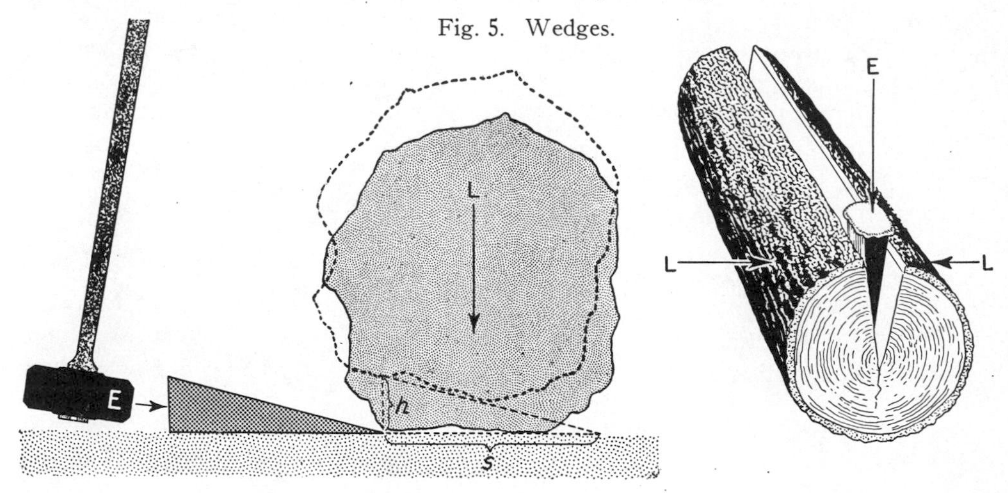

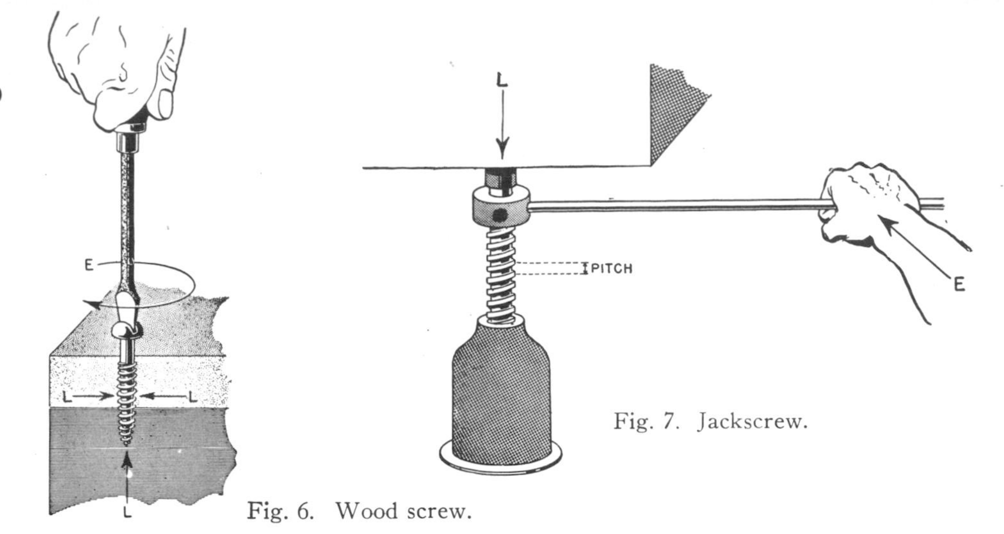

Fig. 6. Wood screw.

Fig. 7. Jackscrew.

When using a wedge, the effort is applied to the head, or thick end, and the load acts against the long sloping side. If the hammer drives the wedge completely under the load, its effort has moved through a distance *s,* equal to the length of the wedge; at the same time, the load will be raised through the distance *h,* the thickness of the head. Dividing the effort distance by the load distance gives us the mechanical advantage; thus, the mechanical advantage of a wedge is its length divided by its greatest thickness. A wedge 6 inches long with a 1-inch head would have a mechanical advantage of 6; it would have the same mechanical advantage if it were 12 inches long and had a 2-inch head.

**The screw provides tremendous mechanical advantage**

It may seem strange to think of a screw as an inclined plane, but you can prove this for yourself. If you follow the thread of a screw with a pointed pencil, you will see that it is a plane, constantly curving upward around a central shaft. In wood screws (Figure 6) the central shaft is tapered; in most machine screws and bolts the shaft is a straight cylinder. Screws are used widely to apply tremendous force with very little effort.

Figure 7 shows a jackscrew lifting a load; the screw is turned by applying effort to the end of a long handle. Each time the handle makes one revolution, the screw does likewise; thus the effort moves through the circumference of a circle with a radius equal to the length of the handle, while the screw, with the load, moves upward through the distance from one thread to the next (the pitch of the screw). If the handle were 30 inches long and the pitch were .5 inch, this would be equivalent to an effort distance of about 188.5 inches and a load distance of .5 inch, or a mechanical advantage of about 377.

We can easily see how much greater this advantage is than that of any practical lever or wedge. With this jackscrew, a 5-pound effort could lift a weight of 1,885 pounds; screws are designed with even higher mechanical advantages. The screw principle is applied in all types of vises and machine tools to exert maximum force.

**The wheel and axle**

In the old-fashioned well shown in Figure 8, the heavy water bucket is raised

Fig. 8. Wheel and axle.

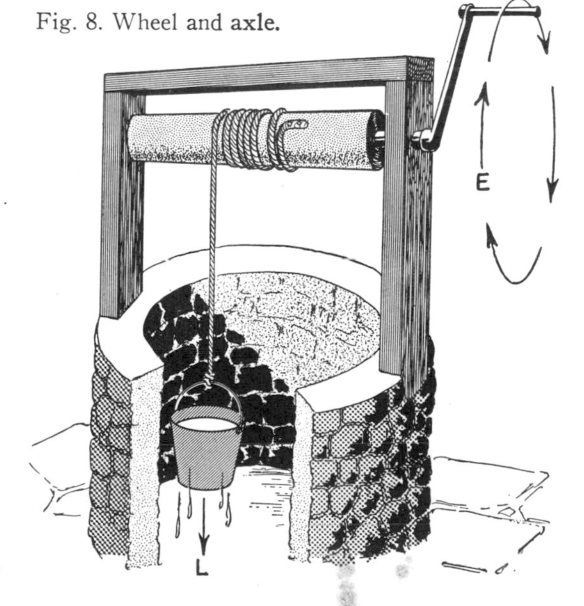

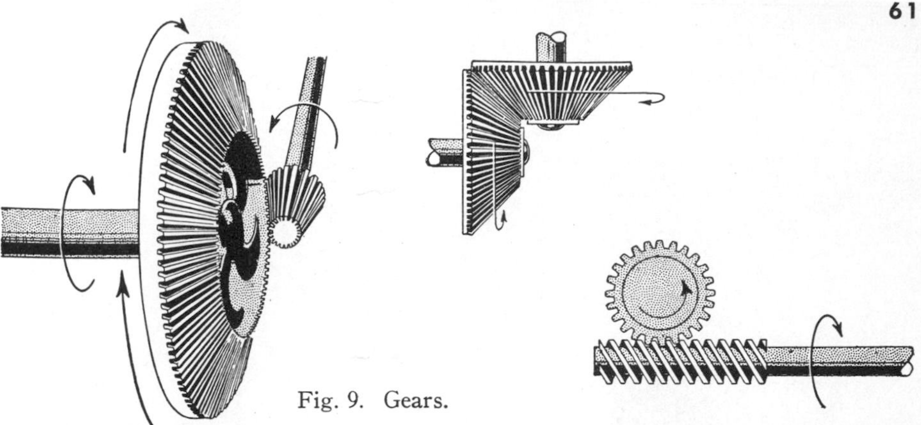
Fig. 9. Gears.

by turning a large crank attached to a wooden axle. As the axle turns, the rope winds around it and thus lifts the bucket. This is an example of another simple machine — the wheel and axle. As the arm of the crank turns, it moves in a full circle; a wheel, turned by a rope that winds around its outer rim, could easily be used in place of the crank.

When we apply effort to the wheel and turn it through a complete circle, the axle also makes one revolution, taking up a length of rope equal to its circumference. The axle radius is small compared to the radius of the wheel — let us say 3 inches as against 12 — and the circumferences of the axle and wheel are in the same proportion as their radii. The circumference of the wheel is the effort distance and that of the axle is the load distance. Thus the mechanical advantage is the radius of the wheel divided by the radius of the axle. In this case, the mechanical advantage would be 4; we could lift a bucket of water weighing 60 pounds with an effort of 15 pounds on the crank.

A wheel with teeth cut into its outer rim is called a gear. When the teeth of two gears are meshed together, one gear can turn the other. Each of the gears is connected to a shaft, one supplying effort and the other moving a load. Thus, a pair of gears accomplishes the same thing as a wheel and axle.

The mechanical advantage of a pair of gears depends on the number of teeth in each gear; this is called the gear ratio. If a gear with 4 teeth is connected to one with 20 teeth, the gear ratio is 5 to 1. The smaller gear turns 5 times for every revolution of the larger one. If effort is applied to the smaller gear the mechanical advantage of the pair will be 5; a 20-pound force on the smaller gear will move a load of 100 pounds on the larger one. If we apply the effort to the larger gear, the mechanical advantage becomes $\frac{1}{5}$; a 20-pound force now would move a load of only 4 pounds on the smaller gear, but the load would move 5 times as fast as the effort.

Gears are vital elements in almost all modern machinery. As we have seen, a set of gears can be used to increase mechanical advantage, to increase speed or to do both at different times. Teeth can be cut in a variety of shapes for different purposes (see Figure 9), and the gears can be arranged to change the direction of motion. An automobile uses gears in almost every conceivable way. The steering mechanism uses a worm gear to apply force for angling the front wheels. The transmission and differential carry the turning force of the engine's flywheel to the rear wheels by means of gears. By shifting these gears we alter the ratio of engine speed to gear speed and thus supply more or less mechanical advantage for starting, climbing hills or speeding over level ground.

#### Pulleys, used singly or in combination

Another simple machine with many applications is the pulley. The housewife pulls on a clothesline looped around two fixed pulleys; the dentist uses a system

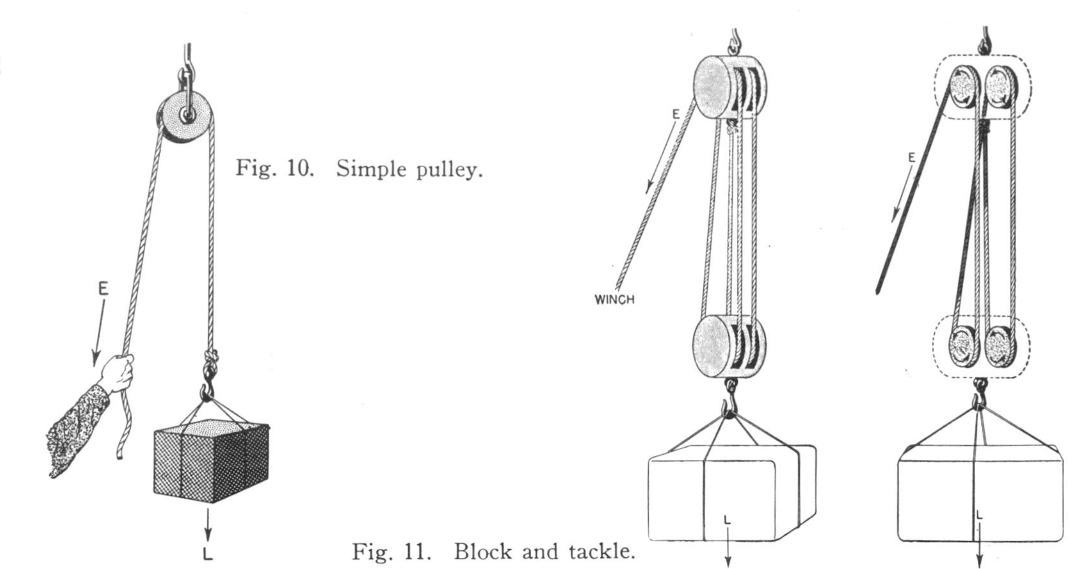


Fig. 10. Simple pulley.

Fig. 11. Block and tackle.

of pulleys to transmit the motion of an electric motor to the drill that he applies to your teeth. A derrick employs a tackle (a combination of ropes and pulleys) rigged at the end of a spar or beam.

The simplest kind of pulley is shown in Figure 10 — just a rope slung over a grooved wheel and attached to a load. When the rope is pulled, the wheel turns but does not move up or down. This single, fixed pulley is useful because it is more convenient to pull *down* a rope than to pull *up* a load; the pulley allows us to change the direction of our effort. However, this kind of pulley has no mechanical advantage. To raise the load one foot, we have to pull down the rope one foot.

In order to have a mechanical advantage there must be at least one pulley that can move with the load. The block and tackle shown in Figure 11 is an example of this kind of pulley system. Here, the rope passes around four pulley wheels; two of them are in the upper block, which is fixed, and two are in the lower block, which moves up and down with the load. One end of the rope is anchored to the upper block. The load is supported, then, by four lengths of rope; if we were to raise the load 1 foot, each length of rope would have to be shortened by 1 foot, making a total of 4 feet for the whole rope. In other words, the effort end of the rope must be pulled through a distance of 4 feet, and the effort will be 4 times smaller than the load — the mechanical advantage is 4.

For all pulley systems the mechanical advantage is equal to the number of ropes (that is, lengths of a single rope) that are supporting the load. Theoretically, there is no limit to the load that can be moved by a block and tackle. We could undoubtedly make a block with a thousand pulley wheels; but such a device would be too heavy and clumsy to be practical.

### How simple machines are combined

In complex machines, we find almost every kind of simple machine. In the automobile, for example, a complicated linkage of levers with many joints and pivot points allows us to move the heavy gears in the transmission by making slight motions with the gear shift. The hand brake, clutch pedal and foot-brake pedal are also parts of lever systems that supply great mechanical advantage. The heart of the engine is the crankshaft, which is simply an axle with a number of cranks built into it; it not only converts the up-and-down force of the pistons into turning force, but also adds a mechanical advantage for moving the entire load of the automobile. Gears, of course, are used in the transmission; they are also used to synchronize the crankshaft with the camshaft and the distributor. By means of belts and pulleys, the crankshaft turns the water pump, fan and electrical generator. These are but a few examples of the simple machines that form part of the automobile.

*See also Vol. 10, p. 280:* "General Works."

# THE BACTERIA

## Microscopic Enemies and Allies of Man

**BY HARRY J. FULLER**

FOR years the harmful part played by bacteria in the transmission of diseases such as tuberculosis, tetanus, diphtheria and cholera has been thoroughly publicized. The result is that many people think of all bacteria as enemies of man, lurking in our food and drink, on the ground and in the air, waiting to pounce upon their victims. The fact is that, as we shall see, the benefits that bacteria confer upon mankind outweigh any harm some of them may do.

The earlier investigators of bacteria thought of them as tiny animals; they were generally grouped together with the microscopic animals called protozoans. Practically all biologists now regard bacteria as plants, belonging to the phylum, or major group, Schizomycophyta. The name means "fungus plants that divide," and refers to the most common method of reproduction among bacteria. The exact relationship of bacteria to the fungi and to other plant groups is uncertain.

Various bacteria show undoubted similarities with true fungi. Certain members of both groups have a substance called chitin in their cell walls. Both bacteria and fungi can obtain nourishment directly from organic matter. Bacteria are microscopic in size; some of the fungi are also extremely small. Some bacteria, however, seem to be more closely related to the lower algae, particularly the blue-green algae. (See the article The Algae, in Volume 3.) Both the bacteria and the blue-green algae reproduce mainly by fission; the cells of both lack definitely organized nuclei. Some blue-green algae have shapes resembling those of bacteria; some are colorless, like bacteria. Probably the bacteria are made up of various kinds of organisms, some related to algae, others to fungi. Future research will doubtless shed more light on such relationships.

General Biological Supply House

Spherical, or coccus, form of bacteria.

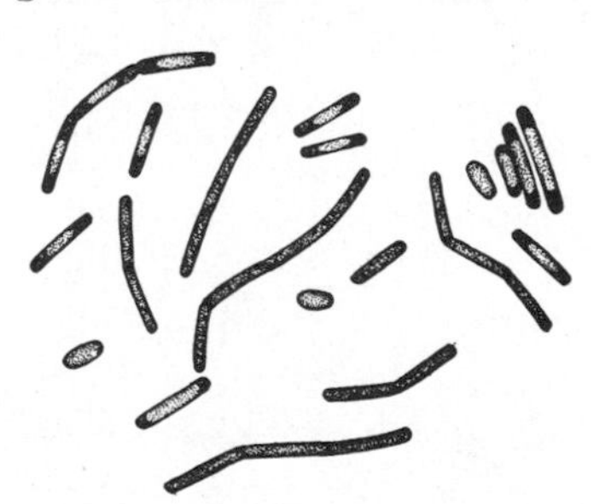

Rod-shaped, or bacillus, form of bacteria.

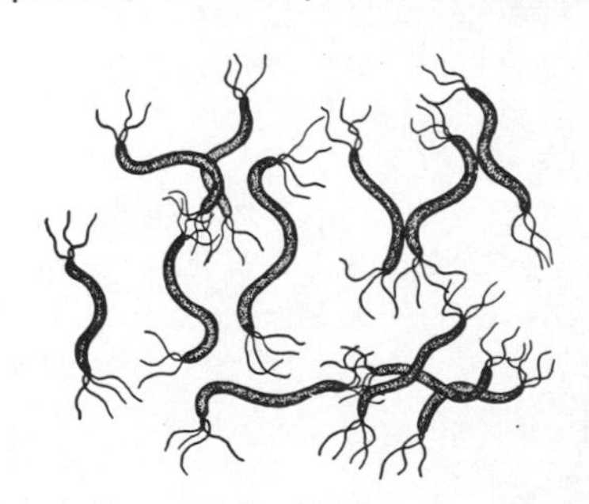

Spiral, or spirillum, form of bacteria.

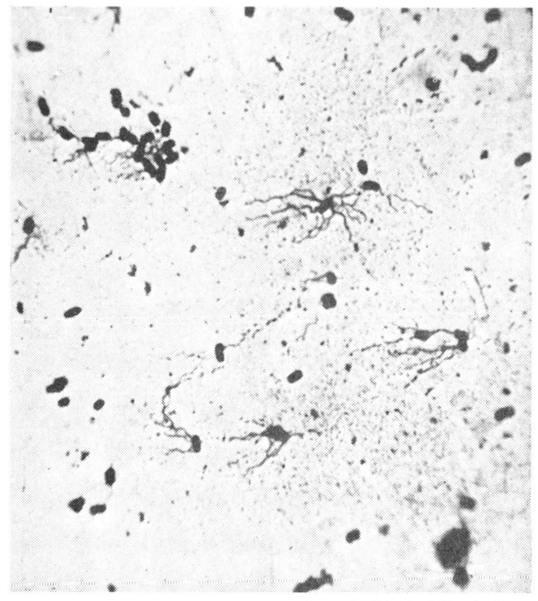

General Biological Supply House

Above are shown the bacteria — called *Eberthella typhi*, or *Bacillus typhosus* — that cause typhoid fever.

Most bacteria are one-celled plants of extremely small size, rarely exceeding 1/5,000 of an inch in their greatest dimension; some average only 1/165,000 of an inch. These plants are the smallest known living organisms (excluding the viruses, which may or may not be living organisms); they can be observed only under microscopes of very great magnifying power. There are three common bacterial body forms: spherical or ovoid (coccus forms), rod-shaped or cylindrical (bacillus forms) and spiral or screw (spirillum forms). Some species of bacteria have much-branched, threadlike bodies; but these species are few in comparison with the sphere, rod and spiral species.

A bacterium has a very thin cell wall, consisting chiefly of cellulose or chitin; this is often surrounded by a slimy, transparent capsule, secreted by the bacterium itself. Within the cell wall is a mass of the living material called protoplasm. When observed under the optical microscope (using beams of ordinary light), bacterial protoplasm appears simple in structure. However, research with the electron microscope (using beams of electrons) indicates that this protoplasm is as complex as that found in the other plants.

Bacteria that live in liquids often have long, threadlike processes called flagella. Rhythmic movements of the flagella propel the bacteria through the liquid, usually in a twisting fashion. The numbers and arrangement of flagella vary greatly in the different species of bacteria and are used as a basis for identification. A few species of mobile bacteria lack flagella; they move about by snakelike, twisting movements of the entire cell.

Reproduction in bacteria is largely, and possibly entirely, asexual. Recent research suggests that a type of sexual reproduction may occur in some species, but the evidence is still inconclusive. Most bacteria reproduce by fission. This is a simple process of cell division, in which one bacterium splits into two new ones. Fission may occur with incredible rapidity — as often as every fifteen or twenty minutes under particularly favorable conditions of temperature, moisture and food supply.

It is estimated that within the space of twenty-four hours a cholera bacterium, reproducing by fission at its most rapid rate, could produce offspring numbering 4,700,000,000,000,000,000,000 and weighing nearly 2,000 tons! This extraordinary rate of reproduction is only theoretically possible. For one thing, there would not be nearly enough food and moisture supplies in any one place for such enormous numbers.

There are other asexual methods of bacterial reproduction. For example, some species form extremely minute structures called spores (one spore per bacterium). This method of reproduction is often, but not always, the consequence of certain environmental conditions, such as temperature extremes or inadequate food supply; such conditions do not favor the active, vigorous growth and fission of bacteria. A spore is usually formed by the condensation of protoplasm within a bacterial cell into a spherical or ovoid (egg-shaped) body. Spores have a lower water content than active bacterial cells; because of this fact, they are more resistant to unfavorable environmental conditions than are active bacteria. In the pasteurization of milk, for example, most of the active bacteria present are killed by the heat treatment; but the bacterial spores in the milk survive.

### The germination of bacterial spores

When bacterial spores encounter favorable conditions of temperature and food supply, they germinate (sprout); each spore then grows into an active bacterium. Fortunately, most bacteria causing serious diseases of man do not form resistant spores. Of course, as only one spore is produced per bacterium, no increase in the total number of bacteria results from formation and germination of spores.

Most bacteria are unable to manufacture their own foods from simple inorganic substances, such as carbon dioxide and water, as green plants can; they feed on organic compounds manufactured by other organisms. Such bacteria are given the name "heterotrophic," which means "other nourishment" in Greek. Those heterotrophic bacteria that derive their organic foods from dead plant and animal bodies or from dung and other waste products of organisms are known as saprophytes. Other heterotrophic bacteria, called parasites, obtain food directly from the tissues of living plants or living animals. (This is really another way of saying that the plants or animals in question have a bacterial disease.) Many bacterial species are exclusively saprophytes; others are exclusively parasites; still others may live saprophytically or parasitically, depending upon the nature of the environment and the organic materials available to them.

Many heterotrophic bacteria are cosmopolitan in their tastes, feeding upon a variety of organic compounds. Other species are highly selective. Certain bacteria, for example, feed mainly on cellulose; others, on sugars, proteins or other foods. Some parasitic bacteria live only upon blood; others, upon specific tissues of animal bodies or host plants. Like other organisms, bacteria produce regulatory chemicals called enzymes. These promote digestion: that is, the conversion of complex or water-insoluble foods to simple or water-soluble foods.

A few bacterial species are able to manufacture foods from simple inorganic substances, such as carbon dioxide. Such bacteria are called autotrophic (self-nourishing). There are two kinds of autotrophic bacteria: chemosynthetic and photosynthetic. The chemosynthetic bacteria obtain the energy required for food manufacture by oxidizing (combining with oxygen) various chemicals. Photosynthetic bacteria contain purple or greenish pigments that enable them to absorb and utilize light energy in food manufacture. This process of photosynthesis (putting food together in light) involves pigments somewhat different from the chlorophyll of higher plants.

The chemosynthetic bacteria are particularly important in the scheme of nature. Sulfur bacteria, for example, convert hydrogen sulfide, a product of protein decay, into sulfur and then to sulfuric acid. The acid undergoes chemical reactions in soils to form sulfates; these substances form the principal source of the sulfur that higher plants require for normal growth and reproduction. Iron bacteria oxidize certain types of iron compounds into other iron compounds. The energy they derive from this oxidation is used in building foods from carbon dioxide. The compounds formed by iron bacteria are deposited in ropy, rust-colored masses; these often appear in springs, water pipes and reservoirs. The chemosynthetic bacteria include the nitrifying bacteria that live in soils and that make nitrogen available to higher plants, as we shall see.

### Bacteria carry on respiration

Bacteria resemble all other living organisms in their ability to carry on respiration: that is, the chemical breakdown of foods and the release from these foods of the energy used in growth, reproduction and various other vital activities. Most bacteria, like most plants and animals, use free oxygen from the atmosphere in respiration and produce carbon dioxide and water as a result of the process. Such bacteria are called aerobic (living in the presence of oxygen). They can live only when they have access to free oxygen, as for example in aerated soil and water, on the surface of other living organisms or on the surface of foodstuffs.

Other bacteria maintain respiration in the absence of free oxygen; they are known as anaerobic (living away from oxygen). They thrive within sealed, imperfectly sterilized cans of food, within the bodies of other organisms and in poorly aerated soils and water. The respiration of such bacteria is commonly called fermentation. Some bacteria are able to respire either aerobically or anaerobically, depending upon whether or not free oxygen is available.

Bacilli of anthrax, from a colony.

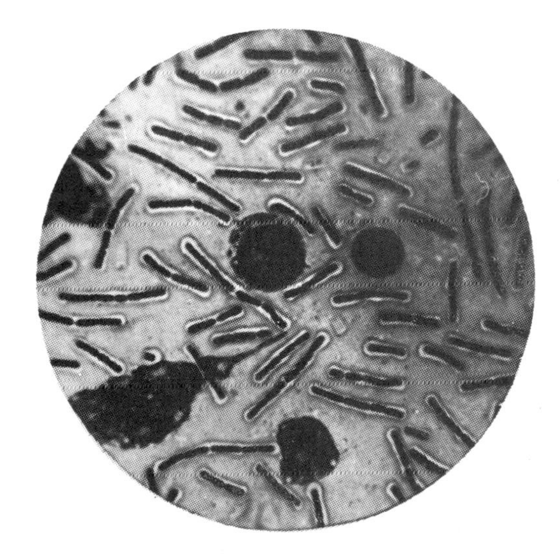

Bacilli of anthrax, from the blood.

The products of anaerobic respiration, or fermentation, are carbon dioxide or other gases and a variety of organic compounds. These compounds include ethyl alcohol, lactic acid (produced, for example, in sour milk), formic acid (responsible for various types of food poisoning) and butyric acid (which causes the disagreeable odor of rancid butter). The swelling of cans containing imperfectly sterilized foods is the result of the pressure of gases formed by bacterial fermentation. Foods in such swollen cans should never be eaten.

The energy released from foods in both aerobic and anaerobic respiration is chiefly in the form of chemical energy; but a certain amount of heat energy is also released. When damp hay is stored in a poorly ventilated barn, the respiration of the bacteria growing on the hay may cause heat energy to accumulate to a dangerous extent. Under such circumstances the hay may suddenly burst into flames. Many apparently mysterious fires may be traced to this cause.

Bacteria have a wider distribution than any other living organisms. They occur in large numbers in air, in soils, in water, in and on the bodies of other living organisms, and in dead and nonliving organic materials, such as cadavers, dung, garbage, humus and milk. Bacteria have been found at depths of many feet in soils and also in the ooze of ocean beds, far below the surface of the sea.

### Bacteria that are harmful to man

Certain bacteria rank high among our deadliest enemies. They cause a disconcerting variety of diseases that attack man, his domestic animals and his crops. They are responsible for such human diseases as tuberculosis, typhoid fever, some types of pneumonia, meningitis, tetanus, cholera, diphtheria, leprosy, several types of dysentery and various wound infections. They cause diseases among domesticated animals

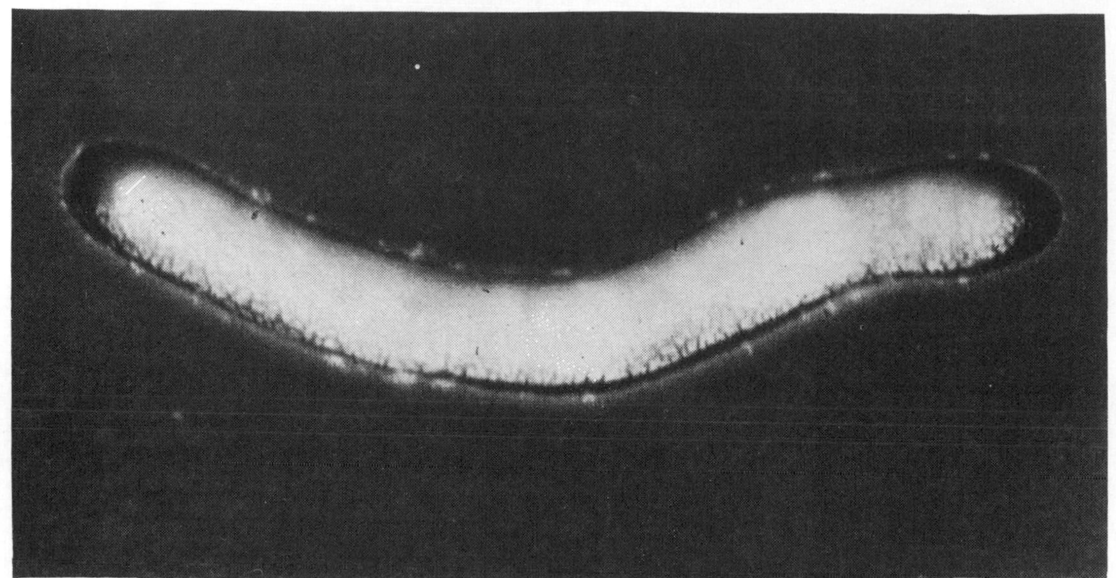

E. R. Squibb and Sons

*Mycobacterium tuberculosis,* the tubercle bacillus. It is magnified here something like 17,000 times.

— tuberculosis, anthrax, fowl cholera, pneumonia and glanders, among others. They are responsible for various diseases of crop plants, such as fire blight of pears, citrus canker, tomato and potato wilts, potato blackleg and soft rot of celery. These diseases result in tremendous crop losses throughout the world's agricultural areas.

Bacteria ruin great quantities of foodstuffs. They cause souring of milk, rancidity of butter and spoilage of both fresh and canned fruits and vegetables. The toxins (poisons) of bacteria in certain spoiled foods may result in ptomaine poisoning, botulism and other types of food poisoning. Fortunately there are various methods of preventing or at least reducing bacterial activity in foods. Heat sterilization, refrigeration and deep-freezing are all effective. So is desiccation, or the drying of foods; since bacteria usually require considerable quantities of water for their activity, they cannot grow, reproduce and respire in dried foods. Food may also be protected against bacteria by adding chemicals that are harmless to man but poisonous to bacteria.

Bacteria cause the decomposition of fabrics, wood and other products of organic origin. This takes place particularly in climates in which excessive heat is combined with high relative humidity; it is much more frequent and obvious, therefore, in tropical regions than in the temperate zones.

### Bacteria that are beneficial to man

The harmful activities of bacteria (harmful, that is, from our viewpoint) are more than offset by other activities that are directly or indirectly beneficial to man.

Bacteria play an important part in the production of many foodstuffs useful to man or beast; among those foods are vinegar, certain types of cheese, sauerkraut and ensilage (fermented plant tissues produced through bacterial action in silos and fed to livestock). Various valuable organic chemicals are also the product, in part at least, of bacterial action. Among these are acetone (widely used in industry, as in the manufacture of acetate rayon), butanol (which serves as a solvent for lacquers) and vitamins.

The activities of certain bacteria result in the retting (rotting) of pectins and other carbohydrate materials that cement together the fibers in the stems of flax and hemp and in coconut husks. The retting process causes these valuable fibers to separate from the tissues that surround them.

The fibers are later dried, spun and woven. The retting is accomplished by submerging bundles of cut stems or coconut husks in water, or by allowing bacterial action to proceed in piles of dew-soaked stems in the fields.

Bacteria cause the decomposition of dead plant and animal bodies and of the wastes of living organisms. They break down the proteins, fats, carbohydrates and other complex organic substances of these materials and transform them into carbon dioxide, ammonia and other simple inorganic compounds. In this way, bacteria rid the earth of organic debris. They also restore to soil and air the simpler compounds that green plants require for food manufacture and, hence, for ther existence.

Green plants make organic foods out of these inorganic substances of air and soil; they build these foods into their own protoplasm and cell walls. Animals obtain such foods by eating plant tissues, or by devouring plant-eating animals or carnivores that prey on plant-eating animals. Without bacteria, the supplies of inorganic raw materials required for the synthesis of organic foods would soon become exhausted, for these raw materials would be locked up in plant and animal bodies. Both plant and animal life would come to an end.

Certain exceedingly important nitrogen transformations are due to the bacteria that live in the soil. These transformations insure a continuing supply of nitrogen, in the form of nitrates, to plants. This is a vital matter, since nitrogen is essential to plant life. Ammonifying, nitrifying and nitrogen-fixing bacteria are involved in nitrogen transformations.

Ammonifying bacteria transform various organic compounds (derived from dead plants and animals and their wastes) into ammonia in the soil. There are two kinds of nitrifying bacteria: nitrite bacteria, which convert ammonia into nitrites, and nitrate bacteria, which transform nitrites into nitrates — the nitrogenous, or nitrogen-containing, compounds most readily absorbed and utilized by most green plants. Nitrogen-fixing bacteria convert nitrogen gas, which makes up 78 per cent of the atmosphere, into compounds that the plants can absorb. The atmospheric nitrogen transformed in this way is found in the soil, together with the other gases of which air consists.

The nitrogen-fixing bacteria include the so-called nodule bacteria, which inhabit the roots of legumes (clovers, alfalfa, soybeans, peas, cowpeas and so on) and a few nonleguminous plants. The bacteria enter the roots of the plants through root hairs in the soil. As they grow and develop, nodules are formed; these are more or less globular lumps, which are large enough to be seen with the unaided eye when the roots of legumes are carefully freed of adhering soil.

#### A fine example of symbiosis

A nodule consists in part of bacteria, in part of root tissues. The association between roots and nodule-forming bacteria is an example of symbiosis: that is, a state in which two organisms, living together, are mutually beneficial. The roots of legumes supply carbohydrates and other foods to the bacteria. These organisms, in turn, through their ability to fix nitrogen gas, provide the roots with nitrogenous compounds. If the bacteria are separated from the roots, they cannot fix nitrogen at all, or else fix it ineffectively.

Root nodules excrete organic nitrogen compounds into the soils in which nodule-bearing plants grow; thus they increase the nitrogen content of the soil and increase its fertility. The practice of crop rotation (one of the crops being legumes) is followed in part to increase the nitrogen content of agricultural soils.

The different varieties of nodule bacteria seem to be widely distributed in the soil. When a farmer plants alfalfa, or sweet clover or cowpeas in a field, the roots of these plants usually develop nodules, because their roots come in contact with the naturally occurring nodule bacteria of the proper variety.

Commercial bacteriological laboratories sell preparations containing different

USDA

Retting kenaf fibers in a Cuban agricultural station. Bacterial action is involved in this process.

varieties of nodule bacteria. Using such preparations, farmers inoculate seeds of leguminous crops before planting them. They can then be reasonably sure that there will be a high degree of nodule formation in the roots of these crops.

Several types of free-living bacteria (bacteria not occurring in roots) also can absorb nitrogen gas from the soil air and convert it into organic nitrogen compounds in soils. Such nitrogen-fixing bacteria enrich the soils just as the nodule bacteria do.

The organic nitrogenous compounds developed in soils as a result of the activities of nodule and free-living bacteria are later transformed by other bacteria into nitrates. The nitrates are then absorbed by the roots of higher plants.

Bacteria play a vital part in the breakdown of foods in the digestive organs of various animals. Cattle and horses, whose diet consists entirely of plant tissue rich in cellulose, do not produce cellulose-digesting enzymes; therefore they cannot digest cellulose independently of bacteria. The anaerobic bacteria that live in the intestinal tracts of such animals digest cellulose material (which may be as much as one third the total mass of the grasses eaten) and produce simpler substances. These can be absorbed through the intestinal walls and thus used as a source of nourishment. Bacteria occur in enormous numbers in the intestinal tracts of cattle, horses and sheep, and are dropped in great numbers with the feces.

Certain intestinal bacteria are important to the animals in which they live for another reason. They produce and secrete certain essential vitamins, notably vitamin K and vitamin $B_{12}$. (See the article The Vitamins, in Volume 5.)

Bacteria also occur in the human digestive tract, promoting the digestion of certain foods and furnishing vitamins. On the average, about 40 per cent of the mass of human feces consists of these bacteria, chiefly *Escherichia coli*. When *Escherichia coli* appears in water, milk and foodstuffs, it is usually assumed that these substances have been contaminated with human feces or sewage.

Certain bacteria produce antibiotic drugs, which are extremely valuable in treating various diseases of human beings and other mammals. Among the important antibiotics produced by bacteria (chiefly soil bacteria) are streptomycin, especially effective in treating tuberculosis and tularemia; chloromycetin, widely used in the treatment of typhus, Rocky Mountain spotted fever and gonorrhea; and aureomycin and terramycin, both highly effective in cases of intestinal, urinary and other internal infections and certain types of pneumonia and influenza. It should be noted that the most widely publicized antibiotic, penicillin, is produced by a fungus, not by a bacterium.

Some kinds of pathogenic (disease-causing) bacteria are used in the preparation of vaccines and serums; these are used to prevent or treat diseases caused by these same bacteria.

### How bacterial vaccines work

A bacterial vaccine is a preparation of dead or weakened bacteria or bacterial products and is injected into an animal body. The bacteria or their products stimulate the animal to produce substances called antibodies in its blood. If active bacteria of the same kind as the injected bacteria enter the body at a later date, they are held in check or destroyed by the antibodies. The antibodies may persist in the blood for long periods of time, conferring upon the animal a type of disease resistance called immunity. This action is similar to that which occurs when the animal has had a disease and has recovered from it; during its diseased state, it produces and accumulates antibodies, which then prevent the later development of bacteria of the same type in its body. Immunity following an attack of typhoid fever or cholera, for example, usually persists during the life of the individual.

Bacterial vaccines are used chiefly in immunizing human beings and domesticated animals against such diseases as diphtheria, cholera and typhoid fever. When a vaccine is injected into an animal it gives it a very mild form of the disease that these same bacteria would cause if they entered the animal's body in an active and unweakened condition. For example, the fever, digestive upsets and headaches that often result when typhoid vaccine is given are actually symptoms of a very light attack of typhoid fever, during which the sufferer produces antibodies.

A serum is a preparation from the blood of an animal that has been inoculated with bacteria (or other disease-producing agents) and has recovered from the disease that these bacteria or other agents cause. Blood is removed from the animal and is cleaned and sterilized; the serum is then separated. This contains antibodies which the animal formed as a consequence of the disease that attacked it. The serum, injected into another animal, confers immunity upon that animal, should disease bacteria enter it. Serums are especially effective in treating or preventing tetanus, diphtheria, meningitis and some forms of pneumonia.

### Active and passive immunity

Since the antibodies that develop in the blood following inoculation with a vaccine are produced by the animal thus inoculated, the type of immunity that results is called active immunity. In the use of serums, the animal that receives the serum does not make its own antibodies but acquires its immunity from the antibodies produced in the blood of another animal; hence immunity of this type is called passive immunity.

Disease-producing bacteria, of course, may be held in check by substances other than vaccines and serums. Sulfa drugs and antibiotics destroy many disease-carrying bacteria within animal bodies. Certain highly toxic chemicals, such as mercuric chloride, iodine and carbolic acid, may be used on body surfaces to kill bacteria; such chemicals cannot be taken internally be-

## BACTERIA THAT SPREAD DISEASE

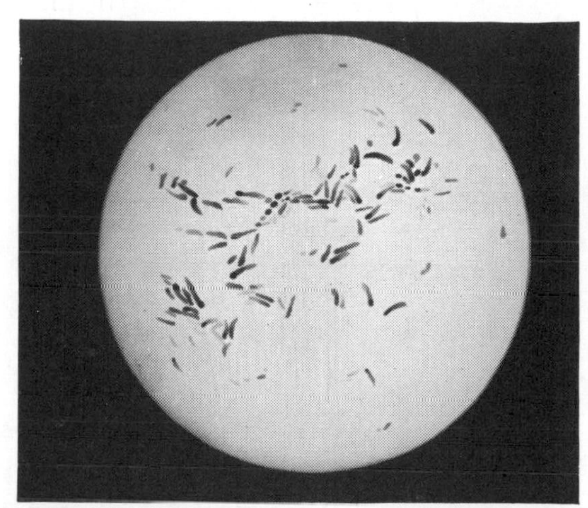

World Health Organization

Colony of the tubercle bacilli that are the cause of tuberculosis. A single bacillus is shown on a previous page.

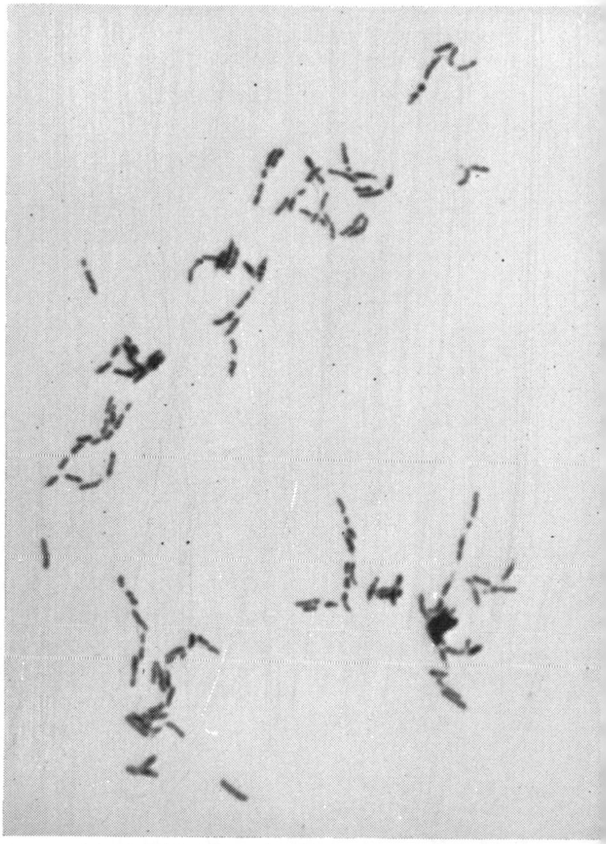

The bacteria (above) that cause diphtheria. The technical name of the organisms is *Corynebacterium diphtheriae*.

Lederle Laboratories Division, American Cyanamid Co.

Charles Pfizer & Co.

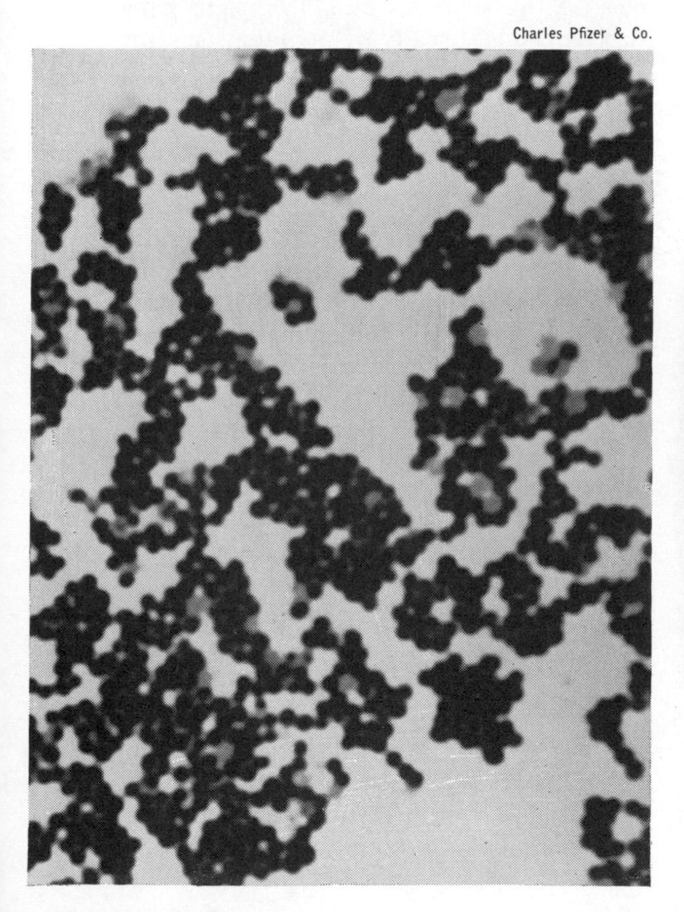

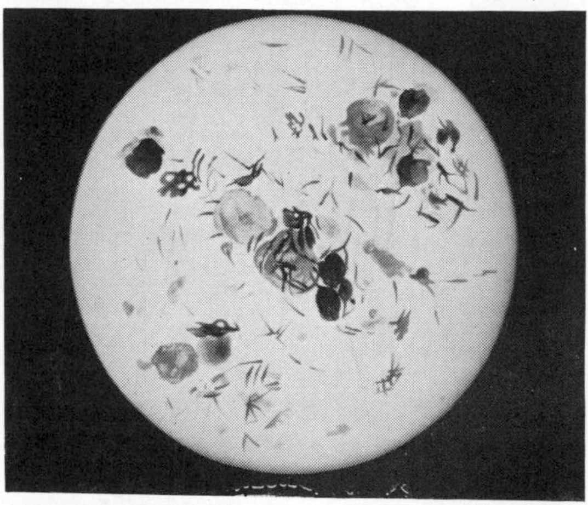

World Health Organization

Leprosy, a chronic and mildly contagious disease, results from infection by *Mycobacterium leprae* (shown above).

Boils are caused by the bacterium *Staphylococcus aureus* (left). It is also known as *Staphylococcus pyogenes*.

cause they are poisonous to body tissues as well as to bacteria.

**How bacteria are cultivated**

Bacteria are extensively cultivated in the laboratories of medical schools, hospitals, medical research institutes and universities, as well as in commercial laboratories and factories that produce antibiotics, useful organic chemicals, cheese, preparations of nodule bacteria and other economically valuable bacterial products. Great care must be taken in cultivating bacteria, partly because of the danger involved in handling disease-producing species, partly because, for many purposes, bacterial cultures must be kept pure. (A pure culture contains only the desired species of bacterium that is being studied or utilized, and is not contaminated by other types of bacteria.) Most bacteria that are important in medicine and commerce are heterotrophic and therefore cannot synthesize their own food. They must be supplied with organic nutrients, such as carbohydrates, proteins, fats and vitamins.

The solid-medium technique is used more widely than any other in cultivating bacteria. In this method, the requisite foods are mixed with gelatin or agar (a gelatinlike material produced by some species of red algae), which has been stirred into boiling water to form a viscous liquid mixture. The mixture is sometimes poured into shallow dishes, called Petri dishes; these are equipped with covers that admit air but prevent entry of dust and bacteria from the atmosphere. Sometimes the mixture is put into glass flasks or other types of containers.

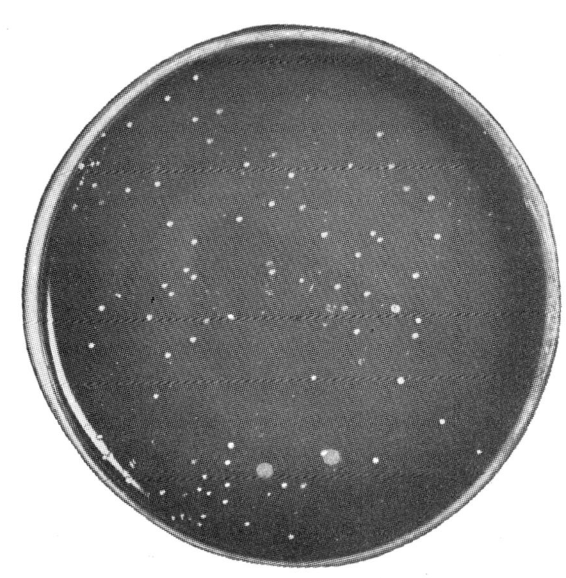

Society of American Bacteriologists

**A photograph of colonies of *Lactobacilli,* which are derived from human saliva. They grow in a shallow Petri dish, in which their food is mixed with gelatin or agar.**

The containers are sterilized for fifteen to thirty minutes under pressure in order to kill any bacteria or bacterial spores that may have fallen into the mixture or the containers during the preparation of the medium. The containers are then removed from the sterilizers and are allowed to cool. During the cooling process, the gelatin or agar becomes firm in the same way as a gelatin dessert; it is then ready to serve as a "garden" for the growth of bacteria.

The bacteria are placed on the surface of the solid medium by means of a sterile needle (usually of platinum or some other metal). The bacteria are obtained from a variety of sources: older cultures, soil, foods, infected plant or animal tissues and the like. Despite all precautions, foreign, unwanted bacteria sometimes enter a culture. In that case the culture is useless because it is not pure; it has to be destroyed.

Different species of bacteria require different cultural treatment. For example, anaerobic bacteria must be grown in containers from which oxygen can be excluded, while aerobic species must have ready access to oxygen. Some bacteria require cellulose in the growth medium; others, blood proteins; still others, amino acids, vitamins or other organic compounds. Certain bacteria are grown in liquid solutions of nutrients, rather than in semisolid media of the gelatin type.

The careful cultivation of bacteria makes it possible for researchers to carry on the fight against bacterial diseases and for manufacturers to make the products in which bacteria play a part.

*See Vol. 10, p. 272:* "General Works."

# THE BONES OF THE BODY

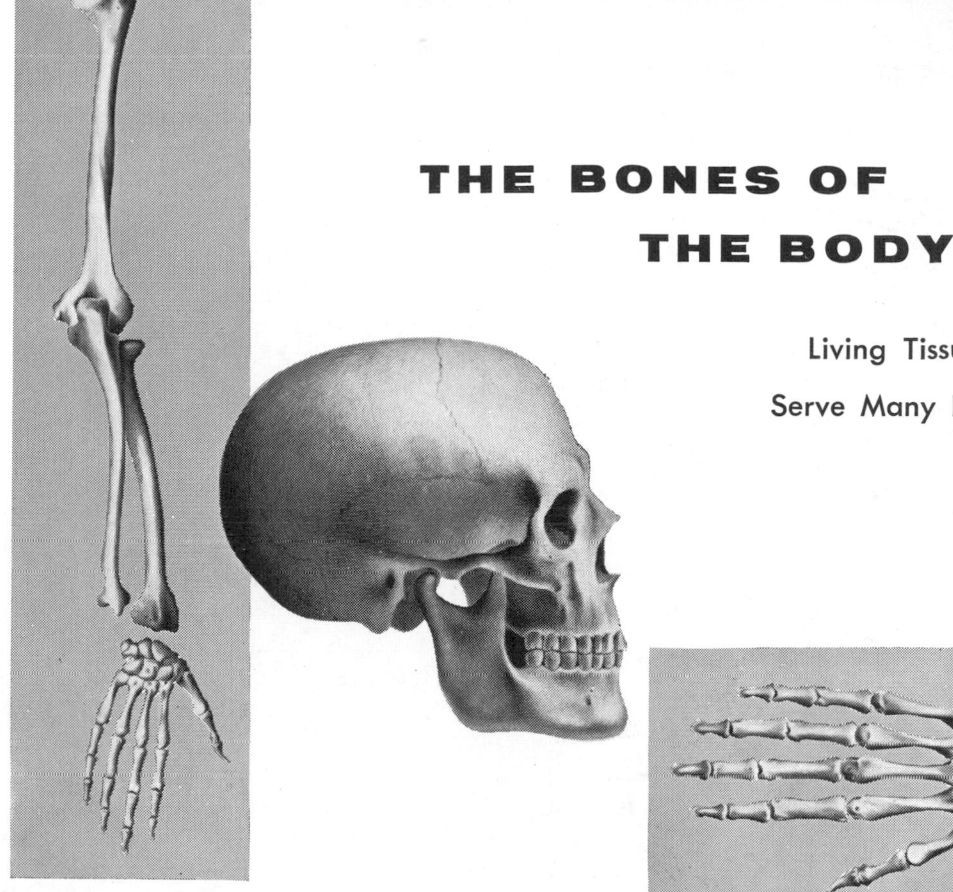

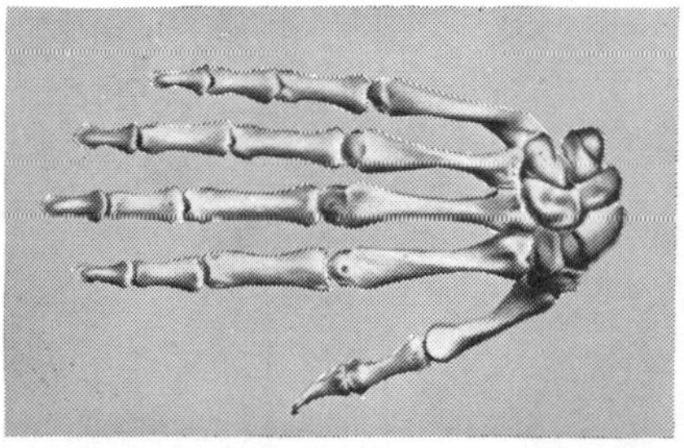

## Living Tissues That Serve Many Purposes

BY H. A. CATES

BONE is one of the most complex and most versatile of all tissues. It serves as a framework, giving form to the body, protecting the vital organs and bearing weight and strain. It is also a chemical laboratory, where red blood cells are manufactured, and a storehouse for the vital mineral calcium, which is supplied to the blood as it is required. A bone resists pressure and accepts strain; in other words it is hard, yet flexible. It owes these properties to its composition. It is two-thirds mineral matter — largely calcium phosphate — and one-third animal matter — chiefly a gelatinlike, elastic substance known as collagen. The hardness of bone is due to its mineral matter; its flexibility, to its animal matter.

Bones are living structures like the other tissues of our body. They generally maintain themselves very efficiently; when they are broken (fractured) they mend themselves. They are generously supplied with blood for their nourishment — witness the pink hue of fresh bone in the butcher shop. The sensitiveness of bone is familiar to all who have suffered a fracture or a bruise.

During man's early years, while a bone is still growing, it is not a single unit but is in three pieces. Each end, or cap, is separated from its long shaft by a plate of gristle or cartilage known as the growth plate. As long as this plate persists, the bone continues to grow and to get longer. As it develops, it often changes its shape and appearance; the thighbone of a child, for example, is quite different from that of an adult. During the growing stage, a cap is sometimes knocked off its shaft as a result of a fall or a hard blow. Unless the cap is accurately restored, a serious deformity may result.

The outer surface of all bone is made up of a dense, ivorylike substance known as compact bone. It is interlaced with numerous small canals that provide passage-

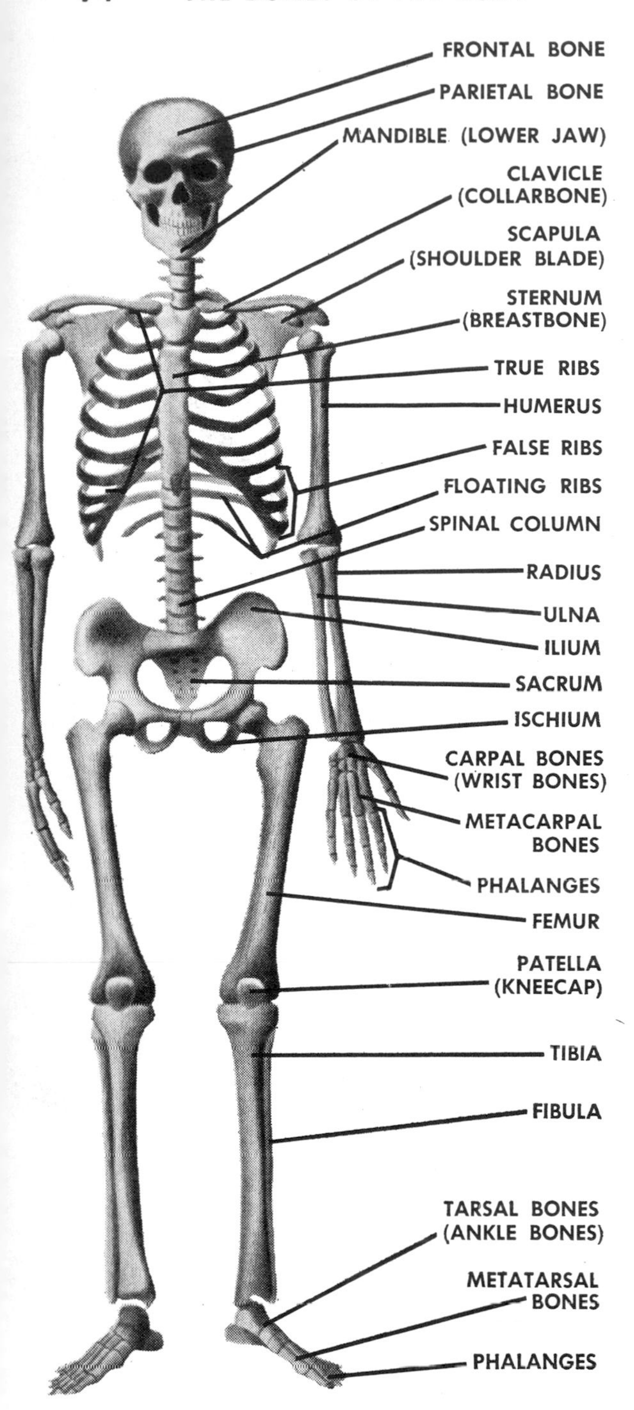


The major bones of the human skeleton: front view.

ways for the nerves and blood vessels that nourish it. Long bones consist entirely of a hollow shaft of compact bone; the central cavity is filled with fatty tissue called bone marrow. In some cases there is an outer covering of compact bone, and inside this outer covering a porous, trellislike structure called cancellous bone. The spaces left within cancellous bone are filled with bone marrow.

In the ends of long bones and scattered throughout the interior of flat bones and ribs, we see streaks of reddish material. These are manufacturing centers for red blood cells, which transport oxygen to the tissues. If the bone marrow that contains these manufacturing centers is damaged, the production of the cells is inadequate and the sufferer becomes anemic. Lead and benzol are injurious to bone marrow; workers who use these materials to any great extent are particularly likely to develop anemia. This condition may also result when a tumor of the bone marrow destroys some or all of the tissues that form red blood cells.

There are over two hundred bones in the human body, each with its own particular function. They are divided into two main groups — axial and appendicular. Axial bones receive and transmit weight and protect the body cavities; they include the skull, the backbone and the bones of the chest. The appendicular bones hang from the main or axial skeleton, as the name indicates; for appendicular comes from the Latin *appendere*: to hang. The bones of the arms and legs are appendicular bones.

The bones are not firmly welded together; they are connected at their ends by flexible joints. Muscles run from one bone to another across the joint between them. When the muscles contract, the bones approach each other; they are drawn farther apart when the muscles relax.

## THE AXIAL BONES

*The Backbone, or Vertebral Column.* The skeleton of a man's body is built around a backbone, just as the skeleton of a ship is built around a keel. So important is the backbone, or vertebral column, that it serves to divide the animal kingdom into two important groups — the vertebrates, who possess it, and the invertebrates, who do not. Fishes, frogs, reptiles, birds, beasts and man himself are vertebrates; they all have backbones constructed on the same basic plan. The invertebrates include worms, snails, lobsters, insects and a multitude of other living things.

The keel of a ship must be straight and rigid. It would be most unsatisfactory, however, if man's "keel" were a ramrod of bone running down his back. The vertebral column must have a certain amount of rigidity, indeed, but it must also be flexible so that our backs can twist, turn and bend. (As a matter of fact, vertebral comes from the Latin word *vertere*: to turn.) Instead of being a single rod of bone, then, the backbone is made up of a number of small cylindrical blocks. Each of these is fastened to its neighbor by a disc of resilient cartilage, which absorbs shocks and strains. The blocks are called vertebral bodies, or vertebrae; the discs of cartilage are called intervertebral discs. The backbone can sway and bend; at the same time it is rigid enough to support the head and to serve as a place of attachment for the ribs and the hip bones.

Behind our sturdy column of bony blocks lies the delicate spinal cord. It runs through a series of arches of bone — the neural arches — each forming part of one of the vertebrae. Within its bony housing, the spinal cord is protected by three layers of membranes. At the sides, between adjacent arches, are large openings called foramina through which the spinal nerves make their way to the organs of the body.

The appearance of bone under the microscope. Like the other tissues of the body, bone is a living substance.

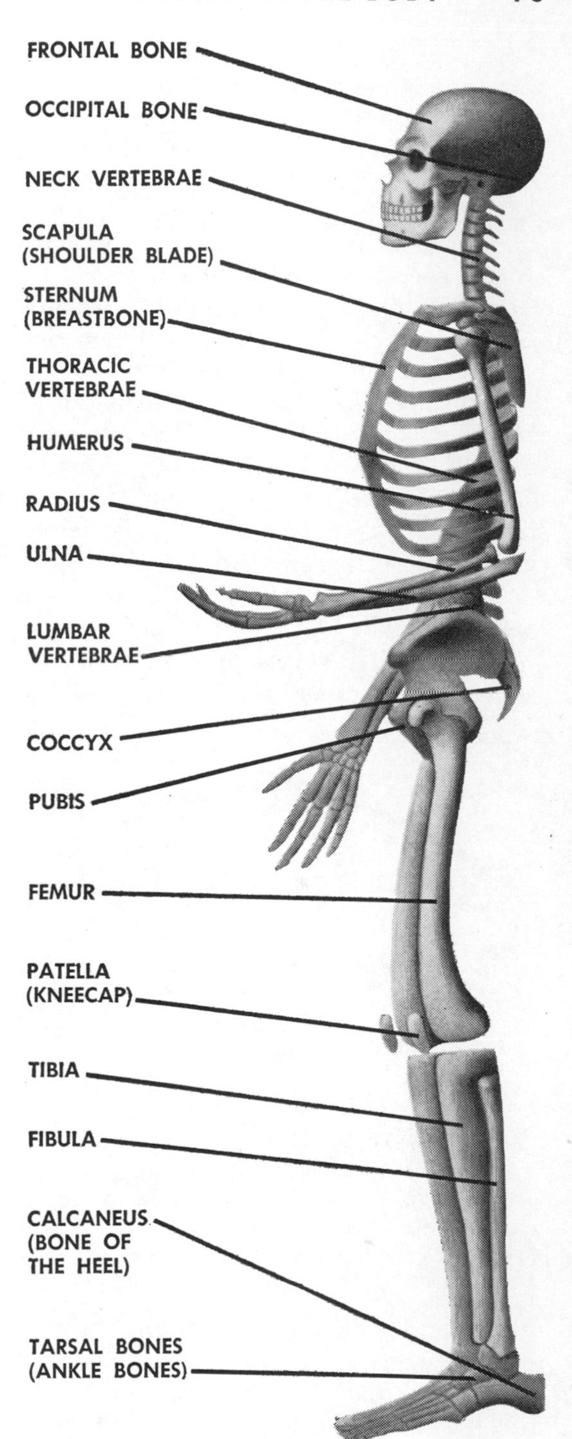


Side view of the important bones in the human skeleton.

The backbone is made up of twenty-four vertebrae, which fall into three main groups. First we have the seven cervical vertebrae in the neck (*cervix*, in Latin). The twelve vertebrae in the chest are called thoracic, from *thorax*, the Greek word for chest. Finally we have the five lumbar vertebrae in the loin (*lumbus*, in Latin), or small of the back. In general the vertebrae become larger as we go down the spinal column.

According to Greek legend, Atlas was a mythical giant who supported the earth on his shoulders. Because the first cervical vertebra supports the head in like fashion, it is called the atlas. This is merely a ring of bone broadened at the sides into two concave surfaces on which the head is poised and on which it can rock back and forth in nodding.

The second cervical vertebra is called the axis because it has a short, axislike peg of bone projecting upward into the ring of the atlas. Around this peg the head and atlas pivot when the head is shaken from side to side. The cervical vertebrae all have thick intervertebral discs, allowing the neck to move freely on its bony axis.

In the thoracic or chest region, the discs of cartilage are quite thin, and therefore there cannot be any considerable movement between adjacent vertebrae. This is very fortunate, since each thoracic vertebra carries a pair of ribs and undue movement would buckle and break them. The place where the last thoracic vertebra meets the first lumbar vertebra is a critical area, for a broken back most commonly occurs here.

The cartilage discs between the vertebrae of the lumbar region are very thick. Hence the spinal column is most flexible in this region, as anyone who has seen a contortionist can testify. A displaced, or slipped, disc has a crippling effect.

Below the lumbar vertebrae is the large os sacrum ("holy bone" in Latin), more commonly known as the sacrum; it really represents five vertebrae that have been fused. Why should this particular bone be called "holy"? According to one explanation, the Greeks also called the bone the "holy bone" (*osteon hieratikon*). But the word *hieratikon* means not only "holy" but also "unusually large." In this case, therefore, *osteon hieratikon* really meant "large bone." The Romans translated the Greek name into the literal Latin equivalent — that is, os sacrum.

The sacrum is a triangularly shaped bone, which tapers sharply downward; it gives solidity to the lower end of the spine. Along its upper sides, the sacrum is attached to the hip bones; through the place of attachment — the sacroiliac joints — the weight of the body is transmitted to the legs. These joints are relieved of strain only when we are lying down; consequently it is no wonder that they frequently cause trouble. Man pays dearly for venturing to stand on two legs.

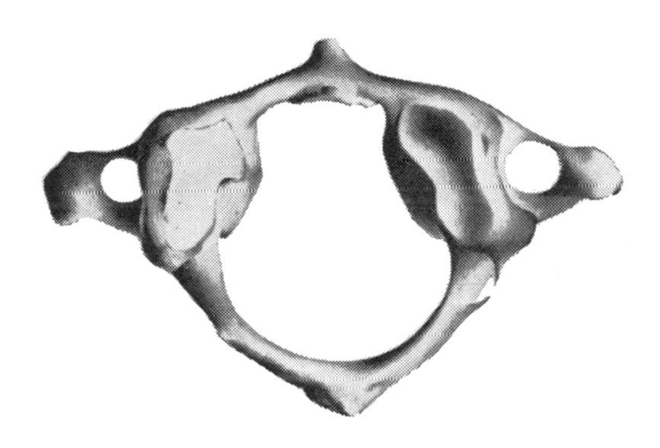

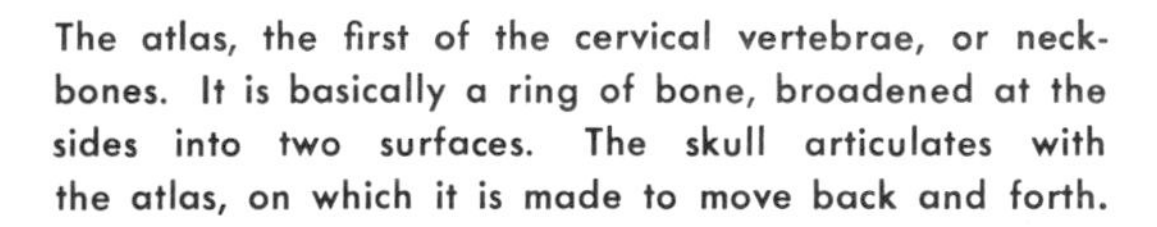

The atlas, the first of the cervical vertebrae, or neckbones. It is basically a ring of bone, broadened at the sides into two surfaces. The skull articulates with the atlas, on which it is made to move back and forth.

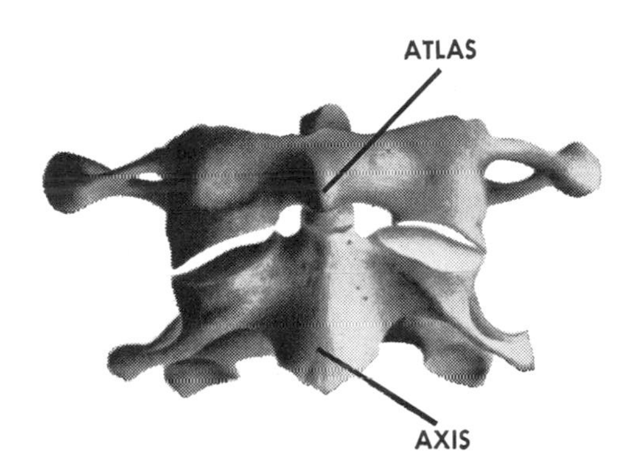


The axis, or second vertebra of the neck, articulates with the atlas above. The latter surrounds the upward-projecting peg of the axis; the entire skull, pivoted on this combination, is thus enabled to turn sideways.

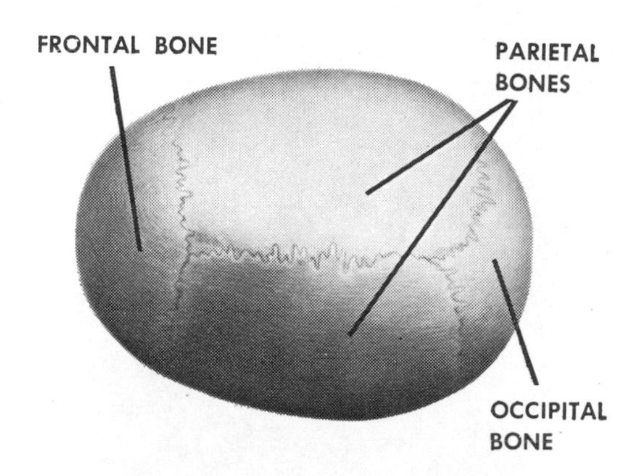


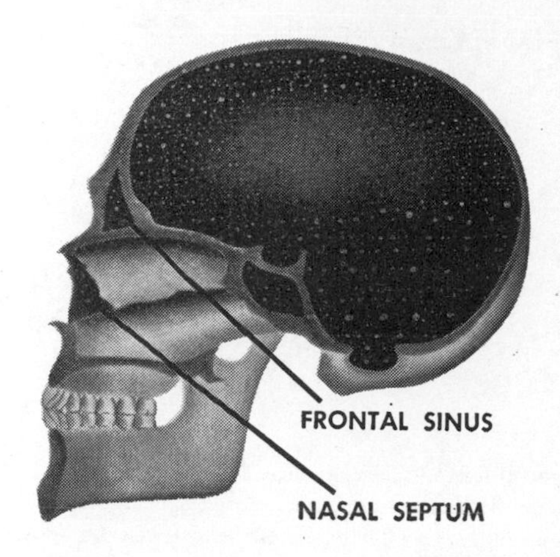


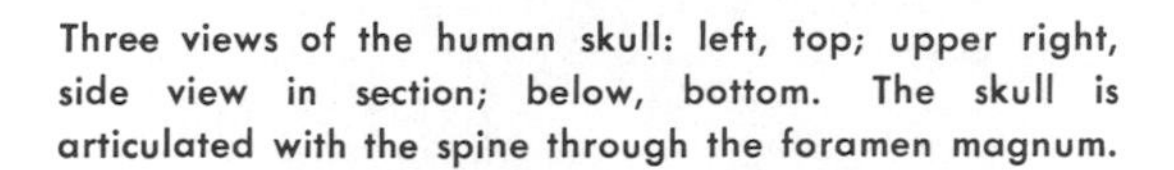
Three views of the human skull: left, top; upper right, side view in section; below, bottom. The skull is articulated with the spine through the foramen magnum.

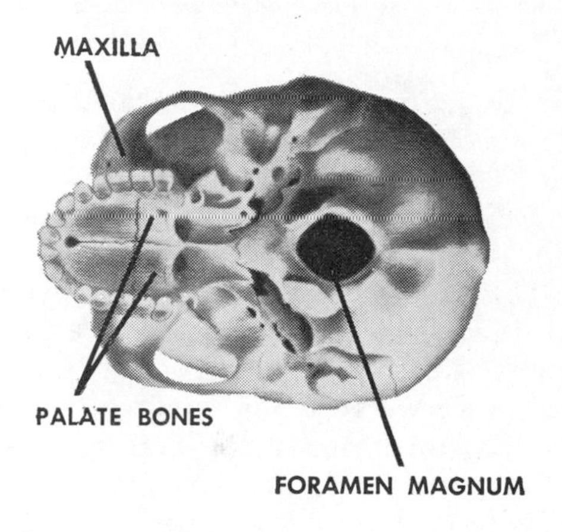


The lower end of the sacrum carries a series of small bones called the coccyx, representing the human tail. The coccyx serves no useful function. It may be broken if one sits down violently on hard ground; such an accident may be extremely painful and may call for the removal of the coccyx.

*The Ribs.* The ribs consist of twelve paired bows of bone; one pair is attached to each of the twelve thoracic vertebrae. They swing toward the front of the body; to their ends are attached elastic bars of gristle known as rib cartilages. The first seven pairs of these cartilages reach the breastbone, at the front of the chest, and the ribs to which they are attached are known as the true ribs. The next five pairs of ribs are called the false ribs. The cartilages of the first three pairs of false ribs are attached in each case to the ribs immediately above. In the case of the last two pair — also known as floating ribs — the cartilages have been reduced to short, pointed tips that are not attached to anything. The floating ribs extend only halfway around the body.

When we take a breath, the ribs are lifted upward and swing out, and the lower part of the breastbone swings forward too. All this increases the girth of the chest. At the same time the diaphragm, the domed muscular partition separating the thorax above from the abdomen below, contracts. This lessens its curvature and lengthens the up-and-down dimension of the chest. The increase in chest size allows air to be drawn into the lungs through the trachea, or windpipe.

Expiration (breathing out) takes place automatically and mechanically — unless of course we consciously try to hold our breath. Normally the muscles that move the ribs relax, and the elastic recoil of the rib cartilages brings the ribs back to their former position; at the same time the diaphragm also relaxes and again arches upward. All this diminishes the chest size. The lungs contract, air is forced out through the windpipe, and we have completed another of our numberless breathing cycles.

Not only do the ribs play an important part in breathing but they also shield the heart and lungs. Some of the abdominal organs, including the stomach, liver and kidneys, are protected by the lower ribs.

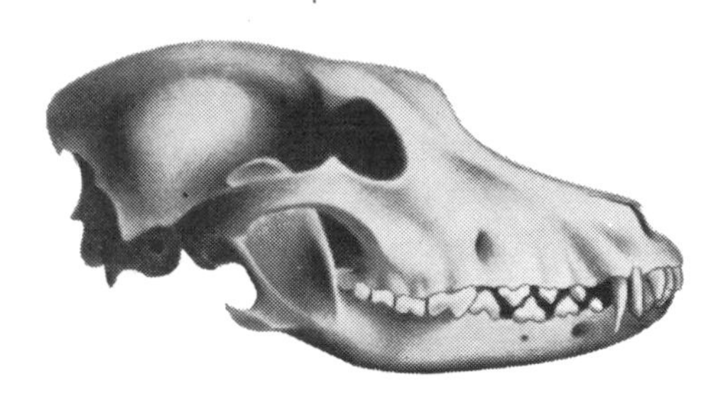

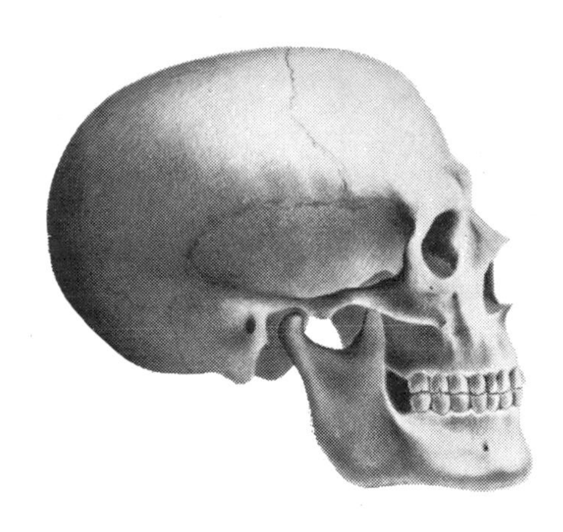

The skull of a dog and of a human being. Note the far larger brain case in man. The dog's long jaws are, in the living animal, worked by strong muscles attached to the sides of the skull. The reduction of these jaw muscles in man may have allowed his brain case to expand.

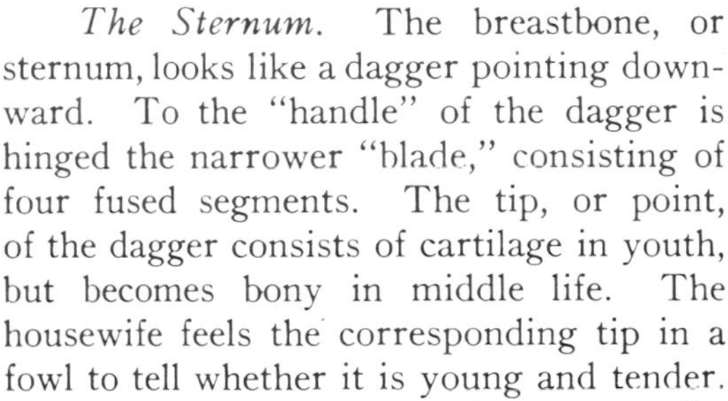

*The Sternum.* The breastbone, or sternum, looks like a dagger pointing downward. To the "handle" of the dagger is hinged the narrower "blade," consisting of four fused segments. The tip, or point, of the dagger consists of cartilage in youth, but becomes bony in middle life. The housewife feels the corresponding tip in a fowl to tell whether it is young and tender.

The cartilages of the first seven ribs are attached to the sides of the sternum; the two clavicles, or collarbones, are joined to its upper part. The wide "handle" of the sternum protects the great blood vessels diverging from the heart. The "blade" and the adjacent rib cartilages guard the heart itself.

*The Skull.* The last group of axial bones are those making up the bony skeleton of the head and face. There are twenty-two bones in the skull, twenty-one of them so firmly united that movement between them is impossible. One bone, the mandible, or lower jaw, is hinged to the rest of the skull just in front of the ear opening and is independent and movable.

The skull is conveniently divided into two parts: the brain box (case), or cranial cavity, and the bones of the face.

In lower animals, such as fishes or rabbits, the brain box is small and relatively insignificant; it lies behind the face, which, projecting forward, is very prominent. The brain box in man has expanded to such an extent that if we look at a human skull from above, no part of the face is visible. The forward expansion of the brain box has resulted in very deep sockets for the eyes. In lower animals the sockets are quite shallow and the eye is less protected than the human eye.

The walls, roof (vault) and back part of the floor of the brain box are formed by six curved plates of bone whose saw-tooth edges fit into one another like the pieces of a jigsaw puzzle. At birth these edges are not yet interlocked; that is why there are soft spots, called fontanelles, on the head of a new-born baby. The floor of the brain box is high in front, coming above the eyes, but it slopes rapidly downward. At its lowest point, some distance forward from the back of the head, there is a great hole called, appropriately enough, the foramen magnum (Latin for "big hole"). Through it passes the spinal cord, as well as various arteries and nerves.

In the middle section of the floor there lies a bone so exceedingly hard that it is called the petrous ("rocky") bone. Within it we find a complex system of minute caverns and tunnels in which the delicate organs of hearing and balance are sheltered. In the middle section of the floor we find also numerous small openings for the passage of nerves and blood vessels; their presence weakens the floor and makes fractures exceedingly serious. In this area the floor roofs the pharynx — the air and food passage at the back of the nose and mouth.

Below the very thin forepart of the floor lie not only the eye cavities but also the narrow slitlike upper part of the nasal cavities between them.

The paired nasal cavities are separated from the mouth by the palate. This is divided into two sections: the hard palate, in front, which is bony and rigid (we can feel it with the tip of our tongue just behind the upper teeth), and the soft palate, thin, soft and movable, lying directly behind it. The movable palate projects backward into the pharynx and by pressing against it shuts off the passageway to the nose when we swallow.

Each nasal cavity is separated from its fellow by a partition called the septum. The septum is mainly bony in structure, but in front it is made up partly of cartilage; that is why we can twist the tip of the nose. The side wall of each nasal cavity has scroll-like bones projecting from it. These are known as the shells (conchae) or whirlers (turbinates). Incoming air must make its way through the long curved passages of these bones; in the course of this journey the air is warmed and filtered. Mouth breathers, of course, lose these advantages. In front, the nasal cavities lead to the face by way of the nostrils. Behind, their much larger openings (internal nares or nostrils) lead into the pharynx; from here, the incoming air enters the larynx, or voice box.

The bones of the face are thin, delicate and often transparent. Those surrounding the nasal cavities contain sinuses (cavities); they are hollowed out till they are but thin shells of bone. Because they all open into the nose and are lined with a continuation of its mucous membrane, they are easily infected from the nose. The largest of the sinuses extends along the nose, above the upper teeth and below the eyes. Another sinus lies deep in the bone above each eye. There are a number of little sinuses in the bone between the eyes and a very inaccessible one at the back of the nose behind the eyes. Infections can spread rather easily from the sinuses to the eyes, where they may do irreparable damage. The hollowness of the sinuses increases the resonance of the voice and lightens the face.

The lower jaw, or mandible, is horseshoe-shaped and forms the bony boundary of the mouth; along the upper edge of the horseshoe are set the lower teeth. A sheet of muscle fills in the horseshoe and forms the floor of the mouth. From the open ends of the horseshoe there springs a pair of vertical plates of bone. They are called rami and they extend the mandible up to the joint.

The lower jaw is hinged to the skull just in front of the ear opening. You can feel it moving in its socket if you place a finger about an inch in front of the ear and open and close the jaw.

*The Hyoid Bone.* In the angle where the floor of the mouth meets the neck is a slender U-shaped bone known as the hyoid bone. "Hyoid" means "upsilon-shaped" in Greek; the hyoid bone is so called because the corresponding bone in animals with a forward-pointing snout looks something like the Greek letter ϒ, or upsilon. Ligaments suspend the hyoid bone from the skull, and from the bone there hangs the Adam's apple, or thyroid ("shieldlike") cartilage. This is the chief cartilage of the larynx (voice box). In swallowing, the larynx is pulled up to the hyoid bone.

## THE APPENDICULAR BONES

*The Clavicle.* The clavicle, or collarbone, is a slightly curved horizontal structure; it gets its name from its resemblance to a type of key common in ancient Rome. (*Clavis* is the Latin word for key.) At its inner end the clavicle is united to the breastbone; at its outer end, about an inch from the point of the shoulder, it is attached to the shoulder blade. The clavicle is really a horizontal strut, forcing the shoulder joint to keep its distance from the breastbone no matter how varied its movements. The swing of the clavicles can be felt when the shoulders are shrugged, depressed, brought forward or thrown backward. When the collarbone breaks, the shoulder collapses inward, and these movements are no longer possible.

*The Scapula.* The scapula, or shoulder blade, is a thin, triangular bone suspended from the outer end of the clavicle.

It is buried in powerful, heavy muscles and moves over a considerable distance as it slides over the upper and back part of the chest. At the armpit the outside border of the scapula forms the socket portion of the flexible ball-and-socket joint between the shoulder blade and upper arm. Because of this joint, the upper arm can move in all directions. Ordinarily bones can move in only two directions at the joints — that is, forward and backward.

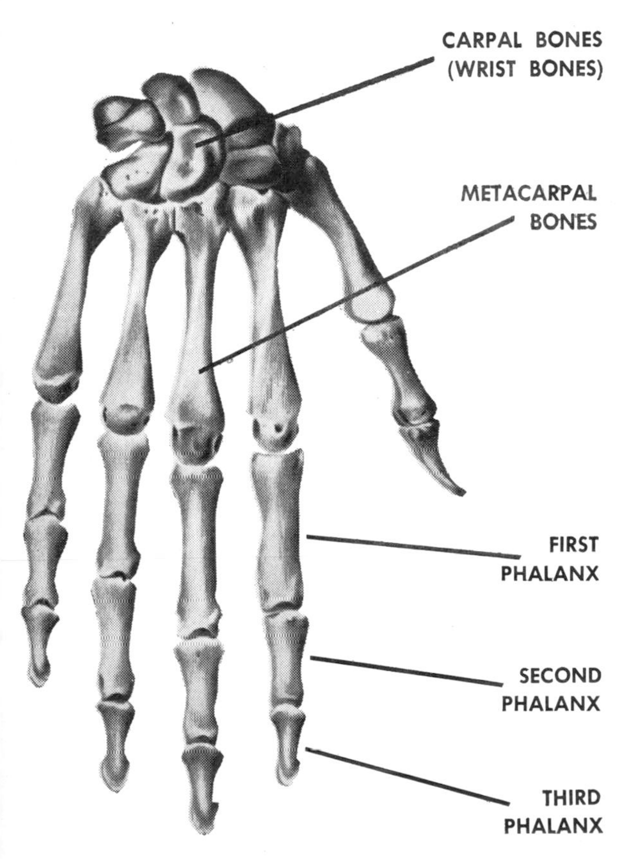


Bones of the hand; compare with diagram on next page.

Running across the back of the shoulder blade is a projecting ledge of bone: the spine of the scapula. As it passes toward the shoulder joint, it becomes more and more prominent until it finally expands into a large bony hood (the acromion process) that protects the vital ball-and-socket joint of the shoulder. The hood ends in a projection; the clavicle is attached to this.

Another strong piece of bone projects forward from the upper border of the shoulder blade just beside the shoulder joint. It is about the size and shape of a bent finger or crow's beak, pointing downward. It is called the crowlike (coracoid) process, and from its tip muscles stream down the front of the arm. The clavicle is firmly bound to the coracoid process by a strong ligament.

*The Humerus.* In the upper arm there is but one bone, the humerus. Its knoblike upper end forms the ball portion of the ball-and-socket shoulder joint. The four muscles that immediately surround the shoulder joint are attached to the outer part of the knob. Since the scapula and the humerus fit rather poorly, it is these muscles that hold the bones in close contact.

The upper part of the shaft is at first cylindrical; as it approaches the elbow it flattens out and expands sideways. There are two joint surfaces, side by side, at the expanded end of the humerus. The inner one resembles a spool, pulley or hourglass laid on its side; it is known as the trochlea, the Greek word for pulley. The inner bone of the forearm, the ulna, hooks on to the trochlea from behind.

*The Radius and Ulna.* Because the ulna is primarily concerned with bending and straightening the elbow, it is thick and massive at the elbow and tapers to a disclike head at the wrist. The radius carries the hand at the wrist and so is thick and heavy there; it narrows at the elbow to the same kind of disclike head that the ulna displays at the wrist.

The upper end of the ulna looks like a monkey wrench whose jaws grasp the lower end of the humerus. The part of the ulna that bends around the trochlea of the humerus is called the funny bone. It is the projection we feel when we run a finger along the elbow. The nerve running to the ulna passes close by it and when we strike the funny bone, this nerve tingles.

When the palm is directed forward and the thumb is pointing away from the body, the radius and ulna lie side by side. As we turn the palm so that the thumb is directed toward the body, the radius turns; its shaft and lower end roll across the ulna. Thus, with the thumbs pointing toward the body, the forearm bones are crossed.

*The Carpus.* The wrists consist of eight little bones that together make up the carpus (Greek for wrist). These bones are arranged in two rows of four bones each, the lower row supporting the bases of the bones that lie in the palm. The varied movements of the carpal bones contribute a good deal to the flexibility of the wrist.

*The Metacarpus.* The bones in the palm of the hand, called metacarpals, are arrayed beside one another like the sticks of a fan. Their square bases are in contact with one another and can move but little. Their rounded heads form the knuckles, and they are not in contact; between them run little tendons which reach the free fingers. Because the metacarpals diverge from one another as they come off the wrist, the fingers tend to spread when they are straightened and to come together when one clenches the fist. The metacarpal of the thumb moves freely and is opposable to the rest of the fingers. This feature causes the hand of man to be a remarkably useful organ.

*The Phalanges.* The thumb has two bones in line; each of the other fingers has three. These bones are called phalanges; they are flat in front and rounded behind. The bases of the first phalanges rest on the rounded knuckles; here the fingers can be bent and straightened and also spread apart and brought together. The other joints serve as hinges, permitting only bending and straightening. The terminal phalanges — those farthest from the knuckles — have triangular tips that support the fingernails.

*The Hip Bone.* The hip, or flank, bone is so irregular in shape that it deserves its scientific name of innominate bone (from the Latin *os innominatum*: "bone without a name"). It consists of two symmetrical parts attached above to the sacrum and uniting below at the crotch. Three sets of bones make up the hip bone: the ilia, the ischia and the pubes (plural, respectively, of ilium, ischium and pubis).

The two ilia are the topmost bones. Each ilium is a large curved structure; it is attached to the sacrum at the sacroiliac joint, which we have already described. The ilium is the bone that we feel when we

The most important bones of the human arm and hand.

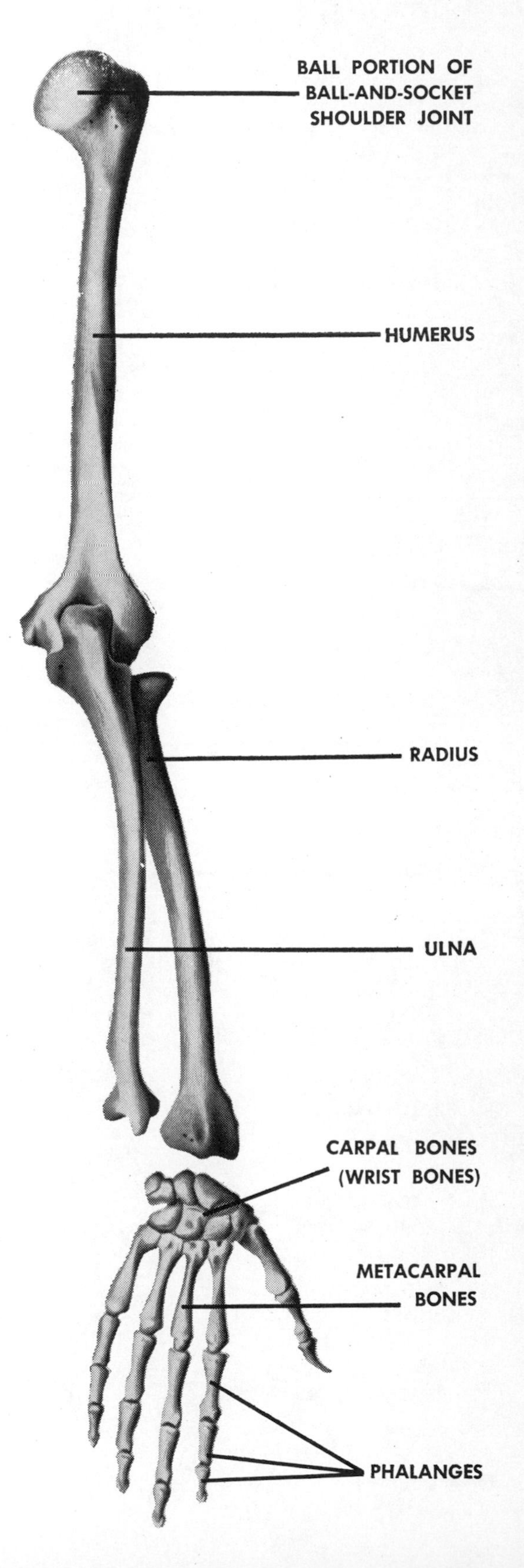

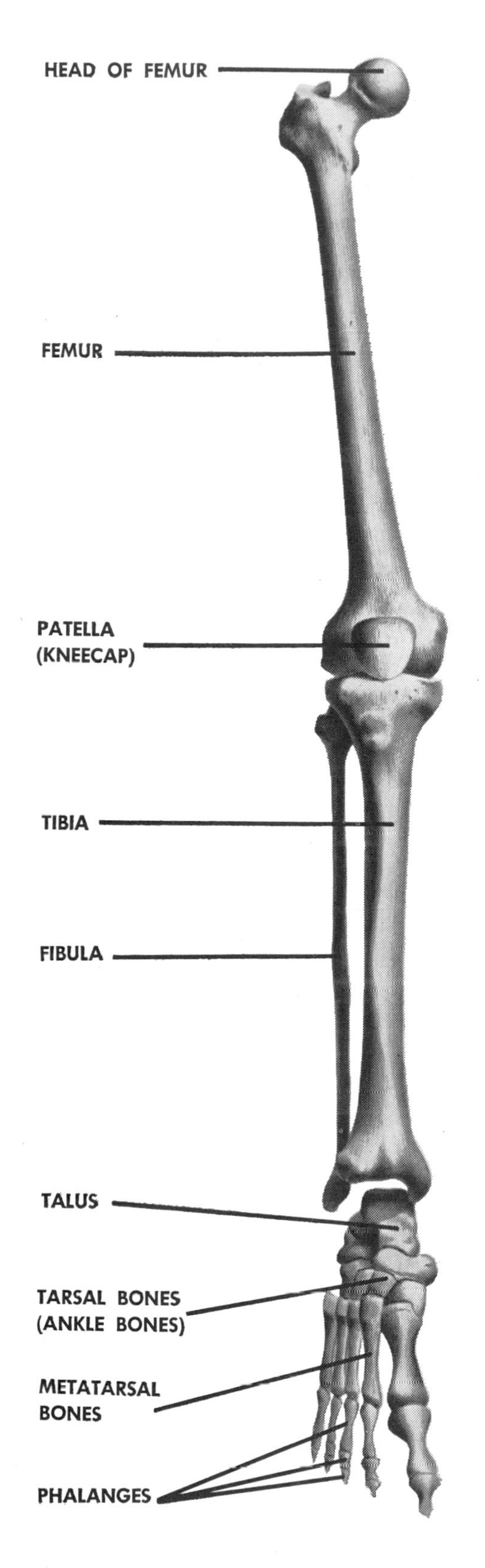


The major bones of the human leg and foot: front view.

place the hand on the hip. The large surface of the two ilia gives protection to the lower abdominal organs.

Each ilium narrows as it descends; at its lower end it is hollowed out into a cup-like structure about two inches in diameter. This is the circular socket known as the acetabulum ("vinegar cup" in Latin), because of its fancied resemblance to a small Roman cup that held vinegar or a sauce. The acetabulum receives the rounded head of the thigh bone, or femur, and with it forms the hip joint, the most secure ball-and-socket joint in the body. There are, of course, two acetabula, one on each side of the body. The ilia are very strong; they transmit the weight of the body to the legs.

A stout column of bone, called the ischium, runs almost vertically downward from the acetabulum; it is the bone we rest on when we sit up straight. The ischium merges with a bony ring known as the pubis. The three sets of bones of the hip region — the ilia, ischia and pubes — together with the sacrum form a bony basin, with the bottom knocked out, known as the pelvis (Latin for basin). Its deep cavity houses the reproductive organs as well as the lower parts of the digestive and urinary systems. At birth all infants must pass from the womb down through the bony circle of the pelvis, which is wider in woman than in man.

*The Femur and Patella.* The thighbone, or femur, is the longest bone of the body. Its neck, representing the upper three inches of its cylindrical shaft, is bent inward and is surmounted with a bony ball, or head, which fits snugly in the acetabulum. As the femur approaches the knee it expands sideways into a pair of prominences known as condyles. The condyles rest and move on the flat top of the tibia, the stout inner bone of the leg.

The kneecap, or patella ("little pan"), is a small triangular bone lying in front of the lower end of the femur; it is imbedded in the thick tendon that passes over the knee joint from the femur to the lower leg. The kneecap slides down the femur when the leg is bent to protect the widely gaping knee joint.

*The Tibia and Fibula.* There are two bones in the leg just as there are two in the forearm. The inner one, called the tibia, is very stout and strong; it alone receives and transmits weight. The outer one, the fibula, is thin and delicate. At the ankle joint the fibula rests in a hollow on the side of the tibia and is bound very firmly to it. Beyond this place the fibula projects downward as a flange — the bump we feel on the outside of our ankles — where it forms an important side support for the ankle joint. The fibula takes no part in the knee joint, which is formed by the articulation of the femur and the tibia. A severe twist of the ankle may break either flange, usually that of the fibula.

*The Foot Bones.* The talus, the uppermost bone of the foot, rests below on the heel bone, or calcaneus, which takes most of the strain when we stand or walk. The talus does not rest squarely on the calcaneus; the front part forms a rounded head that projects beyond the calcaneus toward the inner side of the foot. The talus, calcaneus and the five small bones in front of them are called the tarsals; in front of these again are the metatarsals and phalanges, comparable to the metacarpals and phalanges of the hand.

The human foot, showing the various bones.

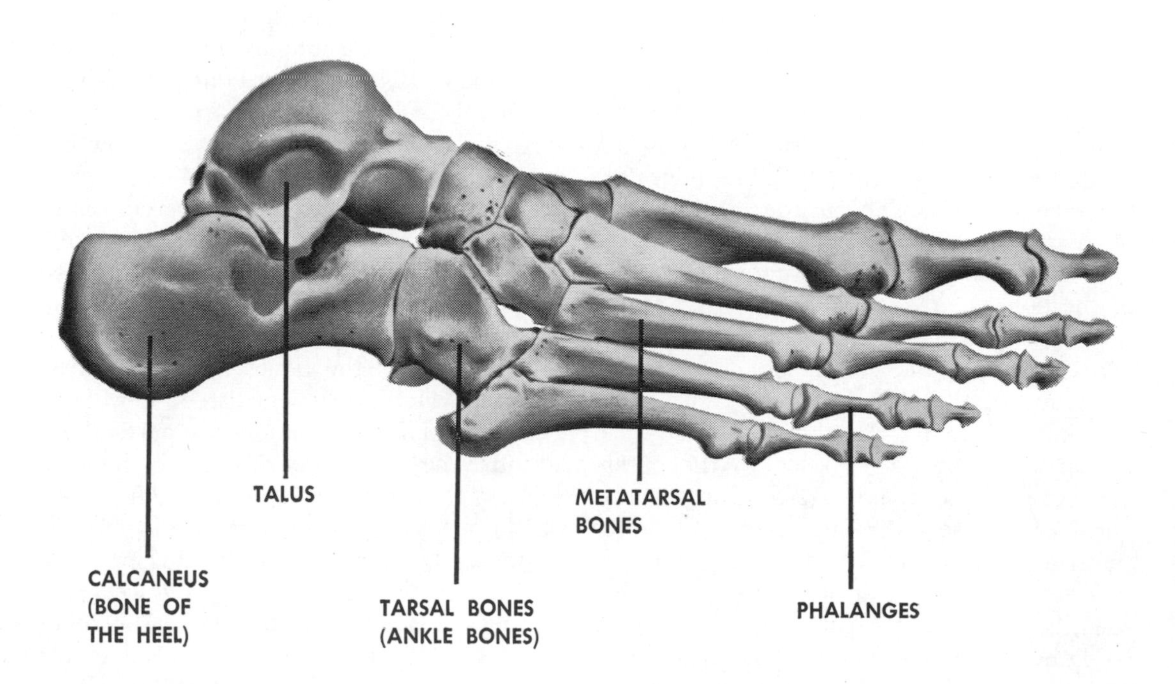


The tibia is flat on top and supports the condyles of the femur; it is firmly held to the condyles by strong ligaments. As the tibia descends, its shaft tapers; below, it expands a little again to take part in the ankle joint. It ends in a flange of bone, forming the inner ankle. The flanges of the fibula and tibia hold the talus — the topmost bone of the foot — as in a vise. At the ankle joint the foot can be bent up or down but there is normally no side play. A

The head of the talus is the keystone in a high arch that gives spring to our walk. Should the head lose the support of ligaments and muscles, it sinks down and the foot becomes painfully flat. This is the well-known condition that is known as fallen arches, or flat feet.

The heel and the little toe of the advanced foot receive the weight of the body; the great toe gives the thrust to propel the body forward. If the arch begins to fall, the effort required to give this forward thrust is a painful one; in order to avoid it, the unfortunate sufferer shuffles along with his feet turned out.

## INJURIES AND DISEASES OF THE BONES

Bruises are the commonest of all bone injuries; who has not bumped his arm or barked his shin? A bruise causes the tearing of blood vessels; as a result blood escapes into the surrounding tissues and there is a swelling. When the blood is absorbed again, the swelling goes down. Unless a bruise has been unusually severe, it has no marked permanent effects.

Far more serious are fractures, or breaks in the bone. The bones of the young are very soft; hence they often splinter only on the convex side instead of breaking completely. This condition is known as a greenstick fracture, because a green twig splinters in much the same way. As one grows older, the bones break more easily. The bones of the aged are particularly brittle; so are those of persons suffering from the rather rare ailment called fragilitas ossium ("fragility of the bones").

There are two general classes of fractures — simple and compound. In a simple fracture the skin is not broken; in a compound fracture the skin is broken and the injured bone is exposed to the outer air. Compound fractures are always dangerous, because germs may penetrate into the wound from the outside.

A fracture is a dramatic thing; in the twinkling of an eye it can transform a perfectly healthy individual into a helpless cripple, writhing in pain. The diseases that attack the bones perform their work more slowly and in a less spectacular way; yet they are often more deadly than fractures.

One of the best-known children's diseases is rickets, which brings about deformity of the bones. Rickets is due to a faulty diet in which calcium, phosphorus and vitamin D are not adequately represented; as a result of this deficiency the bones do not solidify properly. The child becomes knock-kneed or bowlegged, his spine becomes curved and there are other deformities. The treatment consists of supplying an adequate diet and exposing the patient to sunshine or to artificially produced ultraviolet rays.

In some diseases the bones become inflamed. We find this condition in osteomyelitis ("bone-marrow inflammation"), which results when certain microbes of the staphylococcus group infect the inner part of the bone. Pus forms in the stricken area and there is pain and fever. The pressure of the pus may destroy the blood vessels in the marrow and may deprive the bone of its necessary supply of blood; this may cause the death of the bone. Furthermore, if the pus is not drained in time, general blood poisoning may result.

Tuberculosis of the bone is a chronic, or persistent, kind of inflammation, produced by the same bacillus that brings on tuberculosis of the lungs. The bacillus attacks the ends of the long bones or the vertebrae; it causes ulcers to form and makes the bone rot away. Its effects are slow but become increasingly severe in time.

Both osteomyelitis and tuberculosis of the bone are curable if treatment is started at an early stage. Drainage of the affected areas and rest are two important factors in any cure.

Sometimes the bones are attacked by tumors, which really consist of new bone growths. Malignant tumors are particularly serious; they gradually destroy the bone itself and they may also appear in other parts of the body. The only possible cure is removal of the bone; nor is this measure always effective. In some cases cancer originating in the breast, or kidney or other regions of the body may be spread to the bones by way of the blood or lymph. Gradually healthy bone is replaced by cancer tissue, and in time any ordinary exertion, such as raising the arms, may bring about a fracture. Cases of this sort are always fatal, at least in the present state of our medical knowledge.

The injuries and diseases to which bones are subject are a constant reminder that they are living tissue and not inert, stonelike masses. The terms "healthy bone" and "sick bone" are meaningful.

*See also Vol. 10, p. 276*: "Anatomy."

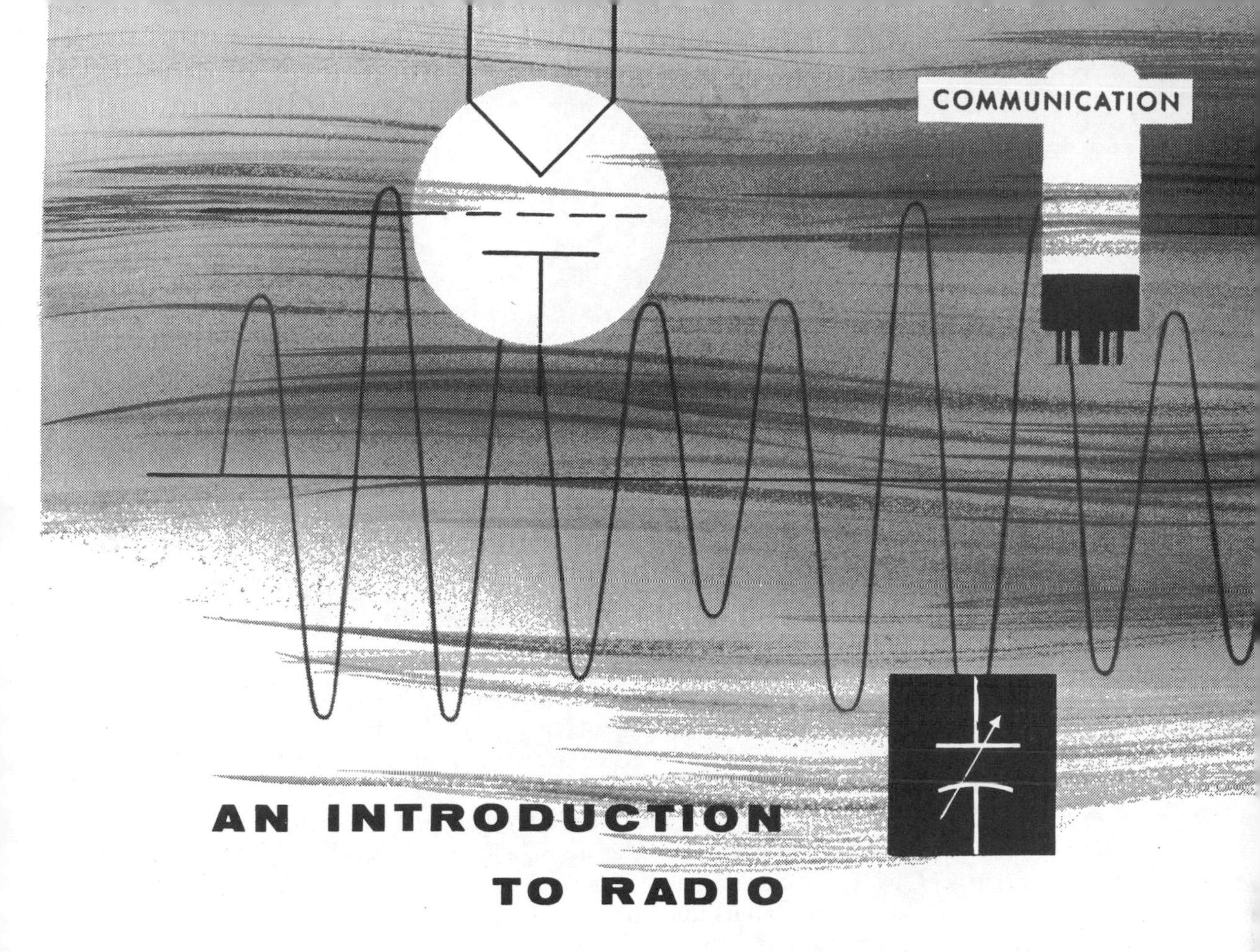

# AN INTRODUCTION TO RADIO

## How Your Receiving Set Brings You Voices and Music from Distant Places

BY GEORGE HOEFER

MODERN science has harnessed certain radiations traveling through space at the speed of light, and has provided methods of communication that would have seemed fantastic a few generations ago. The radiations in question are called radio waves. They make it possible to transmit and receive sounds and pictures over vast distances without the use of connecting wires. In the article on television, in Volume 5, we tell about the pictures that appear on the screens of our television sets. We are going to explain, in this chapter, how sound patterns are sent through space and reproduced in radio receivers.

Radio waves represent one form of the radiant energy called electromagnetic radiation. (See the article on this subject, in Volume 7.) There are other kinds of electromagnetic radiation, including heat rays, light rays, X rays, gamma rays and cosmic rays. They all travel through space at the same speed — approximately 186,300 miles (300,000,000 meters) per second. This speed is several hundred thousand times greater than that of sound waves.

There are various differences between radio waves and light and heat rays. For one thing, light rays will not penetrate many objects, while radio waves will go through any nonmetallic substance. Heat waves are easily absorbed and are not very effective away from their source. Radio waves are not easily absorbed; as a matter of fact, some of these waves keep bouncing back and forth between the earth and the upper atmosphere.

Radio waves, like ocean waves and sound waves, result from a special kind of disturbance. Ocean waves are generated in water by the force of the wind. Sound waves are produced by the vibrations set up in air (or some other medium), as when

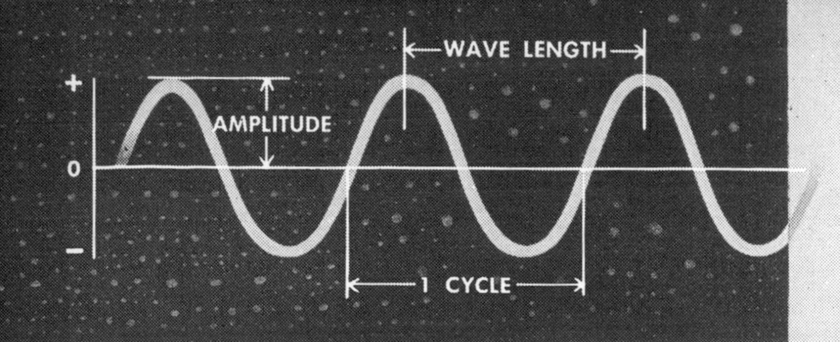


1. Characteristics of an alternating current. From zero, the current builds up to a peak in the positive direction, is reduced to zero, and then reaches a peak in the negative direction.

a tree falls, or a violin plays or a child cries. The disturbances that give rise to radio waves are oscillations — to-and-fro movements — in an electric current.

An electric current is made up of electrons (negatively charged particles) flowing through a conductor. If these electrons move constantly in one direction, the current is direct; it is alternating if the electrons periodically reverse the direction of their flow. It is the back-and-forth movement of electrons in an alternating current that produces radio waves. Conversely, radio waves can bring about an alternating current, as we shall see.

We can represent the characteristics of an alternating current by the diagram in Figure 1. From zero the current builds up to a peak in the positive direction and then is gradually reduced again to zero. Immediately upon reaching zero, the current reverses and builds up to a peak in the negative direction. Again it returns to zero, to repeat the process as long as it continues to flow. The strength of the current at its peak for a given pulse is called its amplitude; the distance from one positive (or negative) peak to the next represents its wave length. A completed back-and-forth movement is a cycle. The number of cycles per second is known as the frequency. Each current cycle produces a single radio wave; consequently, the frequency of the alternating current corresponds exactly to the frequency of the radio waves it generates.

If we know the frequency of a radio wave, we can determine its wave length by dividing its speed by its frequency, as follows:

$$\text{Wave length in meters} = \frac{300{,}000{,}000 \text{ (speed in meters per second)*}}{\text{Frequency (cycles per second)}}$$

* Remember that the speed of a radio wave is always 300,000,000 meters per second.

Obviously, high frequencies go hand in hand with short wave lengths, while low frequencies give long wave lengths. As we shall see, we can vary the frequency, wave length and amplitude of waves by making certain adjustments in the apparatus that is used to create them.

### The generation of radio-frequency currents

To produce radio waves that will bring about the desired sound patterns in our receiving sets, we must first generate radio-frequency electric currents — currents whose frequencies correspond to those of the radio waves themselves. Radio-frequency currents are produced in transmitting stations. These are usually located on the outskirts of cities, where interference is likely to be at a minimum. There are exceptions, however. In New York City, all local stations are serviced by a transmitting tower atop the Empire State Building, in the heart of Manhattan.

A generator such as is used to produce the familiar 60-cycle alternating current supplied to our homes could not possibly generate radio-frequency currents. Such currents oscillate at the rate of hundreds of thousands or even millions of times per second. In order to supply frequencies of this order, we use an oscillating circuit, or oscillator.

Each oscillating circuit has an inductance coil, or inductor, and a capacitor, or condenser. The inductance coil consists of a wire wound around a hollow cardboard tube. Whenever current is passed through the coil, a magnetic field is set up around it. The capacitor is made up of two or more plates of tin foil, brass or aluminum. The plates are separated by a substance (air, glass, paper, mica or oil) that will not conduct electricity; it is known as the dielectric. The capacitor stores electric charges up to a certain capacity when a steady voltage is applied to the terminals of the device.

The simple oscillator circuit shown in Figure 2 shows an inductance coil and a capacitor hooked together. The capacitor has received an electric charge. Plate *a* is negative, plate *b* positive; this means that there is a surplus of electrons at *a* and an electron deficiency at *b*.

When the circuit is closed and the condenser plates are connected, there will be a flow of current from the negative plate, *a,* toward the positive plate, *b*.* As the electrons pass through the inductance coil on their way to *b,* the rate of flow will be retarded at first as a magnetic field is set up around the coil. Then this field will collapse and the current will continue to flow through the circuit in the same direction as before.

There will again be a slight lag as the electrons pile up at plate *b*. The formerly positive plate will become negative, while the formerly negative plate will be positive. When the capacitor is fully charged, current will flow from plate *b* toward plate *a* and again will pass through the inductance coil, setting up a magnetic field. The electrons will flow back and forth between *a* and *b* at an extremely rapid rate. Their oscillations will make up an alternating high-frequency current.

We can produce any desired frequency within certain limits by making adjustments in the induction coil or the capacitor. We can change the length of the inductance coil, or the number of turns of wire or the type of core; we can add more plates to the capacitor or bring the plates into closer contact with each other.

* Electrons always flow from a place where they are in excess to a place where they are deficient.

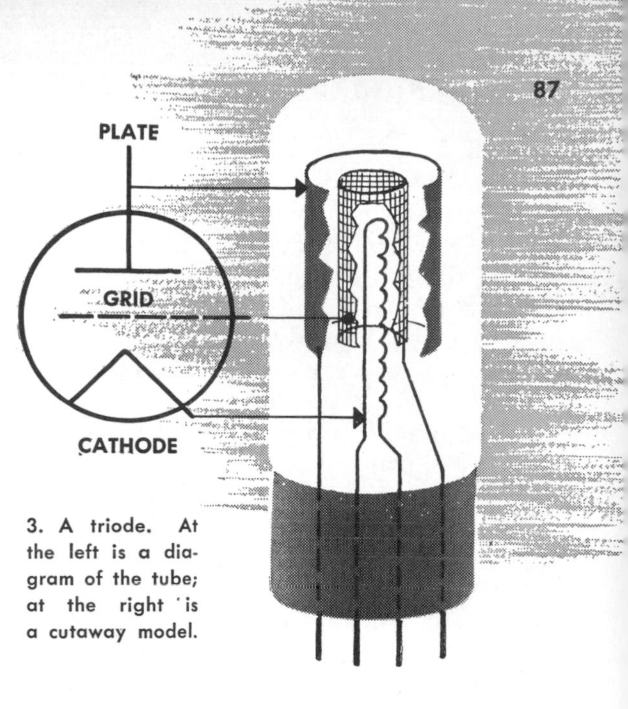


3. A triode. At the left is a diagram of the tube; at the right is a cutaway model.

The to-and-fro movement of current in the circuit we have described would go on indefinitely, were it not for the fact that the oscillating electrons lose energy in overcoming the resistance of the inductance coil winding and the connecting wires. It is necessary, therefore, to furnish a new source of energy that will keep the oscillations from slowing down or stopping. We can do so by inserting a triode, or three-element vacuum tube, in the oscillating circuit we have described.

A triode (Figure 3) has three elements: a negative element, the cathode (*K*); a positive element, the plate (*P*); and a grid (*G*), set between the cathode and the plate.* When the cathode is

* In diagrams, the cathode is represented as being at one side of the tube, while the plate is at the opposite side. Actually, the cathode is in the center of the tube and is enclosed by the grid; the grid, in turn, is enclosed by the plate. See Figure 3.

2. Simple oscillator circuit at work. (A) The capacitor has received a negative charge; current flows toward plate *b*. (B) As current flows through the inductance coil, a magnetic field is set up. (C) The magnetic field collapses, and current continues to flow toward *b*. (D) Electrons have piled up at plate *b*; current now begins to flow toward *a*.

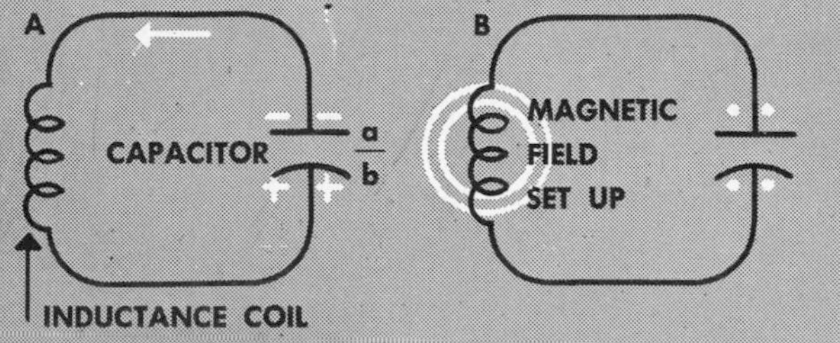


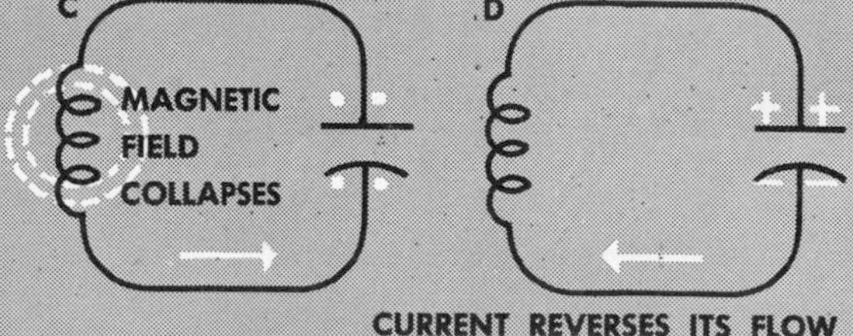

heated, it throws off electrons which are attracted to the positive plate. If the grid is given a strong enough negative charge, it will repel most of the electrons thrown off by the cathode and will prevent them from reaching the positive plate. If the grid is given a positive charge it will add to the attraction exercised by the positively charged plate. In either case, a charge placed upon the grid will be amplified many fold in the plate circuit.

If we put a triode in the oscillating system, enough energy will be added to the oscillations to make up for the energy that has been lost. We show one type of oscillating circuit, complete with triode, in Figure 4. The voltage used to drive current through the grid is extracted from the oscillating circuit by means of a secondary winding, *S*, on the same core as the inductance coil, *L*. Voltage is induced in this secondary winding by the action of the transformer, *T*.

It is highly desirable in radio transmission to have a very steady frequency output, in order to insure that a transmitting station will keep to the frequency that has been assigned to it. In modern circuits, this is often done by using a crystal-type oscillator. A quartz crystal shows the piezoelectric effect that we discuss in another article (see Index, under Piezoelectricity): that is, when subjected to electric stress, a mechanical force is generated within the crystal, and vice versa. If a high-frequency voltage is applied to a cut slice of quartz, it will start vibrating rapidly at its natural frequency. It is possible to cut quartz crystals whose natural frequency is millions of cycles per second.

The oscillating current output from an electronic oscillator is called the carrier current. It must be amplified up to the power level that the transmitting station is authorized to use. For this purpose we use a series of radio-frequency amplifiers, employing large vacuum tubes.

At some point in the series of amplifications, the carrier current undergoes an all-important modification. It is combined with a fluctuating electric current flowing from a microphone at a broadcast.

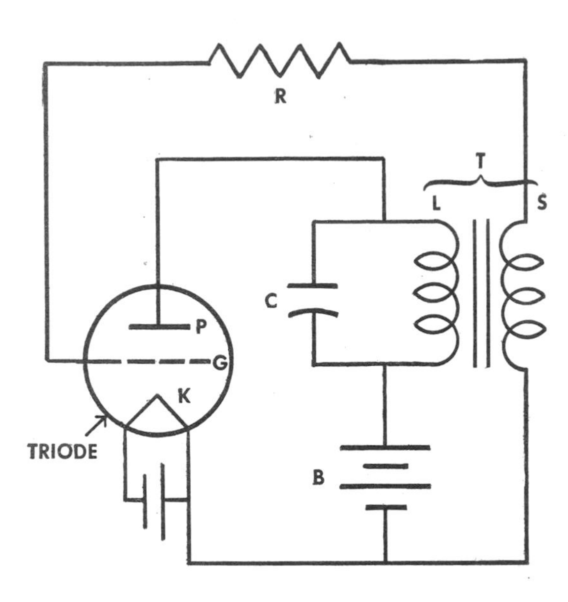


4. The above diagram shows a simple oscillating circuit containing a triode. In the diagram, *B* is the battery; *C*, the capacitor; *G*, the grid; *K*, the cathode; *L*, the inductance coil; *P*, the plate; *R*, the resistance; *S*, the secondary winding; and *T*, the transformer.

### What happens when a program is broadcast

Radio programs usually originate in a studio located in the central area of a city. However, it is possible to pick up on-the-spot action (at a ball park, say, or a football oval) if the proper equipment for remote control is set up. One or more microphones are used, depending on the type of program.

A microphone is a device that converts the mechanical energy released by sound waves into electrical energy. Sound waves are pressure vibrations; their amplitudes vary as loud notes produce peaks and soft notes produce slight bumps in the wave form. The waves also vary greatly in their frequency.

In the microphone, the sound waves strike a diaphragm — a thin disk capable of a small amount of back-and-forth movement. The motions of the disk correspond to the oscillations of the sound waves. The diaphragm is connected to an electric circuit; as it moves to and fro, it modifies the electric current flowing through the circuit so as to produce a fluctuating pattern. This is called an audio-frequency current, because its fluctuations have the same frequency as the sound waves at the broadcast source. ("Audio" is a Latin word, meaning "I hear.")

The current output of the microphone is very feeble and could not be transmitted more than a few feet from the microphone. It is necessary, therefore, to amplify the current by the use of triode or diode electron tubes. (See Index, under Diodes.) Each microphone that is used to pick up a program has its own separate preamplifier tube * located in the control booth adjacent to the studio.

From the preamplifier tube, the current is led to a fader, or microphone volume control, located on a panel in the control booth. The various microphone circuits are mixed: that is, they are combined into one over-all program circuit. Further amplification puts the current through as many stages as are necessary. The sounds from all the microphones used on a program are united in one clear, balanced channel. The current is then carried from the studio to the transmitting station through specially constructed telephone circuits.

The current transmitted from the broadcasting studio is combined with the carrier current generated in the transmitting station. The carrier current is now said to be modulated (modified); it will produce a modulated carrier wave.

* It is called the preamplifier tube to distinguish it from other stages of amplification that come later in the broadcasting process.

Modulation can be brought about in two ways. In amplitude modulation (AM), the frequency of the modulated wave remains constant, while its amplitude varies according to the degree of loudness and the pitch of the original sound (Figure 5). Amplitude modulation is used in standard broadcasting. The programs can be sent over long distances; however, they are affected by static and various other kinds of interference.

In frequency modulation (FM), the amplitude of the wave remains constant, while its frequency varies according to the degree of loudness or the pitch of the original sound (Figure 6). Very-high-frequency waves are used in FM and special circuits are required. A great many receiving sets have a combined circuit, so that one can change from AM to FM by throwing a switch. Frequency modulation is comparatively free from the interference that often mars amplitude-modulated signals; it offers more natural tone reception than does AM.

#### The propagation of radio waves

After the program current from the broadcasting site has been combined with the carrier current generated at the transmitting station, the resulting modulated

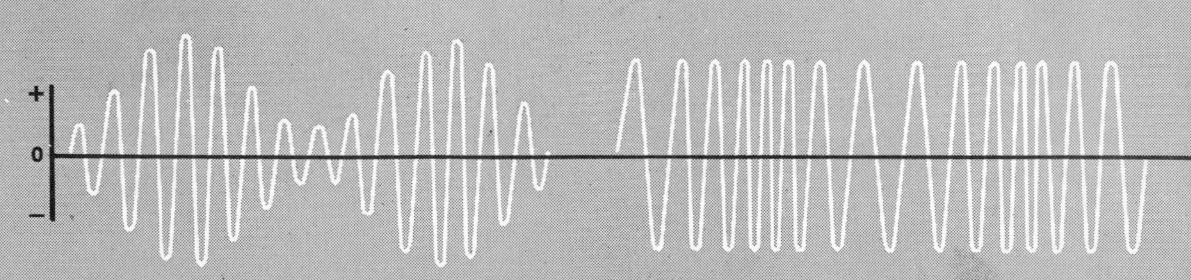


5. Amplitude modulation (AM). Note that the frequency of the wave remains constant, while the amplitude varies.

6. Frequency modulation (FM). In this case, the amplitude of the wave remains constant, while its frequency varies.

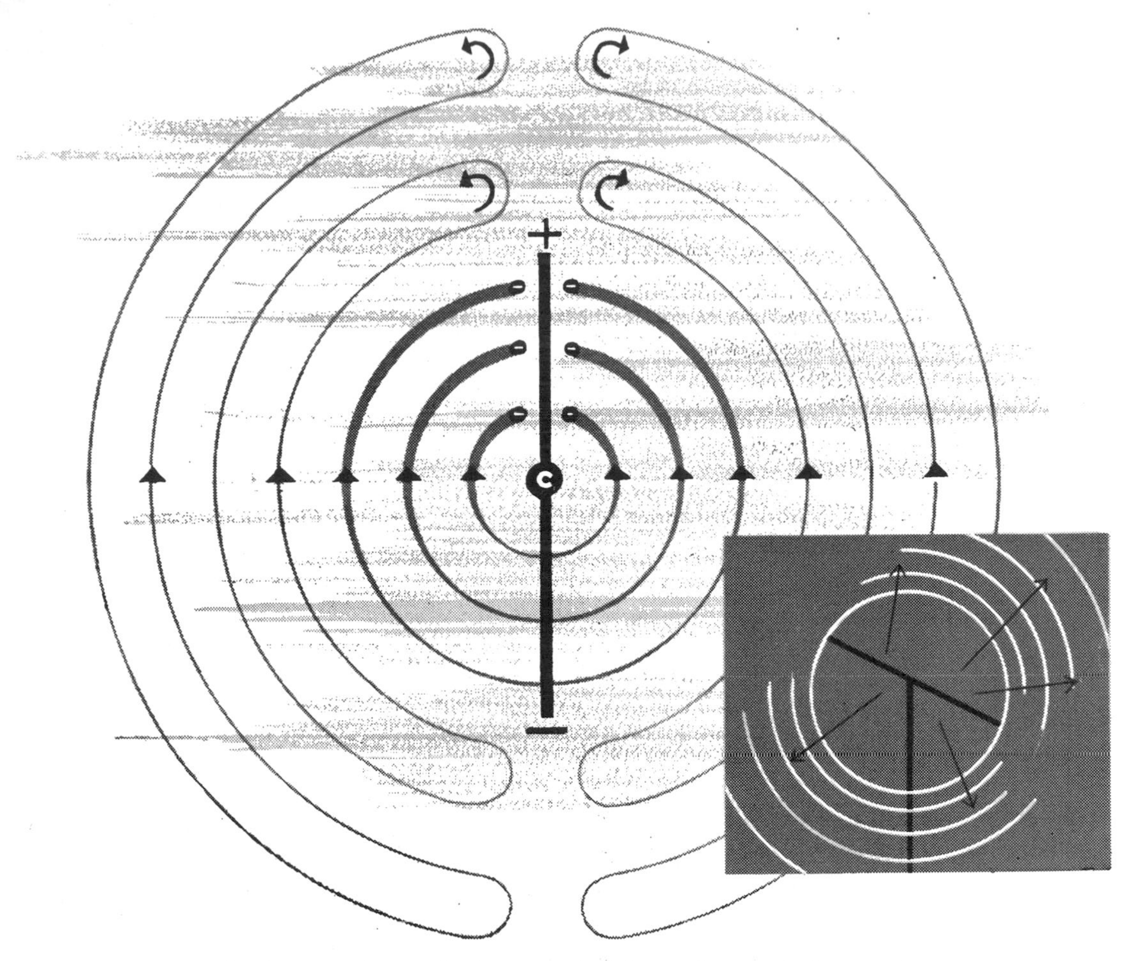


7. The propagation of radio waves from a dipole antenna. C, in the diagram, is the current source.

carrier current — AM or FM — is fed to the antenna wire of the transmitting tower. Here it will be converted into radio waves — modulated carrier waves — that will be sent out into space. The transmitting antenna is a wire (or wires) stretched between two points at some distance above the ground. Radio waves emanate from this antenna in the same way as heat waves do when they spread out from a radiator.

Figure 7 shows a dipole transmitting antenna at work. This is one of the most common types; it consists essentially of two vertical wires with the current source (*C* in the diagram) set between them. Alternating current is flowing through the antenna. During the first half-cycle of current flow, the electrons stream out of one wire (marked with a negative charge in the diagram) and into the other wire (marked with a positive charge). Since the two wires have opposite charges, they may be considered as two plates of a capacitor, with the surrounding air as the dielectric. As the current flows, a field is set up between the opposite plates, as shown by the lines of force on the diagram. The current then dies down to zero, and the field collapses back into the wires. Before the last line of force can reach the wire, the current has begun to go through its second half cycle in the opposite direction and another field has formed. The remaining lines of force from the first field are then pushed out by the second field and become detached from the wire. This process is repeated over and over again. The result is that a radiating stream of lines of force is moved out into space at a speed of 300,000,000 meters per second. This stream — a moving dielectric field — produces a magnetic field, which is at right angles to it. Together they make up a radio wave.

The frequencies of the radio waves that are transmitted into space from the transmitting tower often come to hundreds of thousands or even millions of cycles,* as we have seen. It would be inconvenient to deal constantly with large numbers such as these; hence the units called kilocycles (kc) and megacycles (mc) are generally used. A kilocycle is equal to a thousand cycles; a megacycle, to a million cycles. A hundred kilocycles (100 kc) corresponds to 100,000 cycles; a hundred megacycles (100 mc), to 100,000,000 cycles.

Radio waves used for communication purposes are divided into five distinct groups on the basis of frequency. We list them here, together with the frequency allocation given them by the United States Federal Communication Commission in conformity with international treaties.

Very long waves (30,000 to 2,000 meters) are not used so much as formerly. They used to serve for ground transmission over distances up to 1,000 miles. Transmitting antennas a mile or more in length were employed; a great deal of power was required. The waves also served for transoceanic communication. Today we have found out how to make good use of the shorter waves. The wave lengths used for most of our present radio communication are of 600 meters or less.

The waves with a length of 600 to 200 meters and a frequency of 550 to 1,500 kc are used in standard AM broadcasting. Technically they are considered to be long waves, though of course much shorter than the very long ones we mentioned above.

* That is, cycles per second. The "per second" is generally dropped.

Short waves (50 meters or less) were considered to be useless at first; it was thought they would be too easily absorbed because of their high frequency. Radio amateurs, or "hams," used them, however, and found that they had definite advantages. Not only could they be sent long distances, but much less power was required to transmit them than was needed for long waves. Today short waves are employed in international long-distance communication, police calls and steamship communication. "Hams" still use them. Short waves also serve for television and FM radio.

### Call letters of broadcasting stations

In the United States, the Federal Communications Commission assigns the frequency at which a given AM or FM station will broadcast. In Canada, frequencies are allotted by the Telecommunication Division of the Department of Transport. Radio stations have distinctive call letters. Practically all stations in the United States use the letter "W" as the first of their call letters; the first letter of Canadian stations is "C." Usually, the choice of the other call letters is left to the broadcasting station. These additional letters sometimes stand for the initials of the company that runs the station. For example, in New York City, WRCA is operated by the Radio Corporation of America; WCBS, by the Columbia Broadcasting System; WABC, by the American Broadcasting Company.

| Type of Broadcasting | Frequency Band | Wave Length |
|---|---|---|
| Long Wave Radio | 10 to 150 kc | 30,000 to 2,000 meters |
| Standard | 550 to 1,600 kc | 600 to 200 meters |
| Short Wave and International | 6 to 22 mc | 50 to 14 meters |
| Television | 44 to 216 mc | 7 to 1.4 meters |
| Frequency Modulation | 88 to 108 mc | 3.4 to 2.8 meters |

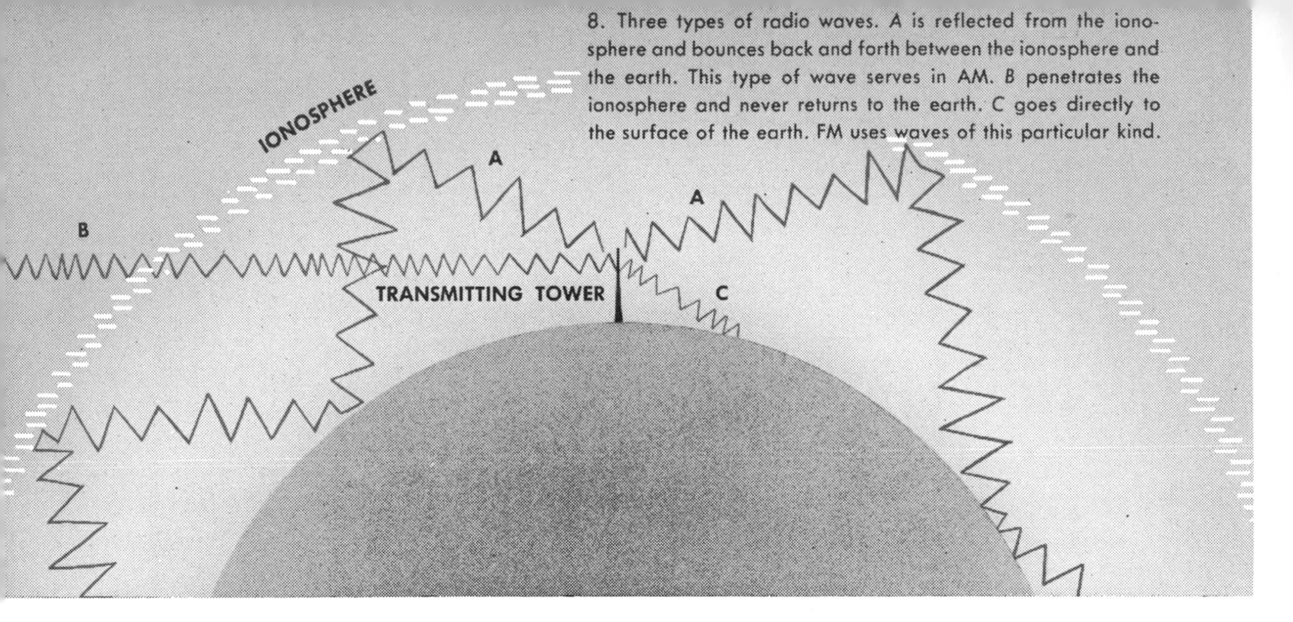

8. Three types of radio waves. A is reflected from the ionosphere and bounces back and forth between the ionosphere and the earth. This type of wave serves in AM. B penetrates the ionosphere and never returns to the earth. C goes directly to the surface of the earth. FM uses waves of this particular kind.

### Radio waves in space

Radio waves spread out from their source in all directions like the water ripples you see after you drop a stone into a pond. As the waves get farther away from the transmitting tower, they become weaker. Some of them — ground waves — travel along the earth's surface; others — sky waves — are radiated straight out into space.

As ground waves travel, they set up induced currents that flow through the ground. In this way, they lose energy and are gradually absorbed. Waves of the higher frequencies lose more power in this way than do the waves of the lower frequencies. Intervening hills, buildings and trees increase the absorption and prevent the signals from being received strongly.

The sky waves that radiate into the upper atmosphere penetrate beyond the stratosphere to the extensive layer known as the ionosphere (see Index, under Stratosphere; Ionosphere). In the ionosphere, the gases of the air are ionized by the action of ultraviolet rays from the sun. It is well known that the higher up one goes, the fewer molecules there are per unit volume of air. Because of this reduction in the number of molecules, there is comparatively little ionization beyond the 200-mile mark. However, it is not entirely absent. In fact, recent data supplied by artificial earth satellites have led scientists to believe that the ionosphere may extend out into space to a distance of 2,000 miles.

The ionosphere is a very important factor in radio transmission and reception. A certain percentage of the radio waves penetrates the ionosphere and never returns to earth. Waves that do not enter the ionized regions are reflected to the earth and then bounce back to the ionosphere. The waves will travel in this way back and forth between the earth and the ionosphere until they are completely absorbed. (See Figure 8.)

The ionosphere changes height from day to night and during different seasons of the year. Radio reception is better at night and during the winter months because the absorption of the waves is reduced at such times. It has been found that absorption reaches its maximum at a frequency of about 1,400 kc.

Ground waves and sky waves may combine their effects in such a way as to cause undesirable changes in the intensity of the sound coming from the loud-speaker. Sometimes the program fades away; at other times there is a sudden increase in loudness. With the constant improvement in radio receivers, this condition is not so prevalent as it used to be; in the early days of radio it occurred frequently and marred radio reception.

Undesirable changes in intensity are due to several factors. Sometimes both the ground wave and the sky wave reach a receiving set. They may both come in to-

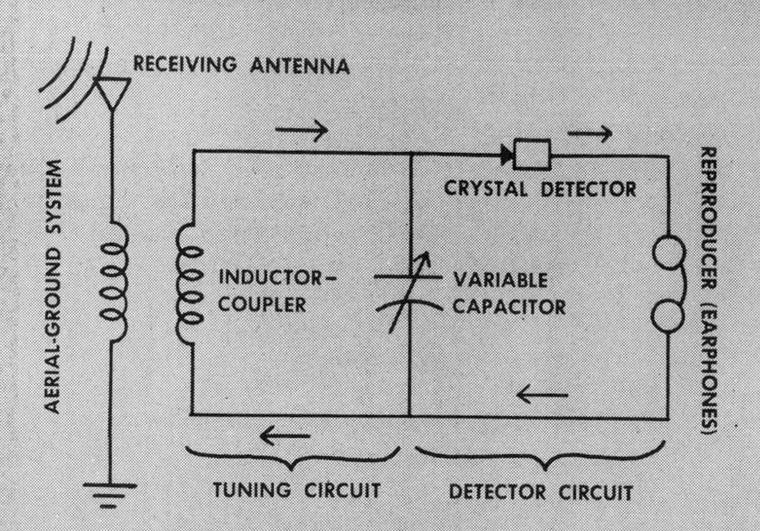


9. Simple crystal radio circuit, showing the four essential parts of a receiver: the aerial-ground system; the tuning circuit; the detector circuit; and the reproducer. Note that earphones are used in this circuit.

gether and reinforce each other; in that case the signal increases in intensity. On the other hand, they may be slightly out of step and may have a tendency to neutralize each other. In this way they bring about a loss in intensity and sometimes even a complete fading away of the signal.

Fading and increase in intensity may also be due to conditions arising in the ionosphere. Changes in this portion of the atmosphere can cause undesirable variations in the sky wave. Such variations may also be due to the interaction between two sky waves. One of these may reach the receiver in one hop; another may hit the ionosphere at a different angle and may arrive at the receiving set in two hops. Here again, if the two waves arrive in step with each other, the signal will be loud; if out of step, they may cancel each other.

### Radio reception

We are constantly surrounded by radio waves — modulated carrier waves — which we can neither see nor hear. The radio receiving set transforms these waves into replicas of the original sound waves that struck the microphone at the source of the broadcast. All receivers, no matter how complex, consist of four essential parts:

(1) An aerial-ground system collects the waves from the air.

(2) A tuner selects the wave or station to be received and rejects all others.

(3) A detector changes the energy of the wave to a form in which it can operate the reproducer.

(4) A reproducer changes the wave energy to a form that our senses can perceive.

In Figure 9, we show these four elements in a simple crystal radio set.

*The aerial-ground system.* Suppose a wave radiating from the transmitting antenna strikes a receiving antenna. A radio-frequency alternating current, with the modulation imposed at the transmitting station, is set up in the antenna wire, and flows to the receiver through a lead-in wire. Formerly, antenna wires consisted of bulky outside aerials. Nowadays, however, the aerial system of most sets is housed inside the cabinet.

The lead-in wire connects the aerial wire to ground through the receiving set. The term "ground" is used in radio work to denote the parts of the circuit directly connected to the earth or to the metallic base of the receiver. Those of you who have worked with simple crystal sets will recall that such sets always have a ground wire that is to be connected to a pipe or a radiator.

If the antenna were connected by wire directly to the tuning circuit, there would be considerable resistance as the current coursed through the wire. The result would be that you would hear several sta-

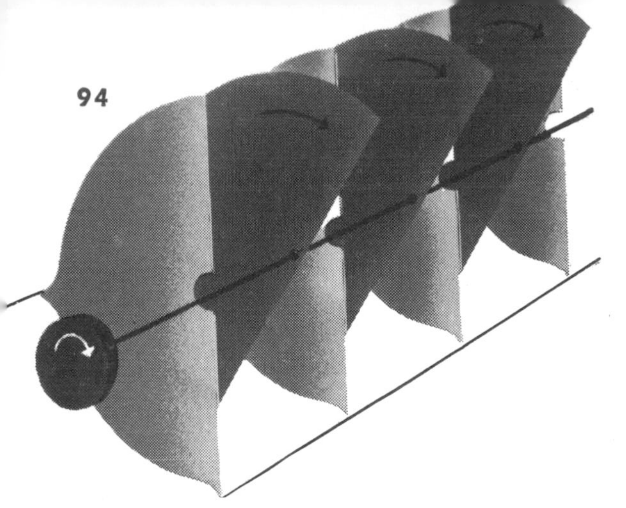

**10. Two views of a variable condenser. One set of plates is stationary; the other can be rotated.**

tions at the same time — an effect known as broad tuning. It is possible to reduce broad tuning to the minimum by reducing the resistance in the circuit. For this purpose a device called an inductor-coupler has been developed (see Figure 9). This coupler is an air-core transformer (see Index, under Transformers). The primary coil of the transformer is in the antenna circuit; the secondary coil, in the tuning circuit. The radio frequency current set up in the antenna wire flows through the primary coil, setting up a magnetic field that cuts across the turns of wire on the secondary coil. This brings about an electric pressure in the secondary coil, causing current to flow through the tuning circuit. In other words, the antenna circuit and the tuner circuit are connected magnetically and not by means of wiring. This cuts down resistance and improves receiver selectivity considerably.

Various transformer coils are used in radio circuits. The magnetic fields set up by these coils are sometimes strong enough to interfere with each other. The best way to prevent this is by shielding. Metal shields, generally of copper or aluminum, are placed around the transformers; they keep the individual fields from spreading into undesirable parts of the circuits.

*The tuning circuit.* Each transmitting station sends out waves of a different frequency within local areas. The tuning circuit, or tuner, must select the frequency of the station whose program we desire to hear. The receiver must be put in resonance with the frequency of the sending station: that is, it must be adjusted so that it will respond only to this frequency.

We can illustrate the principle of resonance by considering the familiar example of the pendulum. A pendulum has a natural frequency: that is, it will swing to and fro so many times per second. Suppose that, when a pendulum is swinging, we give it a series of pushes at the beginning of each downward swing, timing the pushes so that the number per second corresponds to the frequency of the pendulum. The pushes will then be in resonance with the natural frequency of the pendulum.

In a tuning circuit, we match the frequency of the incoming signal by means of an inductor and a capacitor, hooked together. The inductor is fixed; the capacitor, variable. One set of plates is stationary; the other can be rotated (Figure 10).

**11. BLOCK DIAGRAM OF A SUPER-HETERODYNE CIRCUIT**

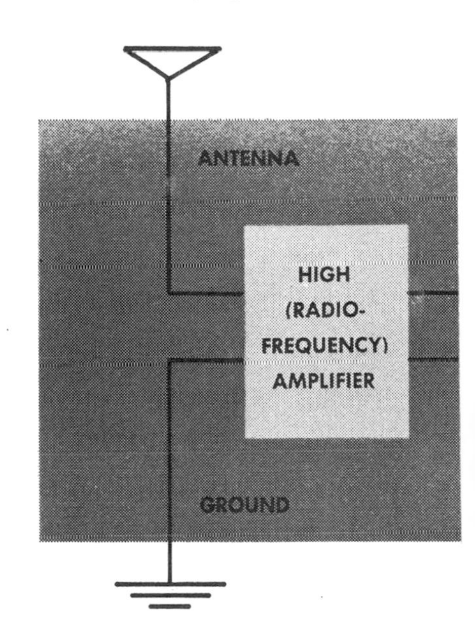

In practically all modern radios, the superheterodyne principle is used in the tuning circuit to assure sharp tuning. It would be just as accurate to refer to the heterodyne principle. The prefix "super" is used because English engineers called this type of circuit a "supersonic heterodyne" in the early days of radio.

The heterodyne, or superheterodyne principle, as applied in radio, involves changing the frequency of the alternating current set up in the receiving antenna to a lower one, called the intermediate frequency. The current can be amplified much more effectively at this frequency. It is not so high as the original radio frequency and it is higher than the audio frequency.

To produce the intermediate frequency, two different ones are combined. The resulting frequency will be equal to the difference between the two original ones. Here is how we apply the heterodyne principle to radio receivers. The radio-frequency transformers, which step up the voltage of the alternating current, are set so that their natural frequency is, say, 200 kilocycles. Suppose a signal is brought in from a station that is broadcasting at a frequency of 1,000 kc. We generate in our set a radio-frequency current of 1,200 kc by means of an oscillating circuit, called a local oscillator. We mix this current with the incoming signal (1,000 kc) in the first tube of the detector circuit. The resulting frequency will be 200 kc (1,200 minus 1,000). This intermediate frequency is then fed into the radio-frequency transformers, which, as we have seen, have also been set at 200 kc. We can tune the local oscillator by turning a dial of the receiving set, thus moving the rotary element of the variable capacitor in the oscillator circuit and changing the frequency. In Figure 11, we show the block diagram of a superheterodyne circuit.

The radio-frequency signal that is trapped by the receiving antenna and that flows through the tuner circuit would be too weak, in most cases, to pass to the next circuit — the detector — unless it were made stronger. We do this by adding two or more stages of radio frequency amplifiers to the tuner circuit.

*The detector circuit.* An intermediate frequency current changes direction thousands of times per second. It could not operate the reproducer mechanism, which requires a fluctuating direct current. It is necessary, therefore, to convert the alternating current to a fluctuating type of direct current.

In the early days of radio, this was done by means of a crystal detector circuit. It was given the name "detector" because, by changing alternating current to direct current, it made it possible to detect radio waves. Crystals of galena,

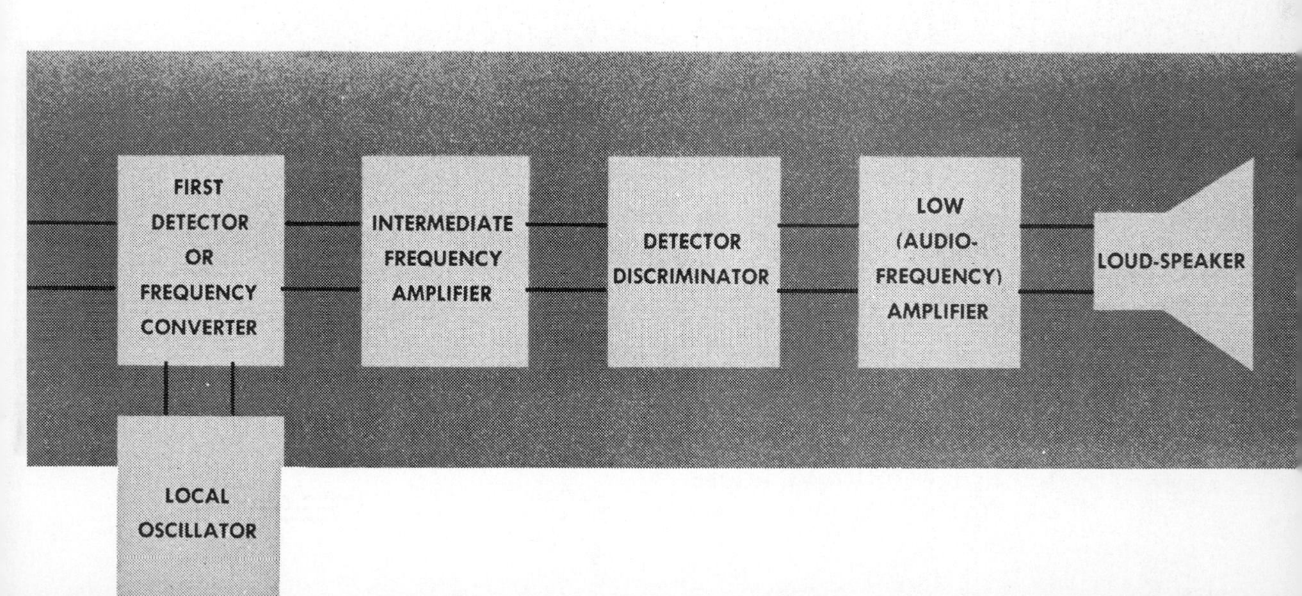

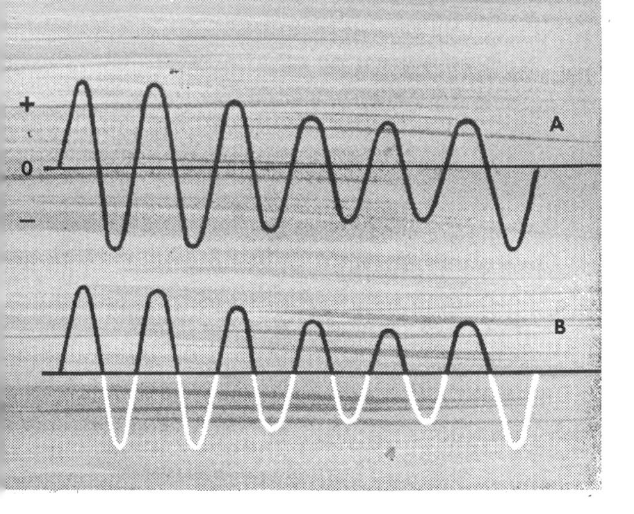


12. Diagram showing how a radio-frequency signal is detected. *A* represents the original alternating current. In *B*, the negative pulses have been eliminated.

a compound of lead and sulfur, were used for this purpose. The galena crystal was capable of acting like an electrical valve; it permitted the flow of electrons in one direction only. The alternating current fed to this crystal would pass through it, therefore, on one half-cycle and would stop flowing on the following half-cycle.

Figure 12 shows how the current would be transformed. In *A* we see the original alternating current, with its hills and valleys — one set of hills and valleys representing the positive pulses, the other set, the negative pulses. In *B*, the negative pulses have been eliminated. The positive ones follow each other so closely that the effect is that of a direct current — a current with a definite fluctuating pattern, reproducing the original pattern sent out from the microphone.

The output of a crystal detector circuit, then, is a pulsating direct current — a current flowing in one direction and periodically changing strength. In early crystal sets a thin wire was set in contact with a sensitive spot on the crystal, where the valvular action was the strongest. The wire was known as the cat whisker.

The crystal detector has been almost entirely replaced by vacuum-tube detectors. Both diode (two-element) and triode (three-element) tubes can be used for this purpose. In the article An Introduction to Electronics, in Volume 6, we explain how these tubes function as detectors. Like crystal detectors, they change the alternating current flowing through the aerial and the tuning circuit into direct fluctuating current. The fluctuations follow the same pattern as in the microphone. We now have an audio-frequency current: one in which the frequency of the fluctuations will be the same as that of the sound waves produced at the site of the broadcast.

*The reproducer circuit.* In this circuit, audio-frequency current is converted into sound waves, which create the sensation of sound when they strike the ear.* In the early days of radio, earphones were used to reproduce the original sound. They proved to be entirely inadequate, and so the loud-speaker was developed. There are two chief types of loud-speakers: magnetic and dynamic.

In Figure 13, we show how a magnetic speaker works. The current pulses that are fed into the reproducer circuit flow to an electromagnet. This consists of a coil of wire surrounding an iron core, which is a permanent magnet. The core exercises a constant pull on a diaphragm set at one end of it. When the fluctuating current is passed through the coil, a magnetic field is set up around it. The diaphragm will then be subjected to the steady pull of the permanent magnet and also to the pull, of varying strength, of the electromagnet. It will vibrate in step with the fluctuating current. In the center of the diaphragm is fastened a stiff wire. The other end of the wire is connected to a large paper cone. As the cone moves rapidly to and fro, it sets a large amount of air vibrating and creates a loud sound.

The dynamic speaker works on a somewhat different principle. In this device, the two poles of the electromagnet consist of a hollow cylinder with thick walls and a small solid cylinder set within the larger one in such a way that there is a circular gap between them (Figure 14). The cone of the loud-speaker has at its smaller end a ring around which is wound fine wire. This ring fits in the gap between the poles of the electromagnet (Figure 15). The fluctuating current passes through the coil of the ring. The ring vibrates as it is

* We tell how this is brought about in the article The Senses, in Volume 7.

attracted more or less strongly by the electromagnet. The cone to which the ring is attached follows the same pattern of vibration and produces sound.

### The use of batteries in receiving sets

When electronic engineers began using vacuum tubes in radio circuits, it became necessary to add sources of power to operate the different elements and circuits of the tubes. A battery called the A battery was used to heat the filament (cathode) of the diode or triode. Another battery — the B battery — served to place a positive charge on the plate of the diode or triode. To put the necessary charge on the third element in the triode — the grid — it was necessary to use a third battery, called the C battery.

All these batteries have been eliminated in modern radio receivers, except for portable types. The house circuit is used to heat the cathode of the tubes. The B battery is replaced by a circuit called the B-battery eliminator. In it, a diode, used as a rectifier tube, converts the alternating current of the house circuit to direct current and uses it to place the required charge on the plate. A C-battery eliminator replaces the C battery; it obtains the current that is necessary for reception from the B-battery eliminator.

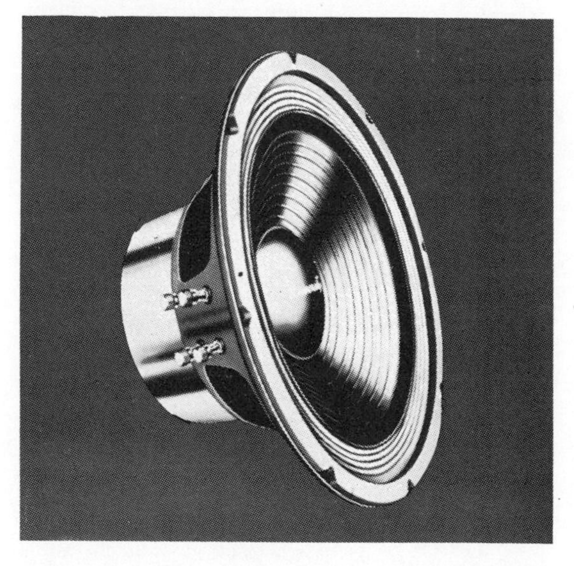

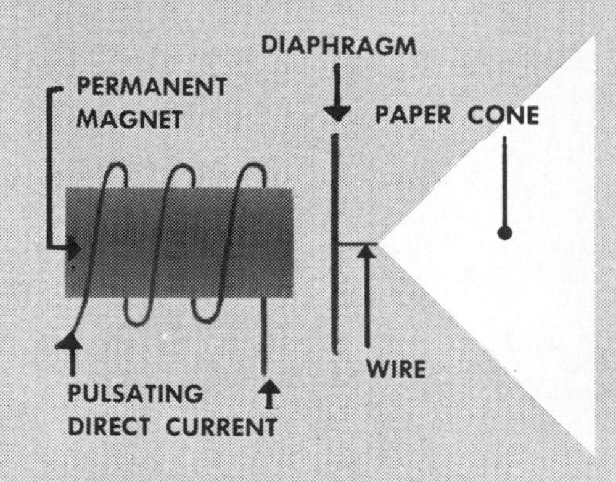


13. Diagram of a magnetic speaker.

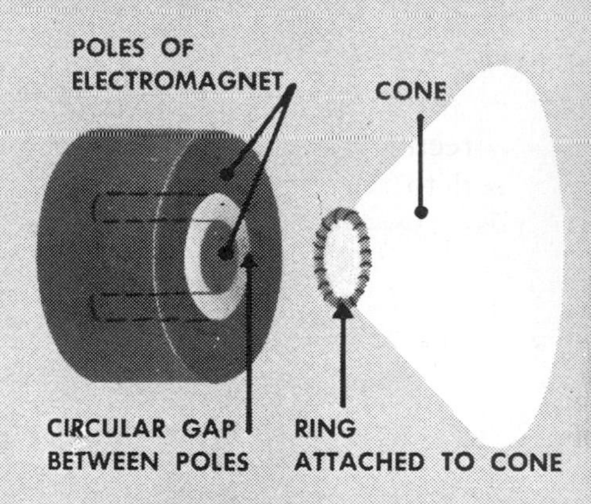


14. Poles, ring and cone of a dynamic speaker.

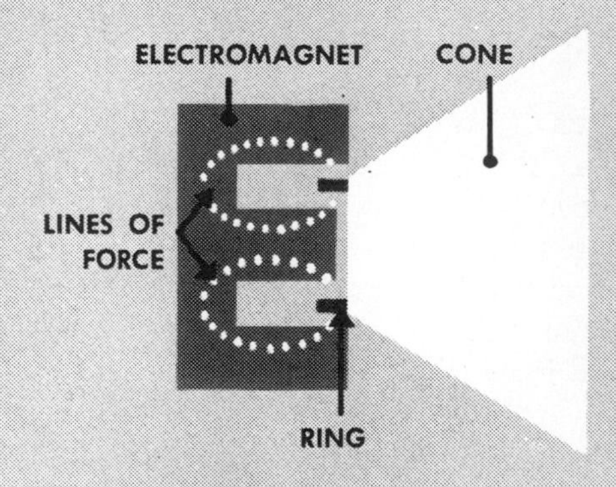


15. Dynamic speaker in cross section. The ring is now in position in the gap between the poles of the electromagnet.

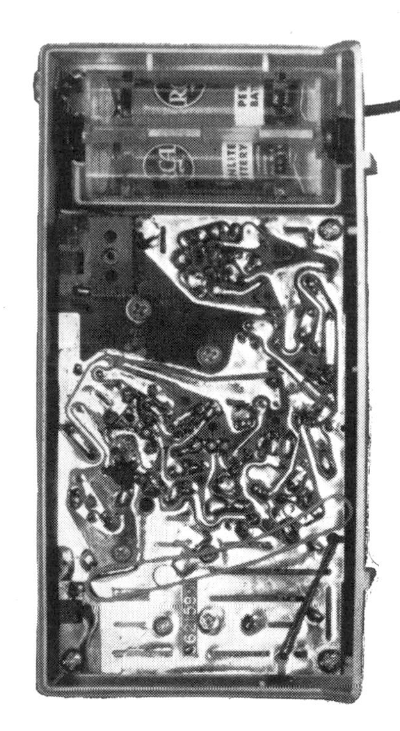

16. Battery-powered portable radio, with transistors, printed circuits and miniature earphone.

RCA-Victor

## The control of volume and tone

The control of the volume of sound emanating from the receiver was a comparatively easy matter in the days of battery-operated sets. All that was needed was a rheostat (a resistance control device), connected in series with the A battery and the cathodes of the tubes. The strength of the current could be controlled by increasing or decreasing the resistance. This would control the flow of electrons emitted by the cathode and would in turn control the plate current. The loudness of the signal would depend upon the strength of the plate current. A knob in the exterior of the receiver could be turned, thus changing the setting of a variable resistor. Once the desired volume of sound had been obtained, however, it would not remain steady but would alternately fade out and become too strong.

Modern receiving sets still have manually operated knobs on the outside of the set to control the strength of current in the audio-frequency amplifier. Once a desired setting has been made, the volume is kept steady by an automatic volume-control method. This is accomplished by devices that control the charges on the grids of the radio-frequency amplifier tubes. These charges are made to vary with the strength of the signal at the antenna. If the signal is strong, the amplification is reduced; if it is weak, it is amplified. This arrangement eliminates or minimizes poor reception due to fading.

The listener can adjust manually the relative intensity of the high and low tones by turning dials on the outside of the set. The circuit used for this purpose has a variable resistance in series with a fixed capacitor. When the higher frequencies (high notes) are decreased in strength, the bass notes come out more clearly, and vice versa. This control is applied in the audio-frequency stages of reception.

## The use of transistors in radio circuits

Some radio receiving sets have transistors instead of vacuum tubes; some have both transistors and vacuum tubes. (See the Mighty Midget of Electronics, in Volume 10.) Transistors serve the same purposes as vacuum tubes; they can act as detectors, rectifiers and amplifiers. They have certain definite advantages. For one thing, they are not so fragile as vacuum tubes. They require less power and occupy less space; hence they are appropriate for portable sets (Figure 16).

In this short article, we have dealt with the basic principles of the transmission and reception of radio broadcasts. We have not mentioned many of the modern refinements of receiving sets: automatic selectivity, push-button remote tuning, tuning indicators, squelching circuits and a host of other features. You will find information about such developments in the books that we have listed under the heading "Radio Broadcasting" in Volume 10, page 287.

# THE SUN, OUR STAR

## Its Disk, Atmosphere, Rotation and Other Features

**BY WALTER ORR ROBERTS**

TO THE astronomer, as to the layman, the most important fact about the sun — the heart of our solar system — is that it is so close to the earth. It is a mere 93,000,000 miles or so away — a trifling distance, astronomically speaking, when we bear in mind that the next nearest star, Alpha Centauri, is 4.3 light-years or almost 26,000,000,000,000 miles distant. When we examine any other star but the sun with the unaided eye, or for that matter with the most powerful telescope, it appears as a mere pinpoint of light. But the sun, even to the unaided eye, shows as a disk of about the same size as the disk of the full moon. (Because of the sun's extraordinary brilliance we must look at it through darkened glass or exposed photographic film to avoid damaging the eyes.)

Because the sun is so close to us, we can make out various details that we can never hope to see on the other stars. This is tremendously important. From a careful telescopic study of the sun's visible features astronomers have learned a great deal about the complex and mysterious behavior of this fascinating heavenly body.

Since the earth's orbit around the sun is in the form of an ellipse, the actual distance of the sun varies according to the position of the earth at a particular part of its orbit. For example, the sun is about

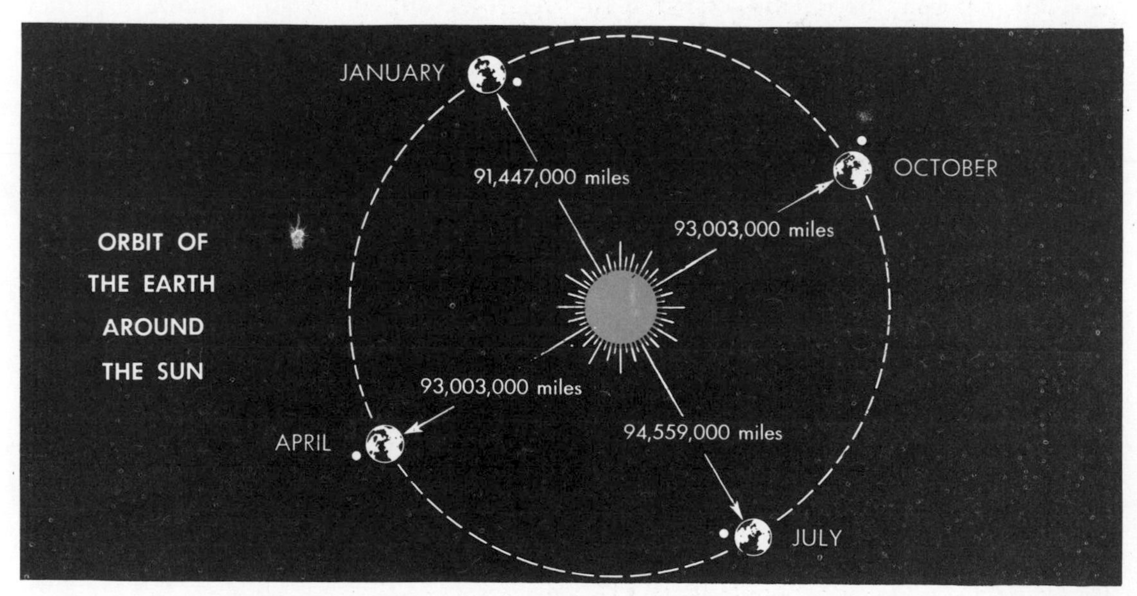

3,000,000 miles nearer to us in January than it is in July. Its mean, or average, distance from our planet is about 93,003,000 miles. Its diameter is approximately 865,600 miles — about 109 times greater than the earth's diameter.

The sun is a large mass of self-luminous gas, glowing with the intense brilliance of its internal energy furnace. Its temperature ranges from about 10,000° F. at the surface to about 27,000,000° F. at the center. Because of such extreme temperatures, not only are all the materials in the sun in the gaseous state but, except for a few stubborn chemical compounds, they have all been reduced to elementary atomic forms. Yet, though it is made up of gases, the average density of the sun is much greater than that of water; it is about a quarter that of the earth. The reason is that the solar gases undergo tremendous compression, because of the sun's great mass — some 333,420 times greater than the mass of the earth. The pressure increases steadily from the surface toward the center; at the center it is something like six million tons per square inch.

Because of its great mass, the force of gravity of the sun is very powerful; it is about 28 times stronger than on the earth. A 10-pound object would weigh about 280 pounds on the sun, provided that we could accomplish the impossible feat of protecting it from the intense solar heat so that it would not instantly vaporize.

Judged by terrestrial standards, then, everything about the sun is awe-inspiring. Its mass is incomparably greater than that of even the largest of the planets that revolve around it; its temperature and the pressures deep below its surface are almost unbelievably high. Yet as a star the sun is not at all outstanding. It is only of average size; it is considerably below average in mass and surface temperature.

#### Conspicuous features of the sun's surface

The bright disk of the sun is called the photosphere ("light sphere"). It has a sharply defined border and represents the accepted limit of the solar surface, as opposed to its atmosphere. Since the photosphere is opaque, we cannot see the part of the sun that lies beneath it.

At one time observers thought that the sun's glowing disk was a uniform orb of light. But when the renowned Italian scientist Galileo Galilei (see Index) began to study the sun with the newly invented telescope in 1611, he was astounded to find that its surface showed a number of dark spots. Galileo was so doubtful about his discovery that he did not announce it at once, but waited until several other scientists had confirmed it. These spots on the sun are by all odds the most conspicuous features of its surface. We know a great deal now about their size, shape and general behavior; we know that they are comparatively cool areas on the sun's surface. In many respects, however, they are still a profound mystery. We shall have more to say about the sunspots later in this article.

Another prominent feature of the sun's disk are the bright faculae (from a Latin word meaning "little torches"). They show up as glowing areas of irregular outline, brighter than the regular surface of the sun in their vicinity. They vary from tiny bright markings to enormous splotched areas completely surrounding large sunspot groups. In general the faculae are most conspicuous in the vicinity of sunspot groups; however, they sometimes occur where there are no sunspots at all. In addition to the sunspots and faculae, which are found only in certain parts of the sun's disk, there are certain markings, known as granulations, that are scattered over the entire surface. The granulations are sometimes called "rice grains"; they look somewhat like grains of rice sprinkled over a rather dark surface.

Above the photosphere lies the solar atmosphere. This is most conspicuous during total eclipses of the sun, when the moon's disk is interposed between the earth and the disk of the sun. The atmosphere is made up of several layers.

The lowest portion of the atmosphere is called the reversing layer. It is because

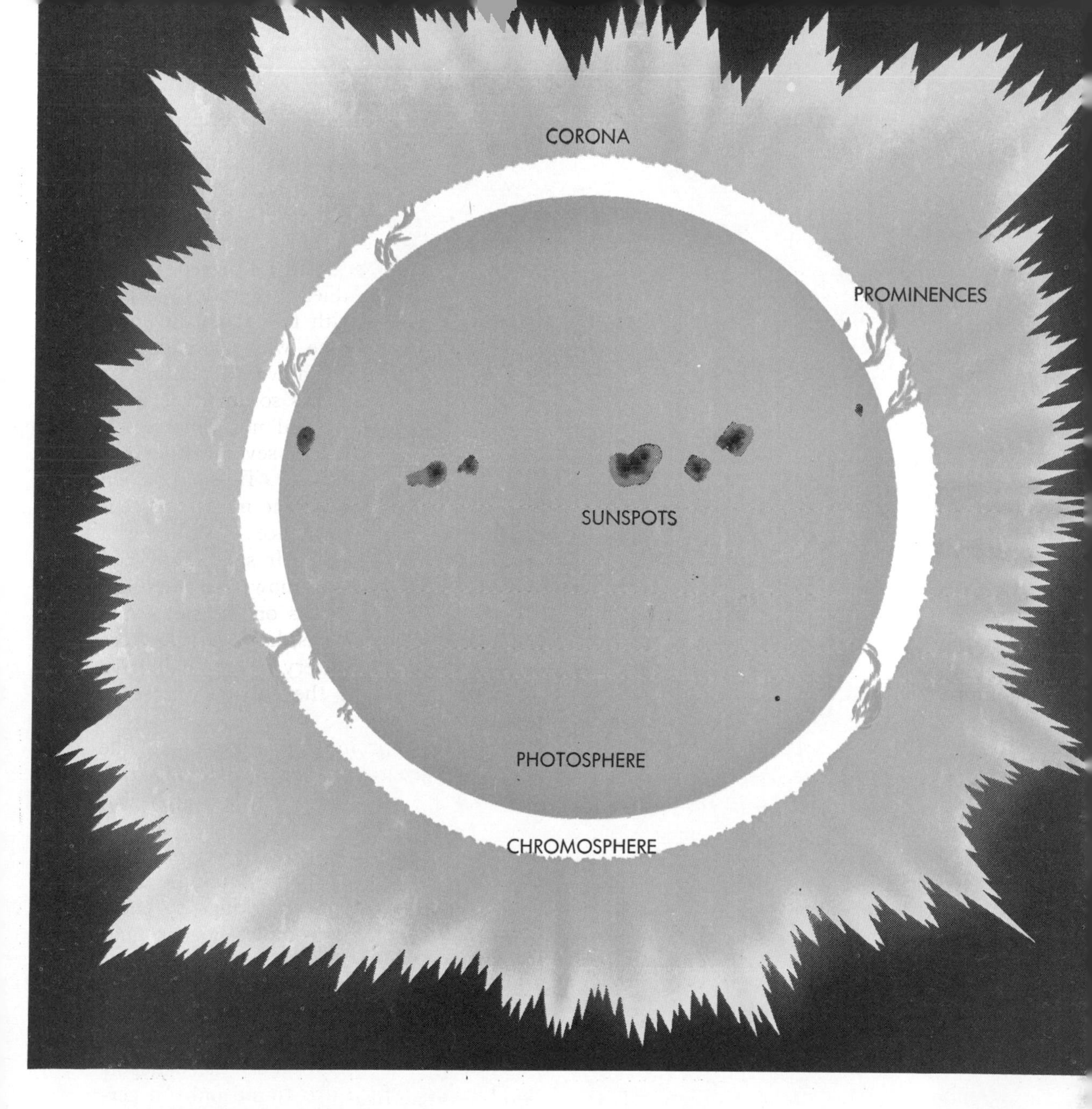


**Some of the features of the sun's surface and atmosphere. The photosphere is the bright disk of the sun; upon it appear the sunspots, which are made up of glowing gases but which seem dark because they are somewhat cooler than the surrounding areas. The sun's prominences are irregularly shaped clouds, made up of gases in motion. The chromosphere and the corona are two layers of the atmosphere that surrounds the sun's disk.**

of this layer that the spectrum of the photosphere does not show a continuous spectrum (see Index, under Spectrum) — that is, one in which all wave lengths, or colors, are represented from violet to red. Instead, the progression of colors is interrupted by certain dark lines and bands, known as absorption lines, because the colors that would otherwise have occupied the space of these lines have been absorbed. Astronomers have assumed that the missing colors in the spectrum have been absorbed in what they call the reversing layer. It has never been seen in any telescope. The reversing layer is supposed to be several hundred miles thick.

Above the reversing layer is the chromosphere; as a matter of fact the reversing layer is sometimes considered to be the lowest part of the chromosphere. The

The solar prominences, shown above, are gaseous clouds that sometimes extend hundreds of thousands of miles above the surface of the sun. After a time they spiral off into space. Astronomers know little about the origin of the prominences. According to some authorities they may be due to the action of local magnetic fields.

chromosphere is made up of gaseous filaments, apparently consisting of hairlike jets of gas originating in the photosphere and extending a few thousand miles upward. The name "chromosphere" ("color sphere") is derived from the ruddy color this layer displays when it is seen around the moon's disk as a total eclipse of the sun takes place. The spectrum of the chromosphere is almost the exact opposite of that of the photosphere; it is made up of the relatively limited number of colors that are absent from the photospheric spectrum.

Rising far above the chromosphere are the giant prominences — irregularly shaped clouds made up of sharply defined filaments of gas in motion. Some of the prominences extend hundreds of thousands of miles beyond the chromosphere. These gaseous clouds are generally much fainter than the chromosphere. Their light is dominated by just a few of the most brilliant lines of the chromospheric spectrum — especially the red line of the spectrum of hydrogen gas, the violet lines of calcium and the yellow spectrum line of helium. Helium was first discovered in solar prominences during a total eclipse of the sun in 1868, though it was not isolated on the earth until 1895. The prominences move under the influence of forces not yet satisfactorily explained.

A particularly interesting feature of the sun's atmosphere is its faintly luminous corona. This is a pale, greenish white halo that completely surrounds the sun and that was formerly invisible except during a total eclipse. (As we shall see, it can now be viewed at other times by means of a device called the coronagraph.) For many years the spectrum of the corona completely baffled observers; it was thought to result from a chemical element — coronium — that was unknown on the earth. In 1941, however, the distinguished Swedish astronomer Bengt Edlén solved the mystery of the coronal spectrum. He found that it resulted from the vapors of iron, nickel and calcium in a highly ionized state. As a result of Edlén's researches, we now realize that there is no such element as coronium.

### How astronomers observe the sun

Since the days of Galileo and his pioneer observations with the telescope, many special telescopic instruments have been devised for the study of the sun. One of the earliest and simplest of these is a projection lens and white screen, used together with an ordinary telescope. With such a device, it is easy to examine sunspots and other solar features, since it provides a large-scale image that is not excessively bright. As we have pointed out, it is dangerous to attempt to view the sun with an ordinary telescope without taking careful precautions to decrease the brightness of the image, so that it cannot injure the eye.

From "Classical and Modern Physics," H. E. White, copyright 1940, D. Van Nostrand Co., Inc., Princeton, N. J.

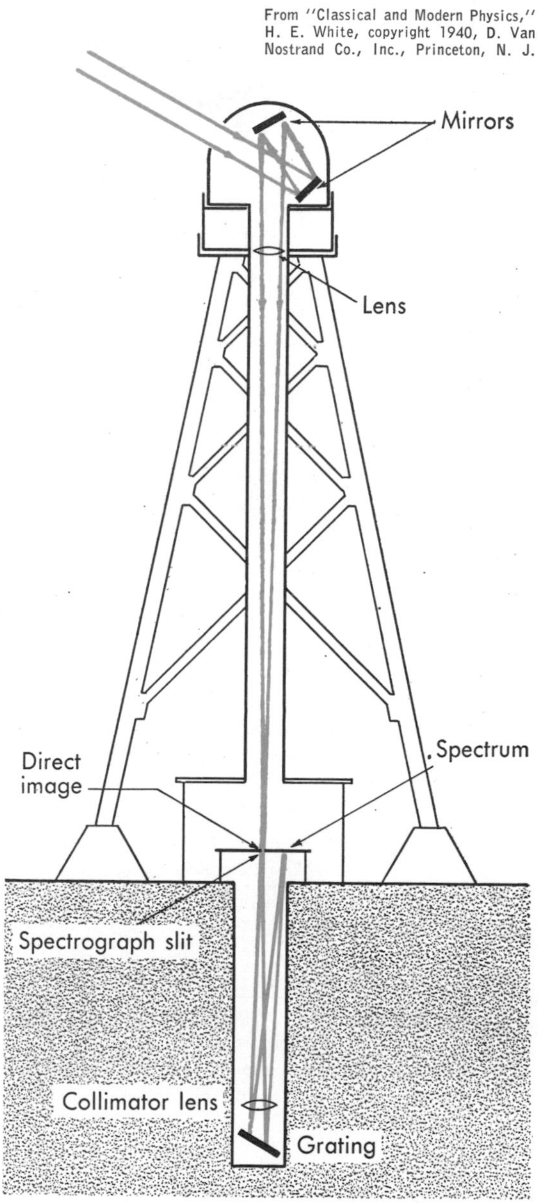


Tower telescope at Mount Wilson Observatory, in California. Sunlight, reflected from two mirrors at the top of the tower, passes downward through a lens, which forms an image of the sun on a table at ground level. The image may be studied directly. Light from a selected part of the image may be passed through a spectrograph slit. The light descends to a collimating lens below, falls on a diffraction grating and returns through the same lens to the table on the ground level. There the spectrum may be observed directly or else photographed.

Observing the sun is such an important part of modern astronomy that its study constitutes a separate science. Very large and powerful telescopes have been designed for solar study; such are the tower telescopes first devised by the distinguished American astronomer George Ellery Hale at Mount Wilson Observatory, in California. Extremely precise spectrographs have been fashioned for studying the spectrum of sunlight. Certain telescopes and even observatories have been exclusively devoted to the examination of the sun's atmosphere.

A modern solar observatory possesses not only telescopes for viewing the bright face of the sun in full sunlight, but also specialized equipment for isolating the monochromatic (one-color) light of a single line of the sun's spectrum. An instrument of this sort is like a filter that transmits the light of only one color and excludes the light of all the rest.

Two types of devices produce this effect. The older one is the spectroheliograph, which was invented independently by George Ellery Hale and the French astronomer Henri Deslandres. The spectroheliograph consists of a telescope combined with a spectroscope that has two slits. Light from a narrow section of the sun's disk passes through the first slit to the grating of the spectroscope. The spectrum produced by the grating passes through the second slit, which permits only a single line of the spectrum to be transmitted to a photographic plate. This gives a monochromatic image of a single narrow section of the sun's disk. By moving the first slit across the sun's disk, a series of images, each showing a single section of the sun, is built up on the plate. The combined images give a complete monochromatic picture of the sun's disk. A photograph taken with the spectroheliograph is called a spectroheliogram. The device called the spectrohelioscope is quite similar to the spectroheliograph; it differs from the latter in that it allows visual observation.

Right: a spectroheliograph, generally used with a tower telescope. An image of the sun is focused on the screen, on which there are two slits a fixed distance apart. One slit (the right-hand one) receives a small segment of the sun's image. Light passes through the slit and the lens and strikes the diffraction grating, producing a spectrum. Each spectrum line is an image of the segment intercepted by the first slit. By tilting the grating, any particular spectrum line can be made to pass through the second slit, where it strikes a photographic plate. As the screen is moved horizontally, the first slit passes across the image of the sun and intercepts a continuous succession of segments. The images of the segments are photographed side by side on the plate.

From "Classical and Modern Physics," H. E. White, copyright 1940, D. Van Nostrand Co., Inc., Princeton, N. J.

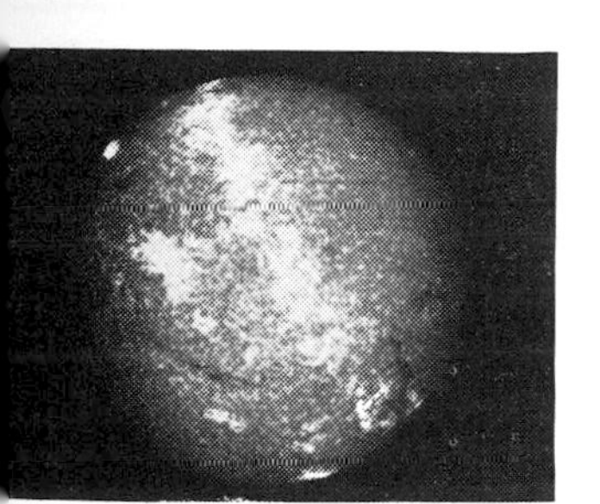

Mount Wilson and Palomar Observatories

Sun photographs. Upper left: ordinary photograph. Lower left: calcium spectroheliogram. Upper right: hydrogen (Hα) spectroheliogram. Lower right: enlarged hydrogen spectroheliogram.

Another device for producing monochromatic images is the polarizing monochromator. It has a filter consisting of alternating layers of quartz crystals and polaroid sheets. Because of optical interference (see Index), only one color is transmitted; any spectral line can be selected for study.

The spectroheliograph has been coupled with rapid-sequence motion-picture cameras at the McMath-Hulbert Observatory, in Michigan, in an instrument called a spectroheliokinematograph. The films taken with such equipment illustrate dramatically the high-speed motions of the gases of the sun's surface and atmosphere.

Until the year 1868 astronomers were able to observe the corona and the prominences of the sun only during total solar eclipses. In that year the Frenchman Pierre-Jules-César Janssen and the Englishman Sir Joseph Norman Lockyer, working independently, directed telescopes fitted with spectroscopes toward the edge of the sun rather than toward its bright surface. They found that it was possible to see the brilliant, irregularly-shaped clouds of gas of the prominences, so familiar to observers of eclipses. Janssen and Lockyer, however, failed to detect the fainter gases of the sun's nebulous corona, discovered a few years later.

In the year 1930, the French astronomer Bernard Lyot developed a completely new type of telescope known as the coronagraph, which made it possible to see the corona in full sunlight. This instrument produced artificial eclipses of the sun by setting an artificial moon of the appropriate size between the lens of the telescope and the disk of the sun. The light scattered inside the apparatus is eliminated by treating the optical surface suitably. Lyot set up the first coronagraph at Pic du Midi, a lofty mountain in the Pyrenees. Various other coronagraphs are now in use, including the well-known one in the high-altitude observatory at Climax, Colorado. With the coronagraph, it is possible to study the behavior of the corona from day to day, instead of only during the brief and rare moments of total solar eclipse.

**The mysterious spots on the face of the sun**

The sunspots are among the most interesting features of the sun's disk. They appear dark because they are somewhat cooler and therefore less bright than the areas that surround them. Yet their temperature is only about 3,600° F. less than the 10,000° mark of the photosphere. The sunspots would look very bright against a dark background. Each spot has two clearly defined areas — a darker region, the umbra, at the center, and a somewhat brighter area, the penumbra, surrounding the umbra. At times there are as many as a hundred spots on the sun's disk; sometimes there are none at all.

According to one theory, the sunspots may be compared to tornadoes. It is held that they are caused by powerful electric currents that circle about the central core of the sunspots, producing whirling effects like those of a tornado. Certain astronomers, however, believe that sunspots are quiet and relatively cool regions in the hot, turbulent solar atmosphere.

Sunspots do not occur at random on the sun's bright face; they show certain strange and unexplained regularities. For example, every 10 to 12 years they are comparatively numerous and large; in the intervening years there are fewer spots and they are smaller. Thus there is a series of sunspot cycles, each cycle averaging about 11.2 years in length.

At the beginning of a cycle, when there are few spots, they are found at

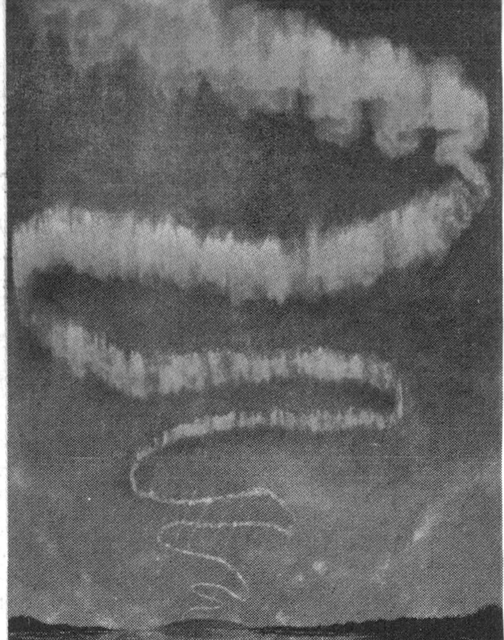

Three forms of auroras, spectacular displays brought about as charged particles emitted from the sun strike atoms in the upper atmosphere of the earth in high latitudes. The aurora borealis, also known as the northern lights, occurs in the Northern Hemisphere; the aurora australis (southern lights), in the Southern Hemisphere.

comparatively high solar latitudes (up to about 40°) north and south. As the number of spots increases, the zone where they chiefly occur draws nearer the equator. Finally, as the cycle is near its end, there are only a few spots relatively near the equator. The first spots of the next cycle then form in higher latitudes again. If a graph is made of the spots throughout the cycle, based on the number of spots as well as the year and latitude in which they occur, the graph will show a curious butterflylike shape, shown on the next page. This effect was first discovered by the Maunders, a famous husband-and-wife team of English astronomers, early in the century.

Other features of the sun seem to be related to the sunspot cycle. Faculae appear at the places where sunspots later break out, apparently acting as forerunners; the faculae remain at the site of the sunspots after the latter have disappeared. The solar prominences are larger and more frequent at the time of sunspot maximum; the corona also changes in shape and brightness in accordance with the sunspot cycle.

Sunspot activity has certain direct effects upon our planet, the earth. These effects are due in large part to the stream of electrically charged and high-speed particles that are emitted from the sun when sunspot activity is at its maximum. The particles are made up of hydrogen nuclei (protons) and certain heavier nuclei, as well as the negative particles called electrons. They are strongly affected by the magnetic field of the earth when they approach our planet, and they spiral around the lines of force created by the earth's magnetic poles. (See Index, under Poles, magnetic.)

The effects of the charged particles emitted from the sun are felt most strongly near the poles and least of all in low-latitude regions. The impact of the particles upon the earth's upper atmosphere in high latitudes brings about spectacular displays — the aurora borealis, or northern lights, occurring in the Northern Hemisphere and the aurora australis, or southern lights, in the Southern Hemisphere. They take various forms. An aurora may look like a waving, many-colored curtain moving across the sky. It may take the form of a series of long rays pointing straight upward or aslant from the horizon. It may look like a crown of light, from which streamers extend to the horizon.

The charged particles affect radio and television reception as they react upon the ionization layers of the atmosphere (see the article Earth's Airy Envelope, in Volume 1). The magnetic storms they bring about may be strong enough to disrupt telegraph service and may lead to electric power failure. These storms may be violent enough to cause a compass needle to move through an arc of several degrees in a single hour.

The sun's ultraviolet and X-ray emissions also affect the earth. When violent, sudden flares occur, for example, a burst of solar radiation produces changes in the earth's ionosphere. These, in turn, cause interference with normal radio communication. The solar bursts may also heat local regions of the earth's highest atmosphere.

The American astronomer Andrew E. Douglass showed that there was a direct relationship between the growth of trees and the solar activity cycle. Annual rings (see Index), which appear in cross section, say, in a felled tree, fluctuate markedly in width in ten-year to twelve-year cycles, corresponding to sunspot cycles.

Graph providing an analysis of the sunspots that occurred during two cycles, extending from 1878 to 1901. The figures at the left-hand margin represent degrees north and south solar latitude; 0 represents the solar equator. As the number of spots increases, they approach the equator. Note the butterfly patterns that are produced.

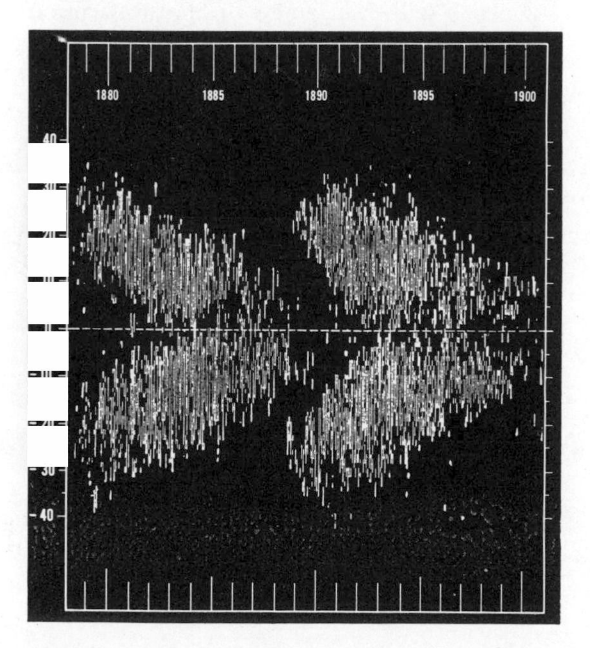


The amount of heat received from the sun seems to vary according to the cycles. A Smithsonian Institution team, headed by Charles G. Abbot, suggested that fluctuations in solar radiation, never exceeding 3 per cent, accompany the various phases of sunspot activity that have been noted by astronomers. Other investigators, however, are rather inclined to doubt the conclusions reached by Abbot.

Sunspot cycles undoubtedly have subtle effects upon the weather, which is controlled to a great extent by the radiation emanating from the sun. Any changes in the nature or the quantity of this radiation — changes such as are associated with sunspot activity — are bound to have a vital effect upon the weather. Attempts have been made to present long-range weather predictions based on sunspot activity. These attempts have not been particularly successful so far; they have been complicated by the fact that weather represents the interaction of a good many different factors — winds, ocean currents, large land masses, lofty mountain ranges and the like. The years ahead, however, may see great advances in research on the different ways in which the sun affects our weather.

### The rotation of the sun

The sunspots persisting for days, weeks or even months, seem to move steadily across the face of the sun. This makes it evident that our star rotates on its axis. The sunspots have provided us with an excellent means of determining the speed of rotation in the latitudes where they occur. The faculae also move across the sun's disk and also serve to measure the speed of rotation of our star. In latitudes where sunspots and faculae do not occur, astronomers must rely on spectroscopic analysis.

The visible colors of the spectrum range from violet to red. If a heavenly body is approaching the earth, all of the lines of its spectrum shift toward the violet end; if it is speeding away from the earth, the lines of the spectrum shift toward the red. This is called the Doppler

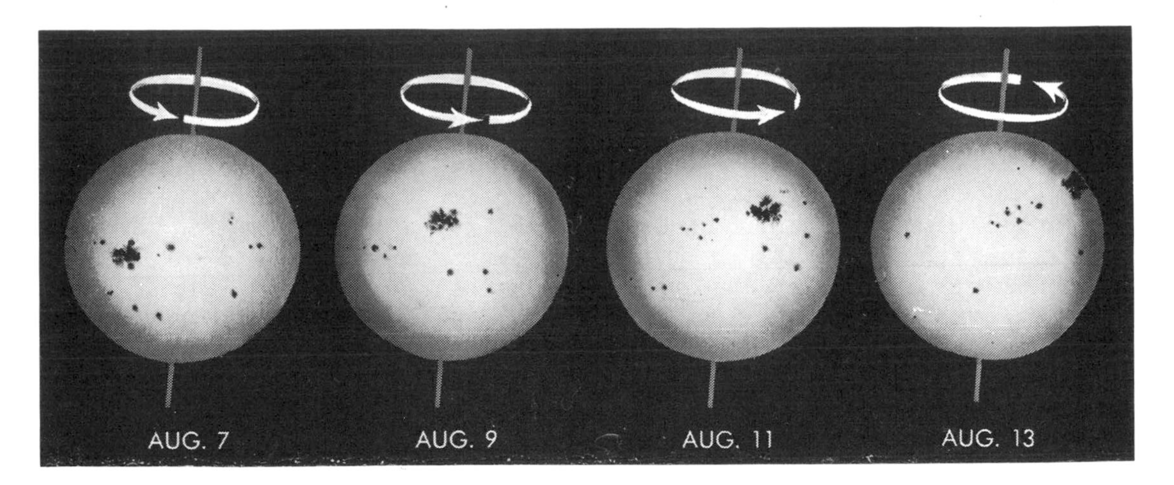


These four drawings show the rotation of the sun, as indicated by the passing of sunspots across the face of the solar disk during a seven-day period — August 7 to August 13, inclusive. Note that on the 7th the sunspots were near the extreme left of the sun's disk. They had moved appreciably toward the center by the 9th; by the 13th they had reached the right edge of the disk. The arrows, above, show the direction of the sun's spin.

effect (see Index). As the sun rotates, each part of it, except the central axis, is either coming nearer to the earth or moving away from it. By studying the shift in the spectrum involved in either case, astronomers can determine the speed of rotation.

Like the earth, the sun rotates from west to east. As it rotates, its axis is inclined at an angle of 82 degrees 49.5 minutes to the ecliptic — the plane of the earth's orbit. The speed of rotation is not the same in all regions. At the equator (latitude 0°), the sun has a period of rotation of 24.65 days — that is, it takes 24.65 days to make a complete turn. The speed of rotation becomes progressively slower at higher latitudes. At latitude 30° it is 25.85 days; at latitude 60°, 30.93 days; at the north and south poles (latitude 90°), 34 days. There is no generally accepted explanation for such differences in rotational speed.

**The chemical composition of the sun**

Our knowledge of the sun's chemical composition is based on the study of the spectrum of the sun as a whole, as well as the spectra of different layers of the atmosphere. As we have pointed out, in the photospheric spectrum the progression of colors from violet to red is interrupted by the absorption lines, often called the Fraunhofer lines, after their discoverer, Joseph von Fraunhofer. We now realize that most of these lines are due to the fact that certain wave lengths of light — that is, certain colors — have been absorbed by the atoms of various gases contained in the sun's atmosphere. Each type of atom absorbs certain definite colors; hence the absorption lines make it possible to identify the atoms that produce them.

It is true that some of the absorption lines in the solar spectrum are caused by atoms existing in the earth's atmosphere, which sunlight must traverse before it can be registered on a photographic plate. About 6,000 lines have been traced to such absorption effects. Fortunately, however, these lines, known as telluric lines (*tellus* means "earth" in Latin), can be readily distinguished from the true solar lines, produced by absorption effects in the sun's atmosphere.

As we analyze the solar spectra, we note that the great majority of the chemical elements known on earth are also found in the sun. It is believed that *all* the elements existing on the earth occur in the sun. If we have not been able to find some of them thus far, it may be because they are as rare on the sun as they are on the earth and, therefore, cannot be detected by the spectroscope.

Hydrogen is by far the most abundant chemical element in the sun. Carbon, oxygen and helium are also found in great quantities. The metals, which of course exist only in the form of gases, are much less common; the most abundant are calcium, iron, magnesium, sodium and nickel.

### The energy released by the sun

The sun radiates energy at the astounding rate of $3.79 \times 10^{33}$ ergs a second. (For a definition of "erg" and an explanation of $10^{33}$, see the Index, under Ergs; Exponents.) What is the ultimate source of this energy? Is it due to the combustion of various materials within the sun? In ordinary combustion, the combustible material is oxidized: that is, it combines with oxygen. There is plenty of hydrogen, a highly combustible material, in the sun, as we have seen; there is also enough oxygen to bring about combustion on a vast scale. We know that when hydrogen and oxygen combine chemically, forming $H_2O$, or plain water, considerable energy is released. But the intense heat of the sun would make it absolutely impossible for the hydrogen and the oxygen atoms it contains to combine; it would serve rather to keep them apart. Even if it were possible for the hydrogen and oxygen atoms to unite chemically, the energy released in this way would be entirely inadequate to maintain the brilliance of the sun for even a fraction of the several billion years that have elapsed since it came into being.

In the nineteenth century the celebrated German scientist Hermann von Helmholtz proposed a "contraction" theory that was widely accepted for many years. According to this theory, solar energy is generated by the work of gravitation in compressing the gases of the sun, thus causing it to shrink. Helmholtz calculated that a change of diameter of only 280 feet per year would suffice to maintain the sun's energy-production rate. This small amount of contraction would have caused no perceptible change in the sun's diameter over the entire span of recorded history. However, the few thousand years of recorded history represent a ridiculously small fraction of the sun's age. The amount of energy liberated by the contraction of the sun would support its radiation for only about two million years.

According to another theory, the heat of the sun is derived from meteors falling into it. The energy released in this way, it was suggested, could maintain the heat of the sun. Nobody takes this theory seriously nowadays. Even if meteors plunged

The center strip, below, shows a region of the sun's spectrum. The black lines in the upper and lower strips represent the bright-line spectra observed when an iron arc and a calcium arc are placed one after the other in front of a spectrograph slit. Each bright line corresponds to an absorption line in the spectrum of the sun.

From "Classical and Modern Physics," H. E. White, copyright 1940, D. Van Nostrand Co., Inc., Princeton, N. J.

into the sun at a fantastic rate, the energy they would emit by so doing could not possibly account for its heat. Such meteor collisions may, however, contribute to the heat of the sun's atmospheric corona.

It has been suggested that the sun's energy is due to the radioactive substances it contains. This explanation would hold true only if the sun were composed almost entirely of radium and other radioactive elements. However, from spectroscopic evidence we know that if these elements are present in the sun, they exist in such minute concentrations that they could account for only a tiny fraction of the total energy release.

Another theory of solar-energy production is based on changes that take place in the nuclei of some of the lighter atoms. The British physicist Ernest Rutherford put scientists on the track of such a theory when he discovered that it was possible to transform atomic nuclei. Ten years later Robert d'Escourt Atkinson and F. G. Houtermans suggested that such nuclear transformation, involving some of the lighter atoms, might account for the energy liberated by the sun and the other stars. Finally, in 1939, a German-born American physicist, Hans A. Bethe, worked out a detailed theory to explain how solar energy could be derived from the transformations of atomic nuclei.

This so-called carbon-cycle theory is generally accepted today. It holds that solar energy is generated as a result of the following six-step reaction:

(1) The cycle starts when a free hydrogen nucleus — a proton — crashes into a carbon nucleus, $C^{12}$, with six protons and six neutrons.* A gamma ray is emitted. The added proton makes the atom a light isotope of nitrogen — $N^{13}$, with seven protons and six neutrons.

(2) A proton is transformed into a neutron; as this takes place a positron, or positive electron, is emitted. The nucleus

**NUCLEAR "CARBON CYCLE" REACTIONS RESPONSIBLE FOR SOLAR ENERGY**

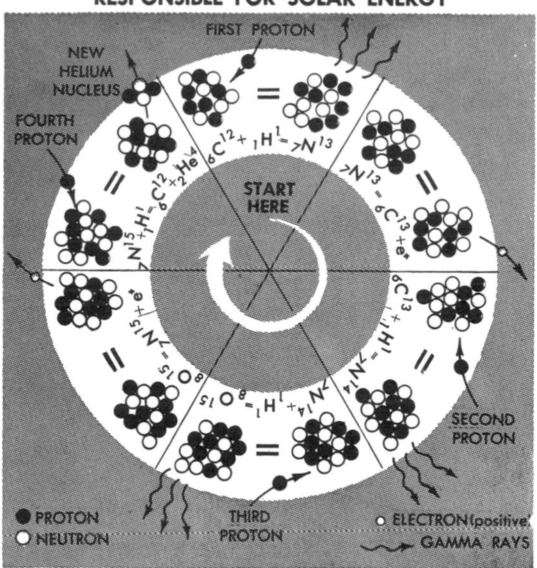


Diagram of the carbon cycle described on this page. According to the carbon-cycle theory, solar energy is generated by the six-step nuclear reaction illustrated above. The letters C, H, He, N and O in this diagram are chemical symbols standing for the elements carbon, hydrogen, helium, nitrogen and oxygen, respectively. The small figure in front of each symbol represents the atomic number; the figure after each symbol, the atomic weight, which is also known as the mass number of the atom.

has one more neutron than before and one less proton. The atom is now a heavy isotope of carbon — $C^{13}$, made up of six protons and seven neutrons.

(3) Another proton comes hurtling into the nucleus of the atom, and another gamma ray is emitted. The atom now has seven protons and seven neutrons; it is ordinary nitrogen — $N^{14}$.

(4) A third proton crashes into the nucleus, releasing more gamma radiation. Now the atom is $O^{15}$, a very unstable isotope of oxygen, with eight protons and seven neutrons.

(5) A positron is emitted from a proton, which is transformed into a neutron. The atom changes into $N^{15}$, consisting of seven protons and eight neutrons.

(6) A fourth proton smashes into the $N^{15}$ nucleus and splits it up into an ordinary carbon nucleus, $C^{12}$, and a helium nucleus, $He^4$, containing two protons and two neutrons.

* The commonest isotope of hydrogen, with atomic weight 1, normally has one proton and one electron. Of course when it loses its electron (in other words, when it is free), it consists only of a proton. The figure 12 in $C^{12}$ represents the atomic weight, also called the mass number, of the carbon atom; the atomic weight represents the sum of the protons (positive particles) and neutrons (neutral particles) at the core of the atom.

As a result of this six-step reaction, four hydrogen nuclei, or protons, have fused to form a helium nucleus. The carbon atom involved in the chain reaction is not changed in any way; it can go through the same cycle again and again as it is struck by one hydrogen nucleus after another. Vast energy is released in the course of the six-stage carbon cycle. It has been calculated that the conversion of a hundred tons of hydrogen into helium would release more energy than all of mankind could use up in a single year. The present brilliance of the sun could be maintained for thirty billion years if its hydrogen continued to be transformed into helium at the present rate.

According to another theory, the sun's energy is derived, in part at least, from a "proton-proton reaction." In this, helium nuclei are built up directly from hydrogen nuclei, or protons. As in the carbon cycle, four hydrogen nuclei fuse to form a helium nucleus.

Since the earth is comparatively small and is far away from the sun, it receives only about 1/2,000,000,000 of the outpouring of solar energy; besides, about 30 per cent of this small fraction is absorbed or scattered as it passes through the earth's atmosphere. Yet the amount that actually strikes the earth's surface is almost unbelievably great; it has been calculated that it is equivalent to 4,690,000 horsepower per square mile.

Solar energy provides the light and warmth necessary for all animal and plant life upon the earth. The sun's light is essential in the all-important process of photosynthesis (see Index). In this process, water, derived from the soil is split by light energy into hydrogen and oxygen. The oxygen is released to the atmosphere, while the hydrogen combines with carbon

How the sun's energy serves us. The remains of certain plants and animals, which have absorbed solar energy, are transformed to petroleum and coal — stored solar energy. The sun's heat causes water to evaporate. The vapor condenses in clouds and falls to earth as rain or snow. It replenishes rivers; their waters turn the blades of turbines, which generate electricity. The sun gives us light and heat; it powers engines and batteries.

dioxide, derived from the atmosphere, to form a simple sugar. The sugar serves as a basic material from which the plant manufactures various essential compounds. These compounds are as necessary to animals as they are to plants; animals obtain them either by eating plants or by devouring animals that have eaten plants. Photosynthesis provides much of the oxygen that we require for respiration and other kinds of combustion.

The energy derived from the sun serves in many other ways. The fuel coal, which provides us with heat and power, represents stored solar energy of ages long past. It is the final stage in the transformation of vegetation buried deep within the earth and compressed for hundreds of centuries — vegetation that thrived in sunlight in prehistoric times. Petroleum is another form of stored solar energy; it is derived from marine animal and vegetable life of prehistoric epochs.

Hydroelectric plants that use the force of running water to generate electricity also depend ultimately upon the energy supplied by the sun. The sun's heat causes water to evaporate from lakes and oceans. The water vapor condenses in clouds and later falls upon mountains and hills in the form of rain, snow and other forms of precipitation. Part of the precipitation makes its way to rivers and other streams. As these course down valleys, their waters are made to turn the blades of turbines; the turbines are connected to dynamos that generate electricity. The winds that turn the blades of windmills are also derived from solar energy; for all winds result from the uneven heating of different parts of the earth's surface.

Solar energy has been used by men directly, though to a limited extent, to supply heat and power. Our homes are heated in some degree by sunlight entering through the windows. Some houses are specially designed so that all or a considerable part of their heat is derived from coils set up behind specially prepared windows and subjected to sunlight. Solar ovens have been operated successfully. Solar engines using sunlight as their power source have been devised; in these engines, the sun's rays cause water to be heated and the steam is made to perform work. In another article we show how sunlight serves to generate electricity (see Index, under Solar battery).

### The past and future of the sun

The age of the sun is still a matter of speculation. Modern astronomers tend to believe that the sun and the earth came into being as a result of the same general process of development perhaps five billion years ago. In the course of its long life, the radiation of the sun has never fluctuated drastically. We know that plant and animal life have existed for long geologic ages in an unbroken span; and all life on earth depends upon the maintenance of a delicate balance between solar heating and terrestrial cooling. If we accept the theory that the ice ages were due to long-sustained minor changes in solar radiation, we know that the sun occasionally departed from strictly normal behavior; we also know quite well when these "lapses" took place.

If the sun was to transform its hydrogen into helium at its present rate, it would continue to provide us with light and heat for some thirty billion years, as we have pointed out. But it is generally believed that as it grows older, the sun will consume more and more hydrogen and will grow hotter. As it becomes hotter, the carbon-cycle nuclear reaction will proceed more rapidly; our star will lavishly radiate more light as its store of energy will grow smaller. That means that its total span of life will be cut down to about 10,000,000,000 years. By the end of this time the sun will probably have become a white dwarf star (see Index, under White dwarf stars), far denser than it is now and emitting very little radiation.

Long before its death, the increase in radiation will cause the temperature of the earth to rise above the boiling point of water and, as far as we can see, life upon our planet will be wiped out. But all this will happen billions of years from now, and we should not worry unduly!

*See also Vol. 10, p. 269:* "Sun."

# FROM CAVE TO SKYSCRAPER

## How Man the Builder Has Triumphed over Space and Matter

THE story of building construction bridges the immense gap between primitive man, seeking a makeshift shelter, to the highly civilized man of today, dwelling amid the mighty piles of steel and concrete and brick that we call cities. It is a fascinating story; its beginnings go back to the Old Stone Age, when man first appeared upon the earth.

Some of the early peoples of that remote era dwelt in open-air sites near their water supply or near the animals that they hunted. In time they built shelters of branches, or they draped skins over poles in order to make rude tents. Other Stone Age peoples dwelt in rock shelters and caves, which they sometimes adorned with astonishingly lifelike paintings. Certain primitive tribes lived in rude huts, thatched with grass or leaves. Others became lake dwellers; they drove piles in lake beds offshore and constructed houses upon these piles.

In the course of the ages more effective building techniques were developed; ingenious tools were invented; new materials were utilized. A genuine art of building construction came into being. The knowledge of this art was passed on from father to son or became the possession of certain groups of people. As time went on, the art became more and more highly specialized; there were now masons and carpenters and plasterers.

By this time, building construction had become extremely complicated. Someone had to devise beforehand the size and shape of structures; he had to indicate the materials that were to be used; he had to decide how these materials could be obtained. This was the province of the architect, who planned so that others might build. The engineer also came to the fore. His task was to determine how building materials would stand up under stress; he was also concerned with the thousand and one problems of construction; he analyzed the equipment for new structures.

In this modern age of vast building projects, the engineer and the architect work hand in hand. They have brought about veritable miracles in building construction. They have had the advantage of working with modern materials such as structural steel, reinforced concrete, cinder blocks, insulating substances (like asbestos and diatomaceous earth), plywood and plastics. (See the article The Builder's Materials, in Volume 7.)

Perhaps the culmination of the modern builder's art is the giant skyscraper, which is as much a product of this industrial age as Europe's Gothic cathedrals were a product of the devout Middle Ages. In a matter of months, a comparatively small force of men can construct a mighty edifice such as ancient builders, with armies of slaves at their command, could never have dreamed of erecting.

The skyscraper, which originated in the United States in the last decades of the nineteenth century, is a wonder of architectural and engineering skill — fireproof, sanitary and built to endure. No tyrant erected these immense edifices as lasting monuments of his greatness; they came into being as the result of economic conditions in the large cities of the United States. A great volume of business had to be conducted in districts of limited area; large floor space had to be provided on a comparatively small plot of ground.

The yurt, the primitive Mongol dwelling shown in this photo, consists of a lattice framework that is covered with thick felt.

Am. Mus. of Nat. Hist.

Under the circumstances, buildings constructed in the traditional way were out of the question. The lower walls would have to be made so thick in order to support the crushing weight of the upper stories that a great deal of valuable space would be wasted and the cost would be prohibitive. American building constructers decided to strike out on a new path.

To provide ample office space in a building erected on a comparatively small lot, they erected a huge up-ended steel bridge to serve as a framework. A series of disastrous fires taught that when steel is subjected to great heat, it becomes a weak building material. They decided, therefore, to sheathe their steel columns, girders and beams with some fireproof material such as concrete or terra cotta. Other problems had to be solved. Builders had to investigate the strength of established materials and to develop new materials. They had to create new techniques in order to reduce to a minimum the time and manpower required for the construction of many-storied buildings.

The skyscraper is original in both conception and execution. It reflects some of the best qualities of Americans — their self-reliance, their willingness to blaze new trails, their mechanical genius — and also certain less desirable traits, such as their impatient haste, their chronic desire for something novel and their preoccupation with mere bigness.

As a matter of fact, the salient traits of many peoples have been summed up in their architecture. The massive and rather monotonous lines of the edifices built by the ancient Egyptians reflect the stability and traditionalism that have always been associated with the Egypt of the Pharaohs. By way of contrast, the lively, versatile and highly imaginative spirit of the ancient Greeks was displayed in architectural contrasts, such as plain surfaces alternating with exquisite decoration, as well as in infinite variations of detail. The unimaginative, matter-of-fact Romans were not so much concerned with charming little details and delightful surprises as they were in building up a succession of infinitely repeated motifs in such a way as to produce an effect of underlying unity. The famous building known as the Pantheon, with its simple divisions, its comparatively plain surfaces and its immense dome brooding over all sums up admirably the Romans' attitude toward architecture.

The skyscrapers include some of the world's most immense edifices. The Empire State Building, in New York City, is the loftiest building ever reared by man; it stands 1,472 feet high, including the television sending station that was erected in 1950. The building was completed on May 1, 1931, on the site of the original Waldorf-Astoria Hotel, at a total cost of something like $43,000,000. Only 25 weeks were needed for the erection of the 57,000 tons of steel in the framework. The total weight of the structure is estimated at 600,000,000 pounds. It has 102 stories, 74 elevators, 2 basements and can accommodate 25,000 tenants. The Chrysler Building, also in New York City, is the second tallest building in the world; it stands 1,046 feet in height.

Both the Empire State Building and the Chrysler Building overtop Paris' famed Eiffel Tower — an architectural freak of iron latticework that was constructed for the Paris International Exhibition of 1889. Of the other skyscrapers in the United States, no fewer than thirty-four are more than 500 feet high. By way of comparison, the Pyramid of Cheops in Gizeh, Egypt — one of the Seven Wonders of the World — stands 450 feet in height.

It would be possible to construct a steel-frame building that would soar far above the impressive height of the Empire State Building. Steel is strong enough so that an architect could fashion a framework of this metal that would spread over a fairly large foundation the weight of an edifice 2,000 feet high or, perhaps, even higher. However, it is very doubtful whether any such superstructure will be erected in the near future. One reason is that it is not considered sound policy, from the dollars-and-cents viewpoint, to construct excessively high buildings.

The foundations of skyscrapers are not laid according to a set pattern; they must be built according to the lay of the land. If the soil is not stable enough at reasonable depths, piles must be driven and footings constructed. (Footings are enlargements at the lower ends of piles and similar structures.) In this way, the load is transmitted from each column down to a number of piles and their footings. Tapered columns of wood, concrete or concrete-filled shells may be used instead of piles.

Occasionally builders take other measures in order to provide adequate support. In the marshy areas of Chicago, for example, skyscrapers are erected upon huge mats of concrete and steel grillage. In New York, the foundations of big buildings are usually carried down to bedrock.

Builders of skyscrapers must also take winds into account. The force of gravity tends to slow down winds; this dragging effect is less noticeable as distance from the ground increases. At great heights strong winds exert tremendous pressure.

Ewing Galloway

**Here is one of the finest examples of ancient Roman architecture — the imposing Pantheon at Rome.**

Caisse Nationale des Monuments Historiques

The famed Gothic cathedral at Rheims, France, restored after the ravages of the first World War.

It took about a hundred years to build the magnificent Gothic cathedral at Rheims. Today a huge edifice can be constructed and ready for use within a period of two years, or even less. The Socony Mobil Oil Building in New York City is an excellent example of such an engineering feat. The ground was broken in March 1954; the first occupant arrived twenty-six months later, in May 1956. The building is forty-five stories high. It is rectangular, measuring 400 feet by 200 feet. Within the building, there are 21,000,000 cubic feet of space, and over 1,500,000 square feet of floor area. The entire building is air-conditioned to provide added comfort for the employees. Special high-voltage circuits (265 to 460 volts) are supplied so that electronic calculators may be operated on the premises.

The most interesting feature of this large edifice is its outer shell. This consists of 7,000 prefabricated stainless-steel panels, each stamped with a uniform design. Since there are no horizontal lines in the individual panels, they are easily washed by the rain. About fifty-eight seconds were required to put each one in its place. The entire skin was constructed in six months. A steel-panel facing has certain advantages over the ordinary concrete facing. The panels are easily maintained, and they allow thinner walls so that greater office space may be made available. There are other metal-clad structures; the Socony Mobil Oil Building is the largest in the world.

The general trend in present-day construction is for eye appeal in addition to usefulness. Buildings are designed so that their function can be emphasized artistically. Glass is used extensively in many of the newer office structures for maximum light penetration.

Lever House, in New York City, embodies some of the newest ideas in modern architecture. It is a twenty-four-story glass structure which seems to float on pillars of stainless steel. Except for the entrance, there is no enclosed space on the ground floor. Part of the building has only three floors. A terrace occupies the roof of this section. The windows and the spaces between the windows consist of a special kind of glass, called Solex glass. It is tinted green and can filter out 35 per cent of the heat rays coming through the windows. Since the entire building is air-conditioned, the windows are permanently sealed. Around the rim of the roof, there is a narrow track. A scaffold, which extends down the side of the building from the roof, moves along this track. An outdoor elevator, installed in the scaffolding, enables a cleaning crew to clean the building's glass exterior in perfect safety.

The building has been designed to give the illusion that it is open on all sides. It is a novel and architecturally pleasing type of modern skyscraper, and it is well worth its cost of $6,000,000.

Charles Phelps Cushing

**Chrysler Building, in New York City.**

Owen Moore from Black Star

The Rockefeller Center, in New York, is an outstanding modern development, covering twelve acres.

The United States Naval Hospital in Washington, D.C., is an example of fine concrete construction.

U. S. Navy

Skyscrapers have a distinctive character of their own — a grandeur that is born of confident strength. Some of them, it is true, may strike us as quite monotonous with their boxlike design and their stark lack of adornment. Yet many of these titanic, clifflike steel towers are models of soaring gracefulness and compare very favorably with the outstanding architectural achievements of the past.

The modern tall building is a marvel of technological achievement. Amazingly efficient elevators whisk passengers to the highest floors in a brief minute or two. Labyrinthine plumbing, heating, lighting and air-conditioning systems assure both safety and comfort. Acres and acres of big windows offer natural lighting for the interior of many of the big buildings. This is possible because the walls of the modern skyscraper are mere skins, comparatively speaking. It is the steel framework that actually supports the building; the walls have become mere curtains for protection against the elements. For example, aluminum sheet has been used for the spandrels of the towering Empire State Building. (A spandrel is the portion of a wall between the top of one window and the sill of the window above it.) In the big McGraw-Hill Building in New York, the spandrels are continuous horizontal bands of terra cotta, and the windows are nearly continuous between the columns.

People have been attracted to the larger office buildings, hotels and apartment houses, not so much because they are large as because of their many conveniences and the almost complete safety from fire that they afford. In the typical steel-frame building, with the frame properly encased in concrete or other protective material, there is less risk of fire than in almost any other kind of structure.

Reinforced concrete has become widely used in the present century. In this type of construction, metal (generally steel) is embedded in concrete in such a manner that the two materials act together in resisting forces. This material can be made fireproof, watertight and durable for use in many different types of buildings. It was first developed many years ago. In 1850, a boat of reinforced concrete was built by the Frenchman Joseph-Louis Lambot. Four years later, W. B. Wilkinson patented his method of reinforced slab construction in England. The theory of reinforced concrete had been pretty well developed by the end of the nineteenth century.

Reinforced concrete will probably never replace steel-frame construction in buildings above twenty stories or so tall, because of the increased size of the columns required to carry the weight of the buildings. However, for factories, warehouses, bridges, private dwellings and the like, reinforced concrete has much in its favor.

The pictures on these two pages show various stages in the erection of a skyscraper framework. Above: preparing to "jump a guy derrick" — that is, to raise it to a higher floor.

All photos, Bethlehem Steel Co.

This big column is being landed on the appropriate floor; soon it will be upended.

Men working on the framework of an upper floor, hundreds of feet above the street.

Two guy derricks are shown in place atop the framework of a huge building under construction. The boom of one of the derricks is lifting a load of steel from a truck on the street below.

Buildings made of reinforced concrete are firesafe, verminproof and resistant to dampness and sudden changes of temperature. Considering the lasting qualities of concrete, it is relatively inexpensive.

Reinforced concrete is made by mixing the correct proportions of portland cement, sand, gravel and water. Steel rods are inserted in the mixture before it has a chance to set. When hardened, the concrete grips the steel rods firmly in place. They supply a resistance to tension, while the concrete resists the effects of compression. This means that concrete is hard to crush, while steel rods are hard to pull apart. If, for example, a horizontal beam is to be made of reinforced concrete, the rods are set in such a position that when the beam is in place, they will be located toward its bottom. As the beam is loaded, its top fibers will tend to be compressed and its bottom fibers will tend to stretch. The compressive strength of the concrete will resist the first of these effects; the tensile strength of the steel will resist the second. The result is that the beam will not flex much. If an unreinforced beam were used, cracks would appear along its bottom surface. Where concrete is subject to compression only, as in the case of the foot of a bridge pier, steel reinforcing is not necessary.

A recent development, related to reinforced concrete, is pre-stressed concrete. The steel rods are placed under tension (stretched) while the concrete sets. Thus they exert a longitudinal compressing force on the concrete. When the concrete is used in the manner of the beam mentioned above, the downward forces on it exert a tension where the steel rods are exerting their compression. If the tension of the load is the same as the compression of the rods, the forces (which are opposing forces, much as pulling and pushing are opposing) cancel. Thus the beam, which may carry many tons, is, if properly designed, relatively free of internal stresses. Pre-stressed beams have found wide use in building construction. They also serve to replace steel in such applications as the horizontal supporting beams of highway bridges.

Reinforced and pre-stressed concrete beams are often made up in the proper size before construction starts. When work begins, the beam is simply lifted into place and fastened there. A great deal of reinforced concrete construction is done by means of molds. A mold is set where the concrete will be utilized in the structure. The rods are then placed in the proper position and the mixed concrete poured in. After several days, the mold is removed. Construction companies that specialize in this type of work maintain a large number of standard-size molds which they use again and again. The molds are frequently heavily built structures, since they have to withstand the weight of the hardening concrete.

A large part of the work in building a reinforced concrete structure is in setting up and taking down molds. A wooden mold is not nearly so strong as hardened concrete and cannot bear nearly so much weight. Hence, only after the concrete of one story has set is the next story erected.

The basic ingredient of concrete (whether stressed, reinforced or simply poured) is portland cement, invented by Joseph Aspdin, an English bricklayer, in 1824. The name has been derived from the resemblance of the substance to the limestone quarried at the Isle of Portland, England. The cement is composed of lime (calcium oxide), silica, iron oxide, gypsum and traces of other elements. Lime and silica comprise 80 to 90 per cent of the mixture. The limestone is first crushed to the size of small stones. Then it is fed into a grinding mill, together with the other materials. The resulting powder is then burned in huge revolving kilns, some of which are as much as 400 feet long. The clinker (fused material) produced by this process is sent to the final grinding mills, where it is finely powdered and mixed with gypsum. After this, it is ready for use in making concrete.

Depending on the hardness desired, the cement comprises from a tenth to a fourth of the total weight. A mixture of one part cement to three parts sand and six parts gravel is standard for ordinary purposes. A harder 1:2:4 mixture is generally used in reinforced concrete construction.

United Nations

The beautiful Secretariat Building, part of the United Nations center in the heart of New York City.

The Price Tower in Bartlesville, Oklahoma, is a fine example of daring new design by the American architect Frank Lloyd Wright. It was completed in February 1956, and has been hailed as an architectural landmark in the Southwest. The exterior walls of the 19-story tower are non-structural, being supported entirely by the cantilever floors that project from four interior concrete shafts.

H. C. Price Co.

When concrete is poured into forms, it hardens in such a way as to form a single mass without joints. It is capable, therefore, of supporting great loads; that is why it is used so extensively for heavy construction — for foundations, piers, dams and so on. Unfortunately, however, it requires a good deal of material and labor to make the forms; it takes time, too, to wait for the concrete to harden. Several ways have been found to overcome these difficulties. There are now new types of concrete that develop high strengths very quickly; the forms can be removed a day or so after pouring. Sometimes, too, the concrete is precast in blocks or slabs before it is placed in the building.

Concrete houses are durable, watertight and fireproof. They are moderate in first cost and require little maintenance over the years. Concrete can be made in various colors and can be molded into harmonious lines and exquisite forms. It is verminproof; no rats or mice or insects can burrow into it. The fungus growth that is sometimes seen on old buildings does not form on concrete; this material offers no lodging place for disease germs. By means of a hose, a building of reinforced concrete can be cleaned inside and out with ease and rapidity. Reasonable in price and beautiful, sanitary and enduring, modern concrete is a superb building material.

*See also Vol. 10, p. 282:* "Construction."

**This concrete apartment house, in Marseilles, France, was designed by the famous French architect Le Corbusier. The seventh and eighth floors of the building contain a shopping center, gymnasium, library and laundry.**

French Government Tourist Office

# THE FIXATION OF NITROGEN

## Putting to Work an Abundant Atmospheric Gas

THE chemical element we call nitrogen makes up almost four fifths of the atmosphere, in the form of a free, or uncombined, gas. As far as we know, it serves very little purpose other than to increase the density of the atmosphere and to dilute the oxygen that air contains. Nitrogen is drawn into the lungs together with oxygen and, like oxygen, is dissolved in the blood. However, it takes no part in the important chemical changes that take place in the tissues; ultimately it diffuses into the lungs, from which it is exhaled. It is a prospective source of trouble, because if a person breathes air that is under excessive pressure, the nitrogen dissolved in his blood may form bubbles, with disastrous results. This condition is called caisson disease, or the bends. (See Index, under Caisson disease.)

The reason for the negative role played by free nitrogen is that it does not readily enter into combination with other elements. But such compounds do exist, nevertheless, and they are extremely important in many respects. Nitrogen, combined with other elements, is an essential constituent of protoplasm; it makes up about 16 per cent of all living tissue. Plants obtain nitrogen from the soil in the form of compounds called nitrates. Animals obtain it by eating plants or else by devouring animals that have eaten plants.

Most compounds of nitrogen are very unstable; they react easily and rapidly and sometimes with explosive force. For this reason, nitroglycerine, $C_3H_5(NO_3)_3$, and certain other nitrogen compounds are used widely in the manufacture of explosives. Nitrogen is a constituent of ammonia ($NH_3$) and nitric acid ($HNO_3$), two very important industrial chemicals. Nitrogen compounds such as sodium nitrate ($NaNO_3$), called Chile saltpeter, and potassium nitrate ($KNO_3$) are excellent fertilizers; they provide plants with adequate quantities of essential nitrogen. There are many other useful nitrogen compounds.

Nitrogen compounds are produced in large quantities by natural processes. When the remains of plants and animals, with their nitrogen-containing proteins and amino acids, are returned to the soil, certain microbes decompose them, forming ammonium salts. Various bacteria then cause the ammonium to combine with oxygen in such a way as to produce nitrates. The rainless provinces of Tarapacá and Antofagasta, in northern Chile, contain vast amounts of sodium nitrate, formed in this way. For years these deposits met the needs of industry and agriculture.

As the demand for nitrates increased in the last decades of the nineteenth century, particularly because of the great expansion of the chemical industry, men began to realize that the supplies of Chile saltpeter were not inexhaustible and that there might be serious shortages in the future. Chemists then sought to supplement the natural supplies of nitrogen compounds. They succeeded in extracting considerable quantities from coal. When this familiar mineral is burned in a so-called by-product coke oven, from which air has been excluded, it yields not only coke but also a variety of by-products, one of which is the nitrogen compound ammonia.

It was evident to chemical pioneers that the free nitrogen in the atmosphere offered even greater possibilities than coal as a source of nitrogen compounds. Of course, to make the boundless supplies of atmospheric nitrogen available, chemists would have to fix it — that is, cause it to combine with other elements.

The fixation of nitrogen takes place in nature on a limited scale. Certain bacteria thrive in knoblike growths on the roots of plants such as peas, beans and clover; these bacteria can effectively fix the nitrogen contained in the air that is found in the soil. Farmers can enrich their lands with nitrogen by planting crops of peas, or beans or clover and then plowing them under. Lightning also fixes a certain amount of nitrogen in the atmosphere. Every flash causes some of this nitrogen to combine with atmospheric oxygen. After several further reactions, the nitrogen becomes available to plants in the form of nitrates. But the compounds produced by such natural fixation could not begin to meet the world's increased needs.

#### The development of the Haber process

As early as the last decade of the nineteenth century, various chemists sought to fix the nitrogen in the atmosphere by causing it to combine with hydrogen, thus producing ammonia, whose molecule is made up of nitrogen and hydrogen atoms. Among the best-known of these researchers was the German chemist Fritz Haber. He found that ammonia could be formed from nitrogen and hydrogen under high temperatures and pressures. A large German chemical firm, the Badische Anilin- und Sodafabrik (Analine and Soda Manufactory of Baden), was impressed by Haber's work and backed him with its great resources. In 1910, the German company set up a pilot plant for the manufacture of ammonia by the Haber process. It was not until 1913, however, that production on a commercial scale began. About 7,000 tons of fixed nitrogen were produced yearly.

#### Improvements in the Haber process

Since that time the process has been greatly improved. There are various modifications of it, known under different names, such as the Casale process in Italy and the Claude process in France. The output of ammonia produced by these processes comes to more than a million tons a year.

The basic Haber process consists of heating a mixture of nitrogen and hydrogen to a temperature of about 500° C.; the mixture is kept at a pressure of from 500 to 1,000 atmospheres. (An atmosphere is a pressure of 14.7 pounds to the square inch.) In the presence of a catalyst — generally iron — about a fourth of the nitrogen and hydrogen is converted into ammonia. The mixture of nitrogen, hydrogen and ammonia is led from the heavy steel bomb where the reaction takes place. The ammonia is drawn off; then the nitrogen and hydrogen are heated and compressed anew in order to form more ammonia. The ammonia can be processed to yield the compounds required for use as explosives or fertilizers or other essential products. In the two following pages we describe the synthesis of ammonia in greater detail, illustrating it with a diagram and photographs.

#### Other methods of fixing nitrogen

The Haber process is the most economical one for the fixation of nitrogen. It is not the only successful method; various others have been developed. In the electric-arc process, air is passed through an electric arc providing a temperature of about 3,000° C. Nitrogen and oxygen in the air unite at high temperatures to form nitric oxide (NO), from which nitrates and nitric acids can be formed. In the cyanamide process, calcium carbide reacts with nitrogen at a temperature of about 900° C. to yield calcium cyanamide ($CaCN_2$). This is a useful fertilizer, and it yields nitric acid and other compounds that serve in industry.

The problem of meeting the mounting demands for nitrogen compounds has therefore been definitely solved. At one time the exhaustion of the Chile sodium nitrate deposits would have had far-reaching consequences in many parts of the world; today it would have little or no effect upon our industry and our agriculture.

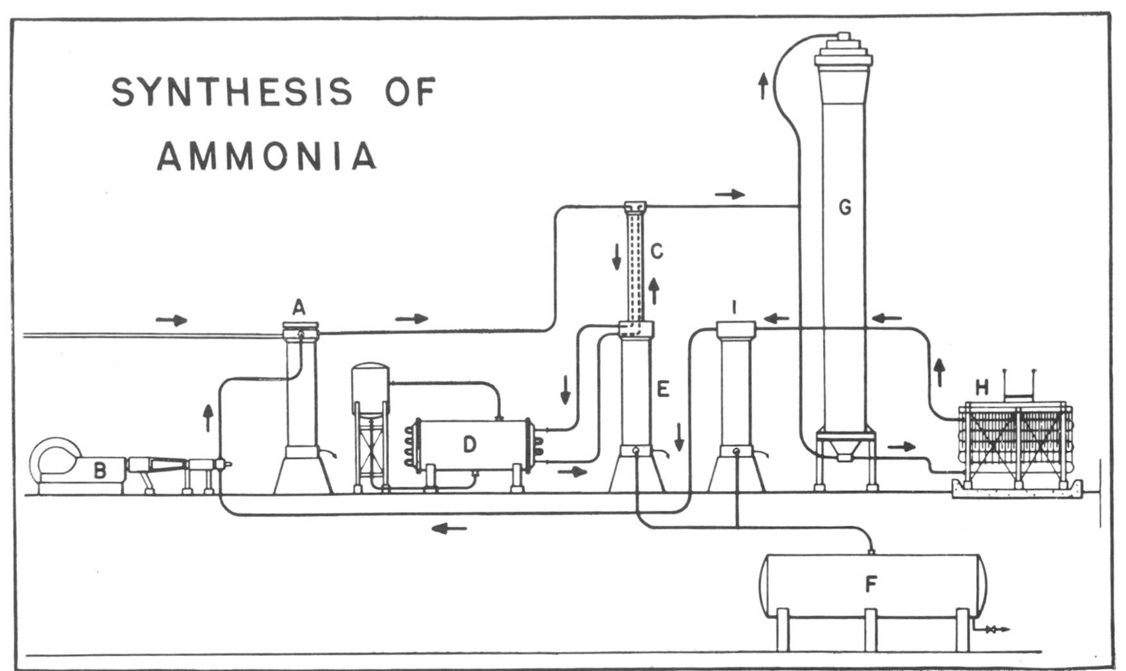


Diagram and photos, Chemical Construction Corporation

The above diagram shows how ammonia is synthesized (combined chemically) from a mixture of two gases, nitrogen and hydrogen. After this mixture has been compressed, it enters the synthesis system through the pipe that is shown at the extreme left. Uncondensed ammonia and unconverted nitrogen and hydrogen, circulated by compressor B, are added to the mixture. (We explain below how the ammonia, nitrogen and hydrogen reach the condenser.) The mixture now goes to a cold exchanger (C) for precooling, before entering condenser D. Here the ammonia contained in the mixture is condensed. The mixture again passes through cold exchanger C. The condensed ammonia is separated from the rest of the mixture in E and is drawn off into an ammonia storage tank (F). After the removal of the ammonia, the nitrogen and hydrogen enter an ammonia converter (G). In this converter the gases are subjected to great heat and high pressure; in the presence of a catalyst most of the mixture is changed into ammonia. Leaving the converter, the gases are cooled in a

# FORM A VALUABLE INDUSTRIAL MATERIAL

water-cooled condenser (H), where most of the newly formed ammonia is condensed. It reaches the separator (I), and from there it passes into the ammonia storage tank (F). The uncondensed ammonia and the unconverted nitrogen and hydrogen pass to compressor B, from which they are recirculated through the system.

In the photograph at the left we see the condensers through which nitrogen and hydrogen pass before they enter the synthesis system. The upper picture on this page shows the building where ammonia is synthesized; in the lower picture we see a synthetic gas recirculator (indicated by the letter D in the diagram). The process described above is a refinement of the Haber Process, which was discovered by the eminent German chemist Fritz Haber in 1906.

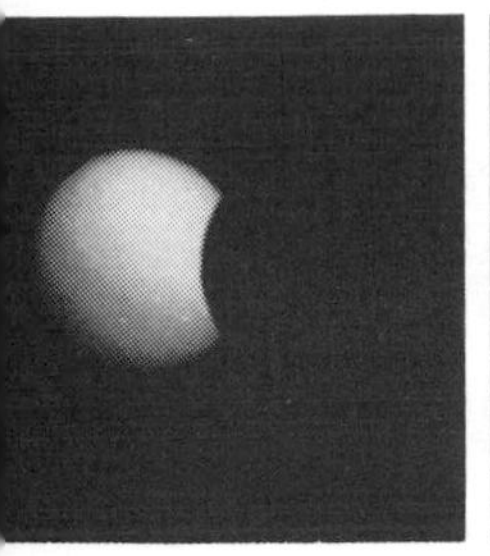

4:27 A.M.

4:41 A.M.

4:54 A.M.

# A TOTAL ECLIPSE OF THE SUN

A total eclipse of the sun takes place when the moon, in its orbit around the earth, passes between the sun and the earth and blots out the sun's disk. The three upper photos show the start of the eclipse; the middle one, total eclipse; the three lower ones, the end of the eclipse.

Photos, Amer. Mus.-Hayden Planetarium

5:24 A.M.

5:37 A.M.

5:50 A.M.

# SHADOWS IN THE HEAVENS

## Eclipses of the Sun and of the Moon

AN ECLIPSE of the sun or the moon is truly an awe-inspiring spectacle. The very word "eclipse" (derived from the Greek *ekleipsis,* meaning "forsaking") mirrors the dread with which the ancients regarded these celestial dramas. As they watched the sun or moon apparently disappear in time of eclipse, it seemed to them that these celestial beacons were indeed deserting mankind. Eclipses, like comets, were held to be portents of war, pestilence, the death of princes or even the end of the world. To this day, certain primitive peoples come to the aid of the sun or moon, as they are being eclipsed, with solemn rites and loud entreaties.

We know now that there is a perfectly logical explanation of eclipses: they are caused by the enormous shadows of the earth and of the moon. Both of these bodies are opaque; hence, when they are illuminated by the sun, each has a shadow extending out into space away from the sun. Since both the earth and the moon are smaller than the sun, the shadow of each is conical and diminishes in diameter as it extends farther out in space until finally it comes to a point.

As the moon travels around the earth — a journey that takes approximately a month — the earth sometimes enters the moon's shadow; in that case, an eclipse of the sun takes place. Sometimes the moon enters the shadow of the earth, and in that case there is an eclipse of the moon. A solar eclipse can take place only at the time of new moon, when the moon is between the sun and the earth; a lunar eclipse can take place only at full moon, when the earth is between the sun and the moon.

There would be an eclipse of the sun at every new moon and an eclipse of the moon at every full moon if the moon's orbit were in exactly the same plane as the earth's orbit around the sun. However, that is not the case; the moon's orbit is slightly inclined to that of the earth.

The moon passes through the plane of the earth's orbit twice every month. Generally it is on one side or the other of the earth's plane at new moon or at full moon. If it is not in the earth's plane at new moon, the moon's shadow passes above or below the globe of the earth and fails to eclipse the sun. If the moon is not in the plane of the earth at full moon, it passes above or below the earth's shadow and it is not eclipsed itself.

From time to time, however, the moon is at full moon or at new moon at the moment when it crosses the plane of the earth's orbit. When that happens, there is a lunar eclipse at full moon and a solar eclipse at new moon.

The plane of the earth's orbit around the sun, or — to put it more exactly — the great circle in which this plane cuts the celestial sphere, is called the ecliptic. The points in the orbit of the moon where it crosses the plane of the ecliptic are called the nodes of the moon's orbit. Naturally, eclipses, whether of the sun or of the moon, can take place only when the moon is at or near the nodes of its orbit around the earth.

The shadow cast by the earth or by the moon consists of several parts. We can show what these parts are by means of a diagram representing an opaque and nonluminous globe lighted up by a larger and luminous one. The diagram that is given on the next page does not, of course, represent the relative sizes or distances of the sun and the earth, or of the sun and moon. However, it shows well enough the various regions of complete and partial shadow which are actually projected into space by the earth and the moon. *A* represents a region of complete shadow, which is known as the umbra (the Latin word for "shadow"). No light comes directly from the luminous body to an object within *A*, which is a cone. An observer within it cannot see the source of light at all.

Surrounding the cone of complete shadow, there is a region of partial shadow, marked *B*. This is known as the penumbra (Latin for "almost a shadow"). Any object within it receives light from a portion of the luminous body; an observer within it can see a part of the source of light.

If the lines bounding the conical region of complete shadow are extended outward beyond the point of the cone, an inverted cone will be formed (marked *C* in the diagram). This is called the negative shadow. An observer within it will see the source of light as a luminous ring around the opaque body that is interposed.

It is not difficult to calculate the length of the umbrae of the earth and of the moon. It is evident from the diagram that the length of the cone of complete shadow depends upon three factors: the diameter of the source of light; the diameter of the opaque body; and the distance between the two bodies. We have pretty accurate figures for these three factors. It is important to bear in mind that while the diameters of the earth, sun and moon are constant factors, the distances between the earth and the sun and between the moon and the sun are variable factors.

For this reason, the umbrae of the earth and of the moon vary in length. The length of the earth's umbra at its maximum is about 871,000 miles and at its minimum about 843,000 miles; its average length is some 857,000 miles. The maximum length of the moon's umbra is about 236,000 miles; its minimum length about 228,000 miles. The average length of the umbra is in the neighborhood of 232,000 miles.

Since the length of the earth's umbra is about 857,000 miles and the average distance of the moon from the earth is about 239,000 miles, it is obvious that when the moon plunges into the cone of complete shadow, it is much nearer to the base of the cone than to its tip. The diameter of the cone, where the moon passes through it, is something like two and a half times the diameter of the moon.

If the path of the moon happens to be right through the center of the shadow, the moon may remain totally eclipsed for about an hour; shadow may cover part of it for about two hours. A lunar eclipse begins when the moon enters the penumbra and ends when it leaves the penumbra. There is little significant darkening, however, until the moon enters the umbra.

If the path of the moon takes it near the edge of the shadow, the total phase of its eclipse may last only a few minutes. If the moon's path is such that only a portion of its disk, and not the whole of it, enters into the conical shadow, the eclipse is partial and not total. Sometimes the moon, in its path, avoids the cone of complete shadow but passes through only the penumbra. Under these circumstances, so much light is still received from a portion of the sun's disk that there will be no marked obscuring of the moon unless it passes very close to the true shadow.

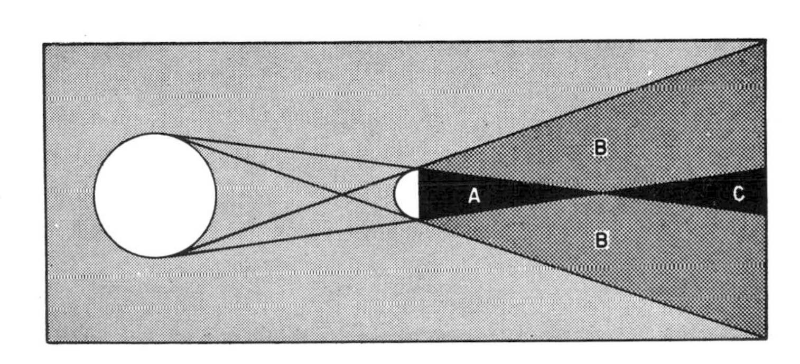


**Degrees of darkness in eclipses, as explained on this page. The big circle is the light source; *A*, the umbra; *B*, the penumbra; *C*, the negative shadow.**

The moon is usually not altogether lost to view even in the midst of a total eclipse; it shines with a strange, lurid copper-colored glow. Although the globe of the earth, so much larger than the moon, is interposed directly between the moon and the sun, enough sunlight reaches the surface of the moon so that the main lunar features may be clearly seen. The reason is that the terrestrial atmosphere acts as a kind of lens, bending some of the sunlight that passes through it into the shadow cast by the earth. The lurid, ruddy color of the moon during an eclipse is due to the scattering of light in the atmosphere — the phenomenon to which we also owe the gorgeous tints of sunset. The light that passes from the sun to the eclipsed moon obviously has to travel a far greater distance through the earth's atmosphere than the rays of the setting sun must traverse before they reach our eyes. Consequently the tinting effects of the atmosphere are greatly enhanced.

Solar eclipse, as viewed from outer space.

### The moon may be entirely invisible in an eclipse

If the atmosphere is heavily laden with clouds and is consequently comparatively opaque, it fails to deflect sunlight into the earth's shadow. Under such conditions the moon's surface may be so obscured as to be altogether invisible.

Eclipses of the moon are not so frequent as solar eclipses. There are at least two eclipses of the sun every year, and there may be as many as five. On the other hand, there are years when there is no eclipse of the moon at all; generally speaking, there are no more than two lunar eclipses in any one year. If, however, there is an eclipse of the moon on one of the first days of the year, there may be a third eclipse in December.

### Relative frequency of solar and lunar eclipses

The statement that eclipses of the sun are more frequent than eclipses of the moon may seem to run counter to our experience, since people living in a given region will see more eclipses of the moon than eclipses of the sun within a specific period of time. But, as a matter of fact, no contradiction is involved in this case. The moon's shadow in a solar eclipse covers a small part only of the earth's surface, whereas the earth's shadow in a lunar eclipse covers more than twice the moon's diameter. Every eclipse of the moon, therefore, is visible over all of the half of the earth that is in darkness (that is, where it is night); but the regions from which any particular eclipse of the sun can be seen lie in a comparatively narrow track across the globe.

Inasmuch as eclipses of both the sun and the moon depend upon the regular movements of the sun, moon and earth, the occurrence of eclipses, both in the past and in the future, can be calculated with great accuracy. In the 1880's, for example, the Austrian astronomer Theodor Oppolzer published a book called the CATALOGUE OF ECLIPSES in which he gave tables of 8,000 solar eclipses and 5,200 lunar eclipses taking place between 1207 B.C. and 2162 A.D. In the case of solar eclipses, he indicated the areas of the earth from which the eclipses would be visible. By consulting Oppolzer's tables, we find, for instance, that on February 26, 1979, a total eclipse of the sun lasting two minutes and forty-two seconds will take place and that it will be visible in various areas in the United States, Canada and the North Polar Sea. To make his predictions, Oppolzer had to be able to calculate the relative positions of the sun, moon and earth for any given time in the period that extended from 1207 B.C. to 2162 A.D.

#### The famous prediction of the philosopher Thales

Predictions of eclipses go back to antiquity. Perhaps the most famous of these early predictions was that of Thales, a philosopher of Miletus, who died in the year 546 B.C. The Greek historian Herodotus has given the following dramatic account of the prediction: "There was war between the Lydians and the Medes . . . In an encounter which happened in the sixth year [of the war] it chanced that the day was turned into night. Thales of Miletus had foretold this loss of daylight to the Ionians [Miletus was in the ancient district of Ionia], fixing it within the year in which the change did indeed happen. So when the Lydians and Medes saw the day turned to night they ceased from fighting, and both were the more zealous to make peace." This eclipse has been identified, after careful research, with the one that took place on May 28, 585 B.C.

#### The discovery that eclipses occur in series

Thales' prediction was not really so surprising as it might appear, though it greatly impressed his contemporaries. Many years before his time, Chaldean astronomers had made a remarkable discovery — one of the most memorable achievements of ancient astronomy. They had noted that eclipses of the sun and moon occur in series and that a definite period of time elapses between one eclipse of a series and the following eclipse. By calculating the time that had passed since the last solar eclipse in a series of this sort, Thales had no difficulty in predicting the time when the next eclipse of the series would take place.

The interval between an eclipse of the sun or moon and the next one in a given series is called a saros (a Greek word derived from the Assyrian-Babylonian word *sharu*). Each saros is 18 years and 11⅓ days long (or 18 years and 10⅓ days long if there are five leap years instead of four in a saros). A given series of solar eclipses may run through as many as 70 saroses and may last for 1,250 years. There are 48 or 49 saroses in the average series of lunar eclipses, and the lapse of time is somewhat under 900 years.

There are many series of eclipses, both solar and lunar, running their course at the same time. If an eclipse in Series A takes place, say, in the first week of May, 1960, the following eclipse of the series will be in May, 1978. In the case of another series, which we shall call Series B, one eclipse may occur, say, in the last week of June, 1961; the next one will take place in July, 1979.

#### A typical series of solar eclipses

The first in a series of solar eclipses is a partial one, in which the moon encroaches but slightly on the sun's disk. At the next eclipse the moon obscures a somewhat larger area of the sun. Next time, the eclipse, though still a partial one, is more pronounced; and it becomes more extensive with each passing saros. Ultimately the partial eclipses are followed by a series of annular and total eclipses. In these, as the moon passes across the sun's disk, it either covers all but the outer rim, as in an annular eclipse, or it obscures the sun altogether, as in a total eclipse. Then there follow a succession of ever diminishing partial eclipses. In the last eclipse of this particular series the bright disk of the sun is only slightly obscured.

#### The three kinds of solar eclipses

Eclipses of the sun are of three kinds: partial, annular and total. An observer stationed in the penumbra of the moon, marked B on the diagram on page 132, sees a partial eclipse; one in the negative shadow, C, an annular eclipse; one in the true shadow, A, a total eclipse.

We have seen that the mean length of the moon's shadow is about 232,000 miles and its greatest length about 236,000 miles. The mean distance of the earth from the moon, however, is about 239,000 miles, so that in general the moon's true shadow is not long enough to reach the earth. At times, however, the moon is only about 221,500 miles from the earth's surface. The true shadow then falls upon a small part of the surface, causing a total eclipse of the sun over an area that cannot exceed 167 miles in diameter. At other times the moon may be as much as 252,000 miles from the earth; if it is then interposed between the sun and the earth, its negative shadow will partially obscure a small area of the earth's surface, causing an annular eclipse of the sun at that point.

**The areas of partial eclipse**

Around the area where there is a total eclipse or an annular eclipse of the sun, there is always a much larger area where there is a partial eclipse. This area of partial eclipse generally extends for about two thousand miles of the earth's surface on each side of the path that is followed by the area of total eclipse. Sometimes the area of partial eclipse extends as much as three thousand miles on each side of the path of totality.

The moon's shadow passes along this path at great speed — about one thousand miles an hour. The longest period of total eclipse, under the most favorable conditions, is about seven and a half minutes.

**A solar eclipse is an impressive spectacle**

The approach of a total eclipse of the sun is impressive, and, to some persons, alarming. The sky darkens, the air becomes chill and a murky gloom begins to prevail. Birds fly to shelter; other animals may show signs of alarm. The darkness rapidly increases. Finally the dark shadow of the moon, like a vast thundercloud, advances with awe-inspiring rapidity from the western horizon and covers the land. Usually, just before the last rays of the sun are obscured, swiftly moving bands of light and shade are observed. (They are probably due to uneven refraction in the atmosphere.) Then the day becomes like night. As the eye becomes accustomed to the darkness, surrounding objects seem to have an eerie appearance; some dreadful calamity appears to be in the offing. No wonder that in former times eclipses were regarded with superstitious dread.

**The phenomenon known as Baily's beads**

As the eclipse approaches totality, the sun is seen as a very narrow crescent of brilliant light. The crescent then becomes a curved line and finally breaks off into irregular beads of light known as Baily's beads (after the English astronomer Francis Baily). The beads are caused by irregularities in the outline of the moon as seen in silhouette — irregularities due to lunar mountains and valleys. The sun's rays pass through the valleys, but are obscured by the mountains.

Total solar eclipses offer astronomers exceptional opportunities to study the atmosphere of the sun: the distribution of its material; the depth of the chromosphere and the reversing layer; the breath-taking splendor of the corona. Furthermore, as we shall see, at the time of total eclipse astronomers can photograph the space close to the sun; it is impossible for them to photograph this region at any other time.

**Organizing expeditions to observe solar eclipses**

Unfortunately few total eclipses of the sun take place in areas where well-equipped astronomical observatories are located; astronomers must generally travel to far-off lands if they are to observe a total eclipse of the sun. Modern eclipse expeditions are generally very elaborate and costly undertakings; temporary villages must be set up to house members of these expeditions, and much costly scientific apparatus must be installed. Parties of this kind go out from each of the lands that are

Official U. S. Navy photo

On a flat desert plain, near Khartoum, in the Sudan, instruments have been set up to study the sun's atmosphere during a total solar eclipse. Note the radio telescope, set up at the right of the tent.

advanced in scientific research. Sometimes, two or three parties may be sent out from a single country. The track of the moon's shadow is dotted in advance of the actual eclipse with considerable numbers of temporary observatories, which are manned by careful watchers. The work must be carefully planned and timed, for an eclipse lasts for only a few minutes.

#### Determining the times of the four contacts

An important point to be determined during a total eclipse of the sun is the precise moment of each of the so-called contacts. The first contact takes place when the moon first encroaches on the sun's disc. In the second contact, the sun disappears behind the moon and the solar corona becomes visible. In the third, the sun's rim is seen again; in the fourth, the moon passes completely off the sun's disc.

Astronomers have noted that the actual eclipse tracks are somewhat farther to the east than their calculations would indicate — calculations based on solar and lunar motions. The phenomenon is due, in part at least, to the gradual slowing up of the earth's rotation, because of the braking action of the tides. (See the article The Tides, in Volume 6.)

The spectroscopic study of the sun during an eclipse is extremely important. It has contributed greatly to the knowledge of our star. It was during a solar eclipse in the year 1868 that the hitherto unknown element helium was discovered by Lockyer and Janssen (see Index, under Helium).

#### Putting the theory of relativity to the test

Observation of the heavens during solar eclipses has made it possible to put to the test one of the most important hypotheses of Einstein's theory of relativity. Einstein had maintained that as light approaches a sizable accumulation of matter, such as a star, it is deflected from its path. The star will then seem to occupy a different position in the sky.

The Royal Astronomical Society of Great Britain decided to put the theory to the test on the occasion of the total solar eclipse of May 29, 1919. It sent out two expeditions, one to Sobral, in Brazil, and the other to the island of Principe, in the Gulf of Guinea. At the time of total eclipse, the stars in the region of the sun were photographed. The plates were then compared with photographs of the same region taken at night when the sun was in a different part of the heavens.

Careful measurements showed a distinct displacement of the star images in the eclipse photographs. Of the two sets of plates taken by the Sobral expedition, one showed a deflection of 1.98 seconds of arc; the other gave a value of .99 seconds. The plates taken at Principe showed a deflection of 1.61 seconds. A number of other measurements of light deflection during solar eclipses have been taken since that time. While the results do not tally exactly with the figure predicted by Einstein (1.75 seconds), they show that a very definite displacement of light occurs.

The study of the sun's corona during times of total solar eclipse was even more important in the past than it is now. Formerly, the corona was visible only during a total solar eclipse. Now with the aid of the coronagraph, an astronomer can examine the corona at all times. In the coronagraph, an artificial moon is arranged in such a way as to block out the solar disc, thus producing an artificial eclipse.

#### Discovering new comets in the total-eclipse period

In the period of total eclipse, certain observers search for hitherto undiscovered planets within the orbit of Mercury, which is nearer to the sun than any of the other planets. At any other time, the brilliance of the sun's light would make it impossible to spot a planet near the sun. Thus far, no new planet within the orbit of Mercury has ever been discovered. The search for hitherto unknown comets during solar eclipses has been more rewarding. A number of new comets have been tracked at perihelion — the part of their orbit in which they pass closest to the sun.

*See also Vol. 10, p. 268:* "Eclipses."

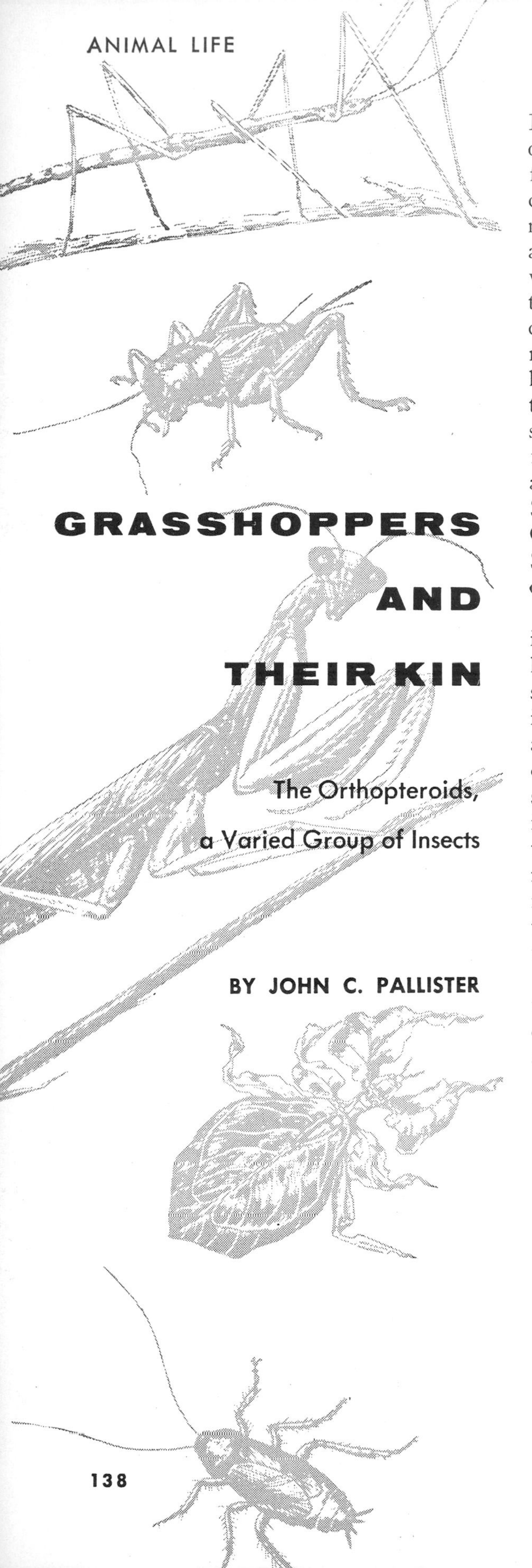

# GRASSHOPPERS AND THEIR KIN

## The Orthopteroids, a Varied Group of Insects

BY JOHN C. PALLISTER

IF ONE wanted to show the great variety of insect life, one might name such diverse forms as grasshoppers, katydids, crickets, cockroaches, walking sticks, leaf insects, mantids and earwigs. Included in this list are loud and mute insects, winged and wingless forms, flesh eaters and vegetarians. There are agile leapers, sluggish crawlers and industrious diggers. Some ravage man's crops and devour his household supplies. Others are allies (unwitting, of course), devouring harmful insects. The insects range in length from a fraction of an inch to over a foot. There are many odd creatures among them. Some hold up their forelegs as if in prayer. Others look amazingly like leaves or twigs. Still others have a formidable looking pair of pincers at the tip of the abdomen.

All these insects, showing such striking differences in form and habit, were lumped together until quite recently in a single order — the Orthoptera. The name is taken from the Greek *orthos:* "straight" and *ptera:* "wings." It refers to the shape of the front wings, which are straight in some of these insects but not at all straight in others. Actually, there is another feature that more of the insects have in common — that is, the rather large, pleated rear wings. These close like a fan along the pleats and disappear under the front wings when at rest.

Some zoologists continue to group all the insects mentioned above in the single order Orthoptera. The tendency nowadays, however, is to divide this up into

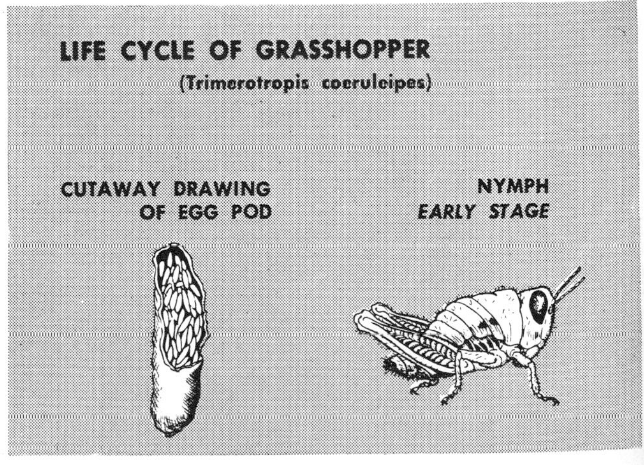

six separate orders: Orthoptera, Grylloblattodea, Blattaria, Phasmida, Mantodea and Dermaptera. The name "orthopteroid" is used to refer to all the insects in these six orders.

The orthopteroids have developed from an old and somewhat primitive group. The cockroaches, belonging to one of the divisions of this group, go back to the Carboniferous period. During the many millions of years that have elapsed from that time to this, the orthopteroids have come to differ considerably from one another. They are more or less alike in at least two respects, however. For one thing, no true water lovers have developed among them. A few will dive into water or even swim under the surface for a short time; but they quickly return to the land. Again, practically all orthopteroids pass through the same type of life cycle. There are three stages in the development from egg to adult. The egg represents the first stage; the nymph, the second; the adult, the third.

The egg is deposited by the female in a place that is suitable for her particular species. From it emerges the nymph. It resembles the adult, except that it is smaller and wingless. The skin or outer covering of the nymph is tough and does not stretch as the young insect grows. Therefore it must be shed, or molted, from time to time. The stage between moltings is called an instar. Nymphs pass through three to twenty instars, according to their species; the majority have five or six. In species that are destined to have wings, tiny wing pads appear at about the middle instar. With each successive molt, the nymph increases in size; at last the adult form appears. This adult is a sexually perfect form, whether it is winged or wingless. Male and female adults mate; the females then deposit their eggs to start the next life cycle.

### The order Orthoptera

The largest of the orthopteroid groups retains the old order name "Orthoptera." This order is made up of the grasshoppers, locusts, katydids, crickets, grouse locusts, tree crickets, cave crickets, mole crickets and others. There are over 28,000 species in all. They are widely distributed throughout the world, wherever vegetation is found, for most of the species consume large quantities of plant material. A very few prey on other animals. Even the plant feeders frequently eat animal wastes; they may devour their own kind if plant food becomes scarce.

Most species of the Orthoptera are able to leap by means of well-developed hind legs. Many are good fliers. The rear wings of the flying species can be folded in pleats under the forewings, which are usually known as tegmina. Not all flying species have straight tegmina; there are a variety of shapes and sizes. The antennae are either long and slender, or short and comparatively thick. Most species of the Orthoptera are voracious feeders; they include many insect pests.

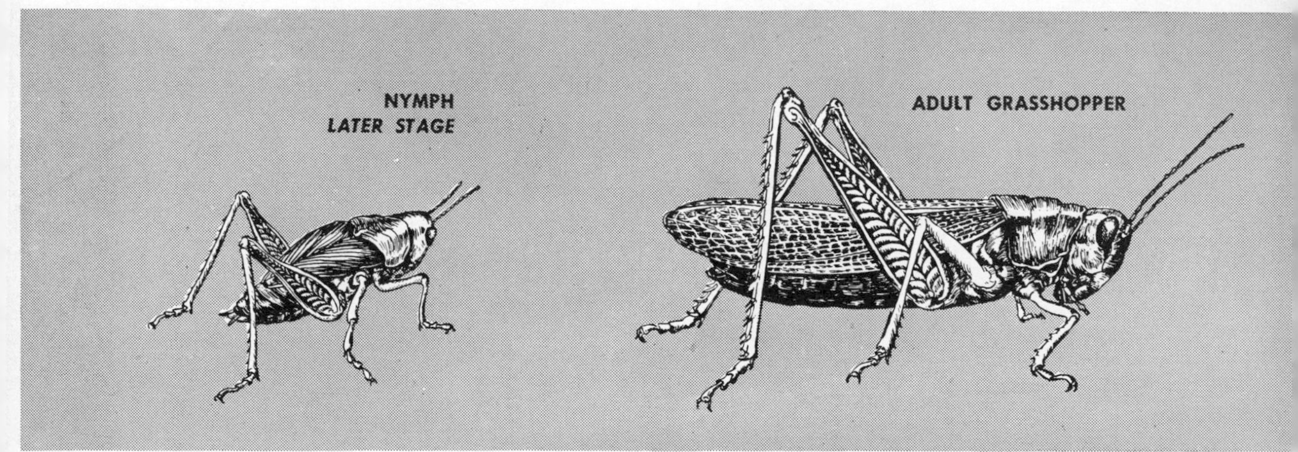

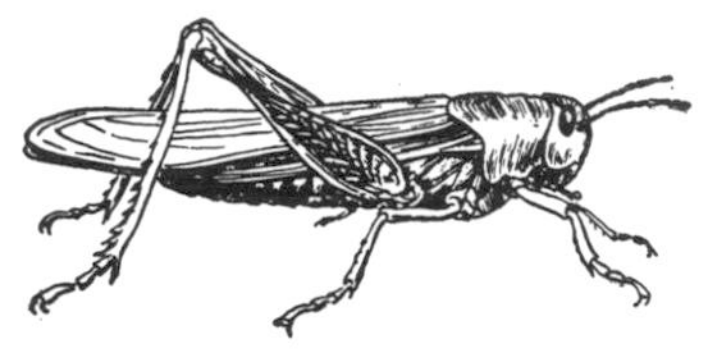

SHORT-HORNED GRASSHOPPER
(Trimerotropis coeruleipes)

LONG-HORNED GRASSHOPPER
(Orchelimum vulgare)

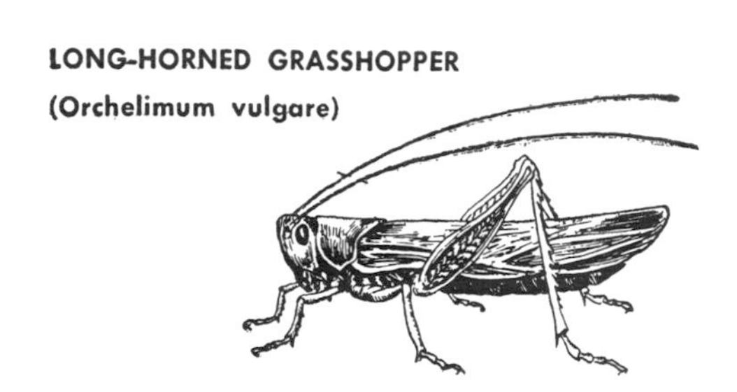

The order Orthoptera is divided into two suborders — the Acridodea, represented by the short-horned grasshoppers, and the Tettigoniodea, represented by the long-horned grasshoppers.*

### The short-horned grasshoppers (Acridodea)

The suborder Acridodea is made up of two families: the Locustidae and the Tettigidae. The familiar grasshoppers found in fields and along roads are members of the family Locustidae. They are sometimes called locusts. The name is correct enough; unfortunately, it is often incorrectly applied to cicadas, which belong to an entirely different insect order. There is a tendency nowadays to use the word "grasshopper" for the species that do not become too destructive. The term "locusts" is reserved for the migratory forms that, because of their vast hordes, cause tremendous damage.

Practically everybody knows what a grasshopper looks like. The more primitive races of mankind are even more familiar with these insects than we are, because they play a far more important part in their lives. Many primitive farming populations find grasshoppers good eating. When a huge horde of these insects descends and devours the crops, farmers may have nothing to eat but the invaders.

* The "horns" in this case are the antennae. The antennae of the Acridodea are much shorter than those of the Tettigoniodea.

I myself have eaten grasshoppers prepared in three ways — fried, toasted and cooked in oil. I must confess that I was not particularly delighted with grasshopper served in any one of these ways. Never could I bring myself to partake of the raw insect. Yet some people consider this a delicacy. In the Peruvian Andes I have seen Incas catch big grasshoppers, four inches in length. They would seize an insect by the head, tear off and discard the wings and legs, bite off the body just behind the head and munch with apparent relish.

Grasshopper species are distributed in a wide belt around the world, extending from the cold regions of the north to those of the south. They range up to the snow fringes of mountains and have reached many isolated oceanic islands. These insects are generally large, averaging about two inches in length; the range is from a half inch to six inches. About 18,000 species are known.

The head is rather large and solidly built; the forewings are usually strong. There are two rather large compound eyes, three ocelli (simple eyes), two short antennae — less than half the length of the body — and a powerful set of jaws. The front wings cover the more delicate rear wings when the insect is at rest. The front and middle pair of legs are short and are used for walking. The rear legs, which serve for leaping, are large and strongly muscled. They catapult the insect several feet into the air. Here the powerful wings take over, sometimes for extended flights of many miles. The legs are armed with rows of spines. These aid the insect in pushing through thick vegetation and also serve as defensive armament against a pursuing enemy.

Along the inner surface of each upper hind leg of many grasshoppers is an elevated, sharp ridge. This may be smooth or notched. There is also a raised vein on each tegmen, or forewing. When the grasshopper scrapes his legs against his tegmina, he produces the familiar rasping noise we associate with his kind. It is amusing to watch immature grasshoppers

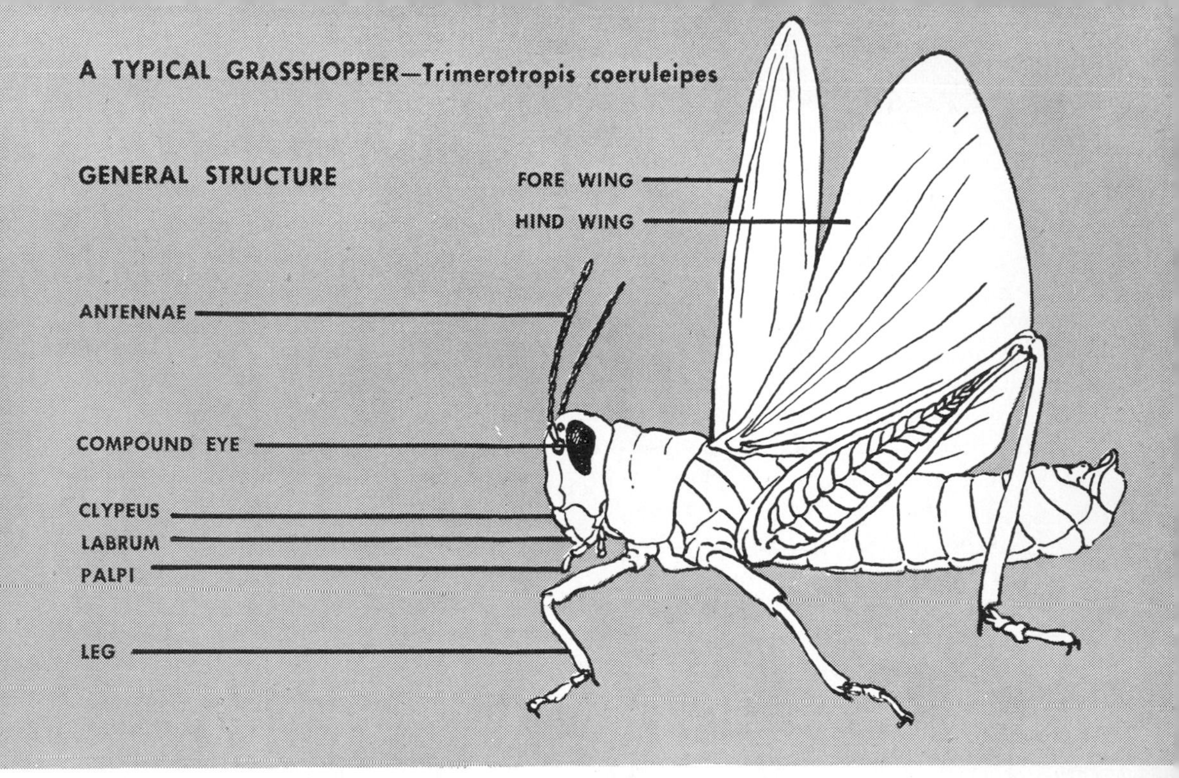
A TYPICAL GRASSHOPPER—Trimerotropis coeruleipes
GENERAL STRUCTURE
FORE WING
HIND WING
ANTENNAE
COMPOUND EYE
CLYPEUS
LABRUM
PALPI
LEG

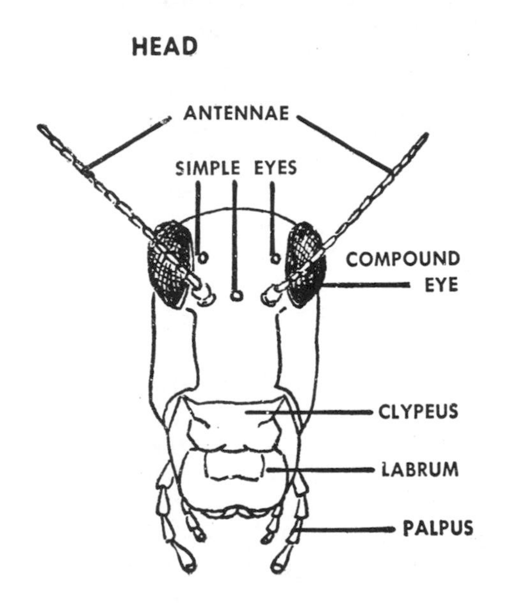
HEAD
ANTENNAE
SIMPLE EYES
COMPOUND
EYE
CLYPEUS
LABRUM
PALPUS

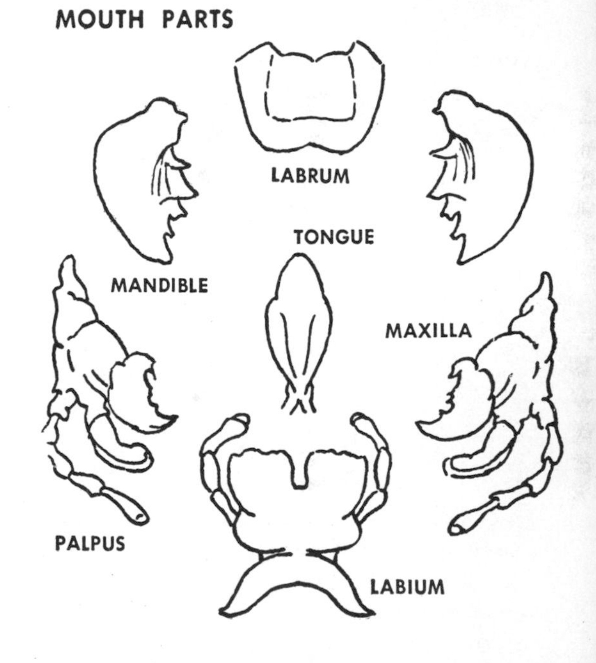
MOUTH PARTS
LABRUM
TONGUE
MANDIBLE
MAXILLA
PALPUS
LABIUM

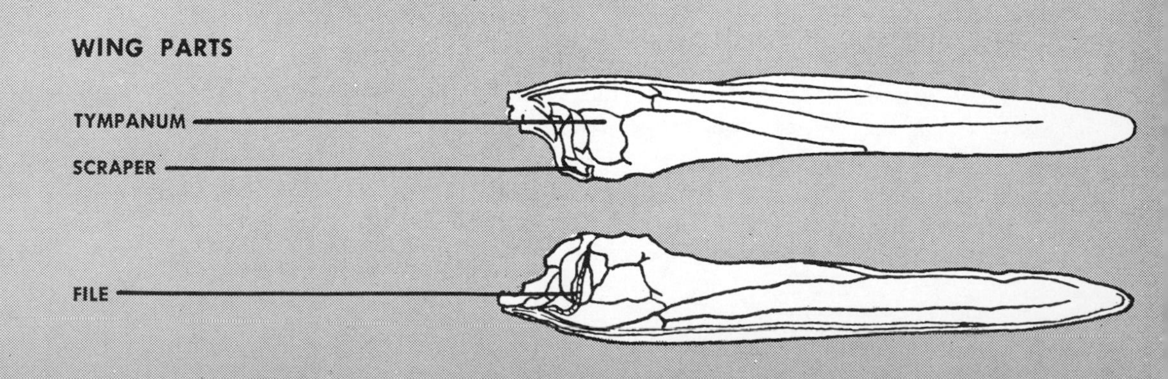
WING PARTS
TYMPANUM
SCRAPER
FILE

working their legs up and down but not making a sound because as yet they have no wings against which to rub their legs.

Some species make a noise by hovering in the air a few feet above the ground and rubbing the front margins of the rear wings against the veins of the tegmina. This action produces a clicking or clacking noise, which is sometimes quite loud and startling. Grasshopper "songs" are mating calls and are produced only by the males. An auditory organ, or ear, usually found in both sexes, is located on each side of the base of the abdomen.

When at rest on the ground, most grasshoppers blend with the color of the soil on which they normally occur. They range in color, therefore, from some shade of tan to dark brown or nearly black. Some, however, are mottled with bright yellows and reds. Grasshoppers that live among thick vegetation take on green shades; those living on sandy beaches are nearly white. In some species the underwings are colored bright yellow, blue or red, or show combinations of these hues. The sunset grasshopper of South America is one of the larger and more brilliantly colored species.

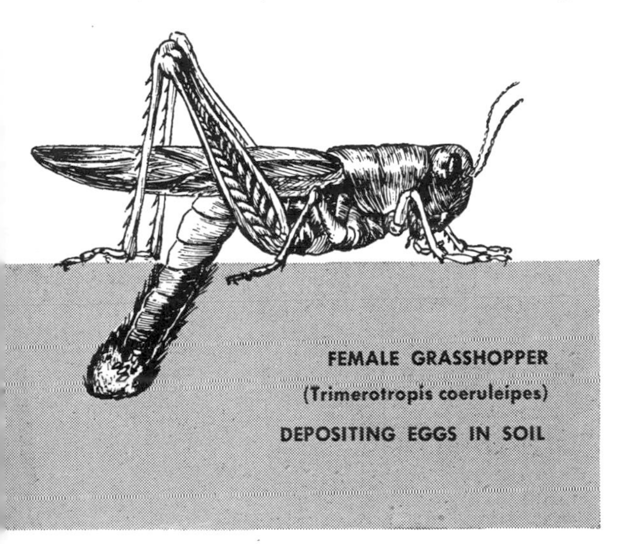


The female grasshopper bores with the tip of her abdomen in the soil; then she deposits from 20 to 100 eggs.

The female grasshopper usually deposits her eggs in the soil. She bores with the tip of her abdomen into the soft dirt as far as it will go. Then she deposits twenty to one hundred eggs, together with a gluelike secretion that cements the eggs and soil particles together. Each female may account for fifteen or twenty egg masses. In the temperate regions the eggs of most species are deposited in the fall and hatch in the spring. The wingless nymphs start eating, and molt their skins as these become too tight. After five to eight molts they become winged adults. In the warmer regions there may be several generations in a year, if conditions are favorable.

Most grasshopper species remain comparatively unimportant in man's economy. They furnish food for birds and smaller mammals, and fish bait for small boys. Only a few species are damaging pests. The curious life cycle of these insects brings about the terrible "plague of the locusts" that has tormented man from the earliest times to the present day. Only in the present century has any relief been attained. This is because we have learned a good deal about the movements and the habits of the various migratory species.

It took a long time for scientists to find out how the locust plagues developed. In general, a migratory species will live quietly in one place for many generations until it has developed an enormous population. Suddenly there appears a generation quite different in color or size from its immediate ancestors. The great majority will fly off for more abundant pastures, eating all leafy vegetation along the way. The few remaining in the old location will breed a new generation that will have the old color and size. The process of building up to a migratory form will then begin again.

One of the most destructive species is the desert locust, *Schistocerca gregaria.* Its ravages have been frequently described. A large species, about two inches long, it ranges from northern Africa and southern Europe east into Asia as far as India. It breeds in sandy areas with sparse vegetation. The nonmigratory form is yellow; the migratory form, pinkish. *Schistocerca gregaria* may fly up to a thousand miles.

The migratory locust, *Locusta migratoria,* has a wide distribution, ranging through most of Africa, Europe, Asia, Oceania, Madagascar, Australia and New Zealand. The insect is about two inches long and yellowish or grayish in its migratory form. It moves in hordes from area to area, doing great damage to all vegetation.

South Africa has two destructive species: the brown locust, *Locustana pardalina,* and the red locust, *Nomadacris septemfasciata.* The South American locust, *Schistocerca paranensis,* nearly three inches long, ranges from southern Mexico into northern Argentina. Until about the turn of the present century, several species of locusts did a good deal of damage throughout the central plains area of North America. The lesser migratory locust, *Melanoplus mexicanus,* and the red-legged locust, *Melanoplus femur-rubrum,* were the worst offenders. Although they are now generally held in check, considerable outbreaks occasionally occur, causing serious damage to crops.

Many nonmigratory grasshoppers are found during the summer and fall in any weedy or grassy area in the United States. The Carolina grasshopper, *Dissosteira carolina,* with its yellow-margined blackish rear wings, is a familiar species. Along the beaches of the Atlantic seaboard and on the shores of the Great Lakes, a grayish white species, *Trimerotropis maritima,* can be found, blending with the white sands. One must be sharp of eye to see one of these creatures and quick of foot to capture it. It can move with great speed and the hotter the weather, the faster it can jump.

In the southern part of the United States, the lubber grasshopper (*Romalea microptera*), a fat, short-winged, flightless species, may be seen walking along sedately. From time to time it stops to feed. The young ones travel together in small herds in early spring.

The other family of the suborder Acridodea — the Tettigidae — is made up of the pygmy, or grouse, locusts. These are small insects, a half inch or less in length, and they are not conspicuous. Some are mottled black and brown; others are pale gray. Although a few live in arid locations, most species prefer rather damp places. Here their favorite foods are algae, mosses, lichens, fungi, low-growing grasses and decaying vegetation. Those that live along the margins of ponds and streams are able to swim; they sometimes dive into the water to escape their enemies.

Pygmy locusts are easily recognized by the prothorax (front part of the thorax), the top part of which extends back over the abdomen and sometimes considerably beyond it, often ending in a sharp point. Under this long prothorax the rear wings lie hidden when not in use. The tegmina, or forewings, are reduced to tiny plates. Pygmy locusts seldom fly unless hard pressed by their enemies. Their well-developed leaping hind legs propel them from place to place in long, rapid jumps. These insects have no rasping apparatus and are therefore silent. They have no organs of hearing.

Over seven hundred species of Tettigidae are known; they range from the tropics to the extreme temperate latitudes. New species are constantly being discovered and added to our lists. The family is particularly well represented in Africa, Malaya and Oceania. Over two hundred species occur in the Americas. In the temperate latitudes the adults hibernate, laying their eggs in the soil in the spring.

### The long-horned grasshoppers (Tettigoniodea)

The members of the suborder Tettigoniodea can generally be identified by their long, slender antennae. Among them are such well-known insects as katydids, crickets, tree crickets and mole crickets. The males of most species are able to produce a noise by opening and closing their tegmina while holding them in a slightly elevated position. One tegmen overlaps the other. On the underside of the overlapping tegmen is a finely toothed vein that rubs across the base of the under tegmen, producing a rasping noise. Each species produces a characteristic sound.

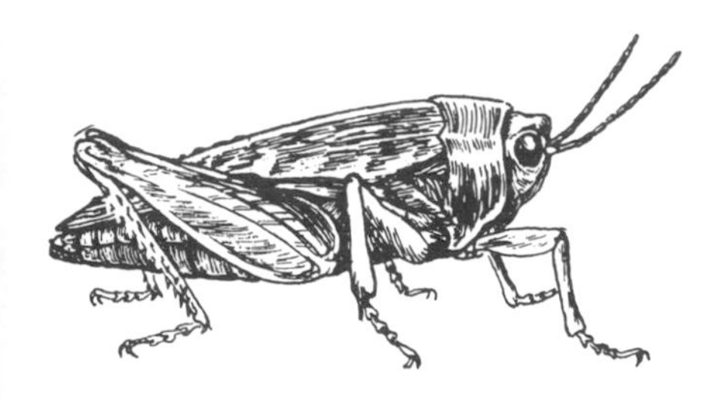

PYGMY LOCUST (Acrydium ornatus)

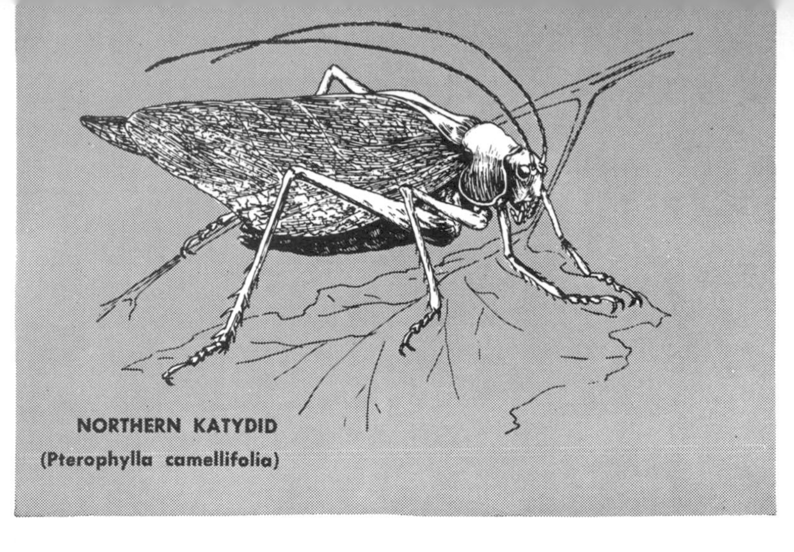

NORTHERN KATYDID (Pterophylla camellifolia)

A person with good ears can soon learn to identify the insects by their calls. The rear wings and probably the abdomen act as a sounding board in spreading the sound. The beat or timing of the call varies with the temperature. When the weather is cool, the call is slow; as the temperature rises, the beat quickens. It is possible to find out the temperature by counting the number of beats of certain species. It has been found that the insects chirp at a rate that increases as the temperature rises.

The females of this suborder usually have long and conspicuous ovipositors (egg-depositing organs). With these they deposit their eggs in the ground or in the leaves or stems of plants. Most species, both males and females, have powerful jumping hind legs. On their front legs they have a set of hearing organs.

The suborder Tettigoniodea forms a large group with some 8,500 known species, making up five generally recognized families. By far the largest of these families is that of the Tettigoniidae. It includes the katydids, the meadow grasshoppers and the green grasshoppers. Many of the species live in trees, shrubs, tall grasses and in the weeds of swamps. These insects are usually green in color, and blend with the foliage so that they are barely noticeable. Those living in trees and shrubs are usually darker than the grass-dwelling species. At night, when most of them are "singing," the beam from a flashlight may reveal numbers of them. The ground-dwelling species are gray or brown; many of them are wingless.

One of the most interesting representatives of this family is the northern katydid, *Pterophylla camellifolia,* which ranges through the eastern and middle parts of North America. There are other katydids; but this is the species giving the call from which the name "katydid" is derived.

The loud, rasping call of the northern katydid is repeated continuously from twilight to dawn throughout the late summer. The interval between notes is about five seconds during the evening; it increases in length as the temperature falls. The call is usually interpreted as "Katy did." If you listen closely, however, you will often hear an intermediate rasp, making the phrase "Katy did did" or "Katy she did." This katydid is broad-winged and yellowish green or, rarely, pink. The insect is about an inch and a half long; its antennae are as long or longer. The female deposits her flat, pointed, slate-gray eggs in crevices in the bark of trees.

There are also narrow-winged katydids, which are a little smaller than the broad-winged kind. They seem to prefer the smaller shrubs and bushes. A common species is *Scudderia furcata,* the fork-tailed katydid, which is widely distributed throughout the United States and southern Canada from coast to coast. This insect can be found in the taller vegetation on marshy ground, along fence rows and at the edge of thickets. About an inch long, it is dark green or brownish green in color. The male's call is not nearly so loud or shrill as that of the northern katydid. It consists of a soft "zeep, zeep," repeated

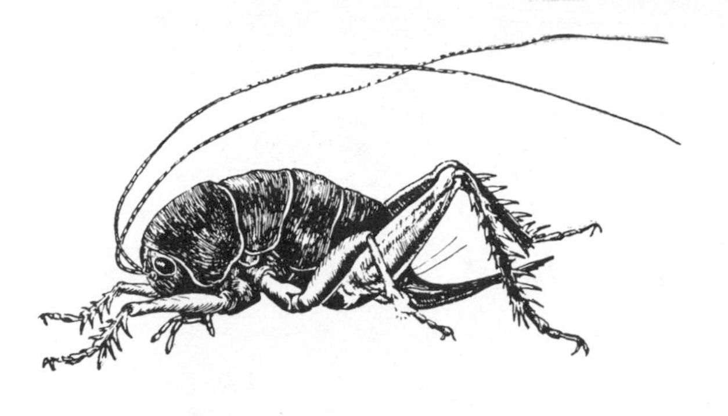

**CAMEL CRICKET (Ceuthophilus)**

**MORMON CRICKET (Anabrus simplex)**

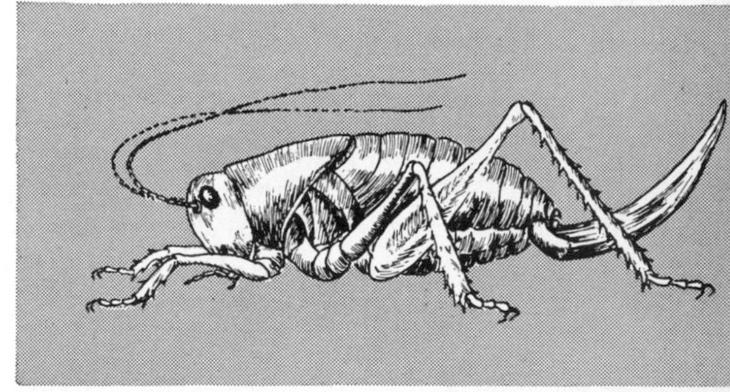

three or four times, followed by a short interval of silence. It is said that the female sometimes responds with a faint "cheep," produced by stretching out the wings, as if for flight. The fork-tailed katydid gives its call in the daytime and early evening. From one to five eggs are deposited by the female along the edge of a leaf. A fluid that comes out with the eggs soon hardens; as a result, the eggs are held firmly in place.

Some members of the family Tettigoniidae have no wings. Among the wingless species is the Mormon cricket, *Anabrus simplex*. Though it cannot fly, it is sometimes as serious a pest as any of the locusts. Nearly two inches long, bulky and dark brown to black in color, it lives in the arid regions of the northwestern United States. Here it can thrive even on sagebrush and cactus. When the Mormon settlers began farming in Utah, about the middle of the last century, the new vegetation brought such vast numbers of the crickets that the life of the colony was threatened. But the cricket hordes attracted flocks of California gulls. These birds disposed of the insects before they could destroy the crops. The Mormons erected a monument to the gulls in Salt Lake City to show their gratitude.

The family Stenopelmatidae is made up of the camel crickets, sand crickets, cave crickets, Jerusalem crickets and the wetas of Australia and New Zealand. Nearly all of the 300 species are wingless. They have stout bodies, very long antennae and large, powerful hind legs. Most species cannot make a noise and have no hearing organs. Usually nocturnal, they live in caves and cellars and under rocks, logs and decaying vegetation. They have a most varied diet.

The true cave crickets, genus *Hadenoecus,* are fragile creatures with slender bodies and very long legs and antennae. They are found throughout the many caves in the central part of the United States. Unlike many other cave creatures, these insects have eyes.

One of the oddest looking members of the family Stenopelmatidae is the camel cricket, genus *Ceuthophilus*. It is a humpbacked insect, with a large head bent downward and backward between its front legs. The hind legs are sometimes more than twice as long as the body. The insect spends its days under rocks at the edges of small streams or under logs in damp woods. At night it comes out to feed upon fresh or decaying plant or animal matter. Camel crickets are quite widely distributed throughout North America; they are also found in other parts of the world, though not in such great numbers. Some Oriental species have come into the United States with various plants and are now common in greenhouses.

The sand crickets, genus *Stenopelmatus,* occur only in the southern and western parts of North America. The most familiar species is *Stenopelmatus longispina;* it is called the Jerusalem cricket in the United States, though no one can tell why. The insect is two inches in length. It is pale yellow to brownish in color; its abdomen is banded with black. The head is large. It looks so startlingly human that the Mexican Indians call the Jerusalem crickets *los niños de la tierra* ("earth children") and look upon them with awe. Sand crickets feed largely upon roots and tubers; they also prey on insects. The female often devours the male after mating.

The land "down under" has some odd crickets, called wetas. One of these, the Australian king cricket, *Anostostoma australasiae,* is a fierce-looking insect, three inches long, with a huge head and mandibles. The even larger *Deinacrida heteracantha* of New Zealand has probably become extinct by now. It was about four inches in length.

The true crickets and tree crickets belong to the family Gryllidae. These insects have long, slender antennae and well-developed rear legs; they produce rasping sounds and have organs of hearing. The rear wings are often poorly developed; hence the insects in this family are more or less earth-bound.

Field and house crickets are common all over the world. In North and Central America and the northern part of South America, the common field cricket is *Gryllus assimilis,* an inch-long black species with brownish wings. It can be found in the fall under logs and old boards and other loose debris. The European cricket, *Gryllus domesticus,* the "cricket on the hearth," has now immigrated to the United States, where it often takes up its residence in greenhouses. It is about as long as *Gryllus assimilis,* but is more slender and is pale brown in color. It devours grain products (especially bread), live cockroaches and wet clothing.

All kinds of crickets and even some katydids have been cherished for their cheerful "songs." The tree crickets undoubtedly emit the clearest, most musical notes. Many people in the Orient keep these insect songsters in little cages. North American tree crickets are commonly slender of body and pale in color. As the name "tree cricket" suggests, many live in trees; they are also found in sunflowers and the taller goldenrods. The beautiful snowy tree cricket, *Oecanthus niveus,* is ivory white, sometimes tinged with pale

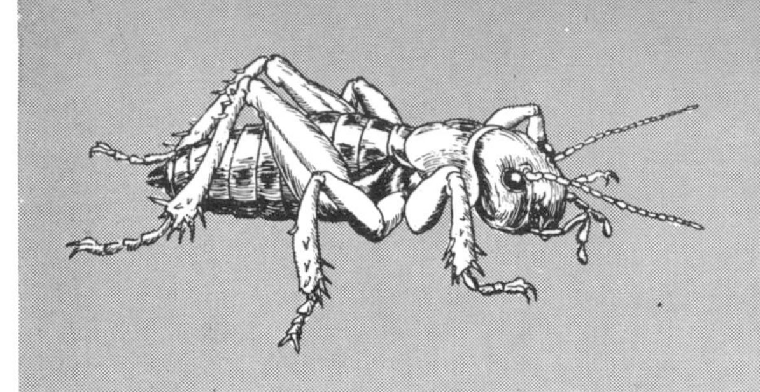

JERUSALEM CRICKET (Stenopelmatus longispina)

COMMON FIELD CRICKET (Gryllus assimilis)

EUROPEAN CRICKET (Gryllus domesticus)

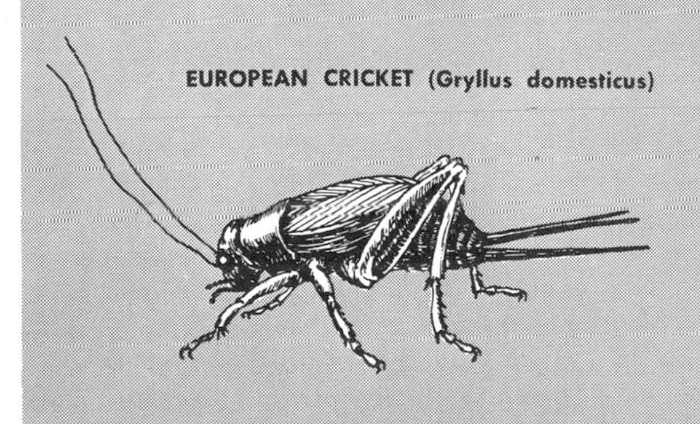

TREE CRICKET (Oecanthus niveus)

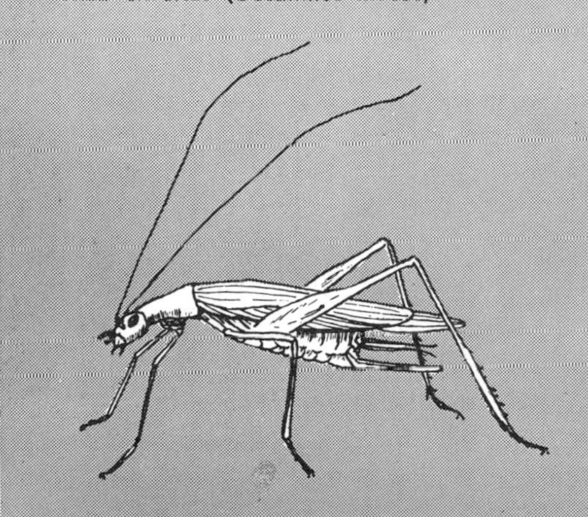

green. It is fairly common throughout the United States. Like many of its relatives, it is not often observed, because it is quiet during the daytime and practically invisible at night.

The mole crickets belong to the small family of the Gryllotalpidae. Though the number of their species is limited, mole crickets are widely distributed throughout the world. About two inches in length, they look like miniature moles. They dig burrows with their enlarged front legs in sand or mud along the banks of streams and ponds. Generally they feed on plant roots. Mole crickets are rarely seen except at night, when they leave their burrows and fly about. Attracted to light, one may land on your porch or come through an open window to make itself at home on your dining room table. When the insect is at rest, the rear wings, folded in fanlike pleats, protrude from the short forewings. The hind legs of these insects are not enlarged for leaping.

Like the mole crickets, the pygmy mole crickets, belonging to the family Tridactylidae, number comparatively few species but are found almost everywhere. These insects are only about half an inch long and are brown or black in color. They could easily be mistaken for beetles, except that the hind legs are built for leaping. Pygmy mole crickets have no sound-producing or hearing organs. They dig their burrows in soft, sandy banks, which are drier than the banks preferred by the larger mole crickets.

### The order Grylloblattodea

Thus far we know only four species of this order; the first of these was not discovered until 1914. The order itself was first set up in 1915. The insects belonging to it are usually less than an inch in length; they are wingless, slender and straw- or amber-colored. They have been found only in mountainous areas of the United States (Montana, California, Washington), Canada (Alberta) and Japan. The insects occur at elevations above 4,000 feet. Here they live under rocks or in moss or debris that is covered with snow during many months of the year. The insects that belong to the order Grylloblattodea have a varied diet.

### The order Blattaria

The cockroaches, belonging to the order Blattaria, were on earth for millions of years before man. If we ever succeed in bombing ourselves out of existence with nuclear weapons, it is likely that some of these hardy insects will remain upon the earth. Nearly all species are medium to large-sized, flattish and oval. They are dull yellow, brown or black in color, though some of the tropical cockroaches sport brilliant colors in striking patterns. About half of the species are winged. Others are wingless; still others have partially developed wings and are called brachypterous (short-winged). In several species the females are brachypterous, while the males are winged.

Almost all cockroaches have a more or less offensive odor. Generally nocturnal in habit, hunger often drives them out into daylight. The insects are extremely active and quickly rush for cover when they are disturbed. Cockroaches can and will eat almost anything, including each other. They are longer-lived than most insects; in fact some of them require five years to complete their life cycle, from egg to adult.

Cockroaches can be found throughout the habitable parts of the earth, but most of them live in the tropics. About 2,300 species have been described. Of this number very few have taken up living quarters with man. However, from year to year other species are discovering the easy living that man can afford them and are moving into his buildings. The great majority remain in their ancestral abodes under loose bark, old logs, stones or piles of decaying vegetation. Some live in ants' nests, where they appear to act as scavengers. Others dig burrows in the soil, or ascend trees and live among the leaves. Cockroaches are sometimes found on banana bunches or clumps of bromelias. A few prefer the margins of streams and ponds and are able to swim under water.

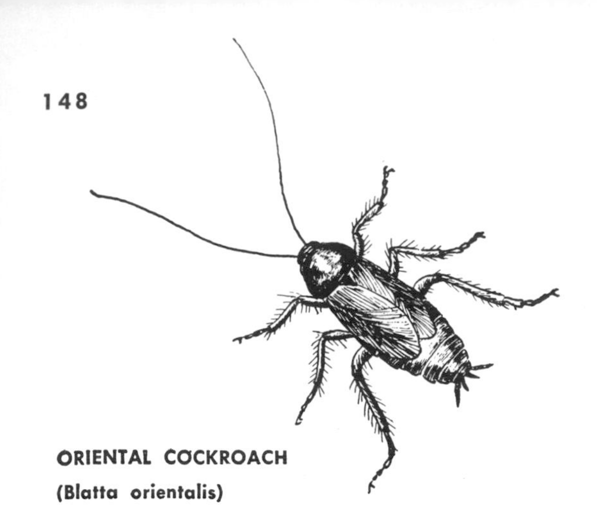

ORIENTAL COCKROACH
(*Blatta orientalis*)

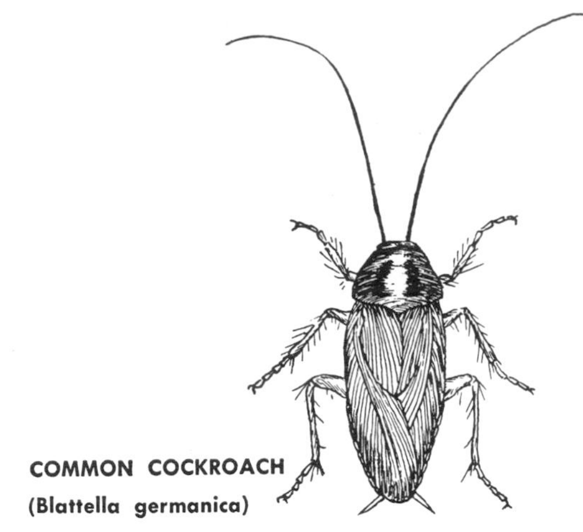

COMMON COCKROACH
(*Blattella germanica*)

The Blattidae are a rather large family of cockroaches, made up mostly of harmless outdoor species. However, at least three species are very annoying to us. The most hated of all is the Oriental, or Asiatic, cockroach, *Blatta orientalis,* also called kitchen cockroach and black beetle. About an inch in length and black or very dark brown in color, it infests houses in crowded districts, hotels, restaurants, garbage containers, ships — in fact any place where food and dirt are allowed to collect. The female has wings, but these are too short to serve as flight organs. The wings of the male are adequate, but the insects seldom use them.

Another annoying member of the family Blattidae is the Australian cockroach, *Periplaneta australasiae.* It is reddish brown with narrow pale yellow margins on the base of the wings. The yellow color is more prominent in the top part of the prothorax; here the basic reddish brown shows as two dark round spots. *Periplaneta australasiae* is slightly larger than the Oriental roach. Both sexes have well-developed wings. This species started its wanderings from its native Malaysia and is now found in many places.

Of late years it seems to be giving way in the United States to the American roach, *Periplaneta americana.* This is a brown or reddish brown insect, about an inch and a half in length. Its long, wide wings make it look larger than it really is. *Periplaneta americana* has wandered northward from its home in Central America and Mexico. It had already become established in the southern part of what is now the United States by the middle of the eighteenth century. Traveling on man's vehicles, living wherever man lives, eating whatever man eats and much that he wears, the American cockroach appears to be increasing in numbers. Unlike many of its related species, it flies readily. Often it startles people on a warm summer evening as it darts about erratically.

The common cockroach, *Blattella germanica,* belongs to the family Phyllodromiidae. It is also called the German cockroach and the croton bug. Only half an inch or less in length, this roach is pale yellow with two longitudinal stripes on the top part of the prothorax. It is believed to have originated in central Europe; it has been cosmopolitan for a long time. Although it is small in size, it more than makes up for this deficiency by the speed with which it can run and its great numbers. Hundreds and sometimes thousands of young and adults may be found in a crack or crevice in an otherwise well-kept apartment. This crowding together accentuates the characteristic odor of the species. Many of the females carry their egg cases, or oöthecae, until they can locate a suitable crevice in which to deposit them. Each case contains thirty-six to forty eggs, arranged side by side in two rows.

Other species of the family Phyllodromiidae vary considerably. Some of them give birth to living young. Certain species are wingless. The giants of the group belong to the genus *Blaberus.* They are tropical and subtropical species, occurring in Central and South America, the West Indies and Florida. A number are four to five inches in length. They like thick, dense vegetation, particularly the tops of palms or the bromelias and orchids clustered on tropical trees.

### The order Phasmida

The walking sticks and the leaf insects, belonging to the order Phasmida, are perhaps the most completely camouflaged of all insects. The walking sticks have slender bodies and legs, which are colored and jointed in such a way that they look like twigs. Most of those found in the United States are wingless; some of the tropical species have the characteristic orthopteroid wings that fold in fanlike pleats. The leaf insects have leaflike longitudinal flat extensions along the legs and body. In some cases the wings also are leaflike.

Most of the species and the most unusual forms of the order Phasmida are found in the tropics. All feed on vegetation; many show a decided preference for a certain kind of foliage. Over 2,000 species are known; they make up five families.

The family Bacillidae is found in many parts of the world, but not in North America. The insects belonging to this family are comparatively heavy-bodied; they are frequently covered with short, heavy spines.

The leaf insects belong to the family Phyllidae. They resemble the surrounding foliage so closely that they can deceive all but the keenest observers. The body is broad and flat. The forewings are flat too; they are green in color and are veined like leaves. The legs have leaflike outgrowths or appendages. These insects are found only in tropical Africa and the East Indies. One of the best examples of leaf mimicry is the female of the species *Phyllium siccifolium,* found in Malaya. The members of this species are about three and a half inches long and one and a half inches wide. The females have leaflike tegmina but no hind wings. The males have hind wings for flying but only short, unleaflike tegmina.

The family Phasmidae, to which many of the walking sticks belong, is distributed throughout the world, but especially in Australia and New Zealand. Some of these insects have stout, heavy bodies. Many — both the slender and the stout kinds — are covered all over with spines.

In the fourth family, Bacunculidae, we find the longest and the thinnest of the North American walking sticks. All are wingless. *Diapheromera femorata* is probably the most common and the most widely distributed in the United States. It is a large species, ranging in length from three and a half to four and a half inches. It may be gray, brown or greenish brown in color. The adults are found from August until cold weather, usually in second-growth areas of oak, wild cherry, hickory, locust and other hardwood trees. At this time the females drop their eggs onto the dry leaves on the ground, producing a sound that suggests the gentle patter of rain. The eggs lie dormant throughout the winter. Some will hatch the following spring; others the following summer; still others in a year or two. The giant of the United States walking sticks is *Megaphasma dentricus,* which occurs on

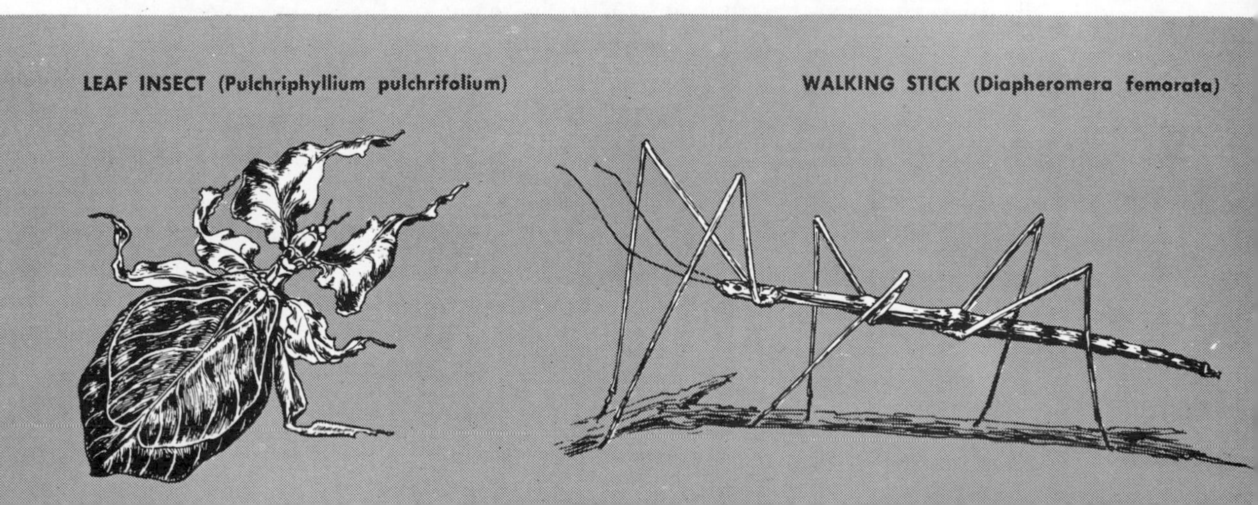

LEAF INSECT (Pulchriphyllium pulchrifolium)

WALKING STICK (Diapheromera femorata)

trees and wild-grape vines along southern and central streams. The female may reach a length of six inches.

The fifth family, Bacteriidae, is found only in tropical regions. It includes many large insects. One species, the giant brown phasmid (*Palophus titan*), found in Australia, is ten inches in length and its wing expanse is also ten inches.

## The order Mantodea

The mantids, belonging to the order Mantodea, are called either praying or preying. Both names are very appropriate. When their forelegs are held up in front of their faces, they appear to be in an attitude of prayer. Let an unwary insect appear on the scene, and the praying mantid becomes a preying mantid with a vengeance!

Many peoples — Arabs, Greeks, central Europeans and Australian aborigines — have regarded the mantids with awe. This is not strange in view of the extraordinary way in which these insects secure their food. Waiting in a favorable spot, the mantid supports itself on its middle and hind pairs of legs. The front pair are held up in front of and close to the upreared body. These legs are strongly developed and armed with heavy spines, or teeth. When an insect passes within reach, the mantid suddenly thrusts out its front legs, seizes its prey and holds it in a pincerlike grasp. Bringing its victim up to its jaws, the mantid feasts upon it, casts the remains away and awaits another insect to satisfy its seemingly insatiable hunger.

Most mantids are winged; the few wingless varieties hunt their prey on the ground. All are carnivorous, feeding on living insects. The winged species lie in wait among the leaves of trees and bushes. Sometimes they sit among flowers, seizing every bee, or butterfly or other insect that comes to gather pollen or sip nectar. A large tropical species is colored in rich reds, purples, blues and white; these colors are arranged in such a way that the insect resembles an orchid. Another species is bright green; the broad and conspicuously veined wings mimic living leaves. Several Brazilian species are dark brown; they look like the old dry leaves among which they live.

The heads of mantids are generally triangular in shape, with two large compound eyes, three simple eyes and fairly long, slender antennae. The head is mounted on a slender neck, which allows the insect to peer in almost any direction. It can even look over its shoulder!

Mantids are usually regarded as beneficial insects, because of the vast quantity of insects they devour. This favorable opinion is open to doubt. The fact is that mantids do not discriminate between the insects we find harmful and those we consider beneficial. In fact, most species of mantids probably destroy more useful insects than harmful ones.

### The Chinese mantid flourishes in the United States

About 1,600 mantid species are known; they dwell chiefly in the tropics. The most common one in the United States is the Chinese mantid, *Paratenodera sinensis,* which was introduced into America toward the end of the nineteenth century. It has spread from Philadelphia, where it was first observed, to a large part of the United States. This has been due chiefly to the sale of egg cases by nursery men and others, who have capitalized on the so-called beneficial qualities of mantids. Actually the Chinese mantid is sometimes a great pest. Beekeepers detest it, for it will establish itself at the entrance to a hive and catch bee after bee going in or out. They are particularly damaging in queen-rearing yards; a single mantid can destroy many of the young queens when they emerge from the hive for their mating flights. However, if you are not keeping bees, you will find that Chinese mantids make excellent pets. They readily learn to accept raw beef or liver from one's hand, and they will drink water from a spoon held out to them.

Another species introduced into the United States is the European mantid, *Mantis religiosa.* It immigrated to the United States on nursery stock imported from Europe in 1899. *Mantis religiosa* has not spread nearly so widely as its Chinese relative. It is only about one half as large as the Chinese mantid. Some European mantids are gray; others greenish, or light brown or dark brown.

The most familiar species native to North America is the Carolina mantid, *Stagmomantis carolina,* found from Mexico north to New Jersey and west to Arizona. From two to two and a half inches long, the males are grayish brown, the females often green. These mantids stalk their prey much as a cat stalks a mouse. The females often devour their mates; they are generally more voracious than the males. One observer reported that a female Carolina mantid devoured eleven Colorado potato beetles during one night, leaving only the wing cases and part of the legs. For all its hunting prowess this mantid is becoming more and more rare. Apparently it is unable to cope with changing conditions. Perhaps, too, it is falling victim to the onslaughts of the Chinese mantid.

**CHINESE MANTID**
**(Paratenodera sinensis)**

### The order Dermaptera

The order Dermaptera, the earwigs, was the first group to be separated from the order Orthoptera by taxonomists (specialists in classification). This was perhaps because the tegmina of the earwigs look like the elytra, or outer wings, of beetles. In fact, many taxonomists thought that the earwigs might be allied to the beetles. One glance, however, at the membranous rear wings will reveal the pleated, folding, fanlike structure, so characteristic of the grasshoppers and other orthopteroids.

Earwigs are elongated insects, small to medium in size and light brown to black in color. The much folded hind wings are tucked under the short tegmina, which are only about a quarter the length of the body. Not all earwigs are winged. In some species the wings are atrophied; in others, they are entirely absent. All earwigs have a pair of forcepslike appendages at the tip of the abdomen. These forceps are found in adults and in most of the

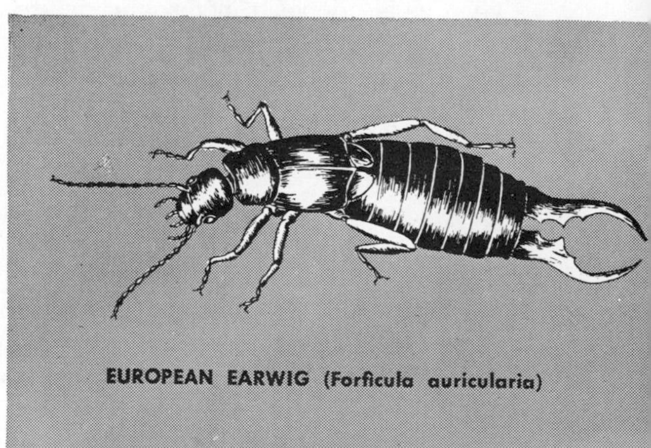

**EUROPEAN EARWIG (Forficula auricularia)**

nymphal stages. They may be long and slender, or short and thin, or short and stout; sometimes they are toothed. In male earwigs, the forceps are generally broadly rounded; those of females are more likely to be long and straight and to lie close together.

Forceps serve in various ways. Earwigs may use them to seize food and drag it to a place of concealment. Forceps also serve to tuck the insect's hind wings under the tegmina. The earwig is quite apt to threaten its enemies with its forceps; sometimes it uses them to pinch the finger of an unwary collector.

It may be because of the forceps that there have been so many erroneous beliefs concerning the earwig. The best-known, perhaps, is that the insect will enter the ear of a sleeping human being, pierce the eardrum and burrow into the brain. This odd idea probably arose in the days when the very poor slept on the ground or in sheds infested with earwigs. These insects like to hide in any small crack or crevice; a human ear would offer such a crevice.

Earwigs spend the daylight hours hiding in damp places under stones, boards, rotting logs, compost heaps and dumps. Some species feed on vegetation, either decaying or living. If they eat growing plants, they may become serious pests, especially in orchards. Certain earwigs are carnivorous. Of these, many are cannibalistic; they are particularly likely to attack immature earwigs while they are still soft and vulnerable.

The female deposits her round, white eggs in a small chamber in the soil or in debris. In a number of species, including *Forficula auricularia,* the female stays with the eggs, taking care of them until they hatch. She then looks after the young until they can forage for themselves.

About 1,100 species of earwigs have been described. Most of the species occur in the tropics; only a few are native to North America. Quite a number of earwigs have become world travelers, hitchhiking on man's possessions to the far corners of the earth.

Although the Dermaptera are among the oldest groups of insects, they have not developed wide variations. The members of the family Arixeniidae form an exception. They have become so distinct in form that one is tempted to include them in a new order. The insects of this family are less than an inch in length; they are wingless and hairy. Only two species are known; these were discovered in bat caves on the Sunda Islands in 1909, and in Java in 1912. They feed on larvae (including their own young) that live in the guano, or dung, of the bats.

One of the most common earwigs in the United States is the seaside, or maritime, earwig, *Anisolabis maritima,* belonging to the family Labiduridae. Dark, brown, wingless and less than an inch long, it came to America from Europe many years ago. It now inhabits the Atlantic, Pacific and Gulf coasts, where it lives as a scavenger just above the high-water mark. If one of these insects is caught by an advancing wave, it is quite capable of swimming back to shore.

The black earwig, *Chelisoches morio,* a native of Asia, has immigrated to California by way of the Hawaiian Islands. It is about an inch long, winged, shining black, with light colored spots on legs and antennae. It is considered an ally of man, since it feeds on the sugar-cane leafhopper, which is a damaging pest.

The European earwig, *Forficula auricularia,* started its wanderings from central Europe and has now reached Australia, New Zealand, Africa and North and South America. This species can be very destructive, attacking many kinds of plants. It does not restrict itself to a vegetable diet, but may feed upon animal material. *Forficula auricularia* seems to prefer locations near human dwellings. Often it gets into houses and stores; it is likely to be found in basements where vegetables are stored. Only about half an inch in length, it is reddish brown to black in color, has wings and flies well. Under favorable conditions enormous populations of these earwigs can develop in garbage dumps or manure piles.

*See also Vol. 10, p. 275:* "Insects."

# COLLECTING INSECTS

## A Brief Account of Equipment and Methods

by CHARLES D. MICHENER and MARY H. MICHENER

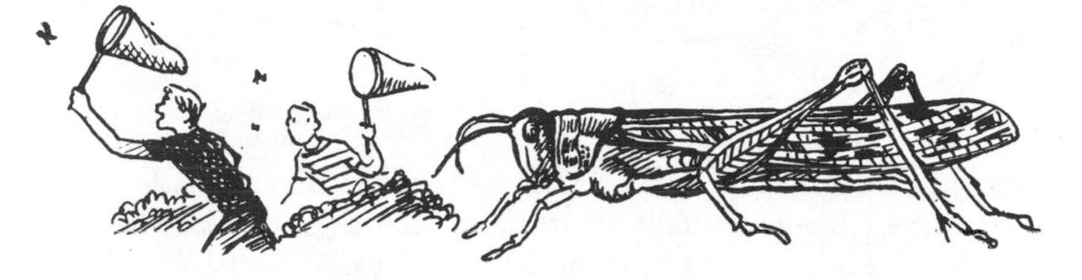

THERE are such endless numbers and varieties of insects and so much about them that we still do not know that even amateur collectors may make worthwhile contributions to our knowledge of these fascinating little creatures. That is one reason why the collecting of insects is such a gratifying hobby. But that is not the whole story by any means. You have the privilege of seeing your specimens alive, observing when and where each insect is found, perhaps how it lives and behaves, what it eats, how it grows and changes, how it finds a mate and breeds. You may notice whether the insect was out in the heat of the day or in the cool of morning or evening, in sun or shade, in a moist or dry location, whether it was lively or sluggish, alone or in a group. If you carry a little notebook to write down your observations, the stories that will go with your specimens will make your collection doubly interesting and will serve as a guide to further collecting.

The equipment you will need for collecting depends upon what kinds of insects interest you most and what methods you want to use. You can make a good deal of equipment yourself with help from some competent person, such as a biology teacher, or from directions in the books listed at the end of this article. A few items will have to be bought from drugstores or biological supply companies.

To collect butterflies and other flying insects you will need a net of light but strong material, such as nylon or best-quality mosquito netting. The mesh must be fine enough to prevent the escape of tiny insects but easy to see through. A short, sturdy handle, such as a three-foot piece of broomstick, is best for the general collector. Long handles are hard to swing with swiftness and accuracy. The bag is rounded at the bottom, not pointed, and it must be deep enough so that when you have captured an insect you can turn the hoop to lie against the side of the net; your prize will then not be able to get away. The rim of the bag, which covers the heavy wire hoop, will get the hardest wear, so it must be made of heavy muslin or light canvas.

Because most insects fly or crawl upward when disturbed, you should generally swing your net sideways and downward. Some kinds dart sideways or drop straight down, however. With experience you will learn what to expect from the various kinds and how to catch them. To take a winged insect resting on the ground, clap the net over it, holding up the end of the bag so that the insect will crawl or fly into it.

After the insect has been caught, it must be put into a killing bottle — a receptacle containing a poisonous vapor that will kill your catch. To secure quick-

Net. Killing bottle. Beating sheet.

moving insects that you have just caught, swing the net rapidly two or three times so as to drive the specimens into the narrow end. Then you can slide your uncorked killing bottle up into the net to complete the capture. In the case of slow-moving insects, you may reach down into the bag with your fingers or forceps.

Killing bottles containing cyanide covered with plaster of Paris are the most satisfactory. (Remember, though, that cyanide is a deadly poison!) Keep strips of paper in the bottle to absorb moisture. Carbon tetrachloride or ethyl acetate poured onto cotton and covered with a disk of cardboard may be used; the fluid must be replaced as it evaporates. Avoid breathing the fumes from any sort of killing bottle.

It is well to have several killing bottles of various sizes so that the insects will not damage one another. Bees and butterflies die almost at once and are damaged if they remain in the jar too long. Beetles, however, have a disconcerting way of reviving after they appear to be dead. As soon as your specimens are quite dead, put them in envelopes or folded paper triangles on which you have written the place and date. You can store insects that have been "papered" in this way in a covered can or carton until you are ready to mount them.

If you sweep your net briskly through grass and other low-lying vegetation, you are likely to be astonished at the number of insects you gather — many of them probably quite new to you. For sweeping in rough or prickly growth you will need a very strong net, such as one made with muslin instead of netting, so that it will not tear or snag too easily.

Sometimes, after sweeping, your net will be so full of creeping, jumping and flying little insects that you will have difficulty capturing them all before they escape. For the very small insects an aspirator is most useful. This is a small bottle whose stopper is fitted with two rubber tubes. You suck on one tube, which is blocked with a fine screen where it enters the bottle, and draw the insects up into the bottle through the other tube. This method is quicker and far less injurious to fragile specimens than picking them up separately. Small insects, incidentally, are just as beautiful, comical or bizarre as the larger, more conspicuous ones. Be sure to carry a small hand lens of about 10X (ten power) with which to look at them. You will need the lens also in identifying many of your big specimens.

"Beating about the bush" is actually a short cut in collecting insects! This is true even in winter when many insects are inactive. To make a beating sheet, sew triangular pockets at the corners of a light-colored canvas square. Long sticks, crossed at the center and thrust into the pockets, will provide a handle and will hold the sheet open. You hold the beating sheet in one hand to catch falling insects while, with the other hand, you give the underbrush a few sharp blows with a stout stick. An opened umbrella held upside down will serve as a beating sheet.

If you collect soft-bodied insects, such as caterpillars and plant lice, and certain

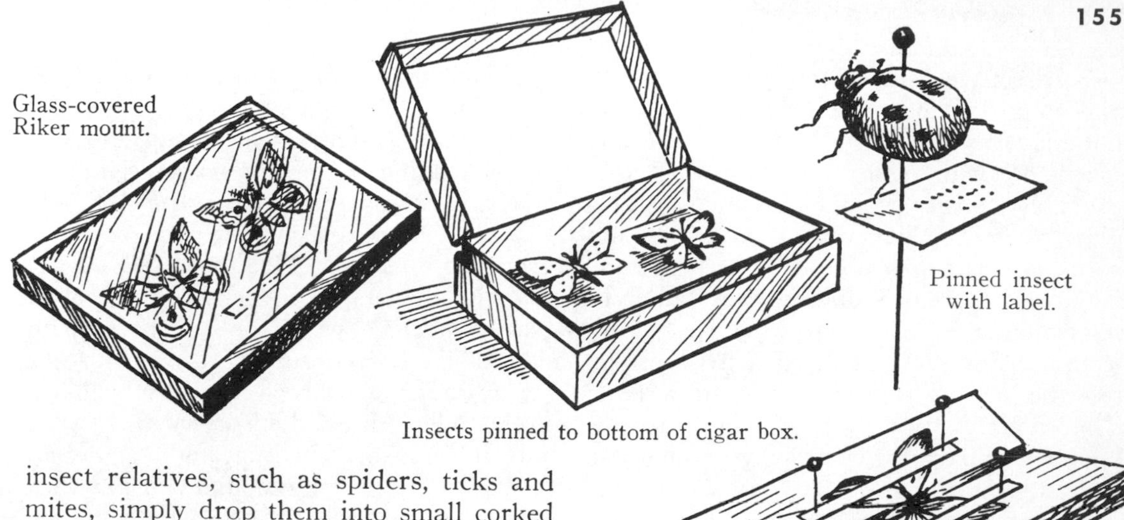

Glass-covered Riker mount.

Insects pinned to bottom of cigar box.

Pinned insect with label.

Preparing a butterfly for display.

insect relatives, such as spiders, ticks and mites, simply drop them into small corked bottles or vials containing 70 per cent alcohol. A fine water-color brush moistened with alcohol will pick up the more delicate specimens. Keep specimens from different localities in separate vials and slip into each vial a penciled label giving the place and date of collection.

Water collecting is another fine way to find insects, especially on a hot summer day. Use a large kitchen sieve for whirligig beetles, water boatmen, mosquito wrigglers and other surface swimmers. A water net of tough, porous material may be used on the same hoop and handle as your aerial, or sweeping, net. For extensive water collecting, a hoop with the side opposite the handle flattened is useful for dragging the bottom of streams and ponds. When you have brought up a small load of mud and stream-bottom litter, spread it out carefully and pick up the beetle larvae, dragonfly and May-fly nymphs and other specimens. To pick up the smaller water inhabitants, such as midge larvae, use a medicine dropper. No special equipment is needed to turn over rocks in shallow running streams to discover caddis worms and stone-fly larvae.

Turn over rocks and rubbish on land, too, to surprise ants, beetles, crickets, centipedes, millepedes and sow bugs. Peel the bark from dead trees, particularly in damp places, to find beetles, roaches, termites and many other insects.

Sifting leaf mold and ground litter is a good method of collecting in winter; so is examining crevices in bark for resting insects. Cocoons and chrysalises may be found among the leaves and bare branches.

You can collect specimens around your porch light or by setting a lantern on top or in front of a white sheet. On a warm evening a light will attract many kinds of moths, as well as May flies, giant water bugs, some kinds of beetles and other night-flying insects.

A good way to get perfect specimens is to rear them yourself. Cocoons and chrysalises may be kept in a box; a jar is not so good. Provide twigs or crumpled paper upon which the adult insect can crawl when it emerges and spreads its wings to dry. You can rear growing caterpillars and plant-feeding bugs if you keep them supplied with fresh food in the form of plants. The food plant can be potted and then, with its feeding insects, surrounded by a glass lamp chimney set into the soil in the pot. Cover the top of the chimney with netting. If you want to rear immature insects from your water collecting, be sure to take some of the pond water, along with algae and other water plants or, for predaceous insects, plenty of smaller living creatures. It is not usually possible to rear running-stream insects in jars or aquaria unless you have some mechanical aerating device.

Insects, especially caterpillars, spiders and so on that have been collected in alcohol may be preserved in the same fluid. Most of your collection, however, will probably consist of dry specimens. The glass-covered Riker mount, in which insects are laid on cotton or cellucotton, makes an attractive display case, but it is not satisfactory for a growing study collection. For such a collection it is best to pin the insects to the bottom of a box. The Schmitt-type box, which may be bought from biological supply companies, is best, but a cigar box will serve well enough. For a pinning bottom use a soft composition wallboard tightly fitted into the bottom of the box. In one corner of the box pin securely a moth ball or, better, a paper container of PDB (paradichlorobenzene) crystals to ward off those arch foes of insect collectors — carpet beetles and mold.

If you do not pin your insects soon after they are freshly killed, you may find them too stiff and brittle to handle without breaking. While they are still in their envelopes or paper folders, place them in a relaxing jar, where they will be softened, for several hours or, if necessary, a day or two. To make a relaxing jar put wet cotton in a wide-mouthed jar, covering it with a disk of cardboard or a wire rack to keep the specimens from getting too wet. Add a moth ball or PDB to retard mold.

For general use, No. 3 insect pins are good, although you may wish to use No. 1 or No. 2 for the smaller specimens. Pin beetles through the base of the right wing cover; pin true bugs through the scutellum, the triangular section between the wing bases; pin butterflies, moths, bees, flies, grasshoppers and most other insects through the thorax. Insects that are too small to be pinned, such as mosquitoes, or that break readily, such as ants, should be pointed. In this method, the pin is put through the broad end of a very small paper triangle, called a point. Tip the triangle with a bit of glue and touch it to the underside of the insect so that the specimen is securely stuck to the point but not too much obscured by glue or paper.

Butterflies, moths, dragonflies and sometimes grasshoppers are displayed with wings spread. The insects should be dried on a spreading, or setting, board — a block or board arrangement that is very slightly **V**-shaped in cross section with a groove in the point of the **V**. Set the insect on the board so that its body occupies the central groove and the wings lie on the sloping sides of the board's surface. With forceps (or an insect pin thrust carefully behind a strong vein) draw the forewing forward until its hind margin is at right angles to the body. Then pull the hindwing forward to overlap it slightly. To hold the wings in place until thoroughly dry (it usually takes several days), pin strips of paper tightly across them or weigh them with pieces of glass. If you cannot buy or make a professional type of spreading board, you may use a flat board, spreading the wings with the insect placed on its back; but the results are not so satisfactory.

Whether pinned, or pointed or preserved in alcohol, every specimen should have a small label giving its name and the date and place of collection. The collector's name may be on the same label or on a separate one. For pinned specimens the label is set on the pin below the insect. Where the same information applies to a number of specimens of the same kind, the label need only be attached to the first in the row or block of specimens. Different kinds should be labeled separately, for as your collection grows you will be moving each order or even each family or genus into a box of its own.

The following books are useful guides in the making of equipment, in collecting techniques and in identifying specimens:

A. B. Klots, *Field Guide to the Butterflies;* Houghton Mifflin, Boston, 1951.

Frank E. Lutz, *Field Book of Insects;* G. P. Putnam's Sons, New York, 1948.

Ralph B. Swain, *Insect Guide;* Doubleday, New York, 1948.

H. S. Zim and Clarence Cottam, *Insects; A Guide to Familiar American Insects;* Simon and Schuster, New York, 1951.

# THE FAMILY OF THE SUN

## A Survey of the Vast Solar System

by FRED L. WHIPPLE

IF you were out in space, billions of miles away from our planet, you would see the earth as a tiny ball moving in a wide path around a star that you might recognize as our sun. You would also see, at various distances from the sun, eight other spherical bodies of different sizes — the other planets — all traveling in the same direction in almost circular paths around the sun. Moving around some of the planets you would see smaller balls — the satellites, or moons, of the planets.

In the space between the orbits of two of the planets — Mars and Jupiter — there would be thousands of little planets, or asteroids, also revolving around the sun. Cutting in, this way and that, across the paths of the planets, you would see comets — starry-headed objects, sometimes with long tails streaming after them as they drew near the sun. You might also catch a glimpse of swarms of even smaller particles — the meteors — swirling through space.

All these bodies — sun, planets, satellites, asteroids and meteors — make up our vast solar system. If you continued to view them for months or for years, you would see that they are moving together through space as a unit, at the rate of some twelve miles a second, in the general direction of the blue star Vega.

The sun is the very heart of the solar system. It is a typical star — one of the several thousand million in our galaxy; like the rest, it is an incandescent body made up of highly compressed gases. It is far closer to us than any other star. The distance from the earth to the sun is 92,900,000 miles; the next nearest star, Alpha Centauri, is more than 260,000 times farther out in space. When Alpha Centauri and the more distant stars are examined with the most powerful telescopes, they seem to be mere brilliant points in the black sky. But even with the naked eye the sun appears as a disc of about the same size as the disc of the full moon; it is so dazzlingly bright that we must look at it through a darkened glass or a film to avoid damaging our eyes.

Compared with the other stars of our galaxy, the sun is of average size; but it is a giant in comparison with even the largest planets. Its diameter of 865,600 miles is 109 times that of the earth; even though it is gaseous, it weighs more than 300,000 times as much as the earth. Its surface temperature is about 10,000° Fahrenheit; at its center the temperature may be as high as 27,000,000° Fahrenheit. The heat energy and light energy radiating from the sun make it possible for life to exist upon the earth. Furthermore, without the reflection of the sun's light, we could not see the other members of the solar system, except for the comets, which are partly self-luminous, and the meteors that leave a fiery trail through friction with the earth's atmosphere.

The planets are the largest bodies in the solar system next to the sun, except for a few satellites that compare in diameter with the smallest planet, Mercury. In the order of their distance from the sun, the planets are Mercury, Venus, the earth, Mars, Jupiter, Saturn, Uranus, Neptune and Pluto. The earth is quite a bit under the average size of the planets. It is much smaller than Jupiter, Saturn, Uranus and

Neptune, which are called giant planets; it is somewhat larger than Mercury, Venus, Mars and Pluto. All the planets, large and small, occupy no more space in the solar system than nine peas would in a huge football stadium.

Each planet travels around the sun in a giant ellipse, which is very nearly a circle. The orbits of all the planets are more or less in the same plane, except for Pluto, the outermost one. The orbit of Pluto is inclined 17 degrees to that of the earth.

In general, as the distance from the sun increases, the planet paths are more and more widely separated. The German astronomer Johann Elert Bode (1747–1826) devised a simple rule of thumb, called Bode's Law, for determining the distances of the planets from the sun. The following table shows how the law works.

HOW TO APPLY BODE'S LAW
(for estimating the relative distances of planets from the sun)

| Planet | Mercury<br>4<br>0 | Venus<br>4<br>3 | Earth<br>4<br>6 | Mars<br>4<br>12 | ———<br>4<br>24 | Jupiter<br>4<br>48 | Saturn<br>4<br>96 | Uranus<br>4<br>192 | Neptune<br>4<br>384 | Pluto<br>4<br>768 |
|---|---|---|---|---|---|---|---|---|---|---|
| Sums | 4 | 7 | 10 | 16 | 28 | 52 | 100 | 196 | 388 | 772 |
| Bode's distance | 0.4 | 0.7 | 1.0 | 1.6 | 2.8 | 5.2 | 10.0 | 19.6 | 38.8 | 77.2 |
| True distance | 0.39 | 0.72 | 1.0 | 1.52 | ——— | 5.20 | 9.54 | 19.19 | 30.07 | 39.46 |

Here is the explanation of the table. We first write down the names of the planets in the proper order of increasing distance from the sun, leaving a gap between Mars and Jupiter. (We shall explain later the probable reason for this gap.) Under each planet and also under the gap, we write a 4; under the 4 we write a zero for Mercury, a 3 for Venus, a 6 for the earth and so on, doubling the preceding number each time. We now add the two numbers in each column. The sum will be 4 for Mercury, 7 for Venus, 10 for the earth and so on. Next we divide each of these sums by 10, by inserting a decimal point before the last figure of each number. This will give us the relative distance of each planet from the sun, the distance of the earth from the sun being given as 1.0. For example, according to Bode's Law, Mercury is only .4 of the earth's distance from the sun; Uranus is 19.6 times farther away from the sun than the earth.

In the last line of the table we give the true distance of the planets from the sun, also in terms of the earth's distance. As you will note, these real distances are remarkably close to those established by Bode's Law, except for Neptune and Pluto, the two outermost planets. Since Bode's Law has no real theoretical basis, we need not be surprised that it breaks down in the case of the two most distant planets.

All of the planets larger than the earth have satellites, or moons, revolving around them. Most of these satellites move around their planets in the same direction as that in which the planets move around the sun; there are exceptions, however, in the satellite systems of Jupiter and Saturn.

Jupiter, the largest of the planets, has the greatest number of satellites; twelve have been discovered thus far. They are especially interesting because, together with Jupiter, they form a sort of miniature solar system. Like the planets, Jupiter's satellites move in what is to all intents and purposes a single plane; and they are spaced somewhat as are the planets.

Saturn has nine satellites; it also has a vast system of tiny moonlets that, re-

Johann Elert Bode, the German astronomer who devised the law named after him for estimating the distance of the planets from the sun.

flecting sunlight, form a magnificent halo around the planet's disc. Uranus has five moons, Neptune two, the earth one and Mars two; Mercury, Venus and Pluto have none. The two moons of Mars are exceedingly small; the larger one, Phobos, is probably less than ten miles in diameter. It revolves around Mars faster than Mars turns on its axis. As a result Phobos is the only moon in the solar system that rises in the west and sets in the east.

We pointed out that according to Bode's Law there is a gap between Mars and Jupiter. Astronomers in Bode's day felt sure that there must be an undiscovered planet between these two bodies, and they eagerly searched the skies for it. On the night of January 1, 1801, the Italian astronomer Giuseppi Piazzi discovered a small celestial body, which he took to be a planet, in the space between the orbits of Mars and Jupiter. This body, which was later called Ceres, was found to have a diameter of only 480 miles. Other small planetlike bodies were found in the course of time in the gap between Mars and Jupiter. Today more than a thousand of these small bodies have been discovered, and it is estimated that there are more than 50,000 in all. They are known as minor planets, or asteroids. In spite of their great numbers, their combined mass is only a fraction of the earth's mass.

#### Theories about the origin of asteroids

Nobody knows how the asteroids originated. According to one theory, they represent the fragments of a big planet whose orbit lay between the orbits of Mars and Jupiter and that exploded for some unknown reason; in support of this theory it is pointed out that at least some of the asteroids are irregular in shape. More probably, there were several smaller planets in this area and they collided, breaking up into many tiny ones.

Among the strangest members of the sun's family are the comets. They abide by none of the rules that govern the nine planets and the thousands of asteroids. Instead of moving in nearly circular orbits in a single direction, the comets revolve around the sun in exceedingly elongated ellipses and in every conceivable direction. Much of the time they are so far away from the sun that they are invisible even in our largest telescopes. It was formerly thought that some comets approach the sun from far beyond the solar system and that, once they withdraw from the sun, they never return. Today it is generally agreed that comets are truly members of the sun's family and that they do not come from beyond the solar system.

When first discovered, comets usually appear as faint, diffused bodies, which are densest at the center. The dense part, which looks like a tiny star, is called the nucleus; the veil-like region that surrounds it is known as the coma. As the comet approaches the sun, the coma becomes brighter. At a distance of some 100,000,000 miles from the sun, some comets show nebulous matter streaming away in the direction opposite the sun and forming a tail. This apparently consists of very thin gases which shine by absorbing and reflecting the sunlight that falls upon them; they are forced away from the sun by the pressure of the sun's light. Many comets never develop a tail; they remain vague and diffuse bodies even when they are close to the sun.

When comets are very old, they break up into the particles known as meteors.

This meteorite was discovered in Canyon Diablo, Arizona. It represents one of the fragments into which a huge meteor was split as it struck the surface of the earth.

Yerkes Observatory

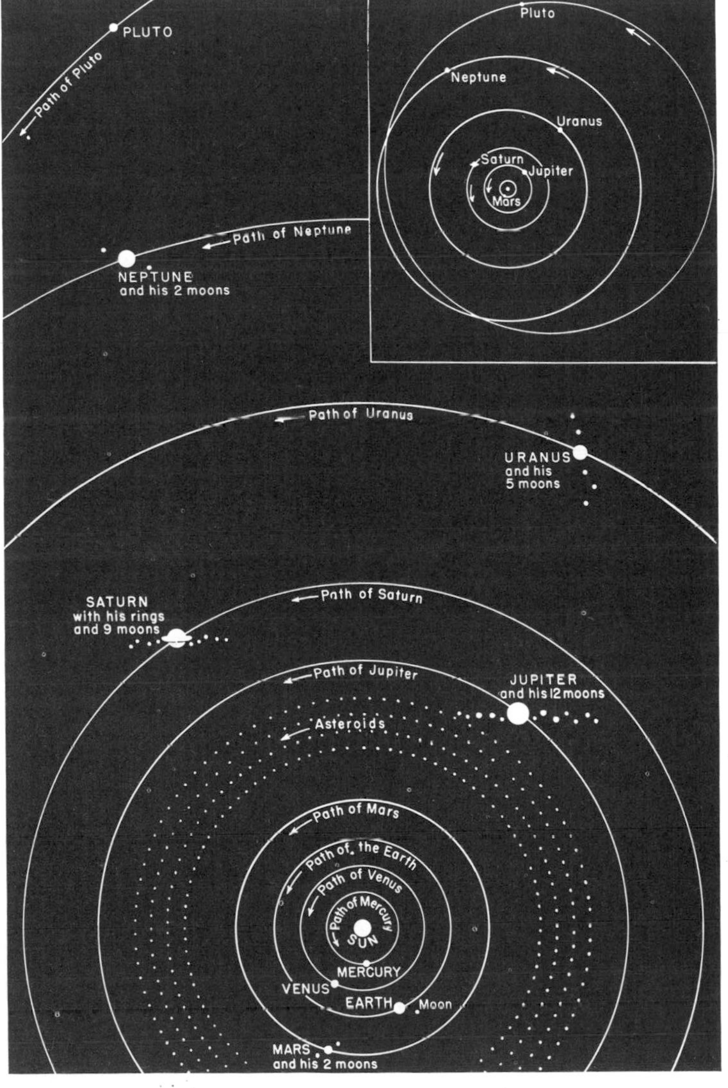


The two diagrams show the order of the planets from the sun outward, though not the comparative sizes, nor the exact shapes of the orbits. Satellites are shown in the larger diagram; so are the asteroids, found between the orbits of Mars and Jupiter. The asteroids are not so symmetrically arranged, in reality, as they appear to be in the large diagram.

International News Photo

Comet first seen in South Africa in 1948. Its tail extended some 25 degrees; its head was as bright as a star of the second magnitude.

Meteors range in size from fragments no larger than pinheads to huge stones weighing many tons. We become aware of meteors only through the bright light produced when they collide with air molecules in our atmosphere. Most meteors disintegrate after striking the atmosphere; some of them, however, land upon the earth. Those that are discovered upon the surface of this planet are called meteorites. They represent the only objects from outer space that we can actually handle and examine at close range. Probably all the meteorites come from broken pieces of asteroids, while the comets contribute most of the small meteors.

The vast majority of meteors are exceedingly small; the bright flash of light seen when a meteor passes overhead is usually caused by an object the size of a pea or smaller. In the heavens these tiny particles form great clouds of dust in certain areas of the sky. It is probable that such clouds, reflecting the light of the sun, cause the zodiacal light — the faint glow in the heavens that may sometimes be seen just before sunrise or after sunset.

In the light of our present knowledge of the solar system, it is hard to realize that only a few centuries ago men had an entirely erroneous conception of the relations between the sun and its family. A few ancient Greek astronomers suggested that the earth moved around the sun, but this idea won scant support. Most scholars in antiquity held an entirely different sort of theory which to us seems fantastic. They did not realize that the earth itself is a planet, or wanderer in the heavens (that is what planet means). The earth, they thought, hung motionless in the very center of the universe. They held that each of the five planets then known (Mercury, Venus, Mars, Jupiter and Saturn) was attached to a great invisible crystal sphere; the moon and the sun were attached to other spheres. The crystalline spheres, set one within the other, revolved around the earth, carrying with them the heavenly bodies that were attached to them. This theory satisfied most people; yet careful observers of the sky found in time that it could not explain certain phenomena.

For one thing, the movement of the

planets in the skies seems to be irregular; apparently they move more rapidly across the heavens at one time than another. Astronomers sought to solve this difficulty by placing the earth somewhat off-center in the universe. But in so doing they left another puzzling phenomenon unexplained. There comes a time when a planet ceases its apparent eastward motion among the stars; it turns about and moves westward for a time. To explain this "reverse motion" of the planets, a complicated system of epicycles was invented. It was held that each planet traveled along the circumference of a small circle, the center of which traveled along the circumference of a larger circle. The earth, it was maintained, was at the center of the larger circle.

For over a thousand years this conception of the universe prevailed; it was taught in the schools of the late Roman Empire and in medieval universities. In the first half of the sixteenth century, however, a Polish astronomer, Nicolaus Copernicus, revived the suggestion that the earth moves around the sun; he held that the other planets also revolve around that heavenly body. Copernicus's system, presented to the world in 1543, was called the heliocentric theory since it placed the sun (*helios,* in Greek) at the center of the universe. The Polish astronomer's ideas were quite sound, on the whole; but they were based on insufficient observation. They were inadequate in some respects and wrong in others.

It required the efforts of several other great astronomers to prove that the heliocentric system of Copernicus had a solid core of truth, for all its defects. The Danish nobleman Tycho Brahe (1546–1601) made a long and accurate series of observations, which later investigators used as a point of departure. Galileo Galilei (1564–1642), a great Italian physicist, staunchly upheld the Copernican system in his influential writings. Johannes Kepler (1571–1630), a German disciple of Brahe, drew up three laws of planetary motion that still hold good today. Copernicus had

Am. Mus. of Nat. Hist.

When Galileo Galilei first looked at Jupiter through the newly invented telescope, he discovered four of Jupiter's moons, "never seen up to our own time." The diagram shows how Jupiter and the four moons appeared to Galileo. The planet has twelve moons in all.

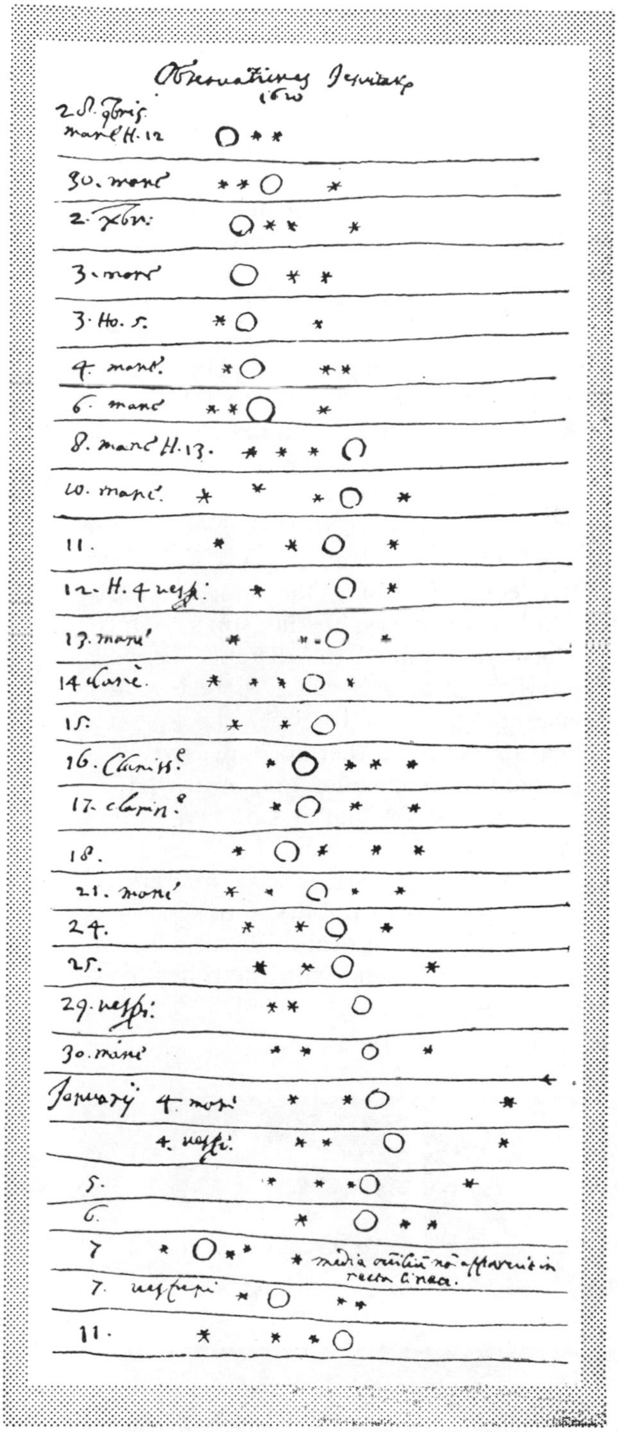

Yerkes Observatory

Galileo's drawings of Jupiter and its moons, as they appeared to him on different nights in 1610. The spheres represent Jupiter; the asterisks, the satellites. At certain times only two moons were visible to Galileo; at other times he could make out three or four.

maintained that the planets move in circular orbits around the sun, and this belief had led to great confusion. Kepler pointed out that the orbits in question are not circles but ellipses. We discuss Kepler's laws of planetary motion in another chapter of THE BOOK OF POPULAR SCIENCE.

### Sir Isaac Newton's law of universal gravitation

Kepler's laws clearly explained the nature of the planets' movements around the sun, but Kepler did not analyze the force that brings about these movements. This force was first revealed in 1687 when the great English scientist Isaac Newton (1642–1727) presented his law of universal gravitation. The law states that every particle of matter in the universe attracts every other particle with a force that varies directly as the product of their masses and inversely as the square of the distance between them. Newton showed mathematically that this is truly a universal law, since it applies not only to objects upon the earth but to heavenly bodies — from meteors to stars — as well.

### How gravitation affects the earth and the heavens

The law of universal gravitation explains why planets and asteroids and meteors keep turning around the sun; this huge body binds its retinue to itself because of its strong powers of attraction. Universal gravitation also explains why we do not fly off the earth even though, at the equator, it is spinning around at the rate of a thousand miles an hour. Men and animals and rocks alike are drawn toward the center of our planet by the force of gravitation (or gravity, as it is called in the case of the earth's attraction). Consequently, the earth has no "bottom side" from which objects can fall off into space. The force of gravitation holds the air and the oceans to the earth. Even the moon, 238,000 miles away, feels the effect of the earth's attraction and because of it continues to revolve around our planet.

By utilizing the law of universal gravitation, we can now analyze the motions of the planets with a very high degree of accuracy; we can account for the small deviations that arise as one planet affects the orbit of another. It was the study of such deviations that led directly to the discovery of the planet Neptune.

After Uranus had been discovered by Sir William Herschel in 1781, careful studies showed that it did not exactly follow the orbit that had been predicted for it in accordance with the law of universal gravitation. This led a young Englishman, John Couch Adams, and a noted French astronomer, Urbain-Jean-Joseph Leverrier, to the conclusion that Uranus was being attracted by another planet even more distant from the sun. Both men calculated the position in the sky of the unknown planet without ever having seen it. On September 23, 1846, on the basis of Leverrier's calculations, the German astronomer Johann Gottfried Galle located Neptune in the heavens almost exactly where Leverrier had placed it, in the constellation Aquarius. This discovery proved once and for all that the law of universal gravitation applies to the heavens as well as to objects upon the earth.

Today we realize that this law is not the last word in the analysis of motion in the universe, for it has been modified by the theory of relativity proposed by Albert Einstein. The modification in question is exceedingly slight, however. Thus far it has been applied to only one case of solar-system motion: it has been shown that the perihelion of the planet Mercury (the point where it is nearest to the sun) moves forward about forty seconds of arc farther in a century than had been predicted on the basis of the law of universal gravitation. But this exceptional case does not invalidate the law; it is still held to be 99.99999 per cent accurate as far as the solar system is concerned. There is, indeed, no other physical law that has been proved so precisely and almost none for which the margin of error is so small.

We have found out many things about the solar system since the discovery of the law of universal gravitation. We have succeeded in weighing the sun, the earth and other members of the sun's family; we have determined the distance between them with great accuracy. Utilizing such devices as the spectroscope, the spectroheliograph and the thermocouple, we have analyzed the composition and measured the temperature of the sun and of many other bodies in the solar system.

We shall discuss these matters in other chapters of THE BOOK OF POPULAR SCIENCE. We shall deal in detail with our star — the sun — and with the other heavenly bodies in the solar system.

*See also Vol. 10, p. 269:* "Solar System."

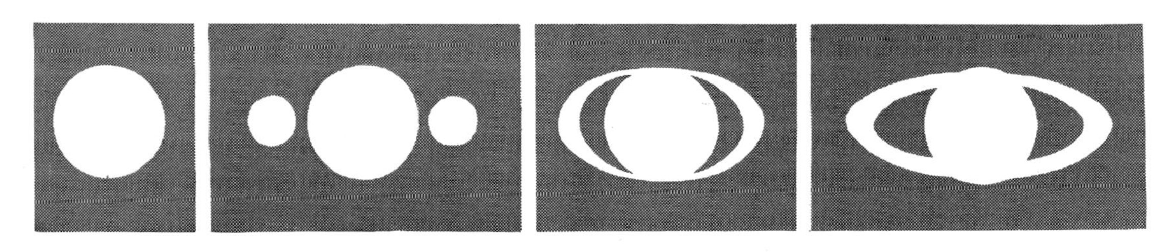

Before Galileo's day, Saturn appeared to be a disklike body like the moon. Galileo discovered what seemed to him a small body on either side of the planet. Later observers thought they saw armlike structures instead of separate bodies. In 1655, Huygens came to the conclusion that the planet was encircled by a ring. We now know that there are three rings in all.

# SUPERACTIVE ELEMENTS

## The Alkali and Alkaline-Earth Metals

IN AN earlier chapter of this volume we discussed the properties and uses of some of the chief metals, such as iron, aluminum, gold, silver and mercury. We now come to two exceedingly active groups of metals: the alkali and alkaline-earth groups. The reactions of these metals are sometimes quite startling. For example, if we put a piece of gold, or silver or iron in a basin of water, nothing much will happen. But if we drop a fragment of potassium, one of the alkali metals, into the basin, the fragment will at once burst into violet-colored flame. Then the burning metal will dart furiously about the surface of the liquid until finally it will explode.

To understand why alkali and alkaline-earth metals are so reactive, we must recall that the atoms of all metals have a tendency to lose their outermost electrons. This tendency is called electropositiveness, since the loss of electrons turns the atoms into positive-charged ions. The alkali and alkaline-earth metals are strongly electropositive. Every alkali metal has only one electron in its outermost electron shell; the alkaline-earth metals have only two such electrons each. When free atoms of the metals are brought in contact with other substances, they show an overwhelmingly strong tendency to lose electrons, to become positive ions and to combine with other ions.

### The Alkali Metals

The alkali metals — lithium, sodium, potassium, rubidium, cesium and francium — form Group Ia of the periodic table (see the Index). The group name, "alkali," is taken from the Arabic *al-qili,* meaning "ashes of saltwort." An alkali was originally a soluble salt that was obtained from the ashes of plants. It has come to mean a substance that has strong basic properties, such as those possessed by plant ashes, when it is dissolved in water. (See the article Acids, Bases and Salts, in Volume 4.) An alkali solution can turn litmus paper blue, has a soapy feeling and bitter taste and can neutralize acids.

The alkali metals have basic properties when dissolved in water. They are generally so soft that they can be kneaded in the hand; they are dull-looking and greasy to the touch. These metals become liquid at comparatively low temperatures: from 28.5° C. (in the case of cesium) to 186° C. (in the case of lithium).

#### Lithium, lightest of all the metals

The first of the alkali metals in the periodic table and the lightest of all the metals (specific gravity, 0.53) is lithium. It was discovered in 1817 by the Swedish chemist J. A. Arfvedson. It is said that the metal was called "lithium" (a Low Latin form of the Greek word *lithos,* meaning "stone") because it was first discovered in a mineral.

Lithium (Li) is a white metal, which tarnishes only slowly in moist air It has a higher melting point (186° C.) and boiling point (about 1,336° C.) than any other alkali metal. It has a hardness of 0.6 on the Mohs' scale (see Index); it can be rolled and welded or drawn into thin wire at ordinary temperatures.

This metal is not abundant in the earth's crust; the principal sources are the minerals spodumene, lepidolite and amblygonite. Important deposits of lithium-containing ores occur in Maine, California, Connecticut and South Dakota. Traces of lithium are widely distributed in other minerals besides those we have named, as well as in plant ashes, soils, seaweed, milk, blood, and muscle and lung tissue.

Small amounts of lithium have a hardening effect on lead alloys. Because of the metal's strong affinity for oxygen, it is often used to deoxidize (remove oxygen from) copper and its alloys; the addition of lithium makes these alloys stronger and tougher.

Compounds of lithium are employed in various ways. Lithium chloride ($LiCl$) and lithium bromide ($LiBr$) are extremely soluble in water and, therefore, serve as moisture absorbers in air-conditioning units. Crystals of lithium fluoride ($LiF$) are efficient transmitters of ultraviolet and infrared light. The compound is used in superior achromatic lenses — those giving images practically free from extraneous colors.

Interest in lithium has been greatly spurred with the discovery that one of its natural isotopes, Li 6, splits into helium and tritium when bombarded with neutrons. Tritium is a major constituent of the thermonuclear (hydrogen) bomb; supplies of tritium are so limited that any new source is of great importance.

At present the softness and extreme reactivity of lithium make it unsuitable for use as a structural metal. Metallurgists hope, however, that some day they may be able to prepare a lithium alloy that will serve this purpose. Certainly lithium would offer many advantages as a structural metal. Its lightness and the ease with which it can be worked would make it particularly valuable in the construction of aircraft.

**Sodium — a highly useful alkali metal**

Sodium and potassium, the most abundant and important of the alkali metals, were first isolated in 1807 within a few days of each other by the famous English scientist Sir Humphry Davy. Pure metallic sodium (Na) is an extremely soft, silvery white substance with a specific gravity of 0.971. It melts at 97.5° C. and boils at about 880° C., giving off a blue-violet vapor. Under high voltage, electric discharges take place in the vapor and brilliant yellow light is emitted Sodium vapor lamps are employed in some places for low-cost, antifog lighting on highways.

Like all the alkali metals, sodium reacts with air and water and must, therefore, be stored under a relatively inert liquid, such as kerosene or naphtha, or in a moisture-proof vacuum. If the metal is placed under water, a violent reaction occurs; if the temperature is high enough, an explosion may result. In air, sodium tarnishes while giving off a greenish phosphorescence; it burns only when heated. The element is essential to living things; it is found in all protoplasm.

Seventh in abundance among the elements, sodium makes up about 2.75 per cent of the earth's crust as a constituent of various minerals. Chile is rich in deposits of sodium nitrate (Chile saltpeter, $NaNO_3$), which is a valuable fertilizer and serves other useful purposes. Sodium carbonate ($Na_2CO_3$) has been mined for thousands of years. The ancient Romans called the substance *natrium,* the word from which the symbol "Na" has been derived.

Sodium is most familiar to us undoubtedly as one of the two elements found in the molecule of sodium chloride, or table salt ($NaCl$). It has been estimated that the world's lakes and oceans contain enough dissolved sodium chloride to cover the land area of the earth to a depth of 400 feet. Vast surface and subterranean deposits, left by the evaporation of ancient seas, add to the reserve. (See the article Where Salt Comes From, in Volume 5.)

For many years sodium was extracted only by the electrolysis of sodium hydroxide. However, increased demand led to the development of an electrolytic process that uses common sodium chloride as a raw material. Molten sodium chloride, with sodium carbonate added to reduce the melting point, is put into a vessel that contains a carbon anode and an iron cathode. When current is passed through the molten salt, sodium is drawn off into a tank; chlorine, which is a valuable gas, is led out through a pipe.

Sodium production soared to a record high of 250,000,000 pounds in a recent

Hamilton Wright

Ore containing sodium nitrate, or Chile saltpeter, in freight cars on the outskirts of Pedro de Valdivia, Chile. This ore has been extracted from extensive beds lying in the northern part of the country. Sodium nitrate serves various useful purposes. It is an excellent fertilizer; it is particularly prized for its quick-acting qualities. It sometimes serves as an ingredient of the batch, or mixture, from which glass is made. Sodium nitrate is also used in the curing of meat products.

Left: beryl ore, from which beryllium is extracted. Below: an X-ray tube, with a base insert of pure beryllium metal (dark area of circular section), which serves as a window through which concentrated X rays can pass.

The Beryllium Corp

year. Modern technology is continually finding new uses for the metal. Sodium and its alloys with mercury and potassium serve the organic chemical industry as powerful reducing and dehydrating agents. Millions of pounds of lead-sodium alloy are consumed annually in treating ethyl bromide so as to produce tetraethyl lead, $Pb(C_2H_5)_4$, an antiknock agent for gasoline. Sodium is sometimes used as a lightweight conductor for exceptionally heavy electric currents of up to 4,000 amperes.

The metal is one of the best conductors of heat. Because of this property and also because of sodium's low melting point, engineers use the metal and its alloys with potassium as a circulating fluid in the heat exchangers of atomic-power plants. (See Index, under Heat exchangers.)

Among the numerous useful sodium compounds employed in the chemical industries, sodium hydroxide (NaOH) holds the most important place. Known as lye, soda lye or caustic soda, it is the strongest of the commonly used bases. Over a million tons of sodium hydroxide are used every year in the manufacture of rayon, soap, dyes and other chemicals, and also in mercerizing cotton, processing wood pulp for paper and removing sulfur from petroleum. Sodium bicarbonate, or baking soda ($NaHCO_3$), is a source of carbon dioxide gas in effervescent mixtures such as baking powder and fire-extinguishing fluids. Sodium cyanide (NaCN) is used to refine gold and silver, caseharden steel, electroplate metals and fumigate citrus trees. The photographic fixing agent known as hypo is sodium thiosulfate ($Na_2S_2O_3 \cdot 5H_2O$).

**Potassium is even more reactive than sodium**

Potassium (K) is even more active chemically than sodium. It sometimes ignites spontaneously in air, burning with a dull flame; we have already noted how it reacts with water. Potassium atoms have so slight a hold on the outermost electron that, when subjected to heat or light energy, clouds of electrons are emitted. Photoelectric cells are often lined with a thin film of potassium in order to take advantage of this property. A soft, waxy, whitish metal, potassium looks a great deal like sodium. It is the second lightest of the metals, with a specific gravity of 0.87. It melts at only 62.3° C. and boils at 760° C., emitting blue-green vapor. The chemical symbol for potassium — K — comes from the Low Latin form *kalium,* which is derived from the same Arabic word as "alkali."

Potassium is found in the cells and blood plasma of animals; it is essential for normal heart function and nervous conduction. It is almost as abundant as sodium in the earth's crust, making up 2.58 per cent of its weight. Potassium is combined with other elements in various micas and feldspars. The orthoclase type of feldspar, containing potassium, aluminum and silica, is abundant in many areas. When it reacts with the carbon dioxide in the atmosphere and water in the soil, one of the products is potassium carbonate ($K_2CO_3$), an essential plant food. This is one of several carbonates to which the name "potash" is given.

Only about 2.46 per cent of the salts in the ocean contain potassium. However, gigantic deposits of potassium salts have been left by evaporation; they are found on and below the surface of the earth. The principal source of these salts are the deep and ancient mines of Strassfurt, Germany, and the deposits near Carlsbad, New Mexico, and Searles Lake, California.

For thousands of years potassium carbonate and potassium hydroxide (KOH) have served in making soft soaps and hard glass. Saltpeter, or potassium nitrate ($KNO_3$), is made by boiling Chile saltpeter (sodium nitrate) with potassium chloride; in gunpowder and fireworks mixtures it provides a source of oxygen. Potassium chlorate ($KClO_3$) is used in quick-burning, detonating powders for percussion caps. About 50,000 pounds of pure potassium metal are produced annually by the electrolysis of fused potassium hydroxide; most of this is used in alloys with sodium of mercury.

### Rubidium and cesium

Rubidium (Rb) and cesium (Cs) were discovered in the early 1860's by two German chemists, Robert W. Bunsen and Gustav R. Kirchhoff, using the spectroscope. This instrument splits up the light emitted by luminous gases into various colors, representing different wave lengths. Since each element emits a different combination of wave lengths of light when heated, the spectroscope helps identify it.

Bunsen and Kirchhoff were analyzing the spectrum of vapor derived from mineral water in 1860 when they found two bright blue lines that they had never seen before. They reasoned that these lines must be due to a new element; they called it cesium, because of the color of the lines in its spectrum. (The Latin word *caesius* means "bluish gray.") A year later a second new element was revealed as the two scientists were examining the spectrum of the mineral lepidolite; its "signature" was two dark red lines. This element was given the name "rubidium" (from the Latin word *rubidus,* meaning "red"). Rubidium was first isolated in 1863; cesium in 1882.

Rubidium has a melting point of 38.5° C. and a boiling point of 700° C. The melting point of cesium is 28.5° C.; its boiling point, 670° C. Both metals are comparatively rare; the annual world production of each is measured in tens of pounds rather than in tons. A rare silicate mineral known as pollucite is the chief source of cesium. Rubidium is derived from the minerals lepidolite and carnallite; traces are found in certain plants, particularly coffee, tea and tobacco plants.

### Industrial applications of rubidium and cesium

Rubidium and cesium have few industrial applications. Because of their effectiveness in combining with oxygen, small quantities of these metals are sometimes added to electron tubes to remove the last traces of oxygen and thus keep the filaments from burning. Rubidium, cesium and their oxides are used in electron-tube filaments and in photoelectric cells as heat-sensitive or light-sensitive emitters of electrons.

The last of the alkali elements in the periodic table was discovered by Marguerite Perey of the Curie Institute in Paris. To honor her and her country for her achievement the metal received the name "francium." This metal is a radiation product of the element actinium and has a half life of twenty-one minutes. (The half life of an element is the time required for half the atoms in a given quantity to disintegrate.) Since francium has been obtained only in very minute quantities, very little is known about its chemical and physical properties except that it gives typical alkali-metal reactions.

## The Alkaline-Earth Metals

The alkaline-earth metals are beryllium, magnesium, calcium, strontium, barium and radium; they belong to Group IIa of the periodic table. They are called alkaline metals because they have alkaline, or basic, properties. The name "earths" was used by medieval alchemists to denote certain insoluble substances. The second part of the name "alkaline earth" indicates that the metals of this group have a low degree of solubility as well as other "earthy" characteristics. These metals are not so reactive as the alkali metals; they are harder.

### Beryllium, least reactive of alkaline-earth metals

The first of the alkaline-earth metals in the periodic table is beryllium (Be). It was discovered in the mineral beryl in 1798 by the French chemist Louis-Nicolas Vauquelin; thirty years later it was isolated by the Frenchman Antoine Bussy and the German Friedrich Woehler. They found that the salts prepared from the oxide of the new element had a sweetish taste; therefore they called the element glucinum (from the Greek word *glykys,* meaning "sweet"). This name was not universally accepted; many chemists preferred to call the element beryllium, since it had originally been discovered in the mineral beryl. In 1949,

Truck body of magnesium metal. The metal is light, and unusually thick sheets can be used. As a result the sheets are so rigid that they do not need a supporting framework.

Dow Chemical Co.

Aluminum Co. of America

Wing spar of aluminum, alloyed with magnesium, on its way to a press for further forging. Such alloys are invaluable in the construction of aircraft.

"beryllium" was adopted as the official name by the International Union of Chemistry.

Pure beryllium is a silvery gray metal (its specific gravity is 1.8). It is quite hard and brittle when cold, but can be rolled at high temperatures. At 1,278° C., beryllium melts; the liquid metal boils at 2,970° C. Beryllium is the least reactive of the alkaline-earth metals and does not readily tarnish in air. As a matter of fact, it bears more resemblance in its properties to aluminum than to the heavier metals in its own group.

The principal source of beryllium is the mineral beryl. This mineral is not so well known, perhaps, as its two gem varieties, emerald and aquamarine. Much of the world's supply comes from Brazil and Southern Rhodesia; the United States, South-West Africa, Madagascar and India are also important producers.

A great deal of beryllium is used in alloys with various metals; it makes them harder, tougher and more resistant to corrosion. Copper alloyed with 1 to 2.75 per cent of beryllium and traces of nickel, cobalt and iron can be heat-treated to yield a bronze that will withstand stresses of up to 200,000 pounds per square inch. A nickel alloy containing up to 2.2 per cent beryllium can be made even harder and stronger than beryllium bronze. It is used for parts that must withstand tremendous stress at high temperatures, such as aircraft engine parts and diamond-core drills.

Beryllium absorbs X rays to a lesser extent than any other metal; for this reason it is used for the windows of X-ray tubes. The beryllium window covers a small opening in the wall of the tube; the concentrated X-ray beam is directed through this opening and the beryllium cover.

### Magnesium, the lightest structural metal

Four of the six alkaline-earth metals — magnesium, calcium, strontium and barium — were isolated by Sir Humphry Davy in 1808. Until that time various oxides known as alkaline earths — lime, magnesia, baryta and strontia — were supposed to be true elements, as much so as gold and silver and copper. Davy had already shown that many supposedly simple elements were really chemical compounds, from which oxygen could be driven off. He was convinced that in the alkaline earths, oxygen held its partner too tenaciously to be separated by the ordinary chemical methods of that day. By means of an electric current, Davy succeeded in liberating calcium from its oxide, lime (CaO); using the same method, he also freed magnesium, barium and strontium from their oxides.

Magnesium (Mg) takes its name from magnesia, the alkaline earth from which it is derived. This mineral was called *Magnesia lithos,* or Magnesian stone, by the ancient Greeks because it was found in Magnesia, a district in Thessaly. Magnesium is a beautiful white metal with a melting point of 651° C. and a boiling point of 1,107° C. It is the only metal of the alkali and alkaline-earth groups with important structural and engineering uses. It is the lightest of all the structural metals; its specific gravity — 1.74 — is about one third lower than that of aluminum.

Magnesium is an important constituent of the green pigment chlorophyll, which enables plants to manufacture food, using the energy of the sun, in the process of photosynthesis. (See Index.) Chlorophyll contains a single central atom of magnesium, around which many atoms of carbon and other elements are linked.

About 2.08 per cent of the earth's crust is magnesium. The most common mineral sources of the element are magnesite and dolomite. A wide variety of silicate minerals contain magnesite; they include soapstone, talc, asbestos, serpentine and meerschaum (used in making fine, cool-smoking pipes). Dolomite is abundant in the South Tirolese Alps — the Dolomites.

Most of the magnesium produced today is derived from ocean water and brine wells. Magnesium salts make up almost 16 per cent of the mineral matter in ocean water and are even more concentrated in deep brine wells, such as those in Michigan. In the Dow process, applied in a vast plant at Freeport, Texas, 260,000 gallons of water are treated to yield a single ton of metal; the cheapness of the raw materials makes the method economical. Oyster shells are heated in kilns to form lime. This is then slaked with water, and the slaked lime is added to sea water, causing magnesium hydroxide to precipitate. This hydroxide is treated with hydrochloric acid and converted into magnesium chloride. Magnesium is then obtained by the electrolysis of the fused chloride.

Magnesium is an important industrial metal. During World War II, its production underwent a tremendous expansion, reaching a peak of 283,000 tons annually. In a recent year, world production was about one fourth this figure.

Pure magnesium is rather weak; because of its high reactivity, it tends to corrode in moist air. But alloys of magnesium with aluminum, zinc and manganese are extremely strong and hard and resist corrosion effectively; at the same time they retain the lightness of the pure metal. Magnesium alloys can be cast, drawn, forged and fabricated in every way. Because of the unusual combination of hardness, strength, resistance to corrosion and lightness, these alloys have become invaluable in the construction of aircraft. It is estimated that some large bombers in World War II contained more than a ton of magnesium alloy; it was used for propeller blades, engine parts, fuselages, wings, landing gears and other parts.

When magnesium is added to aluminum alloys, it greatly increases their strength and stability. Duralumin is such an alloy; it consists of 95.5 parts of aluminum to 3 parts of copper, 1 part of manganese and 0.5 part of magnesium. Duralumin is very light but almost as strong as steel; it is used in airplanes, boats, trains, trucks and other transportation equipment.

Magnesium is one of the less active alkaline-earth metals; but in the finely divided state (either as a powder or a wire), it can easily be ignited. It burns with a blinding white brilliance in air. Magnesium powders were once used for flashlight photography; the metal still serves for flares, incendiary bombs and fireworks. Its affinity for oxygen makes magnesium an ideal agent for reducing other active metals from their ores.

Certain magnesium compounds are used as medicines. Magnesia (magnesium oxide, $MgO$) is an ingredient in tooth powder; it helps to whiten teeth. Suspended in water, magnesia is known as milk of magnesia, a familiar antacid and laxative. The cathartic Epsom salts is magnesium sulfate heptahydrate ($MgSO_4 \cdot 7H_2O$).

U. S. Steel Corp.

Power shovels loading limestone on trucks at Iron Mountain, Utah. Limestone, which is mostly calcium carbonate, is very important in steel manufacture. Under intense heat limestone fuses with impurities in iron or steel and forms slag, which is then removed.

**Calcium is abundant in the earth's crust**

The alkaline-earth metal calcium (Ca) is the fifth most abundant element in nature, forming 3.65 per cent of the earth's crust. Its name is derived from the Latin word *calx,* meaning "lime." As far back as the days of ancient Egypt, lime (calcium oxide, CaO) was one of the ingredients of the building material called mortar.

Calcium is pale yellow in color; it melts at 842° C. and boils at 1,240° C. Its chemical activity and weakness prevent its being used as a structural metal. It is attacked by moisture in the atmosphere but is fairly stable in dry air. When heated in air, it burns vigorously, combining with both the oxygen and nitrogen in the atmosphere.

Calcium is an important constituent of bones and teeth. The compound calcium carbonate ($CaCO_3$) occurs widely in nature. Limestone consists mostly of this compound; so does marble, which is a crystalline form of limestone. Coral, pearls, egg shells and sea shells are made up chiefly of calcium carbonate. Calcium sulfate ($CaSO_4$) is another well-known compound of calcium; gypsum and alabaster are varieties of it. Calcium phosphate, $Ca_3(PO_4)_2$, is an important fertilizer.

Chemists prefer calcium to sodium to remove the last traces of water from organic liquids. Both metals dehydrate (remove water from) such liquids by reacting with the water they contain to form hydroxides (combinations of an element with hydrogen and oxygen). When sodium hydroxide is formed, it is so strong and so soluble that it may interfere chemically with the product that is being purified. Calcium hydroxide is weaker and less soluble; therefore it is less troublesome.

Calcium is used in certain metallurgical processes. Lead-bearing alloys usually contain a small percentage of calcium, which makes them harder. The metal also serves to remove oxygen from copper, nickel and stainless steel.

Millions of tons of limestone — a rock consisting chiefly of calcium carbonate, as we have seen — are used every year by the steel industry. Under intense heat the limestone fuses with various impurities in iron or steel and carries them off in the form of slag. Limestone also serves in the production of lime, employed in great quantities in the construction and chemical industries. Lime is an important ingredient in plaster, cement and mortar. When lime is melted with carbon in an electric furnace, it produces calcium carbide ($CaC_2$). This compound reacts with water to form acetylene ($C_2H_2$), used in welding. When calcium carbide reacts with nitrogen, it forms calcium cyanamide ($CaCN_2$), an excellent fertilizer and an important source of nitrogen compounds.

**Compounds of barium may be poisons or medicines**

The name of the alkaline-earth metal "barium" comes from the Greek word *barys:* "heavy." Since its specific gravity is only 3.5, barium is not among the heavier metals. The adjective really applies to the most abundant barium-containing mineral, barite (originally called barytes, or heavy spar). Most of the world's supply of this mineral comes from Canada, Germany and the United States deposits in Georgia and Tennessee.

Barium is a silvery white metal, slightly harder than lead; it melts at 850° C. and boils at about 1,140° C. It is so reactive that when it is in the form of finely divided powder, it may ignite spontaneously in air. The affinity of barium for oxygen makes it useful as a getter — that is, a substance that removes the smallest trace of oxygen from electron tubes. Since barium is an excellent emitter of electrons when heated, an alloy of barium and nickel is sometimes used for the filaments of these tubes. Barium is often alloyed with bearing metals; it has a hardening effect.

All soluble compounds of barium are highly poisonous, causing muscle spasm and increasing the force of the heart beat. One of these compounds, barium carbonate ($BaCO_3$), is an effective rat poison. Barium fluosilicate ($BaSiF_6$) is used as an insecticidal spray for plants.

Certain insoluble barium compounds are employed in medicine. Stokes-Adams disease, in which the heartbeat is slow and weak, is sometimes treated with barium salts; they strengthen the response of the heart muscle. Barium sulfate ($BaSO_4$) is extremely opaque to X rays. Mixed with flour, sugar, cocoa and water, it is drunk by patients whose intestines are to be X-rayed. Since barium sulfate is insoluble, it cannot be absorbed by the body and is therefore harmless.

Barium sulfate is used as a permanent white pigment, under the name of "blanc fixe"; it also serves as a filler in paper, rubber and linoleum. Certain barium salts produce a beautiful green color in firework displays. Barium sulfide (BaS) is strongly phosphorescent; it is used as an ingredient in luminous paints.

**Strontium, named after Strontian, Scotland**

Strontium (Sr) is a hard, silvery white metal; it sometimes shows a yellowish tint. It has a specific gravity of 2.54; its melting point is 800° C. and its boiling point 1,150° C. Strontium tarnishes quickly in air; it catches fire when it is subjected to friction. The principal ores from which the metal is derived are celestite and strontianite. The name "strontium" comes from the town of Strontian, in Scotland, where the mineral strontianite was first discovered.

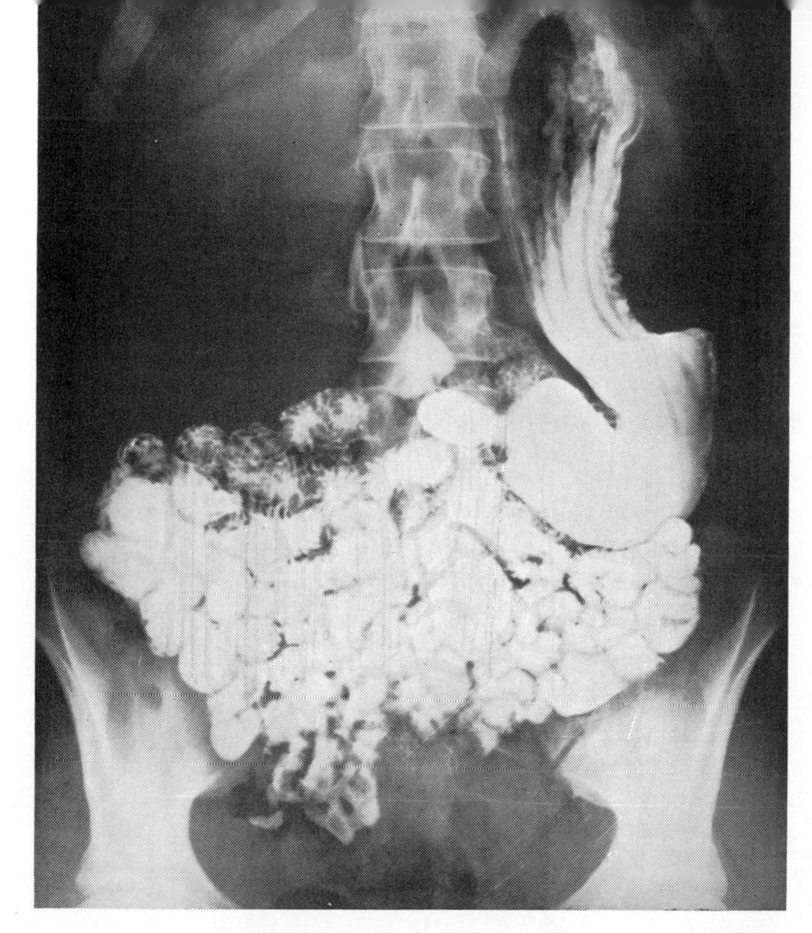

Radiograph of the digestive tract of a patient who has swallowed barium sulfate, mixed with other ingredients. Barium sulfate is extremely opaque to X rays; therefore, when the patient's digestive tract is X-rayed, the portion containing the barium sulfate is clearly indicated.

The compounds of strontium serve certain useful purposes. Most of its salts give a brilliant red color when added to flame; for this reason they are used to produce red light in fireworks, signal lights and flares. Quantities of strontium hydroxide, $Sr(OH)_2$, are employed to remove sugar from molasses. The crystalline powder called strontium lactate, $Sr(C_3H_5O_3)_2 \cdot 3H_2O$, is used as a mild antiseptic and analgesic (pain reliever). Strontium bromide ($SrBr_2 \cdot 6H_2O$) forms colorless crystals that can easily be dissolved in water; this compound has proved useful in the treatment of epilepsy, convulsions, hysteria and insomnia.

### Radium — a long-lived radioactive element

In another chapter (The Road to Modern Alchemy, Volume 8) we tell the fascinating story of the discovery of radium by the French husband-and-wife team of Pierre and Marie Curie. Radium is really one stage in the transformation of the radioactive element uranium into a nonradioactive form of lead — a chain of radioactive changes known as the uranium series.

The element uranium, of atomic weight 238, has a half life of about 5,000,000,000 years. An atom of uranium, in breaking up, emits an alpha particle (the nucleus of a helium atom) and becomes uranium $X_1$, of atomic weight 234. When an atom of uranium $X_1$ breaks up, it shoots off a beta particle (an electron) and becomes uranium $X_2$. After several more transformations, we reach the radium stage of the uranium series. Radium, atomic weight 226, has a half life of 1,622 years. When a radium atom disintegrates, it emits an alpha particle and becomes radon, or radium emanation, a very heavy gas. The half life of radon is short — only 3.825 days. There follows a series of transformations of short-lived isotopes, until we come to a radioactive form of lead, of atomic weight 210. Finally, after a few more steps, we end with a stable (nonradioactive) form of lead, atomic weight 206.

Radium, with its half life of 1,622 years, ranks high among the long-lived radioactive elements. In its pure form it is a shiny white metal, with a melting point of 700° C. and a boiling point of 1,140° C.

It is extremely reactive; it tarnishes rapidly in air and causes water to decompose. This metal is generally available only in compounds, mixed with a barium salt.

Radium is always found in ores of uranium. The chief ores are pitchblende, sometimes known as uraninite, and carnotite. Uranium ores are found in many areas, including Canada, the Belgian Congo, the Union of South Africa, Czechoslovakia and several western regions of the United States. Nowadays these ores are chiefly sought after not for their radium content but for their uranium, used in atomic piles. Radium will always be rare and costly because it is found in such minute amount in its ores. It is estimated that seven tons of uranium will yield one gram (0.035 ounce) of radium.

Overexposure to radium is extremely dangerous, since this highly radioactive element can destroy living tissue. Before the nature of the element was fully understood, many workers in the field of radium research suffered severe injury or death because of their contact with the dangerous substance. The destructiveness of radium has made it valuable in the treatment of cancerous growths, particularly because abnormal cells are even more sensitive to radiation than normal cells.

Radium has been applied in various ways in the treatment of cancer. It has been set at a distance from the body of the patient. Tubes of radium have been fixed in a paste and this has been applied to the surface of the body. Capsules containing the metal have been put in body cavities at or near the site of the cancer. Radium has been put in needles or minute capsules and inserted directly within the cancerous tissue. At present the metal is being replaced in the treatment of cancer by synthetic radioisotopes and high-voltage X rays. (See the articles The Wonder-working X Rays, Volume 7; Atomic Medicines, Volume 10.)

Radium has also found various uses in industry. It has been mixed with a paste of zinc sulfide ($ZnS$) to form a luminous paint. This has been applied to the hands or dials of watches or to other surfaces that are to be made self-luminous in the dark. Radium has also been used to detect flaws in metal castings and other materials by radiographs. In these fields, too, substitutes have proved satisfactory. Luminous paints are now often made of fluorescent materials activated by ultraviolet radiation; X rays and ultrasonic devices are used increasingly to test industrial materials.

*See also Vol. 10, p. 279:* "Elements."

American Cancer Society

Container with radium capsules used in treating cancer. The container, with its lead-lined cover and walls, protects technicians who carry the capsules from room to room.

This type of vegetation flourished in the Carboniferous period, which had a warm, moist climate.

# CLIMATES OF THE PAST

## Methods of Analysis and Results

BY CHARLES MERRICK NEVIN

CLIMATE plays a vital part in the drama of life. Every species of animal and plant develops best under certain limited conditions of sunshine, humidity and temperature. It is reasonable to assume that living things in the past must have been influenced by climate in the same way.

We know something of the living things that dwelt in prehistoric times through the fossil record. There are many different kinds of fossils. They may be the remains of animals and plants, preserved through some accident of nature. They may represent the imprints of once living things — plants or animals — in rock. The track of a worm clearly marked on sandstone or shale is a fossil; so is a skeleton of a mammoth, frozen with flesh intact; so are insects preserved in amber.

The fossil record of life goes back hundreds of millions of years. The fact that these records are more or less continuous for that period of time shows that climate must have been *comparatively* uniform. There have been numerous variations, indeed, as we shall see, but these have been within a relatively moderate range, compared with temperatures found elsewhere in the universe. Life has not been subjected to such extremely high temperatures as exist in the sun, nor to extremely low temperatures, such as space travelers in the future will undoubtedly find on the moon when it is in darkness.

To account for the comparative uniformity of climate, we must assume that from a very early time, the earth has had an atmosphere to shield its living things from excessive amounts of heat and also to prevent heat from escaping freely into outer space. We must also assume that since the time when life appeared on the earth, the

National Coal Association

**Fern-leaf fossil, formed by a plant that grew in the Carboniferous period. Ferns are a sign of a damp climate.**

sun has been emitting radiation at about the same rate as today. For plant and animal life can exist on the earth only because of the warmth and light supplied by the sun. Any radical change in these respects would have catastrophic effects upon all living things.*

Yet, as we have already noted, if climate has been uniform, *on the whole,* there have been innumerable variations in its patterns. There have been periods when high temperatures prevailed and other periods when ice sheets covered the earth. We can prove that such conditions occurred by a study of rock formations, by the fossil records of animals and plants and by other methods of analysis. The more pronounced departures from the normal pattern of climate are of particular interest. It was during these critical periods that the pulse of evolution was quickened and many new forms of life developed.

**Comparing past and present climates**

There are various ways of analyzing the nature and extent of climatic variations. For one thing, if we study present-day climates and their effects on living and nonliving things, we shall have a key to the analysis of the climates of the past. If we see that rocks and animal and plant life are affected in a certain way by this or that type of climate, we may assume that the same type of climate must have produced the same effects in past ages. This is the method of analogy, or comparison. It has to be used with care, because there are numerous exceptions; but it is decidedly helpful.

We assume that certain types of climates existing today have had their counterparts in the past. Among the most significant of these climates are (1) the dry climate-desert type; (2) warm-moist climate; and (3) cold climates.

*The dry climate-desert type.* In a desert climate, wind is the chief agent that transports particles of sand and rock. We should expect to find the smaller grains of sand rounded, because of the action of the wind. This causes stones to rub constantly against each other. The fact that water is so scarce prevents the formation of a protecting film of water, which usually prevents the rounding of small particles in sediments laid down by the work of water. Another effect of dry climate is the frosting of sand grains, resulting from the continual bumping together of the dry surfaces. This contrasts sharply with water-laid deposits, in which the surfaces of sand grains are usually bright and glistening. Still another effect of dry climate is the formation called cross-bedding, in which the different strata, or layers, show a curious wedgelike effect, characteristic of sand dunes. Cross-bedding develops over large areas in climates that are dry.

If thick deposits of salt and gypsum are present in an area, it is likely that they were laid down in an arid climate. We may assume that the rate of evaporation exceeded the rate at which the deposits were precipitated.

Red soil deposits were formerly thought to be evidence of an arid or semiarid climate. This is true for many such deposits. However, very deep-red soils are now being produced by the weathering of certain rocks under a covering of tropical and subtropical vegetation. If the soil that results is transported and deposited in other re-

* In the last two pages of the article The Sun, Our Star, in Volume 3, we list some of the ways in which solar energy affects life.

Illinois State Geological Survey

**Cross-bedding in sandstone of the lower Pennsylvanian period. This effect is usually caused by a dry climate.**

gions, under certain conditions, a red-bed sediment will be formed. Obviously, this particular sediment will have been due to climatic conditions very different from those of an arid environment.

*Warm-moist climate.* The chemical weathering of rocks and minerals is at its maximum in a warm, moist climate. For example, under such conditions feldspar, a common mineral in the earth's crust, is changed to clay. In arid and frigid climates, the feldspar remains unaltered for a much longer period. This distinction serves as a valuable guide.

Vegetation flourishes in a warm, moist climate. As a result, the colors of sediments formed under such conditions are usually various shades of gray to black. Coal beds grade into deposits that give every indication of plenty of moisture. While growth is very rapid in a warm, moist climate, decay is also at its maximum. Hence, though growth is slower in a cold climate, the arrest of decay may cause coal deposits to accumulate more rapidly.

Fossilized forms of warmth-loving types of life give a significant clue to the existence of a warm climate. Among such organisms are butterflies, reptiles, earthworms, certain highly specialized mammals, sponges and snails, palm trees, cycads and tree ferns. However, this particular yardstick does not apply in every case. A warmth-loving species may adapt to slowly changing climatic conditions; ultimately it may come to flourish in an environment entirely different from the original one. For example, the elephants and rhinoceroses of today thrive particularly where it is warm. We know, however, that the mammoth (an elephant) and the woolly rhinoceros, both now extinct, flourished in a cold climate.

Thick beds of limestone formed by reef-building corals would seem to indicate a warm climate, because all present-day coral colonies live in warm seas. We must bear in mind, however, that all of the species of reef-building corals that made up the ancient limestones are now extinct. We cannot therefore be positive that they were as definitely restricted to warm seas as their living descendants.

*Cold climates.* The existence of glacial deposits are reliable indicators of a cold climate. In such deposits we find rock fragments existing in all sizes and shapes, which have not been sorted out. The boulders, too, are striated (marked with striae, or grooves); this provides

**White clay formed by the chemical weathering of feldspar. Such weathering occurs in warm, moist climates.**

British Information Services

**Great numbers of peat bogs were laid down in warm, moist climates in the Carboniferous period. One bog would often be formed atop another, as shown here. Later, the peat was transformed into coal. Actually, peat represents the preliminary stage in the formation of coal.**

evidence of the glacial masses that deeply scored the rocks as they passed over the land. Likewise, the finding of fossilized forms of certain hardy vegetation, such as *Glossopteris,* indicates the existence of a climate such as we find today on the frozen tundras of Siberia.

It is clear that this method of analogy must be applied with caution. Many of the climatic yardsticks we have mentioned are quite uncertain when considered separately. However, when we combine them with other types of evidence, we can obtain a fairly reliable picture of past climates.

### Other methods for studying past climates

Vegetation may serve as an indication of temperature changes. The advances and retreats of the ice sheets in various areas are reflected in changes in forest cover — changes denoted by the different types of pollen that have been preserved. Growth rings in ancient wood clearly mark former wet and dry, cold and warm seasons.

To make an analysis of past climates, it is often very helpful to be able to date various organic remains more or less precisely. The method of radiocarbon dating has proved useful in this connection; it is discussed elsewhere (see Index).

The laying down of sediment in the form of varves may also serve as a criterion. Varves are alternating layers of finer and coarser sediment, formed on the beds of lakes in a seasonal climate. Fine, thin layers are laid down when the lake is frozen over; coarser layers, when spring streams carry in loose materials after the thaw has set in. These alternating strata indicate seasons and perhaps longer cycles.

One method of past-climate analysis is based on the fact that the ice sheets of Greenland and the Antarctic consist of annual accumulations of snow. Scientists taking part in the International Geophysical Year program (see Index) developed a deep-drilling technique for bringing up ice cores from ice sheets. These cores show the annual snow accumulations and provide clues to the climates of the polar regions in past ages.

A fascinating new technique is the use of the so-called "geologic thermometer" developed by the famous American chemist Harold C. Urey, of the University of Chicago. It is based on the analysis of isotopes of oxygen. The isotopes of a chemical element represent the different forms in which the element may occur. All these forms have similar chemical properties, but they differ in atomic weight (also called mass number). For example, the three chief isotopes of oxygen are oxygen 16, 17 and 18 (also written $O^{16}$, $O^{17}$ and $O^{18}$ — that is, oxygen with atomic weight 16, 17 and 18, respectively. (See Index under Atomic weight.)

Urey had made a study of the oxygen isotopes in water. The water molecule is made up of two hydrogen atoms and one oxygen atom; it has the chemical formula $H_2O$. Over 99.7 per cent of the oxygen atoms in the water molecules in a glass of water are $O^{16}$ — oxygen with atomic weight 16. But certain water molecules are made up of hydrogen plus $O^{17}$, or hydrogen plus $O^{18}$.*

Urey noted, in a lecture he gave in Switzerland in December 1946, that when a glassful of water evaporates, the three isotopes of oxygen — $O^{16}$, $O^{17}$ and $O^{18}$ — do not all leave the liquid at the same time. The evaporation process will carry off a slightly higher proportion of the $O^{16}$. As a result the water will have a slightly greater concentration of the heavier isotopes — $O^{17}$ and $O^{18}$ (particularly $O^{18}$). Urey noted that the water in the oceans had been subjected to the process of evaporation longer than fresh water. Hence they should have a somewhat higher proportion of the heavier isotopes of oxygen than fresh water.

A Swiss scientist, Paul Niggli, observed that if this were true, it should give us valuable information about carbonate deposits, such as limestone. Carbonates all contain oxygen in their molecules; if they had been precipitated in salt water they would have a higher percentage of $O^{17}$ and $O^{18}$ atoms than if they had originated in fresh water.

Urey set out to analyze the difference in the isotope ratios in salt-water and fresh-water carbonates. He found that the comparative abundance of the oxygen isotopes other than $O^{16}$ in the carbonate would increase with decrease in the temperature of the water at the time the carbonate was deposited. He now realized that, as he put it, he had a "geologic thermometer" in his hands. By analyzing the temperatures at which carbonate fossils had been laid down, he would be able to find out something about the climate that prevailed during that period.

Urey and his associates at the University of Chicago made their first analysis of "fossil temperatures" in 1950. They chose for this purpose the fossil of a belemnite, a creature resembling the modern squids. It lived about 140,000,000 or 150,000,000 years ago in the shallow sea that covered what is now Scotland. In cross section, the fossil showed rings like the growth rings of trees. Urey and his fellow-researchers shaved off the concentric layers one by one and analyzed the ratios of the oxygen isotopes in each one. The analysis showed seasonal changes of 15°C to 21°C during the growth of the animal's skeleton. It revealed that this particular belemnite had been born in the summer, lived almost four years and then died in the spring.

The oxygen "geologic thermometer" is still in its infancy as a tool of science. It has been used, among other things, to analyze the variations in temperature during the latter part of the Age of Reptiles — the Upper Cretaceous period. The oxygen-isotope analysis of a large number of fossils from North America and Europe has shown that temperatures rose during the first part of the Upper Cretaceous and fell during the second part. This reinforces the commonly accepted belief that the great dinosaurs of the Age of Reptiles became extinct as a result of the cooling of the

* The hydrogen atoms in the water molecules may also differ in atomic weight. 99.98 per cent of water molecules have hydrogen atoms with atomic weight 1 ($H^1$). But in rare instances, the hydrogen atoms will be $H^2$ or $H^3$ isotopes. Water containing an unusually high proportion of the heavier hydrogen isotopes is called heavy water. Urey won the Nobel prize in chemistry for his researches in this field.

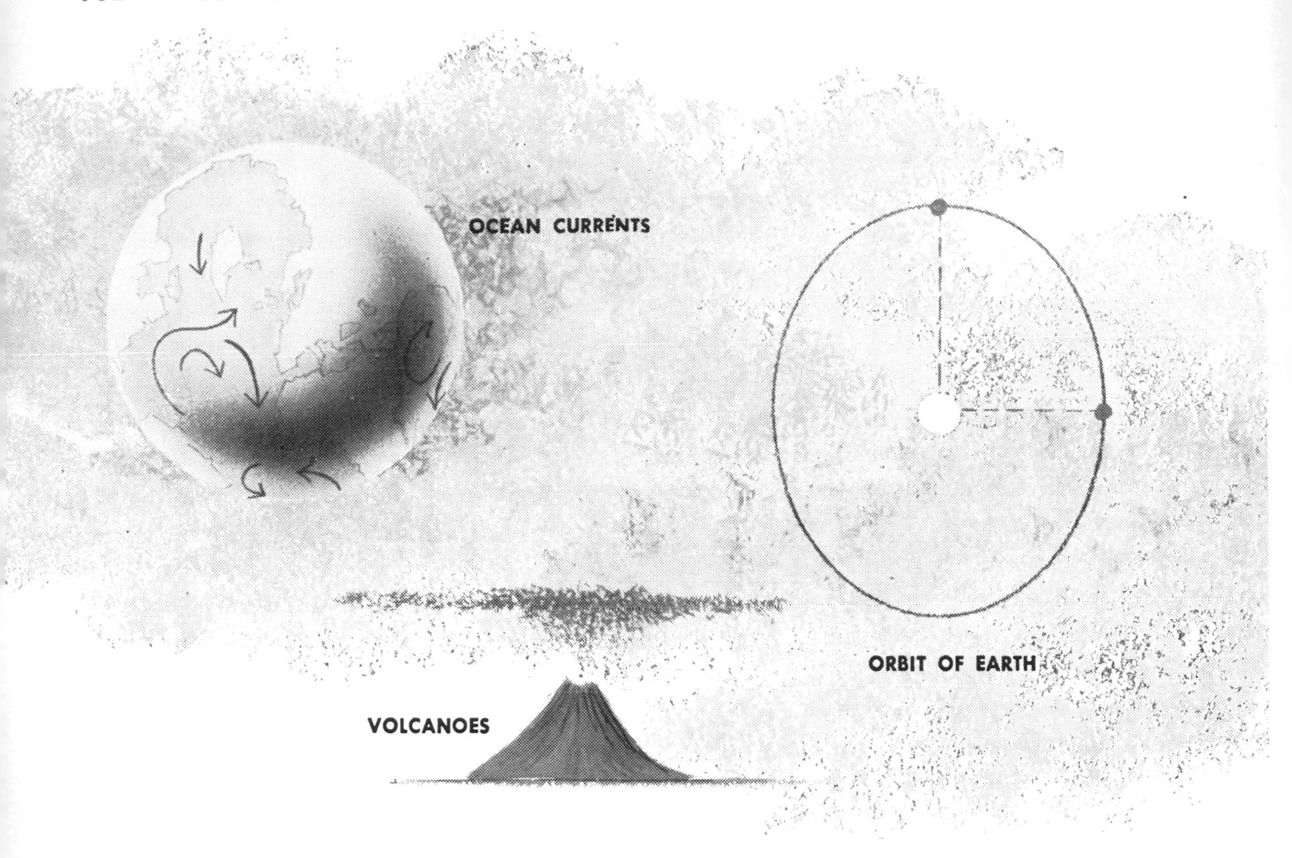


FACTORS THAT INFLUENCE CLIMATE

earth.* The temperature kept dropping on into the Pleistocene.

**The climates of the past**

What have these methods of analysis revealed about the climates of the past? They show, for one thing, that there have been periods during which a warm climate extended from pole to pole. As a matter of fact, there have been such warm periods during the greater part of the earth's history. In many, glaciers were unknown.

Without exception, warm climates were associated with low-lying land. The continental areas were greatly reduced as a result of flooding by shallow seas; broadly rounded hills, instead of the mountainous topography of today, were the rule. Immense oceans extended through wide channels from pole to pole. What is now New York had a subtropical climate; the climate of Greenland was warm to temperate.

Deserts have always existed. During the warm periods of the earth's history, however, they were greatly enlarged, extending from the equator far into the present temperate zones. As we have seen, different forms of life developed as the great climate changes came about. One of the most important occurred in this widespread desert environment — for the air-breathing vertebrates originated under these climatic conditions.

Scientists cannot account for a warm climate extending from pole to pole. The earth's axis is inclined about 23½ degrees to the plane of its orbit. As a result, the tropics receive more heat than the mid-

* For an excellent account of the oxygen-isotope geologic thermometer, see C. Emiliani's "Ancient Temperatures," in the Scientific American, February 1958, pp. 54-63.

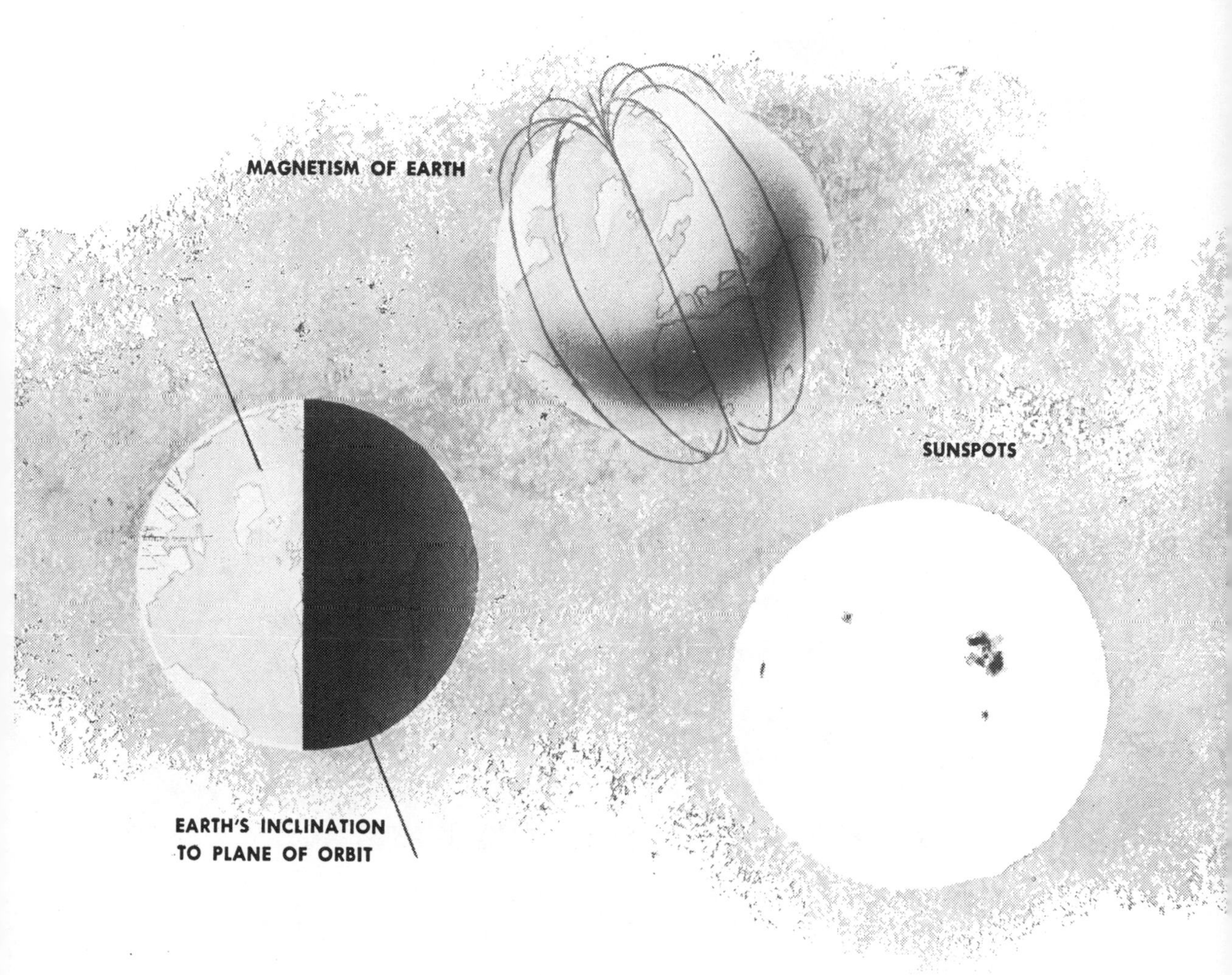


dle latitudes, while the middle latitudes receive more than the poles.* The inclination was probably the same in the past.

Warm periods have alternated with cold; there have been eight major ice ages. They started in Pre-Cambrian times and were spaced about 250,000,000 years apart. Each occurred following a period of mountain-building and the raising of the land into high continents. It seems likely that the land-building activities were an important factor in the coming of the glacial period. For one thing, they brought about an increase in snowfall. Moist air would be forced up mountain slopes. Condensing as it struck the colder air at the mountain top, it would form clouds and the clouds would release their moisture in the form of snow. Then, too, as the continents increased in area, the oceans became smaller and were divided into separate seas. This reduced or even put an end to the flow of warm ocean currents to the polar seas; these seas then cooled very rapidly and became frozen over. Some geologists hold that any such accumulation of ice in the polar regions would have a tremendous chilling effect upon the rest of the world.

There were other possible factors. It is thought that volcanic dust may have been one of them. Volcanic activity was at its peak during the mountain-building periods that preceded the different ice ages. The dust resulting from eruptions consisted of very small particles, which might well have floated about in the air for years after an eruption. They may have been plentiful enough to deflect a considerable portion of the sun's rays, so that the earth would be deprived of its heat.

* We explain why in the article The Planet Earth, in Volume 4.

U. S. Coast Guard

**Two Greenland glaciers merge as they travel to a fiord. Glaciers play an important part in climatic variations.**

It has also been suggested that the decrease in the quantity of carbon dioxide in the air * might have had something to do with bringing on the ice ages. Carbon dioxide serves as a blanket, absorbing some of the heat that is radiated away from the earth and thus preventing it from passing on into space. If, for some reason or other, there were a deficiency in the quantity of carbon dioxide, more heat than usual would be lost to outer space, and as a consequence the climate would become colder. Unfortunately, we have no way of knowing whether there was actually a deficiency in carbon dioxide in the atmosphere, in the period preceding the ice ages.

At one time, the so-called continental-drift theory was supposed to account for the coming of the ice ages. According to this theory, all the continents formed a single land mass in the Paleozoic era. In the Mesozoic, however, they started to break apart and to drift away. The positions of the poles shifted together with those of the land masses. At different times, the South Pole was to be found in various lands of the Southern Hemisphere (South America, South Africa, perhaps Australia), now tropical or mild in climate. The North Pole was farther south than it is now. The coming of the different ice ages, according to the continental-drift theory, would depend on where the poles happened to be. This theory has been largely abandoned.

Some authorities think that the succession of cold and warm eras may have been due to fluctuations in the radiation of the sun. According to one theory, as the heat output of the sun decreased, it would

* Carbon dioxide is found in the atmosphere at all times; it makes up about .03 per cent by volume, on the average.

cause temperatures on the earth to drop. There would be a slowing up of air circulation and less precipitation. The polar seas would freeze over, but no land icecaps would form because of the comparative lack of snowfall.

Suppose now that the heat output of the sun would begin to rise again, though not very much. Clouds would form; precipitation would increase, causing snow and ice to accumulate. Ice sheets would then form on the land. As solar radiation would increase still more in intensity, temperatures would rise notably on the earth. The ice sheet would melt and an interglacial period would result. Then the solar heat output would drop again, and the cycle would start anew. This theory is not universally accepted by any means. Some scientists object to its "hot sun-cold earth" phase.

Still other factors may have been at work. Variations in ocean circulation may have had some influence on long-term climatic changes. Changes in the earth's magnetic field may have influenced the climate, particularly since this field would have had an effect on the radiation from the sun. According to some authorities, the waxing and waning of glaciers has been affected by variations in sunspots.

Each of the major ice ages consisted of several stages, during which the ice fields alternately advanced and retreated. The last great ice age, that of the Pleistocene, consisted of four distinct phases. M. Milankovitch held that these ice ages within ice ages were due to lower summer temperatures. He pointed out that ice sheets in the higher latitudes could melt only during the summer; and if the summer was very cool, they might not melt at all. That would mean that the following year's accumulation of snow would be added to all of the previous year's accumulation and would result in the spreading of the ice sheet.

Milankovitch held that the cooler summers would be due to various astronomical factors. The first would be a lengthening and narrowing of the ellipse formed by the earth as it revolves around the sun. Then the earth may also find itself at aphelion — the part of its orbit when it is farthest from the sun — in the summer. Finally, as it wobbled around its axis, the earth may also be in a position at a somewhat smaller angle to the plane of its orbit during the summer, instead of at the average inclination of 23.5°. However, as the orbit would become shorter and wider, as the earth would reach aphelion in seasons other than the summer, or as it would become more inclined to the plane of its orbit again, the summers would become shorter but hotter, and the ice sheets would recede. Though Milankovitch's theory seems plausible enough, it has not been widely accepted by geologists.

Certain authorities attribute the waning and waxing of ice sheets in the Pleistocene to variations in snowfall. They argue that an ice sheet would press down on the land as it advanced, causing the land level to become lower. There would be less cloud formation and less snowfall.* As the land would sink, it would let in the sea. The glaciers would melt and they would gradually break up. Relieved from the pressure of the glaciers, the land would rise again. There would be more cloud formation and more snowfall. Glaciers would begin to form again; and as ice sheets would encroach upon the land, they would press down heavily upon it and the cycle would be renewed.

According to some authorities, the last of the four ice ages in the Pleistocene period came to an end only about 10,000 years ago; there are still traces of it in the icecaps of Greenland, the Antarctic and other places. Weather records and the analysis of glacier formation indicate a distinct warming trend up to about 1940. This trend has been slowed up since that time, however, and may even have stopped. Is another ice age coming? It is certainly not beyond the bounds of probability that the glaciers will again begin to advance over the earth's surface — in perhaps a hundred thousand years or so!

* As we have pointed out, mountains contribute to cloud formation and precipitation.

*See also Vol. 10, p. 271:* "Weather and Climate."

Three familiar kinds of moss plants. Left: *Mnium punctatum.* Above, left: Peat moss (*Sphagnum acutifolium*); right: *Polytrichum.*

Liverworts. Left: *Riccia.* Below: *Marchantia;* note the gemmae cups and the stalks bearing the plant's sex organs.

All photos, Hugh Spencer

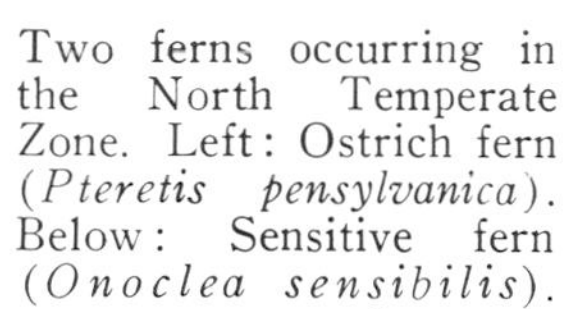

Two ferns occurring in the North Temperate Zone. Left: Ostrich fern (*Pteretis pensylvanica*). Below: Sensitive fern (*Onoclea sensibilis*).

All drawings, Gen. Bio. Supply House

# Mosses, Ferns and Their Allies

## Plants with Unlike Alternating Generations

by THOMAS GORDON LAWRENCE

THE mosses are among the most widely distributed and best known of all plants. A delicate, intensely green carpeting of moss is nearly always to be found in woodlands, along streams and in many meadows. Even in well kept lawns there are frequently glistening moss colonies, more or less shaded by grass blades. Moss often thrives in the cracks of cement walks and old stumps, damp fences and the bark of old trees. It sometimes grows on thatched roofs, which look as if they were covered with green velvet.

The mosses belong to the group of the plant kingdom known as the bryophytes ("moss plants," in Greek). This group also includes the plants called liverworts.

The moss plants with which we are most familiar develop from spores. Each spore gives rise to a green, branching filament, called a protonema, which looks much like a green alga. (See Index, under Algae.) After a time, the protonema produces buds; each bud grows into an erect shoot, with leaflike structures branching off from it. Instead of roots, the shoot has small threadlike bodies, called rhizoids, which draw water and mineral salts from the soil. The leaflike structures of the plant contain chlorophyll and are capable of photosynthesis: that is, manufacturing glucose from carbon dioxide and water in the presence of light.

In the course of time, sex organs develop at the tips of the shoots. The male sex organs, which produce sperms, are called antheridia (singular: antheridium); they are spherical or egg-shaped. The flask-shaped female organs are known as archegonia (singular: archegonium). A single egg is formed at the base of each female organ; in time a canal extends from the base to the open mouth of the organ. In some mosses the antheridia and archegonia are produced at the tip of the same shoot; in others, they are produced on entirely separate plants. Mosses that bear sex organs such as these are called gametophytes ("sex-cell plants").

Fertilization can take place only in water, which may be in the form of rain water, or running water or even a thin film of dew. As the antheridium becomes thoroughly wet, the sperms are released. They swim through the water by means of long whiplike structures called flagella and in due time make their way to the archegonium. Several sperms may go down the canal to the egg at the base of the female organ; but only one of the sperms fertilizes the egg.

Once the egg is fertilized, an entirely different generation is launched. A new plant, called a sporophyte ("spore plant") grows out of the tip of the old plant. This sporophyte consists of a foot, the base of which is implanted into the shoot of the gametophyte, and a long, thin stalk with a capsule, containing spores, at the end; it receives its nourishment from the old plant. When the capsule is mature, it splits open; as the plant sways in the wind or is shaken by animals, spores are thrown from the capsule like salt from a shaker.

Each spore gives rise to a new gametophyte, and the cycle begins anew. Thus there is what the botanist calls alternation of generations: from gametophyte to sporophyte, to gametophyte again and so on through countless generations of mosses.

There are many different kinds of mosses; they are all small plants, never standing more than a few inches high. Some are so low and hug the ground so closely that they look like green turf. The plume moss (genus *Hypnum*) and the fern moss (*Thuidium*), which looks somewhat like a midget fern, are more conspicuous. They thrive in shady woods, often completely concealing the logs and stones on which they grow.

Perhaps the most beautiful of all the mosses are *Bryum* and *Mnium,* with their broad, thin, translucent leaves. More than any of the others, these are the mosses that sparkle like gems when a shaft of sunlight finds its way to the ground through the forest canopy.

A few mosses are dark brown or black. The pincushion moss (*Leucobryum*), on the other hand, is whitish green; its cushionlike masses look ghostly at the bases of trees in damp woods. A moss that makes a satisfactory inhabitant for a home aquarium is the water moss (*Fontinalis*), which lives entirely submerged. In an aquarium it soon forms a dense underwater jungle, where young guppies can escape the eager jaws of their parents and other fish.

Some mosses grow in caves where the light is only $1/500$ as bright as full sunlight. Luminous moss (*Schistostega*) is found in dark places in woods; it reflects light rays from its lenslike cells. The greenish-golden glow of this curious plant may have been responsible for fairy tales concerning goblin gold. One species of moss grows at a depth of 180 feet in Lake Geneva, Switzerland.

The peat mosses (*Sphagnum*) are quite different from the others; they are far larger, for one thing. They grow in boggy places, especially in temperate and subarctic regions. A peat moss leaf is generally only one cell thick and contains many dead cells. These absorb great quantities of water and give a pallid appearance to the plant.

The bogs in which peat mosses flourish are called peat bogs; in addition to peat mosses, they often contain other mosses, as well as reeds, sedges and rushes. The tannins and other organic acids and the salts

Above: Leaves of spinulose shield fern (*Dryopteris spinulosa*). Right: Prothallus of the cinnamon fern (*Osmunda cinnamomea*).

Some young fronds of the Christmas fern (*Polystichum acrostichoides*).

Common bladder fern (*Cystopteris fragilis*).

Common polypody (*Polypodium vulgare*).

that are found in abundance in peat bogs serve as natural preservatives. Consequently, the remains of animals and trees buried in bogs for hundreds of years have been dug up in excellent condition. The peat bogs of Ireland are the most famous, perhaps; there are also extensive formations in Great Britain, Russia, Canada, Finland, Sweden and the United States.

Florists use dried peat moss as packing for flowers, since the leaves absorb and hold moisture. Peat moss that has been ground up is often added to the soil of gardens; not only does it hold water effectively but it prevents the soil from caking. Sterilized peat moss has been used for surgical dressings.

Dead plants in a peat bog sink to the bottom. As one layer piles up on another, the lower layers are gradually compressed and form the substance called peat. In this, the proportion of carbon to other chemical elements (about 60 per cent) is much greater than it was in the living plants. Peat is really the first step in nature's production of coal.

Peat is used chiefly as a fuel. To dig it out, the bog is first partly drained. The loosened surface is then removed, and rectangular sections of peat, called turfs, are cut with specially shaped spades or by machines. The turfs are stacked in the open air for drying — a process that takes about six weeks. Dried turfs are used as fuel in Ireland and other countries. Sometimes peat is pressed into the form of briquettes. These make better fuel than the turfs, since they are more compact and contain less moisture.

The liverworts, which are related to the mosses, are among the simplest of all land plants. Most of them have a simple type of plant body called a thallus; this is a flattened, ribbonlike structure with no stem, or leaves or roots. Rhizoids grow from the underside of the thallus; they anchor the plant and supply it with water and minerals. Many liverworts have small, cuplike growths, known as gemmae cups, on their upper surface. Budlike bodies, called gemmae, are produced in the cups;

Left: Maidenhair fern (*Adiantum*). Right: Christmas-fern leaves.

Common grape fern (*Botrychium obliquum*).

Boulder fern (*Dicksonia pilosiuscula*).

Below: Bracken fern (*Pteridium aquilinum*).

*Lycopodium complanatum,* a club moss. It looks like a very big moss plant.

Right: The glossy leaves of the shining club moss (*Lycopodium lucidulum*).

when the gemmae are separated from the plant, they are capable of growing into new thalli.

You will find thalloid liverworts, as they are called, in the French quarter of New Orleans; here they grow beneath the magnolia trees or along the edges of the walls — a solid, creeping greenness. You will see them, too, in the woods of Canada and the northern part of the United States.

One of the commonest and most interesting of the thalloid liverworts is *Marchantia.* As it grows, it splits again and again at the tip, so that presently it is as if green ribbons were growing in different directions from the original starting point. If the plant prospers, it soon sends up a group of what look like miniature crudely designed palm trees. These growths bear the sex organs. When the egg is fertilized it grows into a spore plant so small that only a specialist would notice it.

Another familiar liverwort is *Riccia,* frequently sold to tropical fish breeders for their aquaria. Grown in water, it repeatedly splits into very thin branches. Grown on land, it divides in much the same way; the branches, however, are thicker.

In the so-called leafy liverworts, the thallus is so divided as to suggest leaves. These plants have a strikingly delicate appearance. They are at their best on dripping rocks, although they are most likely to be found in the dimmest recesses of woods. They are especially abundant in moist tropical forests; here they frequently cover with luxurious growth not only the ground but also the stems and leaves of other plants.

In the ferns and their distant allies — the horsetails and the club mosses — there are alternating generations of sex-cell-bearing plants and spore-bearing plants, as in the mosses and liverworts. There is one important difference, however. In the

ferns and their kin, it is the sporophyte, or spore plant, that is the conspicuous form; the gametophyte, or sex-cell-bearing plant, is much simpler in structure. Ferns, horsetails and club mosses are the more primitive of the vascular plants, or tracheophytes. They were formerly classified together as the pteridophytes (fern plants). Nowadays, however, the ferns are considered by botanists to be more nearly akin to the conifers and flowering plants.

The ferns are more widespread and better-known than the club mosses and horsetails. The roots, stems and leaves of the fern sporophytes are like those of seed plants. Ferns absorb water and dissolved mineral salts by means of root hairs, just as seed plants do. The stems of most species remain underground; these hidden growths are known as rhizomes. The tips of the rhizomes slowly push their way forward through the soil as they grow; thus they constantly tend to invade more and more territory. In certain cases, the stems of ferns grow vertically, forming the lofty plants called tree ferns.

The leaves of the spore-bearing fern plants are called fronds; they carry on photosynthesis in exactly the same way as seed-plant leaves do. When young, the fronds of certain species are called fiddleheads, because the tips are curled up and resemble the scroll, or curved head, of a violin. As the frond grows longer, the tip uncoils. Most of our common ferns have large and usually very decorative leaves.

The underside of some fern leaves is dotted with structures called sori (singular: sorus). Each sorus is made up of a cluster of spore-containing cases known as sporangia, each mounted on a flexible stalk. When the sporangia dry out, the spores are released. If a fern leaf containing sori is kept in position over a glass slide, the slide will be covered after a time with a fine dust, consisting of innumerable spores. Fern spores are provided with a strong protective covering; they are so light that the wind may carry them for miles.

Many millions of spores may be produced by a fern plant in a single growing season; comparatively few of them, however, develop into new plants. When a spore germinates, it gives rise to a small, flat, heart-shaped structure called a prothallus, which looks a good deal like the thallus of a liverwort. The undersurface of the prothallus has rhizoids, in place of roots. The sex organs also develop on the undersurface of the plant. As in the case of the mosses, fertilization can take place only when water is present. The fertilized egg grows into a spore-bearing fern plant of the familiar type.

The best place to look for ferns is on the floor of a forest; here they are often more numerous than any other low-growing plants. They reach their fullest development in the tropics. The richest fern growth occurs, not in dank, lowland woods nor in impassable jungles, but in mountain forests where moisture is abundant.

In some regions, as in Hawaii, New Zealand and parts of Australia, tree ferns, ranging up to forty feet and more in height and three feet in diameter, are a familiar feature of the landscape. Many tropical ferns grow on the trunks or branches of trees; they are not parasites, but simply use trees as convenient anchorages. A few ferns are water plants. One of these, *Marsilea,* has floating leaves which suggest the four-leaf clover; it is sometimes called the clover-leaf water fern.

Among the best-known of all ferns is the brake, or bracken (*Pteridium*), which is particularly conspicuous in the heaths and moors of Great Britain and Ireland

The odd fruiting heads of a typical horsetail plant (*Equisetum arvense*).

Leafy shoots of a horsetail. Most horsetails are relatively small plants.

though found in almost every part of the world. In Ireland the leaves of this plant reach a height of thirteen feet.

The United States and Canada have many interesting types of ferns. One of them is the Christmas fern (*Polystichum*), whose tough leaves shine bright and green even when covered with snow and ice. The long, simple leaves of the walking fern (*Camptosorus*) take root at the tip and bud into new plants. The beautiful maidenhair fern (*Adiantum*) has small, lobed leaflets and shining, purplish brown leaf stalks. There are large ferns, like the royal fern, the interrupted fern and the cinnamon fern (*Osmunda*), each of which bears leaves that may be six feet high. At the other end of the scale there is the tiny curly grass (*Schizaea*) of the New Jersey pine barrens; it grows only two or three inches in height.

**Certain ferns serve mankind**

The leaves of certain ferns are used as packing for fruits and vegetables. The Japanese use the young fronds of some varieties as food. The trunks of some of the sturdier tree ferns serve for construction purposes, particularly in tropical lands. In the Hawaiian Islands the soft hairlike scales of certain tree ferns (*Cibotium*) are used as a filling for mattresses. The male fern (*Dryopteris*) is known to pharmacists as the source of an effective vermifuge — a remedy used to expel or destroy intestinal worms.

If you wander through deep pine or spruce forests, you will often find the ground covered with the dense, bright growths of creeping ground pine. These plants are not related to pines but are club mosses (*Lycopodium*). The name is apt enough; these plants look very much like extremely large mosses and in many species the spores are produced in clublike cones. Most club mosses reproduce much as the ferns do. They are more abundant in the tropics than in temperate regions.

These plants are not especially useful to mankind. Certain varieties are used in wreaths and other decorations. Lycopodium powder, consisting of the spores of a club moss, is used in fireworks; it is sometimes set off in general-science classes to illustrate rapid combustion. It has also been employed in the treatment of certain skin diseases.

Horsetails (*Equisetum*) grow thickly in the gravel of railway roadbeds, or in clumps along the banks of streams and in swamps. Their jointed structure suggests that of sugar cane; the leaves have been reduced to small scales. In some species, the main stem bears a circle of smaller branches at each joint. In others, the stem usually does not branch, so that a thicket of this type resembles a fanciful green colonnade.

The horsetails contain a considerable amount of the hard substance called silica, or silicon dioxide ($SiO_2$), most familiar to us in the form of quartz. As a result the texture of these plants is very rough. The dried stems have been used to scour pots and pans; the plants are sometimes known, therefore, as scouring rushes. Hay containing an abundance of horsetail plants is harmful to livestock.

**Most species of horsetails are small**

Most species of horsetails are small plants, from one to six feet or so high. However, a vinelike tropical variety, found in South America, sometimes reaches a height of forty feet. Most species send up both unbranched, fertile stems, each bearing a cone at the tip, and green, bushy stems. The cones contain the spore cases, or sporangia; green gametophytes arising from the spores are small and ribbonlike.

Today the horsetails and club mosses form only a small part of the earth's vegetation. Even the ferns, which are far more widespread, are rarely the dominant plant in any locality. Yet millions of years ago, in the Carboniferous period, the forests of the world consisted largely of horsetails, club mosses and tree ferns, all of enormous size. The remains of these plants, completely transformed into a dark mineral, are now widely used in the form of coal.

*See also Vol. 10, p. 272:* "General Works."

# BUTTERFLIES AND MOTHS

## Graceful Scale-Winged Insects

by JOHN C. PALLISTER

THE beautiful insects we call butterflies appeal to almost everyone, even to those who fear or dislike other kinds of insects, such as wasps, or bees or beetles. Small boys and girls often start their insect collections with butterflies; many adults also delight in collecting them, sometimes exploring remote and isolated areas in search of unknown species. As a result, we have come to know a great deal about butterflies, and we are constantly adding to our knowledge as new species turn up.

In this article we are going to deal not only with butterflies but with the closely allied insects known as moths. Butterflies and moths belong to the order Lepidoptera, second only to the beetles in the number of their species. So far, nearly 150,000 Lepidoptera species have been described; they represent over 1,000 genera and about 200 families. Moths and butterflies are found throughout the world. A few species inhabit the subpolar regions and ascend to snow line on mountains. There are far more species in temperate regions; in the tropics the insects attain their greatest variety, largest size and most brilliant coloring.

Hugh Spencer

The name "Lepidoptera" comes from two Greek words, *lepis* ("scale") and *ptera* ("wings"). The reference is to the minute scales that generally cover the rather broad, usually opaque and membranous wings of these insects. The scales are usually triangular or elongated modified hairs, each fastened by a stemlike base to the wing. They are laid on the wing in regular rows, each overlapping the row below, like shingles on a roof. Since the scales are held only by the tiny stemlike base, they can be very easily dislodged. Perhaps you have noticed the dustlike particles that coat your fingers when you try to hold a butterfly. Usually scales cover not only the wing surfaces but the rest of the body

Edwin Way Teale

The monarch butterfly (above) is one of the few true migrants among insects.

The luna moth (left) has two delicate tails and antennae that look like ferns.

The four stages in the metamorphosis of the Cecropia moth are shown on these two pages. Above: the eggs. Right: a larva feeding.

as well. They are variously colored and form attractive patterns in most butterflies and many of the moths.

In some cases, however, for instance in the so-called glass-winged butterflies of the genera *Haetera* and *Cithaerias,* the scales are extremely limited in number; the few that are present are very fine and scarcely visible except under magnification. The membrane of the wings is so transparent that printing can be easily read through it. A number of other butterfly species and some moths (including the hummingbird moths and clear-winged moths) show scaleless transparent areas.

The tiny scales on the wings of a Cecropia moth form a striking pattern.

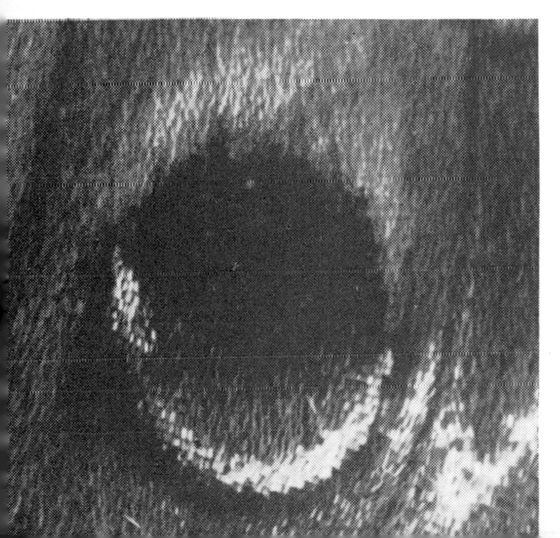

Many members of the family Saturniidae have rounded clear spots margined with rows of variously colored scales of blue, red, pink, white and black, which give an eyelike appearance to the transparent area. Some other butterflies and moths have eyelike spots, without transparent centers. The opaque spots frequently occur on only one surface of the wing, the other surface having a different pattern. This type of design probably serves some protective purpose, which we cannot explain satisfactorily at the present time.

The coloring of butterflies and moths is due in most cases to pigments imbedded in the scales. In some of the Lepidoptera different colors are produced as light is diffracted from minute, closely spaced parallel lines, called striae, on the scales of the wings. This kind of coloration is called structural. It results in a variety of iridescent hues — violet, blue-green, copper, silver and gold; the color changes as the surface is tilted.

Most lepidopterous scales are striated (that is, possess striae); but in only a few cases are the striae fine enough and close enough together to produce iridescent structural color. In such cases, the striae may

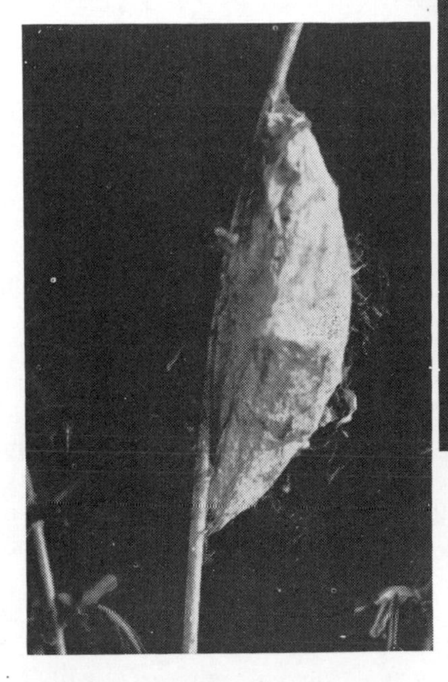

The silken cocoon in which the Cecropia moth pupa is enclosed.

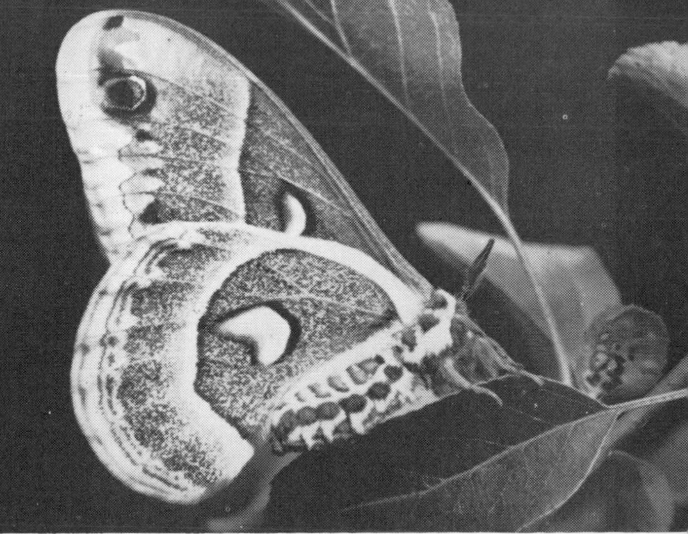

Photos, Hugh Spencer

The adult Cecropia moth is a handsome insect, belonging to the family Saturniidae.

be astonishingly fine. In a Brazilian species of the genus *Apatura,* there are 1,050 striae to a millimeter, or about 26,700 to the inch. In another species, belonging to the South American genus *Morpho,* the striae are even closer together; there are about 1,400 to the millimeter, or 35,600 to the inch.

Butterflies and moths rank among the more highly developed insects, those having a complete metamorphosis. This means that they go through four stages of development: egg, larva, pupa and adult. The eggs are small, round, oval or somewhat elongate; they are variously colored and are delicately sculptured with ridges or pits, according to the species. The female insect usually lays her eggs on or close to a plant whose vegetation will provide the young insects with food. With some exceptions, each species feeds only on one particular plant species or on a closely related group.

The young of most butterflies and moths are herbivorous; a few species are carnivorous, feeding on aphids and scale insects. The creatures that hatch from moth or butterfly eggs are the caterpillars, scientifically known as larvae (singular: larva). Of the four stages this is the only one in which the insect can grow. The caterpillar feeds as voraciously as its food supply will permit. As the skin, or covering, of the animal grows tight, it is shed, or molted. The covering that has formed beneath the old one is soft and expands to accommodate the increased growth. There are five or six molts on the average; some species may molt as many as twenty times.

Caterpillars have three pairs of true legs, one pair to each segment of the thorax. (The thorax is the part of the body between the head and the abdomen.) On the abdomen there are from one to five pairs of prolegs, fleshy protuberances armed with hooks for grasping twigs or leaves of the food plant. Most caterpillars have five pairs of prolegs, one pair each on the third, fourth, fifth, sixth and last, or terminal, abdominal segment. Caterpillars travel by stretching out the front part of the body, taking hold with the true legs and then drawing forward the rear part of the abdomen. The eyes are simple and arranged in pairs; there are from two to six on each side of the head.

After a caterpillar has gone through the required number of molts, it becomes a pupa (plural pupae). In butterflies and moths the pupa is frequently called a chrysalis. It does not in the least resemble a caterpillar; in most cases the appendages are "glued down" to the body (though frequently visible through the pupal skin), giving the insect a compact appearance.

The pupal stage is a quiescent period; it may last from a few days to several months (in some species much longer). Most butterfly pupae are found fastened to some stable object a little distance from the ground. Some hang head downward from a pad of silk, which the caterpillar spun on a sheltered object before molting. The swallowtail butterflies, family Papilionidae, and the Pieridae pupate in a more or less upright position with the tip of the abdomen in a pad of silk; a silken strand, looped around the middle of the body, is fastened at each end to the support.

Moth pupae are sheltered in various ways. Most of them pupate on or near the ground under leaves, old logs or loose bark or in hollow trees. Many moth caterpillars burrow into the ground, where they form a smooth cell in which to pupate. Some line this cell with silk. Others, such as the woolly bears, family Arctiidae, use the spiny hairs from their bodies with a few strands of silk to form a rough cocoon in a sheltered place near the ground. The most conspicuous cocoons are the beautiful silk ones spun by the larvae of a number of the giant silkworms, family Saturniidae. The true Chinese silkworm, *Bombyx mori,* family Bombycidae, spins the cocoon used in commercial silk production.

#### From inert pupa to adult butterfly or moth

After a certain period of inactivity as a pupa (a period that varies according to the species), life begins to stir in the insect; with the final molt the pupal case splits down the back and out crawls a butterfly or moth. It is a poor bedraggled creature at this time; its wings are limp and its body is swollen. But soon the body fluids begin to flow into the veins of the wings, which start to expand and spread out. The spreading of the wings must proceed rapidly; if the air is too dry, the wings may dry out before they are properly spread and the insect will be imperfectly formed. The last transformation of the butterfly or moth usually takes place at night. All the appendages now function; the body has attained its normal size. Soon the lovely creature will be ready to fly off, seek food and mate.

Geologically butterflies and moths are comparatively recent. In North America butterflies occurred in the Eocene and the Oligocene periods of the Cenozoic era, while in Europe small moths were trapped in amber in the Baltic regions during the Oligocene period. It is thought that some European finds may go back to the Jurassic period in the Mesozoic era.

Many different systems of classification have been proposed for the Lepidoptera and even at the present time no system is satisfactory to all workers. Even the division of the Lepidoptera into the butterflies, Rhopalocera, and the moths, Heterocera, is much disputed and considered by some to be artificial. However, this division is a practical one and still serves. The antennae furnish the main point of difference between butterflies and moths. In butterflies the tips of the antennae are distinctly knobbed or enlarged. In moths the antennae assume a variety of forms. They may be slender, tapering to very fine points; they may be feathery, or fernlike or more or less pectinate (provided with teethlike projections or divisions).

There are certain exceptional cases. The skippers (family Hesperiidae), the giant skippers (family Megathymidae) and a few other small families are all definitely butterflies, with knobbed antennae. In most of these insects, however, the knob has a very fine, tapering, pointed tip, which is set at a very sharp angle. The antennae of the Sphingidae, which are moths, are tapering and very finely pectinate, as we might expect of moths; but the tip is curved much as in the skippers.

Another characteristic that distinguishes butterflies from moths is that practically all butterflies fly only in the daytime, unless they are disturbed, while most moths fly only at night, unless disturbed. However, a few moth species, largely of the family Uraniidae, are diurnal. In this family the insects greatly resemble butterflies in their coloring, their broad butterfly-like wings and, in some species, their "tails," or wing extensions.

There is sometimes considerable variation in the coloring of butterflies of the same species. Where there are two or more broods a season, the spring forms may differ radically in the shade of coloring from those in the summer brood; they are usually much lighter. If there is a fall brood, it may differ from both the spring and summer forms. Variation of this kind is called seasonal dimorphism. The two sexes of a given species are often differently colored. The male, which is usually somewhat smaller than the female, is quite apt to show more brilliant coloring. This differentiation is known as sexual dimorphism.

The larvae, or caterpillars, of the Lepidoptera, particularly some of the moths, are often serious pests because, in order to satisfy their voracious appetites, they feed on various plants and products that are useful to man. We can only mention a few of these pests here.

The larva of the gypsy moth (*Porthetria dispar*) is one of the worst; it devours the leaves of the apple, oak, gray birch, alder, willow and various other deciduous trees. When cankerworms — the caterpil-

Above: the egg.

Below: pupa in cocoon.

Photos, Hugh Spencer

Larval form.

Metamorphosis of a typical butterfly, the black swallowtail butterfly. The swallowtails (family Papilionidae) are large; they are excellent fliers.

Adult butterfly.

*Eiroboea endamippus* is a member of the family Nymphalidae.

*Neorina crishna* belongs to the family Satyridae.

*Neurosigma douledaii* is a beautiful nymphalid.

lars of the Geometridae — are very numerous, they can strip the leaves of practically all the trees in extensive wooded areas. The codling moth (*Carpocapsa pomonella*) caterpillar feeds on apples and other pome crops. The wild cherry and other food trees are attacked by the larvae of several species belonging to the family Lasiocampidae. The cutworms — the larvae of various genera of noctuid moths — are so called because they cut the tender stems and leaves of various young plants as they feed. The army worm (*Cirphis unipuncta*) destroys grass, grain and other crops. The tobacco worm (genus *Protoparce*) attacks tobacco plants; the cabbage webworm (*Hellula undalis*), cabbages and other vegetables. The larvae of several moth species, belonging to the family Tineidae, feed on woolens, furs and feathers and cause millions of dollars worth of damage annually.

**Some important families of butterflies and moths**

As we have already pointed out, the Lepidoptera number about 200 families. In the following pages we shall take up some of the larger and more conspicuous ones.

The swallowtails, family Papilionidae, form a large group. Many of the species have "tails," or extensions, on their hind wings. A few even have two "tails" — a feature that accounts for the name "swallowtails." The "tails" of the Papilionidae are not a positive identifying character. Certain members of the family have no "tails" at all; certain butterflies and moths belonging to other families also have extensions on the hind wings.

In general, swallowtail species are medium to large, vividly colored or showing contrasting colors of black, yellow, red and white. There are nearly one thousand described species of papilionids. Most of the species are found in the tropics, but many have invaded the subtropical and temperate regions.

Dimorphism, both seasonal and sexual, is common among the swallowtails. Many of the caterpillars have eye spots on the thoracic region; many have a Y-shaped scent organ that can be extended from a slit on the dorsal (back) part.

Among the most magnificent members of this family are the bird-wing butterflies of the genus *Ornithoptera*. (Some entomologists consider them a separate family.) Most of the bird-wing butterflies are gorgeously colored. The males, which are generally smaller than the females, have brighter colors. The front wings are very large and elongate; the hind wings are much smaller. These butterflies are very strong fliers. They usually prefer the region of the treetops, and for this reason they are difficult to collect. Professional collectors seeking bird-wing-butterfly specimens use long-handled nets and work from platforms built high in the trees. These insects are restricted to portions of the East Indies; they abound particularly in New Guinea, Java, Sumatra and the Malay Peninsula.

The whites, yellows, sulphurs and several other groups make up the family Pieridae. These insects are small to medium in size and world-wide in distribution, ranging from the tropics to the temperate regions. The cabbage butterfly hovering over our garden cabbages or the common sulphur that children pursue over clover fields are familiar representatives of this family. The sulphurs of the tropics are medium-sized; they sometimes migrate in vast flocks to some unknown spot many miles away.

In the tropics of the Americas, from central Mexico south to southern Brazil, occur the gorgeous butterflies known as Morphos, of the family Morphoidae. Tropical species show iridescent blues, greens and purples, for it is in this family that structural coloring is at its best. Over one hundred species of Morphoidae are known. They are much sought after by collectors; thousands are used for decorative purposes in the manufacture of trays and jewelry. So great has the trade in these colorful insects become that some countries have resorted to strict regulations to prevent the Morpho from being exterminated.

It is exciting to see a blue Morpho flitting along a jungle pathway. With its slow easy flight, it looks as if it would fly directly into an out-held butterfly net. Appearances are deceptive, however, for the Morpho is an excellent dodger. It will dash under the net, or sweep upward and over a treetop or backtrack down the path and lose itself in the foliage. You have to be extremely alert to collect a flying specimen.

The Nymphalidae, or brush-footed butterflies, are a very large family. The front legs are greatly reduced in size and are often hairy and brushlike. This character cannot be entirely depended upon to identify the Nymphalidae, because several other families, including the Satyridae, Danaidae and Heliconidae, also have much the same kind of brushlike legs. Nymphalidae are generally medium to large butterflies, garbed in varied patterns and colors. They occupy a wide belt around the world, extending well into the temperate zones. Because so many genera of nymphalids are easily available to butterfly lovers in the United States and Europe, common names have been given to several large groups, including the fritillaries, peacocks, emperors, tortoiseshells and anglewings. Certain species of nymphalids have also received common names, such as the mourning cloak, the red admiral and the painted lady.

The well-known monarch butterfly, famous for its long migrations (see Index, under Monarch butterflies), belongs to the Danaidae. This family has relatively few species but these number many individuals, which are found all over the world. The Heliconidae dwell in the tropics. They flit gracefully from flower to flower along jungle trails on their very long, narrow, brightly colored wings.

The butterflies of the family Brassolidae are found in South America, the West

Photos, Amer. Mus. of Nat. Hist.

Upper left: yellow female tiger swallowtail.

Lower left: dark female tiger swallowtail.

*Kallima paraletka*, shown at the right, is one of the nymphalids.

A moth (genus *Rothschildia*) belonging to the Saturniidae.

Some woolly bear caterpillars of the common Isabella moth.

Indies and tropical North America. These large butterflies are brightly colored on the upperside; the underside is brown, with lines and spots. There is a large eyelike spot in the center of each hind wing. When the wings are spread out, the underside, with its brown markings and two eyelike spots, suggests the head of an owl; hence the common name owl butterflies. These insects are very difficult to collect. When disturbed, they hide in a thicket with their wings folded over their backs, thus covering the bright topside colors; the brown underside markings blend with the leaves and tendrils. If disturbed, they will dash from their retreat, circle a few times, flashing their bright colors, then make for cover in another thicket.

The swiftly flying Hesperiidae, known as skippers, are a large family of small to medium size; their color patterns generally show combinations of brown, yellow and blue. These insects form a connecting link between butterflies and moths, though they are classed with the butterflies. The bodies are heavier and the scales more hairlike than those of most butterflies, giving them a mothlike appearance. However, like most butterflies, they are diurnal, flying at night only when disturbed and then merely to find another hiding place. The antennae are knobbed as in the butterflies, but on the end of the knob is a tiny tapering extension as in the moths.

Among the moths, one of the large and sometimes conspicuous families is the Sphingidae, also known as sphinx moths. The name refers to the sphinxlike position many of the larvae assume when disturbed. Many of the adults are swift of flight, dashing on their long slender wings from flower to flower; they are known as hawk moths. A few hover like hummingbirds in front of flowers, probing into the depths of the corolla with the long, coiled, springlike proboscis; these are called hummingbird moths. Various species of the Sphingidae pollenize flowers with a long corolla, such as the angel's-trumpet and Jimson weed.

The Saturniidae, or giant silk-spinning moths, are an outstanding family; their generally large size and attractive appearance make them a popular group with collectors. The largest known moths occur in this family. The Atlas moth, *Attacus atlas,* of southern Asia and Malaysia has a wing spread of nearly eleven inches. Two other moths of this family — *Attacus edwardsi* of Australia and *Coscinoscera hercules* of Australia and Papua — are just as large but are more rarely found in collections. The Cecropia moth (*Samia cecropia*), spicebush moth (*Callosamia promethea*), luna moth (*Tropaea luna*), polyphemus moth (*Telea polyphemus*) and ailanthus moth (*Philosamia cynthia,* introduced from China) are outstanding representatives of the family in the United States. The Saturniidae are particularly well represented in India, Ceylon, China and Japan.

The Arctiidae, or tiger moths, are a large family of over 4,000 species, worldwide in distribution. They are also known as woolly bears because the larvae of most of the species are covered with a long and hairlike or thick and woolly covering. They combine this hairy covering with a few strands of silk to construct their rough cocoons when they are ready to pupate. One species in the United States, reddish brown in the middle and black at the ends, has acquired a reputation among the superstitious as a weather forecaster. The width of the reddish section is said to indicate a mild or severe winter. The moths of this family are generally small to medium in

A hummingbird moth in a glass cage, above, is puzzled by an artificial flower, to which is attached a tube of honey. The flower does not seem quite like other flowers, but the honey is a powerful lure. After three days of indecision, the moth finds the opening in the flower and projects its long proboscis into the tube, below, drinking deeply. The moth is named after the hummingbird because, by rapid movements of its wings, it also can stand still in flight.

Lilo Hess, from Three Lions, Inc

Photos, Amer. Mus. of Nat. Hist.

A detested household pest: a female clothes moth laying eggs.

size, black or white and decorated with contrasting black, red, brown or white spots, lines or blotches. The Arctiidae are an attractive group and many collectors have specialized in this family.

One of the largest moth families, if not the largest, in number of species is the Noctuidae, or owlet moths. Well over 10,000 species have been described and new ones are constantly being added to our lists. Many destructive lepidopterous pests, including the army worms and cutworms, belong to this family. The noctuids are medium in size, and many of them are monotonously similar in their brown and gray markings. However, the hind wings of some, especially those belonging to the genus *Catacola,* are most strikingly colored; they are jet black or alternately banded with red, black, orange or yellow. They delight the collector because they are easily attracted to his lights or his sugar bait. Their eyes gleam in the dark as they reflect the light from a torch or lantern.

The Geometridae make up another very large family. The family name means "earth measurers"; the larvae are known as measuring worms, inchworms, spanworms and loopers. This family, like the Noctuidae, includes many destructive larva pests; the spring and fall cankerworms are familiar examples. Geometer moths rest with the wings spread out flat.

Up to now we have dealt chiefly with the largest species. Older collectors referred to them as macrolepidoptera ("large Lepidoptera"), or macros. They gave the name "microlepidoptera" ("small Lepidoptera"), or micros, to the families made up largely of very small species. This classification is convenient in certain respects, but modern entomologists consider that it is based on an unnatural division. Many macros are smaller than some micros; many micros are larger than certain macros.

Here are a few families of micros included in the traditional classification. The Psychidae, or bagworms, are interesting because the larvae construct cases, or bags, of silk, interwoven with leaves, bark and other debris. The entire life of both the caterpillar and pupa is spent in the bag. The female adult, which is usually wingless, also remains in her bag and lays her eggs in it. The adult male, however, has wings in order to hunt out the female. The sluglike caterpillars of the Limacodidae, or Cochlidiidae, have no distinct legs. Some of these larvae, such as the saddleback, are covered with spines which cause severe irritation to the skin.

In the Cossidae, the moths are quite large. The small larvae bore into the wood or stems of trees or large plants. The adult females are sluggish and may frequently be found even in daylight, resting near the light that had attracted them during the previous night. In the Aegeriidae, or clearwing moths, the wings are largely transparent; scales occur chiefly along the margins. They are rather small; the larvae are borers in certain plants and shrubs.

The Pyralididae are a very large family of very small moths. Many of these are destructive to hay, grain and other crops. In the Tortricidae many of the larvae roll the leaves on which they are feeding into a small case, in which they can hide and be protected. The Tineidae are chiefly famous because the three species of clothes moths belong here. Most species of tineids are well-behaved little creatures (of course from our viewpoint!); their larvae feed on rotten fungi and other waste products.

The female of *Pronuba yuccasella,* family Incurvariidae, collects pollen in her mouth and places it on the stigma of the flowers of *Yucca filamentosa.* After pollinizing the flower, the moth deposits her eggs in the ovary. The flowers always set enough seeds to provide food for the larvae and to ensure the plant's reproduction. The yucca plant and yucca moth are interdependent; one cannot exist without the other.

*See also Vol. 10, p. 275:* "Insects."

# BUTTERFLIES AND MOTHS

Butterflies and moths belong to the order Lepidoptera; they are second only to the beetles in the number of their species. The name Lepidoptera comes from two Greek words meaning "scale wings"; it refers to the tiny scales that usually cover the wings of butterflies and moths. These insects are often remarkable for their coloring, which shows many striking hues and patterns. In these four pages we show examples of the exquisite colors of some members of the Lepidoptera.

At the right we see the larva, or caterpillar, of the spicebush swallowtail butterfly. In time the caterpillar becomes an inert pupa; then the pupa is transformed into the beautiful creature shown below. Note the "tail," or extension, on the hind wing. Swallowtails form a large group.

Both photos, Ray J. Glover—Nat. Audubon Soc.

202-b

Photos on these two pages, Theo Bandi and Emil Schulthess

The colors of some butterflies and moths are produced as light is diffracted from striae (minute parallel lines), as in the moth *Chrysiridia ripheus* (left). Usually the coloring is due to pigments imbedded in the scales, as in the imperial swallowtail, shown below.

Above we see the complete color pattern of *Chrysiridia ripheus* (left) and the imperial swallowtail, shown in close-up in the other pictures on these two pages. *Chrysiridia ripheus* is found in Madagascar; the imperial swallowtail, in the Himalayan foothills.

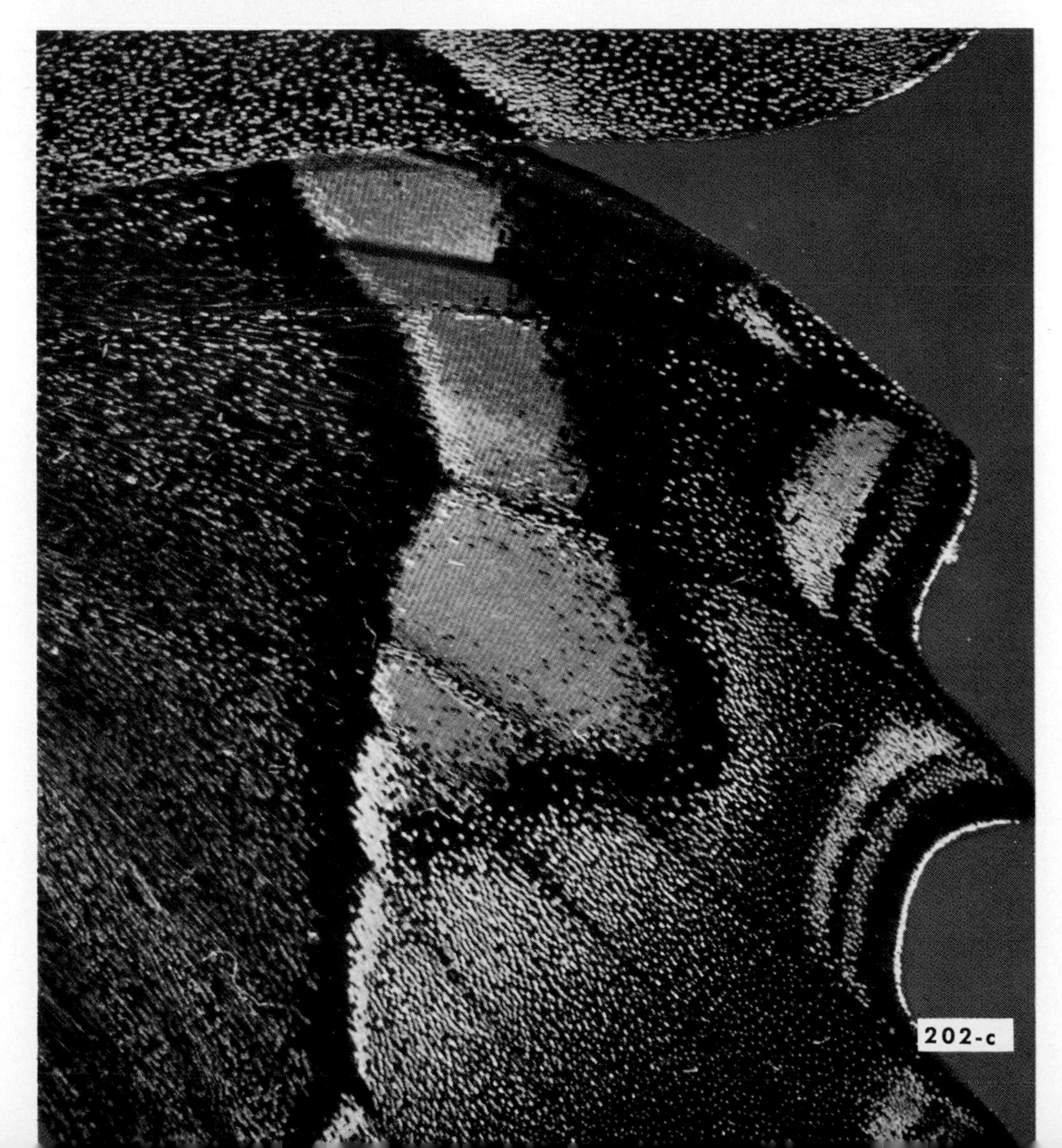

Above: the Lola giant silkworm moth (left) and *Pantherodes pardalaria,* a moth belonging to the family Geometridae. The former is a hill-dweller in northern India. Curiously enough, it has no mouth parts; when it uses up the food it had stored up in its body as a caterpillar, it dies. *Pantherodes pardalaria* flits near the ground in mountainous regions of Central and South America.

Photos on this page, Theo Bandi and Emil Schulthess

Above: two beautiful tropical butterflies. When seen in flight from above, the regal hairstreak (left) is blue-green in color; *Catagramma cynosura* (right) is red and black.

Below: how the regal hairstreak and *Catagramma cynosura* look when seen from below. The underside of the regal hairstreak (left) is greenish; that of *Catagramma* is yellowish.

# THE MUSCULAR SYSTEM OF MAN

## What Muscles Are and How They Work

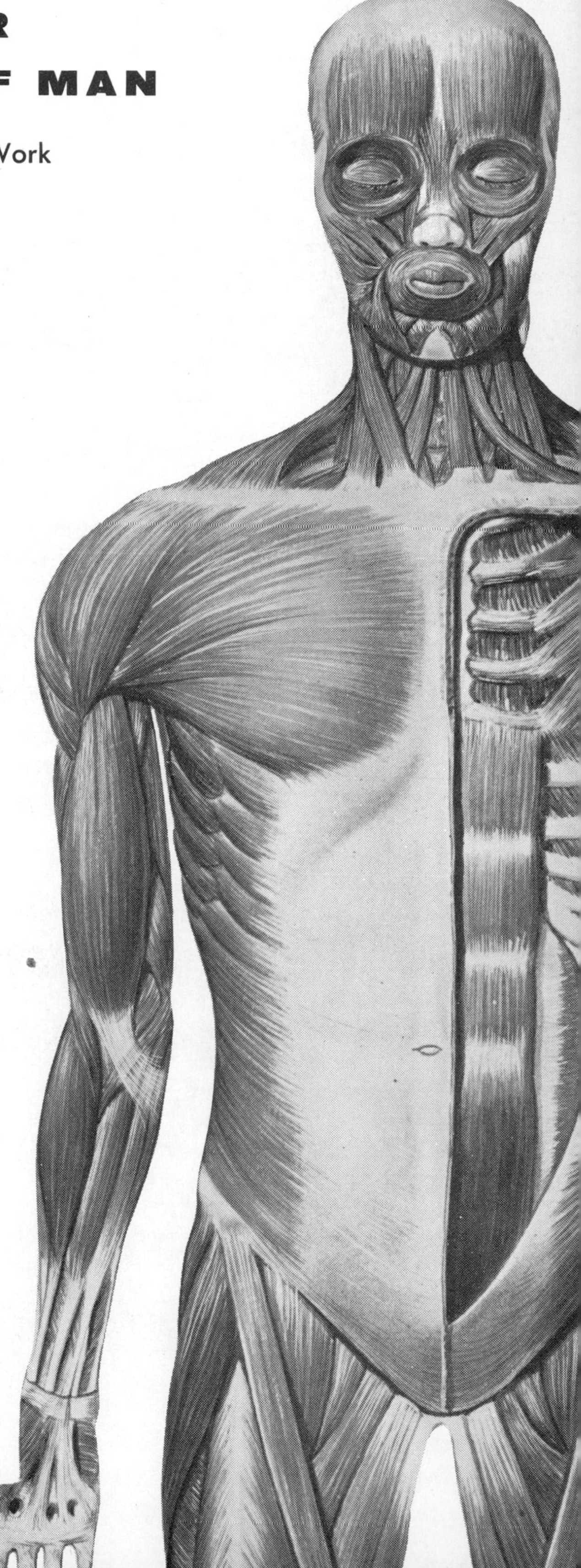

WHEN a small boy offers to show you his muscle, the chances are that he will proudly display his biceps — the upper arm muscle that you can feel hardening, swelling and shortening when you bend your elbow. But the biceps, after all, is only one of more than six hundred different muscles in the human body.

The muscles are essential to life because, by alternately contracting and relaxing, they account for all body motion. Some muscles move the limbs; others open and close the eyelids; still others help us digest our food. The heart is a mass of muscle which keeps working tirelessly day after day. Muscles vary greatly in size and shape; they range from the great latissimus dorsi (the broadest muscle of the back) to a threadlike ear muscle that is only about a fifth of an inch long.

All human muscle is flesh; it corresponds to the red animal meat that we see in the butcher shop. To get a rough idea of the structure of muscle, let us examine, say, a leg of lamb. This cut of meat usually contains several muscles separated from each other by a tough whitish sheath called connective tissue. This is sometimes loose and thin, sometimes dense and thick. Some connective tissue forms tendons, or sinews — cords or straps that attach the ends of muscles to the bones. Parts of tendons are often seen in meat; no matter how long the meat is boiled, these tendons defy vigorous chewing.

Human muscle shows connective tissue like that which we have just described. A good microscope will reveal many other interesting details. We discover, for one thing, that the masses of red flesh, sepa-

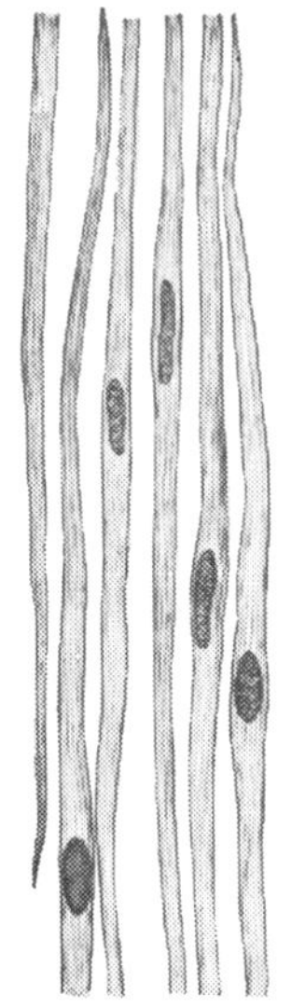

Smooth muscle cells. Each one is elongated and contains a single nucleus.

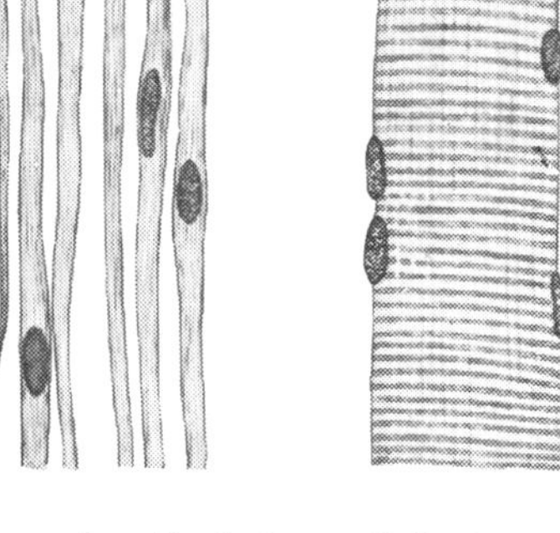

Striped muscle cells. Each one is elongated and has a great number of nuclei.

rated by connective tissue, are really bundles of very firm threads, or fibers. In length, muscle fibers range from about a twenty-fifth of an inch to more than thirteen inches. Some of them run from one end of a muscle to the other; some are joined by connective tissue so as to form long chains.

There are three kinds of muscular tissue in the human body: striped, smooth and cardiac. The name "striped muscle" is derived from the cross stripes, light and dark, of the muscle fibers. Striped muscles are by far the most numerous of the three basic types. Since they are generally attached to the skeleton, they are often called skeletal muscles. Still another name for them is voluntary muscles; they can be controlled by the will, while smooth muscles cannot. For example, you can move your arm, by means of striped muscles, whenever you wish; but did you ever try to control the smooth muscles of the stomach when this organ churns away in the process of digesting a meal?

Smooth muscles do not show any cross stripes. Since we do not have any control over their movements, they are also called involuntary muscles. They are found, among other places, in the walls of the stomach and the intestines, in the bladder, in the blood vessels, around the pupils of the eyes and attached to the roots of the hair. These smooth muscles contract more slowly than striped muscles.

Cardiac muscle (heart muscle) resembles smooth muscle in that its activity is not controlled by the will. It also has a certain similarity to striped muscle since its fibers have a cross-striped pattern.

Each striped muscle cell has a large number of nuclei (as many as several hundred), which lie in the outer part of the

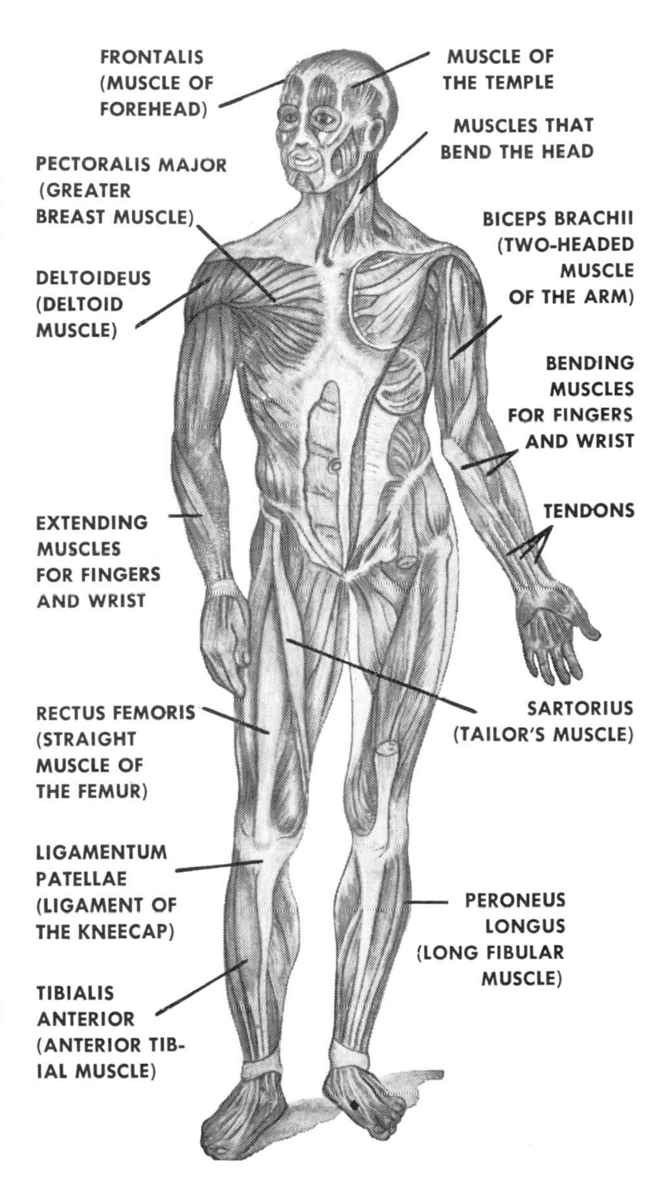


Important muscles — front view. They are usually referred to by their Latin names, which we translate here.

cell. In this respect striped muscle cells differ from most other cells (including smooth muscle cells), which have a single nucleus located more or less in the center of the cell.

What happens inside a striped muscle fiber when it contracts? The ordinary microscope does not help much in this case, and so we call on the chemist to supply us with information. Chemical analysis shows that the striped part of the muscle is made up of the complex substances called proteins. Research chemists have learned much about the arrangement of protein molecules. In the light of this knowledge, we can picture the dark bands of a fiber as made up of rodlike molecules, lying side by side lengthwise in the fiber, while the light bands consist of bent or curved molecules. During a contraction the straight rods would bend or curve, and as this would happen in every dark strip throughout the fiber, the whole fiber would have to shorten.

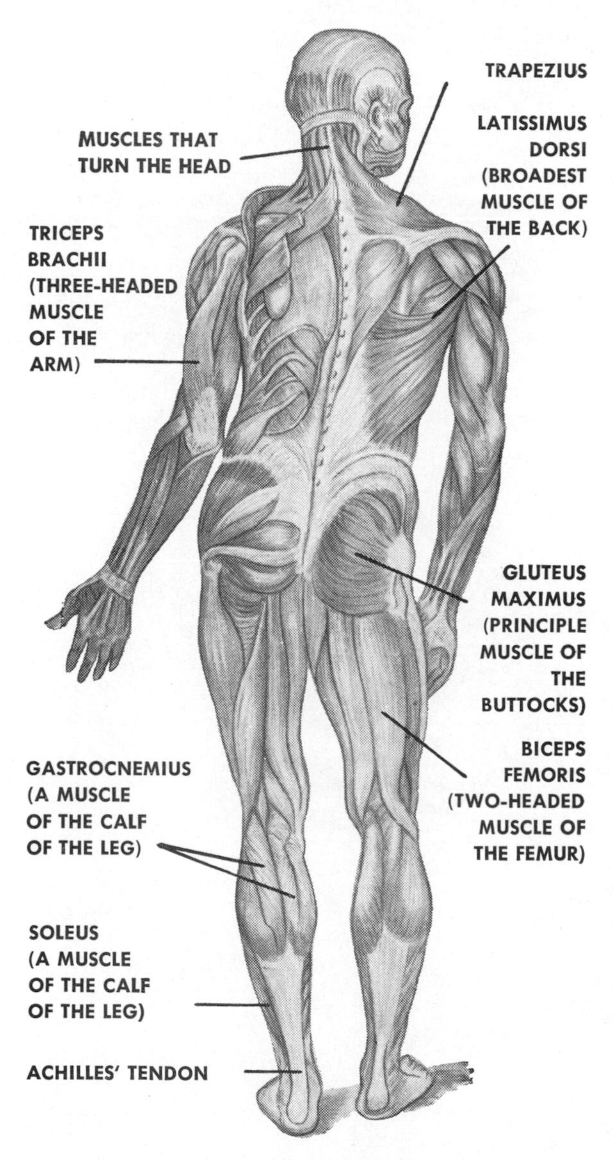


Important muscles — rear view. There are over 600 in all; latissimus dorsi, shown here, is the largest one.

The total effect of contraction is remarkable because, although muscle is soft, it develops great force. If *all* the fibers within a square-inch cross section were to contract at the same time, they would raise from fifty to nearly a hundred and fifty pounds. The muscles used in chewing could lift a man of average weight.

Such force requires abundant energy; to produce this, the body consumes various energy-giving foods. Among the most important of these are sugars, found in sugar cane, sugar beets, fruits, honey, milk and other substances, and starches, derived from foods like potatoes and bread. During digestion and assimilation, sugar and starch are changed into glycogen, which is often called animal starch. Most glycogen is stored in the liver and is given out as needed. Some is found in the muscles and other tissues.

Fat is another body fuel. We probably draw upon it even during light muscular activity; we certainly use it when our stores of glycogen are greatly reduced during long, exhausting exercises or when we go on a reducing diet containing very little starch or sugar.

Fuels such as coal, wood and gasoline contain carbon; so do the foods that we have just mentioned. When fuels are burned, the carbon they contain combines with the oxygen of the air to form a gas called carbon dioxide. During this process heat is produced and is used in various ways; in a locomotive, for example, it causes water to be transformed into steam and this steam pushes pistons to and fro in cylinders. The carbon of foods combines with the oxygen carried by the blood; in this process, too, carbon dioxide is formed and heat is produced. The carbon dioxide is carried by the blood to the lungs and there it is breathed out.

Fuel consumption is much more complicated in the human body than in a locomotive or automobile. The body does not burn glycogen directly but breaks it down into lactic acid; then some of the lactic acid (about one-fifth) is burned — that is, it combines with oxygen. The remaining lactic acid is reconverted into glycogen.

When we want to contract a muscle, we must have energy ready for immediate use. It would take too long to wait for the breakdown of glycogen and the burning of lactic acid. Apparently this is what happens. A phosphorus compound in the muscles — *not* glycogen — breaks down and gives the energy necessary for contraction. A second phosphorus compound also breaks down and gives the energy required to build up phosphorus compound number one again so that it will be ready for the next contraction. Glycogen then decomposes into lactic acid and provides the energy to build up phosphorus compound number two. Finally the burning of some of the lactic acid gives the energy necessary to build up the rest of the lactic acid into glycogen again.

This may seem to be a roundabout way to get results, but it really is most efficient. If muscles contracted only as a result of the energy supplied by burning, we could not engage in sudden spurts of activity. For example, a sprinter may require only ten seconds or less to run a hundred yards. He would not have time, in that short period, to inhale all the oxygen required for combustion within the muscles.

For a time, therefore, the oxygen supply will lag behind the demand. The go-betweens that we have mentioned — glycogen, lactic acid and the two phosphorus compounds — will keep the muscles functioning for a considerable period. After the race the sprinter can gulp in oxygen at his leisure, burn up the accumulated lactic acid and build up anew his supply of glycogen.

If too much lactic acid accumulates within a muscle, it becomes tired and finally stops contracting. If muscles are to act effectively, lactic acid must be broken down into carbon dioxide and water or built up into glycogen. The waste products resulting from the activity of muscle cells also produce fatigue. If they become too abundant in the blood, they may even affect organs which have not been active. Rest after exercise is good because it allows fatigue-producers to be destroyed or passed out of the body.

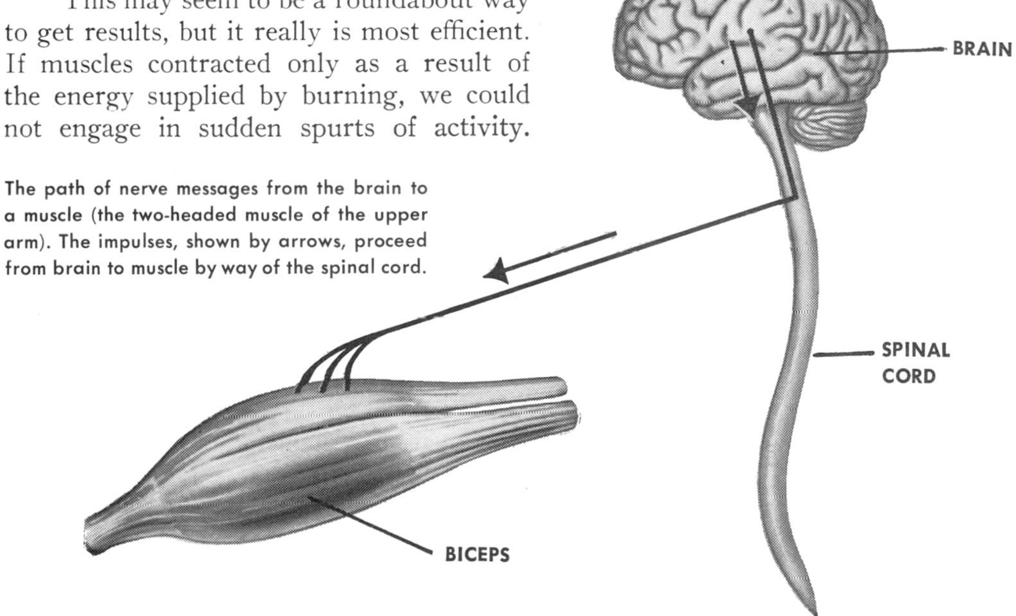


**The path of nerve messages from the brain to a muscle (the two-headed muscle of the upper arm). The impulses, shown by arrows, proceed from brain to muscle by way of the spinal cord.**

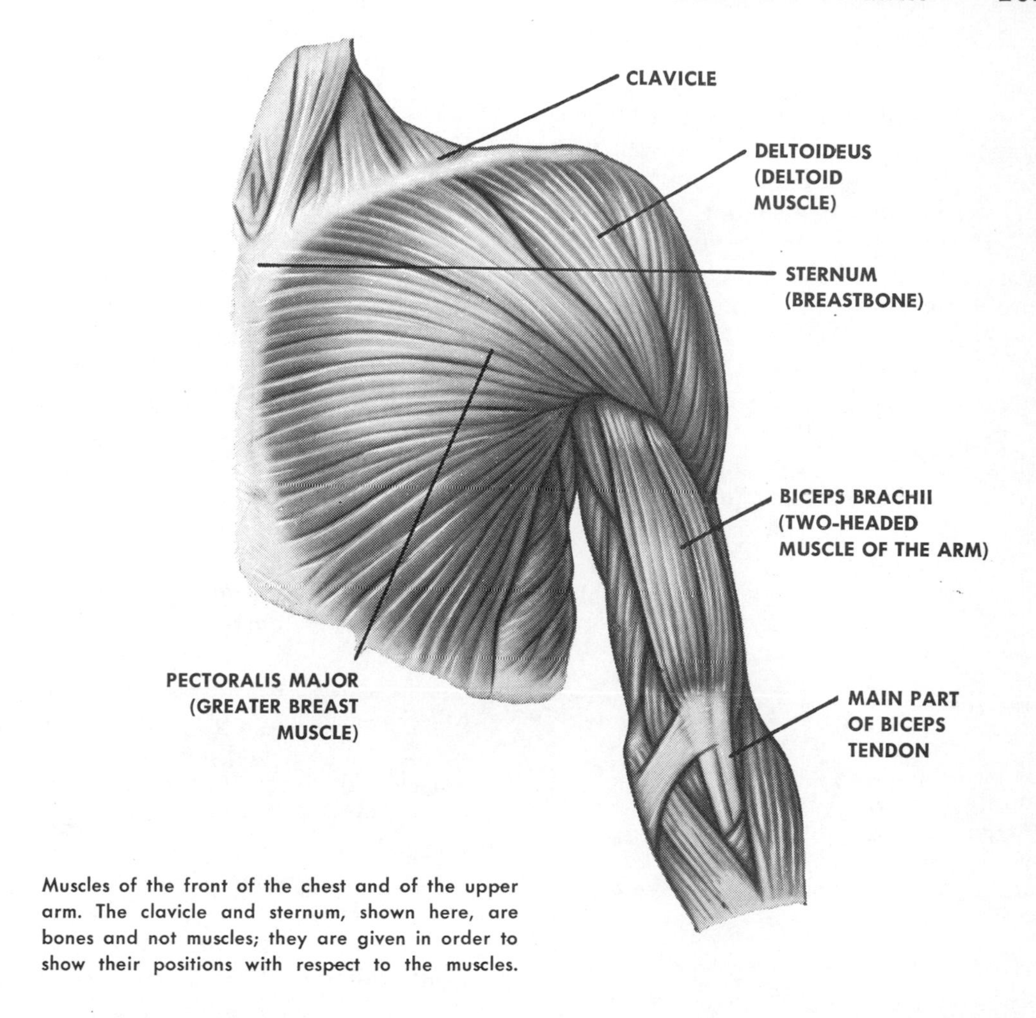


Muscles of the front of the chest and of the upper arm. The clavicle and sternum, shown here, are bones and not muscles; they are given in order to show their positions with respect to the muscles.

We have pointed out that muscles must be well supplied with glycogen and oxygen and that they must get rid of the waste products of combustion. They have an excellent transportation system for this purpose. Blood, bearing the materials necessary for combustion, courses through the arteries and finally reaches the capillaries. The latter have such thin walls that some of the fluid part of the blood seeps through them; it forms a watery fluid — the lymph — which bathes the muscle fibers. The waste products of muscular action dissolve in this fluid; they slowly make their way into the capillaries and, from there, into the veins.

If all the arteries leading to a muscle and the veins leading away from it were to be shut off, the results would be serious. Within a short time, the muscle would suffer great damage; it would die if the condition lasted for more than a few hours. Oxygen would not be brought to the muscle and lactic acid would not be burned; the waste products accumulating in the muscles would not be carried away. First-aiders should bear this in mind when fastening a tourniquet around a limb. Since such a device stops almost all blood flow, it should be applied only when absolutely necessary, and it should be loosened every twenty minutes or so.

Considered as a machine — something that uses energy to do work — the muscles compare favorably with man-made machines. They are more efficient in one important respect, for while machines waste a good deal of heat, the heat produced by muscle contraction is not wasted. It serves to warm the blood, which in turn warms other tissues. This is a most necessary function, because unless body temperature is maintained all cell activities slow down and finally death results. When we shiver in cold weather, we are really calling upon our muscles to produce extra heat in order to maintain body temperature.

How do we start our muscles working? A number of body structures are involved — not only the muscles but also the brain, the spinal cord and the nerves. The nerves connect the brain or spinal cord with the skin, muscles and other tissues. They contain fine long threads, called nerve fibers, which serve the same purpose as telephone lines. If, for example, I wish to bend my elbow, I send messages from my brain down the spinal cord and out along the nerve to the biceps muscle in the upper arm. Beyond the first place of contact between a nerve and its muscle, each nerve fiber divides into many branches and each branch goes to a muscle fiber. A message along one nerve fiber may cause as many as a hundred muscle fibers or even more to contract.

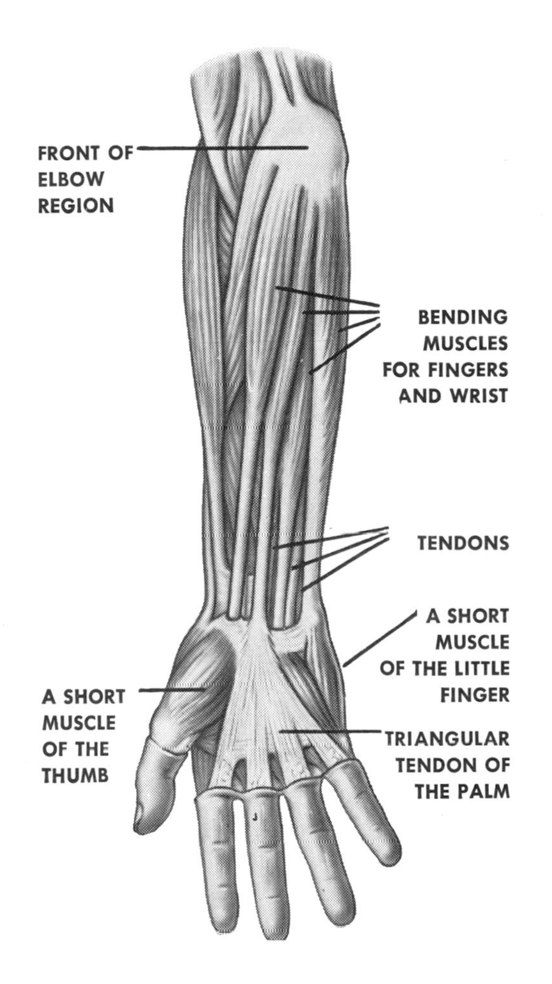


Muscles of the front of the forearm. The hand and fingers are moved by numerous muscles that often begin in the upper part of the forearm and extend into the hand.

When messages are sent by way of nerve fibers to a muscle, only a part of the muscle fibers are called on to contract. They always contract completely; therefore, when more force is required for a given task, messages must be sent along more nerve fibers. Moreover, if a muscle fiber receives only one message, it contracts for only a fraction of a second. Therefore in the case of an ordinary muscle movement we send messages rapidly one after another in order to catch the muscle fibers before they relax.

If necessary, another muscle may be called on to help perform a given task; for at almost every joint there are several muscles that can bring about the same movement. In certain cases muscles that normally oppose a given motion may be called upon either to relax or to offer greater resistance.

Muscles perform an amazing variety of movements. At the elbow, for instance, there are bending muscles, such as the biceps, and straightening muscles, like the triceps at the back of the arm. There are muscles for side-to-side movements, such as those of the wrist; there are also muscles for rotation, as at the shoulder and the hip joints.

In all these cases the bones act as levers and the muscles exert the forces that cause motion. The biceps muscle, to give one example, is attached in front of the elbow joint to the radius, the outer bone of the forearm, which together with the inner bone forms a lever. It is at the place of attachment of the biceps to the radius that effort is applied when we lift a weight; the elbow is the fulcrum.

Muscles often have more than one function. Thus the biceps is not only an elbow-bender; it can also bring about rotating motion. When your right hand goes through the movement of turning a screw,

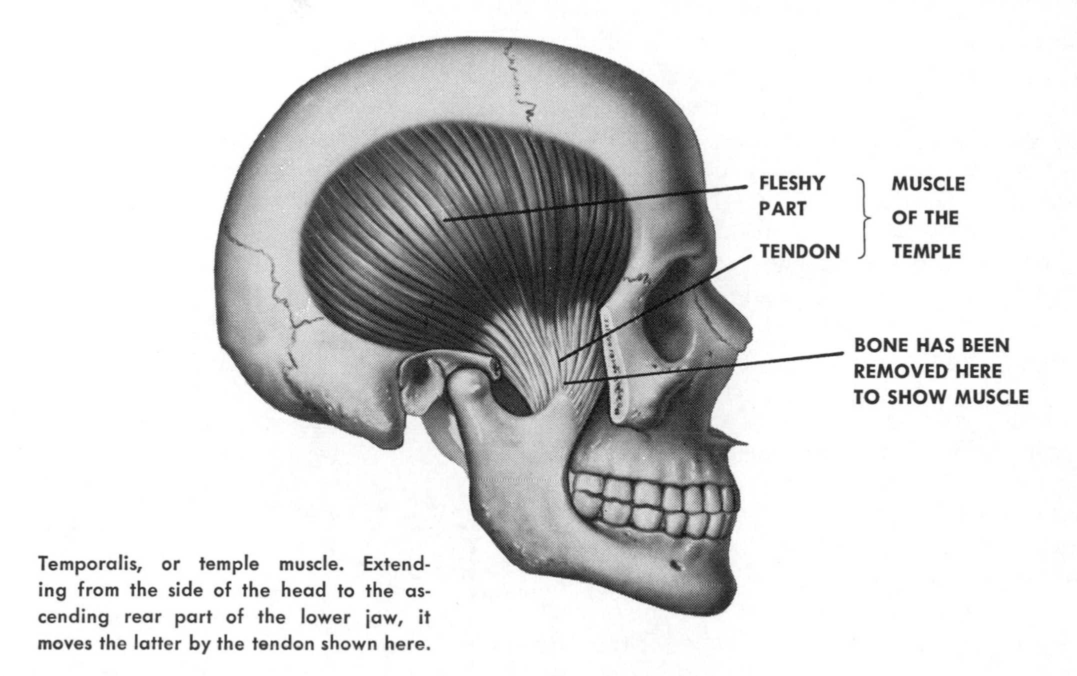


Temporalis, or temple muscle. Extending from the side of the head to the ascending rear part of the lower jaw, it moves the latter by the tendon shown here.

you can feel the biceps contracting to rotate the outer bone of the forearm, which carries the hand around with it. As this takes place, the triceps at the back of the arm contracts so as to prevent the biceps from bending the elbow.

We can study the action of individual muscles, but it is important to remember that muscles act not singly but in groups. We saw that both the biceps and the triceps come into play when we turn a screw. A simple experiment will reveal another case of co-operation between muscles. Move your right arm out from the body and then bring it down slowly while feeling the deltoid muscle with your left hand. (The deltoid muscle, shaped like a reversed Greek letter Δ, or *delta,* covers the shoulders.) Gravity keeps pulling your arm down; the deltoid muscle gradually relaxes (lengthens). Repeat the movement; but this time have someone resist the downward sweep of your arm by holding up your elbow. Now other muscles, such as the pectoralis major (greater breast muscle), come into play; as they meet continued resistance, the deltoid becomes more and more relaxed.

This experiment reveals that gravity often serves to lessen the wear and tear on muscles. It also shows that proper muscle relaxation is as important as contraction. Suppose, for example, that a football player or a dancer wants to do a high kick, bending the hip while keeping the knee straight. He must first learn to relax the muscles of the back of the thigh which cross the hip joint and knee.

Even widely separated muscles may play a part in exerting a given force. When we push or pull a heavy load, the abdominal muscles are rigid. Often in such a case we stop breathing because the muscles of the voice box in the neck have contracted, thus preventing air from passing up or down. The chest and abdomen have become a rigid mass, giving strong support to the arms.

The abdominal, or body-wall, muscles oppose the action of the diaphragm, the dome-shaped muscular sheet between the chest and the abdomen. When we breathe in, the diaphragm contracts, deepening the chest cavity, and air rushes into the lungs. In doing so, the diaphragm pushes the abdominal organs downward; the abdominal wall muscles relax somewhat. When we breathe out, the diaphragm relaxes and the muscles of the abdomen contract.

Other opposing muscles are arranged in such a way as to open and close the mouth and the eyelids. Circular muscles cause the mouth to close; it is opened by muscles than run toward it like the spokes of a wheel. For eye-opening there is the forehead muscle, which wrinkles the forehead when we open the eyes widely, and a muscle that lifts the upper lid.

The face muscles around the mouth and the eye are interesting because they are attached to the skin and not to the face bones, although they are striped muscles and therefore skeletal. We use them to express our feelings — anger, joy, sorrow, fear — and so they are called muscles of expression. They are developed from the same sheet that produces the platelike muscle attached to the skin of the neck. Certain animals have a big sheet of such muscle over the body. If a horse, for example, is annoyed by flies, it will cause its skin to quiver; thus it will shake off the insects.

Muscles are often made far more effective by being combined with tendons. As we have already pointed out, tendons are bands of connective tissue attaching the ends of muscles to the bones. They are immensely strong; a tendon only a quarter of an inch in diameter can support the weight of from three to six men. Some tendons are very conspicuous. You can easily feel those which, like tense cords, pass behind the knees; you can also feel the Achilles' tendon, running from the muscle of the calf of the leg down to the projecting heel bone. People often confuse tendons with ligaments, which are also found at joints and are also made up of connective tissue. The difference is that ligaments do not attach muscles to bones but hold one bone to another; that is, they serve to articulate the skeleton.

Tendons serve to transmit the force exerted by muscle fibers. Suppose that a given muscle is called upon to perform a strenuous task. We have already seen that each muscle fiber exerts a definite amount of force — never more, never less; therefore, in order to get more force we must have more fibers. But if we set all the fibers side by side, there would not be room enough on the bones for all the muscular attachments that are involved.

Therefore, in many muscles the fibers are set at an angle and their force is carried by a tendon running the length of the muscle. This arrangement gives the muscle the appearance of a feather, the tendon being the quill and the fibers the branches. An example of this type of arrangement is the temple muscle; you can feel it when you grit your teeth.

Tendons, then, serve to economize space that would otherwise be occupied by muscle fibers. Here is another example.

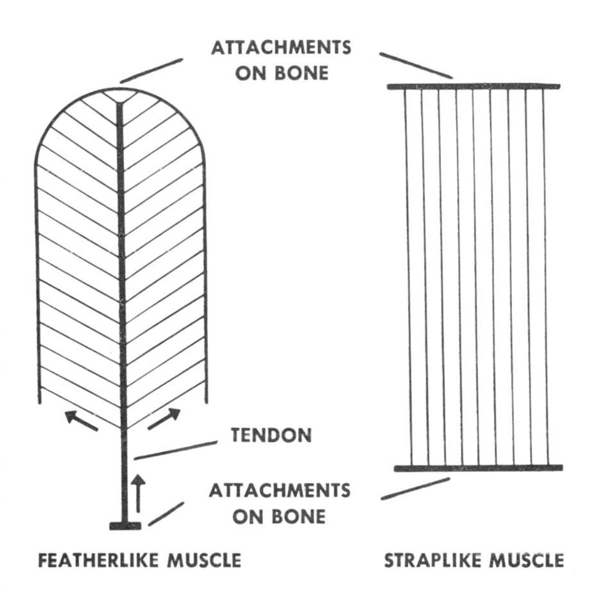


Featherlike and straplike muscles. The former have a central tendon to which the fibers are attached at an angle; the latter have longitudinally arranged fibers.

Our hands and fingers can perform numerous and infinitely varied movements. We can strike a hard blow with the clenched fist, apply light pressure to a glass we are washing or delicately poise the fingers to thread a needle. For such a variety of movements we need many muscles; there would not be room for all the necessary muscular attachments to the bones of the hands. The problem has been solved very ingeniously. Many of the muscles involved in moving the hands and fingers start in the upper part of the forearm; you can feel them when you bend and straighten the wrists and fingers. Near the wrist a great number of fibers are attached to a very few tendons. The forearm muscles manipulate the fingers of the hand by means of these intervening tendons.

Tendons also help muscles to withstand sidewise pressure and friction. Where muscles rub against each other or against bone, they develop protective tendon patches. There are always tendons where muscles have to turn corners, as in passing from the leg to the foot. Here the muscles are shielded by the tendons and the tendons in turn are protected against friction by bags of connective tissue, called tendon sheaths.

These sheaths contain a fluid that acts as a lubricant as the tendon slides back and forth. When tendons are moved quickly for long periods of time, the sheath becomes hot and sore and may fill with extra fluid. As a result the joint may swell, and any movement will result in great pain. The treatment in such cases is rest at the earliest possible moment.

When there is friction on only one surface of a tendon, the connective tissue bags do not surround the tendon but lie between it and the bone or other structure against which the tendon rubs. Simple flattened bags of this kind are called bursae (or bursas); we find them between the kneecap and the skin and in other places where friction occurs. A bursa reacts in the same way as a tendon sheath when there is excessive friction or infection; its cavity fills with fluid and the walls thicken. This condition is known as bursitis, or inflammation of a bursa. You are familiar, perhaps, with the kind of bursitis known as housemaid's knee — the inflammation of the bursa between the kneecap and the skin.

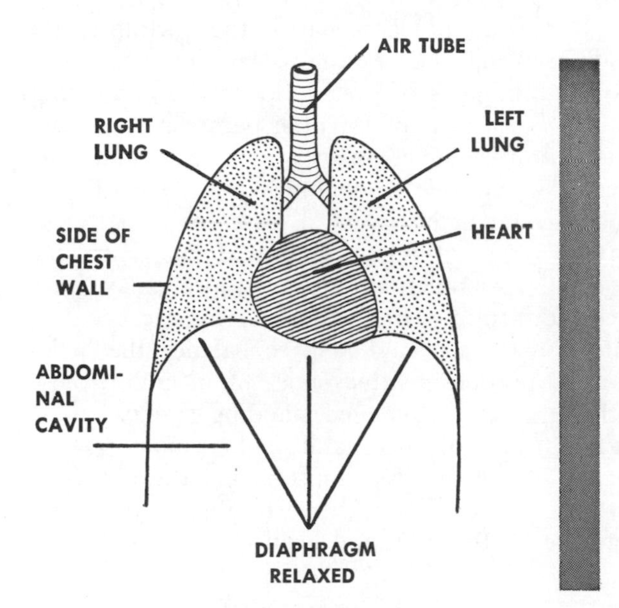


How we breathe. Whenever we exhale, the diaphragm is relaxed and the abdominal muscles (not shown) are contracted. The lungs shrink and gas is forced out of them.

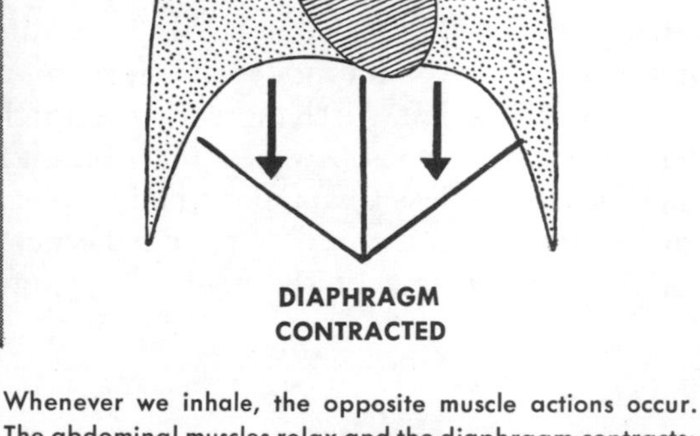


Whenever we inhale, the opposite muscle actions occur. The abdominal muscles relax and the diaphragm contracts; the chest and the lungs expand, drawing in a mass of air.

Muscles are used not only to cause movement but also to restrain movement by holding the bones firmly together. It is true that the bones are attached to one another by means of the bands of tough connective tissue that we call ligaments. But the muscles play the chief part in preventing unwanted movement; the ligaments are only a secondary line of defense, when the strain on joints is too great or too quick for the muscles to control. Even when the muscles seem to be relaxed, some of their fibers are contracting. This partial contraction is called muscle tone or tonus; it serves not only to hold the bones in place but to keep the muscles taut and ready for use.

Muscle tone helps us to maintain upright position or posture. When we stand still we can feel the firmness of the muscles in front of the hip joint and in the back of the thigh; we can feel them relaxing when we sway backward and forward. When we walk, the straightening muscle of the backbone keeps us from falling forward. Whether we stand or walk, the earth's pull always tends to make us slump, with bent head and backbone, sagging belly and bent knees. To prevent this, the brain and spinal cord send myriads of messages to muscles to keep up their tone.

We should always consciously aim at good posture when we stand or walk — or sit, for that matter. Otherwise we shall throw too much strain on some of the muscles involved in maintaining muscle tone; we shall not be able to breathe deeply and, if our bones are still growing, they may become deformed.

It is commonly known that muscles enlarge when they are exercised. The difference in size is due not to the formation of new fibers but to changes in the girth of the existing ones. A new-born infant's muscles weigh less than a pound and a half; an adult's weigh about fifty pounds because of the enlargement of the fibers. If exercise increases the size of the muscles, disuse causes them to become smaller, as many a bedridden patient can testify.

After injury, muscle fibers do not divide, as do skin cells, in order to repair the wound. The end of cut fibers grow to replace some of the lost tissue; but most of the repair work is done by connective tissue, which forms a scar. If the entire nerve path to a muscle has been severed by injury or destroyed by disease, the muscle fibers shrink and die. However, in such cases other muscles can often be trained to perform at least a part of the work formerly accomplished by the useless muscle. Such training is of great importance in treating the victims of infantile paralysis; it sometimes enables them to live a normal life.

Muscle trouble may not always be the result of injury or disease; as a matter of fact, anything that interferes with the proper functioning of the muscle may cause difficulty. For example, muscles control foot bones and help to balance the body. Improperly fitted shoes often cramp these muscles, prolonged standing throws undue strain on them and foot trouble results. This may affect muscles in other areas of the body, such as the thigh, the back and even the neck and head.

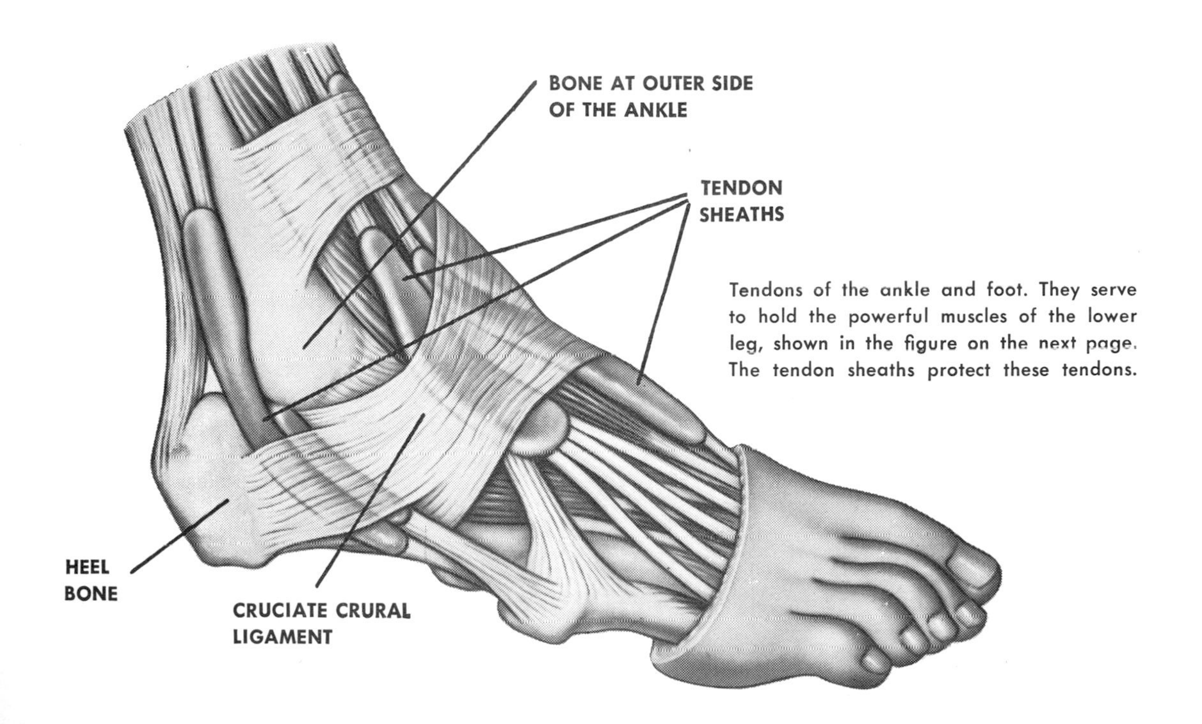


Tendons of the ankle and foot. They serve to hold the powerful muscles of the lower leg, shown in the figure on the next page. The tendon sheaths protect these tendons.

Excessive use of a muscle and too much strain upon it may cause stiffness and pain, especially when one tries to use the muscle. This is a rather common ailment among sprinters and baseball players; Americans call it "Charley horse." We do not know just what happens inside a muscle in such cases, since we ordinarily do not remove sections of living muscle for examination under the microscope. However, experiments with animals have shown that when muscle fibers are greatly fatigued, they are more fragile and inelastic than usual; they lose the power to contract.

Muscles are sometimes bruised as the result of a blow or fall. In such cases the fibers are injured and numerous small blood vessels are broken. The blood then passes into the tissue, which becomes swollen and painful.

Muscles in the vicinity of joints are particularly likely to suffer the disabling injuries known as sprains. These occur when the joints are overstrained or violently wrenched. There is tearing and displacement of tissues inside and outside the joint — the tissues of ligaments, muscles, tendons, tendon sheaths and bursae; often there is bleeding into the tissues. Sometimes the connective tissue of muscles becomes inflamed and gives rise to stiff neck and to lumbago (pain and stiffness of muscles low in the back). Cramps, which occur most often in the foot and leg muscles, are prolonged, painful contractions. Rheumatism is a name often applied to pains not only of the muscles but of the joints and various other areas of the body. Such pains may be due to a variety of causes.

Muscles are often affected by diseases of the nerves, blood vessels, bones, joints and other tissues. If a muscular disorder does not soon disappear, a doctor should be consulted, for it may be a warning signal of serious infection.

*See also Vol. 10, p. 276*: "Anatomy"; "Physiology."

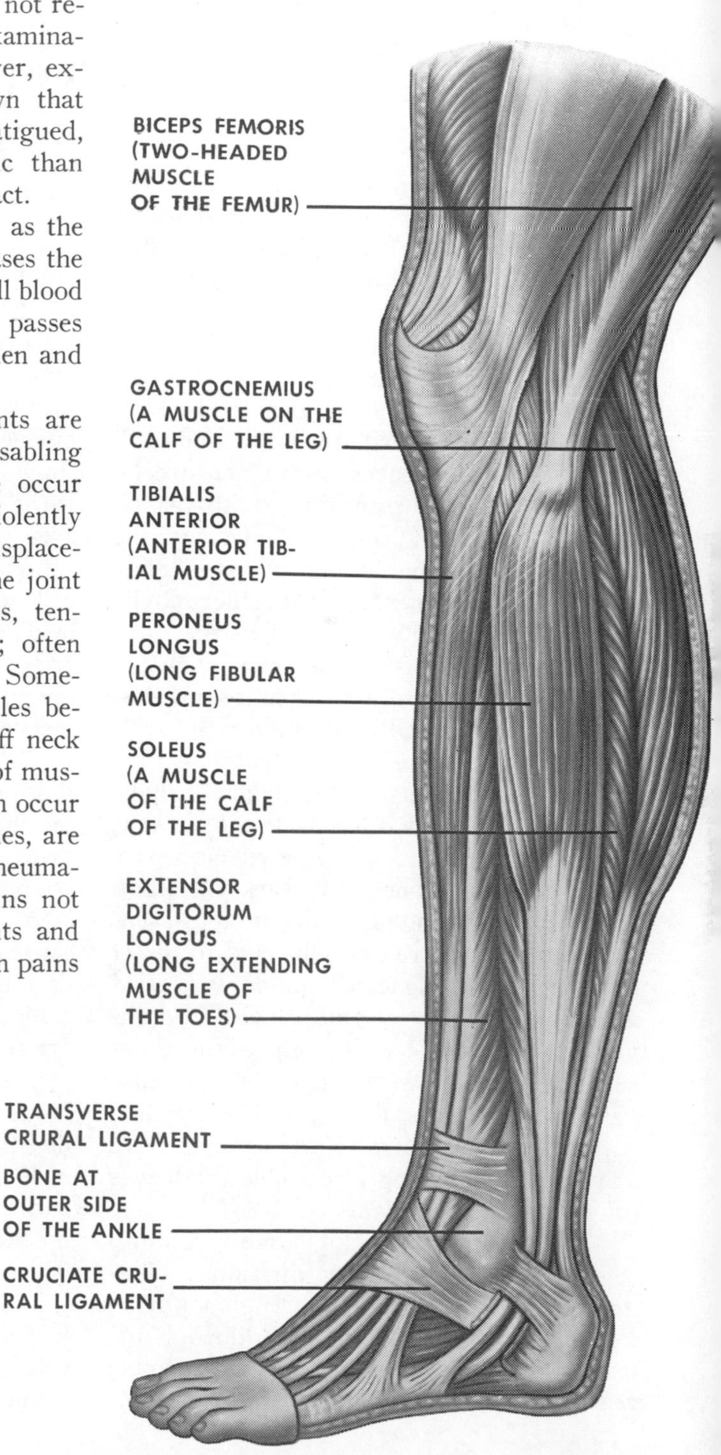


Muscles, tendons and ligaments of the leg and foot. The powerful muscles of the calf, such as the soleus and the gastrocnemius, help maintain foot posture and enable us to walk. The biceps femoris helps flex the knee; the peroneus longus and the tibialis anterior affect motions of the foot.

# THE ALGAE

## Primitive Plants, Infinitely Varied in Size and Shape

by M. H. BERRY

THE microscopic glasslike cells found in almost all the waters of the world; the scum that forms on the surface of stagnant pools; the bluish-green streamers attached to rocks in the vicinity of waterfalls; the immense ribbons of seaweed found on the rocks near the high-tide mark — these are all representative forms of the lowly plants known as algae. They belong to the subkingdom of the Thallophyta, which lack roots, stems and leaves and which are by all odds the most primitive of all plants. In one important respect, however, the algae resemble the higher plants. They possess the pigment called chlorophyll; with this pigment, which absorbs the radiant energy of the sun, they can manufacture food by the process of photosynthesis. (See Index, under Photosynthesis.)

The algae are chiefly water plants, dwelling in oceans, seas, lakes, ponds, rivers, ditches and other bodies of water, large and small. Some species, however, are found on stones, the bark of trees, fences and so on, generally in moist environments that are not subjected to direct sunlight. The algae are infinitely varied in size and shape. Some consist of individual microscopic cells; others form flat sheets, or narrow filaments or immense stemlike structures, that may be more than a hundred feet long. Certain algae have growths that strikingly resemble the leaves of various higher plants.

These primitive plants are divided into seven phyla, or primary divisions: blue-green algae, or Cyanophyta; Euglenophyta; Pyrrophyta; green algae, or Chlorophyta; brown algae, or Phaeophyta; Chrysophyta; red algae, or Rhodophyta.

### The Blue-Green Algae, or Cyanophyta

These slimy algae contain a blue pigment (phycocyanin) in addition to chlorophyll and other pigments. The blue masks out the other hues and gives most species a dark blue color; there are other colors, however, ranging from orange to black. Many species of the blue-green algae are one-celled; the cells are very simple and show no definite nuclei. Other species form colonies; in these, one-celled plants are joined together to form filaments (threadlike growths), sheets or balls. The cells of the colonial plants are generally held together in a mass of sticky slime secreted by the individual cells. The blue-green algae reproduce only by asexual means.

Blue-green algae are found wherever there is ample moisture, in almost all parts of the world. They often contaminate drinking water, causing a very disagreeable odor and taste. The hot springs of Yellowstone National Park are highly colored because of their presence; curiously enough, they bring about the characteristic reddish color of the Red Sea. Blue-green algae are found in the snows of the Arctic; some thrive in the digestive tract of man and the lower animals, apparently without causing ill effects. Certain members of the group, including *Gloeocapsa* and *Nostoc,* have formed a partnership with fungi, making up the separate group of plants known as lichens.

The blue-green alga called *Gloeocapsa* is generally found in the form of single cells. It sometimes forms colonies consisting of three or four cells enclosed in

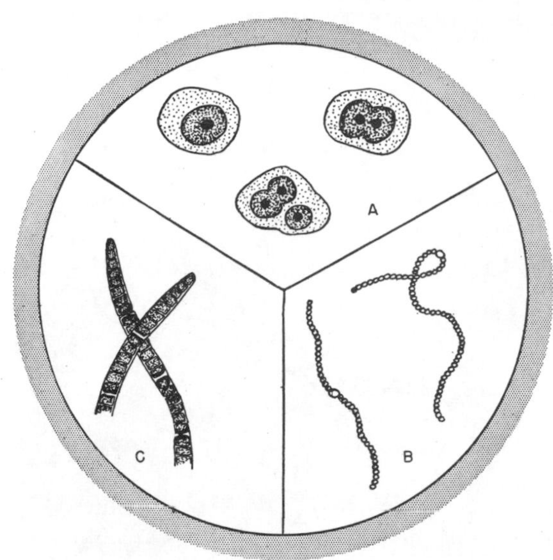


All drawings by the author

Three interesting members of the large blue-green algae group. A. *Gloeocapsa.* B. *Nostoc.* C. *Oscillatoria.*

a gelatinous sheet; eventually, however, these colonies break up into single individuals. *Gloeocapsa* usually occurs as a slimy coating on damp walls and rocks.

*Oscillatoria* is perhaps the commonest of the filamentous blue-green algae. It consists of a long, threadlike filament made up of separate rectangular cells, like so many dominoes placed side by side. Cell division at either end of the filament causes it to become longer. Often, gelatinous partitions are formed at several places within the filament, and the chain is ultimately broken up into several shorter ones. These continue to produce new cells until the chain reaches its former proportions.

*Oscillatoria* is one of the few plants that can move about; it has a wavelike, gliding motion. Botanists cannot account for this phenomenon, since the plant does not have the structures that often bring about motion in primitive plants and animals: flagella, or whiplike structures, and cilia, or hairlike structures.

*Oscillatoria* is found wherever there is ample moisture. Often it is to be seen floating in great profusion on the surface of lakes and ponds, forming the so-called dog-day waterblooms. One species is extremely abundant in moist soil and on flower pots in greenhouses.

Like *Oscillatoria, Anabaena* is a blue-green filamentous alga; it differs from the former in that it cannot move. It is perhaps the most abundant alga in North America. Its cells are beadlike in form; under the microscope the plant looks like an irregular strand of beautiful greenish beads. *Anabaena* is almost wholly aquatic and is generally restricted to fresh water; it is a favorite food of fishes. This alga is often the chief offender when the water supply of small villages becomes foul, with an unpleasant taste and a disagreeable odor.

### The Euglenophyta

These one-celled organisms, which are generally green in color, swim about in the water by means of flagella. (One group has no flagella.) They vary in form; they may be spherical, or egg-shaped or pear-shaped. Reproduction is generally by cell division. Sometimes considerable numbers of one-celled Euglenophyta form colonies. These algae, which are considered by botanists to be plants, are classified as Protozoa — one-celled animals — by zoologists; they are included in the class of the flagellates — animals possessing flagella.

The best-known of the Euglenophyta is *Euglena,* which is found in stagnant ponds, swimming pools and aquaria. This minute green organism has a single flagellum; it also has a minute red "eyespot," which seems to be sensitive to light. *Euglena* causes water to become greenish and cloudy, and often imparts an unpleasant flavor. Water containing excessive quantities of this alga is considered to be undesirable for drinking or swimming purposes.

### The Pyrrophyta

The members of this group, which have pigments ranging from yellowish to brownish, not only manufacture food in the presence of sunlight but also store reserve supplies of it in the form of starch or compounds similar to starch. Pyrrophyta are found in ocean waters and fresh waters. Most members of the group consist of a single cell and move about by means of one or two flagella; some one-celled species have no flagella and do not move. A certain number of the Pyrrophyta are colonial, forming filaments.

The blue-green algae belonging to the genus *Nostoc* (right) form a partnership with certain fungi, making up the plants known as lichens.

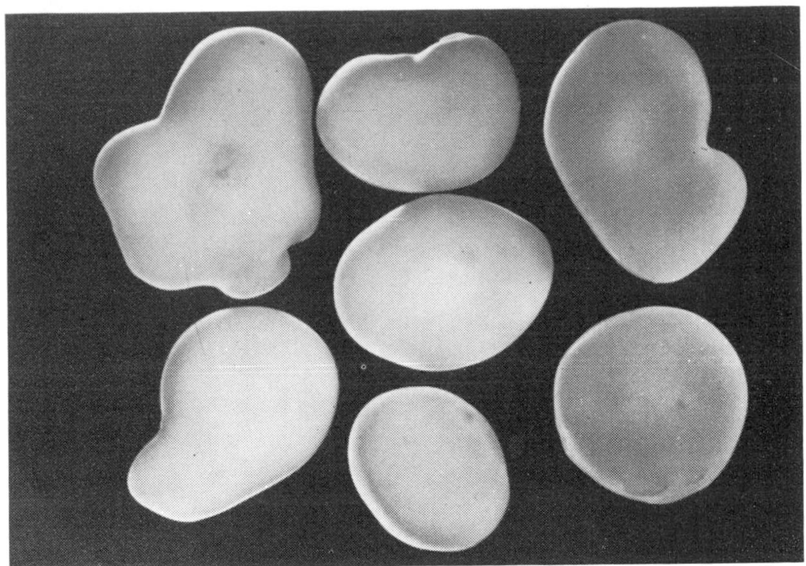

Gen. Biol. Supply House

Like the Euglenophyta, the Pyrrophyta are included by zoologists among the animal group of the flagellates.

### The Green Algae, or Chlorophyta

The green algae are chiefly to be found in fresh water, though there are some marine representatives of the group. Certain forms have adapted themselves to land life. They are to be found particularly in places where the environment is not too dry; they grow attached to moss, rocks, trees and soil. Occasionally these algae are found at high altitudes in patches of snow. Some species are one-celled; others form colonies. Still others are multicellular; they are in the form of filaments (sometimes with numerous branches) or flat sheets. They reproduce in various ways: by cell division, by fragmentation (breaking off into fragments) or by sexual reproduction. Certain species combine with fungi to form lichens.

Four representative types of green algae. A. *Protococcus.* B. *Chlamydomonas.* C. *Oedogonium.* D. *Spirogyra.*

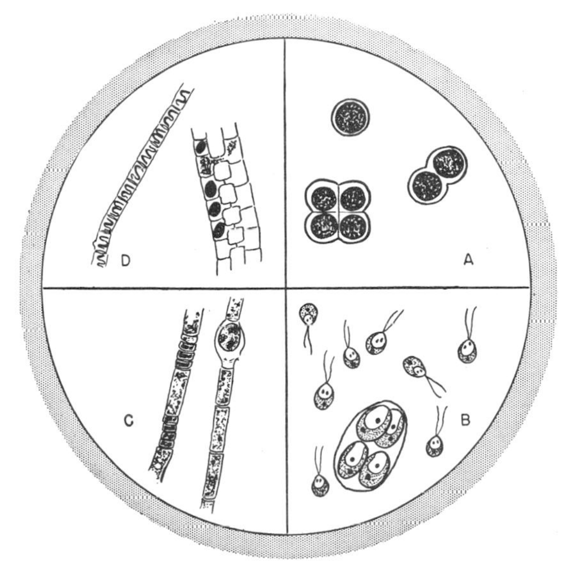


The green algae have an important bearing upon our lives, both for good and for evil. In the course of their food-making activity they add oxygen to the water, thus making more of this essential gas available for the fishes and other organisms that form an important part of our food supply; they also serve as food for these living creatures. On the debit side of the ledger, the green algae often cause pollution of the water in lakes, tanks, aquaria and the like; they may impart unpleasant flavors and odors. If they grow too thickly, their nocturnal respiration may cause the oxygen content of the water to be seriously lowered; as a consequence the fish in the area may die of suffocation. Fortunately, it is possible to eradicate unwanted green algae from swimming pools and tanks by adding minute quantities of copper sulfate ($CuSO_4$) to the water; generally one part of $CuSO_4$ to several million parts of water suffices.

Among the most interesting of the green algae is the form called *Chlamydomonas*. It is very common in ditches, pools and lakes, and is often so abundant that the water appears to be green. It has also been found in the Alps and the Arctic, where it covers entire snowbanks. *Chlamydomonas* is a very primitive plant; like so many other lowly forms of life, it can stand long periods of unfavorable environment.

*Chlamydomonas* often occurs as a single cell which moves about by means of flagella at the anterior, or front, end. The motion of this alga seems to be a pretty definite response to stimuli; for example, the plant moves toward moderate light and away from too much light.

Single individuals of *Chlamydomonas* sometimes withdraw their flagella and divide into two cells; these remain within a single cell wall. In many instances this cell division will go on until there are a number of cells, all included within the same cell wall. The arrangement is temporary, however; the individual cells ultimately break out of the wall, develop flagella and become free swimmers.

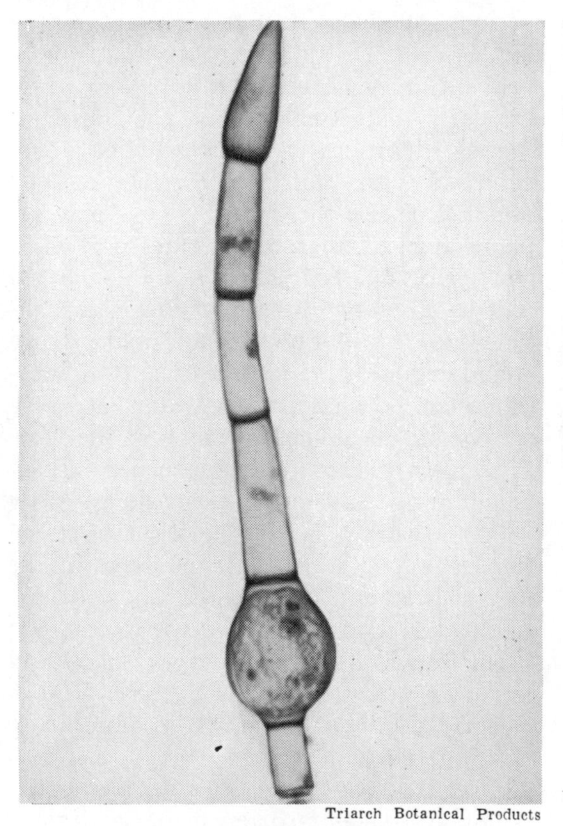

Triarch Botanical Products

*Oedogonium,* a fresh-water green alga that forms filaments. Some of these are attached; others are free-floating.

Each flagellated cell then swims about until it comes in contact with a free-swimming cell from another colony. The two of them fuse and their flagella disappear. We now have the cell called a zygote, representing a union of two gametes. If the environment is favorable (for example, following a spring shower), the zygote will divide and form four free-swimming cells.

*Ulothrix* is a green alga that thrives in fresh-water streams, particularly in places where the water does not get too warm. This plant forms colonies; its cylindrical cells are set end to end in a green filament. At the bottom of each filament there is a specialized cell called a holdfast. As the name indicates, this helps to anchor the plant to sticks and stones in the water. Sometimes the filaments are unable to attach themselves to anything, or else they fail to remain attached; in that case they are free floating.

Reproduction is often carried on by the asexual reproduction bodies called spores. Under certain conditions the protoplasm of the cell divides and forms spores; these are later freed from the cell, and swim about in the water by means of flagella. After a time they become anchored to solid objects in the water and the flagella are lost. Each single-celled spore then divides and divides again, until a new filament is formed.

In *Ulothrix* we also find alternation of generations, in which one generation of plants produces spores and the following one, sex cells. One or more of the cells along a filament may produce gametes, or sex cells, in much the same way as spores are produced. After they are freed from the cell, the gametes swim about in the water. When one of them comes into contact with a gamete from another colony, the two fuse and form a zygote. This later produces spores, each of which is capable of starting a new colony.

*Ulva,* or the sea lettuce, reproduces in the same way as *Ulothrix*. It differs from the latter chiefly in the size of its colony; it forms a sheet two cells thick, a few

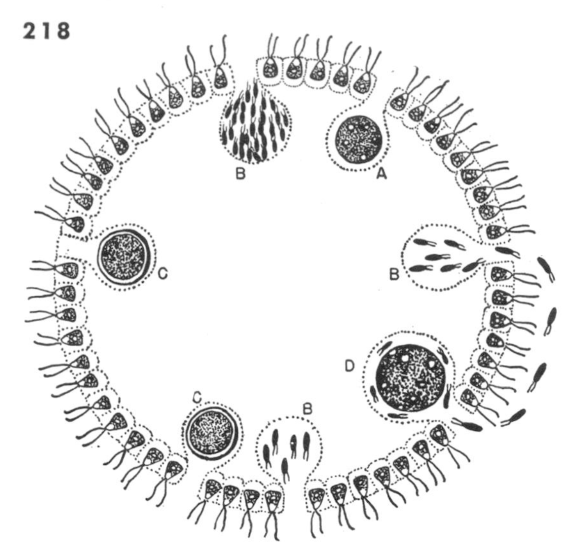


Above: Cross section of *Volvox* colony. A. Female gamete. B. Male gametes. C. Zygote. D. Fertilization taking place. Below: Photo of colonies.

inches wide and several inches long. *Ulva* is usually found in salt water or on piling and rocks along the shore. It often forms part of the debris along the shore after a storm. It is used for food in the Orient.

*Oedogonium* is a fresh-water green alga that forms filaments. Some of these filaments are free floating; others are attached to solid objects in the water. *Oedogonium* reproduces asexually or sexually. In sexual reproduction, certain cells produce large egg cells; others give rise to lively little male cells, provided with many cilia. These male cells ultimately find their way to the egg cells, which they fertilize.

The green alga known as *Volvox* forms a curious sort of colony — a hollow sphere made up of thousands of cells arranged side by side. The tiny cells are provided with flagella; these project beyond the outer surface of the sphere like fuzz on a peach. As the flagella whip about in the water, they give the whole colony a rather definite rotating movement. There are often many smaller colonies within the parent colony; they remain enclosed until the older colony ruptures. *Volvox* is a common alga of fresh-water ponds; it often serves as a host to the microscopic animals called rotifers.

The *Protococcus,* often called *Pleurococcus,* forms striking green layers on bark, on the shady side of trees. It is definitely a land plant; it is often called an aerial alga. *Protococcus* has marvelous powers of resistance; it can survive prolonged periods of drought and temperatures as low as —40°C. The cells are tiny, round, brilliant green objects, with one or two irregularly shaped chloroplasts (bodies containing chlorophyll). *Protococcus* is not equipped with flagella. It reproduces by asexual means; the daughter cells may remain attached, to form colonies of two to four cells.

In *Cladophora,* the filaments are divided to form branches not unlike those found in some of the higher plants. *Cladophora* can be found in salt water and fresh water alike at almost any time of the year. It seems to thrive under especially difficult conditions. It can be seen along rocky shores where it is constantly battered by the waves, or attached to boulders at the base of waterfalls, with the plant body extending several feet downstream; it has been found at the bottom of lakes and under the ice that forms on rivers. The cells of *Cladophora* usually contain more than one nucleus.

The green scum seen on quiet pools, ponds and lakes is often made up largely of the green alga known as *Spirogyra.* This plant gets its name from the peculiar arrangement of the chlorophyll, which extends like a spirally twisted ribbon from one end of the cell to the other. *Spirogyra* is the genus of green algae commonly studied in elementary botany classes; the plant is easy to obtain, and its cells are fairly large. It is used for food in some parts of the world; it is a constituent of Japanese lens paper.

*Spirogyra* is a filamentous plant, in which the cells are quite elongated. The colony is enlarged by cell division, but the number of colonies is increased by a rudimentary form of sexual reproduction. When the filaments are lying close together, the adjacent cells develop protuberances. As these come into contact, the cell walls dissolve. The cell material from one cell (sometimes called the male cell) flows into the other cell (usually called the female cell). The two nuclei merge, and a zygote is formed; this secretes a thick wall and drops to the bottom of the pool. The zygote later develops into a new filament. Sometimes a cell produces a cell similar to a zygote without uniting with another cell; this zygotelike cell can produce a new filament in much the same way as a normal zygote.

The desmids are generally considered to be one-celled relatives of *Spirogyra,* because they reproduce in the same way. They are among the most beautiful of all microscopic plants. To the unaided eye they are only tiny green specks, but under the microscope they rival snowflakes in their infinite variety and their beauty of design. There are many different kinds of desmids; perhaps the most striking of all is the jewellike *Micrasterias.*

Desmids are found in almost any sunny area in fresh water. Fine-meshed towing nets are often used to collect them; it is often possible to obtain them by scraping the bottom mud and debris of almost any pool. Each desmid cell has a constriction in the middle which makes it look at all times like a cell on the point of dividing. Some desmids are united to form chainlike colonies; each cell of these colonies is sharply differentiated from all the others. Certain desmids can move, after a fashion.

An unusual type of green alga is *Vaucheria,* which is found in shallow water or in clumps of damp soil. This plant has no distinct cell walls; many nuclei are scattered throughout the protoplasm, just as in the case of certain molds. It differs from these molds, however, in that it has chlorophyll. Like so many other green algae, *Vaucheria* reproduces both asexually and sexually.

### The Brown Algae, or Phaeophyta

The characteristic brown color of these algae is due to a brown pigment, fucoxanthin, which under normal conditions masks the green color of the chlorophyll that is present in the tissues. The brown algae, which are almost all marine

Living strands of a giant species of *Spirogyra,* a genus of green algae. The cells of this plant are elongated.

Photos, Gen. Biol. Supply House

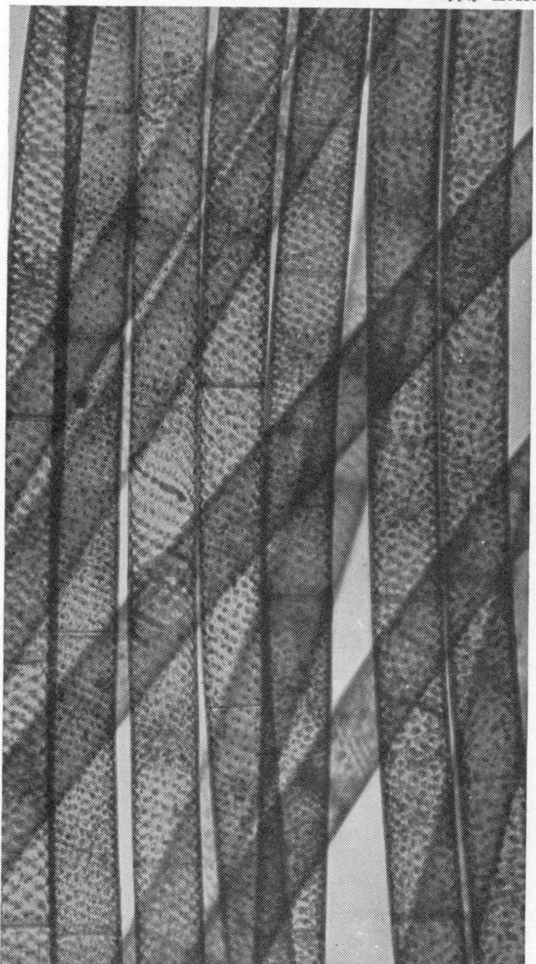

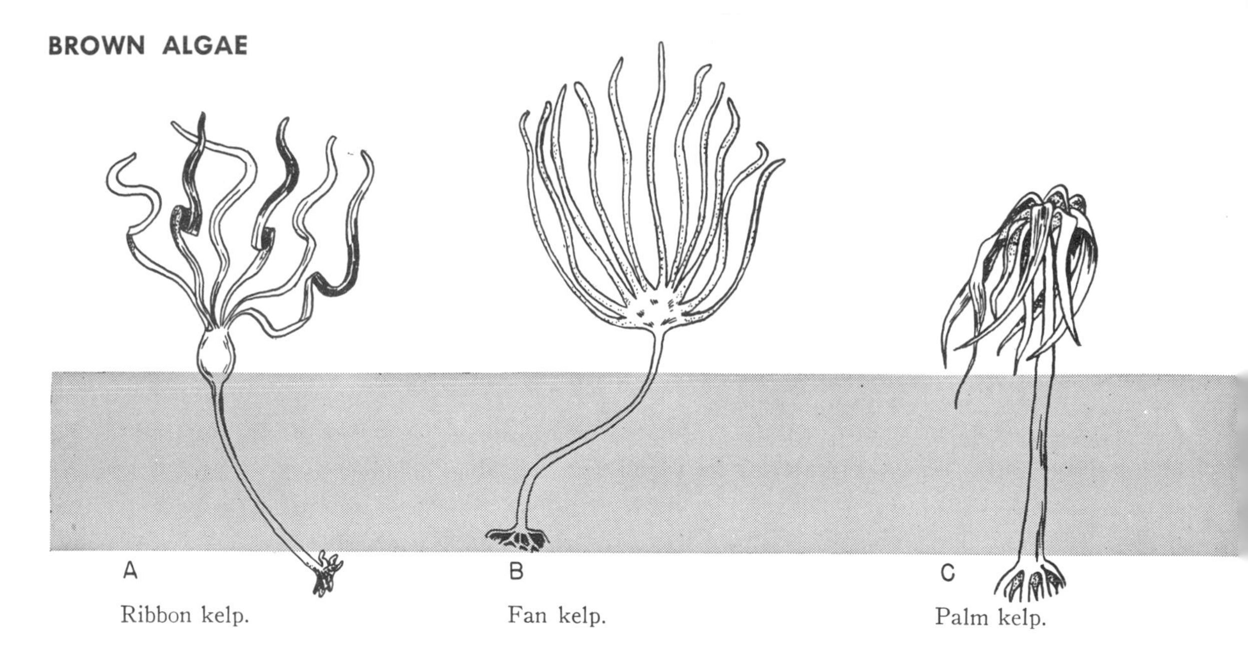

Ribbon kelp.

Fan kelp.

Palm kelp.

plants, show considerable variety of structure. Some are in the form of filaments; others are sheetlike or ribbonlike. Certain members of the group have structures resembling the leaves and stems of higher plants. They are generally attached by means of holdfasts to rocks and other solid objects in the water. Many of them have hollow structures that hold air and that help the upper part of the plant to float on the water or not far below the surface. Some brown algae are found only on the seashore; they are alternately submerged and exposed to the air. These species are generally covered with a jellylike substance that holds water and prevents the plants from drying out while exposed to the air.

In various brown algae there is alternation of generations. At certain times of the year parts of the plant develop spore-producing organs, called sporangia. The spores that develop from these organs develop into plants that produce sex cells; these unite to form spore-producing plants. Other brown algae form spores which function as sex cells.

The brown algae are of considerable commercial importance to man. They are a source of food for fish and other animals living in the sea; when removed from the sea they are sometimes used as cattle feed. In the Orient and Europe they often serve as food for human beings; they are eaten fresh, or dried, or made into soups and broths. Some species yield iodine; others make excellent fertilizer. A British chemist perfected a process for utilizing these algae as a source of rayon.

The brown algae known as kelps include the largest members of the group. The giant kelp (*Macrocystis pyrifera*) is reported to be the longest plant in the world. Thriving at depths of fifty feet or more, it often grows to be many feet wide and well over a hundred feet in length. It is harvested in California as a source of algin, a gelatinous material often used to give body to ice cream and other desserts.

Ribbon kelp (*Nereocystis lütkeana*) is often found along rocky shores and in shoals, and it is sometimes bothersome to small craft. The sea palm (*Postelsia palmaeformis*) is confined to the northern part of the Pacific coast. *Laminaria,* or the common kelp, is a comparatively small member of the group; it generally does not exceed eight feet in length.

The brown algae that are alternately submerged and exposed to the air are called rockweeds. One of the best-known is *Fucus.* The members of this genus are flat, much-branched plants, which are attached to rocks by means of holdfasts and are buoyed up by bladderlike structures. Because of these structures, the plants are sometimes called bladder wracks. *Fucus* is widely presented in elementary biology classes as a typical representative of the

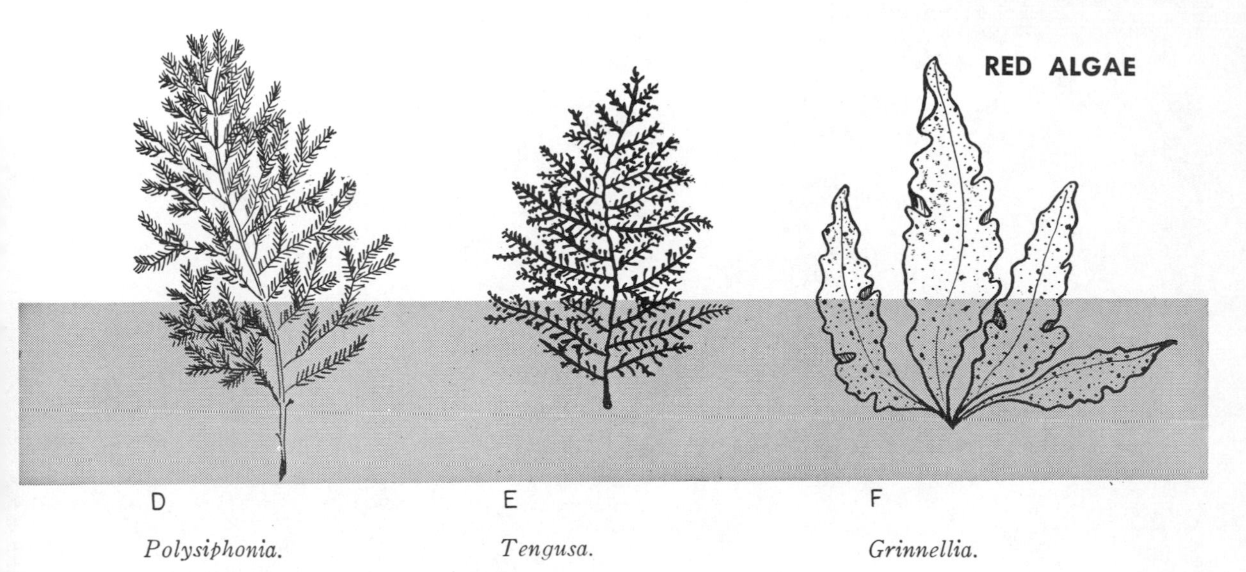


*Polysiphonia.* *Tengusa.* *Grinnellia.*

brown algae. It does not have alternation of generations; gametes develop at the ends of the plant, and these gametes reproduce their kind.

Gulfweed, or *Sargassum,* is a brown alga provided with berrylike bladders; it has leaflike growths set on stemlike structures that sometimes reach great length. Masses of *Sargassum,* torn away from their moorings, are often carried along by ocean currents and collect in floating mats. They are found particularly in the Atlantic area called the Sargasso Sea, extending from the West Indies to the Azores. Columbus was the first to report the existence of the floating seaweed formation in this area. It was widely believed at one time that ships were sometimes so enmeshed in the seaweed that they could not work their way free; but this is now held to be one of the innumerable tall yarns of the sea.

### The Chrysophyta

These algae have color cells ranging from yellowish green to yellowish brown; that is how the name Chrysophyta is derived. (*Chrysos* means "gold" in Greek.) The best known members of the group are the minute plants known as diatoms, which belong to the class of the Bacillariophyceae.

The diatoms are generally one-celled organisms, though they are sometimes found in colonies. They are brownish in color, because of the presence of the pigment diatomin, which resembles the pigment of the brown algae. The cell wall of the diatoms contains silica; it is a glasslike covering, consisting of two sections that fit together like the top and bottom parts of a box. The cell walls are beautifully and delicately sculptured. Some are round; some, oval; some, triangular; some are simple bars with hundreds of crosswise markings. The cell walls, after death, become skeletons which retain their shape for amazingly long periods of time. Some species of diatoms are able to move by swimming, gliding or twisting. Diatoms

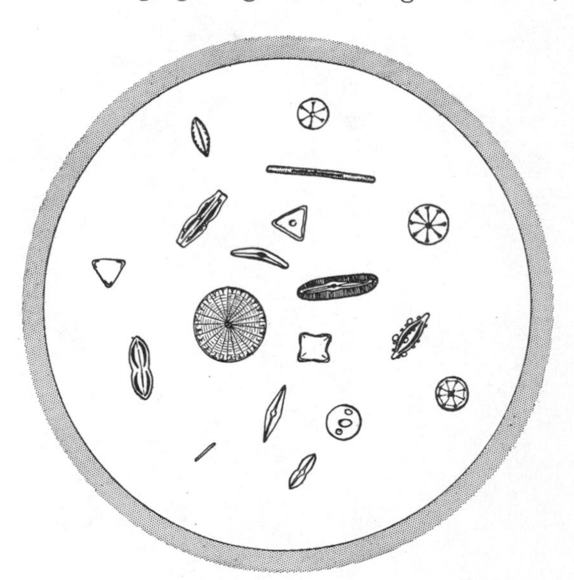

Wonder world of the diatoms, whose cell walls form a glasslike covering.

Amer. Mus. of Nat. Hist.

The microscope reveals microorganisms galore swimming amid strands of filamentous algae.

Right: rockweeds and various other marine algae at low tide. Rockweeds flourish particularly on rocky shores like the one that is shown below.

Hugh Spencer

Standard Oil Co. (N. J.)

generally reproduce by cell division; in some forms a simple form of sexual reproduction takes place.

These tiny plants are found almost everywhere where there is moisture and light: in the sea, in lakes and ponds, in flowing streams, in pools, on moist rocks and in cultivated soil. They are sometimes found even in purified drinking water; fortunately, they are harmless to man. They often make up the bulk of the plankton, the passively floating or weakly swimming plant and animal life of the ocean. The diatoms form thus a sort of ocean pasture, on which countless sea animals feed.

When diatoms die, their skeletons drop to the bottom of the sea or the lake where they have lived. The constant rain of these minute particles over countless centuries has resulted in the formation of large deposits of diatomaceous (diatom-containing) earth, ranging up to hundreds of feet in thickness. Diatomaceous earth is used in different ways. It serves to filter and clarify many liquids. It is an excellent insulating material for boilers, blast furnaces and refrigerators. It is also used as a mild abrasive in metal polishes, scouring powders and tooth pastes.

N. Y. Botanical Garden

The red alga shown above — *Gelidium cartilagineum* — yields the gelatinous material known as agar-agar.

### The Red Algae, or Rhodophyta

The red algae, which derive their color from the red pigment phycoerythrin, are chiefly to be found in the ocean waters of the earth; those growing in fresh water occur principally in cold, swiftly flowing streams. These plants are multicellular; they occur in the form of filaments, or ribbons, or sheets of fernlike or featherlike growths. Generally they range up to a foot or so in length. Almost all of the marine species grow attached to solid objects in the water by means of holdfasts or by special filaments. In practically all species there is alternation of generations. Neither the spores nor the sex cells are provided with flagella; the male sex cells float passively in the water until they come into contact with an egg cell, which they proceed to fertilize.

Like other algae groups, the red algae supply abundant food for fish and other animals living in the sea. They also serve as food for humans, particularly in Europe and the Orient. Among these edible algae are the varieties known as Irish moss (*Chondrus crispus*) and laver (several species of the genus *Porphyra*). Irish moss is also used for curing leather and for shoe polish, as well as an ingredient in the manufacture of creams and shampoos. Certain red algae, including Ceylon moss (*Gracilaria lichenoides*), yield a gelatinous material known as agar-agar. This substance absorbs a great deal of water; when it sets, it has a consistency like that of gelatin. It is used by researchers as growth material for bacteria. It also serves to thicken soups and broths, as a sizing material for textiles, as a mild laxative and to provide body for puddings, pastries, ice creams and other preparations.

Some of the red algae secrete lime; they have helped to build numerous coral reefs in the Indian Ocean and in the waters of various other parts of the world.

*See also Vol. 10, p. 272:* "General Works."

# POPULATION PROBLEMS

## The Mighty Surge and Ebb of Human Tides

FROM a very early age, human beings tend to work together and play together in groups. John Jones is born into the group that we call the family. When he is five or six, he joins a little "club" made up of a half-dozen neighborhood children. He becomes a member in turn of a grammar-school class, a high-school class and a college class. On certain Saturdays in the fall, he and fifty thousand other persons form a group that for two or three hours intently watches twenty-two young men playing a game. During the course of his lifetime, he will be associated with still other groups —fellow employees; a fraternal order; perhaps an amateur orchestra.

The most important group of all is that to which we give the name "population." A population group may be that of a town, a country or a major area on the globe. It is made up of men, women and children, bound together by such ties as language, customs and beliefs. The ancient Egyptians formed a population; so did the Jews of old; so did the Visigoths who helped to overthrow the Roman Empire. The French, the Germans, the Italians, the Danes, the Iranians (Persians), the Irish are all modern populations.

The first human population had to compete for food with other species of living things of the animal and vegetable kingdoms. Animals eat plants or else animals that have eaten plants. Plants themselves need animal products. This grim struggle for existence is, therefore, one of give-and-take.

It is a fact that every species of animal can multiply almost infinitely if left unchecked. The actual reproduction rate of many species seems almost incredible. It is estimated, for example, that the microscopic one-celled animal called the ameba, if unchecked, would in thirty days produce a mass a million times larger than the sun. The average American oyster produces about 16,000,000 young a year. At this rate the fifth generation — the great-great-grandchildren of a single pair of oysters — would form a mass eight times the size of the earth. It is true, indeed, that, as a general rule, the higher we go up the scale of evolution, the lower the birth rate is. Highly developed species like the lion, the elephant and the apes produce only a few young in a lifetime. But even species like these would crowd the globe in time if there were no natural checks.

In the struggle between species, sometimes a species will succumb. More often, it will succeed in maintaining itself on a certain level. This happens even if the reproduction rate is exceedingly high. In such cases, the vast majority of the individuals belonging to the species will die before birth, or in infancy or before having had a chance to reproduce. The over-all deaths in question will just about keep pace with the fantastic birth rate. This is also generally true of species that ordinarily have relatively few young. However, it sometimes happens that because of various factors, so many individuals of a given species will survive that this species may ultimately expand over a vast geographical area. Man is the best example.

Some animals, including toads, have a high reproduction rate. A typical toad egg mass is shown in the upper drawing. The cubs shown in the lower picture are the entire litter of an animal—the lioness—that produces comparatively few young in the course of a lifetime.

In the struggle for survival, early human populations enjoyed certain advantages over their competitors. Man possessed an exceptionally competent brain, capable of great development; he had speech, by means of which he could communicate effectively with his fellows. He had an "opposable" thumb, which could meet squarely the tip of each finger on the same hand; this enabled him to acquire a high degree of manual skill. The apes and monkeys also had opposable thumbs; but their brains were not so highly developed and they lacked the faculty of speech.

By taking full advantage of his special gifts, man was able to survive and, very slowly at first, to increase his numbers. It then became a matter of necessity for him to extend his living space. He was brilliantly successful in doing so. He invaded other areas which had an environment similar to that in which he had lived; he also adapted himself to quite different environments without evolving into a different species.* In the course of time, he became increasingly successful in creating his own environment.

#### Human origins and population movements

For a long time, Central Asia was regarded as man's ancestral birthplace. Since the discovery of very primitive manlike skeletal remains in Africa, authorities are no longer sure just where our birthplace was. At any rate, fossil evidence has shown that various human stocks, from a very early time, were to be found in a number of areas in the Old World — in Africa, Asia and Europe. On one point there has been general agreement so far: the Western Hemisphere did not see the rise of any na-

*Some authorities believe that, though men did not evolve into different species, they became differentiated in the course of their development into the existing more or less dissimilar physical races. According to another theory, modern races developed from separate species of prehistoric men and pro-men in the continents of the Old World. These prehistoric stocks followed a parallel course of evolution, resulting in the different physical races of today. See the Races of Mankind, Vol. 8.

tive human stocks. Very probably, *Homo sapiens* (modern man) migrated into the New World from Asia by way of a broad land connection between Alaska and Siberia more than 10,000 years ago.

The earliest men did not stay in one place, but moved around in search of food and shelter. This flow of population was much like that of a liquid, moving in the direction of least resistance.* It was exceedingly slow; it went on for hundreds of thousands of years.

From a very early time, various human stocks were to be found in a number of areas in the Old World—in Africa, Asia and Europe. Above is shown a representative of one of these stocks—Java Man. The first specimens of this stock were discovered in Java in the 1890's by Eugene Dubois, a Dutch paleontologist and anatomist.

**Factors accounting for the vast increase in human population**

It is true that human migration permitted great increases in population by extending the range of mankind. Yet population movement alone, no matter how extensive, could never account for the human world population of today. Man requires a vast amount of land for his living if he is dependent solely on the supplies that nature provides unaided. A primitive population must always be a sparse population; the number of individuals who can find support in a given area of even rich and fertile land is exceedingly small. But, fortunately for man, while he was gradually spreading over the world, he was developing special tools, like the stone hammer, the bow and arrow, the plow and the wheel. He was also developing special techniques — that is, methods for using these tools.

With these advances in tools and techniques, more human beings found it possible to live on a given area of land. A comparatively modern example illustrates this point. It is estimated that the hunting tribes of eastern Canada, long before the white man came, required from 250 to 400 square miles to furnish food enough for a single family. Gradually, the Indians added to their resources by means of agriculture, fishing and stock-raising. A tribe that used all of these resources found it possible to support as many as 500 people on an area that had been enough for only one person before.

Another important factor in the increase of population was the rise of trade. Primitive groups need all their manpower to provide food and other simple essentials; they have no time to seek relations with other groups. But as tools and techniques become more efficient, different populations learn to know each other. They discover that there are certain commodities that one or the other can produce best. It becomes profitable, therefore, to barter.

Cities gradually arise at the places where exchanges of commodities take place. Many of the greatest cities of antiquity, probably the very earliest of them, grew up in this way. Sometimes, they were located at places where natural trade routes intersected. In other cases, cities came into

* Not until much later did migration become a deliberate movement of peoples en masse from one locality to another.

An important factor in the increase of population was the growth of trade and the rise of cities at the places where exchanges of commodities would take place. One of these cities was Tyre, whose bustling port is shown at the right. This Phoenician city was an important trading center in antiquity.

being at points where land-borne goods had to be transferred to vessels, or where human carriers shifted their loads to camels or horses.

The exchange of products permitted increased population because a tribe would no longer have to obtain all its essential foods and materials from its own area. At first, the exchanged products were essential commodities, like salt, meat, furs for clothing and the like. As time went on, the products included not only essential goods but also comforts and even luxuries. By this time, a great number of persons could live in a comparatively small area because their needs and their luxuries could be supplied by imports from the far corners of the earth. The city of Rome in the days of the Empire received its corn from Egypt, its silk and spices from the Far East, its tin from Britain, its silver from Spain, its pottery from Greece.

Of course, as population increased, population pressures increased also. They varied widely from one age to another. Sometimes natural calamities like the plague and drought would depopulate considerable areas of the world's surface; sometimes man-made scourges, like wars and massacres, would produce the same effect. Yet in time the losses were made up, and population pressures began to mount again. Then the surge to "green fields and pastures new" once more began.

#### The Malthusian theory of population growth

Men must have been dimly aware of these problems from very early times. It was not, however, until the dawn of the nineteenth century that anyone worked out a theory of population growth. The man who did this was Thomas R. Malthus (1766–1834), an English clergyman.

In the closing years of the eighteenth century, men were very much interested in the possibility of unlimited progress for mankind. Malthus's father believed firmly that such progress was possible. He maintained that if men could only perfect their political systems, life would be forevermore happy, prosperous and peaceful throughout the world. The younger Malthus denied that this was possible; he maintained that the pressure of population would inevitably result in misery for mankind. Father and son took part in heated discussions over the matter. The son eagerly sought support for the arguments that he sometimes drew out of thin air; thus he was led to examine the whole problem of population size and growth. His famous ESSAY ON THE PRINCIPLE OF POPULATION appeared in 1798. An immediate sensation, it brought about heated controversy, which has not yet died down. Malthus devoted five years to research and travel and then brought out a second edition of his work (1803).

The Malthusian theory is based to a large extent on the difference between geometric and arithmetic progression. These formidable-looking terms are really easy to understand. Let us take two series of numbers:

(1) 2, 4, 6, 8, 10, 12.
(2) 2, 4, 8, 16, 32, 64.

In both series, we start with the number 2. In the first series, we add 2 to 2 to make 4; we add 2 to 4 to make 6; we add 2 to 6 to make 8 and so on. That is, we go forward by small steps, each step higher by 2 than the one preceding it. This is an arithmetic progression.

In the second series, we multiply 2 by 2 to make 4; we multiply 4 by 2 to make 8 and so on. In other words, we produce a given number in the series by multiplying by 2 the figure preceding it. This is a geometric progression.

Arithmetic progression is comparatively slow. If we take 2, 4, 6, 8 and so on up ten steps, the number is only 20; the twentieth step only 40; the thirtieth is only 60. But in the geometric progression, the results are really startling. The tenth step of 2, 4, 8, 16 and so on is 1,024; the twentieth step is 1,048,576; the thirtieth is 1,073,741,824!

Now Malthus took the position that human beings can produce their young in what is practically a geometric progression. His study of population problems in the American colonies convinced him that under favorable conditions a human group could double itself every twenty-five years. Such a progression would lead to enormous numbers in a very few centuries. But, of course, all these individuals would have to be fed. Malthus, therefore, turned his attention to the practical possibilities of increasing food production. He came to the conclusion that mankind could not hope to increase its subsistence by more than an arithmetic ratio. That is, man would add to food production every twenty-five years an amount equal to that which was being produced at the time that Malthus wrote.

Under such conditions, the geometrical increase in population would have to be limited by the arithmetical increase in food. Population would always tend to press upon food supplies; and in Malthus's opinion, this pressure would be so great that food supplies would always run short, and much of the population would be condemned to a life of misery.

Malthus realized that food supplies can be increased by advances in farming and other techniques. But he claimed that no matter how advanced its methods might be, society would never succeed in providing for all the offspring that man could produce. Therefore there must be some checks. Malthus divided these into two groups. In the first group, he included the preventive checks that limit the birth rate, such as celibacy, deferred marriage and vice. The second group consisted of positive checks that increase the death rate; here he included war, famine, pestilence and, again, vice. All of these checks involved misery,

The English clergyman Thomas R. Malthus held that food production could be increased only in an arithmetic progression, whereas human population, if unchecked, increased in geometric progression. The diagrams show five stages in such progressions, the starting number being 2. Note that we keep adding by 2 in the arithmetic progression and multiplying by 2 in the geometric progression.

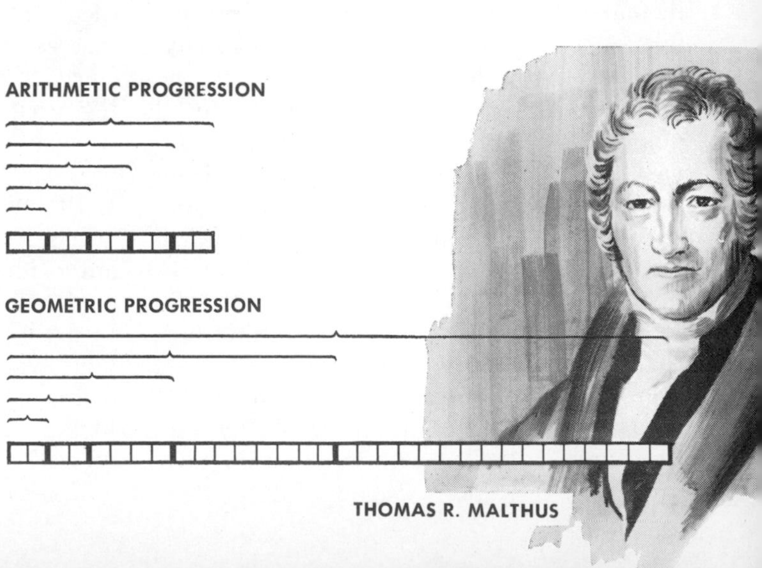

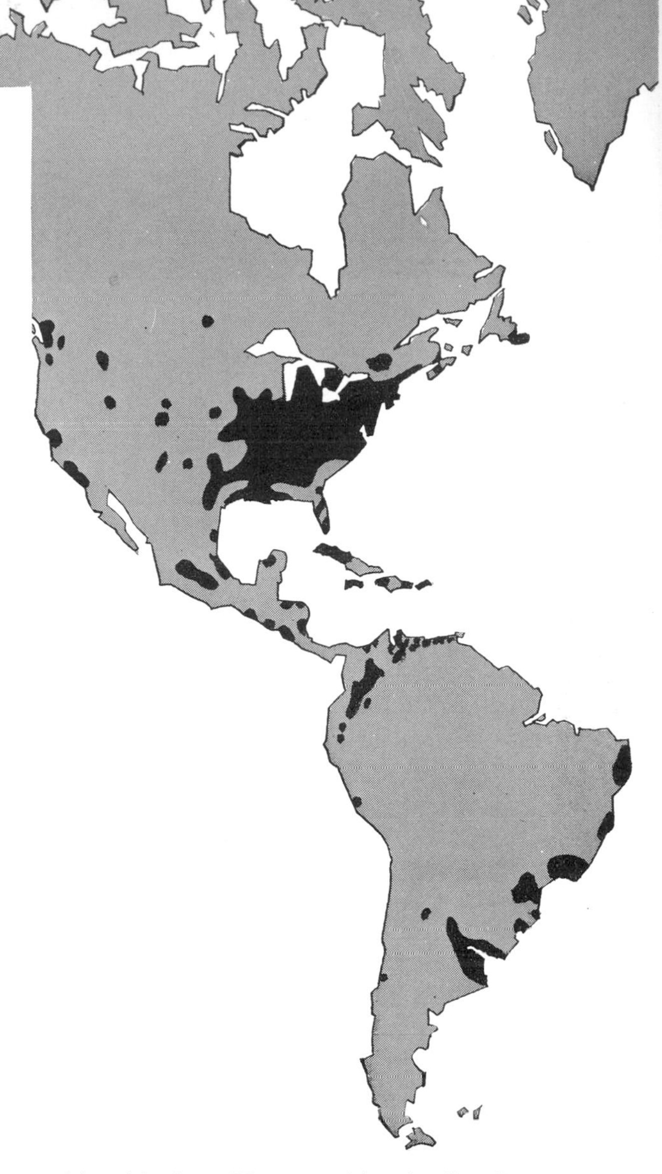

said Malthus. But he insisted that the minor misery of the preventive checks was preferable to the major misery of the positive checks. He advised young people to postpone marriage until they felt reasonably sure that they could provide satisfactorily for their children. In this way, they would be serving both themselves and society.

The Malthusian theory is still being argued. Its opponents insist that history has disproved it. The population of the world, they say, has more than tripled since the publication of Malthus's essay. Indeed, the human race has added more to its total numbers in the century and a half since Malthus wrote than in the whole previous span of its existence. In 1800, there were only about 900,000,000 people in the world. At the present time, the world population is about 3,000,000,000; yet the general level of living of many peoples has gone up amazingly. Where, then, has been the pressure; where the terrible misery?

The supporters of Malthus claim that the English clergyman's theory has not yet been refuted. It is true that the population of the world has increased as never before and that there has been a seemingly marked improvement in the general level of living. But certainly there has been no dearth of war and vice, nor any lack of pestilence and famine, if we consider the world as a whole. As for misery, it is probable that there are actually more people in the world living in destitution and wretchedness than there were in 1798.

Malthus said that men could not hope to increase food production every twenty-five years by an amount greater than that which was produced at the time that he wrote. In 1800, enough food was produced to provide (rather badly) for 900,000,000 people. If that amount of food had been added to the world's total supply each successive quarter century, by the middle of the twentieth century there would have been food enough for over 6,000,000,000 people. Actually, say the Malthusians, there are only about 3,000,000,000 people and no surplus food. Not so, say the opponents of the Malthusian theory. The world, they say, is plagued with local surpluses — and local lacks. If we could only distribute available food supplies effectively, there would be no shortages.

Renewed interest in the Malthusian theory has been brought about by a terrific increase in the rate of population growth since the end of World War II and a resulting great increase in the number of the world's inhabitants. This "population explosion" is discussed elsewhere in THE BOOK OF POPULAR SCIENCE (see Index). It has caused a good deal of concern. Many authorities fear that a grave crisis may confront the world in the not-too-distant future if present population trends continue. The available land, they say, will become overcrowded; natural resources will no longer suffice to support the greatly increased number of inhabitants. Human

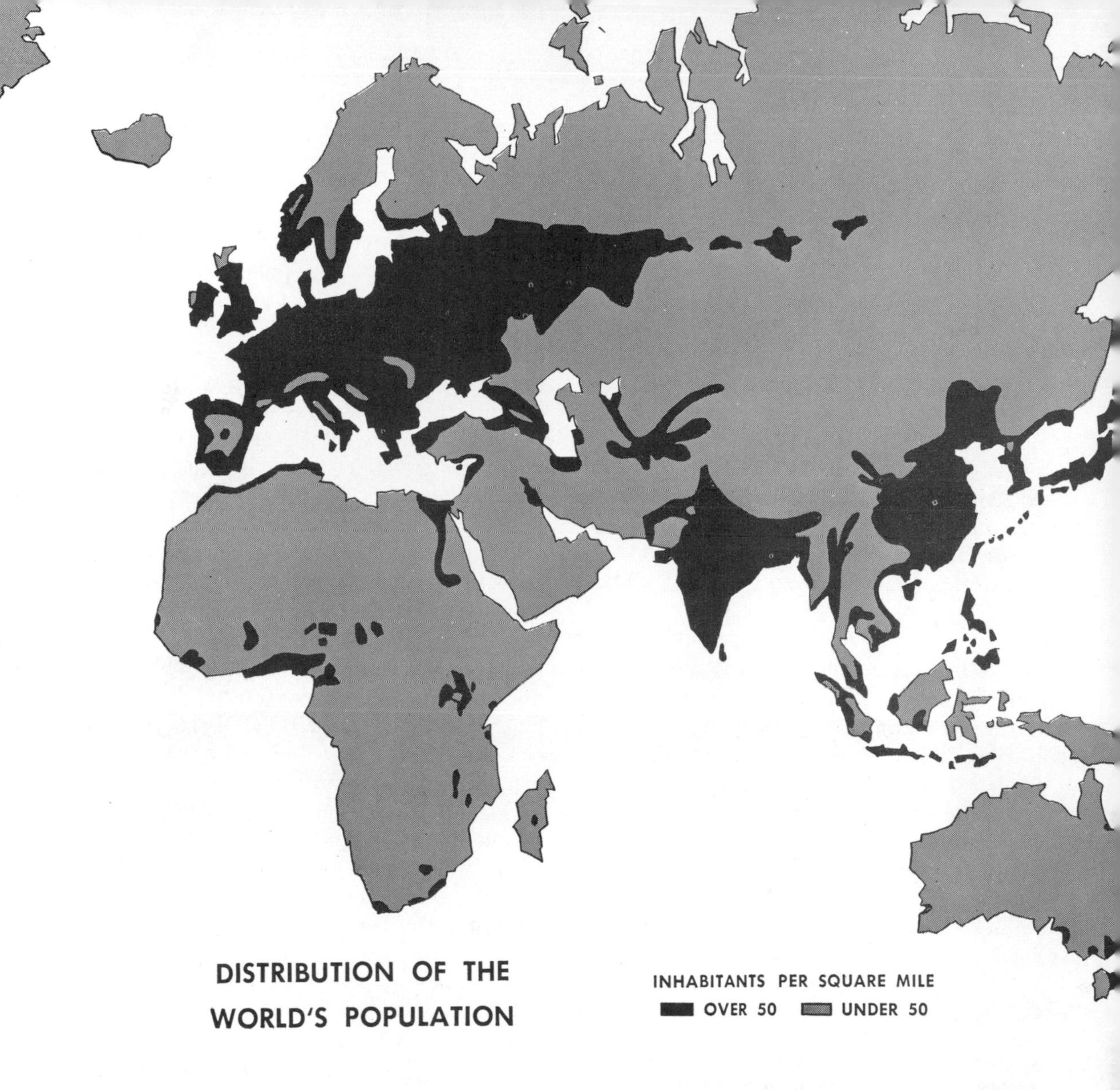


DISTRIBUTION OF THE WORLD'S POPULATION

tensions and discontent may build up to such a degree that another world war may become inevitable. The gloomy predictions of Malthus may well become grim reality.

#### An analysis of the world population

As we have pointed out, the population of the world today is about 3,000,000,000. Its distribution is very uneven. More than half the world's inhabitants are crowded into the single continent of Asia, whose population is 1,665,000,000 (not including the Asiatic portion of the Soviet Union). Of this vast total, about 672,000,000 dwell in China and 408,000,000 in the peninsula of India. Something like 427,000,000 people live in Europe, not counting the Soviet Union. The latter has a total population of about 214,400,000 in its European and Asiatic domains; North America has some 265,000,000 inhabitants; South America, 140,000,000; Africa, 244,000,000. Australia, one of the most sparsely inhabited continents in the world, has not much more than 10,000,000 human beings. About 6,400,000 persons dwell in the islands of the Pacific (Oceania).

Such figures do not tell the whole story. It is also extremely important to know the "density of population" figures. We obtain these figures by dividing the total number of persons in a given population by the number of units of area (for example, square miles) in the section of the world's surface occupied by that population. The following table shows how wide the range of population density is:

| *Country* | *Persons per square mile* |
|---|---|
| Netherlands | 916 |
| Belgium | 773 |
| Japan | 656 |
| Great Britain | 556 |
| India | 301 |
| China | 178+ |
| United States | 51 |
| Republic of South Africa | 32 |
| Soviet Union | 25 |
| Brazil | 20 |
| Argentina | 19 |
| Canada | 4+ |
| Australia | 3+ |
| Antarctica (continent) | 0 * |

* Only inhabitants are a few scientists and technicians.

The above table shows that the most densely populated countries in the world — the Netherlands and Belgium — are in Europe. They enjoy a relatively high standard of living because of their industrial development and international trade. By way of contrast, many of the nations of South America are underpopulated. Yet we often found an extreme degree of misery among many of their inhabitants, due in part at least to insufficiently developed industry and trade and outmoded farming techniques.

In considering the wide range of population density in the world, one must realize that density of population is not necessarily an evil and that there is such a thing as too sparse population. In every case, we must consider the relationship between population and the material level of living. The larger the number of people that has to be supported on a given area of land, the smaller the share of each individual or family in the total production, *other things being equal.* But the situation may be modified by different factors. Suppose that a given people has reached an advanced state of industrialization, with mass production, high division of labor and world markets. If the population is too scanty, that particular people may not be as prosperous as it could be. There will not be enough population to absorb the huge domestic production; there will not be enough workers to meet the world-wide demand for the country's exports. Up to a certain point, therefore, an increase in population may tend to raise the level of living rather than depress it.

We must, therefore, realize the existence of both overpopulation and underpopulation. A society is overpopulated when the number of people is too large for the maximum per capita supply of necessary

An invasion represents one form of outward population movement. In a typical movement of this sort, such as the invasion of Europe by Attila the Hun (below) in the fifth century A.D., a barbarian people would move toward the land of a more highly civilized and wealthy people. Attila's invasion of Europe was turned back when his hordes were defeated at the Battle of Châlons in the year 451 A.D.

commodities, comforts and luxuries. A society is underpopulated when there are not enough people to take full advantage of available land and technical resources. In between these two extremes, there is the optimum population (ideal population), in which the number of people is just right to bring about the maximum level of living for everyone. The optimum population is not always the same, of course, from one generation to another. As circumstances change, as industrialization is introduced, as farm techniques improve, the optimum will change accordingly. We must point out, however, that industrial progress may bring complications of its own. Thus the rapid growth of automation (see Index) has resulted in greatly decreased employment in certain industries. If the unemployed victims of automation cannot be put to work, they will constitute an increasing drag on the economy; the ultimate result of industrial progress, in that case, would be overpopulation.

**Immigration is an effective form of population movement. It accounts in large part for the vast increase in population in the United States. Above, we see immigrants arriving at Ellis Island, New York, which was formerly the chief immigration station in the United States. Its facilities were shut down in the year 1954.**

**How the human race meets its population problems**

The great population problem will probably always be overpopulation, since where underpopulation exists, natural increase in the number of offspring will probably bring the population up to the optimum mark in time. Up to now, the human race has been able to meet its population problems more or less successfully by spreading over the earth's surface. Even today there are wide underpopulated areas — Australia, New Zealand, large parts of Canada, Central and South America — which would be available for this purpose.

Outward population movements have taken various forms. Sometimes invasion has taken place. There were many spectacular movements of this sort in the past. A rude and vigorous people would send out large sections of its population, or would move as a whole, toward the land of a more highly civilized and wealthy people. They would seek more and better land; they would also have a vague longing to share in the benefits of a more advanced and luxurious culture. Among such population movements were the inroads of the Goths and Vandals in the days of the Roman Empire, the invasion of Europe by the Huns in the Dark Ages and the mass forays led by Jenghiz Khan and Tamerlane in the Middle Ages. There have been practically no similar invasions in the last few centuries.

Conquest represents a different kind of expansion. It involves both military and political domination; its logical result is empire. Such ancient empires as those of the Assyrians, the Macedonians under Alexander the Great, and the Romans were based on conquests; so was the empire of Charlemagne; so, to a certain extent, were the empires established by the British and the French in modern times. Fascist Italy, Nazi Germany, Imperial Japan and Communist Russia overran various nations before and during World War II. Italy, Germany and Japan, losers in the war, had to give up their conquests. The Russians maintained their grip on most of the nations they had occupied; these nations became satellites, taking their orders from Moscow. Communist China overwhelmed Tibet in

1950–51; with its powerful armies, it now threatens its neighbors. Apparently, conquest is still a pretty effective means of expansion. It should be pointed out that the conquering powers of today profess, not to conquer neighboring countries, but to "liberate" the people of these countries.

Colonization, as a form of migration, has gone on from very early times. It takes place when a relatively advanced nation settles either upon uninhabited land or land occupied by a backward people. The latter may offer ineffective resistance or perhaps may not resist at all. The Chinese and the Greeks were great colonizers in ancient days. In more recent centuries, Britain, France, Holland and Spain, among other nations, have developed great colonial systems. A great many new colonies were set up in the nineteenth century, particularly in Africa. But colonization, or at least avowed colonization, has gone out of fashion. The American colonies won their independence in the eighteenth century; the Latin-American states, in the nineteenth; India, Burma, Indonesia and many African principalities, in the twentieth.

The most effective type of population movement at the present time is immigration. It differs from the foregoing types of expansion in its peaceful character. Under immigration, the citizens of one state are permitted or invited to move into the territory of another; they generally act on their own initiative and they use their own resources. The populations of the United States and other new countries have been recruited very largely from immigrants from foreign lands. After World War II, considerable numbers of displaced persons joined the ranks of immigrants. Many countries have now imposed more or less strict limits on newcomers, but immigration is still a force to reckon with.

Immigration is not necessarily an ideal solution of the overpopulation problem. The crowded areas of the world cannot always solve their difficulties by shipping surplus population out of the country; the natural rate of population increase within the country may soon make population pressures as grievous as before. On the other hand, if the more favored areas of the world receive an endless flow of immigrants, the result may be the depression of the level of living in these areas as soon as the optimum population figure has been reached.

There are certain areas in the world that are sparsely populated because present living conditions are unfavorable. There are immense desert tracts, like the Sahara and the Gobi; dense jungles like those of the Congo and the Amazon Basin; extensive areas, like certain regions of central Africa, where disease makes prospective immigrants hold their distance. If we could succeed in making these regions habitable, they would provide living space for billions of immigrants. Such projects are not beyond the power of man. He has succeeded through irrigation in making the desert bloom in Arizona and in Israel; he has converted the Panama Canal Zone from a pest-hole into one of the most healthful areas in the world; he has at least begun the task of making the Amazon jungle a fit abode for man.

Even in the well-populated areas of the earth, there would be room for many more inhabitants if men used nature's resources wisely. It would mean selecting the best varieties of crops for a particular kind of soil. It would mean using the correct sort of fertilizers to restore fertility to the soil. It would mean planting thick-growing vegetation and building dams to prevent excessive loss of water. It would mean new irrigation canals for dry lands. It would mean a better use of scientific knowledge in nutrition and animal husbandry. It would mean a vastly greater use of our plant and animal resources from the sea. It would mean breaking the logjam of distribution.

There is a sobering thought, however. If we succeeded in multiplying the food resources available to man, the population might well increase by leaps and bounds because of these very advances. Unless some effective method of controlling population were devised, within a few generations the population pressures might become just as great as before.

# THE STRUGGLE FOR LIFE

## Adaptations That Serve in the War for Survival

BY F. L. FITZPATRICK

WE ARE all more or less familiar with the kind of war that is fought with weapons such as howitzers, rifles, flame throwers, planes and submarines. However ruinous such a war may be, it comes to an end in the course of time, and it is followed by a period of peace, which may last for a good many years. There is another sort of war that is being fought to a finish every single day in the vast arena of nature — a grim, unending war for survival among the numberless species of living things. Billions upon billions of individuals succumb each day in the fateful struggle. Sometimes entire species may be ruthlessly wiped out in the course of a few years.

The species to which we belong — *Homo sapiens* — has played an important part in this never ending war. The advance of man's civilization has caused the extermination of some species and has endangered the existence of others. On the other hand, man himself has fought a losing battle in certain regions of the world. The terrible tsetse fly, the carrier of a parasite that causes sleeping sickness in man, has made large areas of Africa almost uninhabitable. The plants of the jungle have overwhelmed human settlements again and again in such widely separated places as Yucatan and Indochina.

In the continuous struggle for existence, each species is faced with three critical problems. (1) It must obtain food; (2) it must defend itself against its natural enemies; (3) it must provide for its offspring. To solve these problems it must possess certain special structures, such as teeth, or wings or thorns. It must also display certain forms of behavior, such as the building of nests or the making of long migrations in search of food.

We give the name of special adaptations to the structures and behaviorisms that enable a species to survive. There are literally thousands of such adaptations in nature, and it would take many pages simply to give a list of them. In this article we shall tell you about a few of the more significant or striking ones.

### Adaptations for obtaining food

We find in both plants and animals special adaptations for obtaining food. Among the most striking are the teeth of animals. The teeth of carnivores, or flesh eaters — dogs, wolves, cats, tigers, lions, hyenas, bears, seals and the like — are especially adapted to seizing and rending prey. Not all carnivores, it is true, live by the hunt. Most bears, for example, are vegetarians; and it is not too often that a domestic dog has occasion to seize and devour prey. But most of the carnivores are hunters, and the possession of suitable teeth is for them a matter of life and death.

The four canine, or eye, teeth of carnivores are large and pointed; they are the fangs that seize a victim. Together with the sharp claws that are found in most species, the canines are the chief offensive weapons of the carnivores. Behind the canines are the premolars and molars, provided with sharp cutting edges, well suited to shredding meat, or with broad crushing and grinding surfaces. As for the carnivores' incisors, or cutting teeth, they are so insignificant as to be almost useless.

We find adaptations of a completely different kind when we examine the teeth

N. Y. Zool. Soc.

has no surface to wear against, and continues to grow unchecked. Such an upper incisor does not grow straight downward, but curves back into the mouth cavity and finally punctures the brain case at the base of the skull.

The great anteater, a mammal that lives in the jungle country of Central and South America, shows some remarkable adaptations for obtaining food. In its native haunts this animal feeds almost entirely upon ants, termites and other crawling insects. An anteater's front feet are provided with powerful claws, which can be used to tear apart ant hills. The animal has a long snout, at the end of which is a small, toothless mouth with a tiny slit as an opening. The tongue is a tubular affair about eighteen inches long, and it can be used to pick up ants on the surface of the ground or in the passageways of ant colonies. There is no truth in the popular belief that an anteater's tongue is covered with a sticky saliva that will entrap ants like so much flypaper. The animal's tongue is quite effective enough as it is.

Two very effective adaptations for obtaining food: the sharp fangs of the puma (upper photo) and the long snout of the anteater (lower photo).

Chicago Nat. Hist. Mus.

of the rodents, or gnawing animals, which include beavers, muskrats, rats and squirrels. The large, chisel-like incisors in the upper and lower jaws are fine for biting or cutting off food. The rodents grind their food with their formidable broad-surfaced premolars and molars in the back of the mouth cavity. They have no canine teeth at all—no great loss, since they are vegetarians and never hunt prey.

A rodent's incisor teeth continue to grow throughout life. Thus, a beaver that depends upon incisors to chisel off the bark that it eats need not fear that these teeth will wear down to useless stumps. As a matter of fact, it is probable that some rodents gnaw a good many things that they do not eat or use, and that this keeps the incisors worn down to a comfortable length. The unchecked growth of such teeth would be disastrous. If a rodent's lower incisor is broken off by accident, the upper incisor

The bills or beaks of birds are very efficient devices for obtaining food. A beak consists of an upper and a lower mandible, or jaw. In the heron both mandibles are long and pointed, and the beak is especially suitable for spearing small fish and frogs in shallow water. A pelican is also a fish-eater, but its beak is constructed quite differently. The upper mandible is curved abruptly downward at its tip, and it is with this portion of the bill that the pelican seizes a fish. A fleshy, pouch-like sac extends between the two sides of the lower mandible. When a pelican has captured a fish, it tilts its head backward and works the fish down into the pouch. The fish is now unlikely to escape, and the pelican can swallow its captive at leisure.

The powerful beak of a hawk, with its curving and sharply pointed upper mandible, is very effective in tearing up the flesh of prey. Hawks also have powerful talons, or claws, with which they secure a firm hold on their victims. The bills and claws of owls are similar to those of hawks. The woodpecker has a strong chisel-like bill with which it cuts into the decaying wood of trees; it can thus get at boring insects and their worm-like larvae. Woodpeckers also have horny, lance-like tongues, which can be used to spear larvae when their tunnels beneath the bark have been opened.

The powerfully muscled boa or python kills by constriction the birds and mammals upon which it feeds. The snake wraps several coils around its prey and begins to squeeze. As the coils tighten, fresh air cannot be breathed in by the victim and

The pelican's long bill is a very powerful weapon.

Photos, N. Y. Zool. Soc.

The heron spears its prey with its pointed beak.

action of the heart is stopped; death comes quickly. Because of the snake's mouth structure, prey several times larger in girth can be swallowed whole.

Other snakes, such as the rattlesnake, the copperhead, the water moccasin and the cobra, subdue their prey by means of venom. The rattlesnakes of North and South America, for example, develop venom glands on the two sides of the head above the roof of the mouth cavity. Tube-like structures known as ducts extend from each gland to a pair of hollow fangs in the upper jaw. The fangs have openings a short distance from their tips and the venom spurts through these openings.

The fangs are attached to small bones hinged to long, rodlike bones of the upper jaw. Normally, the fangs lie flat against the roof of the mouth. When the mouth is opened in striking prey, the upper jaw bones slide forward rotating the fangs into a vertical position. Muscles in contact with the venom glands contract so that venom, forced from the glands, passes through the fangs into the wound.

Certain fishes secure prey by literally shocking it. The members of the torpedo-ray, or electric-ray, family, related to the sharks, store up electricity in organs on each side of the head. The discharge of the stored electric energy paralyzes or kills fishes in the vicinity and the ray pounces upon them. Even humans are affected; many a swimmer has had his arms and legs benumbed by the electrical discharges. The electric eel, found in the streams of tropical South America, has a similar shocking apparatus.

The mouth parts of the female mosquito represent a rather complicated adaptation. In the center of the mouth parts is a sucking tube; around it are grouped a series of sharp, piercing structures that literally cut a disc out of a victim's skin. The mosquito inserts its sucking tube into the wound and drinks its fill of blood.

Striking adaptations for obtaining food are by no means limited to the animal world. There are plants that trap insects and obtain necessary proteins from their victims. One of the best-known plants of this type is the Venus's fly-trap, which grows in moist places around fresh-water ponds. The leaves of the Venus's fly-trap bear double-bladed structures with tooth-like edges; they close together to hold an insect that alights in the space between the blades. The victim is then partially dissolved by digestive juices, and the dissolved portions are absorbed by the plant tissue.

### Pitcher plants are well-known insect-catchers

The pitcher plants form another famous group of insect-catchers. In this case the leaves are in the shape of pitchers, which hold the water of rains and form tiny ponds in which insects are drowned. The pitchers of one South American species, thriving in an area of very high annual rainfall, have small holes near their tops. These holes permit excess water to be drained off. Hence the pitchers cannot overflow and the drowned insects contained in them cannot be floated off.

The adaptations of certain desert plants enable them to obtain food and to conserve water. Normally a higher plant bears many green leaves, which are food-making centers and through which water vapor is given off to the air. This sort of structure, however, is not at all suited to life in the desert. Here water comes only from the rare showers or the melting of scanty winter snows, and it must be carefully hoarded throughout the dry months if the plants are to survive.

The spray discharged from the scent glands of the skunk serves to keep off possible attackers.

Members of the cactus group, such as the giant cactus of the North American southwest, are well adapted to desert life. They have no broad leaves through which water may be lost to the atmosphere; their food-making activities are carried on entirely in the green stems. They are therefore able to retain much of the water that is absorbed during or after a rain, and to conserve it throughout long periods of drought. At the same time, these plants are able to carry on normal food-making processes.

### Adaptations for defense

Some adaptations serve for defense as well as for providing food. The rat often uses its teeth to defend itself when it is cornered by a dog or a man; it can bite viciously. The electrical discharge of the torpedo ray is sometimes directed against larger fish which are pursuing it. The venom of snakes helps to ward off the attacks of stronger enemies, including man. Other adaptations, however, are concerned almost altogether with defense.

The scent glands of skunks are a remarkably effective defensive weapon. Both the striped skunk and the little spotted skunk of North America possess such glands, just beneath the skin on either side of the anal

A foe that tries to harass a porcupine soon acquires an unwanted collection of barbed quills.

opening. Skunks are able to discharge their ill-smelling secretion in the form of a spray to a distance of about ten feet. The odor of this spray is overpowering; it is very hard to remove from clothing. If any of the secretion gets into an animal's eyes, temporary or permanent blindness may result.

Most other animals give skunks a wide berth, and if unmolested the skunks quietly go their way. Apparently they realize that they have little to fear from other animals of the wild; they are apt to adopt a carefree attitude that often has disastrous results. For example, they show little disposition to get out of the way of approaching cars on highways at night, and many of them are run over and killed.

The bombardier beetle is a poison-gas specialist among insects. Within its body it secretes a foul-smelling liquid, which turns into vapor as it is discharged from two glands near the anus. There is a sound like the report of a tiny popgun as the gas attack is launched against a pursuer, and the beetle generally makes good its escape from its startled foe.

The Canada porcupine is well prepared in its normal habitat to defend itself against almost any enemy. This animal, a rodent, lives in Canada and the northern part of the United States. It feeds largely upon bark, buds and other plant materials. It is armed with a large number of sharply barbed quills, which are especially well developed on the animal's back, sides and tail. The quills are loosely inserted in the skin and normally point backward; but they can be raised in moments of excitement.

Contrary to popular belief, a porcupine does not throw its quills; but they are so loosely inserted that they come out readily and fasten themselves upon an attacker. A dog that attempts to harass a porcupine will soon have an unwelcome collection of quills in its face and neck. The presence of barbs makes the quills difficult to pull out. In fact, they are likely to be still more firmly embedded when a victim tries to get rid of them.

The defense mechanism of squids and cuttlefishes is also most effective. These animals are soft-bodied, sea-dwelling mollusks, related to clams and snails. Each squid or cuttlefish develops an internal structure, known as an ink sac, which is filled with a black fluid. When an enemy appears, the mollusk squirts its ink into the surrounding water, and then beats a

Photos, N. Y. Zool. Soc.

Hawks grasp their prey firmly with their claws.

hasty retreat while the attacker is trying to locate its prey again.

Many of the defense mechanisms encountered in nature are effective because they make animals inconspicuous and therefore less likely to attract the attention of their enemies. The walking-leaf insect of Asia, for instance, shows a remarkable resemblance to a green leaf, both in form and coloration. The upper surfaces of its green-colored wings bear a brown pattern that looks very much like the veins of a large leaf. Its legs are flattened out and veined so that they look like smaller leaves.

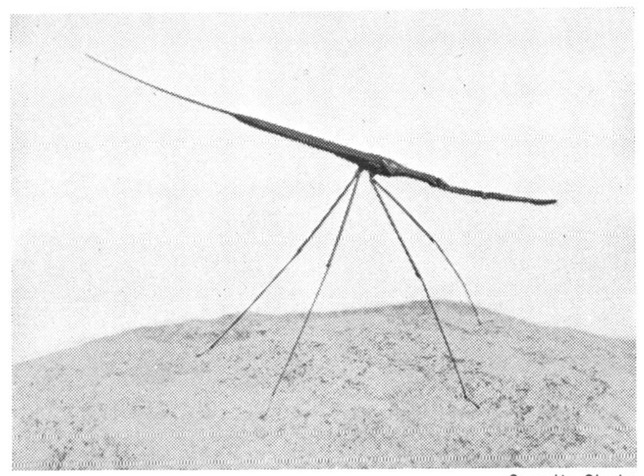

Cornelia Clarke

This *Ranatra* markedly resembles a dried stick.

The dead-leaf butterfly of the East Indies, as its name indicates, looks remarkably like a dead leaf. When this butterfly is at rest on a twig, its wings are folded together and only the lower surfaces can be seen. These are dull brown in color and bear a black, veined pattern. The upper wing surfaces are brightly colored by comparison with the lower surfaces; in flight the dead-leaf butterfly looks like a different insect.

Certain defenseless insects closely resemble insects of other species that are feared by their neighbors. Bees, as you know, are four-winged insects that have an effective sting; they are hairy and show contrasting colors. Bee flies have only two wings and they have no sting. But since they too are hairy and are colored after the fashion of bees, they may easily be mistaken for these formidable insects and it is possible that potential enemies may avoid bee flies for this reason. Certainly the writer has often seen people go out of their way to avoid bee flies.

Some animals feign death in order to try to escape from their foes. Sometimes the performance is striking but not particularly convincing. The writer has often seen a Virginia opossum lying inert, with eyes closed, as a dog approached; but I could never see that the dog was deceived by this performance. The hog-nosed snake of North America also feigns death on occasion; when confronted by an enemy it will turn over on its back and lie without movement. If the snake is turned over on its belly, the chances are that it will assume its former position, as if this were the only proper pose for a dead snake!

### Adaptations for preserving offspring

The third group of adaptations has to do with the preservation of offspring. There are numerous examples among the members of the vegetable kingdom. The seeds of many plants contain stored food for the early growth of the young plants that are destined to sprout from them. A maple tree may bear thousands of seeds in spring or early summer. If all of these seeds fell directly to the ground beneath the parent tree, the resulting offspring would be far too crowded for successful growth. Maple seeds, however, bear wing-like structures, and when they break off from the tree they are carried for some distance by the wind before they come to rest. The seeds of the common dandelion have similar adaptations. Each seed bears an umbrella-like process, which is caught and lifted by air currents, so that the seed may be carried far from the mother plant.

The nut of the coconut palm is a great traveler, too, but it voyages by sea. This fruit is protected by a thick, light, fibrous husk. When such a fruit falls into the water from a palm on a tropical beach, it may be carried for great distances by waves and currents, perhaps to be washed up and to sprout upon the beach of a distant island.

Even animals serve to transport various plant seeds. Thus, the fruits in which the seeds of the burdock are enclosed are covered with prickly envelopes known as burrs that are provided with innumerable "hooks." When an animal brushes against the burrs, they stick to its hair. As a result the seeds go wherever the animal may wander until they finally break out of their protective coverings and drop to the ground.

The result of such special adaptations is that seeds are scattered far and wide by wind, water, animals and gravity. We have already pointed out that this spreading prevents too much crowding around the parent plant. It also enables a species to penetrate into new areas that will favor its growth. This may be important because environments change; a spot may be favorable for certain species today that will not necessarily be favorable for the same species a hundred years from now.

There are many outstanding examples of special adaptations for the care and protection of the young among animals. The eggs of the little marine creatures known as sea horses are protected by a special brood pouch in which they are hatched. It is the male and not the female that develops this pouch, on the lower surface of the body near the tail. The female inserts her eggs into her mate's pouch and here the eggs remain until the young have emerged.

The male of the midwife toad also bears the responsibility for the safety of its mate's eggs. The female of this species lays two strings of eggs, which the male wraps around its hind legs. The male then takes refuge in out-of-the-way places while the eggs are developing; it comes out now and then to moisten the eggs in a pool of water. When the time for hatching is near, the male again takes to the water and well-developed tadpoles emerge from the eggs. The tadpoles complete their development in ponds before becoming adults.

Doremus and Co.

Coconut-palm nuts that fall in the water may be carried to distant beaches, where they may sprout.

After a female Surinam toad lays its eggs, the male spreads them over her back, which in the meantime has become swollen and spongy. The eggs sink down into little cavities in the skin and they are sealed by protective coverings. The female toad carries her eggs with her, and when they hatch the tadpoles remain in the cavities. Eventually, the tadpole stage comes to an end and small but fully formed adult toads emerge.

A well-known adaptation for the protection of the young is found among the kangaroos and wallabies of Australia and the opossums of North and South America. This is the marsupial pouch, a pocket on the surface of the female's abdomen. (The name "marsupial" refers to the order to which these animals belong.) The marsupial pouch serves as a place of refuge for the young, which vary in number from one or two, in the case of kangaroos, to the fourteen or more of a Virginia opossum. The young are small, immature and quite helpless at birth; but they are housed safely in the pouch, where they can get at the milk glands without leaving their shelter.

A kangaroo carries her young around in her pouch until the baby has grown to considerable size. When pursued by an enemy, a kangaroo will sometimes pull a young one from the pouch and throw it to the ground. It is not known whether she does so in order to lighten her own load, or to provide for the escape of the baby kangaroo in case the mother is overtaken by her pursuer.

Some animals safeguard their young by remarkable adaptations of behavior. The Siamese hornbill makes her nest in a cavity in a tree. After she has begun to sit on her eggs the male hornbill seals up the entrance to the hole almost entirely with mud. This dries and hardens, leaving a slit-like opening, through which the male feeds his mate while she is nesting. The opening is not large enough to permit the entrance of animal foes that climb about among the trees in warm countries.

The runs, or migrations, of the king salmon of North America are associated with the finding of a suitable place in which the eggs may hatch and the young may develop. A king salmon spends most of its life in the sea and may grow to a weight of over a hundred pounds. Each year adult salmon four to seven years of age gather in bays along the western coast of North America in preparation for a run up a fresh-water

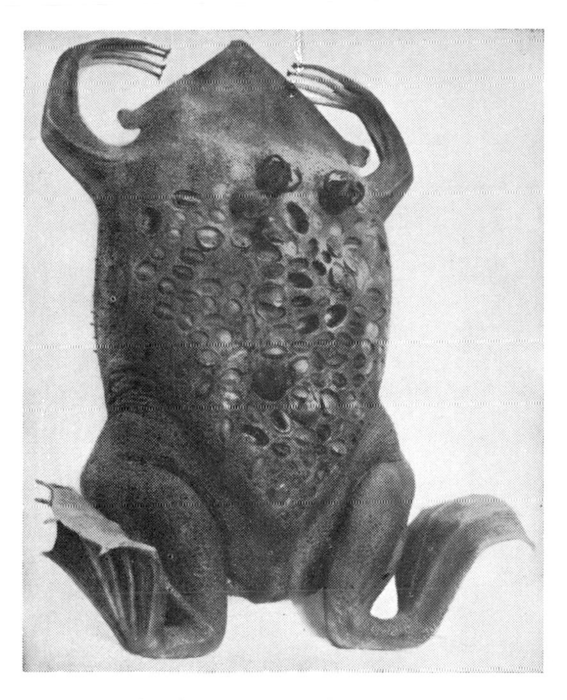

How a female Surinam toad carries its young.

Male hornbill feeding the female through a hole

stream to the quiet pools at its source. The fish do not stop to feed or rest for any length of time on the way. They always press on upstream, and this despite the fact that some of their journeys are hundreds of miles in length. Their courses are made difficult by the presence of numerous rapids and falls. They pass all these barriers, sometimes leaping completely out of the water.

Eventually the salmon reach the headwaters of the stream, often in a sadly battered and worn condition. The females deposit their eggs on the bottoms of the small streams or shallow pools of the headwaters; the male salmon discharge milt, or reproductory fluid, over the eggs. Thereupon, the adult salmon usually die.

After hatching from the eggs, the young remain for a time in the fresh water where the eggs were laid. Here there are not likely to be many large enemies and the young salmon have a good chance to survive. As they become larger, they move downstream and finally enter the sea. Most of them are a year old and four inches in length before they see the ocean.

Another type of fish that has become famous for its spawning migrations is the common eel, which is found in fresh-water ponds, lakes and streams of Europe and eastern North America. This eel has a long, slender, snake-like body; yet it is a bony fish. Europeans have used it as food for centuries.

Until recently the spawning habits of eels were an unsolved mystery. People knew only that fully grown eels went out to sea, never to return, and that young eels came in from the sea. The Greek philosopher Aristotle (384-322 B.C.) thought that these fishes did not lay eggs but that their young arose in some unexplainable way from the sea itself. It was not until modern times

Lilo Hess

Female kangaroo carrying her baby in her pouch.

Both photos, Amer. Mus. of Nat. Hist.

in the mud wall with which he seals up the nest.

that the real facts in the case were revealed and even today a number of questions remain unanswered.

Eels live in their fresh-water homes until they become mature; in some cases they attain a length of between five and six feet and a weight of several pounds. In the autumn of the year in which they become mature, the fishes begin a migration to the sea. When they reach salt water, they continue onward for hundreds of miles to a deep trench in the Atlantic, to the southeast of Bermuda. Here the females deposit their eggs in the spring of the year. It is probable that male and female adults die soon afterward; at least they never return to their former homes.

### A transatlantic crossing that takes three years

When the eel eggs hatch, a vast number of almost transparent, threadlike young rise to the surface of the sea. Some of them turn eastward toward Europe, and others begin the shorter journey to North America. The young usually are three years old before they appear on the European coast; they make the trip from the breeding grounds to North America in two years. They now work their way up fresh-water streams to the places where they will remain until they become adults.

Why should adult eels make this long spawning migration which, as far as we know, ends in their death? According to one theory, at some time in the distant past the coasts of Europe and North America were much closer together than they are today and therefore the eels had a relatively short trip to their ancestral breeding ground. In the course of time the two continents became farther removed from each other, but the eels still obeyed the instinctive urge to return to the place of their origin.

We could extend almost indefinitely this account of the special adaptation of animals and plants. We have already provided enough instances, however, to show the nature of these adaptations and their importance in the struggle to survive.

*See also Vol. 10, p. 271:* "Ecology."

U. S. Fish and Wildlife Service

Salmon leaping up a waterfall on their way upstream to the ancestral spawning ground.

# HOW ATOMS COMBINE

## The Nature of Valence

BY ANTHONY STANDEN

EVERYTHING in this world undergoes change. This is as true of living things, such as men, trees, rodents, mosses, amebas and fungi, as it is of nonliving things — the atmosphere, mountain ranges, sand dunes, the waters of the sea.

All changes may be divided into two classes — physical and chemical. Physical changes do not affect the molecules of which substances consist. For example, the molecule of water is made up of two atoms of hydrogen and one of oxygen. If water is heated and turned into steam or if it is frozen into ice, its molecule still consists of two hydrogen atoms and one oxygen atom. The steam will turn back into water when it cools; the ice will become water again when it melts. If we bend or crush a solid or grind it into powder, it undergoes only a physical change. Its molecules have not been affected.

A chemical change is entirely different. It involves the transformation of the molecules: the combination or the recombination of atoms. A different substance is formed. When the atoms of oxygen and hydrogen, both gases,* combine chemically, they are transformed into water, a liquid. Mercuric oxide is a red solid; it is quite different from the liquid metal, mercury (quicksilver), and the colorless gas, oxygen, which make up its molecules. An even more startling transformation results when sodium, a highly reactive metal, and chlorine, a sharp-smelling, poisonous gas, combine chemically. They form ordinary table salt, an important part of the diet.

In this chapter we shall concern ourselves with chemical changes and particularly with the ways in which atoms combine to bring about such changes. At first glance there does not seem to be much rhyme or reason in these combinations. In a molecule of mercuric oxide ($HgO$),* an atom of oxygen combines with an atom of mercury. The ammonia molecule ($NH_3$) consists of one atom of nitrogen plus three atoms of hydrogen; a single atom of one element has combined with three atoms of another. In the compound

* When we refer to the substances mentioned in this paragraph as "gases," or "liquids" or "solids," we should really add "under ordinary conditions." For example, the chemist can transform both hydrogen and oxygen into liquids.

* The chemical formula $HgO$ is made up of the symbols Hg, standing for "mercury," and O, standing for "oxygen." A small figure after and near the bottom of a symbol indicates the number of atoms; if no such figure is given, the number "one" is understood. For example, the formula for calcium carbonate — $CaCO_3$ — shows that the calcium carbonate molecule is made up of one atom of calcium (Ca), one atom of carbon (C) and three atoms of oxygen (O). The complete list of chemical symbols is given on page 295, Vol. 10. For more details on the formation of molecules, see the chapter Chemical Reactions, in Volume 2.

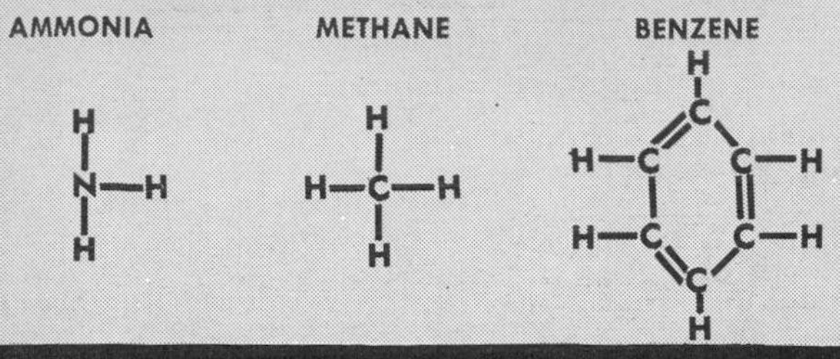

1. Reading from left to right: the structural formulas of ammonia ($NH_3$), methane ($CH_4$) and benzene ($C_6H_6$).

2. The seven concentric shells of the atom.

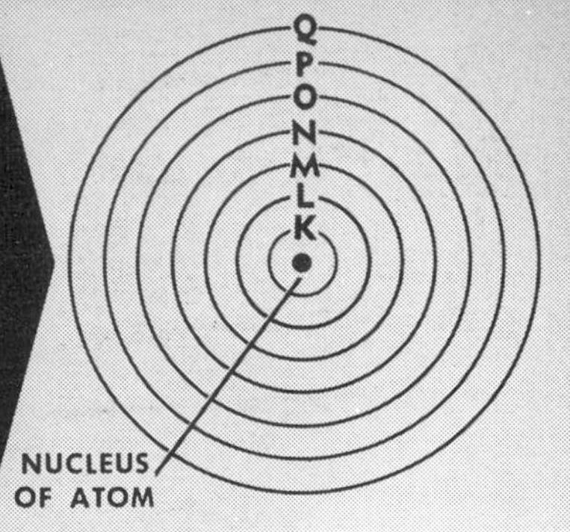

methane ($CH_4$), a carbon atom combines with four atoms of hydrogen. In other words, an atom may be linked, in a molecule, to a single atom or to two or more atoms.

A formula such as $CH_4$, which simply shows the number of atoms of each kind in a molecule, is called an empirical formula. The name "structural formula" is given to any formula, such as those in Figure 1, that shows lines for the bonds connecting one atom with another. In Figure 1, note that a nitrogen atom has three bonds (lines) and carbon four bonds, while hydrogen has only one. The number of bonds leading from an atom is called the valence of that atom. In the compounds shown in Figure 1, nitrogen has a valence of three, carbon four and hydrogen only one. "Valence," therefore, means the "combining power of an atom," expressed in terms of the number of hydrogen atoms with which it can combine. (It is convenient to express valence in terms of hydrogen, since hydrogen always has a valence of one, and never more.)

The ideas of valence as "combining power" and of structural formulas have been known to chemists since about the middle of the last century. But it was not until the present century that chemists learned *why* atoms combine as they do and what there is about some atoms (such as hydrogen) that makes them always have a valence of one, whereas other atoms have valences of two, three, four or even more.

To understand why atoms combine, we must consider the structure of the atom. It consists of a core, or nucleus, made up of positive particles, called protons, neutral particles, known as neutrons, and other particles. The nucleus contains nearly all the weight of the atom. Revolving around the nucleus are a number of negatively charged particles, called electrons. They move around the atomic nucleus in a series of orbits, somewhat like a tiny version of the solar system, in which planets move around the sun.*

Atoms are listed, on the basis of increasing complexity, by their atomic number, which always corresponds to the number of protons in the nucleus. In an electrically neutral atom, there are as many electrons orbiting around the nucleus as there are protons in the nucleus. Hydrogen, the lightest element, has atomic number 1; it has a single proton and a single electron. Oxygen, with atomic number 8, has eight protons and also eight electrons. The number of neutrons varies according to the atom. The commonest form of hydrogen has no neutrons at all; the commonest form of the heavy atom uranium has 146.

Electrons revolving at the same average distance from the nucleus are said to occupy the same shell. A shell may contain from one to thirty-two electrons. The hydrogen atom, which has but one electron, has only one shell. Other atoms have two, or three or more; the most complicated ones have seven. Scientists refer to the seven shells, beginning with the one just outside of the nucleus, as the K, L, M, N, O, P and Q shells (Figure 2). The maximum number of electrons in the K shell is two; in the L shell, eight; in the M shell eighteen,** and so on. The shells are not always filled to capacity.

Let us consider the electron arrangement in a few representative atoms (Figure 3). Hydrogen has one electron, which is in the K shell. Helium has two elec-

* There are some resemblances between the atom and the solar system; but there are also a number of differences, in addition to the obvious difference in size. The comparison with the solar system helps us to understand the atom; but we should not take it too seriously.

** In some cases the full quota of electrons for the M shell is eight.

3. The electron arrangement in various atoms.

K SHELL
NUCLEUS
M SHELL
L SHELL
K SHELL
NUCLEUS

HYDROGEN ATOMIC NUMBER 1 | HELIUM ATOMIC NUMBER 2 | BORON ATOMIC NUMBER 5 | MAGNESIUM ATOMIC NUMBER 12



trons, filling the K shell to capacity. Boron has five electrons. Two of these fill the K shell; the other three are in the next shell — the L shell. This can hold eight electrons in all. Hence it is not filled in the case of boron; it lacks five electrons. Magnesium, with atomic number 12, has twelve electrons. Two of these fill the K shell, and eight the L shell. The remaining two are in the M shell. The electrons of the outer shell, in each case, are the only ones involved in chemical reactions.

### The inert, or noble gases

In certain atoms — helium, neon, argon, krypton, xenon and radon, the so-called inert, or noble gases — the outer shells have their full quota of electrons. This is a very stable arrangement; it accounts for the fact that the inert gases almost never enter into chemical combinations. As for the other atoms, the outer shells are not filled to capacity. Such atoms can fill the full electron quota in their outer shells by gaining or losing electrons, or by sharing them. We say of such atoms that they are chemically active.

### Electrovalence: atoms gain or lose electrons

Let us see what happens when atoms combine by gaining or losing electrons — the type of chemical combination known as electrovalence.

The elements sodium and fluorine join together in this way. In Figure 4, we show an atom of sodium. It has eleven positive charges on its nucleus, balanced by eleven electrons with one negative charge each. The innermost shell — the K shell — holds two electrons; the next, or L, shell holds eight. These shells are filled. The eleventh electron occupies the M shell all by itself.

Now the nucleus attracts each electron with a force varying inversely as the square of the distance. This means that the two K electrons are held most tightly of all, while the lone electron in the M shell is attracted much less strongly than any of the others. It would not take much energy to remove this electron from the atom altogether.

Figure 5 shows the structure of a fluorine atom, with atomic number 9. It has nine electrons in all. The first two, as in the sodium atom, fill the K shell, closest to the nucleus. The other seven electrons go into the L shell; they do not quite fill it, since this shell can hold eight electrons.

Both the sodium and fluorine atoms are electrically balanced, with equal numbers of positive and negative charges. They are capable of chemical activity because their outermost shells — the M shell of the sodium atom and the L shell of the fluorine atom — are not filled.

A chemist has various ways of representing atoms. Instead of showing the

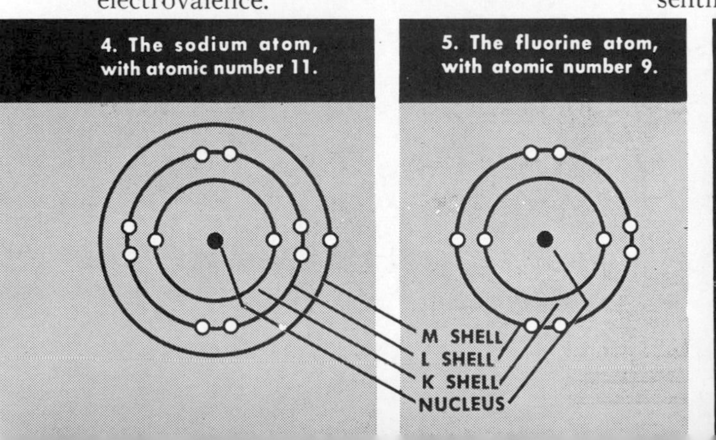

4. The sodium atom, with atomic number 11.

5. The fluorine atom, with atomic number 9.

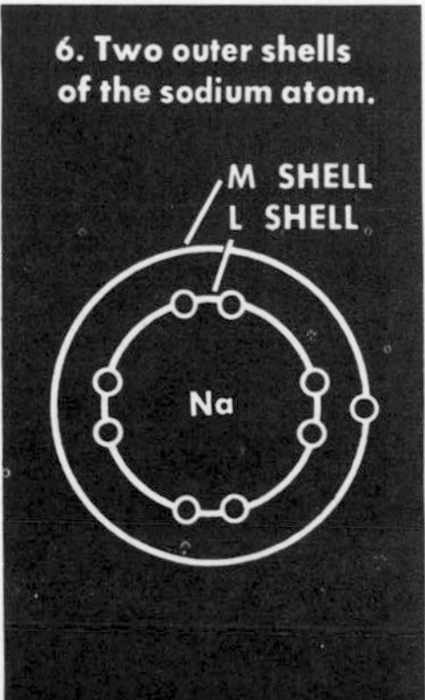

6. Two outer shells of the sodium atom.

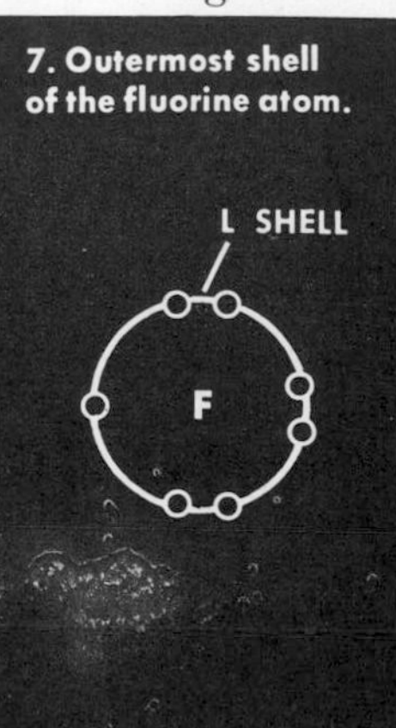

7. Outermost shell of the fluorine atom.

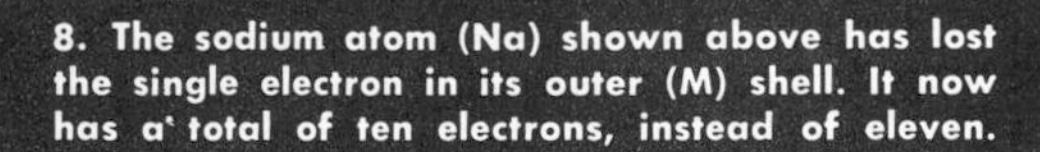


8. The sodium atom (Na) shown above has lost the single electron in its outer (M) shell. It now has a total of ten electrons, instead of eleven.

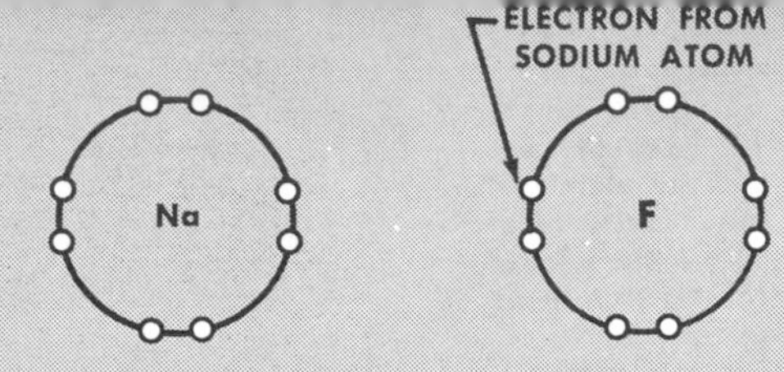


9. The electron from the M shell of the sodium atom has been taken up by the fluorine atom. This atom now has eight electrons in its outer shell.



nucleus and all the shells of the sodium and fluorine atoms, as in Figures 4 and 5, he might show only the two outer shells of the sodium atom and the one outer shell of the fluorine atom, as in Figures 6 and 7. In Figure 6, the dots represent the 8 electrons in the L shell and the one electron that is by itself in the M shell. The symbol for sodium Na,* in the center, stands for the nucleus and the two electrons that are always in the K shell. The drawing of the fluorine atom in Figure 7 shows the seven electrons in the outer (L) shell. The symbol for fluorine, F, in the center represents the nucleus and the two electrons of the K shell.

The electron in the outer (M) shell of the sodium atom can be rather easily dislodged (Figure 8). As we have seen, it is comparatively far from the nucleus and is not attracted greatly by it. If the electron that a sodium electron has lost comes close to the fluorine atom, it will occupy the vacancy in the L shell (Figure 9).

An important transformation in the sodium and fluorine atoms has taken place. The sodium atom was electrically balanced formerly with eleven positive charges in the nucleus and eleven electrons. Now that it has lost an electron, it has eleven positive charges and only ten negative ones. Hence it has one net positive charge, indicated by a plus sign set to the right and near the top of the symbol for sodium: $Na^+$. The fluorine atom was also balanced to begin with, with its nine positive and nine negative charges. Since it has picked up an additional electron, it has ten negative charges as against nine positive ones. It has one net negative charge, indicated by a minus sign: $F^-$. A charged atom, such as $Na^+$ or $F^-$, is called an ion. We discuss ions elsewhere in this set (see Index, under Ions). They differ greatly in their properties from the regular, uncharged atoms from which they have been derived.

We know that like electric charges repel, while unlike charges attract. The positive sodium ion and the negative fluorine ion will, therefore, attract one another. This "electrostatic attraction," or Coulomb attraction, as it is called, is very strong; it is far more powerful than gravitational attraction. Hence, if there are a number of sodium and fluorine ions in a receptacle, say, they will be drawn together. They will arrange themselves so that each sodium ion

* The symbol Na is derived from *natrium*, the Latin name for the substance called sodium carbonate.

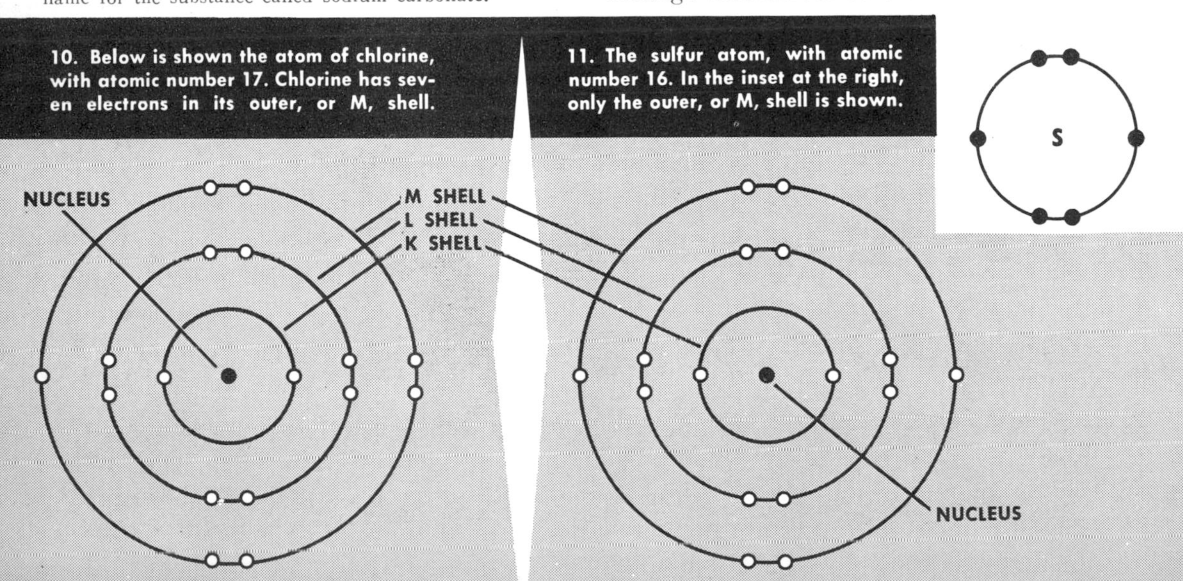


10. Below is shown the atom of chlorine, with atomic number 17. Chlorine has seven electrons in its outer, or M, shell.

11. The sulfur atom, with atomic number 16. In the inset at the right, only the outer, or M, shell is shown.

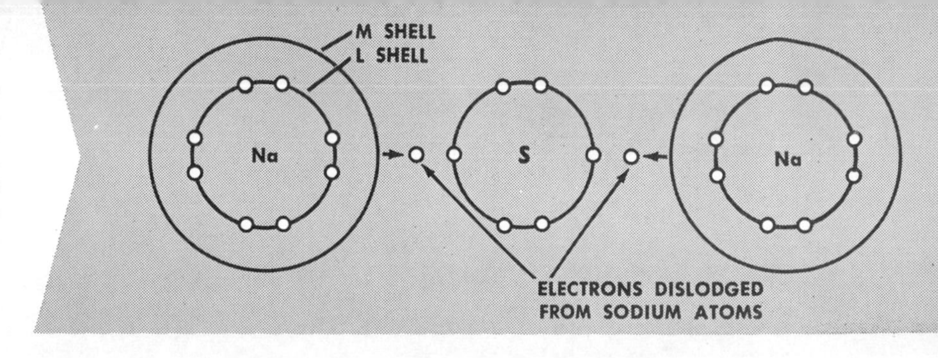


**12. Each of two sodium atoms in the vicinity of a sulfur atom loses the electron in its outer (M) shell.**

will be as close to as many fluorine ions as possible, and each fluorine ion as close to as many sodium ions as possible.

X-ray analysis of crystal structure has indicated that the ions of sodium and fluorine form a cube, with sodium ions and fluorine ions in alternate corners. A great number of cubes may be combined in this way, making up a crystal of sodium fluoride (NaF).

Sodium chloride (NaCl), or table salt, is formed in the same way as sodium fluoride. Chlorine, atomic number 17, has complete K and L shells and a one-less-than-complete M shell (Figure 10). It can take up an electron from a sodium atom to complete the outer shell, just as fluorine does. A chlorine ion, written $Cl^-$, is formed. Sodium and chlorine ions combine, forming cubic crystals.

Lithium, potassium and cesium are metals that, like sodium, lose a single electron in combining chemically. Some metals (such as beryllium, magnesium and cesium) lose two electrons; others (scandium, yttrium, barium), three; still others (titanium, zirconium, hafnium), four. When a magnesium atom, say, loses the two electrons of the outer shell and acquires two net positive charges, becoming an ion, two plus marks are added to its symbol: $Mg^{++}$. Three positive charges are indicated by three plus marks; four positive charges, by four plus marks.

We saw that the nonmetals fluorine and chlorine can each acquire a single electron to fill the outer shell. The nonmetal sulfur can take up two electrons when it combines with another atom. The sulfur atom is shown in Figure 11. You will note that its K and L shells are complete; its M shell contains six electrons and has room for two more. We can also represent the sulfur atom in abbreviated form, as in the inset in Figure 11.



Suppose that each of two sodium atoms in the vicinity of a sulfur atom loses the lone electron in its M shell (Figure 12). The sulfur atom can capture both of the free electrons (Figure 13). Each sodium atom, having lost an electron, now has a single net positive charge; it is now a sodium ion, $Na^+$. The sulfur atom has acquired two electrons. It now has a net negative charge of two; it has become a sulfur ion, $S^{--}$. The two positively charged sodium ions are attracted to the negatively charged sulfur ion, and the compound sodium sulfide, $Na_2S$, is the result. In this compound, sodium has a valence of one; sulfur, a valence of two.

An enormous number of compounds can be formed in the same way as those that we have just described — sodium fluoride (NaFl), sodium chloride (NaCl) and sodium sulfide ($Na_2S$). In each case an electron (or more than one electron) leaves one atom and is picked up by another atom; hence both atoms become electrically charged. Since they have different charges, they attract one another and combine. The resulting compound is called electrovalent.

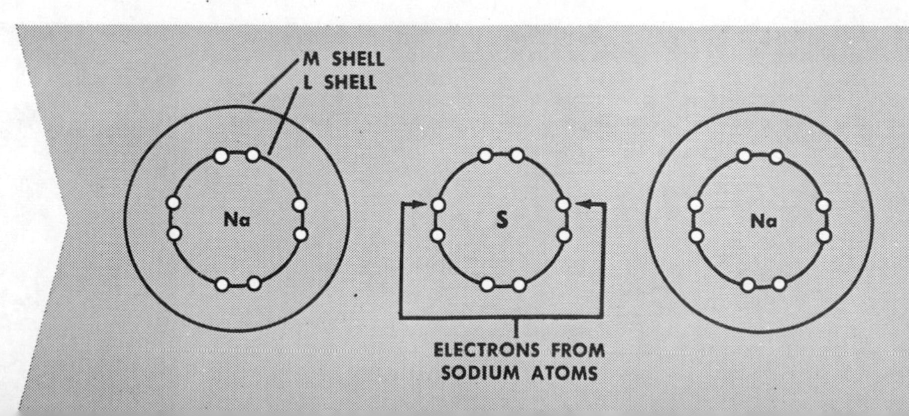


**13. The sulfur atom has captured an electron from each of the two sodium atoms in Figure 12, above.**

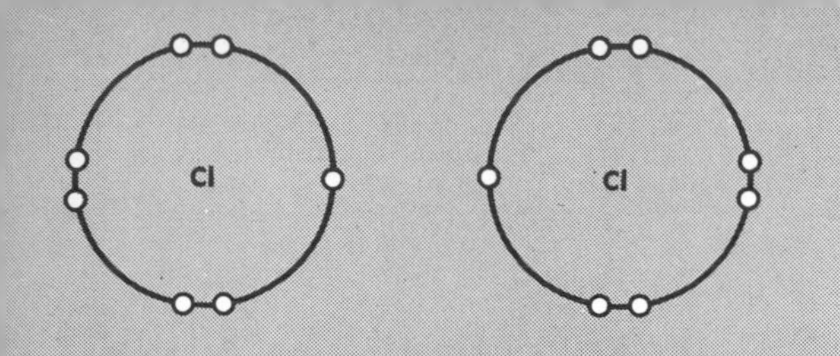


14. Two chlorine atoms, with atomic number 17 and with seven electrons each in the outer (M) shell.

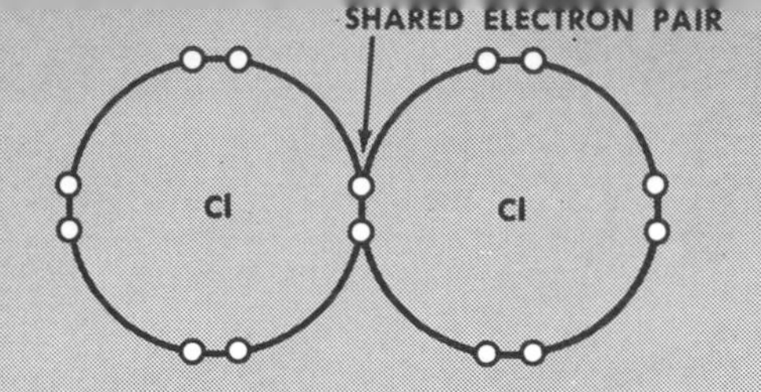


15. Two chlorine atoms have combined, sharing an electron pair and forming a chlorine molecule ($Cl_2$).

The bond that holds the ions together is known as electrovalence or ionic valence. If an atom can lose one electron, thus acquiring a single positive charge, it is said to have an electrovalence of plus one. An atom (such as that of fluorine) that can take up one electron, acquiring a single negative charge, has an electrovalence of minus one. An atom with an electrovalence of plus two can lose two of its electrons; an atom with an electrovalence of minus two can take up two electrons. Other atoms have electrovalences of plus or minus three or four.

### Covalence: atoms share electrons

Electrovalence is not the only kind of bond that unites atoms. Suppose two different atoms, such as those of chlorine or fluorine, have one vacancy in their outer shells. They can fill this vacancy by sharing a pair of electrons.

In Figure 14, we see two chlorine atoms, each with seven electrons in the M shell. They can combine as shown in Figure 15. Each atom now seems to have eight electrons in its outer shell, but only because two of the electrons are allowed to count twice, once for each atom! The orbits of the outer-shell electrons would look something like the one shown in Figure 16. You will note that six electrons are orbiting around only one or the other of the two chlorine nuclei. Two electrons, however, are revolving around the nuclei of *both* of the chlorine atoms.

This type of bond is called covalence. Note that *two* electrons take part in *one* covalent bond. The two chlorine atoms linked together by a covalent bond in Figure 15 form the molecule of the gas chlorine ($Cl_2$). A chlorine atom and a fluorine atom can unite in a similar way. Both of these atoms, as we have seen, have seven atoms, out of a possible eight, in their outer shells. When they combine, as in Figure 17, they form the compound chlorine fluoride (ClF). Both chlorine and fluorine have a covalence of one.

Nitrogen has a covalence of three: that is, it is bound to three atoms of the element with which it unites. Figure 18 shows a nitrogen atom and three hydrogen atoms in the uncombined state. The nitrogen atom has five electrons in its outer shell, while the hydrogen atoms have one electron each. Figure 19 shows the four atoms united in a covalent bond, to form the compound ammonia, $NH_3$. The nitrogen atom shares electron pairs with the three hydrogen atoms. Carbon generally has a covalence of four. When it combines with four chlorine atoms, each with a covalence of one, it forms carbon tetrachloride, $CCl_4$, a familiar dry-cleaning fluid (Figure 20). In all these diagrams, note that the electrons that are indicated as "shared" by two atoms really go around *both* the atomic nuclei.

16. These two chlorine atoms share a pair of electrons. Six electrons revolve around each of the nuclei; two, around the nuclei of both atoms.

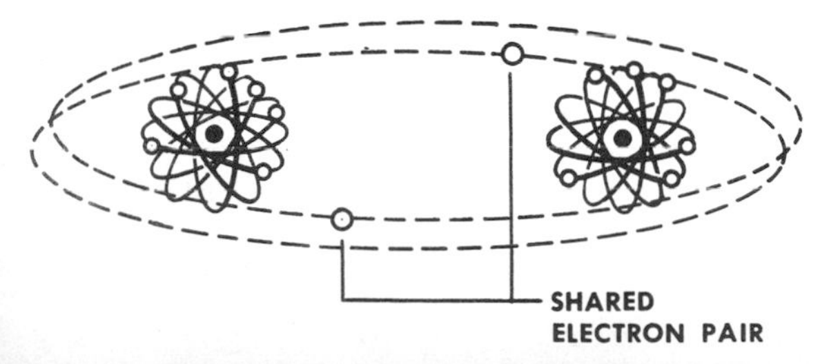


17. Chlorine and fluorine atoms combine, forming chlorine fluoride.

Cl
F
SHARED
ELECTRON PAIR

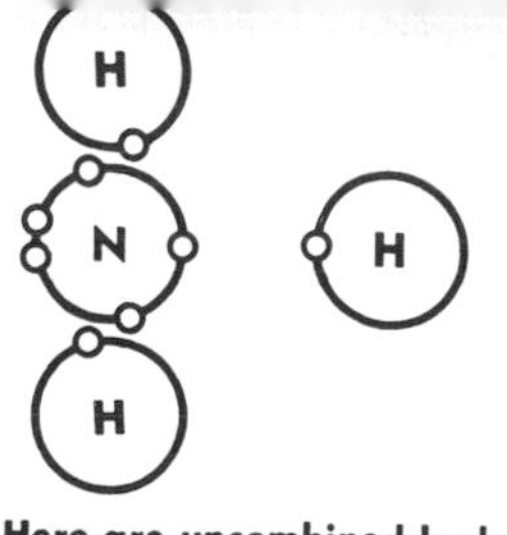


18. Here are uncombined hydrogen (H) and nitrogen (N) atoms.

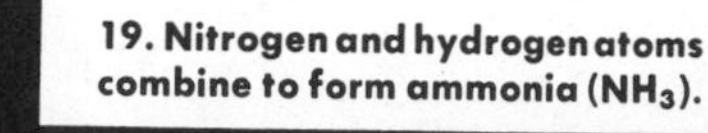

19. Nitrogen and hydrogen atoms combine to form ammonia ($NH_3$).

20. Tetrachloride ($CCl_4$) is made up of carbon and chlorine atoms.

**In-between cases of valence**

The compounds we have described up to now have presented clear-cut cases of electrovalence (sodium fluoride, sodium chloride, sodium sulfide) or covalence (the chlorine molecule, chlorine fluoride, ammonia, carbon tetrachloride). There are also a great number of intermediate cases, falling somewhere in the range between electrovalence and covalence.

For example, two atoms may share a pair of electrons, as in covalence, but as the two electrons orbit around the nuclei of both atoms, they may spend more time in the neighborhood of one of the nuclei. The atom favored in this way will acquire a partial negative charge; the other atom will have a partial positive charge. We would say of such a compound that it has "covalent bonds with partial ionic character." A good example is water.

The water molecule ($H_2O$) is made up, as we have seen, of two atoms of hydrogen and one of oxygen. Hydrogen has only one electron, which is in the K shell. Oxygen, with atomic number 8, has the usual two electrons in the K shell and six electrons in the L shell. There are two vacancies, therefore, in this shell. The vacancies are filled as the oxygen atom combines with two hydrogen atoms. If all these atoms were linked together by covalent bonds, we could show the linkage as in Figure 21, or by the notation H—O—H. On the other hand, if the bonding were completely ionic, there would be two $H^+$ ions and one $O^{--}$ ion, sharing two electrons and kept together by electrostatic attraction only. Neither way of indicating the molecule would be accurate.

For one thing, the hydrogen atoms do not form a straight line with the oxygen atom, but a bent line, making an angle of 105 degrees. We know also that the water molecule is not neutral electrically. The oxygen atom has a slight negative charge; each of the two hydrogen atoms, a slight positive one. Figure 22 would give at least some idea of the water molecule. Note that the two shared electrons are shown nearer to the oxygen atom than to the hydrogen atoms, indicating that they spend more time in the vicinity of the oxygen atom as they orbit. The negative charge of the oxygen atom and the positive charge of the two hydrogen atoms are given in parentheses to show the partial nature of the charges.

These are the principal ways in which atoms combine to form molecules. We have dealt with only a few of the simpler compounds in this brief article. There are vast numbers of different molecules and some of them are exceedingly complex, consisting of a thousand atoms or more. Yet they all follow the basic principles of combination that we have outlined.

*See also Vol. 10, p. 279*: "General Works."

21. How the water molecule ($H_2O$) would look if its hydrogen and oxygen atoms were joined together by covalence.

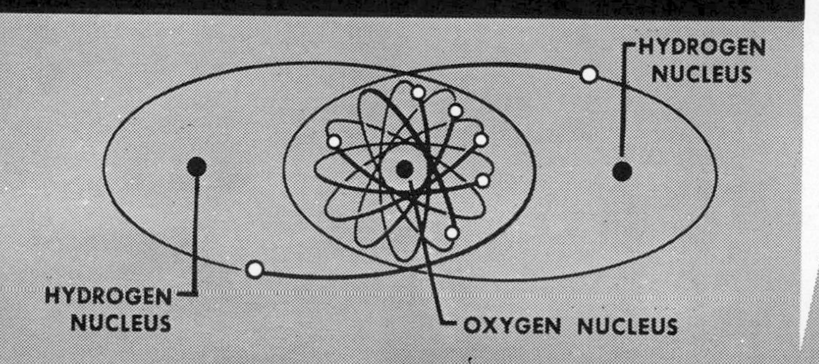


22. Probable combination pattern of hydrogen and nitrogen atoms of water.

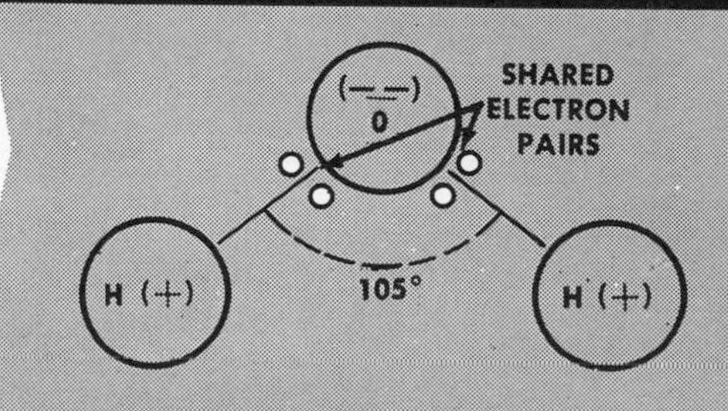

# THE TRUE BUGS

## Stinkbugs, Chinch Bugs, Bedbugs, Water Striders and Other Hemiptera

BY JOHN C. PALLISTER

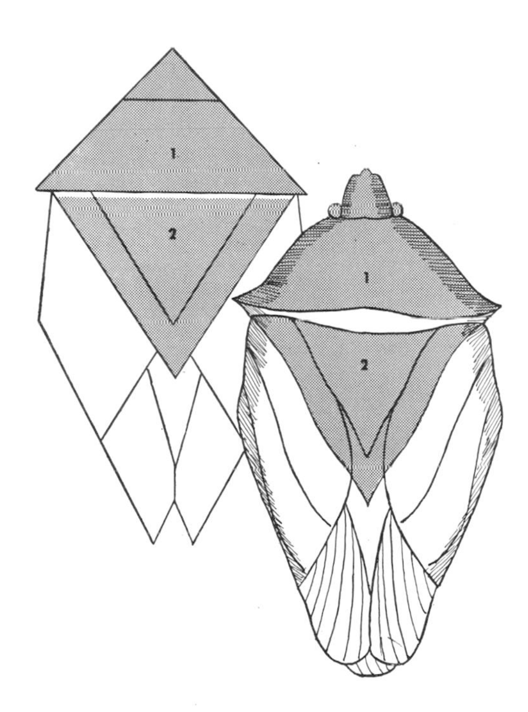


If you look at a true bug (right-hand drawing) from above, you will note two conspicuous triangles — the head and prothorax (1) and the scutellum (2). These triangles are clearly indicated in the left-hand drawing.

THE word "bug" has a variety of meanings. People sometimes apply it to insects in general, particularly to crawling insects. Occasionally it is used as a synonym for bedbug. Bacteriologists and others jokingly refer to microorganisms (particularly those causing disease) as bugs. A bug may also mean a "flaw in a mechanism," a "fanatic," a "flashlight" and other things besides.

In zoology, however, "bug," particularly when the word "true" is put before it, has a much more restricted meaning; it refers only to a particular group of insects. Many entomologists (specialists in the study of insects) put these insects in an order called the Hemiptera. This name, meaning "half-wings," refers to the forewings, which are thick and opaque at the base, and thin and membranous at the tip. Other authorities prefer to call the true bugs Heteroptera ("different-wings"), a name derived from the fact that the basal part of each forewing differs from the tip of the wing.

There is an allied group of insects — the aphids, cicadas, leafhoppers and others — known as the Homoptera ("similar-wings"). Certain entomologists have proposed a classification in which the Homoptera are put in the same order as the true bugs. In this classification, the Homoptera make up a suborder. The true bugs form another suborder, which is called the Heteroptera. The old term Hemiptera is applied to the enlarged order, combining the Homoptera and the Heteroptera.

Each of these methods of classifying the true bugs has its supporters among entomologists. In this article, I shall follow the one that I mentioned first; I shall refer to the true bugs as Hemiptera and shall consider them as a separate order.

As you look at almost any hemipterous insect from above, you will see a pattern of triangles. The head and prothorax form one triangle; the shield, or scutellum, form a second one.* There are other triangle patterns, which are not so obvious as the two we have mentioned.

* The prothorax is the front segment of the thorax (see Index). The scutellum is a plate set between the wings.

If you examine one of these insects closely, you will discover a few other characteristics of the order. The mouth parts are tubular, beaklike and fitted for piercing and sucking. All true bugs have such beaklike parts — long and straight, or short and curved. Some other orders, including the fleas, sucking lice, thrips and Homoptera, also have sucking mouth parts; but in each case the beak is differently formed and placed.

The structure of the Hemiptera forewings, which are called the hemelytra, is particularly characteristic. As we have pointed out, the basal part of the wing is thick and opaque; the tip is thin and membranous. The forewings usually fold flat over the back of the insect; the rear wings are membranous and fold under the front pair. Some Hemiptera have only one pair of wings; the bedbug has none. The forewings of one or two families do not have a thick base.

The head can move freely and often has a necklike portion. The compound eyes of the Hemiptera are unusually large. (One family has no eyes at all.) The antennae are generally long in the land bugs and very short in the aquatic families. The thorax may have various odd adornments. There may be flaps that flare out at the sides; there may be hairs or horns. Nearly always, between the base of the wings, we find the hard triangular plate of the shield, or scutellum.

The legs of the different species show an astonishing variety of forms. All six legs may be rather short; or the forelegs may be quite short while the middle and hind pair are long. This is especially true of the aquatic bugs, which use the long legs for swimming and the short ones for holding a captive or a mate. The legs of some species are as long and slender as those of the daddy longlegs. In some of the preying species, the short forelegs are curved and toothed; they serve as pincers. The middle and hind legs may carry a comblike structure or a feathery fan. Any pair of legs or all of them may be notched or clublike, or both notched and clublike. Some legs have large flat, leaflike projections.

Below are shown some of the details of a typical true bug's mouth parts, wings, legs and prothorax.

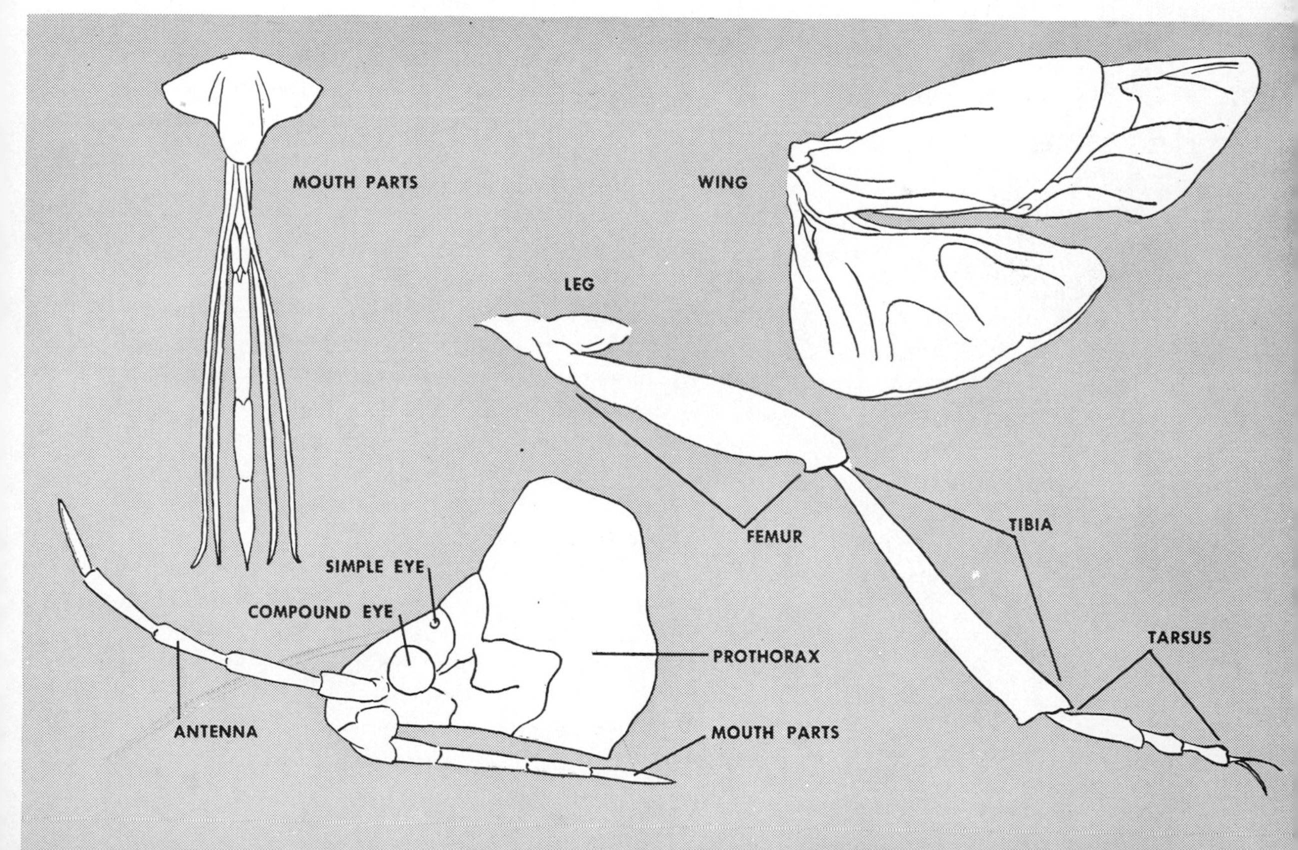

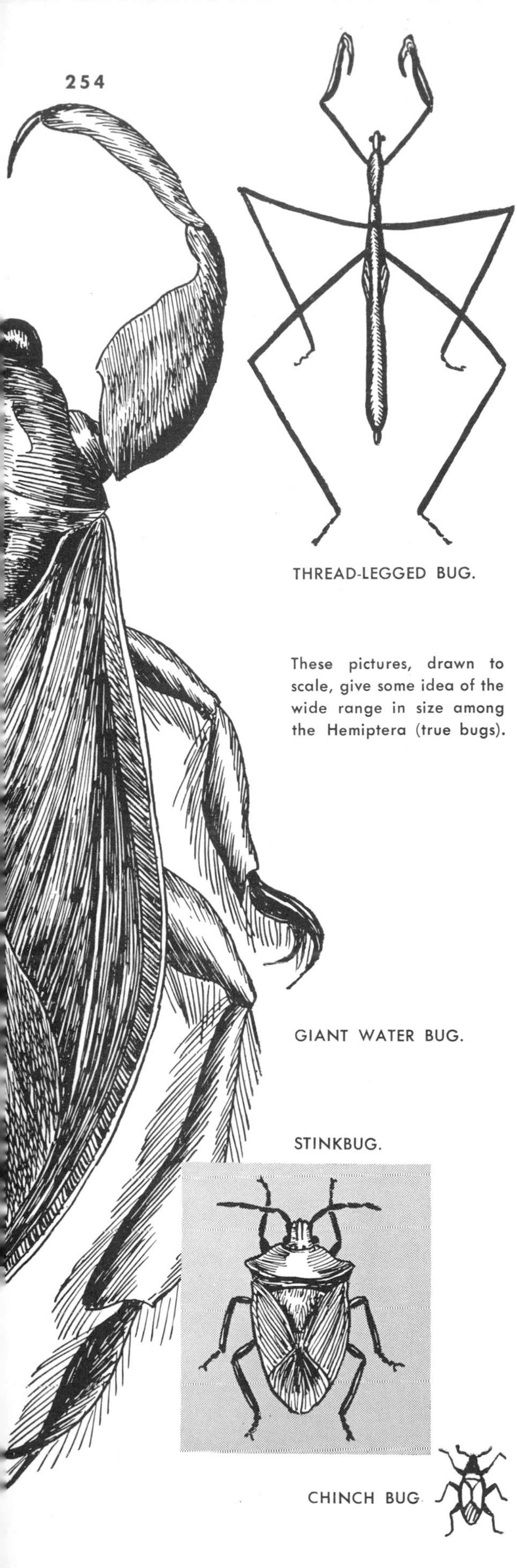


These pictures, drawn to scale, give some idea of the wide range in size among the Hemiptera (true bugs).

They are often fringed or covered with hair, fine or coarse.

In size the Hemiptera range from the tiny garden flea hoppers, about a twelfth of an inch long, to the giant four-inch water bugs, or toe biters. The great majority of species are from a quarter of an inch to three-quarters of an inch in length. Most bugs show plain colors — blackish, brownish, reddish or greenish. In a number of species, however, the insect's small back displays an intricate pattern in many contrasting hues. Tropical species are often large and brilliantly colored. Whether in Torrid or Temperate zones, patchwork coloration serves as camouflage, which causes the bug to blend remarkably well with its background.

Nearly 30,000 species of Hemiptera are known. Most of these dwell in the tropics, but many are found in the polar areas and the Temperate Zones. The great majority of species live on land. Several are found in fresh water and a very few spend their entire lives on the surface of the ocean. There are no true hemipterous cave dwellers.

The phytophagous, or plant-feeding, bugs suck juices from every type of vegetation, including trees, lichens and cacti. The predacious ones (preying bugs) will feed on anything they can attack, including man.

As cold weather approaches, most bugs go into hibernation in whatever state of development they happen to be — adult, immature or egg. If no food is available, many species can live for long periods without it. Because of these features, bugs have survived transportation from one continent to the other. The bedbug originally came from Asia, but is now at home wherever man dwells. Many hemipterous plant pests have a world-wide distribution; their "travel agent" was man.

The earliest hemipterous fossils discovered go back about 150,000,000 years to the Triassic period. They were found in New South Wales, Australia. These fossil forms show such an advanced stage of development that it seems certain the order must have originated earlier, perhaps in the

Permian period, which preceded the Triassic. Zoologists have found what is probably a living representative of the true bugs' Permian ancestors in Tasmania, Tierra del Fuego and other regions in the Southern Hemisphere. This insect is very tiny, green and has no hind wings. On its prothorax are flaps like the ones seen in fossil remains.

True bugs develop from the egg to the adult form by what is called gradual or incomplete metamorphosis. The creature that emerges from the egg resembles the adult, except that it is smaller and has no wings. This immature insect is called a nymph. As it grows, its outer covering becomes tight and is shed; after two or three such sheddings, or molts, little wing pads appear. These grow larger with each succeeding molt until they become functioning wings; the insect is then an adult. It usually takes a bug from five to six molts to reach the adult stage.

In almost all adult Hemiptera, scent glands are located on the underside of the thorax.* These glands secrete odorous oils. To the human sense of smell, the scents emitted by bugs are usually repellent. Some, however, are rather pleasant, suggesting the smell of pineapple, or cinnamon or ether. The purpose of the odors is not known. They may keep enemies away or attract a mate. The odorous oils may serve some purpose that we cannot guess, while the odor itself may be a mere by-product and nothing more.

Many bugs are able to produce sounds by rubbing one part of the body against another. Ordinarily only the males do so; but in some families both sexes and even the nymphs can make a noise. The apparatus used for this purpose is located in various places. In some families, there is a small groove on the front part of the body (the prosternum); the beak can be scraped back and forth in this groove. Both males and females of some stinkbug species have a grooved area on the underside of the abdomen and a set of short pegs on the hind legs. By rubbing the pegs against the grooved area, the insects can produce a weak, rasping sound. In other stinkbugs, the sound-producing apparatus is on the back of the abdomen and along the edge of the forewings.

Although most Hemiptera have two pairs of functioning wings, at least two families are wingless. In several instances, some adults of a species have short, useless wings, while others in the same brood develop normal ones. When in flight, bugs use a curious device to fasten the front and rear wings together. A turned-up portion of the forward edge of the rear wing hooks into a groove along the underside of the forewing's rear margin. As a result, both pairs of wings work together. In more primitive insects, such as the damsel flies, as the front pair of wings goes up, the hind pair goes down.

Land bugs generally deposit their eggs in small groups on the backs of leaves or on tender stems. In some species the bugs watch over the brood until it is hatched. The eggs, which are of various sizes and shapes, are usually quite visible to the human eye; often they are attractively decorated with patches or streaks of color. There are from one to four or five generations a year, depending not only on the species but also on the environment. The rate increases when the weather is warm, or when the season is longer than usual or when there is a bountiful food supply.

As far as we know, the Hemiptera do not migrate (in the sense of traveling back and forth between summer and winter locations), if you except the different species that move a little distance to hibernate in a sheltered spot. But emigration following the pattern of human emigration is common among fresh-water bugs. They will move en masse at night from one body of water to another several miles away, in order to seek a new food supply. We shall discuss this habit later.

Land bugs rarely have to go far to locate more food; they can usually find something in the next field. About half of the bug species live on plant juices. The other half — the predacious bugs — devour the first group. They also feed on

* In the nymph, the glands are near the end of the abdomen.

other insects and, in the case of aquatic bugs, on small fishes, frogs and salamanders. A very few are parasitic on man, birds and bats.

From man's point of view the bugs that live on plant juices are a pest, damaging or destroying his field crops, vegetables and fruits. It is unfortunate that several species feed on almost any kind of herbaceous plant, so that it is not possible to get rid of them by changing from one crop to another. Certain species found in the United States feed only on plants of the grass family. However, since much of the central United States is planted to corn and wheat and other grass-family crops, this is not a restricted diet for the bugs, and they live prosperously on the farmer's labor. It is fortunate for us that practically every bug that feeds on our crops is preyed on by another bug, often a close relative. Yet such preying bugs are not an unmixed blessing, for they also destroy great numbers of honeybees.

Of the few parasitic bugs, the bedbug is the best-known; it is attacked by several species of other bugs. Some members of the genus *Triatoma,* belonging to another family, are disease-carriers, especially in the warmer climates. Members of several bug families can inflict painful wounds when their beaks puncture the hands of careless collectors. This is true not only of the preying bugs, but of any bug with a strong enough beak. The pain from some of these punctures is great and it may persist for a week.

Most Hemiptera are too small to be of much use as human food. Nevertheless, in tropical countries, where the bugs are larger than elsewhere, some species are commonly eaten. Certain bugs belonging to the family Pentatomidae are roasted in southern Rhodesia; the Assamese cook another species of this family with their rice. Still another species is consumed in Mexico. The giant water bug is eaten, roasted or raw, in the American tropics; it is also said to be a favorite food in Laos, Indo-China. In Mexico, the eggs of the water boatmen — the corixids — are collected, dried and used as flavoring.

Besides man, the principal enemies of the Hemiptera are other insects (including bugs), spiders and birds. Dragonflies, cockroaches, robber flies, crickets and spiders do away with great numbers of adults and nymphs. Hymenopterous insects * are parasitic on the eggs.

The Hemiptera may be divided, according to habitat, into three groups: the land-dwelling; the subaquatic, living partly on the surface of the water and partly on the shore; and the aquatic, living in the water. Of the fifty or more families found in the United States, Canada and Mexico, we can discuss only the most important or interesting in this article.

### The land-dwelling Hemiptera

The land-dwellers contain by far the most families and species of the Hemiptera. They include a number of bugs that attack man's crops (harlequin bugs, squash bugs, chinch bugs, tarnished plant bugs and others) as well as the predacious bugs that prey on plant-feeding Hemiptera and other insects. The bedbug feeds on human blood. Several land-dwelling true bugs transmit disease to man.

*Stinkbugs* (Pentatomidae). The pentatomids, or stinkbugs, are probably the best-known of all the true-bug families. Their shape is distinctive; their colors,

* The order of the Hymenoptera includes bees, ants, wasps, ichneumons and true gallflies.

STINKBUG.

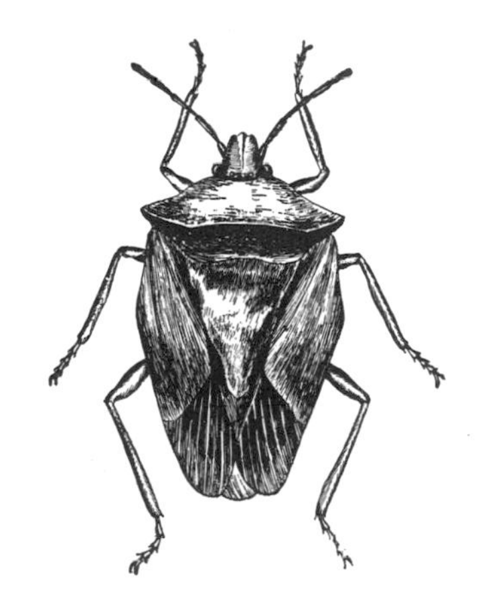

bright and metallic; their odor, vile. A person may know about bedbugs only by hearsay, but the chances are good that he has encountered a stinkbug in a garden or a berry patch, or even in a box of berries.

Most stinkbugs are shaped like a shield or an arrowhead. Once you are familiar with a few species, you will have no difficulty in recognizing members of the family. The prominent eyes bulge out at the base of a rather long head. The antennae are well developed and so are the wings, although you will seldom see a stinkbug flying.

In length the insects range from a quarter of an inch to an inch and a half, and in color from the glittery grays, greens and browns of northern species to the gorgeous and startling hues of southern ones. The inch-and-a-half giant that lives in Australia shows a color pattern of black, maroon, orange and purple.

The pentatomids, numbering 5,000 known species, are one of the largest families in the order of the Hemiptera. Most of them are found in the tropics of the Western Hemisphere, but there are representatives of the family in almost every part of the world. Wherever plants grow, there are some pentatomids, feeding either on the plants or on other insects.

Some species are injurious to truck gardens. Chief of these in the United States is the harlequin bug, or calico bug, *Murgantia histrionica,* a small, gaudy insect, with red, black, white, orange and yellow spots on its tiny back. It prefers cabbage and related plants, but in their absence it will feed on many other garden, orchard and field crops. The harlequin bug came to the southern part of the United States from Mexico shortly after the Civil War. For a long time the Yankees were blamed for its appearance, and for this reason it was called the Yankee bug.

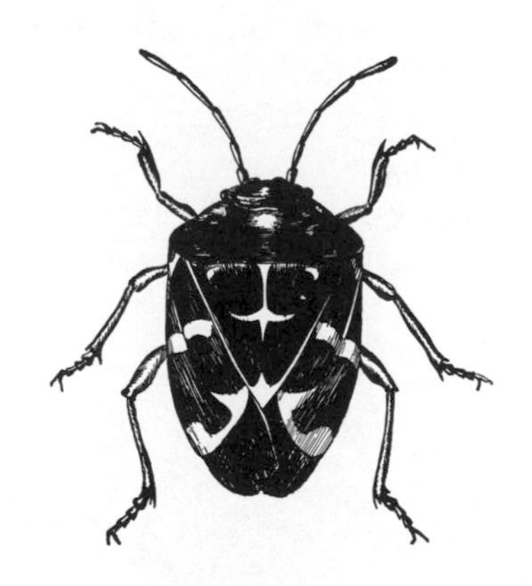

HARLEQUIN BUG.

While many stinkbugs are agricultural pests, others destroy various insect pests. Predacious pentatomid species feed on caterpillars and on the grubs of the Colorado potato beetle and bean beetle.

*Shield bugs* (Scutelleridae). Smaller than the pentatomids, the scutellerids are odd little turtle-shaped bugs, blending so harmoniously with their environment that they usually escape notice. The shield, or scutellum, that triangle at the base of the wings that is generally so conspicuous in the Hemiptera, is especially so in the shield bugs. In some species it covers the entire abdomen; when the wings are not in use, they are concealed under it. Most of the family feed on plants, especially on grasses, but so far have not caused much damage in North America. Some large and beautiful species are found in the tropics and in Australia.

*Negro bugs* (Thyreocoridae). These tiny, shining black bugs often swarm on early summer flowers. They look so much like the round black beetles found at the same time on the same flowers that one must examine them closely to determine which is which. The back of the beetle is covered by two hard elytra,* which spread out when it flies. The back of the negro bug is made up of a single hemispherical piece, the scutellum, from which the wings unfold for flight. There are not many known species, but these are common in the United States, where they are sometimes injurious to gardens. Like the stinkbugs, to which they are closely related, the diminutive negro bugs impart a decidedly bad taste to strawberries.

* The elytra are the first pair of beetle wings. They are hard and shell-like and serve as a covering for the second pair of wings and for the abdomen.

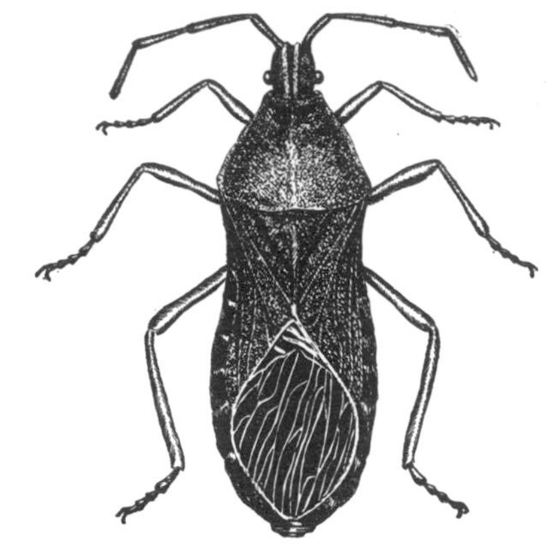

SQUASH BUG.

*Squash bugs and others* (Coreidae). The coreids form a large and variable family with over a thousand species, including the obnoxious squash bug (*Anasa tristis*). Coreids differ from the bugs we have just described. They are longer; their sides are straighter and their backs flatter; the scutellum is not so prominent. Most northern species are brownish, with or without red or yellow markings. Adult squash bugs are about five-eighths of an inch long and are grayish-brown in color. The younger nymphs have a green abdomen and crimson head, thorax, legs and antennae; but they lose these colors gradually through successive moltings. Young and old feed voraciously on leaves of the melon family, particularly squashes and pumpkins. What is worse, each time they puncture a leaf they inject a poison that burns it to a black crisp. Where they are present in any quantity, squash bugs will destroy the entire crop. Another coreid, the rice bug, an evil-smelling pest, injures grain plants in India. The so-called leaf-footed bugs have curious leaflike enlargements on their hind legs (not their feet, as their name implies). Brighter in color and longer than the squash bugs, they are quite common in the southern and western United States. They feed on all kinds of plants, including cactus, cotton and orange.

*Grass bugs* (Corizidae). Suburban residents, especially those living near golf links, often complain in the fall that the grass bugs are moving in with them. These are small, oval, yellowish-green (grass-colored) insects, not more than a quarter of an inch long; they are found in great numbers all over the world wherever there are open fields and pastures. During the winter, they hibernate under the bark of trees or, failing that, in barns and houses. One of the brightest-colored of the northern corizids, the box-elder bug (*Leptocoris trivittatus*), is red and black in varying proportions. It often leaves its normal hosts, the box elder and the maple, to feed on orchard fruits.

*Chinch bugs, milkweed bugs and others* (Lygaeidae). The scientific name Lygaeidae refers to the dull, dark color of many of the species. (The Greek word *lygaios* means "shadowy.") However, some of the most beautiful bugs in Europe and North America — the milkweed bugs — are members of this family. The spotted milkweed bug (*Oncopeltus fasciatus*) is a slender insect ranging in length from a half-inch to five-eighths of an inch. It is orange or bright red in color, with a black thorax and bands across the wings. The eggs and the nymphs are bright red. The insect feeds on milkweed and hibernates on trees and buildings.

Most notorious of the lygaeids is the chinch bug (*Blissus leucopterus*). A quarter of an inch in length, it is dark in color, with reddish or yellowish legs and antennae. Some members of the species have very short, brown wings, barely covering the abdomen and useless for flying. Others in the same brood have normally long, white wings and are able to fly for great distances. Chinch bugs are most injurious in the farm belts of the United States and Canada. Here they gather in huge numbers and suck dry the young plants of corn and other grains; they also feed on hay.

CHINCH BUG.

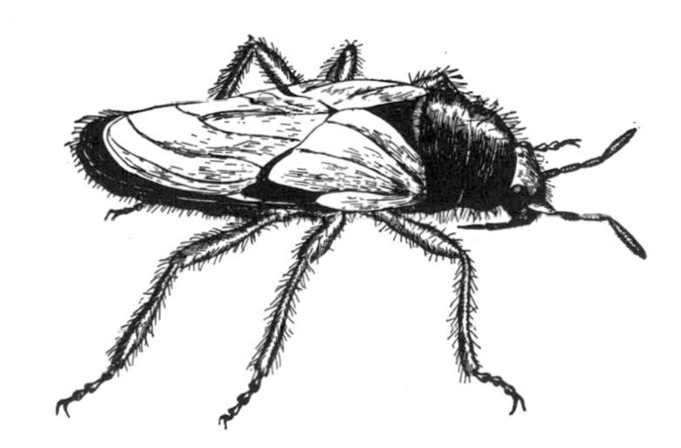

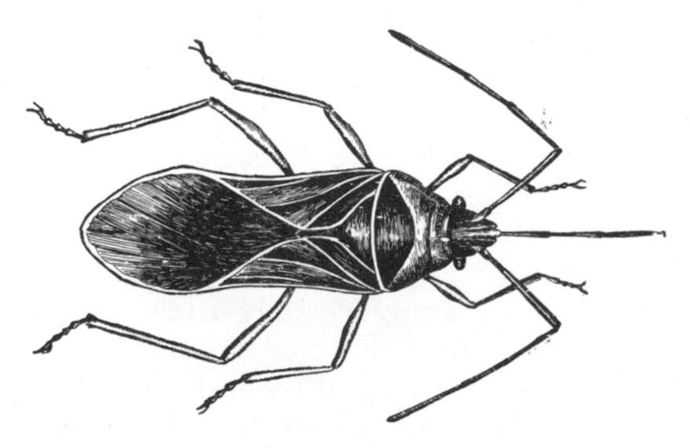

COTTON STAINER.

Not all lygaeids feed on plants. The odd-looking big-eyed bugs of the genus *Geocoris* prey upon the destructive chinch bugs and rank with man's allies. They are very small, flattish and ground-colored insects, with huge popeyes.

*Cotton stainers and others* (Pyrrhocoridae). This is a small family of stout and medium-sized to large insects. Red is the predominating color in most species. Many pyrrhocorids are brightly decorated with contrasting bands and spots. Apparently all are herbivorous and several are injurious to crops. Members of the genus *Dysdercus,* the cotton stainers, suck the juices of young cotton bolls and leave red stains on the fibers. The forty-odd species of *Dysdercus* are found wherever cotton or hibiscus plants grow. The common "stainer" in the United States, *Dysdercus suturellus,* is a slender red bug. Its beak is nearly half an inch in length; the rest of its body, about three-quarters of an inch. Swarms of "stainers" collect at night around street or porch lights.

The bordered plant bug, *Euryophthalmus succinctus,* also common in cotton fields, is more black than red; its forewings and thorax are bordered with orange. The nymphs of these species are brilliant blue, with red legs and a red spot on the tip of the abdomen.

*Lace bugs* (Tingidae). Most dainty and delicate of the Hemiptera, the lace bugs are so small (averaging about an eighth of an inch in length) that one would rarely notice them if they did not collect in such large numbers. Examine one with a hand lens. It looks like a bit of lace or filigree, patterned to suggest some highly imaginary insect. From the fore part of the thorax a high lace cap or hood arises, covering the head and much of the thorax. On each side of the hood there is a fanlike projection, also lacy in appearance. The forewings are wide, spreading far beyond the sides of the abdomen.

All lace bugs are believed to be plant-feeders, sucking sap from the veins on the underside of leaves. They cannot fly very well, but sometimes on a summer evening you may see a small flock of them drifting on a gentle wind.

Lace bugs are very common throughout the United States, and even more so in Europe. Several species produce two broods a year, one in May or June and another in August or September. Both adults and eggs may live through the winter under bark or piles of dead leaves.

*Assassin bugs and others* (Reduviidae). The distinguishing structures of this large bug family are small and not easily recognized, except perhaps for the head, which is long and narrow. These bugs are all predacious. If you watch a flowering bush or plant with several bugs on it, the ones that are devouring other bugs are likely to be reduviids.

The assassin bugs, genus *Apiomerus,* are rather stout and from a half-inch to three-quarters of an inch in length. They are black or brown with red or yellow decorations. Legs, head and thorax are often covered with coarse hairs. These bugs feed on caterpillars, plant lice and anything they can capture. One species, the bee assassin, is a pest around beehives; it devours beetles and flies as well as bees.

ASSASSIN BUG.

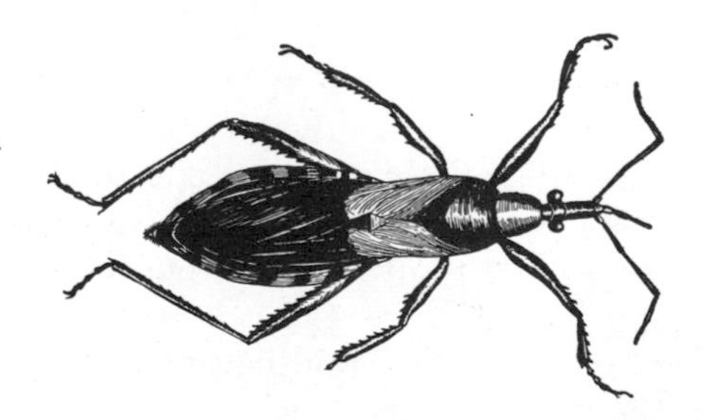

The genus *Triatoma* lives on the blood of mammals. Several species transmit various diseases from rodents to man, including Chagas' disease in South America and kala azar in India. At least one species of *Triatoma* and several belonging to other genera attack bedbugs, especially those that have just feasted on human blood. Some of these bedbug-destroying insects smell like their prey, but the odor is much stronger. They are sometimes called "big bedbugs." Occasionally these insects bite humans on the lips; for this reason they have received the picturesque name of "kissing bugs."

All reduviids with beaks strong enough to pierce the skin can cause extremely painful and long-lasting wounds. With the exception of *Triatoma* and possibly one or two other genera, these bugs usually bite only in self-defense.

*Thread-legged bugs* (Ploiariidae). These bugs look like tiny walking-stick insects. Most of the genera are less than half an inch in length and, although plentiful, are not likely to be observed. However, one species common in North America, *Ploiaria* (or *Emesaya*) *brevipennis,* may exceed an inch and a quarter in length. This bug is brown. Its front legs, much shorter than the others, are not used for walking but as pincers to capture and hold other insects. The middle and hind legs and the antennae are much longer than the rest of the body. The wings are short, covering less than half the abdomen; *Ploiaria brevipennis* can fly, but not very well. All thread-legged bugs are predacious, feeding chiefly on mosquitoes and gnats.

*Bedbugs and others* (Cimicidae). This famous family has only a few species, all of them small, flat and reddish-brown, with short legs, a small tough beak and no wings. They suck the blood of birds, bats and man. Of the two species that attack man, only one, *Cimex lectularius,* is found in North America; it came to this continent from Asia by way of Europe. The other species is restricted to the tropics of the Old World. Certain cimicids attack pigeons, wild birds and chickens. The poultry bug, *Haematosiphon inodorus,* a serious pest in Mexico and the southwestern United States, has recently begun to invade houses. Unlike most cimicids, it has no odor and it is not nocturnal.

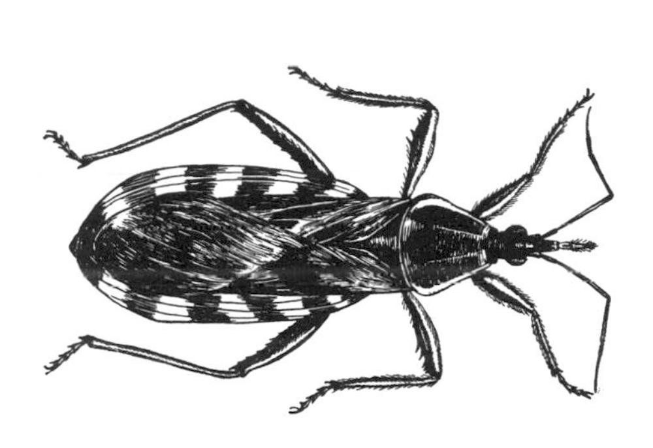

TRIATOMA.

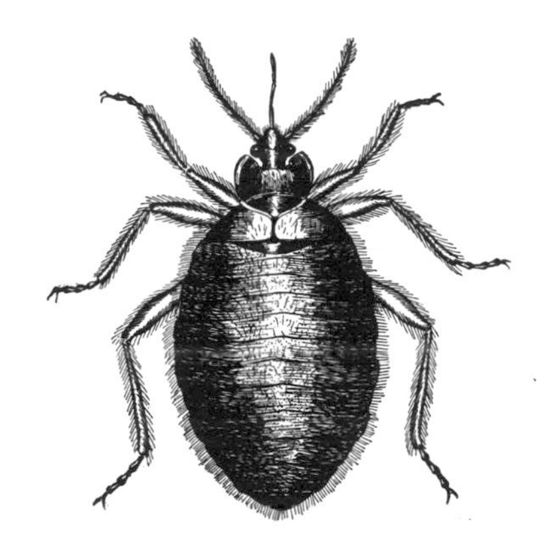

BEDBUG.

*Flat bugs* (Aradidae). Sometimes when you pull off the bark from a dead tree or log, you will find a small group of little, flat brownish bugs that resemble bedbugs. They are often mistaken for those unpleasant insects. A casual examination, however, will show that the flat bugs, unlike the bedbugs, have wings. They are quite innocent of blood-sucking, feeding upon tree fungi and possibly beetle larvae. The flat bugs spend the winter under bark; they are often frozen stiff, only to thaw and become active again with returning warm weather.

*Plant bugs* (Miridae). Like the pentatomids, or stinkbugs, the mirids form a very large family, with some 5,000 species; they include plant-feeding and predacious forms. Their average size is much smaller than that of the stinkbugs. Some mirids have long wings; others have very short ones; still others are wingless. The beak is long, often half as long as the rest of the body. Certain species feed chiefly on plant juices but require an occasional insect meal for their complete development.

One of the smallest known Hemiptera is a mirid — the garden flea hopper, *Halticus citri.* This tiny, black jumping bug, about a twelfth of an inch long, is common in the western United States and Canada. Collecting in vast numbers, it generally damages sugar beets, cowpeas, clover and many other crops.

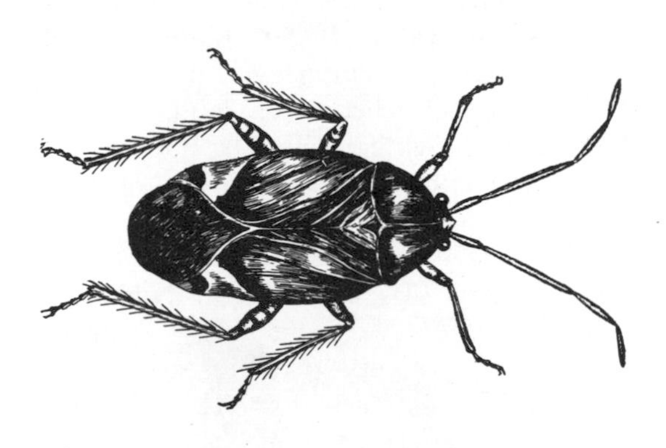

TARNISHED PLANT BUG.

The most serious pest in the entire order of the Hemiptera is probably the tarnished plant bug, *Lygus pratensis,* which is common throughout the Northern Hemisphere. A flat, oval bug, only a quarter of an inch long and less than half as wide, its back is decorated in a mottled pattern of browns, yellows, black and white. In the northern areas this bug hibernates, but in warmer climates it is active throughout the year. The tarnished plant bug is said to attack more than fifty cultivated plants as well as weeds and grasses. Vegetables, cotton, tobacco and fruits are all acceptable foods for this pest. As is usually the case with the mirids, the damage it does is due in part to its huge numbers. Swarms settling on an orchard just setting fruit can spoil the entire yield.

The genus *Deraeocoris* appears to be entirely predacious. *Deraeocoris ruber,* the aphid destroyer, is a broadly oval, shining little bug, about a quarter of an inch long; it is yellowish with red and black trimmings. Of European origin, it is now common throughout North and South America. It feeds not only on aphids but also on insect eggs.

*Shore bugs* (Saldidae). As you walk in summer along any shore, of marsh, stream, pond or ocean, you are likely to stir up a swarm of shore bugs that you had not seen until they rose before your feet. They do not fly high or far, and when they alight they merge with the sand or vegetation and are again invisible. Examined under a lens, they are not very impressive in appearance. They are small, oval, winged bugs, having the general color of their environment; they are mottled in darker shades along marshes and in lighter ones on the beach. They have made your walk a little more pleasant, for they have been destroying gnats and mosquitoes as fast as they could.

### The subaquatic Hemiptera

The amphibious or subaquatic bugs are at home both on shore and on the surface of the water; they do not dive below the surface. There are several families. The majority are extremely small (an eighth of an inch or less); they are sometimes winged and sometimes wingless. They spend most of their lives on floating vegetation and tangled mats of duckweed or algae, feeding on minute crustaceans and on small insects that have fallen into the water. They hibernate on shore under any kind of sheltering debris.

Not much is known about the lives and habits of the subaquatic bugs. Most of them are so minute that they can be studied

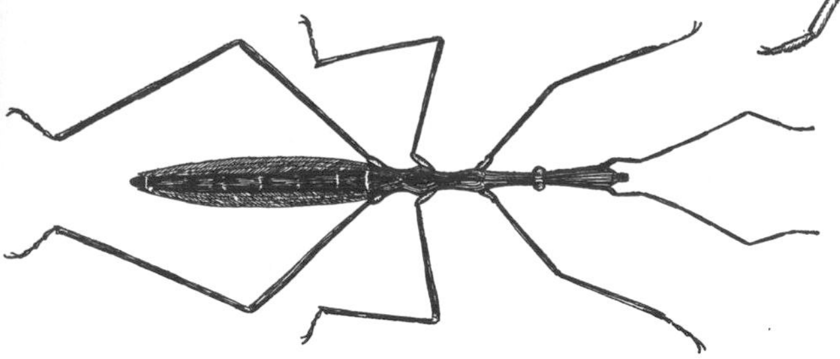

MARSH TREADER.

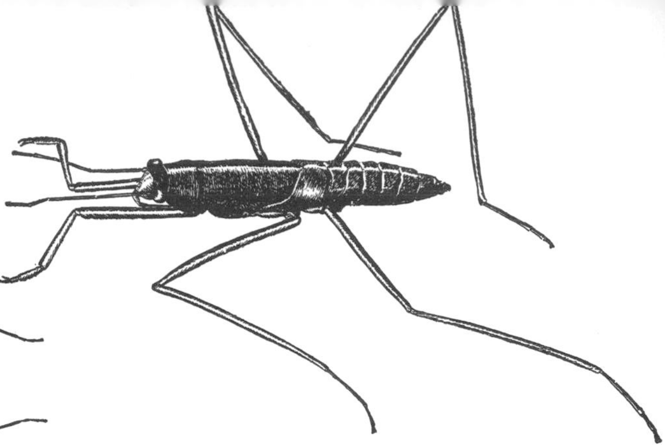

WATER STRIDER.

only under the artificial conditions of an aquarium in a laboratory. The marsh treaders and water striders, however, are large enough to be observed.

*Marsh treaders* (Hydrometridae). Marsh treaders, or water walking-sticks, have very slender bodies, about half an inch long, and equally long and thin legs. They are about the color of the water they patrol: brown with a bluish tinge. If you watch carefully, you may sometimes see them strolling across stagnant patches of water, stopping to examine each fallen leaf or bit of duckweed for small insects and water life. The marsh treaders are too slight to break through the surface film of water, too thin and shadowy to be noticed by their prey. If a large insect falls into the water, a number of these wispish creatures will surround it at once, each inserting its beak into the struggling victim. Most hydrometrids are wingless, but some in each generation will develop wings, with which they will reach a new feeding place. They hibernate on land.

*Water striders* (Gerridae). On a suitable stretch of water there are sure to be water striders: medium-sized, slender, dull-colored bugs, often covered with a velvety pile. Like most subaquatic bugs, they are dark-colored above and light-colored beneath. The front legs of the gerrids are short; they are set just behind the head and are well adapted for capturing other insects. The middle and hind legs are long and arise close together near the base of the thorax. The middle pair serve as oars; the hind pairs, as rudders. Some species have very short wings; many are wingless. As in other aquatic families, when need arises, some individuals will develop long wings and fly off in search of more plentiful food. The gerrids are active until the water begins to freeze over, when they go ashore to spend the winter under snow-covered leaves and rocks. On sunny days in winter, you may see them hopping about in sheltered nooks at the water's edge.

In contrast to the slow-moving marsh treaders, the water striders glide back and forth swiftly, scurrying even more rapidly if they are disturbed. On land, however, they have an awkward kind of hopping motion; frequently they fall over on their backs.

Of all the countless millions of insects, only one genus is truly ocean-dwelling — the bug *Halobates* of the family Gerridae. This short, velvety-gray insect has no wings and almost no abdomen; it lives hundreds of miles from land on great tracts of floating seaweed or on the open water in tropical and subtropical oceans.

### The aquatic Hemiptera

Several families of the Hemiptera spend all their lives in the water, except for occasional nights when they start hunting for a new home. Nearly all of them have well-developed wings. These bugs are not equipped with gills, as are the young of dragonflies and mosquitoes. They must come to the surface to breathe and must carry a supply of air with them when they go below. Some aquatic Hemiptera can hold a little bubble of air between the abdomen and the wings. Others are covered partly or entirely with a thick pile of hair which keeps the water out and the air in. Still others can rest just below the surface and draw in fresh air through a tube that projects from the tip of the abdomen.

*Water scorpions* (Nepidae). Both the common and scientific names of this family * are based on two faint resemblances. The front legs of the nepids look something like the pincers of a scorpion and indeed are used for the same purpose: that is, to capture and hold prey. Short and curved and placed very near the head, they are held straight forward and are not used for walking or swimming. At the tip of the abdomen there is a breathing tube, usually quite long, that looks a little like the tail of a whip scorpion.

Nepid species vary from extremely slender to widely oval, and from one to two inches in length. Middle and hind legs are long and slender and not at all well-adapted for swimming, though the insect can swim after a fashion. It seems to prefer hanging motionless, head downward on some plant, just below the surface of the water, with its respiratory tube projecting into the air like the snorkel of a submarine. Many species can produce a loud, rasping noise with the front legs. The nepids are very aggressive, successfully attacking creatures much larger than themselves. In Florida, the long slender *Ranatra australis,* called the alligator flea, or water dog, is said to bite humans savagely, causing considerable pain.

* *Nepa* is Latin for "scorpion."

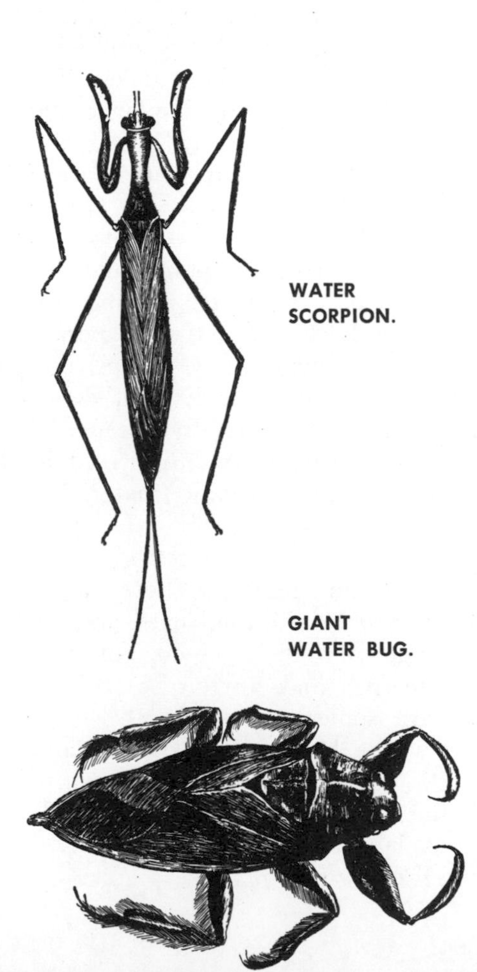

WATER SCORPION.

GIANT WATER BUG.

*Giant water bugs* (Belostomatidae). Other expressive names for these savage creatures are electric light bugs, fish killers and toe biters. They are by far the largest of the true bugs and rank high in size among the larger insects of all orders. Four inches is a common length in the United States; four and a half, in the tropics. The insects are husky and vigorous, with broad, leathery, dull brown bodies and, usually, strong wings.

They live at the bottom of shallow water, feeding on small fish, frogs, and insect larvae — in fact, anything that their powerful forelegs can seize and hold. When a bare-footed human wader steps too near one of these bugs, it will grasp a toe and jab its sharp beak again and again into the flesh, causing exquisite pain.

All belostomatids fly at night, usually in great swarms. If they are fortunate, they may find a fish hatchery and have a great feast for the short time that it takes them to devour all the young fry. Sometimes they mistake the sheen or reflection from a glass greenhouse for water and rush headlong to their death. They are also attracted, to their great peril, to electric lights. When the writer was in Mérida, in Yucatan, the city was invaded night after night for some two weeks by hordes of a belostomatid species, *Lethocerus americanus,* drawn by the bright lights. They would fall dead or dying or merely exhausted in the streets, making driving or even walking hazardous.

The bugs belonging to the species *Lethocerus americanus* are among the largest belostomatids in North America. They are from two to four inches long and from three-quarters of an inch to one and a quarter inches wide. They are also the commonest and most widely distributed forms, growing larger in Mexico than in the United States.

A few genera have an unusual way of assuring that the next generation will at least survive the egg stage. The female glues the eggs to the back of the male. The eggs of one genus, *Belostoma,* hatch in from ten to twelve days, and then the male is able to shake them off. Thereupon the female glues another brood on his back!

In tropical countries, where these bugs are most gigantic, the natives commonly strip them of their legs and wings and eat them, either raw or roasted.

*Back swimmers* (Notonectidae). Members of this family differ from all other insects in several conspicuous ways. For one thing, they swim or dart about on their backs, which in this inverted position are shaped like the keels of boats. The hind legs, which serve as oars, are long, flat and often feathered. The middle pair of legs is placed close to the front pair, which it resembles; it is used to assist the front pair in clinging to objects and grasping prey. The color shades are the reverse of the usual arrangement; the bugs are dark on the underside and lighter above.

Back swimmers are black and brown in color; they range in length from half an inch to three-quarters of an inch. The very large eyes occupy most of the head; the antennae are short and hidden. Two small tanks on the sides of the abdomen supply air, which is replenished through short tubes at the tip of the abdomen. Like their larger relatives, the notonectids are ferocious hunters and long-distance night fliers. Occasionally they descend, probably by mistake, on swimming pools, where they jab viciously at the bathers.

WATER BOATMAN.

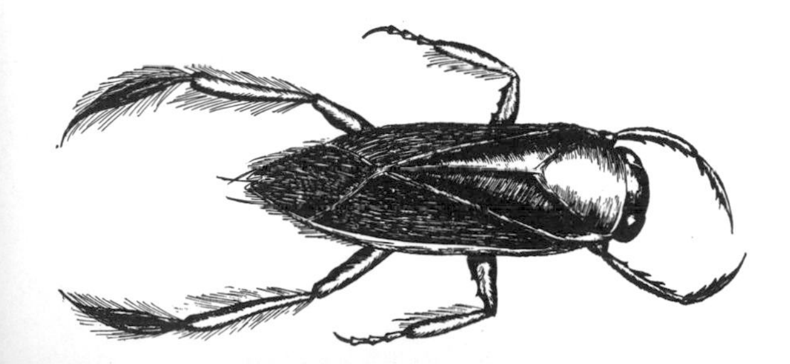

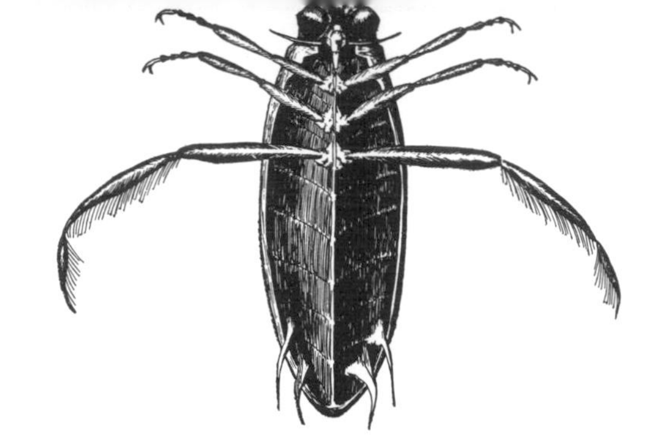

BACK SWIMMER.

*Water boatmen* (Corixidae). In general appearance and behavior, the corixids resemble other water bugs. They are small, mud-colored insects, with a triangular or crescent-shaped head as wide as the thorax, and large triangular eyes. Like the other water bugs, they have strong wings, fly in great swarms and may invade swimming pools and bite the bathers. The beak is not nearly so complicated as in other Hemiptera; the insect can suck small particles into its mouth, instead of being restricted to juices. The males of some species make shrill chirping or twittering sounds by rubbing a front leg against the side of the head. In Europe they are called "water crickets."

Corixids spend most of their time on the bottom of streams, ponds and lagoons of fresh or brackish water. Some breed in summer puddles and retire to permanent waters in the winter. One species lives at the bottom of Lake Erie at depths of from fourteen to thirty-five feet. Wherever water boatmen are, they feed in the muck on minute organic material, both plant and animal.

A few species have considerable economic value in Mexico. There the insects are collected by the ton, dried and sold at home and abroad as food for poultry, aquarium fish and songbirds. Sticks and poles are sometimes set in the water in places where the bugs deposit their myriads of eggs. After the sticks have been pulled out, the eggs are scraped off, dried and used as flavoring in soups and gravies.

*See also Vol. 10, p. 275:* "Insects."

# SCIENCE GROWS UP (1600-1765) II

BY JUSTUS SCHIFFERES

### *INVISIBLE COLLEGES AND SCIENTIFIC SOCIETIES*

THE progress of science depends upon the communication and criticism of ideas. Today, if one scientist wants to communicate urgently with another, he can simply pick up a telephone and call him or jump on a plane and see him in a matter of hours. Phones and planes are themselves the outgrowth of scientific progress. But in the seventeenth century the growing fraternity of scientists — gifted amateurs and lovers of science whose income and support came from other sources — had still to solve the problem of communicating effectively with each other. This was solved by the gradual development of scientific fraternities, or "invisible colleges," as they were sometimes called. Out of them grew the formal scientific societies, many of which are still in existence today and provide, above all, a forum for the clash of ideas out of which reasonable scientific certainty arises.

Probably the first scientific fraternity, which was more like a secret society as we know the term today, was the *Academia Secretorum Naturae* (Academy of the Secrets of Nature). It was founded in Naples in 1560 by Giambattista della Porta but was soon disbanded when it was suspected of practicing magic and "the black arts." A second Italian academy was founded in Rome in 1603, with both Della Porta and Galileo among its members. It took the name of *Accademia dei Lincei* (Society of the Lynxes) and had as its emblem a lynx clawing away the curtains that hid the truths of nature. This society was considered far more respectable, and a modern version of it still exists.

The most important Italian scientific society of the seventeenth century — the *Accademia del Cimento* (Academy of Experiments) — did not arise until nearly half a century later, when such institutions were already in the wind in England and France. In the year 1651 the Grand Duke Ferdinand II of Tuscany and his brother Leopold, members of the powerful Medici family, had set up a well-equipped laboratory in their spacious palace in Florence. The two brothers, who had studied under Galileo, invited scientists of Italy and of foreign countries to make use of their apparatus and to join them in experiments. The leading spirits of the "invisible college" that arose in this way were two pupils of Galileo — Evangelista Torricelli, the inventor of the barometer, and Vincenzo Viviani, a noted mathematician.

In the year 1657, the group acquired a more regular status when it received a charter as the *Accademia del Cimento*. The *Accademia* numbered among its members some of the most renowned scientists of the day. There was Giovanni Alfonso Borelli, an anatomist who applied the principles of mechanics to living things and who was also interested in capillary phenomena, such as the rise of liquids in small tubes and wicks. There was the embryologist Francesco Redi, who attacked the false idea of spontaneous generation by demonstrating that maggots do not arise spontaneously in decaying flesh but hatch from eggs deposited by flies. There was the Danish anatomist Nicolaus Steno (born Niels Stensen), who began to clear up the differences between the lymphatic and the ductless (endocrine) glands in the human body, and who helped found crystallography. There was likewise the astronomer Giovanni Do-

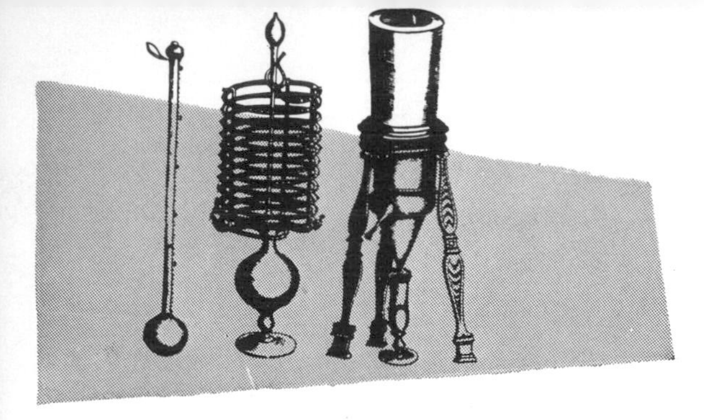

These instruments—two thermometers (left) and a hydrometer (right)—were used by the *Accademia del Cimento.*

menico Cassini, who later became the director of the Paris Observatory.

The Italian academicians were greatly interested in the study of heat, and they developed an improved kind of thermometer — the so-called Florentine thermometer, a miracle of the glass blower's art. They performed numerous experiments on the freezing and compressibility of water (hammered in a silver vessel), on elementary magnetism and electricity and on atmospheric pressure. These experiments, jointly conducted, were reported with restraint in a publication called the ESSAYS OF NATURAL EXPERIMENTS MADE IN THE ACCADEMIA DEL CIMENTO. This publication was suspended when the *Accademia* closed its doors in the year 1667.

The "invisible colleges" as forerunners of formal scientific societies were more numerous in England than elsewhere during the seventeenth century. There was one "invisible college" at Oxford; Robert Boyle frequently referred to it by that name in his letters. It centered in the rooms and around the person of Sir William Petty, a man of many scientific interests — physician, surveyor, businessman and cofounder of the science of vital statistics. There was another "invisible college" in London. These "invisible colleges," of course, had no buildings, no faculty, no students, no resources and they granted no degrees. They were informal associations of brilliant men excited by the new "experimental philosophy" and anxious to share and compare their ideas. This was the great age of amateurs in science.

The letters of the mathematician John Wallis tell us about the humble beginnings of the "invisible college" that was to develop into Britain's famous Royal Society. "About the year 1645," he wrote, "while I lived in London (at a time when by our civil wars academical studies were much interrupted), I had the opportunity of being acquainted with divers [various] worthy persons, inquisitive into Natural Philosophy and other parts of human learning and particularly what has been called *New Philosophy* or *Experimental Philosophy.* We did by agreement, divers of us, meet weekly in London on a certain day . . . The meetings were held sometimes at Dr. Goddard's lodging in Wood Street, on occasion of his keeping an operator [an instrument-maker] in his house for grinding glasses for telescopes and microscopes; sometimes at a convenient place in Cheapside and sometimes at Gresham College or some place near adjoining." Among those present at the meetings were physicians, astronomers, mathematicians and a Palatinate German, residing in London, Theodore Hank (or Haak), who "gave the first occasion and first suggested these meetings."

These "worthy persons" discussed many subjects — the circulation of the blood, the valves of the veins, the Copernican hypothesis, the nature of comets and new stars, the satellites of Jupiter, sunspots, the irregularities in the moon's surface, the improvement of telescopes, the weight of air, the newly invented barometer and the descent of falling bodies.

The meetings were interrupted in 1658 because of the political unrest of those troubled times of civil wars and of the Cromwellian Protectorate. After the Restoration of Charles II, the group began to meet again. The "invisible college" acquired official status in July 1662, when it became the Royal Society under a charter from the king. This renowned Society still flourishes.

At its early meetings it was the custom to assign special research projects to individual members or groups (today we would call them teams), who later made known their findings in reports. The Society investigated theoretical matters such as the production of colors by different combinations of chemicals, the measurement of the density of air and the incompressibility of water. But it got its early popular reputation and royal support because it proposed also to investigate practical problems: the recoil of guns, ship building and navigating, brewing, mining techniques, improvements in existing machines and the like.

In March 1665, the Society began the publication of a periodical called the PHILOSOPHICAL TRANSACTIONS OF THE ROYAL SOCIETY; the secretary of the Society, Henry Oldenburg, paid the expenses of publication. The PHILOSOPHICAL TRANSACTIONS are still in publication. It is the second oldest scientific periodical in the world; the oldest is the French JOURNAL DES SAVANTS (Scholars' Journal), which began its career only a few months before the PHILOSOPHICAL TRANSACTIONS. The establishment of scientific periodicals for the prompt publication, exchange and above all *criticism* of new scientific ideas marked an important step in the advance of science. The journals also provided a means of establishing priorities in scientific discovery and reduced the jealous and bitter controversies that often arose on this point.

In the course of its long and illustrious career the Royal Society has included in its ranks almost all the great men of British science. Among the early members we find such men as the famous architect of London Sir Christopher Wren, who was also interested in blood transfusion, mathematics and astronomy; John Evelyn and Samuel Pepys, both famous for their diaries; Sir Kenelm Digby, who dabbled in a "sympathetic powder for healing wounds." Also on the rolls were Sir Robert Murray, the society's first president (before incorporation); Lord Brouncker, a mathematician; Sir William Petty, co-founder of vital statistics; Robert Boyle, the "father of chemistry"; Robert Hooke, the microscopist; Edmund Halley, the astronomer; and Sir Isaac Newton, the greatest name in English science.

With the founding of the Royal Society, science for the first time became popular and fashionable. It was encouraged by the patronage of many noblemen, who, dressed in the height of fashion, attended its lecture-demonstrations just as they attended the theater. These nonproductive scientists came to be called virtuosi. Indeed the Royal Society was required for a time to "put on a show." The demonstrations were directed by Robert Hooke, who was appointed Curator of Experiments in 1662. His task was to prepare three or four novel or spectacular experiments for each weekly meeting.

In France the headquarters of a scientific fraternity and "invisible college" was for many years in the cell of a Minorite friar, Marin Mersenne (1588–1648). Mersenne was a tireless correspondent; his exchanges of letters with many of the leading scientists of the day served as a substitute (a poor one, indeed) for a scientific journal. The friar himself was particularly interested in research on music and acoustics. He worked out the relationship between the pitch of a tone and the length, thickness and tension of the violin string on which the tone was produced. In 1636 he published a huge work in twelve volumes on UNIVERSAL HARMONY.

During the 1630's and 1640's French scientists and philosophers like Pascal, Fermat, Gassendi and Descartes visited the learned friar in Paris in order to hear the news of the scientific world. Gradually a discussion group was developed; it attracted an increasing number of famous scientists and it continued after the death of Mersenne. Finally Claude Perrault, architect, inventor and mathematician, sug-

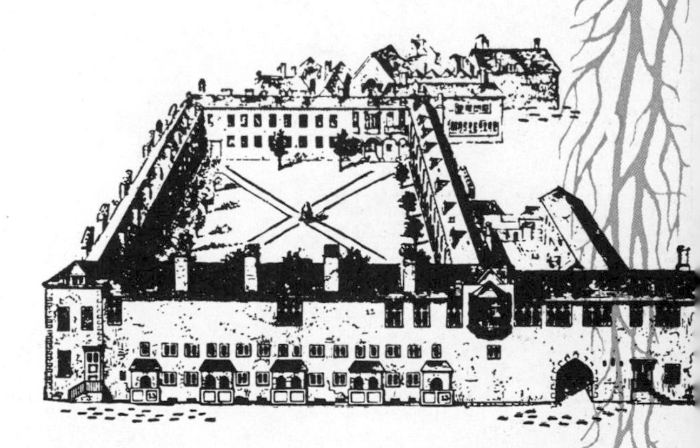

Gresham College, in London. The distinguished members of the "invisible college" that was to develop into Britain's Royal Society often met here in the 1640's.

gested to Jean-Baptiste Colbert, the minister of finance, that it would be a good idea to set up a formal scientific academy in France, with the learned members of the Parisian "invisible college" as a nucleus. Colbert brought the matter to the attention of his royal master, Louis XIV, who agreed to charter and support an *Académie des Sciences* (Academy of Sciences). With twenty-one royally paid scientists on its staff — one of the first examples of support of science with state or public funds — the *Académie* held its first meeting in the Royal Library in December 1666.

The scientists of the *Académie* worked on new inventions, such as automatic saws and comparatively frictionless pulley combinations. They were sincerely interested in the history of animals; they dissected a number of beasts (including a panther and an elephant obtained from the Versailles menagerie) in order to correct current errors in natural history.

They sent expeditions to Tycho Brahe's ruined observatory, Uraniborg, in Denmark, and to Cayenne, in French Guiana, which was close to the equator in South America. In Cayenne, Jean Richer observed that a pendulum, in order to mark the seconds accurately, had to be made shorter than at Paris; this observation reopened inquiry into the exact shape of the earth. Improved measurements of the earth — more specifically, the length of the degree of a great circle — had previously been undertaken with success by the academician Jean Picard.

The academicians occasionally engaged in some rather frivolous experiments, such as making burning glasses of ice. Other experiments of theirs seem to have been pretty silly, at least in the light of our modern knowledge. For example, the academicians distilled plants and even toads in order to discover their chemical nature; they obviously had no idea that they were destroying the very substances that they were trying to analyze.

Many unsuccessful attempts were made to found a satisfactory scientific society in Germany in the seventeenth century. The only one, however, to rank later with the Royal Society and the *Académie des Sciences* was the Berlin *Akademie der Wissenschaften* (Academy of Sciences), established in 1700 under the patronage of the Elector of Prussia. The great German mathematician, philosopher and educator, Gottfried von Leibniz (1646–1716) had labored for many years to bring the *Akade-*

King Louis XIV of France visits the *Académie des Sciences*. Among other activities, the members dissected animals to correct current errors.

*mie* into being, and he was its guiding spirit for a time. The financial support of the *Akademie* was derived from a monopoly on the sale of calendars. (The Gregorian calendar had been adopted in Germany in 1699.)

A Russian society, the St. Petersburg Academy, was founded by Peter the Great in 1724. The American Philosophical Society was established in 1743, with Benjamin Franklin as its guiding light; the American Academy of Arts and Sciences, in 1780.

The "invisible colleges" and the scientific societies that evolved from them had a most beneficial effect. In the give-and-take of open discussion, fanciful speculation had short shrift. To stand up to the acute criticisms of their peers, scientists had to hew closely to the line of careful observation and experimentation. Furthermore they learned the value of continued and unselfish co-operation among scientists as a means of obtaining scientific truth.

The French scientist Jean Richer noted that if a clock was to mark the seconds accurately, its pendulum would have to be shorter in Cayenne than in the city of Paris.

There are thousands of specialized scientific societies in the world today. New ones are still being founded every year and for the same reason that the invisible colleges of the seventeenth century came together: that is, to share and also to communicate exciting scientific ideas.

## THE BEGINNINGS OF TRUE CHEMICAL SCIENCE

In the course of the seventeenth century the pseudo-science of alchemy, with its mystical hokum and its confused symbolism, began to be transformed into the genuine science that we call chemistry. Perhaps the truest measure of its development as a science was the fact that it came to concern itself with fundamental principles as well as with practical applications.

Certainly, at the beginning of the seventeenth century, alchemy, or chemistry (chymistry), as it was also known, was a confused art. Some of its devotees were still trying to transform base metals into gold; others were seeking a universal remedy or an elixir of life, which would restore youth or prolong life indefinitely. But there was also a more practical outlook on what we should now call chemical investigations. There was interest in metallurgical chemistry — the chemical processes connected with mining operations and the refining of ores and metals. Again, partly under the influence of Paracelsus, a system of medicinal chemistry had arisen. It represented the application of chemical products to the treatment of disease and it was called iatrochemistry. (*Iatros* is the Greek word for "physician.")

Jean Rey, a French physician, was also a metallurgical chemist of this period. His chief contribution was his discovery of the role of air in calcination (the process of making a substance powdery by the application of heat). In his ESSAYS ON AN INQUIRY INTO THE REASON WHY TIN AND LEAD BECOME HEAVIER WHEN THEY ARE CALCINATED (1630), Rey pointed out that when a metal such as tin or lead is heated in air, it is converted into an ashlike calx that weighs more than the metal from which it was derived. He reasoned, correctly, that the increase in weight comes from the air in which the metals were heated. But his explanation of this phenomenon was altogether wrong. He thought that when air is heated, its lighter parts are driven off and its denser parts remain; these denser parts then adhere to the calx, or ash, of the metal.

Perhaps the outstanding iatrochemist of the seventeenth century was Jan Baptista van Helmont (1577?–1644?), a warm-hearted Belgian nobleman and physician. He was the first to distinguish different kinds of gases from ordinary air. In fact, he invented the word "gas"; it represented the Flemish version of the Greek word *chaos.*

Helmont talked about "windy gas," "smoky gas," "fat gas," "wild or unrestrainable gas" and "dry gas." He taught that they could all be changed by chemical combination into visible, material bodies. Since he had no way of collecting, examining and weighing gases, his ideas about them remained vague. His concept of a gas as a substance quite distinct from ordinary air did not win favor until the days of the chemical revolution more than a hundred years later. Even the best chemists, for several generations to come, spoke of different kinds of airs, rather than of different gases.

Helmont proposed the curious theory that "all tangible bodies are the product of water only and may be reduced to water again by nature or art." To prove his point, he put a small willow tree weighing five pounds in an earthen vessel containing 200 pounds of dry earth. He moistened the earth with water, as required, over a period of five years. At the end of this time the tree weighed 169 pounds and three ounces; the earth in the pot still weighed about 200 pounds minus about two ounces. Helmont concluded that "164 pounds of wood, bark and roots had been formed out of the water alone." We know today that the growth of the tree was due not only to water but also to the carbon dioxide in the air as well as to various substances (nitrates, phosphates and so on) contained in the ground.

The German Johann Rudolf Glauber (1604–68) was an iatrochemist who earned a fat living by selling secret chemical and medicinal preparations. He won fame for a substance that he called the "wonderful salt" and that is today called Glauber's salt, after him. (It is the chemical compound sodium sulfate decahydrate.) Glauber hailed his salt as a splendid medicine for internal and external use; it also, he said, would serve as a substitute for vinegar and it would impart a wonderful tang to meat and poultry. Glauber's salt is used today as an aperient, or mild laxative; it is also used in glassmaking, in the dye industry and in the manufacture of paper from southern pine.

With Robert Boyle (1627–91) we come at last to true scientific method in chemical research. This gifted man, who was born in Ireland, studied at Eton and then traveled widely on the continent, was one of the first members of the Royal Society and one of its leading lights.

Boyle did much to rid chemistry of the trappings of alchemy that still hampered it. In 1661, in his SCEPTICAL CHYMIST (Chemist), he insisted that chemistry was a science and not simply a practical system for treating metals or preparing medicine. As a science, he said, it should try to explain phenomena rather than seek to put them to immediate practical use. He gave the scornful name of "sooty empirics" to those experimenters who were intent only upon their own little projects and who concerned themselves not at all with general ideas.

In the SCEPTICAL CHYMIST, Boyle introduced the modern concept of a chemical element. "I mean by elements," he said, "certain primitive and simple or perfectly unmingled bodies; which not being made of any other bodies, or of one another, are the ingredients of which all those called perfectly mixed bodies are immediately compounded and into which they are ultimately resolved."

For centuries men had more or less accepted the ancient doctrines of Empedocles that there were only four elements — water, air, fire and earth. The alchemists had reduced the number of elements (they called them principles) to three — sulfur, mercury and salt. Boyle maintained that four (or three) elements could not account for even a tenth of all the phenomena that we behold in nature. We realize today that he was right. We have already identified more than 100 elements.

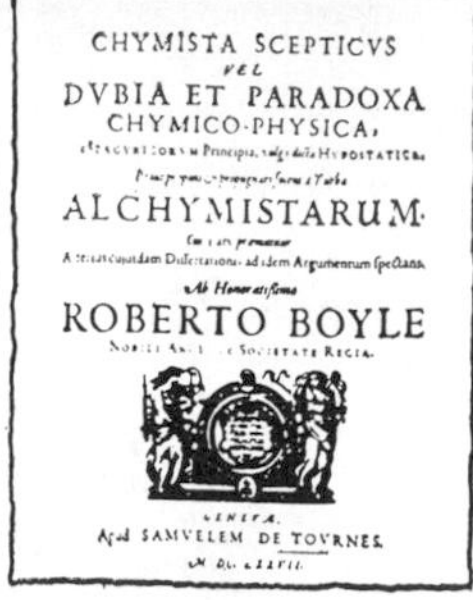

CHYMISTA SCEPTICVS
VEL
DVBIA ET PARADOXA
CHYMICO-PHYSICA,
ALCHYMISTARUM.
Ab Honoratissimo
ROBERTO BOYLE
GENEVÆ,
Apud SAMVELEM DE TOVRNES.

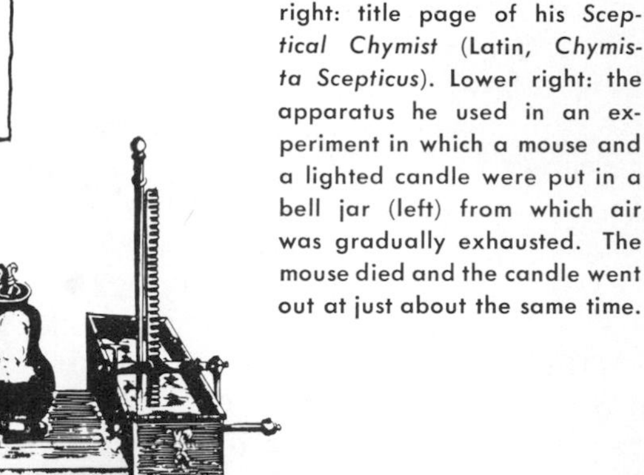

Left: Robert Boyle. Upper right: title page of his *Sceptical Chymist* (Latin, *Chymista Scepticus*). Lower right: the apparatus he used in an experiment in which a mouse and a lighted candle were put in a bell jar (left) from which air was gradually exhausted. The mouse died and the candle went out at just about the same time.

Boyle was greatly interested in the composition of the air. He recognized that only a part of the air is necessary for the combustion that takes place, say, when a candle is burned. We know this "part of the air" to be oxygen, which makes up, roughly, one-fifth of the atmosphere. For Boyle, however, it was "some odd substance, either of a solar, astral or other foreign nature, on account whereof the air is so necessary to the subsistence of flame." It was not until toward the close of the eighteenth century that oxygen was finally identified.

Boyle came to the conclusion that the life of animals is just as dependent upon that mysterious "part of the air" as the flame. To prove this point, he placed a live mouse and a lighted candle under a glass bell jar, from which the air was gradually exhausted with an air pump. The mouse died and the candle went out at just about the same time.

He investigated the spring (elasticity) of air by means of a manometer — a U-shaped tube closed at one end. When mercury was poured in the open end, the air in the closed end was visibly and measurably compressed. He worked out the relationship between the volume of a gas and the degree of pressure upon it in what is now called Boyle's Law: "In a closed chamber, if the temperature remains constant, the volume of a gas is inversely proportional to the pressure exerted on it."

The versatile Robert Hooke, who discovered vegetable cells (see page 273), was also interested in the problem of air. He thought that both combustion and respiration depended on a "volatile nitrous substance" contained in the atmosphere. He believed that plants, not less than animals, require air for growth. He thought he had proved it by sowing some lettuce seed in the open air and other seed in a container from which the air was removed. The seeds planted in the open air yielded flourishing plants; no plants came up in the container.

By the end of the seventeenth century, chemistry was becoming a bona fide science. It is true that alchemy still had its followers, but an increasing number of chemists had given up the search for the philosopher's stone, for the universal remedy, for the elixir of life. They sought, instead, for general laws that governed the various transformations of matter.

This search for general laws was, however, responsible for the rise of the fanciful phlogiston theory. The principal sponsors of the theory were Johann Joachim Becher (1635–82), a physician and alchemist, and

Georg Ernst Stahl (1660–1734), professor of medicine at the University of Halle. The word "phlogiston" itself was first introduced by Stahl in 1702.

According to the phlogiston theory, all substances that can be burned give off an "inflammable element" called phlogiston, or the spirit of fire, while burning; this inflammable element is supposedly absorbed by the air. If, for example (according to the theory), a metal like lead is burned in the presence of air, phlogiston is given off and the metal is reduced to a calx (see page 269). If the calx is heated with charcoal out of the presence of air, the phlogiston in the charcoal unites with the calx and the original metal is produced.

The phlogiston theory persisted until almost the end of the eighteenth century. We know today that it was utterly wrong, yet for a time it provided a working hypothesis that apparently explained many natural phenomena. It was simply one of many scientific hypotheses that have been tried in the balance and that have been found wanting in the course of the years.

## *THE MICROSCOPE REVEALS A NEW WORLD*

In the 1660's the members of the English Royal Society were greatly stirred by a letter from a modest but reliable observer in Holland. It announced that the writer, peering through microscopes fashioned by his own hands, had discovered a vast number of "little animals" in rain water. These "living atoms," or "animalcules," as he called them, were tiny; several thousand would fill the space of a grain of sand.

The Dutchman who thus first spied upon the world of infinitely small creatures was Anton van Leeuwenhoek (1632–1723), an untutored ex-shopkeeper and minor official of the picturesque city of Delft. (Some say that he was the janitor of the city hall.) He built his own microscopes — hundreds of them — and with them he observed anything that aroused his curiosity: the brain of a fly, the legs of a louse, sections of the crystalline lens of an ox's eye, the stinger of a bee and human semen.

A microscopic view of the cells in a section of cork. From the *Micrographia* of Robert Hooke.

Leeuwenhoek wrote almost 400 letters in all to the Royal Society describing his discoveries. These letters were long, rambling affairs, written in Dutch, the only language that he knew. They were translated into English and published in the PHILOSOPHICAL TRANSACTIONS. In 1680 the Royal Society elected him to membership; he repaid the honor by bequeathing to it twenty-six of his microscopes.

Leeuwenhoek has been called the "father of bacteriology" — the study of bacteria — and the "founder of protozoology" — the study of protozoa, or one-celled animals. He first beheld bacteria in "scum scraped from his own teeth"; stagnant water provided him with great numbers of protozoa. He faithfully observed the size, shape and motions of these tiny organisms under a variety of conditions. He saw some tiny animals "put forth two little horns, continually moving"; others were "furnished with extremely thin feet, which moved very nimbly"; still others "swam gently along, moving as gnats do in the air," or else had a "serpentine motion."

In 1688 Leeuwenhoek turned his microscope on the tail of a tadpole and saw the tiny capillaries connecting the veins and the arteries. As we shall see, Malpighi had

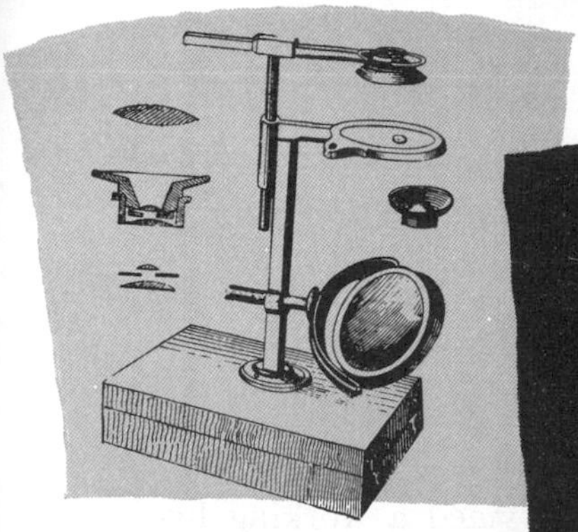

Early simple microscope, complete with mirror.

This compound microscope was developed by Robert Hooke.

The compound screw microscope was useful.

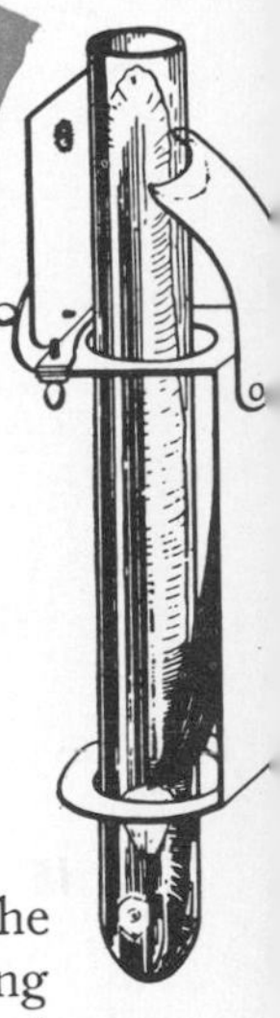

Simple microscope of Leeuwenhoek clamped to a tube.

already discovered the capillaries; but Leeuwenhoek was not aware of the fact, and he reported with great gusto on his amazing "find." "A sight presented itself," he wrote, "more delightful than any mine eyes have ever beheld; for here I discovered more than fifty circulations of the blood in different places while the animal lay quiet in the water and I could bring it before my microscope at my wish. For I saw not only that in many places the blood was conveyed through exceedingly minute vessels from the middle of the tail toward the edges, but that each of the vessels had a curve or turning, and carried the blood back toward the middle of the tail, in order to be conveyed again to the heart . . . And thus it appears that an artery and a vein are one and the same vessel prolonged and extended."

When Leeuwenhoek died in 1723 at the advanced age of ninety-one, he had an amazingly large number of microscopic discoveries to his credit. Among other things, he had discovered spermatozoa — the male elements in human beings and lower animals. He had made out the red blood corpuscles; he had correctly noted that they are circular in the blood of man and mammals, but oval in the case of frogs and fishes. Leeuwenhoek contented himself throughout his long scientific career with being an observer; he took little or no interest in theories or general laws.

Among the members of the Royal Society who showed most interest in the many communications of Leeuwenhoek was the versatile Robert Hooke (1635–1703), who himself made several notable contributions to microscopy. Like Leeuwenhoek, he built his own microscopes, including one of the outstanding early compound microscopes, with a collapsible tube for focusing. With this instrument, Hooke discovered in thinly sliced sections of cork the tiny honeycomb cavities which he was the first to call *cells.* According to his estimate, each cubic inch of cork contained about 1,200,000,000 cells. He gave an account of his discovery in his MICROGRAPHIA (1665).

Hooke was a many-sided man. He developed a theory of the wave motion of light. He pointed out that the motions of the heavenly bodies represented a problem in mechanics. He invented a new kind of barometer, called a wheel barometer, and he made several improvements in clocks and watches. He is known for Hooke's law, which states that in an elastic body, such as a coiled spring, stress is proportional to strain. Hooke had certain most unamiable traits. He was vain and irritable, and he was continually claiming credit for work done by other men of science.

There were other famed microscopists in the second half of the seventeenth century. Among the most notable was Marcello Malpighi (1628–94), a learned Italian physician, who taught at Pisa, Messina and Bologna. During the last years of his life he was the private physician of Pope Innocent XII. His scientific investigations are found chiefly in papers sent to the Royal Society of London. He became a Fellow of that learned body in 1668.

Marcello Malpighi (left) was a great Italian anatomist. He was one of the first to use the microscope in biological study and was the founder of microscopic anatomy.

In 1660 Malpighi discovered the capillaries in the lung of a frog; later he made out the capillary systems in other parts of the body. He discovered the pigments in the skin between the dermis (true skin) and the epidermis (surface skin); the layer containing these pigments is now called the Malpighian layer, after his name. Malpighi also investigated the layers and structures of the liver, kidney and spleen. He carefully studied the anatomy of the brain; he described the distribution of its gray matter accurately. He discovered that the buds on the tongue are organs of taste.

Malpighi carefully studied silkworms under the microscope; he discovered that they breathe through a minute system of air tubes distributed all over the body. He also studied the anatomy of plants. One day, it is said, while walking through the woods, he saw some thin threads hanging from the end of a broken branch. Examining these under the microscope, he observed that they were hollow spirals, much like the air tubes of silkworms. This led him to a number of interesting, though often incorrect, speculations about the manner in which plants breathe and receive nourishment.

The Dutchman Jan Swammerdam (1637–80) was a pathfinder in the study of the minute anatomy of insects. This is the beginning of the science called entomology. Though he received a medical degree, he never practiced; instead, he devoted himself, at great cost to his health and fortune, to the anatomy of bees, May flies, water lice and other small creatures. His writings were collected many years after his death by the Dutch physician Hermann Boerhaave, and published in 1737 under the title of THE BIBLE OF NATURE.

Swammerdam created his own minutely pointed dissecting instruments and his own techniques. He ground, under the magnifying glass, the tiny knives and scissors and scalpels that he used in his work. He drew glass tubes as fine as hairs; with them he injected colored liquids in the tiny blood vessels of insects so as to make them more visible. His descriptions and drawings of bees, May flies and other tiny living creatures were unsurpassed for over a century.

Swammerdam firmly opposed the Aristotelian notion, still widely held in his day, that certain animals arise from slime or decaying matter. He claimed that living things are born only of living things. Not only does decaying matter never produce living organisms, according to Swammerdam, but certain organisms can bring about decay in matter. Swammerdam marks a considerable advance over Harvey in his ideas on the spontaneous generation of living things. As we have already seen, Harvey believed that most animals, including man, are evolved from eggs; but he had maintained that certain lowly forms of animal life arise spontaneously from decaying matter.

A famous experiment by the Italian Francesco Redi (1626?–97?) confirmed the findings of Swammerdam. Redi put a dead snake, some dead fish and a slice of veal in four flasks, which he then closed and sealed. He filled four other flasks in the same way and left them open. Flies constantly entered the open flasks and their contents became wormy; but there were no worms (or rather maggots) in the closed flasks. This proved, he maintained, that "the flesh of dead animals cannot engender worms unless the eggs of the living be deposited therein." Later Redi put meat and fish in a vase which he then covered with gauze. "I never saw any worms in the meat and fish," he says, "although there were many on the gauze."

One of the foremost English microscopists was Nehemiah Grew (1641–1712), a physician who became the secretary of the Royal Society in 1677. With Malpighi he is credited with founding the science of vegetable anatomy. He made a number of original discoveries. He made out pores in the lower surface of leaves and came to the correct conclusion that leaves are the respiratory organs of plants. He held that flowers are the sexual organs of plants — that stamens are the male organs and pistils the female ones.

The German Jesuit priest, mathematician and physician Athanasius Kircher (1601–80) was a many-sided man, who was interested in optics, medicine, magnetism and geology. He was also an accomplished linguist; it was he who revealed Egyptian hieroglyphics to the world. He made one valuable contribution to microscopic research. With a 32-power simple microscope, he studied the blood of victims of the plague and pointed out that this blood contained microscopic organisms. Thus he confirmed the theory of Fracastoro (see Index) that disease is spread by "seeds of contagion." He was the first to investigate disease with a microscope.

Leeuwenhoek, Hooke, Malpighi, Swammerdam, Grew, Kircher — all these great pioneers in the science of microscopy drew the attention of mankind to the incredibly fine structure of many-celled organisms and to the teeming world of tiny animals and plants. In spite of the imperfect microscopes at their disposal, they did such amazingly fine work that their era has been called the "classical period of microscopy." Their achievements were unsurpassed until the nineteenth century.

## ON THE NATURE OF LIGHT

What is light? What is color? How do we see? Questions like these had stirred men's minds from the earliest days. The Pythagoreans theorized that vision was due to particles emanating from the eye and falling on objects. The Platonists believed that vision resulted from the interaction between the rays emitted by the sun or other source of light, the viewed object and the eye. Aristotle taught that colors were mixtures of light and darkness. Alhazen, the Arabic authority on optics, held that the form of the object viewed passes into the eye and is transmuted by the lens.

In the seventeenth century optics (the science of light and vision) began to be studied more intensively than ever before. Men began to form more accurate ideas about the nature of light and about the manner in which vision is brought about. They ferreted out the reason for the iridescent colors on a peacock's tail and for the rainbow. Efforts were made, too, to measure the incredible velocity of light.

In general, two theories concerning the basic nature of light held the field. One theory held that light was propagated in waves (like sound); this was called the undulatory, or wave, theory (*unda* means "wave" in Latin). According to the other theory — the emission theory — light is made up of particles emitted from luminous bodies and reflected or refracted (bent) by other objects.

The idea of the wave theory of light was hinted at by Leonardo da Vinci and by Galileo. It was definitely suggested by a Bologna mathematics professor, Francesco Maria Grimaldi (1613?–63), who thought of light as a fluid capable of moving in waves like any other fluid. Grimaldi made other contributions to the science of optics. He discovered, for example, that rays of light could be deflected as they passed by the edges of thin opaque bodies, producing fringes of light — the phenomenon called diffraction.

The foremost champion of the wave theory of light was the Dutch mathematician, astronomer and physicist Christian Huygens (1629–95). This thoughtful and imaginative man made notable contributions to science. He developed the telescopic eyepiece that has since been known by his name; he discovered the rings of the planet Saturn; he applied the pendulum to the regulation of the movement of clocks. Newton drew upon his contributions to the theory of dynamics in formulating the theory of universal gravitation.

### Huygens and the wave theory of light

Among Huygens' claims to fame was his research work in optics and, particularly, in the wave theory of light. He presented this theory to the public in his TREATISE ON LIGHT (1690), but he had worked out its details long before. He held that the different points of a wave front (or wave crest) of light set up an infinitely numerous series of secondary waves. These secondary waves reinforce each other after a time and produce a new wave front. Other secondary waves arise, and these result ultimately in still another wave front. Thus light is propagated.

Waves have to move in some medium or other. Huygens felt that this medium could not be air, since it was well known that light rays could pass through a vacuum. He therefore assumed the existence of an all-pervading, mysterious "ethereal substance" — the ether — as the medium through which light waves traveled. The ether, he taught, was made up of small, hard and elastic particles, which transmitted the wave impulses without being permanently displaced themselves. This invisible and intangible ether was destined to trouble men of science for several centuries; today, few physicists will concede its existence.

The chief champion of the emission (or corpuscular) theory of light was Sir Isaac Newton, whom we shall discuss more fully in another chapter; it cannot be said, however, that he personally insisted on this single explanation. Sir Isaac believed that light consists of small particles emanating from luminous bodies. At first he was inclined to believe that these particles set up waves in the ether; he thus combined the undulatory and the emission theories. "Assuming," he said, "the rays of light to be small bodies, emitted every way from shining substances, these, when they impinge on any refracting or reflecting superficies [surface] must as necessarily excite vibrations in the ether, as stones do in water when thrown into it." As time went on, however, Newton came to favor the emission theory of light. His chief objection to the wave theory was that he could not reconcile it with the propagation of light in straight lines (rays), which he took for granted.

Nowadays, physicists use both the undulatory and the emission theories of light, depending on the phenomena that are being studied. Thus, certain effects of light — such as refraction and reflection — can be explained most simply by supposing that light, emitted in tiny quantities called quanta, travels in straight lines. In the case of various other phenomena, like polarization or diffraction, the physicist assumes that light is transmitted in the form of electromagnetic waves.

### How Kepler explained vision

Various explanations of vision were offered in the seventeenth century. The astronomer Johannes Kepler proposed a theory that comes astonishingly close to our modern ideas on the subject. He held that rays were refracted (bent) by the crystalline lens of the eye in such a way as to form an inverted image on the retina; when this image is transmitted to the brain, vision takes place. Kepler correctly explained far-sightedness and near-sightedness on the basis that in one case the light rays reach the retina before coming to a focus and that in the other case they come to a focus before reaching the retina.

Newton gave the first satisfactory and generally accepted explanation of color phenomena. He began by experimenting rather idly with a prism — the wedge-shaped glass that causes the light of the sun to be broken up into a spectrum, or

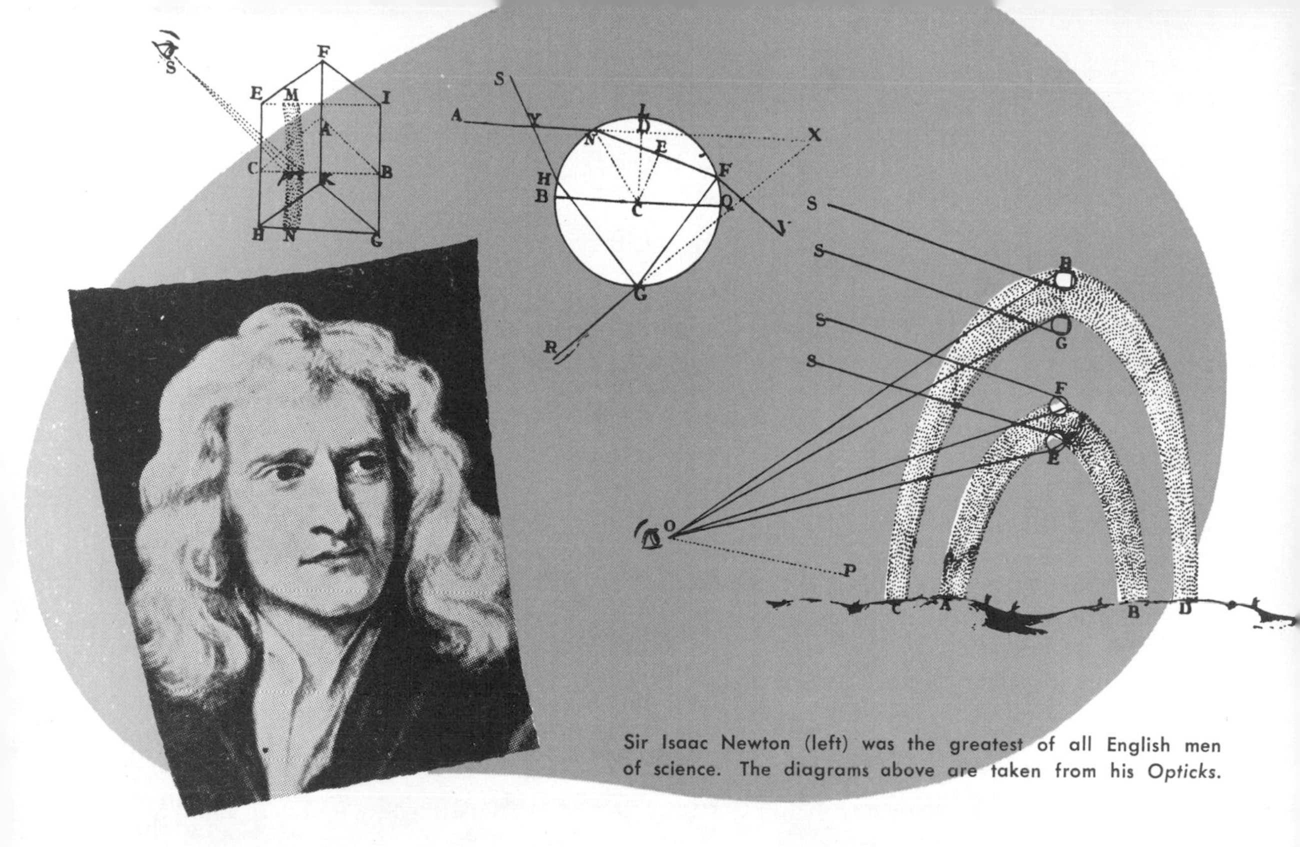

Sir Isaac Newton (left) was the greatest of all English men of science. The diagrams above are taken from his *Opticks*.

series of colors ranging from red to violet, as they are seen in a rainbow. "In the year 1665," he wrote, "I procured a triangular glass prism, to try therewith the celebrated phenomena of colors . . . Having darkened my chamber and made a small hole in my window-shutters, to let in a convenient quantity of the sun's light, I placed my prism at its entrance that it might thereby be refracted to the opposite wall. It was at first a very pleasant divertisement."

Newton then proceeded to make a serious study of color phenomena. By an ingenious arrangement of prisms, lenses and screens, he succeeded in demonstrating that white light, or sunlight, is a "most surprising and wonderful composition" of the colors of the spectrum, mixed in due proportion. He showed that each color of the spectrum, ranging from red to violet, has its own degree of refraction (that is, that it is bent in just one direction). Today we would say that each color has its own individual wave length.

Newton studied the phenomena of colors in many subtle experiments. He dealt, for example, with the phenomena described since as Newton's rings. These are alternate circles of color and darkness produced by pressing the flat side of a lens that is flat on one side and convex on the other against a double convex lens of great thickness. He studied the nature of chromatic aberration (color distortion in refracting telescopes). He gave an explanation of double rainbows. He also developed a theory of the color of different bodies — why one flower is red, for example, and another is blue; he held that the light rays are reflected in different ways according to the thickness of the films of which surfaces are composed.

Other men who shed light on the problem of light in the seventeenth century should be mentioned here. Descartes thought of light as a kind of thrust or pressure from luminous bodies and attempted to explain its refraction in mechanical terms; he used the illustration of a man hitting a tennis ball against a flat surface or through a curtain. Willebrord Snell (1591–1626), Dutch professor of mathematics at Leipzig, formulated but did not publish the law of refraction that bears his name. Pierre Fermat (1601–65), a mathematically minded magistrate of Toulouse, formulated the principle of "least time" for the transmission of light and showed that its velocity would be less in a more highly refracting medium. The Frenchman Edme Mariotte (1620?–84), who explained the halos occasionally seen around the sun and moon, also made the discovery of the blind spot in the eye (where the optic nerve joins the retina).

One of the most significant scientific advances in the seventeenth century was the realization that light does not travel instantaneously as had been formerly assumed, but requires *time* for its propagation. Galileo had advanced the belief that light has a definite velocity, and he had tried to measure it. But the speed of light was too great to be measured by the crude apparatus at his disposal.

It was the Danish astronomer Olaus Roemer (1644–1710) who first definitely demonstrated that light has velocity and who gave the first approximate idea of its high speed of travel. Working at the Paris Observatory in the 1670's, he observed a series of eclipses of one of the satellites of Jupiter. (According to some scholars, he used a series of observations made by the Italian astronomer Giovanni Domenico Cassini several years before.) He noticed that the eclipses succeeded each other more slowly when the earth, in its orbit, was moving away from Jupiter than when it was approaching that huge planet. He came to the conclusion that the farther the earth was from the satellite, the longer it took light rays reflected from the satellite to reach the earth. Using the time of the eclipses as a basis, Roemer calculated the speed of light as about 155,000 miles per second. (Actually it is about 186,280 miles per second.) The fact, announced by Roemer in 1676, that light has a finite and measurable velocity is a foundation stone of modern physics and astronomy.

## NEWTON AND THE FALLING APPLE

"Mortals, congratulate yourselves that so great a man has lived for the honor of the human race." This inscription, on the tombstone of Sir Isaac Newton in Westminster Abbey, is a fitting tribute to the great Englishman who gave man new insight into the nature of the universe. For he demonstrated that the heavenly bodies scattered through space — comets, planets (including the earth) and meteors — are all parts of the same system and that they all obey identical mechanical laws. With almost unmatched mathematical genius, Newton proved that the same natural force that caused an apple to fall from a tree holds planets in their courses. We call this force of mutual attraction gravitation. (The attraction of our earth for terrestrial objects is called gravity.) Newton showed that gravitation acted in the same way throughout the universe. This is his theory of universal gravitation.

Newton was born on December 25, 1642, at Woolsthorpe, in Lincolnshire, the puny posthumous son of a small farmer. At the age of ten he was sent to the free grammar school of Grantham, where he showed marked mechanical ability. In 1661 he entered Trinity College, Cambridge, and obtained his Bachelor of Arts degree in 1665.

The plague raged in England in the years 1665 and 1666, and Newton sought refuge at the family farm in Woolsthorpe. It was during this "lazy period" of his life, as he called it, that he laid the foundations for some of his most momentous later discoveries. "In those years," he wrote later, "I was in the prime of my age for invention, and minded mathematics and philosophy more than at any time since." In those golden young years he worked out the binomial theorem; he discovered calculus; he elaborated his system of colors. At this time, too, his theory of universal gravitation first began to take shape in his mind. The story goes that as he was sitting in the quiet orchard he observed apples falling to the ground. It came to him then or later that the same force that pulled the apple to the earth also drew the moon toward the earth and held the planets in their courses.

After the plague had run its course, Newton returned to Cambridge and obtained his Master of Arts degree. In 1669 he became Lucasian Professor of Mathematics in Trinity College. In the first period of his academic career he did a good deal of research in chemistry. He also completed his studies on the composite nature of white light (see page 277). In 1672 he reported on the subject to the Royal Society, of which he was a member.

Newton returned at intervals, too, to the problem of universal gravitation. His friends and colleagues, recognizing his mathematical genius, urged him to write out his proof of this problem, which none but he could solve. Finally, in 1684, at the urging of the astronomer Edmund Halley, he decided to devote his full attention to it. At first he planned to publish the results of his researches on universal gravitation in the journal PHILOSOPHICAL TRANSACTIONS OF THE ROYAL SOCIETY. But the Society, immensely impressed by Newton's work, decided instead to print it in the form of a book.

Unfortunately, the Society could not go forward with the project because of lack of funds. At this point Halley came to the rescue and arranged to have the work published at his own expense. It finally appeared in July 1687. Like most scientific books of that day, it was written in Latin; it bore the title of PHILOSOPHIAE NATURALIS PRINCIPIA MATHEMATICA (The Mathematical Principles of Natural Philosophy). It is generally referred to as Newton's PRINCIPIA.

During this period of supreme scientific creation, Newton remained a tortured and introspective bachelor, who yearned for quiet and seclusion. He feared to publish his great discoveries lest they draw him into controversy and widen the circle of his friends — a thing he tried to shun as he would the plague. Many stories are told of this recluse Cambridge professor; that he failed to keep certain dinner appointments; that he arrived at others unkempt and ill-clad; that he cut a small hole in his door so that his cat could enter and leave at will; that, when she presented him with kittens, he cut a smaller hole for them; that either he or the cat tipped over a candle and burned up notes on optics that had been accumulating for twenty years.

In 1689 Newton was elected to Parliament; it is said that his only "speech" was a request to an usher to shut a window. He was appointed a Warden of the Mint in 1695 and took charge of the important operation of recoining the silver currency, which had become seriously debased. Four years later he became Master of the Mint, and he held this office until his death. Newton resigned in 1701 from his professorship and removed to London, where he lived with his niece.

In his later years, though he was no longer scientifically productive, Newton was an Olympian figure, honored by men of science at home and abroad. In 1669 he had been elected one of the eight foreign associates of the French Academy of Sciences. He became president of the Royal Society in 1703; in 1705 he was knighted by Queen Anne. His later years, however, were marred by a long and bitter quarrel with the German mathematician and philosopher Leibniz over who had invented calculus. Newton called calculus a

"method of fluxions" for calculating infinitesimal and moving quantities. Evidence seems to show that Newton had come upon the idea of calculus before Leibniz and that he had used it to solve important problems. However, Leibniz' method of writing it down was unquestionably better and it has been adopted. Newton died on March 20, 1727, in his eighty-fifth year.

#### The theory of universal gravitation

Newton's supreme contribution to science was his theory of universal gravitation, as set forth in the PRINCIPIA. Its cornerstone is the law that all bodies attract each other with a force proportional to their mass and inversely proportional to the square of the distance between them. Newton never defined "mass"; on earth, in terms of terrestrial mechanics, it is usually equivalent to weight. Newton's three famous laws of motion also form a part of the theory:

(1) Every body continues in its state of rest, or of uniform motion in a straight line, unless it is compelled to change that state by forces impressed upon it.

(2) The change of motion is proportional to the motive force impressed and is made in the direction of the straight line in which that force is impressed.

(3) To every action there is always opposed an equal reaction; or the mutual actions of two bodies upon each other are always equal, and directed to the contrary parts.

In the PRINCIPIA Newton shows how the law of universal attraction and the laws of motion apply to earthly phenomena: the movement of pendulums, bodies passing through water, sound passing through air. He also shows how the heavenly bodies obey these universal laws. Thus, two mutually attracted heavenly bodies describe similar orbits (paths) about their common center of gravity and also about each other. When three bodies attract each other mutually — such as the earth, moon and sun — certain irregularities in their orbits result. Thus the moon's motion is irregular because it is attracted by (and attracts) the sun at the same time that it is attracted by (and attracts) the earth.

Newton concludes that the celestial bodies making up the solar system move in accordance with the theory of Copernicus and that their orbits are determined by mutual attraction. On the basis of this doctrine he gives an explanation of the tides. He points out that the highest tides occur at times when the sun and moon reinforce each other's attraction for the earth; that the lowest tides occur when the sun and moon "act against each other." Newton showed that comets, which had formerly been regarded as capricious phenomena, also follow the law of universal gravitation. He held that they move in parabolas or ellipses about the sun.

The PRINCIPIA contains Newton's reflections on true scientific method. This is the counsel he gives men of science; these precepts still hold today:

(1) Admit no more causes of natural things than such as are both true and sufficient to explain their appearance.

(2) To the same natural effects, assign so far as possible the same causes: for instance, breathing in man or beast.

(3) Assume properties common to all bodies within reach of the experiment as pertaining to all bodies.

(4) In experimental philosophy [that is, science], look upon propositions inferred by general induction as accurately or very nearly true, notwithstanding any contrary hypotheses that may be imagined, until such time as other phenomena occur by which they may either be made more accurate or liable to exception.

The principles set forth in the PRINCIPIA were not the exclusive product of Newton's own thought and research. Kepler had already suggested the inverse square relationship. Galileo had supplied him with the data upon which he built the first two laws of motion; there are hints of the third law of motion in Huygens, Wren and Wallis. But it was Newton who combined the isolated conjectures and theories of others in a single vast, comprehensive system and who bolstered it up by rigorous mathematical demonstration.

Newton's system of the universe remained uncontested for more than two hundred years; it formed the basis upon which men of science built their own theories. To many scientists the Newtonian concept came to represent a mechanistic and materialistic view of the workings of the universe. They not only held the universe to be a vast machine, blindly following changeless laws; man and the mind of man came to be considered as so many cogs in the machine. Voltaire presented this viewpoint in his IGNORANT PHILOSOPHER: "It would be very singular that all nature, all the planets, should obey eternal laws, and that there should be a little animal, five feet high, who, in contempt of these laws, could act as he pleased." Newton would have been the first to protest against the narrowly materialistic interpretation of his theories, for he was a devout man.

In the twentieth century the sharp lines of the Newtonian mechanics have been blurred by Einstein's theory of relativity and by such modern concepts as quanta, radioactivity and the like. We now realize that the universe does not function in quite the mechanical way that Newton had imagined. Powerful forces of which he had no inkling profoundly affect its workings. Yet these new developments have not overthrown Newton's system; they have served, rather, to supplement it.

#### Newton's other contributions to science

The theory of universal gravitation represented the capstone of Newton's work. He made many other outstanding contributions to human knowledge. His work on optics was of the very highest rank; his achievements in mathematics were no less outstanding. Yet he had the modesty of the true scientist. "I do not know," he said, in a conversation recorded by Joseph Spence, "what I may appear to the world, but to myself I seem to have been like a boy playing on the seashore — now and then finding a smoother pebble or a prettier shell than ordinary, while the great ocean of truth lay all undiscovered before me."

For all the mathematical genius and scientific insight displayed in the PRINCIPIA, it was nevertheless the new information on the motions of tides, offering practical help in the art of navigation, that won Newton a popular reputation and royal favor for his work. The kings of England might not have been able to understand Newton's abstract mathematics but they acutely appreciated all discoveries that would contribute to England's progress as a seafaring nation. Astronomical and mathematical research won support in England because it was an essential aid to practical navigation.

#### The founding of the Greenwich Observatory

The famous Greenwich Observatory — whose "meridian of Greenwich" remains the base line for world time zones — was founded by royal warrant about this time (1675). Charles II is reported to have said that "he must have the star places and the Moon's motion anew observed, examined and corrected for the use of his seamen." He had an observatory erected at Greenwich for the promising young astronomer John Flamsteed (1646–1719), whom he named royal astronomer.

Flamsteed, a tireless observer, determined the position of thousands of fixed stars; but he was determined to withhold his results until they could be presented in complete form. This led to a violent quarrel with Newton and Halley, who held that Flamsteed's observations should be made public without further delay. Over Flamsteed's objections the first Greenwich star catalog was issued in 1712. The royal astronomer denounced the publication and went forward with the preparation of a more complete catalog. The result of his labors — the monumental CELESTIAL HISTORY OF BRITAIN, in three volumes — was presented to the public in 1725, six years after Flamsteed's death; it had been finished by two assistants. Flamsteed was succeeded as royal astronomer by Newton's good friend and devoted disciple, the learned and jolly astronomer, Edmund Halley (1656–1742).

Halley's fame has been overshadowed by that of the great Sir Isaac. Yet Halley, who has been called the "second most illustrious Anglo-Saxon philosopher," was an eminent scientist and won ample recognition in his own day. He became a Fellow of the Royal Society when he was a callow youth of twenty-two; he was named Savilian Professor of Geometry in Oxford University in 1703.

**The prediction of a famous comet's return**

Halley is best remembered for his accurate prediction, in 1704, of the return of the comet that now bears his name. Before that time several men of science, including Newton, had predicted that various comets would return; but their calculations had been based upon wrong principles and had been very far from the mark. Halley's prediction was the result of a careful study of the orbits of certain comets. He noted that the ones that had appeared in the year 1531, 1607 and 1682 had moved along practically the same paths in space. He came to the conclusion that these supposedly different comets were really one and the same. He predicted that it would return to visibility from the earth about 1758; the fulfillment of his prediction, sixteen years after his death, marked a new epoch in the history of "precise astronomy." Halley's comet was seen again, as predicted, in 1835 and in 1910.

A prodigious worker at figures, Halley had many other claims to scientific fame. As a mere lad he journeyed to the island of Saint Helena (later to be Napoleon's place of exile); by patient observation, often under murky conditions, he determined the positions of 341 stars for which no accurate records existed. When he published his results in 1679 in his CATALOGUE OF THE SOUTHERN STARS, he was hailed as the "Southern Tycho." (For Tycho Brahe, see the Index.) Halley found evidence that the so-called fixed stars, such as Sirius and Arcturus, change their position as well as their order of brightness. His work with telescopic sights led him to discover the correct formula for the focal lengths of spheric lenses and mirrors.

One of the few landlubbers ever put in direct command of a British naval vessel, Halley made two trips to South America in order to study the variations of the magnetic compass. Upon his return he published the first map (1701) showing the declination and inclination of the compass. (For declination and inclination, see the appropriate Index entries.) At the same time he published maps giving reasonably accurate information on prevailing trade winds and monsoons.

As a mathematician, Halley's chief contribution, perhaps, was his "estimate of the degree of mortality of mankind," which laid the scientific foundation of the life insurance business. The mathematics problem was to calculate accurately the price of annuities on lives. This first "life table" was drawn from "curious tables of the births and funerals of the city of Breslau, in central Germany." With typical British understatement, Halley informs us that the work cost him "a not ordinary number of arithmetical operations."

Halley was a most jovial fellow and a lover of good food; he founded the Royal Society's dining club at a London coffeehouse. There is a story that he spent a convivial evening with Peter the Great, Tsar of Russia, who was then traveling incognito in England; according to this tale, the Englishman wound up the evening's festivities by wheeling the Russian monarch through a yew hedge in a wheelbarrow. Halley's wife, Mary, was devoted to him, but she sometimes complained that he was dissipating the family fortune on "useless" scientific experiments, expeditions and publications. One can imagine the scene at the breakfast table when Halley announced that he intended to publish Mr. Newton's book at his own expense!

*Continued on page 373.*

# FLIES AND MOSQUITOES

## Two-winged Members of the Vast Insect Group

by HERMAN T. SPIETH

DAMSEL FLIES, stone flies, dragonflies, dobson flies and butterflies are not true flies at all, in spite of their names. They all have four wings — an anterior, or front, pair and a posterior, or hind, pair. The true flies differ from these insects and from almost all other insects, for that matter, because they have only one pair of wings, which correspond to the front wings of other insects. Instead of the hind pair, they have a pair of rodlike structures, with a knoblike expansion on the outer tip of each. These structures, which are called halteres, are actually modified wings; they serve as living gyroscopes, enabling the insect to maintain its balance in flight. Houseflies, mosquitoes, horseflies, botflies and midges are true flies because they possess only one pair of wings and a pair of halteres. The scientific name of such insects is Diptera, which comes from the Greek words meaning "two wings."

The Diptera have a remarkable life cycle, which begins with the tiny eggs. Some females lay only an egg or two at a time; others deposit masses of a hundred eggs or more. (As we shall see, certain true flies do not lay eggs at all, but give birth to living young.) The eggs hatch from within a few hours to a few weeks.

The little wormlike creatures that emerge from the eggs bear no resemblance to their parents, since they are legless, wingless, slender and generally white, or even transparent. They are commonly called maggots; the young of mosquitoes are known as wigglers. Both maggots and wigglers are often referred to as larvae (singular: larva).

Though the true-fly larva is wormlike in appearance, its internal anatomy is that of an insect. As in the case of insects, too, the entire outer surface of the body is covered by a thin layer of material that serves as the skeleton. This outer skeleton, or exoskeleton, is found not only in insects

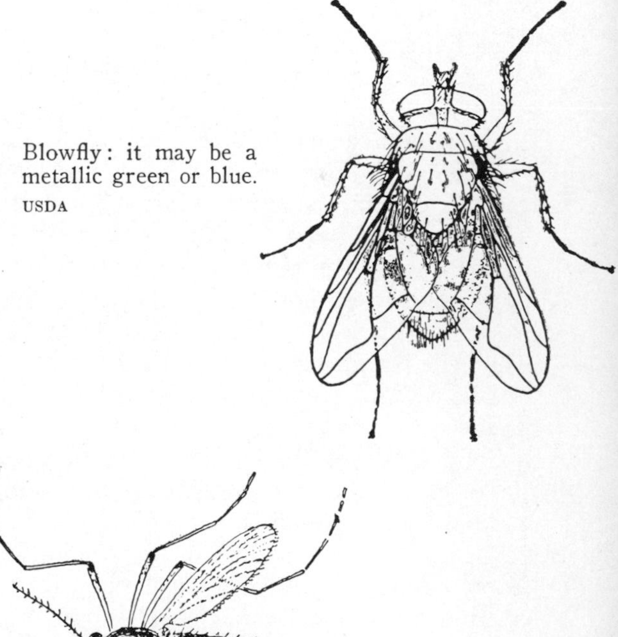

Blowfly: it may be a metallic green or blue. USDA

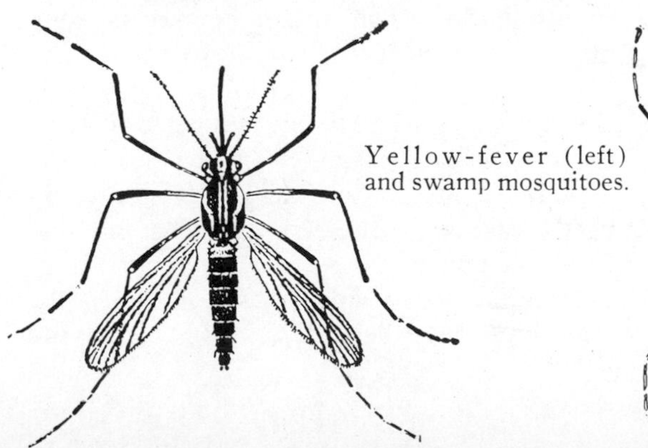

Yellow-fever (left) and swamp mosquitoes.

but also in their arthropod relatives, such as the spiders, the crabs, the lobsters, the mites and the ticks.

Since the exoskeleton covers the entire outer surface of the body as a continuous sheet, the insect must shed the old skeleton and manufacture a new one periodically while it is in the process of growth. The shedding of the exoskeleton is controlled by a series of hormones.

Almost invariably the larvae live in moist surroundings, so that their food is bathed in liquid. The little creatures have jaws; with these they tear and shred food into small pieces that they can swallow. Some larvae, dwelling in ponds and other bodies of water, feed upon tiny animals or plants. Others burrow into the soft parts of plants and feed upon the living tissue; they often cause abnormal growths, called galls, to develop. Certain larvae feed on fungi. The larvae of houseflies and various other Diptera live upon fermenting or rotting materials, such as manure. They feed not upon these materials, but upon the microscopic organisms — bacteria, yeasts and the like — that cause rotting or fermentation. Some larvae are parasitic; they feed upon the tissues of insects or other living animals.

As it greedily eats, the little larva grows apace. Periodically it sheds its exoskeleton, increasing in size until finally it becomes a full-grown larva. It is then transformed into a quite different kind of creature — the pupa. In the great majority of species, the pupa is inert: it cannot move and it cannot feed. It is usually brown or black in color; its exoskeleton, called the pupal case, is quite thick.

In the quiescent pupal stage, the body of the insect undergoes considerable remodeling. When the process is completed, the insect breaks the pupal case and out comes a full-fledged imago, or adult. Adult true flies, as we shall see, are generally hearty eaters. No matter how much they eat, however, they can never grow because they cannot shed the exoskeleton.

The complete life cycle of the true fly, then, consists of four distinct stages. There is, first, the egg stage, which ends when the egg hatches and the larva emerges. In the larval stage, the insect feeds ravenously, grows rapidly and sheds its exoskeleton from time to time. It does not feed nor grow in the pupal stage, but its body is transformed. Finally, in the adult stage, the true fly feeds but does not grow; it mates, lays eggs if it is a female and eventually dies.

There are certain variations in the life cycle of the Diptera. The females of some species do not lay eggs. They retain the eggs in their bodies until the hatching stage; then they give birth to larvae. In some cases, as in the tsetse fly, the larvae remain in the body until they are ready to pupate (that is, to become pupae).

Only during the last few hundred years have men known the correct relationship between the different stages of the life cycle; for example, it was formerly common belief that maggots arose spontaneously from decaying flesh. Even today it is often necessary to rear flies in the laboratory in order to determine what form the larvae of an unfamiliar species will take. There are many species whose larvae have never been discovered.

It generally takes a relatively short time for the life cycle of the true flies to be completed. Low temperatures lengthen the cycle; high temperatures shorten it. Most species produce at least several generations a year. In some species, however, there is only one generation a year; in others, only one in several years.

The body of the adult true fly, like that of all other insects, is divided into three parts: head, thorax and abdomen. The head is relatively large and usually more or less spherical in shape. It bears the antennae (or feelers), the eyes and the mouth parts. The antennae may be long and slender, as in the mosquitoes, or short and thick, as in the housefly. They bear the organs of smell and in certain cases the organs of hearing.

In all insects keenness of vision is directly related to the size of the compound eyes, structures consisting of numbers of eye units placed side by side. Since the true flies have relatively large compound

Photos, Amer. Mus. of Nat. Hist.

Stages in the life cycle of the housefly (*Musca domestica*). Above: Eggs are laid in garbage, manure and decaying matter. Right: The larva, which may mature in 4 or 5 days.

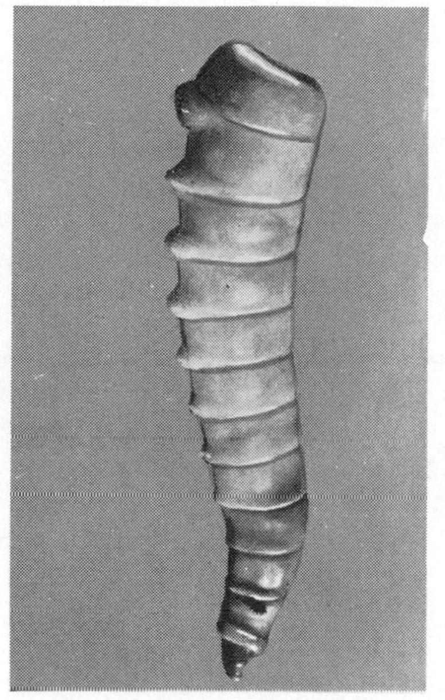

The pupal stage (shown above) is completed in less than a week's time.

eyes, consisting of a great many units, they have excellent vision.

The mouth parts are formed for sucking, or for piercing and sucking; no adult true fly can chew solid food. The housefly feeding upon a lump of sugar pours salivary fluid over it; at the same time the insect rasps at the sugar in order to loosen tiny particles. The resulting fluid, which contains dissolved sugar and tiny particles in suspension, is then sucked up and swallowed. The female of the mosquito and of many other true flies pierces the flesh of its victim and sucks the blood. The males of these species, in almost all cases, must content themselves with nectar or other freely available liquid foods, since the mouth parts are not constructed for biting. As a matter of fact, most true flies, male or female, feed upon such substances.

The thorax of the Diptera consists of three ringlike segments, tightly joined together. Each segment bears a pair of legs on the ventral, or bottom, surface. The second segment carries the wings, which arise from the upper part of the sides. The halteres are borne on the third, or hind, segment.

The wings are thin, flat, sheetlike structures; they are usually transparent,

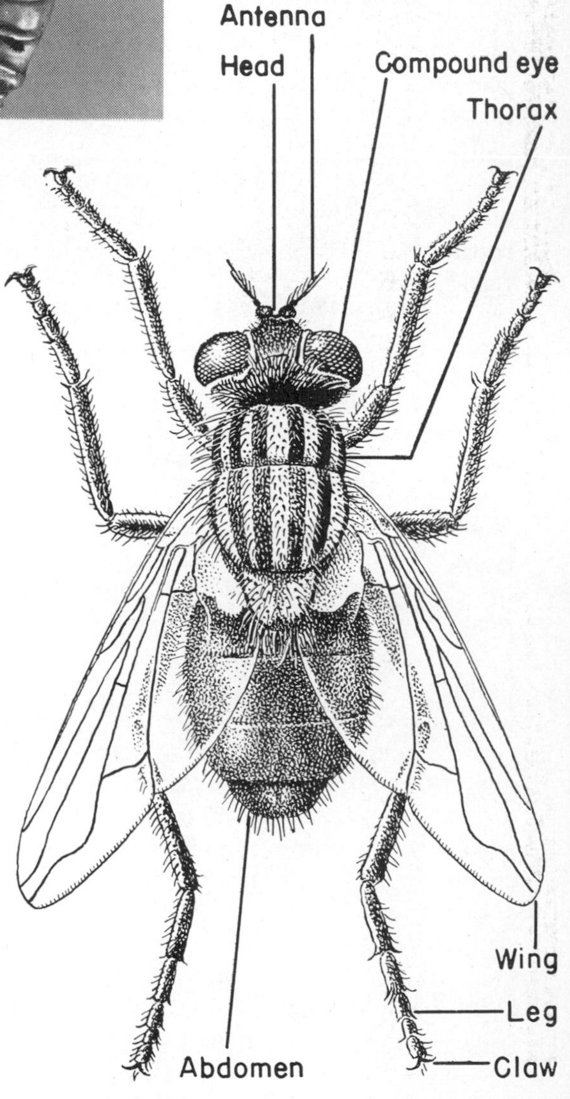


Adult housefly — a cosmopolitan insect.

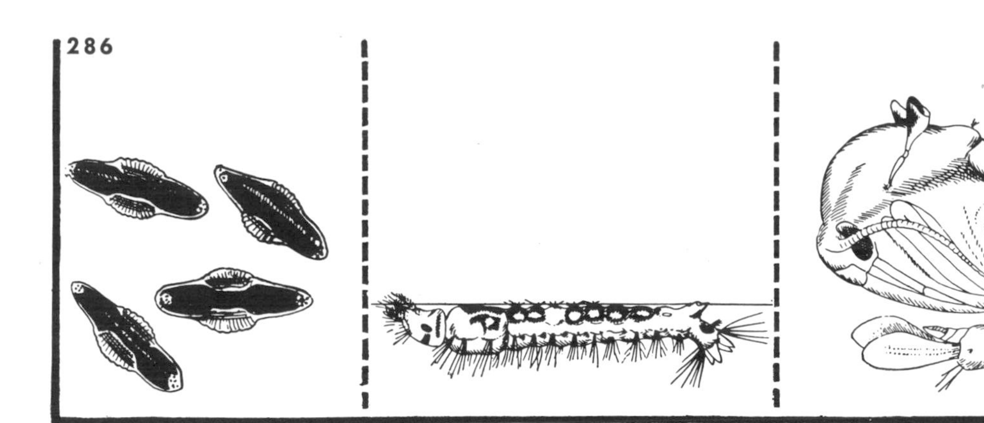
USDA

The eggs, which are furnished with ornate floats.

Larva in resting position at the water's surface.

Respiratory trumpets of pupa pierce the surface.

though in some species they display beautiful color patterns. When true flies are at rest, they usually fold their wings flat over their backs. In flight they commonly beat their wings very rapidly; this is particularly true of the mosquitoes and some of the midges.

The abdomen, the hind section of the fly's body, is made up of a number of segments; the different segments can be made out, however, only under the microscope. The abdomen contains the major parts of the digestive, respiratory, circulatory and reproductive systems.

The adults of most species live for several weeks — perhaps two or three months, on the average. The warmer the temperature, the shorter the adult life span; in the summer, for example, the housefly generally does not live more than two or three weeks. The coming of cold weather is fatal to most true-fly adults. The species is continued in various ways. In certain cases, adults hibernate in sheltered places, including houses, and become active again in the spring. Some of the Diptera, including the horseflies, remain in the earth in larval form until the coming of spring; they then pupate.

True flies are found in almost all parts of the world. They thrive in various cold regions; in the very short summers of the arctic areas, mosquitoes and other bloodsucking Diptera make life almost unbearable. The larvae of a few species live in the sea; the adults of these species, however, are generally winged insects that feed on land. There are a few notable exceptions. A little midge called *Pontomyia* lives in the shallow waters of the South Seas; the adults lack wings and swim about in the ocean.

Most true flies probably never travel more than about two miles from the place of their birth; of course they cover many, many miles as they fly about the area in which they spend their lives. Certain Diptera, however, may travel considerable distances from their breeding sites.

Only a few species of the Diptera are considered to be allies of man. Some of them destroy harmful insects, either by devouring them or by using their living bodies as breeding places for larvae. Others act as scavengers, ridding us of animal and vegetable wastes. The vinegar fly and some of its close relatives have been used extensively in biological research.

A considerable number of the true flies rank among the most formidable insect enemies of mankind. Some of them feed upon the crops and other plants upon which man and his animals depend. Since these true-fly vegetarians are in direct competition with us, we call them destructive. Actually, of course, they are no more destructive than man himself; from their point of view we are human pests!

It is the larvae of the Diptera that damage plants. They like to feed upon moist, relatively soft substances; it is not surprising, therefore, that fruits, vegetables and young plants are the chief objects of attack, particularly in the tropics and semitropics. Certain true flies, such as the Mediterranean fruit fly and its kin,

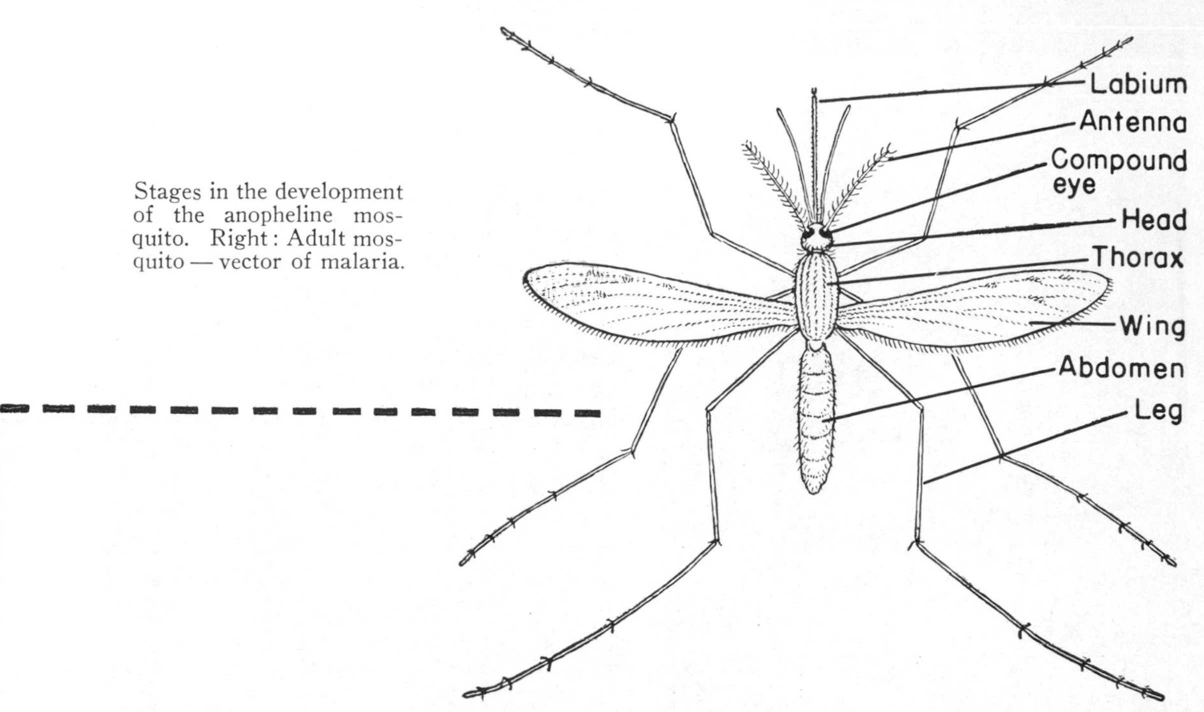


Stages in the development of the anopheline mosquito. Right: Adult mosquito — vector of malaria.

cause great damage. However, the Diptera are not nearly so destructive to plants as are various other insect groups, such as the beetles (Coleoptera), the moths and butterflies (Lepidoptera) and the true bugs (Hemiptera).

The true flies do considerably more damage as pests of domestic and game animals. The adults of mosquitoes, black flies, horseflies and others pierce the skin of animals and suck their blood. In most cases, only the females of the offending species do so. These females can produce no eggs unless they have had a meal of blood; therefore they aggressively seek out a victim upon which they can feed.

As the female bites, it injects a small amount of fluid (saliva), which prevents the victim's blood from clotting; this usually causes the area about the bite to swell and itch. Most animals gradually develop an immunity to the injected salivary secretions; it is the young animals, which have not had the time to develop immunity, that are most seriously affected.

The biting of these Diptera pests is a constant source of annoyance to the animal victim; but it is often more than that. The loss of blood may represent a serious drain upon the animal's vitality. For example, a water buffalo in the Philippines may lose as much as a quart of blood in a single night because of the swarms of female mosquitoes that feed upon it. The combination of blood loss and of the injected saliva may produce most serious effects; in extreme cases it may prove fatal. In some instances, biting flies — including the mosquito, tsetse fly and horsefly — transmit various diseases to animal victims.

The larvae of certain flies, such as the warble flies, the botflies and the screwworms, live in the bodies of various domestic animals. Some of these pests may do little harm; others may cause serious injury or death.

The Diptera that are pests of man are always annoying and in some cases extremely dangerous. Certain true flies, including the housefly, carry disease-causing organisms on their feet and other parts of their bodies; they distribute these organisms upon man's food. Other Diptera, such as the horsefly, tsetse fly, sand fly and punkie, pierce the skin and suck blood. Sometimes these bloodsuckers transmit disease; the tsetse fly and the mosquito are notorious in this respect.

Generally speaking, the larvae of the Diptera do not bite man and drink his blood. The Congo floor maggot (*Auchmeromyia luteola*), it is true, sucks the blood of sleeping natives in central Africa; but this is an exceptional case. A con-

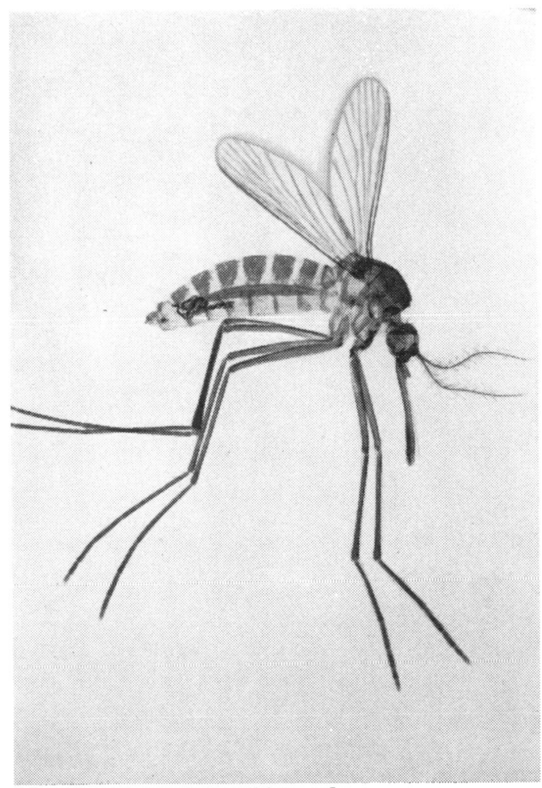

Ward's Natural Science Establishment, Inc.

Rain-barrel mosquito (*Culex pipiens*).

Right: An egg mass of *Culex pipiens*.

Yellow-fever mosquito (*Aëdes aegypti*).

USDA

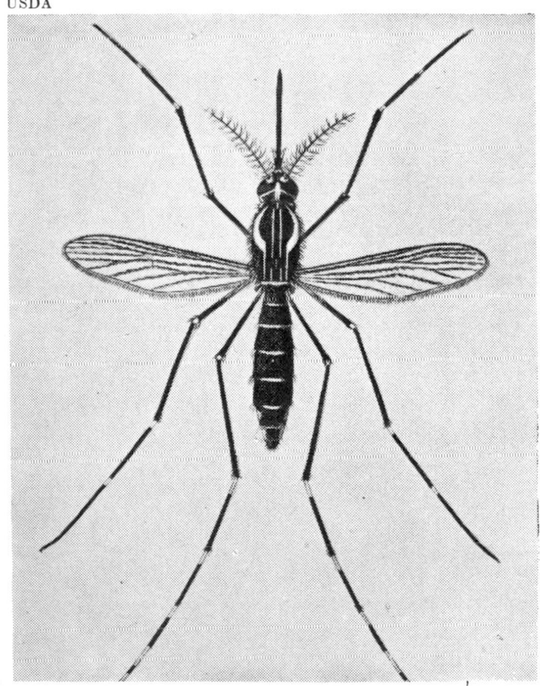

siderable number of true-fly larvae, however, including those of the human botfly (*Dermatobia hominis*) are dangerous to man because they live in human flesh. They often hatch from eggs that have been deposited on the skin, or in unprotected wounds or in the nose and ear. When the larvae hatch, they burrow inward, causing extreme discomfort to their victims and occasionally killing them.

In the following pages of this article we shall have more to say about the Diptera enemies of man. We discuss in other chapters the never-ending warfare that is waged against these insect pests. The prosperity and health of mankind and, in some areas, his very survival, depend upon the effectiveness with which this warfare is waged.

### Important true flies

Something like 85,000 different species of true flies have been described and classified by entomologists (scientists who spe-

USDA

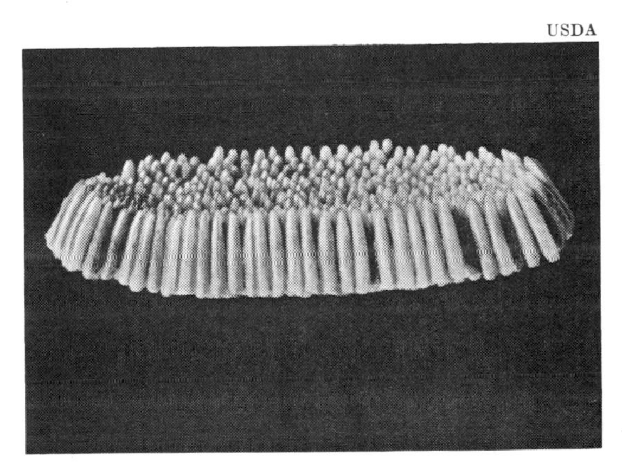

cialize in the study of insects). Many species are still unknown to science; new ones are constantly being discovered. It is estimated that eventually at least 200,000 different species of Diptera will be known to man.

The Diptera are subdivided into various groups, according to the structure and habits of the different species. Here are a few of the more important groups:

*Crane flies, or tipulids.* These are slender-bodied, often relatively large flies with very long legs. They are often seen flying slowly about the edges of streams and in damp woods. They look much like overgrown mosquitoes; fortunately for us they are harmless, since they cannot bite. The larvae are found in the soil, and also in water and mud. Because of their tough skin they are sometimes known as leather-jackets.

*Midges, or chironomids.* They are to be found almost everywhere, but especially in the moist parts of the earth and around bodies of water. They range in size from extremely small creatures to insects as large as big mosquitoes. Certain tiny midges, including the sand flies and punkies, are vicious biters. Most midges, however, cannot bite; they annoy man because they swarm around him. The larvae usually dwell in the water; some of them are bright red and are therefore known as blood-worms. Both larvae and adults are important items of food for fresh-water fishes and many birds.

The midge called the Hessian fly (*Phytophaga destructor*) attacks wheat, rye and barley. Eggs are deposited along a stem; when the larvae hatch, they feed on the stem, sometimes killing the plant. This pest was introduced into North America at the time of the American Revolution. It was secreted in the straw that the Hessian troops, employed by the British, brought with them from the homeland as bedding.

*Black flies and buffalo gnats, or simuliids.* The larvae of these flies live in fresh-water streams, where they attach themselves to stones and sticks by means of a suckerlike structure on the hind end of the body. The adults are small, blackish, humpbacked insects. The females are aggressive bloodsuckers; swarming about their animal or human victims, generally in the daytime, they inflict painful bites, which result in considerable swelling and occasionally prove fatal. The simuliids are strong fliers.

*Mosquitoes, or culicids.* These pests are the most dangerous insect enemies of man. They are found in almost all parts of the world, from arctic regions, such as Alaska and Greenland, to the tropical swamps of Africa and South America.

The eggs are laid on or near the water of ponds, pools or marshes; they are sometimes deposited in the water that has collected in empty tin cans or discarded automobile tires. After a short time the eggs hatch into the aquatic larvae called wigglers. These feed upon minute animals or plants or floating particles in the water; they frequently come to the surface to breathe. The pupae are also active wigglers. Adult males have featherlike antennae; the antennae of the females are slender and hairy. Only the female bites.

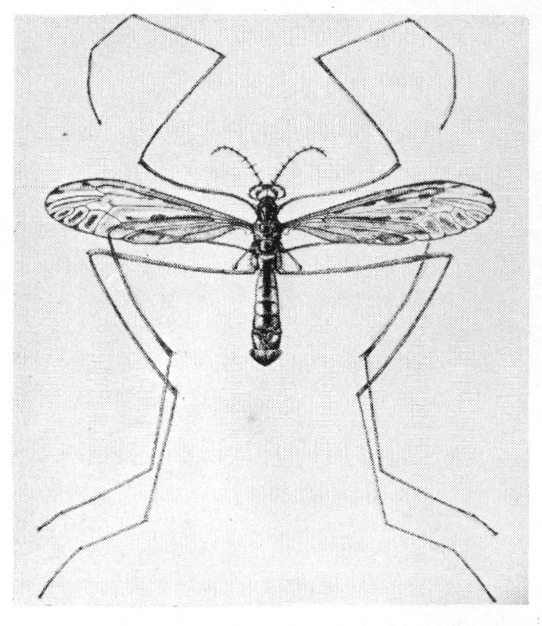

The range crane fly (*Tipula simplex*).

Hessian fly (*Phytophaga destructor*).

USDA

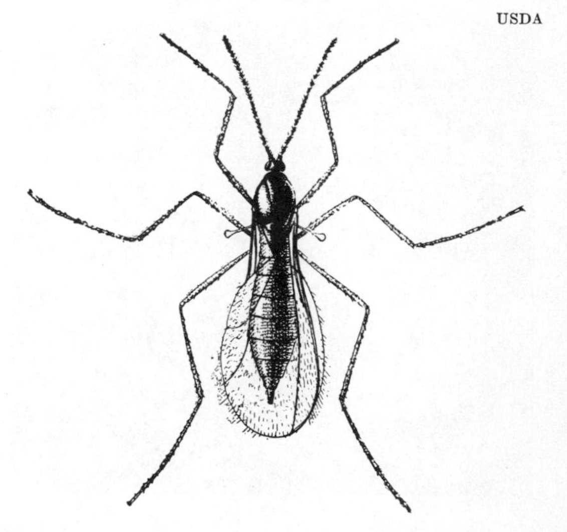

When the female is ready to mate, her supersonic song (one that is beyond the range of human hearing) attracts the male. Man has taken advantage of this fact in his constant warfare upon mosquitoes. The song of the female is recorded and then broadcasted. Males are attracted to the scene of the broadcast and are then electrocuted in cleverly placed traps.

All over the world mosquitoes annoy man by sucking his blood. It is within the tropical and semitropical areas, however, that they cause the greatest amount of human misery. They carry several major diseases that are deadly to man. Malaria is transmitted by certain species of *Anopheles;* yellow fever, primarily by *Aëdes aegypti,* but also by other species of *Aëdes* and some species of *Haemagogus;* dengue, by *Aëdes aegypti* and some of its close relatives; filariasis, by various species of *Culex* and *Aëdes.*

Fortunately nature has provided certain effective checks upon the spread of the mosquito population. Wigglers and pupae are pounced upon by fishes, water newts, aquatic beetles and dragonfly larvae. Adults are the prey of dragonflies, hornets, swallows and swifts by day and of bats and nighthawks by night.

Man's fight against the mosquito is directed chiefly against the larval and pupal stages. Certain shallow pools, swamps and marshes are drained; others are sprayed with oil or dusted with Paris green and other substances. Pyrethrum, rotenone and DDT kill mosquito larvae. These substances should be used with caution, however, since they may also kill fish if the concentration is too strong. Screening and mosquito nets protect houses against adult insects; mosquito veils are used to protect the face out of doors in certain areas. Aerosol bombs containing pyrethrum, rotenone or DDT keep small areas free of the pests for a time. Repellents, such as citronella, are also used.

The small, yellowish-hued vinegar fly.

General Biological Supply House

*Horseflies and their kin — the tabanids.* These are large, stout-bodied flies, which bite various mammals, including man. Only the females suck blood; they sometimes transmit various diseases, especially tularemia and anthrax. The larvae are found mostly in swamps, ponds and damp places, where they feed upon tiny animals, especially the larvae of other insects.

*Flower flies, or syrphids.* Some flower flies look like honeybees, others like bumblebees or wasps; for that reason most people leave them severely alone. They are entirely harmless, however, because they cannot bite. Some of the syrphid larvae are beneficial to man, since they feed exclusively upon the aphids, or plant lice, that suck juice from many of our domesticated plants.

*Fruit flies, or trypetids.* The larvae of these insects are destructive to fruits and vegetables. Among the members of the trypetid group are the apple maggot (*Rhagoletis pomonella*), the Mediterranean fruit fly (*Ceratitis capitata*), the Mexican fruit fly (*Anastrepha ludens*), the cherry fruit fly (*Rhagoletis cingulata*), the olive fly (*Dacus oleae*) and the walnut husk fly (*Rhagoletis completa*). The adults are beautiful little creatures, smaller than houseflies; they have banding or spotting on the wings. Because of the way many of the fruit flies alternately raise and lower their brightly colored wings as they move slowly about on vegetation, they are often called peacock flies.

*Vinegar flies, or drosophilids.* These are tiny creatures, almost always less than an eighth of an inch in length and yellowish to black in color. The adults are attracted to rotting fruits and vegetables as well as to fermenting substances; the eggs are laid in these materials. The life cycle is always short, usually two to three weeks in length. More than six hundred different species have been described. Of these *Drosophila melanogaster* is the most famous because it has been used extensively as an experimental animal in research on the laws of inheritance and evolution.

*Shore flies, or ephydrids.* The adults are small, dark-colored flies, which do not molest man. One species, *Psilopa petrolei,* breeds in pools of crude petroleum. The larvae of *Ephydra gracilis* are found in great numbers in the strongly saline waters of the Great Salt Lake; the popular name of this species is Great Salt Lake shore fly.

*Horse botflies, or gasterophilids.* The adults of these flies lay their eggs upon the living body of a horse. When the larvae hatch from the egg, they are often licked up by the animal. Sometimes they burrow their way into the horse's mouth, travel down the throat and finally attach themselves to some point inside the digestive system. Here they live and grow until they are ready to become pupae. They then loosen their hold and are carried out with the feces. They pupate in the soil and eventually emerge as adults, ready to mate and initiate the life cycle again.

Amer. Mus. of Nat. Hist.

The infamous tsetse fly of Africa.

*Warble flies, or hypodermatids.* Certain species of warble flies are serious pests of cattle and other animals. The adults, which are strong, beelike insects, cannot bite or even feed, since their mouth parts are degenerate. The females of the cattle warbles (*Hypoderma lineata*) lay their eggs on the hairs that cover the legs or flanks of cattle.

The young larvae hatch, crawl down the hair and bore into the flesh. In the space of about four months they work their way through the body of the cow and finally reach the esophagus. Here they stay for three months, feeding and developing. At the end of that time they again bore into the flesh and finally make their way to the skin on the animal's back. They then cut a hole through the skin and form a swelling called a warble; the insect breathes and discharges wastes through the opening. When the larvae are full-grown, they tumble off the cow onto the ground; here they pupate and eventually emerge as adults. There is only one generation a year. The larvae cause annoyance and loss of weight to milch cows and lessen milk

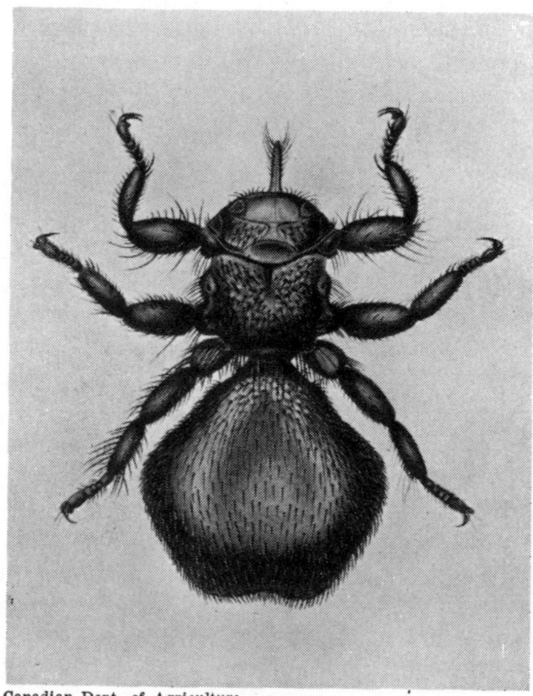

Canadian Dept. of Agriculture

Sheep tick—a wingless bloodsucker.

production. The holes they make in the skin greatly reduce the value of the hide.

*Tachina flies, or tachinids.* The larvae of the tachinids are all parasitic on insects. Often the female tachinid seeks out the insect host — either a larva or an adult — and lays its eggs directly upon it. Many species produce living young upon the host instead of laying eggs. The young tachinid bores its way into the host, which it usually eventually kills but not before it has completed its larval development. The tachinids are the most directly beneficial to man of all the true flies. They help control many of the insect larvae, especially caterpillars, that are destructive to crops and forests. Tachinids are often mistaken for houseflies, since they are quite similar in size and general appearance.

*Blowflies, or calliphorids.* The adults are often a bright metallic blue or green in color. The females lay their eggs in decaying materials, especially decaying meat, the rotting bodies of animals and excrement. They perform a valuable service in helping to dispose of this material; they rank among the leading scavengers of the insect group. Some species of blowflies are reckoned among the harmful insects, because they lay their eggs in the open wounds of animals and man. The larvae later burrow their way into the body of the host and cause serious damage.

*Houseflies, or muscids.* The housefly (*Musca domestica*) is one of the ever present uninvited guests that live in or near the abodes of man. It does not bite but its habit of alighting upon the exposed parts of our bodies is most annoying. Houseflies breed upon manure, garbage and other kinds of decaying matter and feed not only upon such materials but also upon our food. Hence they are efficient transmitters of such human diseases as typhoid fever, dysentery, cholera, yaws and anthrax.

**Houseflies are now better checked than ever before**

Because of improved sanitary conditions and the development of modern insecticides, particularly DDT, man can control the housefly more effectively than before, at least in civilized communities. Unfortunately this pest has developed marked resistance to various insecticides and still represents a definite problem. It is as deadly a menace as ever in the more backward regions of the world.

The stable fly (*Stomoxys calcitrans*) looks much like the housefly. Unlike the housefly it can bite; it sucks the blood of animals and men and is suspected of transmitting disease. It breeds chiefly in decaying vegetation.

*Tsetse flies, or glossinids.* Among the most dreaded of the Diptera is the tsetse fly (genus *Glossina*) of Africa. The female brings forth a single full-grown larva at a time, depositing it in a shady place on the ground where it enters the soil to pupate. Both male and female adults are bloodsuckers, attacking animals and man. They serve as carriers of trypanosomes, parasitic protozoans that cause such diseases as sleeping sickness and nagana.

*The Pupipara.* The Pupipara look more like ticks than insects; they are commonly known as tick flies, and also as louse flies and keds. They live as external parasites upon the bodies of birds and mammals, sucking the blood of their hosts. They have flattened bodies, with leathery, somewhat flexible exoskeletons; most of them are wingless. The Pupipara are found throughout the world, though there are only about 500 different species.

The name Pupipara ("pupa-bearing," in Latin) has been given to this group because the female retains the larvae in her body until they are ready to pupate. When the larvae are finally born, they are immediately glued to the hair or feathers of the host, and then they enter upon the pupal stage. When the adult insect emerges from the pupal case, it finds itself upon a host on which it can feed. One of the most common members of this group is the sheep ked, or sheep tick (*Melophagus ovinus*). It is reddish brown in color; its pupae are often mistaken for eggs. Other species live upon pigeons, grouse, horses, camels, dogs, deer and other animals.

*See also Vol. 10, p. 275:* "Insects."

# ROADS AND ROAD MAKING

## The Avenues of Travel and Trade in Many Times and Lands

FEW countries in the long history of mankind have reached a high state of prosperity without an adequate network of roads. Even today, in spite of the tremendous development of rail and air travel, roads form an essential part of a nation's transportation facilities.

It is true that at one time the railroads seemed destined to supplant the vehicles that traveled upon highways. At the turn of the century, many thoughtful men feared that the competition of the railroads would cause some of the world's greatest roads to fall into disuse and decay. The opposite thing happened; as railroads opened up new areas, they caused new roads to come into being. Today, where railroads compete seriously with the busses and trucks and automobiles that use our highways, it is likely to be the railroads that will succumb.

As for the airplane, it has afforded a providential means of transportation in places where roads are nonexistent or few and far between. It has made accessible various snowbound fastnesses in Canada and Alaska as well as many communities scattered in the jungles of the Amazon or the boundless wastes of the Sahara. But it has not affected existing roads in the slightest degree, nor has it caused any appreciable slackening in the rate of growth of new roads. Even with the advent of the helicopter, the airplane cannot hope to offer the flexible sort of transportation that is provided by the automobile, which threads its way slowly through crowded city streets and speeds along broad superhighways. The airplane provides an exceedingly rapid mode of transportation, but the ever increasing congestion of automobile traffic shows that air travel has not had the slightest effect upon travel on our highways.

Roads offer an excellent means of gauging the prosperity of a community. If it has no roads or only poor roads, with no access to a main highway, the chances are that its people will be retarded socially. The beautiful simplicity of such isolated communities as Oliver Goldsmith described in his DESERTED VILLAGE is apt to have its unfortunate drawbacks, even if the people live in such favorable surroundings as those provided by "Sweet Auburn! loveliest village of the plain." It is true that a life of isolation may produce a certain quota of noble and understanding characters, but a community off the beaten track is in danger of becoming narrow-minded, superstitious and tied down by hampering traditions.

When a highway connects such a community with the vast outside world, the effects are apt to be amazing. Contact with other communities brings new and vital interests. Trade and commerce begin to flourish. The community begins to find new markets for the products of its labor; the importation of new commodities brings about a higher standard of living. The intolerance of the community breaks down as it establishes closer relations with other communities and discovers that strangers can be good people. Its intellectual life quickens as it comes in contact with new ideas. Increased prosperity and higher intellectual standards lead to the building of new schools, new hospitals, new theaters and new concert halls. It becomes increasingly difficult to understand those who long for the "good old days" — that golden age that exists only in men's imaginations.

The historical division of roads into the few that are very old and the many that are modern — a product of the present age of technical progress — is striking, for it illustrates the long period of industrial and mechanical stagnation between ancient and modern times. The Romans did many things as well as the men of the Middle Ages and the Renaissance, and they did some things — including the construction of roads — much better.

When Hannibal was forced to make a road for his army across the Alps into Italy, in 218 B.C., he could use only the methods invented by the ancient Egyptians. Small rocks were cleared away by hand tools and wedges. The fire-setting system was employed for very large stones. The rock was heated by lighting a fire on it; and when very hot it was suddenly chilled with water, the rapid contraction often causing it to split. It was slow work; and when the road makers were confronted by a granite mass too large to be broken by hand tools or fire and water, they had to twist their path around it.

Now all is changed. The making of a modern road is an incident in the dramatic history of the mastery of space and time that has been proceeding so sensationally. It is indeed true that modern road making goes back only about two hundred years. John McAdam was born in 1756, and Thomas Telford in 1757, and with these contemporaries a new era in road construction may be said to have begun. Since their day, road making has transformed the world and made comparisons with other times absurd. Countries that have not built new-type roads are not in the modern age of speedy and comfortable highway travel.

The history of the highway is a story in four chapters. The first concerns the man who walks; the second, the man who rides a horse; the third, the man who uses a wheeled vehicle drawn by animals; the fourth, the man who rides on or in other forms of wheeled conveyances. In modern times, of course, the last-named mode of transportation has become predominant on the roads of civilized countries.

Public Roads Administration

Model showing the method of construction of the Appian Way, in ancient Rome. Careful preparation of the foundation and painstaking shaping and fitting of paving blocks made possible a road that was smooth and durable.

The first roads were tracks, foot paths and trails, designed chiefly to keep the traveler out of wet places. One may still see the pattern in many a winding country road. Perhaps our sentimental attitude toward foot paths, anywhere, is an unwitting reversion to the primitive days when the single-file trail was the only road. Half the charm of the country, whether we wander through the meadows or the woods, or climb the beckoning hills, comes from the faint romance of the foot-road. The instinct of the African following his forest track from the coast to the interior, or the Indian of the New World traversing the wood that is trackless to the stranger, or the mountain guide who walks with eyes that trace a way ten miles off, lives in a dim way in almost every man—a scarcely heard racial echo from the primeval world.

Nor is the bridle-path, the next stage in the evolution of the road, deprived of sentiment. One feels it when one stumbles upon the deserted packhorse trail across the Rockies, and pictures instinctively the cavalcades of the vanished years; nay, one even feels it when in the Pyrenees or the Alps short-cuts on the old mule-paths are taken while the tame road, crowded with speeding cars, winds smoothly and slowly below us. The energetic and educated little countries — such as Switzerland and Norway — have fully realized, as more backward countries have not, that the road for wheeled traffic is indispensable; and the mule-path is becoming a romantic inconvenience, that only satisfies the parts of the world still slumbering under semi-civilizations.

Rather curiously, some of the most prosperous countries on earth were but slowly beginning to understand the true place of the road as late as the first decade of the present century. The most conspicuous examples, perhaps, were the United States and Canada. The prosperous agricultural sections of these two countries got on quite well with very few roads worthy of the name, except in the great cities and long-settled districts, until the automobile came along. The automobilist came to realize through bitter experience what roads should be when he began to ride upon them in the newfangled "horseless buggy." He now began to understand why visitors from the Old World to the New were so unfavorably impressed by the makeshift character of so many of the roads, even those in localities of considerable size.

Myers, from Gendreau

**Primitive transportation — carrying supplies to the chicle gatherers in a Mexican forest.**

Today, of course, the United States and Canada have an abundance of good roads. The European visitor is apt to be bewildered by the profusion of splendid concrete highways that lead into the metropolis of New York. But why did Americans of a past generation accept their abominable roads with no idea that anything was wrong?

The explanation is simple. Until a town is large enough to afford a debt and until problems of heavy traffic arise, a wide strip of cleared land is adequate. When the inevitable ruts appear, traffic simply moves over, making a parallel set of ruts. This situation was common until well into the twentieth century and is still found in the less heavily settled parts of North America, especially parts of western Canada and the northwestern United States, including Alaska. The fact is that road making must always come at a comparatively late time in economic organization and must be a co-operative effort.

**Road building in the Roman Empire**

The Romans, the first great road builders, had a much clearer idea of the value of roads. They had established a great empire by conquering one ancient state after another. When they added a newly acquired area to their realms, one of the first things they would do would be to build a system of roads through the area. These thoroughfares gave their armed forces mobility; a comparatively small number of men were able to control a relatively large area. In time the roads brought about growth in trade and communication and ultimately bound the province to Rome more strongly than military rule could do. Of course it gave the Roman traders access to the region and so helped repay the cost of governing and conquering.

The roads were paid for largely through local taxes. The labor that built them was local, though some road construction was done by the Roman soldiers. The Romans extended their roads to the limits of their empire and built them to last for centuries. These roads surpassed any built up to the present century, and some are still in use.

**The extent of the Roman road program**

In the period from 300 B.C. to 400 A.D., the Romans are known to have constructed 50,000 miles of roads (not all paved). It was the largest system ever built until the twentieth century, when programs such as the Federal highway program of the United States were introduced, marking a rebirth of planning on a national, or even a continental, scale.

In the heyday of the Roman Empire, roads extended from the Italian peninsula into what is now Spain, France and Germany. An extensive road system was built up in England. It was possible to travel all the way from Caledonia (modern Scotland) to southern Italy by road, with the exception, of course, of the boat trip across the English Channel. Other roads extended into Asia Minor (southwestern Asia). The Egnatian way led from Rome through Macedonia to the Hellespont (Dardanelles); it is still in use in some of the more backward parts of the Balkans.

**The famous Appian Way**

The best known of all the Roman roads was the Appian Way, familiar to tourists. Begun in 312 B.C. by Appius Claudius, it has been in use ever since. It has been said that no other human construction has been used for such a long period of time. Two of the main features of this famous thoroughfare are still apparent. These are its straightness and narrowness, as evidenced by the tombs and other ruins of the past that line its way for miles. Roman roads were never wide; except in urban areas the usual width was about fifteen feet, providing about enough room for two wagons abreast. On either side of the paved roadway was a drainage ditch.

**How Roman roads were constructed**

Most Roman road construction followed a pretty definite basic pattern. First a wide, shallow ditch was dug in the ground. The exposed earth was rammed hard to prevent it from settling after the road was laid down. Then several layers of flat stone were set in place. On top of these, there was a layer of rubble or possibly one of concrete (at that time a mixture of lime, water, sand and gravel or stone). Another layer, of concrete, was added. The final layer consisted of blocks of stone, carefully

American Export Lines Collection

Magnificent triumphal arches bend over the famous Appian Way as it enters Rome from the southeast. This durably constructed highway, once paved with lava blocks, attests to the competence of the Roman road builders.

joined and fitted. The final product was usually more than a yard thick. At times, roads were built above the level of the countryside by first digging ditches and then using the dirt from the ditches to form embankments. A highway of this kind was protected from the effects of water in low or rainy country. At the same time the combination of ditch and embankment served as a fortification of sorts.

The Roman roads were massive enough to resist weathering quite effectively; nor were they as much damaged as other roads by the wedging action of tree roots or ice. However, despite their solidity of construction and vast extent, it is doubtful that they were ever quite adequate for the needs of the population.

#### The fall of the Empire and the decline in road building

With the collapse of the Roman Empire in the fifth century A.D., central government disappeared from Europe. Without some group to assume responsibility for the roads, the system gradually decayed. The Empire had been, in many respects, an economic enterprise; it represented a huge trading area, with Rome at the center. Hence the roads had been an economically sound investment. In the Middle Ages, however, the various regions of Europe were more or less self-sufficient and isolated from each other; there was comparatively little travel between them and trading activities were at a low ebb. Few were interested in a far-flung system of roads.

Local roads served mainly for the movement of livestock. Since a good surface was not needed for this purpose, most roads were simply dirt paths. In the Middle Ages, the building of roads and bridges was often a work of piety. The Frères Pontifes, or bridge-building monks, were active in France in this period.

In those days, traveling was hazardous, and prayers were regularly said for those who traveled on land, as well as those who voyaged by sea. So bad were the highways of England that a meeting of Parliament under Edward III was postponed to await the coming of members delayed on the road. There was to be no marked improvement in road conditions for several centuries.

#### The revival in road building

Not until the Middle Ages had passed into history was there a definite revival in road building. In the middle of the seventeenth century, England began to create a new system of roads to meet the demand created by steady growth in population and trade. Trusts were enfranchised to build and maintain highways, and in return they were allowed to collect tolls. Various local communities began to construct roads, taxing their residents and levying tolls on travelers. The system had certain obvious disadvantages. Roads were built solely with the idea of making profits through tolls; the builders were not concerned with furthering the interests of the public at large. Even if they had been inspired by the loftiest of motives, the trusts would not have had the capital to build and then maintain highways in reasonable condition.

#### How the toll-road system furthered road building

The toll-road system, however, had its good points. For one thing, to some degree at least, it fixed the responsibility for building and maintenance. Likewise it created an incentive, on the part of various groups of people, to provide more and better roads. It was the most powerful incentive of all: the hope of large profits.

With the coming of the industrial revolution, popular pressure was so great that abuses were corrected and improvements made. It is interesting to note that England, which was the first nation in Europe to feel the effects of the economic awakening, was also in the van of road-building progress. The toll-road system, despite all its faults, was continued in certain parts of England up to the end of the nineteenth century. (With many modifications, the practice has been reintroduced in the United States with the building of such superhighways as the Pennsylvania Turnpike, the New York State Thruway and several other major toll roads.)

# ROADS

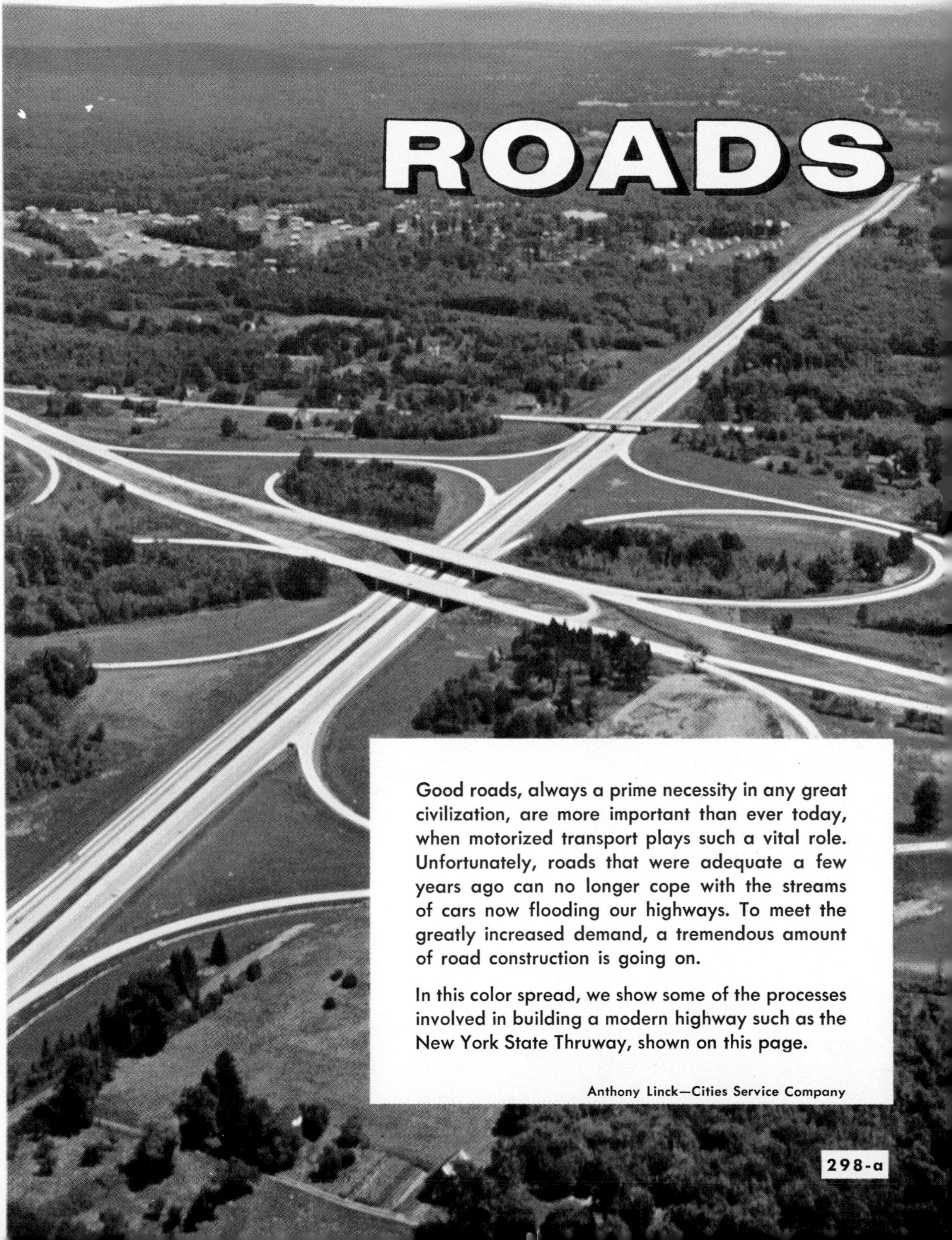

Good roads, always a prime necessity in any great civilization, are more important than ever today, when motorized transport plays such a vital role. Unfortunately, roads that were adequate a few years ago can no longer cope with the streams of cars now flooding our highways. To meet the greatly increased demand, a tremendous amount of road construction is going on.

In this color spread, we show some of the processes involved in building a modern highway such as the New York State Thruway, shown on this page.

Anthony Linck—Cities Service Company

# HOW A MODERN

## I. EXCAVATING AND GRADING

**1.** After the route has been surveyed, bulldozers remove trees and other obstacles.

**2.** Drills operated by compressors bore holes for the explosives that will blast rock.

**3.** Putting drainage pipes in place. Drainage is a vital factor in road construction.

**4.** Grading: giant scoops remove earth from high spots and spread it to low places.

**5.** A sheep's-foot roller compacts the material spread on low places in the road.

# HIGHWAY IS BUILT

## II. PAVING (Two Major Methods)

### ASPHALT (Bituminous)

Dump truck piles crushed rock, followed by asphalt-penetrated crushed rock.

A truck feeding hot material to the spreader, which is shown at the left.

An efficient spreading machine levels the mixture of asphalt and crushed rock.

### CONCRETE

A dump truck piles crushed rock on an embankment. A blade grader levels it.

A paver mixes the ingredients of the concrete and pours the mixture in place.

A finishing machine levels and smooths the concrete mixture dumped in place.

## III. LANDSCAPING AND MARKING

Landscape crew plants foliage to prevent erosion and beautify surroundings.

Mobile crane hoists direction sign on an overhead trestle to guide motorists.

# FINISHED HIGHWAYS IN CROSS SECTION

## ASPHALT

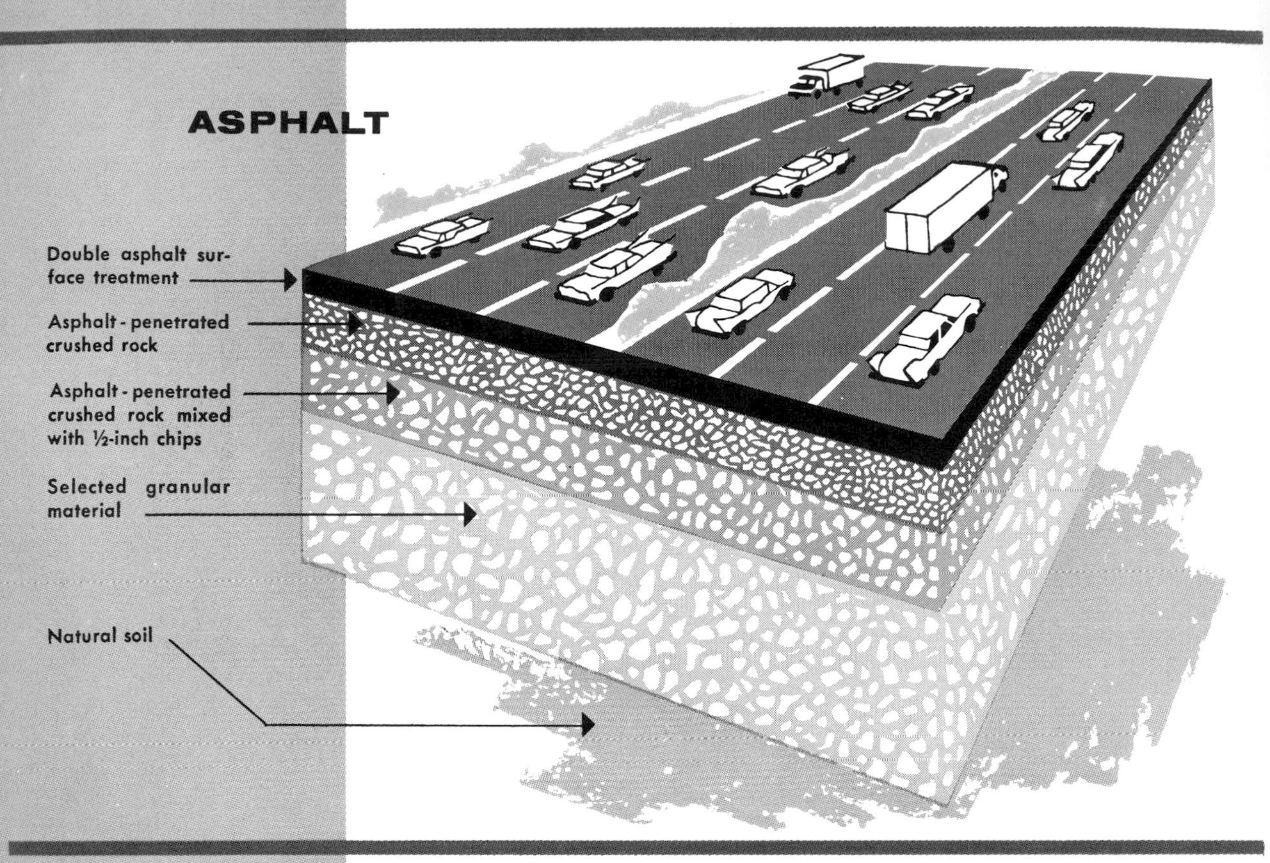


## CONCRETE

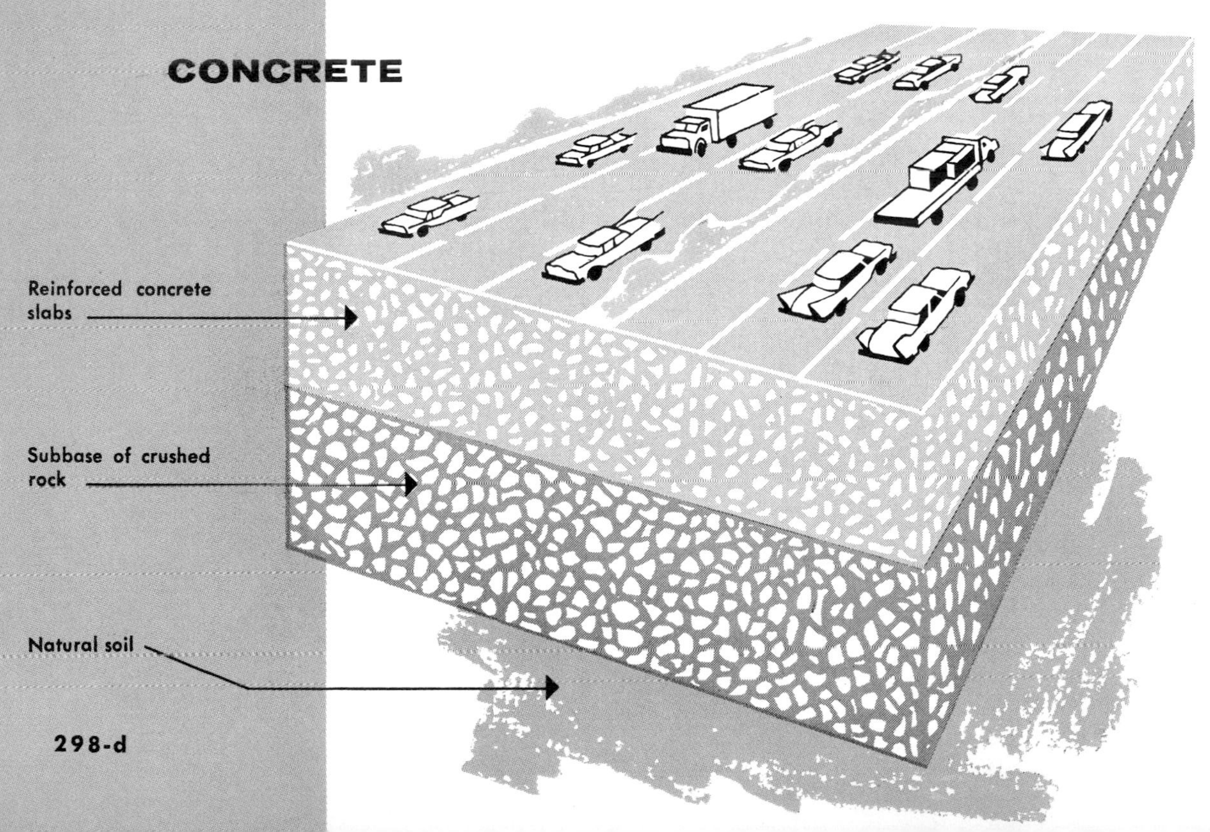

Thomas Telford

Culver Service

John L. McAdam

Toll roads were commonly built with a prominent crown, so that water would drain off toward the sides. Unfortunately, the crown was often made too high, so that vehicles traveling at one side or another of a road would be uncomfortably tilted. Consequently, drivers sought to travel along the crown; ruts would be worn in it and the ruts would become water traps.

The beginning of the nineteenth century marked a new development in the art of road building. A Scotchman, John Loudon McAdam (1756–1836), argued that it was not necessary to have an elaborate stone foundation for roads. He maintained that a good, smooth, hard surface could be maintained by spreading a layer of irregular, small stones a foot or so deep. The stones would gradually settle; they would be cemented together by the action of rain upon the stone dust ground by the heavy wheels of the vehicles passing by. In time the compacted stone would form a solid pavement. Another Scotchman, Thomas Telford (1757–1834), introduced a variation of McAdam's method of road building. He laid a foundation of large stones, with enough space between them to allow for proper drainage. Above this rough and strong foundation he placed broken stone, the smallest stones on top.

Many miles of macadamized roads, based on McAdam's original idea and utilizing improvements by Telford and others, were built in England and in the United States. Up to about 1900 these roads were generally considered to be the best of all.

The introduction of the automobile brought about a great change. Macadamized roads were not suitable for automobile travel. For one thing, the rubber tires of automobiles did not grind enough rock dust to cement the small stones of the road properly. As a matter of fact, the swiftly moving tires were more likely to stir up whatever dust was already on the road; this dust would be blown away by the wind, and only loose, rough stones would be left. The road would no longer shed water; holes and ruts would be formed. Engineers sought to develop new kinds of roads expressly designed for automobile travel. The result has been a complete revolution in the art of road building.

Roads of dirt or gravel are still to be found in various rural areas. Often, however, these roads are specially treated by adding a material that will serve as a waterproofing agent, or as a binding agent to provide added cohesion between the particles of soil or rock. Bituminous materials are often used for this purpose.

Standard Oil Co. (N. J.)

Above: This road crew is applying a new coating of black sheet asphalt to an old cement sidewalk in New York's Central Park.

Below: Two road rollers are shown working on a road in Bardstown, Kentucky. Diesel engines provide power for the rollers.

Caterpillar Tractor Co.

Bituminous surface treatment, without a heavy foundation, suffices for pleasure vehicles and a medium amount of truck traffic; with a heavier foundation, it may serve for heavier traffic. The bitumen used in this treatment penetrates into the wearing surface of stone particles, coating them, binding them together and developing a waterproof surface.

Sheet-asphalt roads also withstand heavy traffic; they provide a smooth surface that is easily cleaned and maintained. In these roads the wearing surface is composed of a mixture of sand and mineral filler with an asphaltic cement. Sheet-asphalt wearing surfaces are laid on good foundations; a concrete foundation of from six inches to nine inches in depth is quite common. An existing pavement may be adapted for use as a base for the sheet asphalt. The asphalt is solid at normal temperatures; it is shipped to the site, heated until it is viscous and laid hot so that it may be easily spread. The road is rolled while hot, immediately after spreading.

Portland-cement concrete is a favorite material for through highways. A concrete road is durable, light in color, easily cleaned and generally skidproof. The initial cost is high, but maintenance requirements are less than with other roads. Concrete roadways for heavy traffic are gen-

Public Roads Administration

Big snowplows, like the one shown above, help keep our highways passable throughout the winter.

erally from seven inches to eight inches thick at the center. At the outer two feet of the paving, where additional strength is required, the concrete is thicker than at the central portion of the road. Concrete roads are generally reinforced with a lightweight steel-mesh reinforcing.

In building concrete roads, a uniform subgrade, or base, is prepared and shaped. The concrete is often mixed on the way to the job in "transit-mix" trucks with portable mixers; sometimes it is transported, ready mixed, to the site in dump trucks. The concrete is placed on the subgrade and spread until its top surface lacks only two inches of the intended height. The reinforcing steel is laid in place and the balance of the concrete is poured. The concrete is then agitated to make sure that it will occupy the entire roadway section and leave no unsightly and perhaps dangerous gaps.

After ten to twenty minutes of standing, excess water and organic impurities will rise to the surface of the concrete; this laitance, as it is called, is removed. Just before the concrete sets, the surface is broomed at right angles to the roadway, either by hand or with mechanical equipment, to produce a rough, nonskid surface. During construction, joints are left to allow for the expansion of the concrete. The joints are generally filled with a mastic compound to keep water from undermining the concrete slab, to prevent chipping at joint edges and to provide a smoother riding surface for automobiles.

Although the concrete will set in a short time, it will not be strong enough to withstand traffic for several days, and it will not reach its designed strength for almost a month. As the correct quantity of water must be available in the concrete for complete reaction with the cement, care must be taken to prevent the loss of that water. This process is known as curing the concrete. In hot weather, the concrete slab may be covered with burlap or waterproof paper and kept damp by spraying; in winter burlap is often used to keep the frost from the concrete. Frequently, in either hot or cold weather, the surface is sprayed with a bitumen compound, which seals in the moisture. After curing is completed, traffic is permitted to use the road.

Block pavements of asphalt block, paving brick and stone blocks, often laid on concrete foundations, have occasionally been built in or near large cities, where traffic concentration is high. Few block-type pavements of this kind have been built in the United States in recent years.

The art of road building is seen at its best, perhaps, in cities and mountainous areas. The paving of many of our cities,

United States Steel

Bridges like this one, at Cuscatlán, El Salvador, form parts of highway systems.

Standard Oil Co. (N. J.)

Workmen laying paving stones on the Champs Elysées, a world-famous avenue in Paris.

constructed to conform to varying conditions of geography and traffic, is an example of masterly engineering applied to a great variety of problems. We are apt to take for granted the uniform and well-maintained roads that run past our homes; yet the least of these would be considered a marvel and a luxury by many people in the world's underdeveloped areas.

Today's road builders are not discouraged by the prospect of carrying their highways across mountainous terrain. All the resources of modern industry and technology can be brought to bear on such projects. Unlimited power, explosives of every form and strength and land-moving and scraping machines of every shape and capacity are now available. As a result, highways can be built in almost every land region. They now span whole continents. It took modern engineers less than two years to complete the 1,523-mile Alaska Highway, from Dawson Creek, British Columbia, Canada, to Fairbanks, Alaska, through swamps, mountains, gorges and subzero temperatures. This all-weather road, with few grades over 4 per cent, is 24 to 36 feet wide and includes about 700 bridges.

#### The great Inter-American Highway

The Alaska Highway forms part of a vast system of roads — the Inter-American Highway — that will extend from the northern to the southern part of the two Americas, traversing Alaska, Canada, the United States, Mexico, Central America and South America, through Argentina. Much of the highway has been completed.

Standard Oil Co. (N. J.)

In the ancient city of Bombay, India, contract laborers are carrying heavy rocks on their heads for the foundation of a new road, which is being built for the refinery of a British oil company. One of the most urgent problems that confronts India at the present time is the improvement of its inadequate system of highways.

Ed Sievers

At San Diego, California, U. S. Highway 101 parallels a picturesque stretch of beach for many miles.

Port of New York Authority

This great plane is taxiing across an underpass at New York's International Airport on Long Island.

Arnold and Kellogg

An intricate maze of roads guarding one approach to New York City. It helps to make driving more safe.

N. J. Turnpike Authority

The New Jersey Turnpike, a modern superhighway, is 118 miles long and traverses ten counties of New Jersey.

The United States has developed an outstanding system of roads, under the leadership of the state and Federal governments; the total mileage has been estimated at 3,000,000. Originally road building was a function of township or county authorities; but the dirt roads that were built under their auspices proved to be entirely inadequate when automobiles began to multiply. The state then had to take over a certain measure of responsibility for the roads. Every state now has a highway department, which supervises the state highways. The cost of building and maintaining highways is met by bond issues, gasoline taxes, registration of motor vehicles and taxes levied on motor carriers.

The Federal Government has also taken an important part in the road-building program. It first assumed responsibility for the construction of roads in 1806, when Congress made appropriations for the construction of the Cumberland Road, which was to extend westward from Washington, D.C. Congress later appropriated money for other roads.

### Federal aid for road construction as a national policy

In 1916 Federal aid for road construction was established as a national policy, and it has been continued ever since. In the past few years Federal-aid funds have amounted to some $500,000,000 a year. The states bear part of the cost of the construction of Federal highways, and they assume the cost of maintenance after the roads have been completed.

It would be an endless task to name all the outstanding highways in the United States. Let us mention here the Lincoln Highway, from New York to San Francisco; the Pacific Highway, from Vancouver, British Columbia, to the Mexican border beyond San Diego, California; the Dixie Highway, from Chicago to Jacksonville, Florida; the Jefferson Highway, from New Orleans to Winnipeg; the Pennsylvania Turnpike, from the Ohio line to Philadelphia; the New Jersey Turnpike, extending to Wilmington, Delaware.

*See also Vol. 10, p. 285:* "Roads."

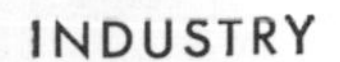


Glass Container Manufacturers Institute

Steuben Glass

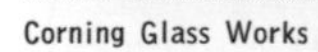
Corning Glass Works

# THE STORY OF GLASS

## A Basic Material of Modern Civilization

BY GEORGE J. BAIR

SAND—ordinary sand—appropriately blended with various other substances has given us the astonishingly versatile material that we call glass. Windowpanes and mirrors are made of glass; so are bottles, fruit jars, drinking glasses and many dishes. Glass tubes and lamps form an essential part of radio and television sets. We use glass for the lenses of still, motion-picture and television cameras. The electrical industry makes wide use of glass insulators. Stable electronic resistors and capacitors for computers and for space exploration are made of glass. The chemist needs vast amounts for the retorts and tubes in which he fashions his modern miracles. Through the big lenses of his telescope, the astronomer peers far out into space. The delicately adjusted lenses of the microscope bring into view living things that are too small to be seen by the naked eye. Glass blocks serve in building construction; glass fibers can be woven into fabrics. Ceramic materials made of glass serve for heat-resistant cooking ware and the nose cones of missiles. Our life would go on without glass, but it would be poorer indeed.

There are many different kinds of glass. Make a list of glass articles that are familiar to you and opposite each item set down the properties that each must possess. Your list will begin something like this:

| *Article* | *Properties* |
|---|---|
| Windowpane | Transparent, translucent or colored, depending on purpose of window. |
| Bottle | Impervious to water; sanitary. |
| Ovenware | Heat-resistant. |
| Cover over key of fire-alarm box | Fragile, so that it may easily be broken. |
| Retort | Resistant to chemicals. |
| Glass door | Hard and durable. |
| Electrical insulator | Nonconductor of electricity. |

To produce these different properties, we must vary in a thousand different ways the composition of glass and the heat treatment to which we subject it. We must also form it in different ways; naturally the shaping operation that will produce a flat sheet of window glass will not do for a milk bottle, or a 40,000-pound telescope-mirror blank or a glass fiber $1/5{,}000$ of an inch in diameter.

### The history of glassmaking

Nature was the first glassmaker, fashioning the natural volcanic glass that we call obsidian. At a very early period, man began to ape nature's production of glass. To sand he added soda and lime or other substances; he fused the mixture under intense heat and then shaped the molten mass. The Egyptians were pioneers in the art of making glass; but the foremost glassmakers of antiquity were the Romans.

Glass served the Romans for personal ornaments and for architectural decorations, sometimes even for windows. They made beautiful artificial gems of colored glass; they decorated glass with gold leaf. When the barbarians overthrew the Western Roman Empire, the art of glassmaking came to an abrupt halt in Europe. It continued to flourish, however, in the East, particularly in the Byzantine Empire.

Glassmaking was revived in Western Europe toward the end of the eleventh century. The Italian city state of Venice became especially noted for its fine colored glass with decorations in enamel and gold tracery and for exquisite filigree patterns, like lace made of glass threads.

By the time Venice had produced these masterpieces, all the fundamental glass-forming processes were known. Until the late 1700's, advances in the glass industry were limited to the discovery of a few new raw materials and to the development of artisans' skills. Then machinery was introduced and came to be widely adopted.

Today, the industry is almost completely mechanized, though hand operations are still used in some small-volume specialized fields. There has been a remarkable advance in glass technology, particularly in providing new ingredients and in modifying the heating treatment so as to bring about a variety of desired properties.

Corning Glass Works

**A lump of obsidian, formed when rock was heated in the interior of the earth and disgorged from a volcano.**

### The chief glassmaking processes

The chief processes used in making glass are much the same as they have been for many years past. (1) Sand and other materials are assembled to form what is called the glass batch. (2) These materials are melted in a furnace at a very high temperature and produce molten glass. (3) While the glass is molten, it is shaped and made to cool so as to form a rigid piece. (4) The glass article is annealed: that is, it is reheated and then gradually cooled.

Proper selection of batch materials is essential. The basic material (50 to 75 per cent) is still sand — the type composed almost entirely of the two chemical elements silicon and oxygen and known as silica sand, or "glass sand." Soda is another important ingredient; so is lime, a compound of calcium and oxygen. Among the elements needed for various kinds of glass are arsenic, barium, boron, lime, lead, magnesium, potash, zinc and manganese.

Glassmakers prefer materials in granular form, because they mix well with sand. Raw materials are usually stored in large

**Two fine examples of Venetian glass, from the collection in the Metropolitan Museum of Art. Venice was a renowned center of glassmaking in the medieval period.**

Metropol. Mus. of Art. Left: gift of J. J. Jarvis, 1881; right, gift of G. Marquand, 1883

bins or silos above the batch house, the chamber where the glass is prepared. The materials are fed by gravity to a scale room, where they are accurately weighed; then they are dry-mixed in a batch mixer, generally a revolving cylindrical drum, like that used in making concrete. When the batch has been thoroughly mixed, it is transferred to the furnace.

Large furnaces, called tank furnaces, many of them holding as much as 200 tons, are used extensively to melt the batch ingredients. A tank furnace consists of a big rectangular chamber, the lower half of which is always kept full of molten glass. The space between the glass and the arched roof is the combustion space, where gas or oil and preheated air are burned, giving temperatures of up to 2,900° F. The furnace is constructed of large blocks made of fire clay or special refractory materials that will resist the corrosive action of molten glass at exceedingly high temperatures. The level of glass is kept constant by feeding the raw batch into one end as the melted glass is removed from the other.

Bubbles that form during melting are allowed to rise to the surface and break or are redissolved chemically into the melt; this is called fining. A tank furnace is divided into melting and fining zones by a wall which extends clear across the middle of the furnace and which is a few inches higher than the level of the molten glass. The wall serves to hold back unmelted batch and scum from the fining chamber. The molten glass flows into the fining end of the tank furnace through a throat near the bottom of the wall.

Certain substances, such as sodium sulfate or sodium chlorate, are added to the glass mixture to form large bubbles; these gather up the tiny bubbles and carry them along to the top of the mass. Sometimes fining agents, such as arsenic and antimony, are used. These materials behave chemically to redissolve any bubbles that are left. Finally, the molten glass, free of bubbles, is worked out of ports at the end of the furnace. In some cases it flows to feeders above automatic forming, or shaping, machines.

Tank furnaces are operated continuously, day and night, week after week. They produce the large volume of glass required for bottles, windows, light bulbs and the like.

For low-volume special glass products, pot furnaces and day tanks are used. A pot furnace consists of a chamber in which one or more fire-clay tubs, called glass pots, are placed. The raw batch is charged into the hot pot and the glass is melted. Sometimes the glass is worked directly from the furnace; a small working door is constructed for this purpose near the top of the pot. In other cases, the pot is removed from the furnace and the molten glass is poured from it on to casting tables. Pots usually hold about a ton of glass.

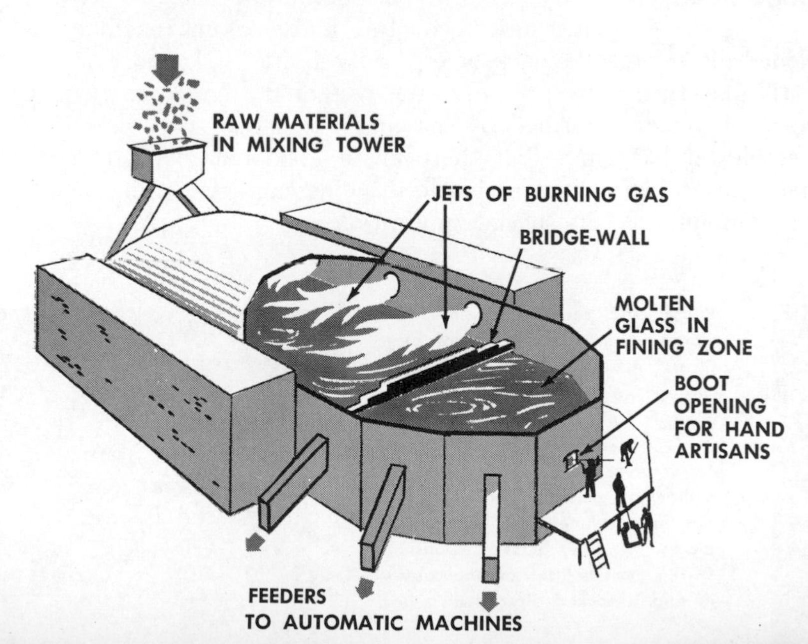


A tank furnace, used in high-volume glass production. Raw materials from a mixing tower are fed into the tank. Here they are subjected to the roaring heat of burning gases, issuing in jets of flame from sidewall ports. The glass is more and more completely melted. Finally, it makes its way through an opening (throat) near the bottom of a bridge-wall into the fining zone. (The wall holds back impurities and unmelted batch.) In the fining zone, bubbles are removed from the molten glass. It then passes from the tanks through feeders, which deliver hot glass to various types of automatic forming machines. Hand artisans gather molten glass from the tank through boot openings just above the glass level.

Adapted from diagram supplied by Corning Glass Works

Right: a jar-making machine at work. A gob of molten glass drops into a blank mold (*A*). The blank passes into a finishing mold; here compressed air blows the jar into the final shape (*B*). The iron jaws of the finishing mold swing open (C), revealing a finished jar. The process takes about six seconds.

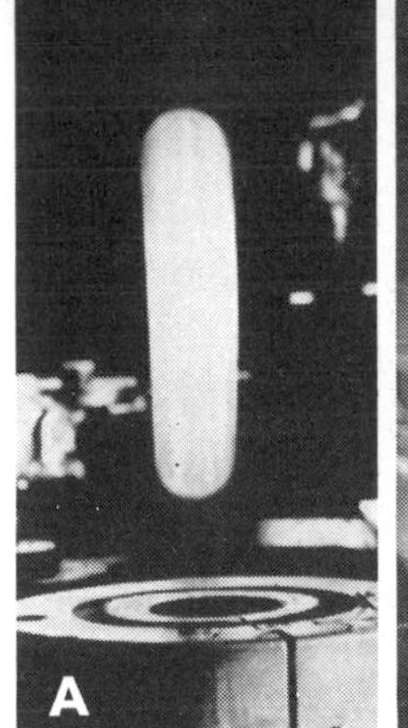

Glass Container Manufacturers Institute

Left: a master glass blower, called a gaffer, blows a bulb for a 75,000-watt lamp. The hand-blowing process is still used for short-run production, for special glass and for large or unusual shapes.

Corning Glass Works

Day tanks are small rectangular furnaces, which are built of high-temperature blocks. The raw batch is charged into the hot tank where it is melted; after the glass is fined, it is worked by hand. The day tank is so called because the glass is usually melted during the night and worked in the daytime.

After the raw batch has been melted in the furnace, it is ready for forming. There are several different ways in which this may be done.

Glass blowers once fashioned glass for the emperors of Rome; they still play their part in the glass-forming process. To make an article like a bottle, a glass blower has a crew of four men. The first man, called a gatherer, collects the proper amount of molten glass on the end of a long tube, called a blowiron. The blowiron is then passed on to the next man, who shapes the gob of hot glass so that it will have the proper distribution, and starts a bubble. Then he turns the iron over to the glass blower, who alternately blows the bubble larger, and swings the iron to elongate the molten shape. At just the right time the glass blower lowers the partly formed hot bulb into a mold at his feet and blows it to fill the mold. It quickly cools and, still attached to the blowiron, becomes rigid. A crack-off boy now breaks the glass from the iron and takes it to the finisher. The finisher reheats the neck and forms it with hand tools.

Most artware, many special kinds of bulbs and a certain amount of glassware used for scientific or industrial purposes are made by the method that we have just described. Some ware is not molded at all; it is shaped entirely by hand.

Automatic bottle machines mechanically perform the operations that we have just described. In a bottlemaking machine, the neck or top of the bottle is shaped first, by pressing; the rest of the bottle is then formed by blowing. It should be pointed out that there are a good many different kinds of bottle machines and that each has its special applications.



Pittsburgh Plate Glass Co.

Photos *A-E* show some important stages in the manufacture of plate glass. *A*. Molten glass is passed through large rolls and a ribbon of glass is formed. *B*. The glass ribbon is annealed and cooled to room temperature. C. The surfaces of the ribbon are ground. After the grinding process, the glass is cut into plates. *D*. The glass is polished. *E*. The glass is lifted off the conveyer and is packed directly in a box.

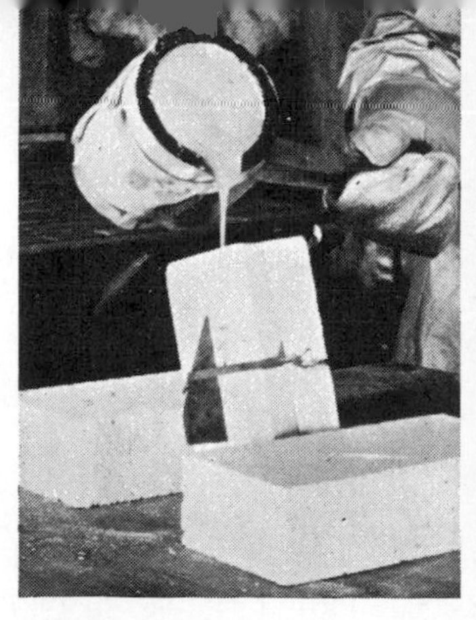

Right: the casting method as applied to glass. (1) Pouring the hot glass into a ceramic mold; (2) shaping a sealed-beam headlight by pressing. The casting method is also employed in forming tableware, ovenware and lenses.

Corning Glass Works

Glass baking ware, dishes, bowls, heavy tumblers, building blocks, insulators and many other kinds of glassware are shaped entirely by pressing. A gob of hot glass of the right size and shape is delivered to a metal mold which is set in position under a plunger. The plunger forces the hot glass into the space between the mold and the plunger and holds it there until it solidifies. After the glass has been removed, the mold is ready for another charge.

Window glass is made by drawing a continuous vertical sheet from a pool of molten glass. This sheet is drawn through a slot in a fire-clay float, set on the glass surface or drawn directly from a chilled surface; special cooling attachments set the glass at the proper time. The sheet is usually pulled upward to the floor above the tank; here it is trimmed and cut to length.

Plate glass is made by rolling. The molten glass leaves the fining end of the furnace through a horizontal slot; it passes from the slot through rolls as a continuous ribbon. It is then annealed. (We shall describe annealing later.)

After the annealing process, the glass ribbon continues through a line of large grinding heads located above and below the sheet so as to grind both surfaces simultaneously. Graded abrasive sand of successively finer grit is fed in a water slurry (a thin, watery mixture) between the revolving grinding heads and the glass ribbon. After the plate is ground to the proper surface, the ribbon is cut into convenient lengths and mounted on an endless table for polishing with revolving felt pads and rouge. The polishing is usually done on one side at a time. The plate must be turned over for the finishing of the second side. In some cases, the ribbon is cut after annealing, and both grinding and polishing is done on one side at a time, on an endless table. Finally, the plate is inspected and cut to proper size.

Some plate glass is melted in open pot furnaces. The pots of molten glass are removed from the furnace by large cranes; the glass is then poured from the pot between two or more rollers. After annealing, the glass is cut into the proper lengths and ground and polished as described above.

The casting method is used to make massive shapes such as large architectural blocks and mirror blanks for big telescopes, like the one at Palomar Mountain. For this purpose, heat-resistant molds are used and molten glass is poured into them. After the glass is annealed in the mold, the mold is broken away and the piece is finished.

Like most materials, glass contracts in cooling. Suppose that, during the forming process, various parts of the glass piece are at different temperatures and will so-

Corning Glass Works

Annealing casseroles in a lehr, a continuous furnace in which adjustments in temperature are made so as to prevent internal stress from developing. The diagram shows how such stress arises. *A*, *B* and *C* are three parts of a single piece of glass. *A* and *C* have set at a higher temperature than *B*. On cooling, they have contracted more than section *B* and hence tend to pull on this section. As a result, an internal stress is set up at *B*.

lidify at different times. Each part will tend to contract a different amount as it cools to room temperature.

Because cold glass is so rigid, even a slight difference in contraction will cause internal stress to develop. The diagram on this page shows how it comes about. To prevent such stresses, glass must be annealed. By this we mean that it is held at a temperature where internal adjustments can be made to release stress; then it is cooled at a rate that is slow enough to prevent the forming of any new stresses. Fortunately, the annealing temperature is below the temperature at which the glass object would begin to lose its shape.

Most glass today is annealed in continuous furnaces called lehrs. Glassware is set upon a metal belt that conveys it through a tunnel. The temperature along the tunnel is graduated so that, after passing through a hot zone, the ware is slowly cooled to a safe handling temperature at the end of the belt. In batch-annealing furnaces the ware is stacked on shelves and in metal baskets. After loading, the furnace is sealed and heated to the annealing temperature, then cooled to room temperature.

During all the foregoing operations, special care is taken to make sure that the glass is perfectly uniform throughout. If the glass in each part of a piece of ware is like that in every other part, we say that it is homogeneous (from the Greek, *homo*, same, and *genos*, kind). Inhomogeneous glass may be "cordy"; it may contain "seeds" or "stones." Glassware with too many blemishes like these is defective.

A cord usually results from a streak of glass of slightly different composition from the rest of the glass. To have a clear idea of what cords are like, stir a teaspoonful of corn syrup in a glass of cold water; when you hold the glass up to the light, you will see cords of syrup. Cords distort light passing through the glass; they may bring about excessive internal stress.

Seeds are small bubbles of gas. The fining operation is designed to eliminate such bubbles; but they will develop in spite of all precautions if the glass is not homogeneous. Stones are particles of unmelted batch material or pieces of furnace material, or else result from the crystallization of one of the materials in the batch.

### Some important varieties of glass

So much, then, for the major processes in the making of glass. Let us now consider some important varieties of glass.

*Tough glass.* We generally think of glass as a brittle material. If a thrown or batted ball strikes a window, the pane is almost sure to break, just as a drinking glass will generally break if we let it fall to the floor. In some cases, it is desirable to have brittle glass. For example, the glass used as a cover over the key in a fire alarm box is so brittle that a slight tap will shatter it; the key will then be accessible to the person who wishes to sound the alarm. For some purposes, however, we want tough glass.

As a material, glass is inherently very strong; freshly formed glass fibers are almost as strong as steel. Why is it easy, then, to shatter a windowpane with a comparatively light blow? To understand this problem, we must recognize the difference between tension and compression stresses. Tension stress is the force induced in an object when it is stretched; compression stress results from a squeezing force. If we push down at the middle of a rubber eraser (*A* in diagram on next page), there will be tension stress at the under surface; it will stretch. There will be compression stress at the upper surface; it will contract.

Much the same thing happens when we apply a load to the center of a bar of glass, supported at its ends, as shown in *B* in the diagram. The under surface of the bar will tend to stretch and as it does so, it causes the upper surface to be compressed. Glass is relatively nonelastic. The under surface of the glass will bend just so much and no more; when the applied force exceeds the glass's tensile strength, a crack will start. It will travel rapidly through the glass and will cause it to break.

The crack usually starts at some scratch or surface defect. Glaziers put this characteristic to good use when they cut window glass. They scratch the surface with a glass cutter and then apply pressure on the surface opposite the scratch; the glass will break along the scratch.

For many years, glass manufacturers obtained maximum strength in glass by carefully annealing it, thus removing internal stress. Today, the strength of glass is increased many times by controlled tempering — suddenly chilling after heating. The object of the chilling operation is to establish a uniform temperature difference between the glass surface and the interior at a high enough temperature so that any temporary stress in the glass may be released. On further cooling to room temperature, the hotter interior must cool more and hence contract more than the surface. The surface, therefore, develops high compressive stress, balanced by tensile stress in the inner part.

Let us again consider our bar of glass. If the underside of the bar is in compression to begin with, the first pressure exerted from above will merely remove some of this compressive stress. As more pressure is applied, the compressive stress in the underside lessens, until finally there is tension stress instead of compression. If pressure still increases, it will exceed the tensile strength of the underside in time, and the glass will break. But a good deal of pressure will have to be applied.

The tempering process has made it possible to manufacture much tougher glass than before. Tempered thin-walled drinking glasses are now more durable than the old heavy-walled tumblers; glass tableware will withstand severe restaurant service. Beautiful plate-glass doors without supporting frames are used in modern store fronts. Automobile side and rear windows will stand powerful blows without breaking; when they do break they shatter into pieces too small to cause fatal injury.

*Laminated glass.* The automobile windshield is an interesting example of a kind of glass that must not be too strong. Automobile manufacturers favor laminated safety glass (glass in layers) for windshields. In this material, a tough, elastic sheet of plastic is sandwiched between thin pieces of plate glass; the three layers are cemented together by heat and pressure.

In case of an accident, if a person is hurled into the windshield, the glass will break and the plastic will stretch to absorb some of the shock. All the pieces of glass will be held to the plastic sheet and serious cutting will be avoided. A tempered single-plate windshield would be so strong that it would not break if a passenger in the car were hurled against it; the result would probably be a fractured skull.

Bullet-resistant glass consists of a number of layers of glass and plastic sheets. The outer layers of glass are relatively thin. With the elastic plastic layers they serve to absorb the force of a projectile, so that the heavier glass backing will resist penetration. Bullet-resistant laminated plate is usually an inch or more in thickness.

**The diagrams below show the difference between tension stress and compression stress. If we push down at the middle of a rubber eraser, as in diagram A, there will be tension stress at the under surface; it will stretch. There will be compression stretch at the upper surface; it will contract. Much the same thing happens when we apply a load to the center of a bar of glass, as in B.**

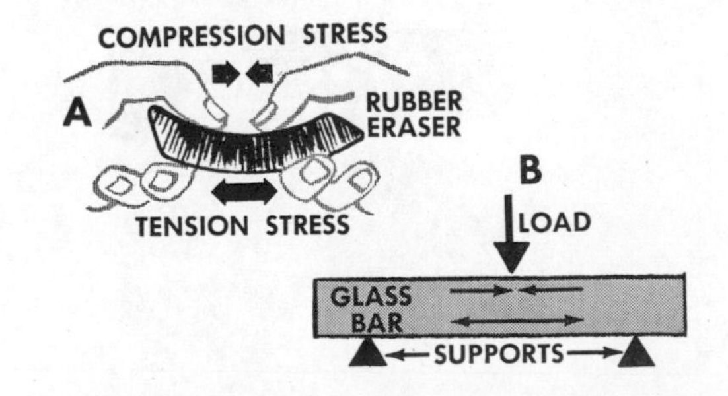

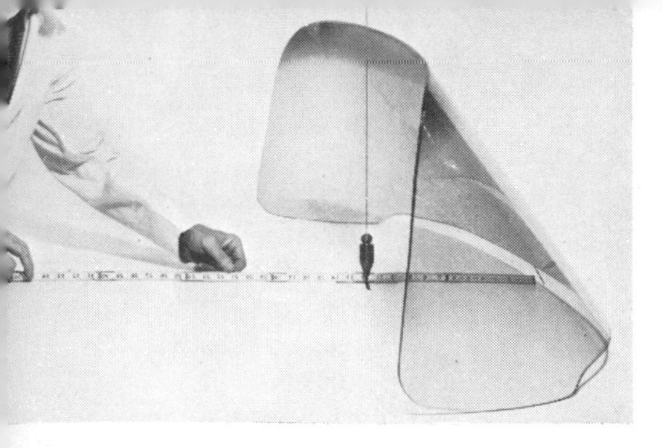

Demonstrating the slope of a panoramic, or "wrap-around," automobile windshield. Curved glass such as this is produced by subjecting the finished glass to just enough heat to make it sag. The glass then adopts the contours of the mold on which it is set.

Libby-Owens-Ford Glass Co.

Below: the 84-inch telescope mirror blank cast by the Corning Glass Works for the Kitt Peak National Observatory, near Tucson, Arizona. It will later be ground and polished. The weight-saving honeycomb pattern is seen through the surface of the glass.

Corning Glass Works

Below: optical quality glass for spectacle lenses and scientific instruments is mass-produced today, making possible high-quality glass at lower cost. The glass is produced from chemicals of the highest purity, and great care is taken in weighing and mixing.

Corning Glass Works

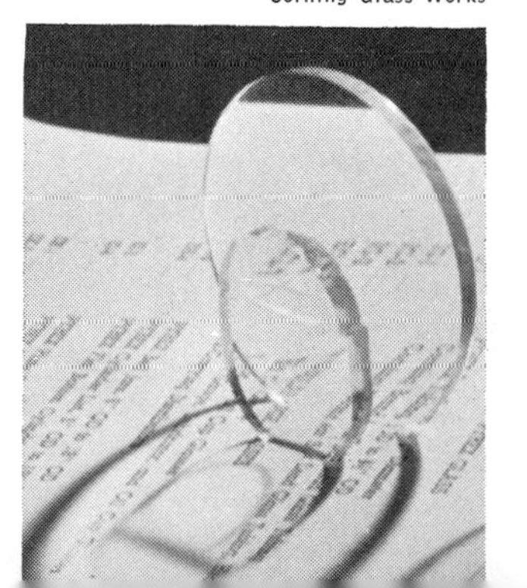

*Curved plate.* Many of the windows and windshields in our modern automobiles have curved surfaces; so has the plate glass of certain store fronts. To produce curved plate, the finished plate glass is first cut to size, then placed on molds of heat-resistant metal and sent through a furnace. The temperature in the furnace is just high enough to allow the plate glass to sag, but not high enough to mar the polished surface. To prevent sticking, the molds are usually dusted with an infusible powder (one that does not fuse or melt when it is heated). In some cases suction is applied in the mold to help make the glass sag.

*Mirrors.* When plate glass is coated with a metal film on one side, it becomes a mirror. Two chemical solutions are made to react at the glass surface, causing metallic silver to be deposited upon it. In one method two solutions are sprayed from a double-nozzled spray gun so that they mix and react upon striking the surface. The silver surface is usually protected from scratching by a coat of paint.

Precision mirrors used in scientific instruments are often made by evaporating aluminum or other metals from a hot filament; the metal vapor is allowed to condense on a clean glass surface. This was the process used in making the 200-inch mirror for the Palomar Mountain telescope. In this case, the glass is the mirror support, not the mirror itself; the metallized surface is the front face.

*Optical glass.* Certain kinds of glasses, used for lenses and prisms, must have specific light-transmitting properties. They are called optical lenses. They must be free from cord, seed and stone to a greater degree than any other glassware. Certain types of optical glasses must withstand high temperatures and pressures, or tropical humidity or the effects of sea water. Because of all these exacting requirements, optical glass must be manufactured by certain special processes.

Optical glass is manufactured from chemicals of the highest purity, and exceptional care is taken in weighing and mixing. Most optical glass is melted in small continuous tanks, usually equipped with a stir-

## GLASS

From antiquity until well into the twentieth century, glass was formed by hand. Although most glass is made by machine today, hand processing is still used for special items for which there is a limited demand and also for high-quality art and table ware. In hand blowing, molten glass is gathered from the furnace at the end of an iron blowpipe. As the glassblower blows through the pipe, he forms the hot glass into the desired shape and size. Sometimes the blob of glass is blown against the wall of a mold. In the hand casting method, molten glass is poured into a mold and permitted to cool. Glass can be drawn into a rod or tubing by hand drawing. In hand pressing, a gob of glass is dropped into a mold. A plunger is then forced into the latter, causing the glass to flow into the mold shape.

## MODERN GLASS MANUFACTURE

### PLATE GLASS

Molten glass flows out of a tank, is picked up between two rollers and is converted into a moving ribbon. This passes through an annealing oven, called a lehr. Then it is ground by twin disks, which use emery and fine sand as abrasives. The ground glass is polished by felt-surfaced disks. It is washed and dried and finally cut into the desired lengths.

### MACHINE-BLOWN GLASS

Molten glass, passing between a pair of rollers, is formed into a ribbon. As the ribbon travels along a steel track, it sags into holes; air-blowing plungers, coming down from above, fill out the gobs. Almost at the same time, molds come up from below and clamp around the gobs. More puffs of air are forced into each one, filling it out into the mold shape. The mold is then opened and the shaped product is passed to an annealing oven.

### MACHINE-DRAWN GLASS

To form glass tubing, molten glass, flowing directly from the furnace, passes around a hollow ceramic spindle. The glass is then pulled rapidly along a series of rollers. Air blowing through the center of the spindle helps maintain the glass as a continuous tube. The dimensions are controlled by the drawing speed, glass temperature and air pressure. Machine drawing can also be used in making rods.

### MACHINE-PRESSED GLASS

From a large tank, gobs of glass are fed continually into the molds of a revolving press. As a mold containing the hot glass is moved beneath a plunger, the latter forces the glass into its final shape, as shown here. When the mold opens, the piece is transferred to a conveyer and then passes through an oven.

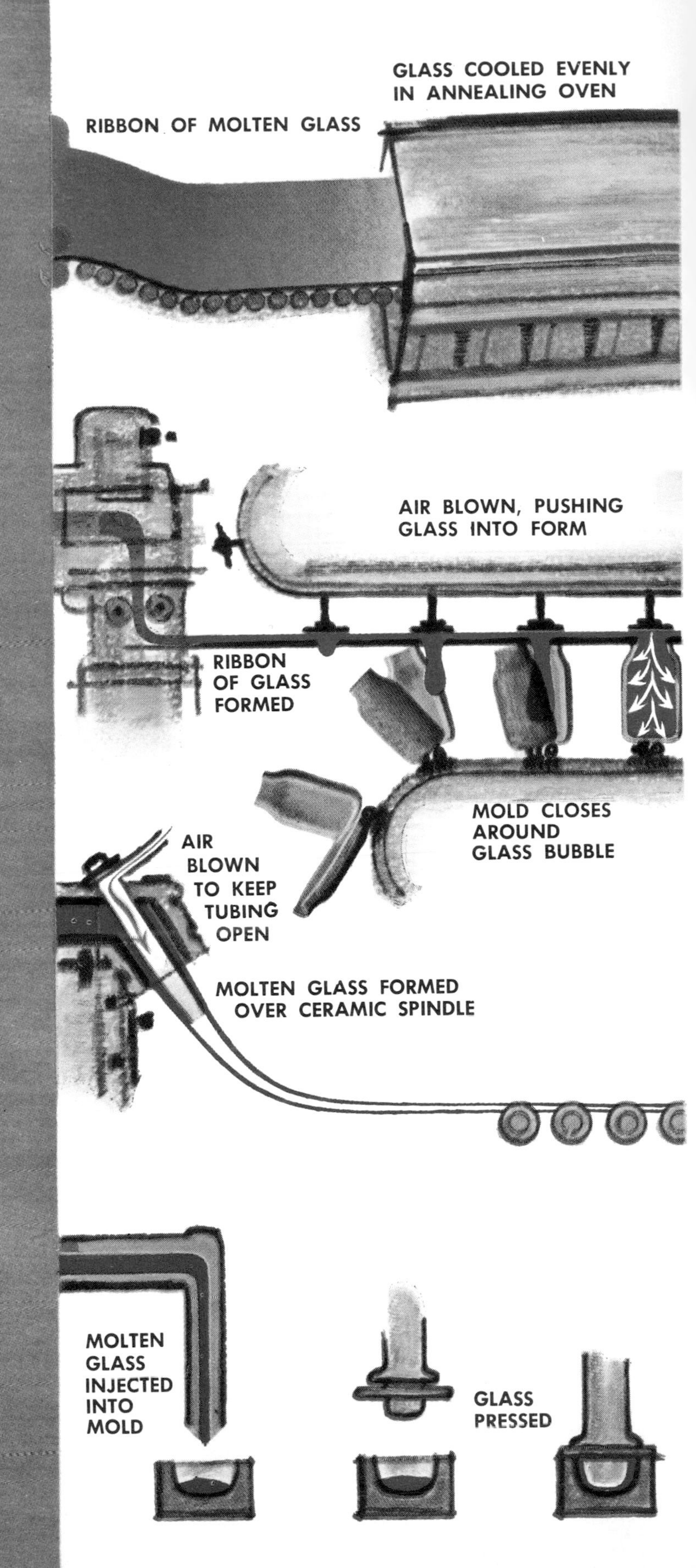

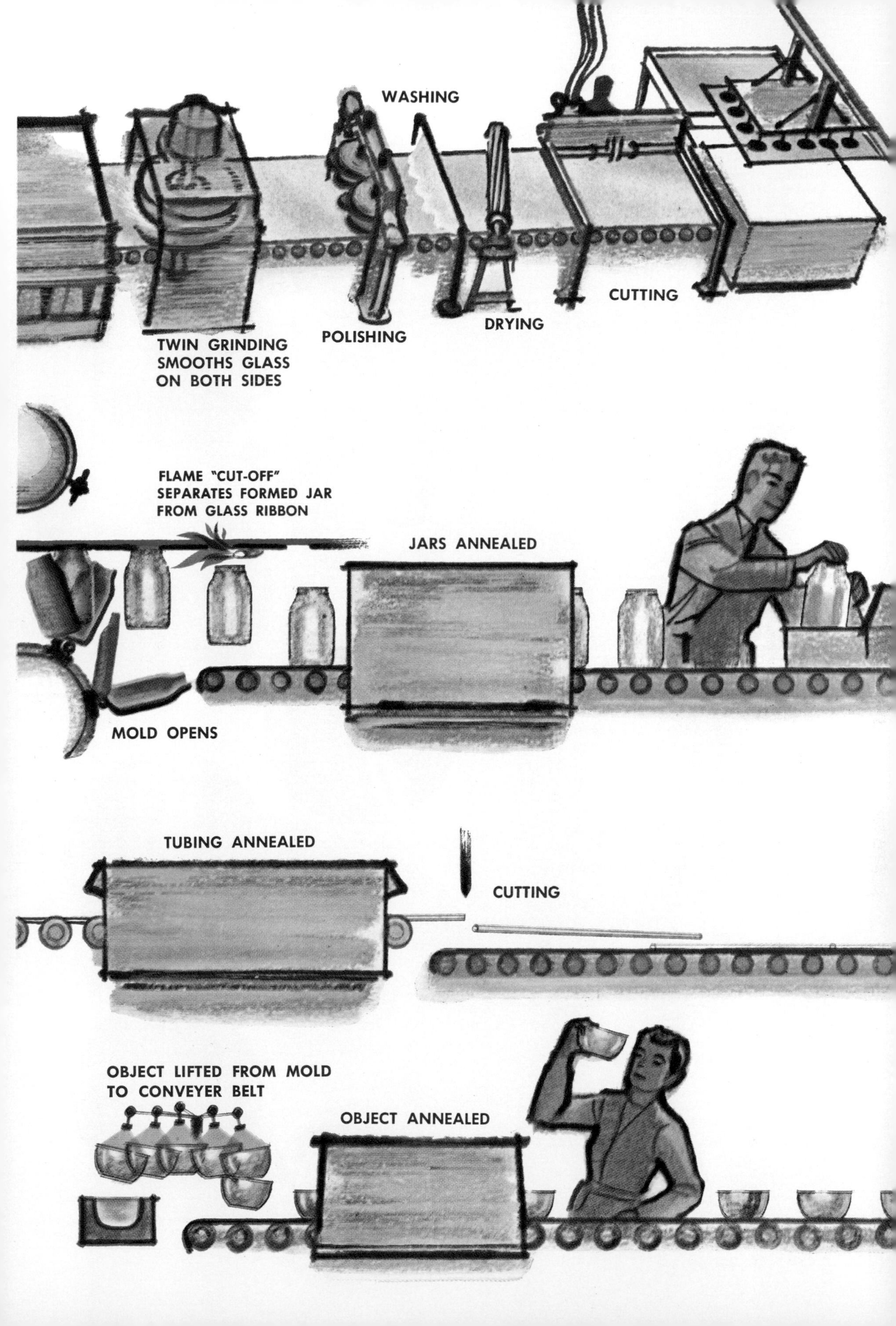
WASHING
TWIN GRINDING
SMOOTHS GLASS
ON BOTH SIDES
POLISHING
DRYING
CUTTING
FLAME "CUT-OFF"
SEPARATES FORMED JAR
FROM GLASS RIBBON
JARS ANNEALED
MOLD OPENS
TUBING ANNEALED
CUTTING
OBJECT LIFTED FROM MOLD
TO CONVEYER BELT
OBJECT ANNEALED

## PROPERTIES OF GLASS

**Glass is ordinary sand, mixed with soda and lime and various other substances and fused to form a molten liquid. By varying the composition of glass and the treatment to which it is subjected, its properties can be controlled quite amazingly. On this page, we indicate some of the properties and their applications.**

## USES OF GLASS

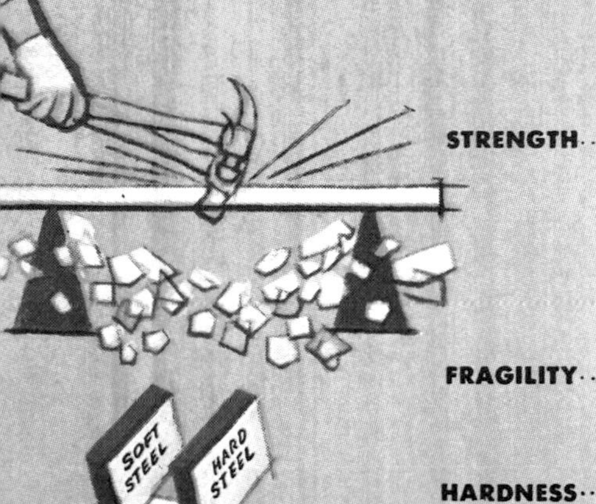


**STRENGTH**

**FRAGILITY**

**HARDNESS**

**PERMANENCE OF SHAPE**

**CORROSION RESISTANCE**

**HEAT RESISTANCE**

**ELECTRICAL PROPERTIES**

**OPTICAL PROPERTIES**

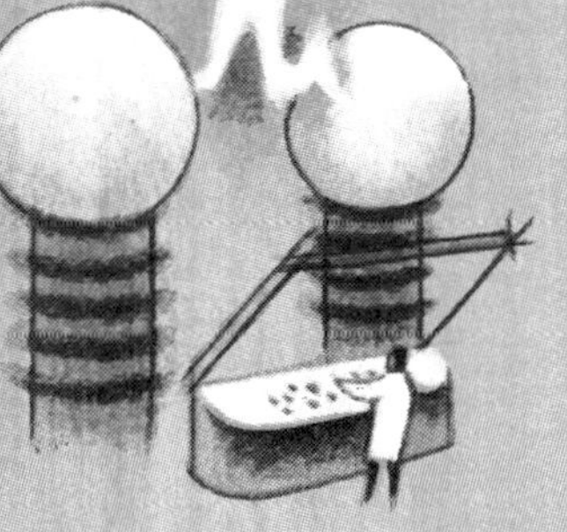

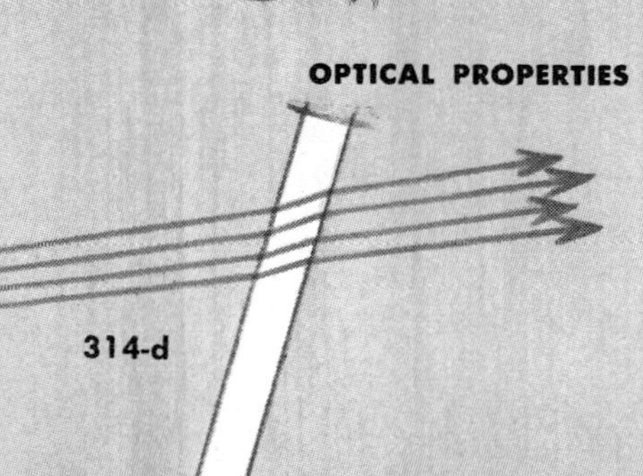

ring chamber and often lined with platinum to avoid contamination. The continuous stream of homogeneous molten glass is cast into bars; later it is reheated and repressed into lens blanks. Spectacle lens blanks are pressed directly from the tank stream on small automatic presses.

Some optical glasses are melted in open pots holding about 500 pounds of glass. After the glass is melted and fined, it is cooled slightly. Then it is stirred to bring about complete homogeneity and to remove cord. After the stirring process, the pot of glass is removed from the furnace. The molten glass is poured from the pots and cast into slabs or is allowed to cool in the pot. As it cools, the pot breaks and the glass cracks into a number of irregularly shaped pieces. These are placed in molds and transformed into slabs in a furnace. In some cases, the glass is remolded to lens or prism form.

Color-filter glasses form a special group of optical glasses. By means of a special glass or combination of glasses, we can filter out any color represented in the light spectrum. (See Index, under Spectrum.) Color-filter glasses are used extensively in photography, in railway, marine and airway lighting and in scientific equipment.

*Heat-resistant glasses.* Glass, like most other materials, expands on heating and contracts on cooling. The change in the dimensions of glass brought about by such variations in temperature is known as thermal expansion. The thermal expansion of glass depends upon its composition. Ordinary glasses, used for making windows and bottles, have relatively high thermal expansion. Such materials are not suitable for ovenware or for chemical-laboratory ware, which must withstand intense heat. For these latter purposes, special glasses, containing boron, are used; they have about one third the thermal expansion of ordinary glass. Fused silica, or "quartz glass," has still lower thermal expansion; it serves in special instruments, such as mirrors for reflecting telescopes used in artificial satellites.

The thermal expansion of lamps and tubes used in radio or television sets must match the thermal expansion of the various metals used for lead wires. If the glass does not match the metal in expansion, the seal will fracture on cooling. A series of glasses with special compositions has been developed for these purposes.

*Glass ceramic.* For years, the glass chemist has developed his compositions so as to avoid the natural tendencies of silica melts to crystallize, because the resulting crystals produce defects in ordinary glassware. In a new family of materials now under manufacture, the glass is made to crystallize. After it is melted, shaped and annealed like ordinary glass, it is heated in a kiln for several hours at the proper temperatures to cause crystallization to take place.* The resulting product is called glass ceramic.

Glass-ceramic materials can be made to provide a wide range of properties. In general, they are stronger and tougher than ordinary glass and can be made with zero thermal expansion. Among other things, they are used for cookingware, tableware and the nose cones of ballistic missiles.

*Fiber glass.* When fiber glass was first introduced, it opened new horizons for glass products. The art of drawing glass into threads is very old; but the modern fiber-glass industry is based on comparatively recent developments. Fiber glass is fireproof, strong and light in weight; it resists fungi and chemicals; it does not conduct electricity; it insulates against heat.

There are two kinds of glass fibers: short length, or staple, fibers and continuous fibers. Staple fibers are made by blasting tiny streams of molten glass from numerous jets. The ejected fibers are caught on a continuous belt, where they intertwine to form a mat or blanket. Continuous fibers are produced by drawing molten glass through tiny orifices at such speeds that single fibers as small as $1/5{,}000$ of an inch are formed. A number of these single fibers are twisted together to form a thread.

For the insulation of new homes, mats made of staple fibers are used. The mats are cut into suitable strips and attached to the paper or metal lath that is to be placed

* The kilns used for this process are similar to those which serve in the pottery industry.

For many centuries, glass was the most fragile substance known to man. Today, some kinds of glass are so strong that they can be used to hammer nails into wood.

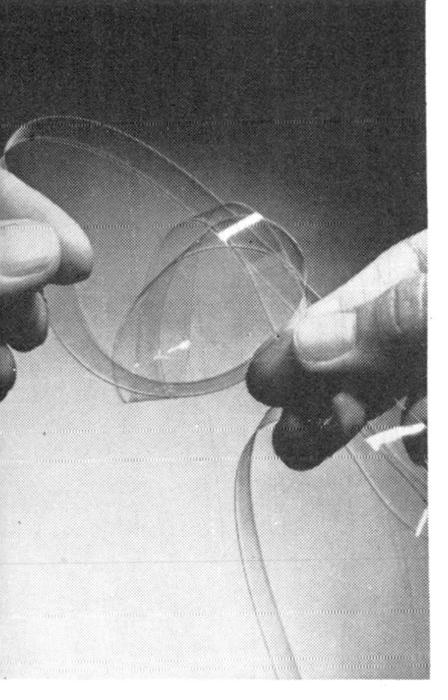

The ribbon glass at the left is only one millimeter thick. It can be bent, twisted and even tied into knots. Ribbon glass is used to replace mica sheets in certain capacitors.

Photos, Corning Glass Works

Glass can be processed in such a way as to have exceptional resistance to both heat and cold. Here, molten bronze is being poured into a glass dish resting on ice.

in the wall or ceiling. A different insulation method is used for homes that have already been constructed. The fiber-glass mat is shredded and chopped to form small balls or granules; these are then poured or blown into the areas to be insulated. Insulating board is made by wetting the fiber-glass mat with an adhesive and pressing it into boards. Acoustical (soundproofing) tile is prepared in the same way.

Fiber-glass threads for textiles can be made from staple fibers by carding, combing and twisting the fibers into thread. The strongest fiber-glass threads, however, are made of continuous fibers. Fabrics are woven from glass threads on conventional looms. They are used for curtains, drapes, bedspreads, lampshades, men's ties and covers for ironing boards.

The combination of fiber glass and plastic produces one of the strongest of lightweight materials. It is used in making boats, automobile bodies, fishing poles and even armor plate. A fiber-glass-plastic heat shield was used on the *Mercury* capsule carrying astronaut John H. Glenn on the first American manned orbital flight around the earth. The shield served to absorb the terrific heat generated on reentry into the earth's atmosphere.

*Foam glass.* Foam glass is another product that is unlike normal glass in appearance and properties. In the manufacture of foam glass, instead of processing the molten glass in order to remove all bubbles, special materials are added to the batch so that it will bubble excessively. This glass froth is formed in slabs and cooled. Foam glass, then, consists of a multitude of thin glasswalled bubbles. It is exceedingly light in weight, strong, impervious to moisture, fireproof and vermin-resistant. It is sawed into blocks for insulating and other uses.

This, then, is the story of glass. Derived from the common and abundant materials of the earth, it will always be plentiful. Made in large part of the rock fragments that we call sand, it is durable; being pliable while it is hot, it can be molded at will. It serves us in thousands of different ways, and new uses are continually being found.

*See also Vol. 10, p. 283*: "Glass."

# NONMETALLIC ELEMENTS

## From Cleansing Oxygen to Versatile Carbon

IN our survey of the chemical elements, we have reached the group generally known as the nonmetallic elements. They are decidedly in the minority, yet they include elements of the very greatest importance. Among these are the gases oxygen, nitrogen, hydrogen and chlorine and the solids silicon, sulfur, phosphorus and carbon.

Oxygen, the most widely distributed of all the elements, is a colorless, odorless and tasteless gas. It forms something like 21 per cent by volume of the atmosphere, eight ninths of the water of the world, about half the material in the earth's crust and about 20 per cent of animal tissue and 40 per cent of vegetable tissue. Oxygen does not burn in the ordinary sense, but it is the greatest supporter of combustion. It is necessary for human life, since respiration is a form of combustion that could not take place without the oxygen that is drawn into the lungs from the outer air with every breath that we draw. Oxygen enters into combination with nearly all the metals.

The different compounds that an element forms with oxygen (that is, with oxygen alone) are called its oxides. Thus, both carbon monoxide ($CO$) and carbon dioxide ($CO_2$) are oxides of the element carbon; both ferrous oxide ($FeO$) and ferric oxide ($Fe_2O_3$) are oxides of the element iron; both cuprous oxide ($Cu_2O$) and cupric oxide ($CuO$) are oxides of copper.

When an element combines with oxygen, it is said to oxidize, or to undergo oxidation. (Oxidation is also used nowadays in the sense of removal of electrons from an atom or group of atoms.) Heat is produced in oxidation; in many cases, as in ordinary combustion, light is also produced. When hydrogen oxidizes, a large amount of heat is evolved. This combustion heat of hydrogen is utilized in the oxyhydrogen blowpipe, which serves to weld iron and steel and to cut iron and steel plates. (The oxyacetylene blowpipe, using the compound acetylene, serves the same purpose.)

Another striking example of the intense heat developed by rapid oxidation is found in the Goldschmidt process. In this process, when a mixture of powdered aluminum and iron oxide is ignited with the aid of a bit of burning magnesium wire, the aluminum steals the oxygen from the iron compound. The aluminum reduces this iron to the metallic state and evolves heat enough, at about 3,000° C., to make the iron run like water. On the other hand, there is no *noticeable* heat in the rusting process, which represents the slow oxidation of iron. Oxygen can be liquefied at the temperature of —118° C. under a pressure of 50 atmospheres; the liquid boils at —183° C. and freezes at —218° C.

Nitrogen occurs free in the atmosphere, of which it forms 78 per cent by volume; like the element oxygen, it enters into the formation of all living tissues. Nitrogen can be liquefied at the temperature of —147.1° C., under a pressure of 33.5 atmospheres; its boiling point is —195.8° C. and it freezes at —209.86° C.

Like oxygen, nitrogen is a gas without color or odor, and it is tasteless. Unlike oxygen, however, it does not enter readily in combination with other elements. For example, nothing short of a lightning flash, the continuous action of a high-power electric arc or the mysterious processes of life will link together nitrogen and the oxygen contained in the atmosphere.

When nitrogen does form compounds, they are usually highly unstable; many of

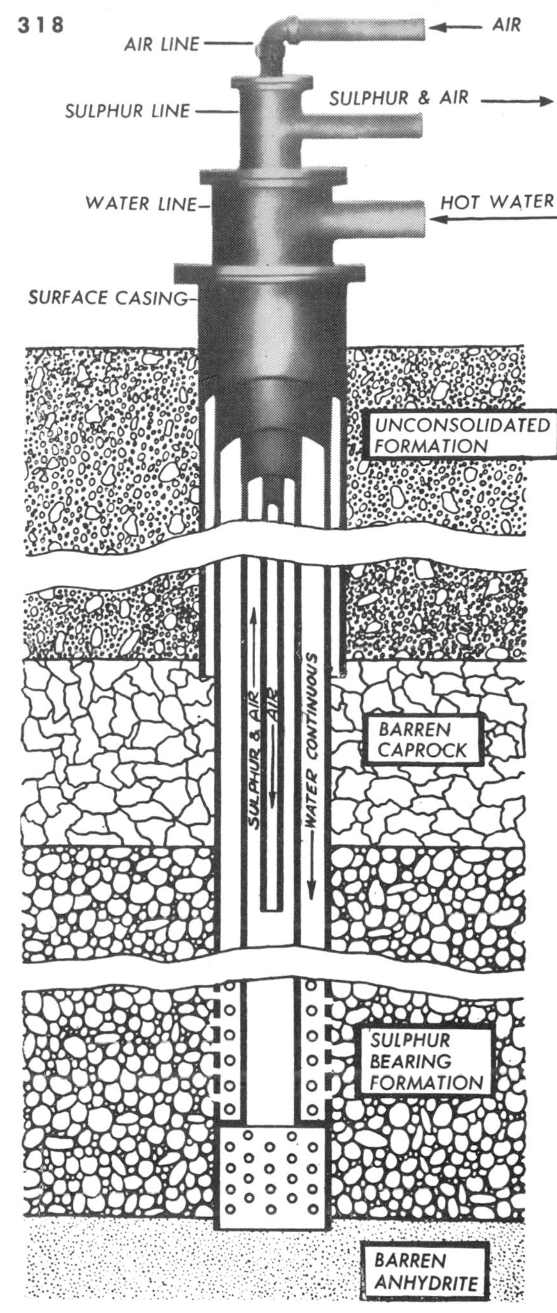


A sulfur well. Hot water flowing down the outer pipe melts the sulfur; the melted sulfur is forced up through another pipe by means of compressed air.

these react so easily and rapidly that an explosion results. This is true of such nitrogen compounds as nitroglycerin ($C_3H_5(NO_3)_3$) and potassium nitrate, or saltpeter ($KNO_3$), used in explosives.

Nitrogen has many uses in the field of industry. It is particularly important in the production of ammonia ($NH_3$), a compound of nitrogen and hydrogen, and nitric acid ($HNO_3$), a compound of nitrogen, oxygen and hydrogen. Ammonia is a raw material in many industries; dissolved in water, it is a common household cleaning agent. Nitric acid is required for the manufacture of organic dyes, drugs and explosives, as well as inorganic nitrate fertilizers. Since nitrogen is relatively inert, nitrogen gas is ideal for filling large electric bulbs; it is also used to harden the surface of steel. The compound of nitrogen and oxygen known as laughing gas, or nitrous oxide ($N_2O$), is an anesthetic.

As we pointed out, nitrogen is a vital element of protoplasm, forming 16 per cent of all living tissue. It is present in the amino acids that make up the long, complicated chains of protein in protoplasm. Plants, upon which all animals depend for food, cannot use pure nitrogen directly in the building of these proteins; they require the salts of nitric acid, or nitrates, for this purpose. However, the boundless resources of nitrogen in the atmosphere are not lost to the cause of life. Certain bacteria, living in the roots of plants such as clover, alfalfa and the other legumes, can combine the free nitrogen of the air with oxygen to form nitrates. Some soil bacteria decompose dead plant and animal tissue and liberate organic nitrogen in the form of ammonia. Other bacteria oxidize this ammonia and thus produce more nitrates for the use of plants.

### The lightest of the elements — hydrogen

Hydrogen is another element that is found in all living tissue. In the free state, it is a colorless, odorless, tasteless gas that liquefies at —252.8° C. and solidifies at —259.1° C. Free hydrogen forms only .001 per cent of our atmosphere, but modern astronomy has revealed that much greater percentages are found in the stars, including our sun. The fusion of hydrogen atoms to form helium is the source of the sun's tremendous energy; it is this process that also accounts for the awesome destructiveness of the hydrogen bomb.

Hydrogen is one of the more abundant elements in our planet, since it makes up 11.1 per cent of the water contained in oceans, lakes, rivers, the soil and the tissues of living things. The process of electrolysis is used to free hydrogen from water. This involves passing an electric current through water containing some acid, base or salt. Hydrogen is a fairly active element and has a strong tendency to unite with oxygen, liberating a great amount of heat. At temperatures above 550° C., hydrogen and oxygen unite so rapidly and completely that an explosion results.

Hydrogen is the lightest of the elements, weighing 14.4 times less than air. A balloon filled with hydrogen will therefore rise in the air. Practically all balloons, until comparatively recent times, were inflated with hydrogen. (The very first balloon to achieve flight was filled with hot air.) Since hydrogen may burn or explode, modern lighter-than-air craft are generally filled with the inert gas helium, which is the second lightest element.

The most common isotope of hydrogen has only one proton and one electron in its atoms (see the article Inside the Atom, in Volume 1). There are two other isotopes of hydrogen, which, though they are very rare, are so important that they have separate names. These are deuterium and tritium; they have one and two neutrons in the nucleus, respectively, in addition to the proton of ordinary hydrogen. Both heavy isotopes play significant roles in atomic-energy research.

More hydrogen is used for the hydrogenation of organic oils than for any other purpose. In this process, hydrogen is chemically added to certain animal and vegetable oils, transforming them into more useful and edible solid fats, such as Crisco. Hydrogenation is also used to improve the quality of petroleum.

In the case of certain compounds, called acids, a hydrogen atom separates from the rest of the compound as a hydrogen ion ($H^+$) when the compound dissolves in water. The hydrogen ion then combines with a molecule of water to form a hydronium ion ($H_3O^+$). (See the article Acid and Bases, in Volume 4.)

**Chlorine, one of the most active nonmetals**

Chlorine is normally a gas, with a yellowish green color and an irritating, pungent odor. The gas can be liquefied at −34.6° C. and solidified at −103° C. Unlike its life-giving neighbor, oxygen, chlorine is a poison if inhaled; it burns and inflames the lungs and upper respiratory tract. It was the first poison gas to be employed in modern warfare, coming into use during World War I.

Luckily for us, chlorine never occurs free in nature. It is the second most active of the nonmetals, being surpassed only by its relative fluorine; it has a very strong tendency to gain electrons from other elements, thereby forming compounds known as chlorides. With active metals such as sodium and calcium, this reaction is explosive; even such inactive metals as iron and antimony react violently with chlorine, becoming vaporized and incandescent. Chlorine can also oxidize some nonmetals; exposure to a little direct sunlight is enough to explode a mixture of hydrogen and chlorine.

Though free chlorine is a rarity in nature, its compounds are plentiful. Sea water provides an inexhaustible source of chlorides, particularly those of sodium (common salt), potassium and magnesium, from which chlorine can be liberated by electrolysis. In many places, huge deposits of pure chlorides have been left by the evaporation of ancient lakes and seas.

USDA

Javelle water, a compound of chlorine, serves as an antiseptic and disinfectant and as a bleaching agent.

The oxidizing power of chlorine has many applications. Since oxidation destroys the color of most dyes, chlorine is widely used as a bleach, in the form of chlorine water. Oxidation also kills bacteria and other microbes; small, nonpoisonous amounts of chlorine are often added to drinking water and swimming pools to serve as an antiseptic. A mixture of hydrochloric acid (HCl) and nitric acid will oxidize and dissolve the noble metals — such as platinum and gold — and is therefore known as *aqua regia,* or royal water. Chlorine is also an important element in many drugs, such as chloral and the anesthetic chloroform. Combined with carbon it forms carbon tetrachloride, an efficient fire extinguisher and solvent.

Chlorine is a member of the family of elements called halogens, in Group VII of the periodic table, which is given in the article The Periodic Table, in Volume 4. This family includes fluorine, bromine and iodine as well as chlorine. Generally speaking, all the members have properties similar to chlorine, but these properties become less pronounced in the heavier members. Fluorine, the first and lightest member, is the most active of the nonmetals and the strongest oxidizing element known; like chlorine, it is ordinarily a gas. One of its compounds, hydrofluoric acid (HF), even attacks and dissolves glass, and is therefore used in etching. The acid must be stored in vessels made of paraffin wax, lead or platinum.

#### Phosphorus — an element with a split personality

At ordinary temperatures, phosphorus is a solid, but it may take two very different forms. One of these is white phosphorus, a waxy substance that melts at 44° C. and boils at 280° C., evaporating steadily at room temperature. This form is superreactive and combines explosively with the oxygen of the air at moderate temperatures. For this reason, and because of its poisonous vapor, white phosphorus must be stored under water. At lower temperatures, it is oxidized slowly, giving off a white phosphorescent light.

If white phosphorus is allowed to stand, it gradually becomes yellowish or reddish. This is due to the gradual transformation of white phosphorus into what is known as red phosphorus. Though this form does not melt below 600° C. and is not volatile or reactive, it is the same pure element as white phosphorus. When a single element has two or more different forms in any one of its physical states — solid, liquid or gas — the different forms are known as allotropes. The differences between allotropes are caused by varying arrangements of the atoms in a molecule or crystal of the element. Many elements other than phosphorus have allotropic

Dmitri Kessel — Fortune Magazine

Conveying matchsticks on a belt as the heads are being tipped with a compound containing red phosphorus.

forms; one of the most interesting is sulfur, which we shall discuss next.

White phosphorus was once used to make matches, which were stored under water, and burst into flame spontaneously when they were taken out. Modern safety matches are much less dangerous since they use the nonpoisonous form, red phosphorus, which will not ignite spontaneously.

A great deal of phosphorus, in the form of phosphate salts, is used as an ingredient in fertilizers. All plants require these salts for proper growth and development, as do animals. Calcium phosphate is vital for the formation of teeth and bones. Other phosphorus compounds play important parts in the functioning of the brain, muscles and nerves.

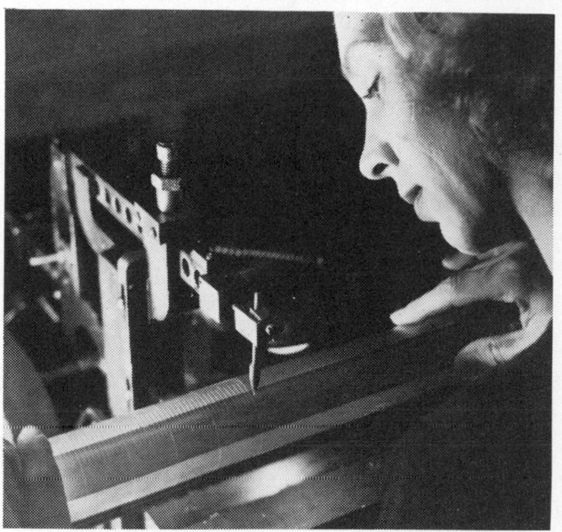

Corning Glass Works

Cutting through the wax applied to a glass before inserting the glass in a hydrofluoric acid bath for etching.

#### Sulfur — an actor with many faces

In the earth's crust the element sulfur is found both free and in compounds called sulfides and sulfates, but it is not one of the most abundant elements. Over 80 per cent of the world's sulfur is produced in the United States; most of this comes from the free deposits in Texas and Louisiana, where it is taken out by the method known as the Frasch process. Superheated steam is piped down to melt the sulfur, which is then forced up through another pipe by compressed air. When the frothy mass cools it yields a yellow, crystalline solid, 99.5 per cent pure sulfur.

The allotropic forms of sulfur are much more varied than those of phosphorus. Below 96° C. sulfur is arranged in diamond-shaped crystals — this is the rhombic form. Above 96° C. it appears in long, needlelike crystals, known as the monoclinic form. This melts at 119° C. to give a straw-colored liquid known as $\lambda$-sulfur ($\lambda$ is the Greek letter *lambda*). It contains molecules of eight atoms linked together in a ring. As the temperature of the liquid reaches 280° C., the liquid thickens and turns dark red as it changes into $\mu$-sulfur ($\mu$ is the Greek letter *mu*). The molecules now consist of long, intertwined chains of sulfur atoms. If the $\mu$-sulfur is rapidly cooled by immersion in cold water, it forms a rubbery, shapeless substance known as amorphous sulfur. This is a supercooled liquid; it changes into the rhombic form upon standing.

Sulfur is fairly active and forms many compounds with other elements. It burns in air to form the pungent, colorless gas sulfur dioxide ($SO_2$), which serves as an antiseptic and fungus-killer, as a preservative for dried fruits and in the manufacture of wood pulp for paper. What is even more important, sulfur dioxide can be combined with oxygen and water to form sulfuric acid $H_2SO_4$. This is perhaps the most versatile and useful acid known to man. When highly concentrated, it is a powerful dehydrating (water-absorbing) agent, and at high temperatures is a strong oxidizing agent; in dilute solutions its acid action is unsurpassed. For all these reasons, sulfuric acid has found thousands of applications in the petroleum, chemical, rubber and metal industries. Sulfuric acid is the electrolyte in wet-cell storage batteries.

Combined with carbon, sulfur forms a valuable organic solvent known as carbon disulfide ($CS_2$). Sulfur is also contained in living tissue. The odor emitted by rotten eggs is due to the poisonous gas hydrogen sulfide ($H_2S$), which is formed by certain bacteria in putrefying animal matter.

### Silicon and carbon — the basic elements of our world

We come now to the two fundamental elements of our world — silicon and carbon. With the exception of oxygen, silicon is the most abundant element in the earth's crust, accounting for one quarter of its weight. Silicon, in the form of its dioxide, silica, is the chief constituent of almost all minerals and rocks. Silicon is the central element around which others group themselves in inorganic nature. Carbon holds the same position in the realm of life.

Carbon and silicon are the first two members of Group IV of the periodic table, and therefore have similar atomic structures. There are differences between them, however, and these explain why one is associated with life and the other with inorganic matter. The chemical union between carbon and other atoms, particularly oxygen, is easily broken and reformed; life processes depend on this unceasing chemical change for their very existence. The chemical bond between silicon and oxygen is quite different; it is one of the strongest in nature. Minerals owe what permanence they have to this bond.

### The role of silicon in nature and industry

Silicon hardly ever plays a part in the metabolism of living creatures, but there are exceptions to this rule. The microscopic plants called diatoms and the protozoans known as radiolarians both secrete intricate shells made of silica.

Because of its affinity for oxygen, silicon is never found free in nature. Pure silicon is a gray, crystalline solid that can be prepared only with great difficulty; ferrosilicon, an alloy with iron, is more easily prepared and is widely used as a deoxidizer in metallurgical work.

The natural varieties in which silica appears form quite an array. Transparent quartz and amethyst are made of large crystals of silica; opal, onyx, agate and flint are also made of silica crystals, sometimes combined with water. Amorphous or fragmentary forms of silica include sand, gravel and sandstone. The presence of metallic impurities gives beautiful colors to some of these minerals.

If quartz is melted and then supercooled it forms a transparent substance much like glass; it can be molded and blown in the same way. It has many advantages over glass. It is much purer chemically and can therefore be used to make apparatus for superfine chemical analysis. Since it has a much higher melting point than glass and hardly expands at all when it is heated, a red-hot quartz vessel may be plunged into ice water with no danger of cracking. Unlike glass, quartz does not absorb ultraviolet light and can thus be used in making ultraviolet lamps, telescopes and microscopes.

Glass itself is actually a mixture of various silicates — compounds of a metal with silicon and oxygen. (See Index, under Glass.) Most rocks are also mixtures of silicate minerals, such as asbestos, mica and kaolin, as is Portland cement.

In recent years, organic chemists have created a wonderfully useful group of substances (oils, greases and plastics) known as silicones. These are made up of long chains, rings and networks of silicon and oxygen atoms attached to organic molecules. The silica skeleton of these silicone molecules makes them more resistant to fire, heat and chemical attack than pure organic plastics. Some of the many uses of silicones are shown in the illustrations on the opposite page. Silicon carbide (SiC) is a well-known compound of silicon; it is marketed under various trade names, including Carborundum. It rivals the diamond in hardness and is used to grind metals and stones.

### Carbon — the supremely important element of life

Carbon is truly the basis of life, for every molecule of living tissue is built around a skeleton of carbon atoms. In some important organic molecules, atoms of oxygen, nitrogen, phosphorus or sulfur are also linked in this chain. Well over half a million carbon compounds have already been identified by chemists, and new

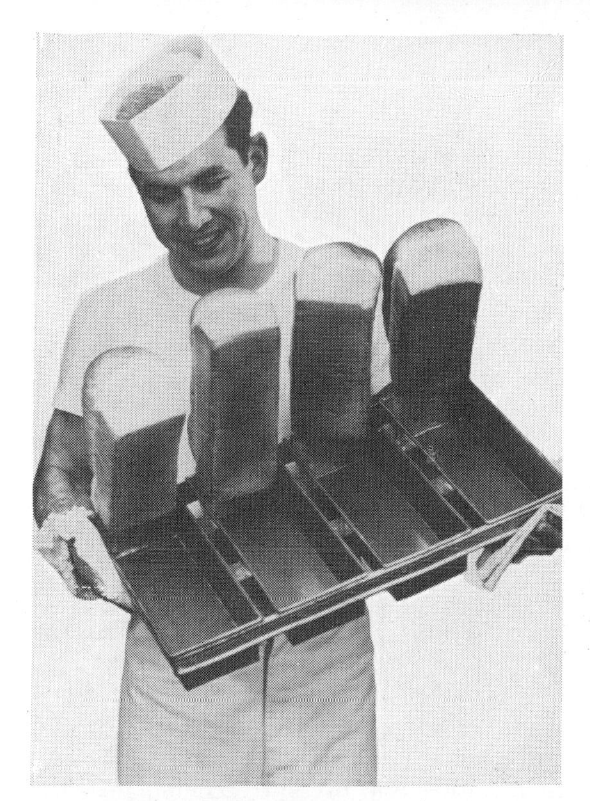

Silicones are synthetic compounds, derived from atoms of silicon, oxygen, carbon and hydrogen; they serve many purposes. Above: bread will not stick to a pan that has been coated with a silicone.

This silicone-insulated test motor has been alternately operated at temperatures up to 310°C. (590°F.) and exposed to 100 per cent humidity for several thousand hours; yet it is still in perfect condition. Without this protection the motor would have broken down long before this.

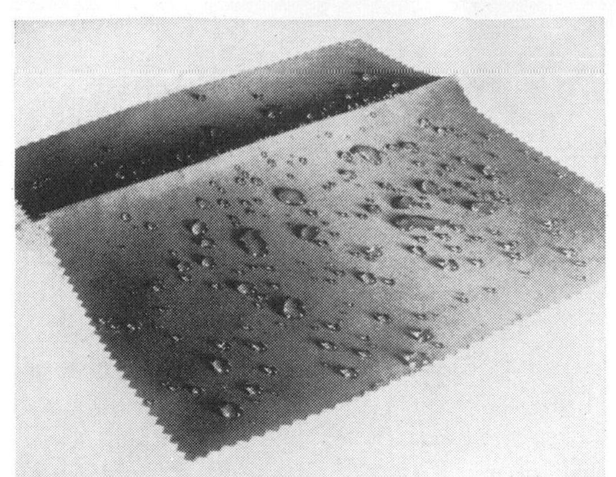

Nylon treated with a silicone continues to shed water even after it is laundered or dry cleaned.

The eyeglass wipers shown above have been treated with some silicone oil. They wipe glasses clean.

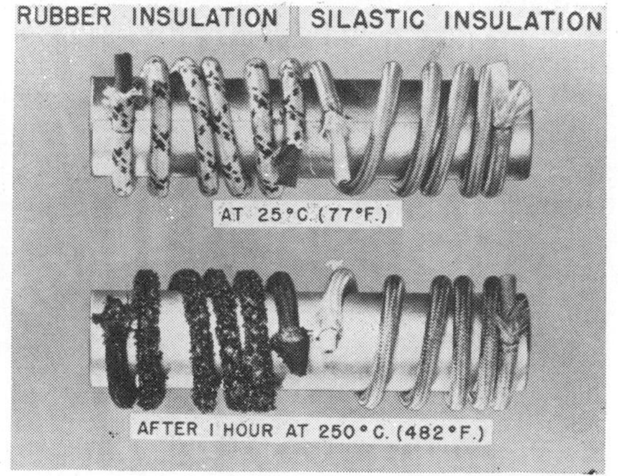


All photos, Dow Corning Corporation

Two wires, one insulated with rubber and the other with a silicone, are shown above as they appeared at room temperature. They were exposed for one hour to 250°C. (482°F.) heat. The lower picture shows what happened. The rubber insulation has melted but the silicone insulation is still intact.

discoveries are being made continually. A special branch of science — organic chemistry — has been created to deal exclusively with carbon compounds.

Among other things, organic chemists have learned that more than a thousand carbon atoms may be linked in a single organic molecule; also, that two separate organic compounds may have identical chemical compositions but may differ in the way their atoms are joined together and arranged in space. The achievements of organic chemistry are too numerous to be taken up here; you can find out more about this fascinating science in the article The Chemist as Creator, in Volume 8.

Carbon is also widely distributed in the mineral world; free carbon is found in the form of the hard gem called diamond and the soft mineral graphite, or plumbago. Graphite and diamond are allotropic forms of pure carbon. In graphite crystals the carbon atoms are arranged in loosely bound layers that can slide over each other and give the mineral a soft, greasy feeling. This property makes graphite an excellent lubricant for heavy machines. Graphite is also used for crucibles, electrodes and as an absorber of radiation in atomic-energy research. The "lead" used in pencils is really a mixture of graphite and clay. Graphite is produced artificially by heating coal or coke in an electric furnace.

In diamond crystals, the carbon atoms are packed and held much more tightly than in graphite; diamond is the hardest substance known to man and is used industrially to cut metals, rocks and other gems. The beautiful, large and clear stones that are used for adornment are found only in natural rock formations, but small, dark diamonds for industrial use can be made by heating graphite crystals under enormous pressure. Natural diamonds were probably formed in a similar way, crystallizing out of cooling masses of lava, deep in the earth.

The name "black diamond" is often given to the rock formation called coal, whose carbon content ranges from 67 per cent to well over 90 per cent. As a matter of fact coal is much more useful to man than the sparkling king of gems. For thousands of years man has used coal as a source of heat; in recent years we have learned to extract a marvelous array of valuable chemicals from coal tar, a substance obtained by distilling bituminous coal. Besides carbon, coal contains hydrogen, oxygen, nitrogen and various minerals such as silica and alumina.

Scientists believe that coal deposits were formed by the slow decomposition of dead organic matter. Most deposits began to form during or after the Carboniferous period, about 300 million years ago. The crust of the earth was then rapidly changing and lush, humid jungles covered the land. Over millions of years, thick sediments of dead plant remains were formed on the swamp bottoms, and were covered with heavy layers of mud and rock. As temperature and pressure accumulated, the organic compounds decomposed and left a residue of carbon. Under the growing weight of various sediments, this carbon was compressed into rocklike formations, now known as coal.

Carbon black and lampblack, which are made by the partial burning of natural gas and petroleum respectively, are both composed of tiny crystals, called microcrystals, of graphite. Carbon black is used in making rubber for automobile tires; large quantities of lampblack go into printer's ink. Charcoal also consists of graphite microcrystals; it is made by the incomplete combustion of wood. It is widely used in the chemical industries as a purifier and catalyst.

Calcium carbonate ($CaCO_3$) is the most widely distributed carbon mineral; it is found almost pure in chalk, pearls, marble and limestone. Limestone is an important raw material for smelting iron and steel, and large quantities are also used in the chemical industries. Carbon dioxide ($CO_2$) is given off when limestone is heated and is a natural product of plant and animal respiration; it is present in our atmosphere and dissolved in water. The bubbles in soda pop are due to this gas. Solid carbon dioxide is a freezing agent.

*See also Vol. 10, p. 279:* "Elements."

MAN

# THE LIFE STREAM

## What Blood Is and What It Does

BY LOUIS FAUGERES BISHOP

THE blood stream that circulates through the body during the lifetime of the individual comes in contact with all the body cells. It serves as a highway over which materials travel to their destinations. It transports oxygen from the lungs to the cells, bears food elements from the alimentary canal to different parts of the body and carries off waste materials from the cells. Defense cells move in the blood to wherever they are needed to fight infection; various hormones are also transported in the blood stream. Besides serving as a highway, the blood helps maintain a more or less constant temperature in the body.

Blood is a red, sticky fluid with a salty taste; it varies in color from a bright scarlet to a bluish red. The adult man's body contains from five to six quarts of this precious fluid. Under the microscope we can see that it is composed of a watery fluid called *plasma,* in which certain *formed elements* are suspended. The formed elements are different types of cells — red blood cells, white blood cells and platelets.

**The plasma — the fluid part of the blood**

The plasma is the watery part of the blood, making up from 50 to 60 per cent of the total. It is a clear yellow fluid, serving as a vehicle for the transportation of red blood cells, white blood cells, platelets and various substances necessary for the vital functioning of the body cells, for clotting and for the defense of the body against disease. After clotting occurs, a straw-colored fluid called serum is left; this retains its liquid form indefinitely.

About 90 per cent of plasma is water, in which a great variety of substances are held in suspension or in solution. These include proteins, such as fibrinogen, albumin and the globulins, and also sugar, fat and inorganic salts derived from food or from the storage depots of the body. Plasma contains urea, uric acid, creatine and other products of the breakdown of proteins. There are enzymes, such as adrenal hormones, thyroxine and insulin, derived from the glands of internal secretion. There are also various gases: oxygen and nitrogen, diffused into the blood from the lungs; and carbon dioxide, diffused into the blood from the tissues.

The proteins in the blood serve several vital functions. For one thing, the cells of the body are made up mostly of proteins. Proteins attract water; if they were present only in the cells and not in the plasma, the cells would draw the liquid into themselves until they would be swollen. The proteins in the plasma serve to balance the pulling effect exerted by the proteins in the cells. Serum albumin is the plasma protein most concerned with this function.

Among the other proteins contained in the plasma are various globulins. Gamma globulins are antibodies that serve to immunize the body against disease. They have been used to give passive immunity to such diseases as measles, infective jaundice and infantile paralysis. The globulin prothrombin, as we shall see, plays a vital part in the blood-clotting mechanism. Fibrinogen is another protein contained in the plasma; it too is involved in clotting.

The various proteins in the plasma can be separated — a development in which the researches of Edwin Cohn and his associates at the Harvard Medical School played an important part. The separation of these proteins is called fractionation; each protein obtained in this way is called a blood-plasma fraction.

The salts in the plasma are useful in various ways. Sodium chloride (common salt) helps proteins dissolve — a vital function, since proteins must be in solution before they can serve their purpose. Other salts in the plasma are known as buffers; they serve to maintain the same degree of alkalinity in the blood regardless of whether acids or bases are added.

### The red blood cells

The red blood cells are the most numerous of the formed elements. Their function is to carry oxygen to the cells throughout the body and to return carbon dioxide to the lungs. There are normally about 5,000,000 in each cubic millimeter of blood in men and about 4,500,000 in women during the childbearing years. Each cell is about 1/3500 of an inch in diameter. Normally the cells are in the form of disks, both sides of which are concave.

Red blood cells are developed in the red marrow, found in the ends of the long bones and throughout the interior of flat bones, such as the vertebrae and ribs. The cells have definite nuclei in the early stages of their formation; in man and the other mammals the nuclei are lost by the time the cells have become mature and before they are released into the blood stream.

Each mature red blood cell has a structural framework called the stroma, which is made up chiefly of proteins and fatty materials. It forms a mesh extending into the interior of the cell; it gives the cell its shape and flexibility. The most important chemical substance in the cell is hemoglobin, which causes blood to have a red color. Hemoglobin is composed of an iron-containing pigment called heme and a protein called globin; there are about four parts of heme to ninety-six parts of globin. In man the normal amount of hemoglobin is 14 to 15.6 grams per 100 cubic centimeters of blood; in woman it is 11 to 14 grams.

Hemoglobin combines with oxygen in the lungs after air has been inhaled; the resulting compound is called oxyhemoglobin. When the red blood cells later make their way to other parts of the body deficient in oxygen, the oxygen in the compound breaks its bonds and makes its way by diffusion to the tissues of the oxygen-poor areas. Thus the red blood cells draw oxygen from the lungs, transport it in the blood stream and release it to the tissues as needed. In high altitudes the rarefied air contains less oxygen so the body needs more red cells to carry the needed amount of oxygen to its cells. Therefore, within limits, the higher the altitude, the more hemoglobin there is.

The average life span of a red blood cell is 110 to 120 days. Worn-out cells are destroyed by the spleen; the remnants are transported to the liver, which stores the iron they contain. Eventually, the iron passes to the marrow of the bones, where new red cells will be formed. In the healthy individual, the rate of formation of red cells in the bone marrow keeps pace with the rate of destruction.

It is sometimes very important in making a diagnosis to determine the number of red blood cells and also the hemoglobin content of the blood. To count cells, the blood is diluted by a carefully measured amount of 0.9 per cent sodium-chloride solution, and then the red cells are counted one by one in a tiny glass compartment within a counting chamber. The dimensions of the compartment are 1/20 millimeter by 1/20 millimeter by 1/10 millimeter; its capacity, therefore, is 1/4000 cubic millimeter. After one finds out the number of red cells in this amount of diluted blood, it is a relatively simple matter to calculate the number of cells in a cubic millimeter of undiluted blood. This test can be misleading unless the diagnostician considers that while the count may be normal, the individual cells may be lacking in an adequate amount of hemoglobin.

Various methods are used to determine the hemoglobin content of the blood. Most of these depend on the color of the hemoglobin. Sometimes a drop of blood is drawn and compared with a chart containing various shades of red or pink. Each of these shades represents a certain color of blood and this color corresponds to a known quantity of hemoglobin.

A more accurate method requires the use of a spectrophotometer. A sample of the blood is mixed with ammonium hydroxide and then shaken, the hemoglobin thus being converted into oxyhemoglobin. This pigment is then measured by means of the spectrophotometer.

Another test used in the study of the red blood cells is the hematocrit reading, which gives an estimate of the volume of packed red cells in relation to the volume of the plasma. First an oxalate (a salt of oxalic acid) is added to the blood to prevent its clotting. A sample of blood is then inserted into a "hematocrit" tube, closed at one end and graduated in millimeters, and the tube is whirled, or centrifuged. The red cells become packed in the bottom, since they are heavier than the plasma. If the level reached by the packed cells is lower or higher than the normal reading, it may indicate disease. From the hematocrit reading, it is possible to figure the volume and

**Looking through a microscope, one can count red blood cells in a glass compartment within a counting chamber.**

hemoglobin content of the average red cell and the concentration of the hemoglobin in the red-cell mass. This test is usually more accurate than measuring the hemoglobin.

A doctor may also want to know the sedimentation rate of the red blood cells. The oxylated blood is placed in a hematocrit tube and left standing upright in a rack for a certain period. The red cells sink because they are heavier than the plasma in which they are suspended. The rate at which they fall may be hastened or slowed by abnormalities. The principal factors influencing the sedimentation rate are the concentrations of the proteins. After sedimentation has taken place, the blood is centrifuged in order to pack the red blood cells tightly; then the hematocrit is read.

The red blood cells may also be examined on a slide. A drop of blood is smeared on a slide, dried and then stained. The technician can tell whether the cells are abnormal in shape, in hemoglobin content and in other respects.

### The white blood cells

The white blood cells are the body's military force, attacking disease organisms such as staphylococci, streptococci and meningococci. These cells are far less numerous than the red variety; the proportion of white to red under normal conditions is 1 to 400 or 500. The white blood cells are semitransparent bodies. They differ from red cells in several important respects; among other things, they contain no hemoglobin and they always have nuclei.

There are several easily distinguished varieties of white cells: neutrophils, lymphocytes, basophils, eosinophils and monocytes. Neutrophils, basophils and eosinophils are formed in the bone marrow. Lymphocytes are made in the lymphatic tissues; monocytes, in the reticulo-endothelial system.*

The neutrophils are by far the most numerous of the white blood cells, making up from 65 to 70 per cent of the total. They derive their name from the fact that they readily take the color of a neutral dye. These cells are about half as large again as red blood cells.

When germs infect the body, they begin to destroy the tissues because of the poisons they liberate. The nearby blood vessels begin to dilate and quantities of blood flow into the area. The neutrophils migrate in great numbers to the affected region and make their way through the walls of the expanded blood vessels. They then attack the invading bacteria and engulf them; they also absorb fragments of body cells that have succumbed to the invading organisms. Many of the neutrophils are killed by bacterial poisons. In dying, they release digestive enzymes, which serve to break up nearby dead cells.

As a result of all this activity, the area of the infection becomes swollen with blood, tissue fluids, dead cells, living and dead bacteria and neutrophils and various kinds of cell fragments; all these elements form a thick semiliquid mass called pus. A large accumulation of pus is called an abscess.

As more and more neutrophils are drawn into the struggle against the invading organisms, the formation of neutrophils in the bone marrow is speeded up. The result is that there is a notable increase in the number of white cells in the blood. This fact gives the physician an important tool for diagnosing pneumonia, appendicitis, peritonitis, mastoiditis and other diseases. In these diseases the white-cell count may rise from the normal 5,000 or 10,000 per cubic millimeter to 15,000 or 20,000 or, in extreme cases, to 50,000. In most cases the additional white cells are neutrophils.

The white blood cells known as lymphocytes are second only to the neutrophils in number; they make up from 20 to 25 per cent of the white cells. They are found in large numbers in the lymph nodes (see Index) and spleen. It is thought that lymphocytes are transformed into connective-tissue elements, which aid in the healing process. Some authorities believe that lymphocytes may release antibodies (see Index, under Antibodies).

The other varieties of white blood cells are far less numerous than the neutrophils and lymphocytes. The basophils (0.5 to 1 per cent of the white-cell total) are so called because they have an affinity for a basic dye called methylene blue. The exact function of the cells is not known; it is evident, however, that they are greatly increased in leukemia. It is believed that they secrete an anticoagulant, called heparin. The eosinophils (2 or 3 per cent), which are easily stained with the dye called eosin, increase in number in some diseases, particularly allergic infections such as hay fever and asthma and in certain parasitic conditions, including hookworm and trichinosis. (See Index, under Trichinosis.) Monocytes (5

* The term "reticulo-endothelial system" is applied to various cells in or associated with certain fine networks (reticula) and inner linings (endothelia) and also found in connective tissue.

WHITE CELLS

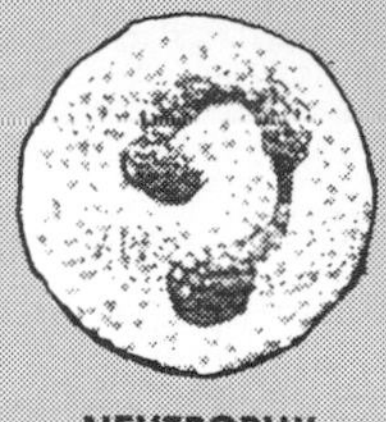

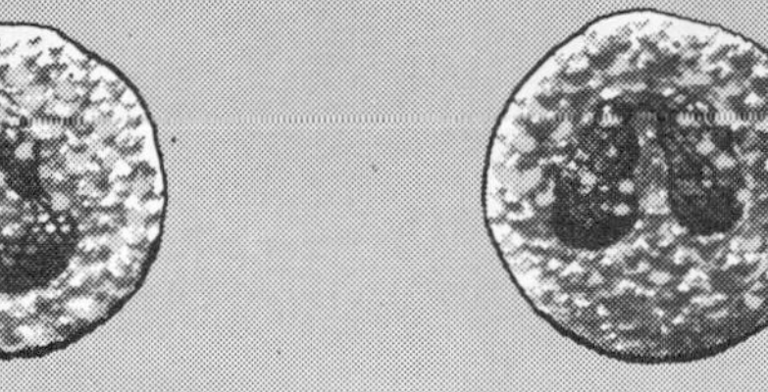

NEUTROPHIL EOSINOPHIL BASOPHIL

to 12 per cent) are active little cells, moving around in the blood stream at a great rate. They act much like amebas, pushing out their pseudopods * and enveloping debris resulting from infection. They seem to form part of the body's defense against microorganisms, and are especially active in tuberculosis, malaria and typhoid. These are the largest of the white blood cells; they are two or three times as large as red blood cells. All white blood cells have a life span of less than two weeks.

Since the proportions of the different white cells in the blood can indicate some abnormality, the doctor may want to know what percentages of each are present. This is determined by what is called a differential white-cell count. A drop of blood is smeared on a slide and is dried and stained. A count is then made of the actual numbers of each type of white cell out of a given total — say, 100. In some reports of differential white-cell counts, the percentage of each type of white blood cell is given; in others, the number of each type per cubic millimeter of blood is calculated.

**The platelets in the blood stream**

Platelets are tiny circular or oval disks, which are derived from certain giant cells in the bone marrow, called megakaryocytes. Their number ranges from 200,000 per cubic millimeter to 500,000 or more. The platelets, which are much smaller than the blood cells, serve several useful purposes. When they disintegrate, they liberate a substance called thrombokinase or thromboplastin, which is vital in the blood-clotting process. They also help to plug leaks in the tiny blood vessels called capillaries.

If the doctor suspects the platelets of being abnormal, he may want to examine the blood for them. This, also, is best done with a stained slide. Exact counts of platelets are difficult to do with accuracy because the little cells tend to clump together. Doctors usually consider it sufficient if several clumps are seen in nearly every field observed under a microscope. Of course this tendency to stick together is one thing that makes the platelets so useful; they act in this way when they plug up little holes in capillaries. Platelets live a few hours.

**The clotting of the blood**

The blood is so vital that the loss of a considerable quantity would have the direst consequences. Fortunately for us, nature has provided a means of stanching the flow from ruptured blood vessels through the process known as coagulation, or clotting. In clotting, the blood is first converted into a jellylike mass. As the process continues, threadlike structures, which are composed of fibrin, form a tangled mesh. The formed elements of the blood — red and white blood cells and platelets — are trapped in the meshes. As the clot solidifies, it shrinks and serum is squeezed out from it. The red color of the clot is due to the red blood cells that have been captured between the interlacing strands of fibrin. These cells

* Pseudopods ("false feet") are formed by a flowing of the ameba's protoplasm.

RED CELL

PLATELETS

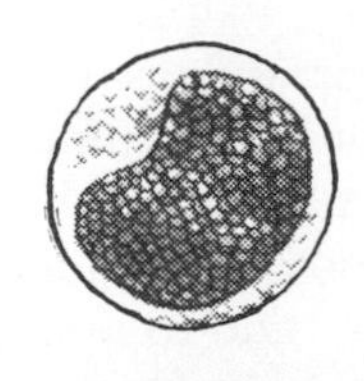

LYMPHOCYTE

In clotting, blood is converted into a jellylike mass, and threadlike structures of fibrin form a tangled mesh.

have nothing whatever to do with the clotting process.

Fibrin is not present in the blood; it is formed through the union of two substances called fibrinogen and thrombin. Fibrinogen is always present in the plasma; thrombin is not, though the plasma contains an inactive form of thrombin called prothrombin. A rather complicated train of events occurs before fibrin appears.

First, the injured tissues liberate a clot-inducing substance called thrombokinase; this is also freed by the disintegration of platelets. Thrombokinase then reacts with the prothrombin present in the plasma, in the presence of calcium ions (electrically charged calcium atoms), to produce thrombin. Finally thrombin and fibrinogen react to produce fibrin, which then enmeshes to form the clot. Blood normally clots in one to five minutes in the case of a superficial flesh wound. Much more time is required if the wound is deep.

### A survey of blood diseases

A number of illnesses are characterized by some abnormality in the blood. In some cases, the red blood cells are seriously reduced in number or else they are deficient in hemoglobin. We then have the condition known as anemia — or, more correctly, the anemias. Because of the hemoglobin deficiency, less oxygen reaches the tissues and the cells cannot function normally. Among the symptoms are pallor, fatigue, headache, general weakness, black spots before the eyes, palpitation of the heart or heart murmur and an increased pulse rate. In severe cases, breathing may be shallow and difficult. Occasionally there is an excessive amount of water in the tissues; this is usually due to poor circulation, since the heart muscle is not getting enough oxygen. In some instances a person may be anemic without displaying any of these symptoms.

There are various types of anemias. The nutritional anemias are due to some malfunctioning of the blood-forming centers. Enough of the materials necessary for blood formation may not be present in the diet, or the body may not be able to utilize those materials. If enough iron is not taken in, for example, an iron-deficiency anemia may develop. The red blood cells will be smaller than is normal, perhaps fewer in number, and they may have less hemoglobin than usual. They will not be able to carry enough oxygen and so an oxygen deficiency will develop in the tissues. Women are particularly likely to develop this kind of anemia. Another nutritional anemia, causing similar symptoms, is achlorhydric anemia. In this disease, not enough hydrochloric acid is secreted in the stomach. This prevents absorption of iron, even though there may be enough in the diet.

In hemolytic anemias, the circulating red blood cells are destroyed at too fast a rate. This may be either an inherited or acquired characteristic. It may be due to overactivity of the spleen, to the effects of certain poisons or to the fragility of the red blood cells. In certain cases, it may be necessary to remove the spleen.

Pernicious anemia is particularly serious. It results from the body's inability to absorb vitamin $B_{12}$, or cyanocobalamin. In order to absorb this vitamin, a certain enzymelike substance must be present in the gastric juice. The vitamin and enzyme react, forming an antianemic factor which is stored in the liver and is necessary for red-blood-cell production. People with pernicious anemia lack the enzymelike substance. They must be given vitamin $B_{12}$ by injection since they cannot absorb it through the normal processes of digestion. Death occurs in two or three years after the disease begins if the treatment is not applied. Once started, the treatment must be continued as long as the patient lives.

Certain anemias are secondary: that is, they result from some underlying condition, which may or may not be clearly defined. Such anemias may be a consequence of hemorrhage, chronic blood loss, acute infection, malignant growths or gastrointestinal disease. Since the liver acts as a storehouse for iron, damage to the liver may interfere with iron metabolism, so that there will not be enough iron for the formation of hemoglobin in adequate amounts.

Various toxic (poisonous) drugs may cripple the bone marrow so that it is unable to produce an adequate number of red blood cells. These drugs include benzol and arsenic. Other factors which injure bone marrow are X rays and radioactive substances, such as thorium compounds.

From what we have just said, you might think that the more red blood cells you have, the better. This is not necessarily so. One serious disorder, polycythemia, is characterized by an overproduction of red blood cells. The blood becomes too thick and moves through the veins sluggishly. The patient complains of headaches and dizziness and he has a tendency to form clots in his blood vessels. If these clots drift around in his veins and arteries, they may cause gangrene to develop; clots that plug up a vessel in the heart or brain may cause death. The patient is bled at regular intervals and may be given a drug to slow down the red-cell production.

In some illnesses there is an overproduction of white blood cells. This is sometimes due to the disease known as leukemia, in which the white-cell count may reach the amazing total of 500,000 per cubic millimeter. The white cells overrun the bone marrow, crowd out the red-blood-cell and platelet-producing material, and cause an accompanying anemia to develop. Lymph nodes and reticulo-endothelial tissue swell with their burden of excessive white cells. As far as we know, leukemia is not due to infection, but to an abnormal condition of the bone marrow or lymphoid tissue, causing white-cell-forming elements to multiply. Overexposure to X rays and radioactive materials seems to be one contributing cause. Treatment can prolong life, but it cannot cure the disease. There are acute leukemias in which life expectancy is only a few months; in chronic cases, the patient may live from two to ten or more years.

In infectious mononucleosis, a virus (as yet unidentified) attacks lymph nodes. As a result, the spleen becomes enlarged and there is a great increase in lymphocytes. The disease is usually found among young people under thirty. It is transmitted only by close personal contact and is often called the "kissing disease" for this reason. Rest seems the best treatment.

Sometimes there are too few white cells — a condition known as agranulocytosis. The body's resistance to disease is lowered and secondary infection may develop. Agranulocytosis may be caused by toxic drugs and diseases of the white-cell-producing tissues. With this disorder, a patient may die from the effects of the infections he cannot resist.

In another group of diseases, the body may be deficient, for one reason or another, in one or more of the essential clotting elements. Platelets may be defective or too few in number, so that not enough thrombokinase is liberated. The fibrinogen content of the plasma may be seriously reduced (as when the liver has been damaged). There may be a prothrombin deficiency, brought about because there is not enough vitamin K in the body — a condition found particularly in infants (see Index, under Vitamin K). If any of these conditions are present, even slight wounds may cause dangerous bleeding.

In thrombocytopenia, a disease in which there are too few platelets, the blood tends to seep out of the circulatory system, making black and blue bruise spots and pe-

**A blood clot from one blood vessel has blocked another vessel. Such an obstruction is called an embolism.**

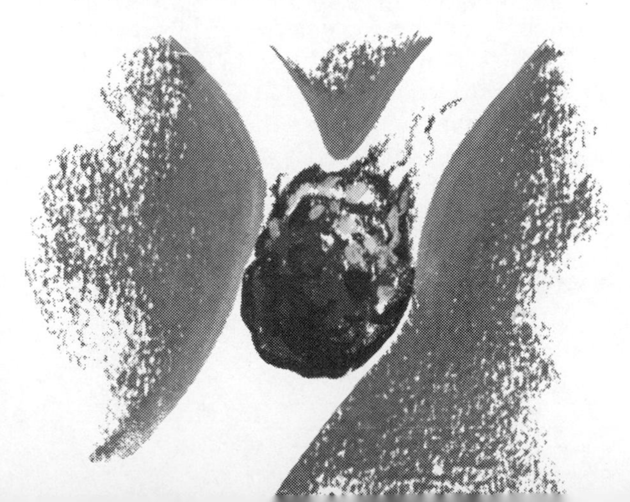

techiae (tiny pinprick-sized blood spots). This happens because there are not enough platelets to plug up small breaks in the capillaries. If the condition is not corrected, fatal bleeding into the brain or other organs may take place. The reduction in platelets may be due to benzol, arsenic or radioactivity, among other things.

The failure of the blood to clot properly has particularly serious consequences in the disease called hemophilia. It is a hereditary disease, occurring almost exclusively in males and transmitted directly only by females. Those suffering from the disease are called hemophiliacs, or bleeders. The most trivial happening — friction on the gums, accidental biting of the gums, a very slight cut on any part of the mucous membrane, the extraction of a tooth — may bring on serious hemorrhages and may result in death. Because of the high mortality rate in youth, comparatively few bleeders live to old age.

Hemophilia still baffles medical men. There does not seem to be any deficiency in any of the essential factors in clotting — fibrinogen, prothrombin, calcium, platelets. Some authorities have pointed out that the platelets in the blood of hemophiliacs are unusually stable. Perhaps they fail to disintegrate quickly when injury takes place, and for this reason an insufficient quantity of thrombokinase is released.

There is no known cure for hemophilia. Every effort is made to stop bleeding by administering clotting agents or calcium salts; repeated blood transfusions are given. Bleeders are urged to avoid exposure to accidents — even the most trivial sort of accidents. Of course, if females capable of transmitting hemophilia did not have children, the disease might be eradicated in time.

There are times when the blood may clot too quickly. The clotting reaction may be triggered too easily and may take place in the veins or arteries. A clot, called a thrombus, may form upon the wall of a ruptured blood vessel; it may extend completely across the vessel in such a way as to shut off the flow of blood. If this vessel supplies blood to a vital body structure, the consequences may be most serious. Death may result unless nearby blood vessels can take over the task of supplying blood to the affected areas.

In some cases a thrombus may break loose from a blood vessel; it may be carried off in the blood stream to another blood vessel, which it may partly or wholly block up. A foreign or abnormal particle of any kind carried in the blood stream in this way is called an embolus (plural: emboli). If an embolus is transported by the blood stream in the veins to the right side of the heart it may block up the pulmonary artery, wholly or in part. Death results if the artery is entirely obstructed; if the stoppage is only partial, gangrene or inflammation of part of the lung commonly takes place. Emboli may block up arteries in other parts of the body and may result in the shutting off of the blood flow in these areas. If the embolus is septic (that is, poisoned by bacteria), it will cause infection wherever it is lodged and may lead to pyemia, a kind of blood poisoning.

### The blood groups in man

Each person's blood possesses certain characteristics which distinguish it from the blood of every other person. They are inherited in accordance with the Mendelian laws of heredity * and remain unchanged throughout life. On the basis of these characteristics, human beings have been divided into different groups. Scientists first came to understand about the blood groups when they studied certain unfortunate consequences of blood transfusion.

In blood transfusion, blood from an outer source is introduced into the veins of a person in order to replace the blood lost through serious hemorrhage, or to prevent shock, or to combat infection or to treat certain conditions such as the anemias. It was formerly a hit-or-miss procedure; anyone willing to contribute his blood was accepted as a donor (giver of blood).

Doctors were puzzled by the serious aftereffects that sometimes occurred. The

* See Index, under Mendel, Gregor Johann — Laws of heredity.

red blood cells of certain donors would form clumps after transfusion into the recipient (person receiving blood); the clumps would sometimes block up the recipient's blood vessels leading to vital organs. After a while the clumped cells would disintegrate, with disastrous consequences.

We now understand, thanks to the researches of Karl Landsteiner (see Index) and others, the why and wherefore of these unhappy secondary effects. We know that they are due to substances called agglutinogens, or antigens, which occur in the red blood cells (See Index, under Antigens), and also to agglutinins, the antibodies that originate in the serum of the blood. A certain variety of agglutinin will cause red blood cells to form clumps if these cells contain a certain variety of agglutinogen. Two important types of agglutinogens are those called A and B; corresponding to these are two kinds of agglutinins, called anti-A (or $\alpha$) and anti-B (or $\beta$). Clumping of red blood cells takes place if blood containing A is mixed with blood containing anti-A, or if blood containing B is mixed with blood containing anti-B. The red blood cells of a given person may contain both A and B, or either A or B or neither A nor B; his serum may contain both anti-A and anti-B, or either of these substances or neither of them. Of course no person could possibly have both A and anti-A or B and anti-B in his blood, because if such were the case, clumping of red blood cells would be going on all the time.

The blood of human beings has been classified in four groups on the basis of the A, B, anti-A and anti-B factors they contain; these groups are known as O, A, B, and AB. The table at the bottom of the page shows the nature of this classification:

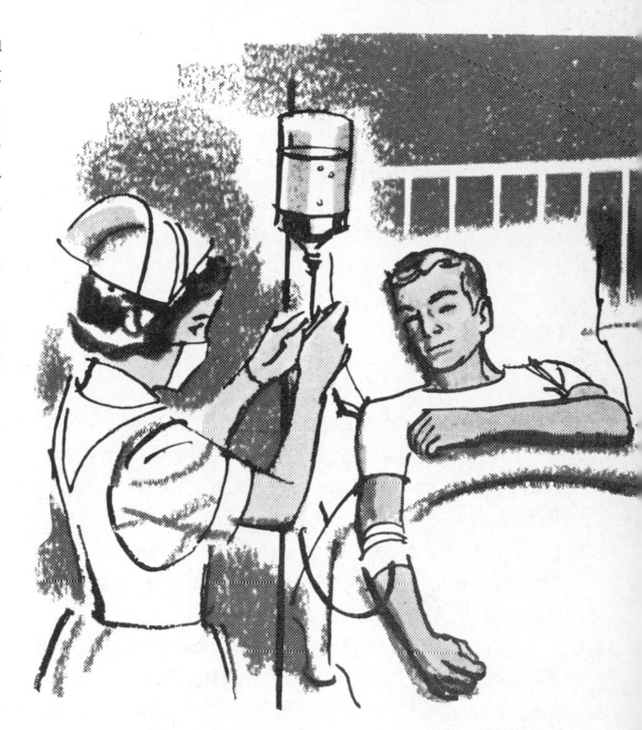

A blood transfusion. The blood passes from a glass container through a tube into the veins of the recipient.

There are also subgroups within the main groups listed in the table.

There are other blood groupings besides the one we have discussed. One that also has an important bearing on blood transfusion is based on the so-called Rh factor, an agglutinogen found in the red blood cells of most people (85 per cent of white Americans, 95 per cent of Negroes and 99 per cent of Japanese, Chinese and American Indians). It is called the Rh factor because it is similar to an agglutinogen found in the red blood cells of the rhesus monkey. Blood possessing the Rh factor is called Rh-positive; blood without this agglutinogen is known as Rh-negative.

Table showing the major human blood groups and their distribution among three races.

| Name of group | Agglutinogens in red blood cells | | Agglutinins in serum | | Percentage of Caucasians belonging to group | Percentage of Negroes belonging to group | Percentage of Japanese belonging to group |
|---|---|---|---|---|---|---|---|
| | A | B | anti-A | anti-B | | | |
| Group O | none | none | present | present | 40% | 42% | 29% |
| Group A | present | none | none | present | 45% | 24% | 42% |
| Group B | none | present | present | none | 10% | 28% | 20% |
| Group AB | present | present | none | none | 5% | 6% | 9% |

If blood from an Rh-positive person (one with Rh-positive blood) is transfused into the veins of an Rh-negative person, the serum of the latter will produce an anti-Rh agglutinin. If the patient receives another transfusion from an Rh-positive donor, the anti-Rh agglutinin will cause the red blood cells contributed by the donor to form clumps as soon as they enter the recipient's veins. Clumping of the cells may occur following a single transfusion in the case of Rh-negative women who are pregnant or have just given birth.

Various other blood groups have become known. Factor P was discovered by Landsteiner and Levine in 1927. There are two types, P-positive and P-negative. In Caucasoids, 75 per cent are P-positive; in Negroes, 97 per cent are P-positive.

Another group of blood factors, named M and N, was discovered by Landsteiner and Levine in 1927. The Caucasian population has three classes under this system — M, N and MN, occurring in 30, 20 and 50 per cent of the population respectively. More recently, a new factor, S, was found. It is linked with the MN system; thus, a person may be MS-positive or MS-negative and so on. Other blood groups discovered in recent years include the Lewis, Kell, Lutheran, Duffy, Levay, GR and Jobbins groups. As you can imagine, many different combinations of all these factors are possible — thus lending to the complexity and individuality of each person's blood.

**Procedures in blood transfusion**

Knowledge of the different blood groups has made blood transfusion a relatively safe procedure. Doctors test for the O, A, B and AB groups and also for the different subgroups under these before allowing a transfusion to take place. The donor and the recipient should belong to the same main group. In case of extreme emergency, blood of the group-O classification may be given to patients of any group; group O, therefore, is known as the universal donor group. In case of emergency, too, patients belonging to group AB may be given blood from donors belonging to any of the three other groups. For this reason, AB is called the universal recipient group. Doctors also carefully examine prospective donors and recipients for the Rh factor before permitting a transfusion to be made.

A donor gives about 500 cubic centimeters of blood (roughly, a pint) at a time; if he is in good health, he will be ready to contribute blood again in about 6 weeks. Blood is first withdrawn from the donor into a citrate solution, which prevents clotting from taking place. It is filtered through sterile gauze and placed in a glass container at a height of three and a half or four feet above the recipient, who is in a lying position. Then it is permitted to run through a tube into the recipient's veins.

Much of the blood used in transfusions nowadays comes from the blood banks of hospitals. These banks consist of stored blood, contributed by voluntary or professional donors. The blood is preserved in containers, in a solution of sodium citrate and dextrose, and is kept at a temperature of 4° C. (about 39° F.).

In many cases human plasma is used for transfusions, particularly to replenish deficiencies in blood proteins, to build up or maintain blood volume, to lessen blood concentration and to treat cases of shock. The great advantage of this type of transfusion is that plasma from any donor may be used. It can be stored under refrigeration for an indefinite period, either in liquid or frozen form. It can also be prepared in powdered form after drying; when it is to be used, enough distilled water is added to make up for the water removed in the drying process. Fluid, frozen or powdered plasma is kept in the blood banks of hospitals. Powdered or frozen plasma can be safely shipped anywhere in the world; powdered plasma is used extensively by the armed forces.

Other substances have been employed for transfusion purposes. The plasma fraction called serum albumin has proved particularly effective in the case of shock. Sodium chloride (table salt), sugar and plasma substitutes, such as gelatin, have been introduced into the circulation.

*See also Vol. 10, p. 276:* "Physiology."

# WING-SHEATH HORDES

## The Beetles, Most Numerous of the Insects

by JOHN C. PALLISTER

TO MANY people a beetle is a dark, scurrying little insect, which arouses disgust and perhaps apprehension. The very name of the insect reflects this attitude; for "beetle" comes from the Anglo-Saxon *bitol,* meaning "creature that bites." Actually, many members of the beetle group, as we shall see, are extremely useful to man and to the world at large.

The word "beetle" is applied to all the members of the great insect order of the Coleoptera. This order includes the lady beetle or ladybug, a children's favorite for many centuries, the scarab, which was venerated by the Egyptians, the fireflies, which delight all of us, the weevils, which only an entomologist can admire, and many, many other kinds. The beetles are the most numerous of all the orders of insects; the insects, in turn, far outnumber the other classes of animals.

A beetle may be no larger than a pinhead; it may be as big as a man's fist. It may be dull in color; it may gleam like a precious jewel or it may be intricately designed like an exquisite brooch. It may be slender and graceful, or antlike in shape. Beetles are a favorite of collectors: amateurs find them easier to catch, mount and keep than butterflies; professional entomologists never tire of studying these fascinating insects.

Like all insects, a beetle is a boneless animal; its vital organs and muscles are protected by a jointed, segmented case of hard material called chitin. Its body is divided into three parts: the head, the thorax and the abdomen. The head carries the eyes, the antennae and the mouth parts, which are very complicated. The thorax, the middle section, bears the six legs and the two pairs of wings, and, within, some of the digestive organs. The abdomen is made up of nine or ten ringlike segments of chitin, connected by a softer tissue; it contains the organs for breathing, digesting and reproducing. The stridulating organs, which produce sound, are also found on the abdomen. Not all beetles possess these organs.

All beetles have one pair of jointed antennae, usually projecting in front of the eyes. They may be so short that you can hardly see them, as in carpet beetles and lady beetles; or they may be two or three times as long as the insect's body, as in the long-horned wood-boring beetles. Under a low-powered microscope, the antennae show an astonishing variety of shapes; they may suggest a brush, or a feather, or a string of beads, or a comb or a club.

The most remarkable structure, which distinguishes the beetle from practically all other insects, is the outer, or first, pair of wings, called elytra. They do not look at all like wings; they are hard and shell-like and serve as a covering for the second or inner pair of wings and for the abdomen. The under wings are thin and membranous, and when not in use are folded and refolded under the elytra. The elytra are of little help in flying.

All beetles have elytra or traces of elytra; the name "Coleoptera" ("sheathed wings") is derived from this wing arrangement. The only other insects that have wing sheaths are the earwigs, some grasshoppers and some Homoptera; but their sheaths are not nearly so firm and shell-like as those of the beetles.

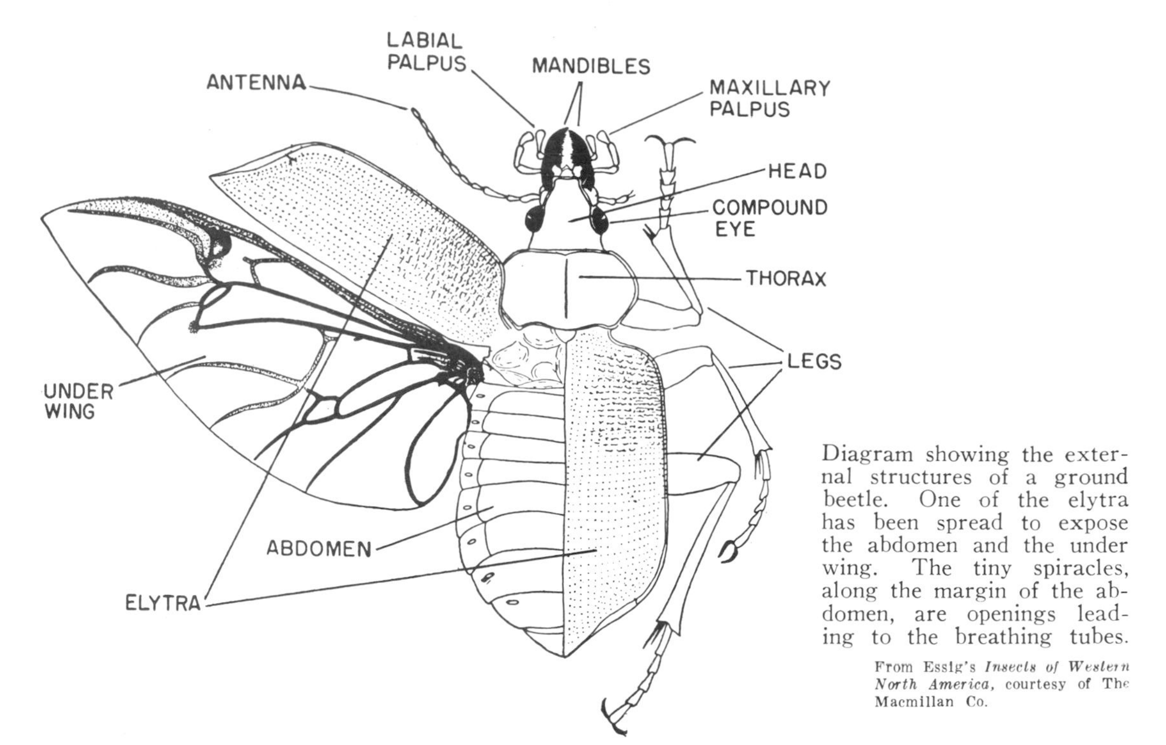


Diagram showing the external structures of a ground beetle. One of the elytra has been spread to expose the abdomen and the under wing. The tiny spiracles, along the margin of the abdomen, are openings leading to the breathing tubes.

From Essig's *Insects of Western North America*, courtesy of The Macmillan Co.

A beetle's thorax is made up of three segments, each of which bears a pair of legs. The elytra are fastened to the top of the middle segment, covering it, the back segment and part or more often all of the abdomen. The top of the front thoracic segment — the prothorax — is as hard and shell-like as the elytra. In some species it fits so neatly against the elytra that when the beetle is resting, its back appears to be in one piece, like the back of a turtle.

During its life a beetle passes through four distinct stages. First it is an egg; then it hatches into a wormlike larva, commonly called a grub. The grub eats voraciously; this is the only stage of the beetle's life during which it increases in size. (Little beetles are never the young of large beetles; they are always of a different species.) As the grub grows too big for its skin, it molts, or sheds the skin; it usually molts five or six times before it is full-grown. Having eaten all it needs, the larva seeks a secure resting place, often in the plant or tree where it has been feeding; or it burrows into the ground, where it builds itself a little cell. Some leaf-eating weevils spin cocoons.

Once it is secure in its shelter, the grub becomes a pupa, eating nothing and remaining motionless while great changes go on in its body. This period may last only from three to four days in the case of lady beetles, when the humidity and temperature are just right; it may last all winter, in the case of beetles that become pupae in the autumn months. Finally the adult beetle emerges from the pupal case. The ordinary beetle lives only from two to six months; certain wood-boring beetles, however, have a life cycle of several years.

Land beetles are found all over the world, except in the extreme northern or southern areas. A few species dwell in fresh-water ponds and streams, but none in the oceans and seas.

Certain beetles are man's allies (unwitting allies, of course), preying upon insect pests. Many are scavengers, burying small dead animals, cleaning up debris and making the earth a more attractive place in which to live. Other members of the Coleoptera are among mankind's most damaging pests, destroying trees, crops, processed foods, clothing and furniture. But though beetles may damage man's belongings, they seldom attack his person. They are rarely parasitic upon man. The only beetles known to carry disease to man or animal are a few that transmit a tapeworm to rats, other animals and man.

Though beetles seldom carry the attack to us, they will sometimes defend themselves when we try to handle them. A few have mandibles (mouth parts) strong enough to draw blood from an unwary finger. When handling a beetle of this kind — a pinching bug, say — it is best to grasp it just back of the head, so that it cannot turn and nip you. The bombardier beetle will shoot out a caustic liquid when it is picked up; this liquid can blister a tender hand. When other beetles are handled, their leg joints ooze juices that are evil-smelling and stain the fingers.

The order of the Coleoptera belongs to the general class of the Arthropoda, which also includes the lobsters, crabs, spiders, scorpions and millepedes. The Coleoptera are subdivided into about 150 families, totaling some 300,000 species. In the following pages we will briefly examine 23 of these families in order to catch a glimpse of the fascinating beetle world, with its infinite variety of forms, sizes and habits.

### Tiger Beetles (Cicindelidae)

"Tiger beetles" is an appropriate name for the fierce and swift cicindelids, which stalk their prey along sandy banks and beaches, on roadsides and woodland paths. They are rather small, slender insects, just about the color of the sand or dirt upon which they run. However, one of the commonest species in the eastern United States is a brilliant metallic green, and several related species are greenish or bronze-colored. The elytra are often spotted or banded in lighter tints. The tiger beetle feeds on other insects.

When you approach a tiger beetle, it will run rapidly for a few feet, then fly close to the ground for several more feet; as it alights it will turn to face you. You will find it very hard to catch.

Tiger-beetle larvae are whitish grubs with big, metallic heads and very large mandibles. They are even more dangerous than the adults to small insects. They lurk in small burrows they have dug, with their heads just protruding and their mandibles wide open, ready to seize any insect that walks over them. Hooks on the back of the abdomen serve to anchor the grub to the sides of its burrow, so the intended prey is not able to launch a counterattack by jerking the larva out of its lair.

### Ground Beetles (Carabidae)

As you walk along a path or sidewalk in the evening, you may encounter small, dark beetles scurrying in front of you. The great majority of these will be ground beetles, or carabids, belonging to the large family of the Carabidae. Carabids are most numerous in areas where there is plenty of rainfall.

Most ground beetles hunt at night, hiding by day under rocks, logs or debris. If you raise an old board from the ground, you may catch a glimpse of some carabids, black or brownish, oblong, medium in size, with long, slender legs. They will try to

Ground beetle of the family Carabidae.

Edwin Way Teale

run away rather than fly, rushing to hide under the nearest cover.

In some species the under wings have atrophied and the elytra have fused together. Other carabids fly very well. One of these, called the searcher (*Calosoma scrutator*), preys upon caterpillars in trees; it is larger than most ground beetles and is colored a beautiful metallic green with red edges and blue legs. It is common in the central United States in the early summer. Sometimes crowds of searchers will fly around street lights.

Most ground beetles feed upon the larvae and adults of other insects, and also upon slugs, snails and every other creature they are able to capture. A few are seed eaters; the grubs of a very small number are said to feed on sprouting corn.

Ground-beetle larvae are long, flattened, almost white grubs. They live in burrows just below the surface of the earth; here they feed on the larvae of leaf-eating insects that have entered the ground to pupate.

### Water Beetles (Dytiscidae, Gyrinidae and Hydrophilidae)

Although most beetles are strictly land insects, there are a few families that spend much or all of their lives in ponds and streams. The diving beetles (Dytiscidae) make up the largest family of the water beetles. They are oval in shape and generally range from medium to large in size; they are brownish or greenish black; they have long, sturdy hind legs which they use as oars. Under their wing covers they carry a supply of air for breathing while they hunt or rest on the bottom. Both adults and larvae feed voraciously on all kinds of aquatic insects, tadpoles and even small fish. The adults sometimes migrate at night in great numbers from one pond to another, stopping on the way to whirl around a street light. One of the larger species is a common food in China and in the Chinatowns of the West.

If you approach a pond or pool quietly, you may surprise a group of whirligig beetles (Gyrinidae) circling around on the surface of the water. Disturb them and they will quickly dive to the bottom. They are small dark insects; their long, slender front legs are equipped to seize prey, while the middle and hind legs are broad and flat for swimming. Their habits and food are much like those of the Dytiscidae.

The third important water-beetle family, the Hydrophilidae, are sometimes called water scavengers, because they feed on decaying vegetable and animal matter; however, they will also eat any living creature they can catch. They somewhat resemble the diving beetles, but they use their legs differently and are not as good divers. The

Robert C. Hermes from Nat. Audubon Soc.

Diving beetle devouring a tadpole.

underside of their bodies is hairy, and enough air clings to the hairs to give the beetle buoyancy; it also carries air under its wings for breathing. Like other water beetles the hydrophilids sometimes fly into town on a summer night. I saw the streets and out-door cafes of Merida, in Yucatan, invaded night after night for weeks by thousands of huge two-and-one-half inch hydrophilids. The bright lights were too much for them and by midnight the city would be littered with their dead bodies.

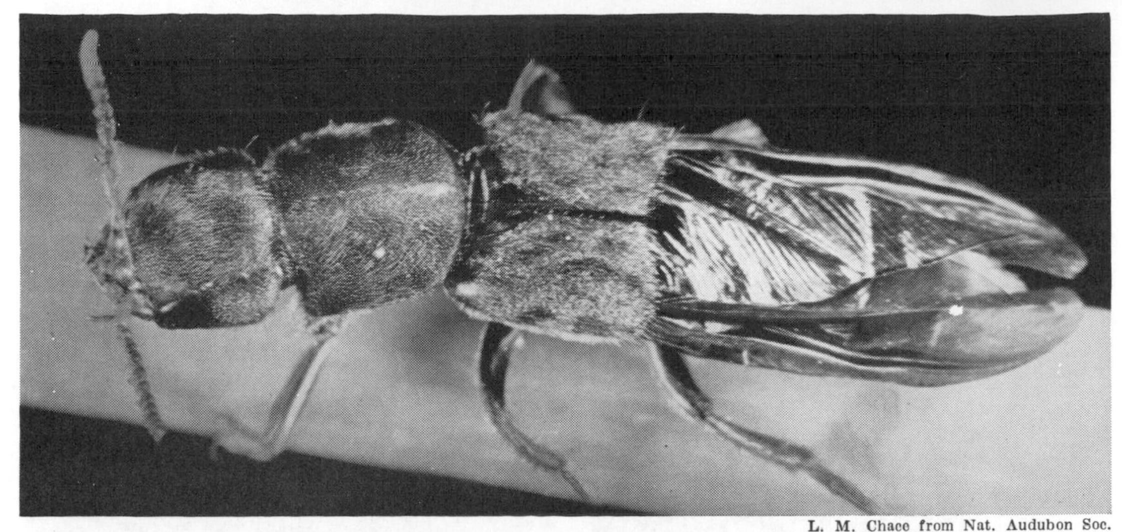

L. M. Chace from Nat. Audubon Soc.

Rove beetle with wings extended.

The insect called the black water beetle, which is found around kitchen sinks and drains and other damp places, is really no beetle at all; it is the black Oriental cockroach, or kitchen cockroach.

### Carrion or Burying Beetles (Silphidae)

Certain members of this beetle family help to keep the surface of the earth clean by burying small dead animals. When a pair of these beetles find a dead mouse or small snake, the female deposits her eggs in it; then the pair quickly bury the body several inches under the ground. When the larvae are hatched, they feed upon the decaying flesh.

There are many species of burying beetles, most of them fairly common throughout the United States. One of the largest and most conspicuous, *Necrophorus americanus,* is about one to one and one-half inches long; it is a shining black, oblong, heavy-bodied insect, with two large reddish spots on each wing cover. The prothorax is hemispherical and red; the head is almost as large as the prothorax. Other species of burying beetle are oval-shaped or almost hemispherical; they may be black or brownish.

Leonard Lee Rue

The carrion beetle, an insect scavenger.

The Silphidae family, to which the burying beetles belong, has other interesting members. Some species dwell in caves and have no eyes; others, very small, live in decaying fungi and even in ants' nests.

### Rove Beetles (Staphylinidae)

The rove beetles (Staphylinidae) make up one of the largest beetle families, numbering more than 30,000 species, which are found all over the world. They are usually slender-bodied and range from minute to medium in size. A good many are black; some are dark red, or dark brown or yellowish; a few shine in metallic greens and blues. The larvae are found with the adults, which they resemble.

The elytra of these beetles are so brief that they do not cover the last three to five segments of the abdomen. The under wings are large and thin. When a rove beetle alights it folds these wings and doubles them back; then, bending up its abdomen, it uses the tip like a hand to poke

The firefly, a carnivorous beetle.

every portion of the wings completely under the tiny elytra.

Rove beetles are also called short-winged scavenger beetles, an appropriate name, for they feed on decaying animal and vegetable materials. They aid in reducing manure to an available plant food form. Some 300 species live in ants' nests, eating dead ants and cleaning the nests. Ants tolerate them and care for the beetle young as carefully as they tend their own young. Some rove beetles are also found in the nests of termites.

### Ladybirds, Lady Beetles (Coccinellidae)

Unlike most beetles, the ladybirds, or coccinellids, have always enjoyed wide popularity. They were long considered omens of good luck; children throughout Europe and North America sing affectionate little verses to them. Nowadays most people know that ladybirds eat the aphids or plant lice that attack our house and garden plants; they have heard the dramatic story of the ladybirds imported from Australia to prey upon orchard-destroying scale insects. There are coccinellids, however, such as the squash beetle and the Mexican bean beetle, which eat plants instead of other insects and do considerable damage to truck gardens.

Both the helpful and harmful varieties of coccinellids are attractive beetles, rather small; they are broadly oval or hemispherical. The many species of the true ladybird are enameled in different shades of red with black spots. Usually these are the aphid eaters. Other species are black with red spots; many of these attack scale insects. The herbivorous lady beetles are likely to be yellow with black spots.

When disturbed, ladybirds discharge an evil-smelling secretion that is believed to protect them from insect-eating birds. During the winter months the beetles tend to congregate in large numbers in crevices on the sunny side of buildings. The aug-

Upper photos, Edwin Way Teale

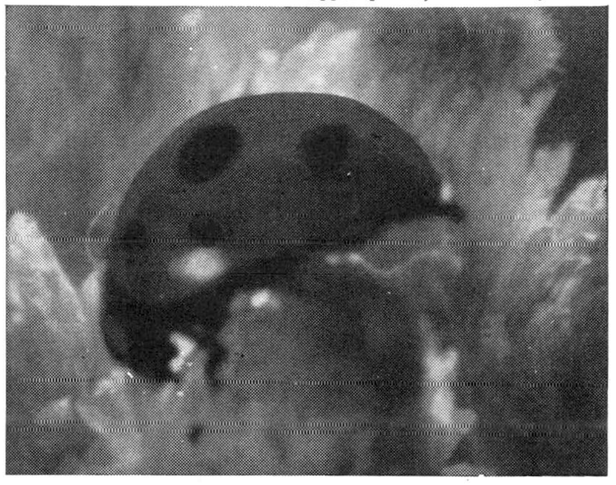

The ladybird, deemed a good-luck omen.

Click beetle with under wings spread.

John C. Pallister

mented odor from the group probably increases their protection. Convergent lady beetles, common in the western states, collect in masses to hibernate during the winter under leaves and debris on mountain slopes. Horticulturists gather these insects and keep them dormant until the beetles are needed for attacking aphids.

### Skin Beetles (Dermestidae)

Everybody hates the skin beetles, or Dermestidae (skin-eaters), and with good reason. They love to feed on every animal product that man has processed for his own use. Fur and feathers, woolen cloth, bacon, ham and cheese, flour and meal, dried insects and stuffed birds and animals in museums — each of these products is a food for one or more dermestid species.

Many of these pests are quite small, under a quarter of an inch in size. Larger ones may be as much as half an inch long. They are usually oval, plump and dark-colored; they are partly covered with fuzz in a lighter tint, or with scales that rub off. The larger dermestids in general feed on hides and dead animals; they are called hide beetles. The smaller ones eat pollen in the fields; they come into our homes to feast on clothes, carpets and dried foods. Many species remain out of doors and never bother us at all.

Dermestid larvae are small and brownish, covered with many black hairs and bristles. They can move very rapidly. If you watch one closely, you may see it do a curious thing — run a short distance, stop, vibrate its hair rapidly, then start running again. The larvae are more destructive than the adults, eating voraciously when food is available; yet they are able to live a long time without any food.

Because the adults are small and have no prominent features, they are rather difficult for anyone but an entomologist to identify. Common names of some species that have attained notoriety as pests are the larder beetle, which infests stored bacon and ham, the black carpet beetle, which destroys carpets, silk, woolen goods and feathers, and the buffalo moth, a name for the hairy grub of another carpet beetle.

### Click Beetles (Elateridae)

A large variety of extraordinary beetles belong to the family of the click beetles, or elaters. Their most amazing characteristic is their ability to hurl themselves up in the air. When disturbed, the elater drops on its back to the ground, feigning death. When it seems safe to move again, the beetle bends its head and thorax back, pushing a special spine on its prothorax almost out of the groove in which it lies. Suddenly the tension is released, the spine snaps back along its groove with a clicking noise, driving the base of the elytra against the ground with such surprising force that the little insect may be shot as much as four or five inches up into the air.

Click beetles are also known as skipjacks and snapping beetles. In the United States the most conspicuous one is the eyed elater (*Alaus oculatus*), a sturdy fellow sometimes two inches long; it is shining black, flecked with silvery scales. On top of its large prothorax two big black spots, outlined by a ring of white scales, imitate two glaring eyes. Eyed elaters are found all summer long around old stumps and logs, where their larvae live, feeding on the rotting wood.

Elater larvae are long, slender, smooth, yellowish grubs, so hard and stiff that they are called wireworms. Many feed on decaying wood; some, however, live in the ground, annoying farmers by burrowing into bulbous roots, tubers and sprouting corn seed. The tunnel you sometimes find in a potato has probably been dug by a wireworm. Some wireworms require several years to complete their growth.

The largest and most beautiful elaters live in the tropics. Among these is the luminous fire beetle (*Pyrophorus*), found from the southern United States down to Argentina. Some of the larger species have a two-way light system. At night, when one of these beetles is resting on a tree trunk, two spots on its prothorax glow with a soft green light; when it takes wing, a bright orange light streams from its abdomen. Indians in the tropics use fire beetles for illumination and decoration, and have developed an extensive folklore about them. Swarming fire beetles so impressed early explorers in these parts that they discussed the insects in detail.

### Metallic Wood Borers (Buprestidae)

Nearly all the metallic wood borers, or buprestids, are striking insects. Their copper, gold, green, blue or red backs shine with a metallic lustre and often are decorated with intricate patterns in contrasting colors. Since their bodies are hard and the colors do not fade, buprestids are often used as decorations, not only by natives in tropical forests but by art workers everywhere. Two of the most gorgeous species are called jewel beetles; Australians set them in mountings to wear as jewels.

Buprestid species may be short and flat, or oblong and cylindrical; they are large in the tropics — a Brazilian giant is over two inches — and small to medium-sized in the United States. Most of them are tropical and nearly all live in forests. The adults like to sun themselves; you may come upon one in some dark woods, glittering in a small patch of sunlight.

The larvae are blind and legless, with a small head and a large flat thorax. The thorax is often mistaken for the head and gives the grub the name of flat-headed wood borer, or hammer head. The adult female deposits her eggs in cracks in the bark of trees that have been injured by fire or overexposure to sunlight, as when a clearing has been made by storm or lumbering. Curiously enough, forest fires and smoke seem to excite some buprestids, so that they will attack persons working around fires, biting the hands or neck or any exposed skin area, often quite severely.

The family does not restrict its boring to tree trunks. Some species attack herbaceous plants, and some very tiny buprestids, called leaf-miners, bore into leaves.

### Fireflies, Lightning Bugs, Glowworms (Lampyridae)

The fireflies add mysterious charm to summer evenings; even the city dweller will see a few sparklers on his little patch of lawn and around his few feet of hedge. In the daytime, with its lamps turned off, the firefly is not a particularly notable insect. It is medium to small-sized, elongated or oblong, black or brownish, edged with red or yellow. The elytra and the thoracic covering are not so hard as in most beetles. The light-giving apparatus, which is on the underside of the abdomen, is composed of a fatty tissue through which air tubes run. The tissue contains a substance called luciferin, which is acted upon by the enzyme luciferase to produce a nearly cold light.

The female of many species are wingless, wormlike creatures; they are sometimes three or four times as long as the male. They glow at night from spots along the sides of the abdomen and thorax. The European glowworm belongs to this group.

Each species of firefly appears to have its own code of signals. It has been assumed that the lights are signals between the sexes; but many larvae also are luminous, and so are a few pupae. In some tropical species the larvae have lights but the adults do not. Not all lampyrids are luminous; a great many species are diurnal and have no need of light.

Adults and larvae of nearly all species are carnivorous, feeding on small worms, snails and the larvae of other insects.

### Deathwatch Beetles, Drugstore Beetles, Spider Beetles (Ptinidae)

The ptinids are a family of small (one-quarter inch or less), dull-colored scavengers, living on old dry vegetable or animal material. The most dreaded member of this family used to be the deathwatch beetle, a tiny brown insect that feeds on decaying wood. It often feasted on the wood of old houses; as it did so it made a ticking noise that, in the silent night, seemed portentous. To untutored people in past generations the insect appeared to be ticking off the minutes until the death of a member of the household. Though it was merely eating its favorite food, the deathwatch beetle literally scared many a sick person to death. It took many genera-

Jack Dermid from Nat. Audubon Soc.

A male stag beetle displaying its antlerlike mandibles. Stag beetles live in forested areas.

tions for doctors, clergymen and scientists to free the minds of Europeans from the superstitious fear of the deathwatch beetle. It is not a harmless creature, indeed; it has eaten into and destroyed many a great oak beam in the ceilings of houses and churches.

In the days when druggists carried large supplies of dried roots and leaves from which to compound their medicines, they were much annoyed by drugstore beetles, which ate their medicaments and also their cigars. Today's medicines are compounded in laboratories and packaged in glass; the drugstore beetle, therefore, has moved to the grocery store, where it can live on dog biscuits, cigarettes and anything else that is dry and available.

You may find in your bathtub what looks like a little red spider, with a globular body, either smooth and shining or partly fuzzy, and long slender legs. This is really a beetle, one or another of a few species of the Ptinidae. It has been living on small forgotten remnants of your food, or on the paste in the wallpaper. It has fallen into the bathtub, in which it is trapped; it cannot climb the slick sides of the tub and it has no wings for flying. Only in the bathtub is it visible; elsewhere it would blend completely with its surroundings and would almost certainly escape detection.

### Darkling Beetles (Tenebrionidae)

Darkling beetles, or tenebrionids, are found mostly in arid regions; they are usually nocturnal scavengers, feeding on dead or decaying vegetable matter. Some western species of the United States devour living plants, and sometimes become so numerous that they denude the sparse natural vegetation and damage cultivated crops. Small black tenebrionids are seen wherever the cactus called the opuntia flourishes. The eastern species are found in fungi and beneath bark.

In general, darkling beetles are small to medium, black and stoutly built. Many species are wingless, and their elytra have been fused together. Their legs are long, but they move rather slowly with an awkward, loose-jointed gait.

The larvae, long and slender, somewhat resemble wireworms; they live in decaying wood and dried vegetable products. Some species, of European and Asiatic origin, have specialized on grain and grain products, and are now spread around the world, doing considerable damage. The meal worm (*Tenebrio molitor*), which attacks our cereal products, has become a commercial product itself; it is raised in large quantities to feed pet birds and can be purchased in any pet shop.

### Stag Beetles, Pinching Bugs (Lucanidae)

The giant male stag beetle, which lives in rotting logs and stumps, is a rather formidable-looking creature. It is a large beetle, ranging in length from an inch and a half to two and a quarter inches. From the head of its highly polished chestnut brown body protrude two antler-shaped

Maslowski and Goodpaster from Nat. Audubon Soc.

The superb male unicorn beetle.

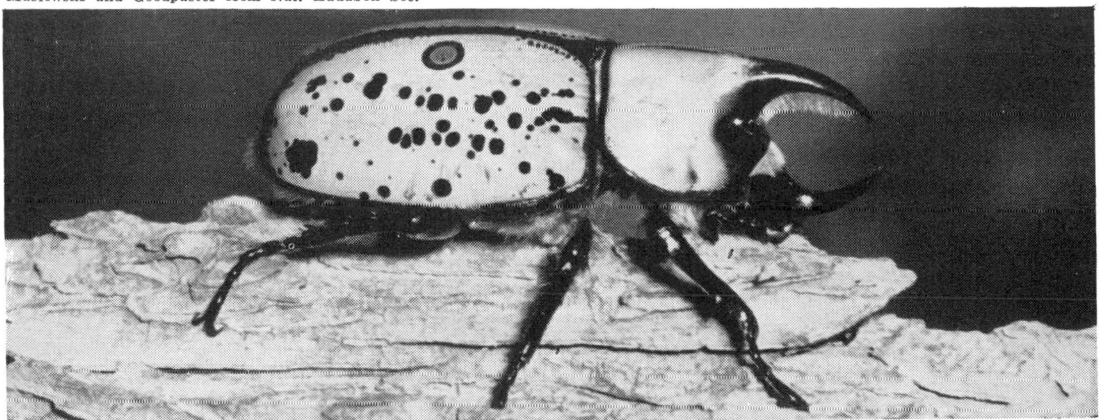

L. M. Chace from Nat. Audubon Soc.

The Hercules beetle of tropical America.

mandibles, or jaws, almost as long as the whole body. The antennae are black, elbowed, and end in small combs. The strong black legs are edged with short spines. The female is smaller than the male; her mandibles are short and stout. It is not known what use the males make of their mandibles. They have been observed fighting each other, not by pinching but by pushing or butting.

The giant stag beetle is particularly common toward the southern part of the United States. Another species, widely distributed east of the Rockies, the pinching bug, is somewhat smaller and lighter brown; its mandibles are much shorter and are sharply curved in. If you push a small stick in front of the pinching bug, it will grasp it with its mandibles and hold on tight while you lift it into the air. Do not offer a finger; the beetle can draw blood.

Like all lucanids, stag-beetle larvae live in old logs and stumps of oak, maple and apple, and require two or more years to mature. The lucanids are not a large family, but they number interesting species in all temperate and tropical forests. A beautiful green species in Chile has a two-inch body adorned with two-inch mandibles. By way of contrast, certain Californian lucanids are less than half an inch in length. Once common in the United States, the lucanids are disappearing as old forests are cut down and logs and stumps are cleared away.

### Scarabs, Dung Beetles, June Bugs (Scarabaeidae)

The scarabs are a numerous and famous family. They are so varied in form, size and habits that about the only characteristics they have in common are antennae which end in a leafy club, large eyes, large and prominent prothorax and strong legs.

They include the largest beetles in the world and also some of the smallest; they may be any color from dull black to brilliant hues on metallic or enameled surfaces. Because of these variations the Scarabaeidae have been divided into six or seven subfamilies. Some entomologists have raised the family to the status of a superfamily, called Scarabaeoidea, and the subfamilies to family rank; but the relationship is the same.

The scarab, the ancient Egyptian sacred beetle, after which the whole family was later named, is one of the dung beetles. These are useful scavengers, which clean up excrement by rolling it into little balls; the females deposit their eggs in the balls and then bury them. The Egyptians held one or two scarab species sacred, as a symbol of resurrection, and placed them in the tombs with their dead. They painted pictures of the scarab on their stone coffins, and made models of it in precious and semiprecious stones.

Another well-known group are the May beetles or June bugs, big-bodied creatures that bumble around in the early evening. Tropical kinds are brilliantly colored; American species are brownish. These beetles are vegetarians, the adults eating leaves and the larvae living underground on roots. The beautiful and destructive Japanese beetle is closely related to the June bug.

Pictures and museum specimens have acquainted many people with the giant tropical scarabs and their grotesque horns. The horn projecting from the prothorax of the male Hercules beetle of the West Indies may be two and a half inches long; the entire insect is about six inches long. The massive elephant beetle, which dwells in the tropical Americas, is the thickest and heaviest of all beetles. It is four inches long; its wing spread of eight inches enables it to fly quite well. The males of the Goliath beetles of Africa and eastern Asia are five inches long. They have no large horns, but the prothorax is enormously swollen and beautifully marked.

### Long-Horned Wood-Boring Beetles (Cerambycidae)

The double-jointed name of the long-horned wood-boring beetles aptly describes the majority of the species in this large family. The term "horn" refers to the insect's antennae and not to the protuberances on the head or prothorax found on some other beetles. The cerambycids range in size from the pygmy beetle of central United States, about one-tenth of an inch long, with antennae of the same length, to the startling *Batocera* of New Guinea, whose three-inch body carries seven-inch antennae. Cerambycids have large eyes and mouth parts; the mandibles of some males are very large and antler-shaped in some tropical species. The legs are long and slender; sometimes the front pair are nearly twice as long as the others. The insects display all colors and color patterns. Usually they have large, powerful

Bernard L. Gluck from Nat. Audubon Soc.

The common milkweed long-horned beetle.

wings; a few are wingless. Many have stridulating organs for producing sound; they make a peculiar squeaking noise. Some species have a pleasant odor; the European musk beetle, a beautiful copper and green insect, smells like attar of roses. Many long-horns are good mimics: some look like bumblebees; others, like wasps. One African species is camouflaged to resemble a piece of velvety moss as it rests on a tree trunk; its antennae appear to be dried twigs.

All cerambycids are vegetarians. The adults feed on fungi, pollen or green leaves; the larvae live inside a plant or tree, where they may spend from one to three or four years. Some larvae grow very large, and furnish a much-desired food for tropical residents, native or European. The larva that lives in the agave, or century plant, in Mexico is a shrimplike creature that is an appetizing addition to a salad.

Cerambycids reach their greatest development in the tropics, where they are avidly sought by collectors; every natural history museum has specimens of the larger and more striking long-horns. Many of the North American species are quite attractive. The common milkweed long-horn, only half an inch long, is bright red with black spots. The *Prionus imbricornis*, about two inches long, is a dark reddish brown, with magnificent, heavy plumed antennae; its larvae infest the roots of orchard trees and grapevines.

### Leaf Beetles (Chrysomelidae)

Wherever you find green leaves you will find leaf beetles, very attractive little insects and very destructive from our point of view. Third largest of the beetle families, the chrysomelids are small to medium-sized and hemispherical, oval or oblong in shape; their mandibles, antennae and legs are usually short. Some are enameled in brilliant colors; many are striped and spotted; others are a dull black or brown.

Most leaf beetles live in the tropics, but there are a thousand species in North America. If you go out on a summer morning after a heavy dew you will find

L. M. Chace from Nat. Audubon Soc.

Head-on view of a long-horned beetle.

leaf beetles drying themselves on the top of sunny leaves. If it is early, you will be able to capture them fairly easily. Later in the morning they will be more alert, dropping to the ground as you approach, or leaping or flying out of reach. Some of the brighter colored leaf beetles may not move so quickly; when you pick one up, a yellow malodorous juice will ooze from its leg joints and will stain your hand. It is believed that birds find this secretion ill-tasting, and so leave the vivid little insects alone.

Each species has its preferred plant food. Many of the larvae feed on the outside of the leaf in company with the adults; they are active, bright-colored, chunky little grubs, in contrast to the pale sluggish larvae that live in the ground or inside a plant stem.

Females of the beautiful genus *Donacia* drop down into the water to deposit their eggs on roots and stems of water lilies, pickerelweed and other aquatic plants, where the larvae will live until they pupate. *Donacia* beetles are gregarious; you can often see numbers of them flying over or resting on the lush vegetation around ponds and swamps.

The chrysomelids number several bad pests, among them the famous yellow and black Colorado potato beetle, which has devastated crops in Mexico, has eaten its way eastward from the southwestern United States and has invaded Europe. Three or four species of brightly marked cucumber beetles do their share of damage to vines. A tiny brown and yellow turnip beetle has migrated from Argentina to become a pest in the southern United States.

### Weevils (Rhynchophora)

The order of the Coleoptera is divided into two suborders. All the beetles described so far belong in the first group. The second contains the weevils, or snout beetles; they comprise several families, most of which have beak-shaped heads. In some species the beak is very long, slender and rigid; other species have spoon- or shovel-shaped beaks. In one

Lee Jenkins from Nat. Audubon Soc.

Snout beetle, or weevil, on a corn stalk.

family, the bark beetles, the beak is so short as to be hardly noticeable or is missing altogether.

All weevils are vegetarians, attacking most trees and cultivated crops. They have been a major pest for so long that now almost any damaging insect is called a weevil, and the term "weevily" is used to describe damaged grains and grain products. However, only a few of this enormous group attack man's possessions. The great majority feed on plants with which man is not at present concerned, or on weeds, which compete with his cultivated crops.

The weevils of the Brentidae family are the oddest-looking; their heads are often as long as their slender bodies, or even longer. This is especially true of the female; she uses her head to bore a deep hole in which to deposit her eggs. Sometimes she gets stuck and cannot withdraw her head; the male, who has been standing guard, then tries to pry her loose by pushing down on the end of her upturned abdomen.

Only a few Brentidae are found outside of the tropics. There they are sometimes two inches long, but never thicker than a match stick. Some live in large colonies under loose bark. One small reddish species is fairly common on oak and maple in the United States; although small, the males of this species are quite as pugnacious as their larger tropical relatives. The tiny one-quarter-inch adult of the sweet-potato borer in the south looks more like a red ant than a beetle.

The curculios, weevils belonging to the Curculionidae family, are the largest family of beetles; there are more than 40,000 described species and probably many others still unknown to man. The curculios include many formidable pests: the cotton-boll weevil, the apple-blossom weevil, the plum curculio, the rose weevil, the granary weevil and the rice weevil, among others.

Curculios range in size from minute to three inches, and all have prominent snouts. A great many are colored dull grey or brown. When alarmed, they fold their legs close to their bodies and remain motionless, looking for all the world like seeds or bits of dirt. Some are extremely beautiful. One of these is the diamond beetle of Brazil, which is covered with scales reflecting brilliant blues and greens from minute grooves. It was in such demand for jewelry at one time that it was almost exterminated.

Bark beetles and ambrosia beetles, belonging to the weevil family of the Scolytidae, have no long snout, are minute in size and dull in color. They are serious forest pests. The female of the bark, or engraver, beetle excavates a passageway along the grain of living trees, just under the bark, and deposits eggs on either side at regular intervals. The larvae work at right angles to the lengthwise passage through the cambium layer, so that the route of one never intercepts that of any other. This makes a pretty pattern, but kills the tree.

Ambrosia beetles penetrate deep into the wood of dead trees. There the female lines tunnels with a yellow fungus, called ambrosia, probably as food for the larvae; the tunnels spoil the value of the trees as lumber. The beetles spend their whole existence in the tunnels, often remaining after the wood has been cut into lumber. The owner of a home constructed with green lumber should not be surprised to find numbers of ambrosia beetles crawling on the walls or floors of his new home.

## Other Beetles

The family of the Meloidae includes the blister beetle, which is well-known to pharmacists. A variety of blister beetle — the Spanish fly (*Lytta vesicatoria*) — is dried and reduced to powdered form to yield the pharmaceutical preparation called cantharides. This is used externally as a blistering preparation for the skin, and internally as a stimulant and diuretic.

The smallest of the beetles are those belonging to the Ptiliidae family, all less than one-sixteenth of an inch in length. They are particularly common under loose bark. To the naked eye they appear to be mites; but the microscope reveals that they are true beetles, with elytra and wings.

*See also Vol. 10, p. 275:* "Insects."

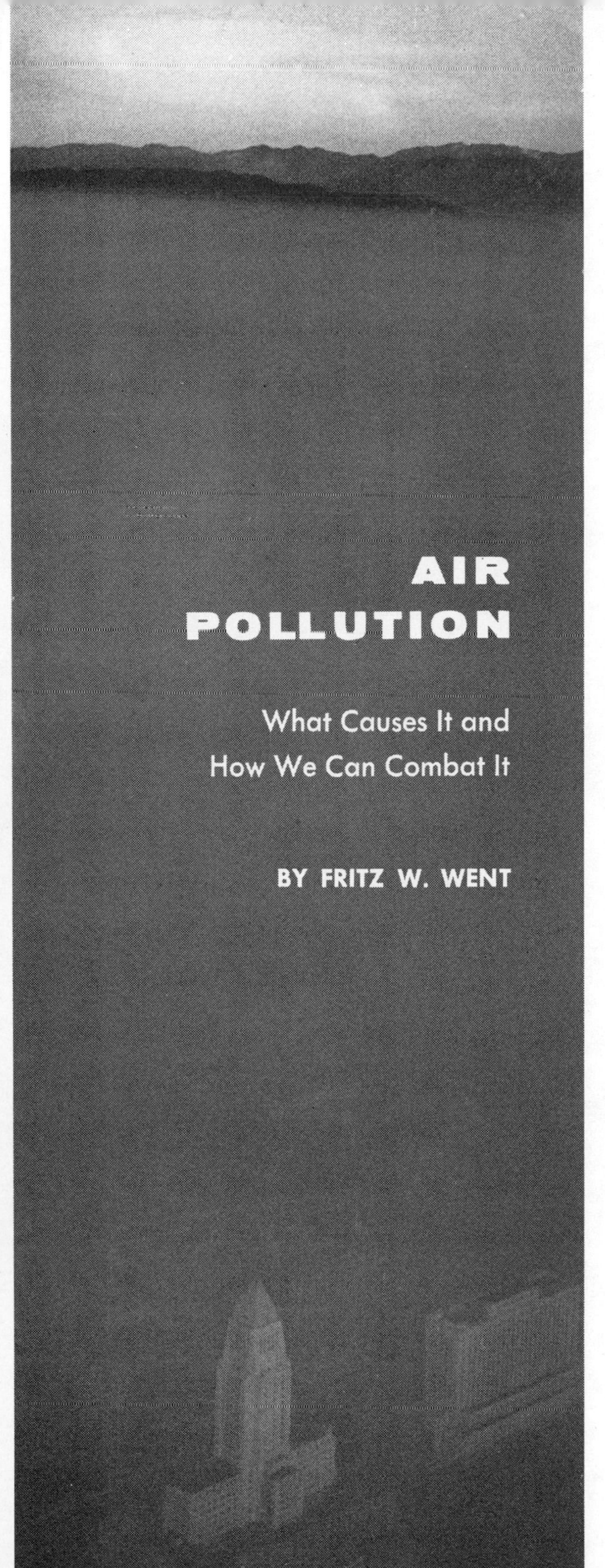

# AIR POLLUTION

## What Causes It and How We Can Combat It

BY FRITZ W. WENT

THE air we breathe can become polluted in various ways. Sand and dust storms and volcanic eruptions introduce into the atmosphere various materials that do not belong there. Salty ocean spray flung into the air along coastal areas is a form of pollution. Forest fires contribute smoke and gases to the atmosphere. The rotting of plant and animal remains produces great quantities of gases. Bacteria, molds and spores are borne aloft; so are wind-blown pollen grains, which are shed from certain plants during their flowering periods. People who are sensitive to these pollens may develop allergic symptoms. (See Index, under Allergies.)

Man himself has contributed considerably to the pollution of the atmosphere. He began to add foreign substances to the air when smoke arose from his first fires. As his civilization has become increasingly industrialized, his contributions to air pollution have increased by leaps and bounds. The smokestacks of factories have added particles and gases to the atmosphere; gases have been emitted from the exhaust pipes of automobiles. The air has also been contaminated in recent years by radioactive fallout from atomic-bomb blasts. In this article we shall deal principally with man-made air pollution.

The solid particles that man's industrial activities contribute to the atmosphere are often in the form of smoke.* Sometimes the smoke is mixed with fog to produce the type of pollution called smog (from "SMoke" plus "fOG").

It is said that the word "smog" was coined in 1905 by a British physician to describe the "pea-soup" fogs of London. United States Weather Bureau officials in Indianapolis used the term in 1926 in their local weather reports. The word is now often used in referring to atmospheric conditions in industrial communities.

In these communities the presence of particles of soot and other polluting substances tends to increase the formation of fogs. The particles provide centers on

* According to a recent definition, smoke consists of small air-borne and gas-borne particles that are numerous enough to be observed.

Allan Grant — Courtesy LIFE Magazine. © 1954, Time, Inc.

Above: diagram showing how temperature inversion is brought about in the Los Angeles basin. In the remarkable photograph below we see how automobile exhaust and other polluting materials are concentrated under a low temperature inversion, causing smog in Los Angeles. The base of the inversion is most clearly defined against the three lofty skyscrapers in the background.

Diagram and photo, Los Angeles County Air Pollution Control District

which the water vapor in the air can collect. As the "smoke-fog," or smog, blankets the community, it holds in suspension the products of pollution and prevents them from escaping to the upper atmosphere. The inhabitants must breathe the polluted air as long as the smog persists.

Smog is particularly liable to develop in areas where temperature inversion develops This is an atmospheric condition that often arises on calm, clear nights, when the soil cools rapidly as heat radiates away from it. The air near the surface of the ground then becomes colder than the air higher up. The name "temperature inversion" is given to this condition because generally the temperature decreases with height above sea level. The cold air near the surface is denser than the warm air above it; therefore it continues to hug the ground. The products of pollution accumulate in it.

Smog is frequent in the Los Angeles area because temperature inversion frequently occurs there and also because the San Gabriel Mountains, to the north, prevent normal air drainage. Temperature inversion causes smog in various other cities.

Large particles that pollute the air — such as carbon soot and fly ash — are retained in the mucous lining of the upper air passages, and few penetrate the inner lungs. However, smaller particles, less than a micron (a thousandth of a millimeter) in diameter, are liable to penetrate deep in the lungs and may cause trouble. For one thing, the victim may become susceptible to pneumonia and other lung ailments. It is believed, however, that the most dangerous type of air pollution is due to toxic, or poisonous, gases, such as sulfur dioxide, hydrogen fluoride and chlorine, which form highly corrosive acids when they are combined with water.

There are various methods of analyzing air pollution. If the air contains particles such as soot or dust, its transparency will be reduced. This will affect the visibility, which can be measured.

A device called the Ringelmann Chart is used in the United States to grade the density of smoke. Black, white and four shades of gray are represented in the chart and are indicated by appropriate numbers. No. 0 is white; Nos. 1-4 are shades of gray ranging from light to dark; No. 5 is black. An observer measuring the density of smoke notes the shade of the Ringelmann Chart that corresponds most closely to the degree of darkening of the sky. No. 0 stands for no smoke; No. 1, 20 per cent density; No. 2, 40 per cent density; No. 3, 60 per cent; No. 4, 80 per cent; No. 5, 100 per cent.

Of course gases are invisible, and so one cannot apply the methods of visual measurement. We can use chemical apparatus to measure the concentration of sulfur dioxide and other gases. Unfortunately analysis with such apparatus is expensive; besides, it tells us only about the gas concentrations that exist while the apparatus is working. However, there is a very clear and simple method by which we can detect the occurrence of most of the toxic gases in the air even several weeks after they first appeared. This is by observing the vegetation.

Many plants, especially leafy vegetables and tender grasses, are injured by industrial gases. Each gas produces different symptoms and generally injures only certain types of plants. If hydrogen chloride gas has been present in the air, the tips of gladiolus leaves will turn yellow or brown and dark brown areas will develop on grape leaves. When ivory-white spots develop between the main veins of alfalfa, spinach or beet leaves, the chances are that the leaves have been injured by sulfur dioxide.

Usually the younger leaves are most sensitive to these gases. It is possible, therefore, to tell how long ago toxic concentrations of the gases occurred by counting the number of uninjured leaves between the growing point and the first leaves that show signs of injury. By taking into account the average length of time it takes a new leaf to develop, one can determine just when these gases were present in the air. From the extent of the injuries, it is also possible to make a rough guess as to the actual concentration of a toxic gas in the atmosphere. In cold climates, where no tender leaves are visible outdoors during winter, one can usually find one or more of the gas-sensitive plants growing in a greenhouse.

Copyright 1954, McGraw-Hill Publ. Co., Inc.

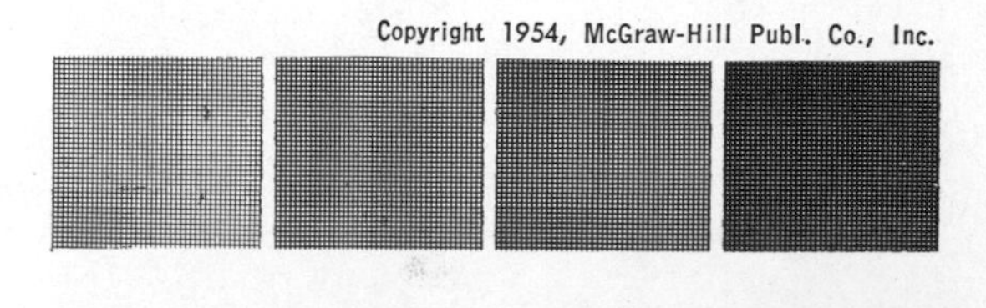

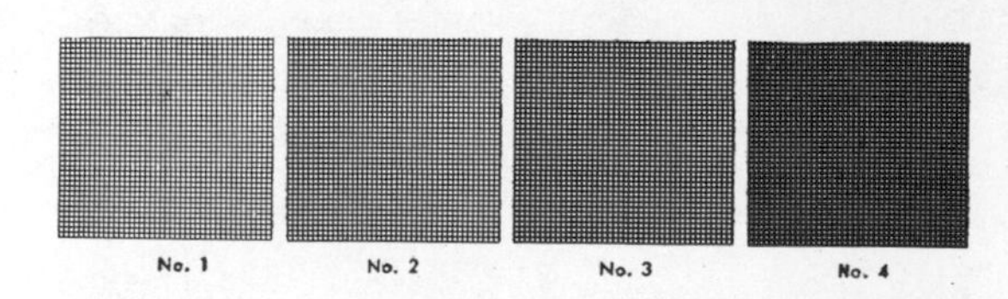


Above is the standard Ringelmann Chart issued by the United States Bureau of Mines. To grade the density of smoke, one notes the shade of the chart that corresponds most closely to the degree of darkening of the sky.

"N. Y. Times" photo

The lowest concentration of sulfur dioxide that can be traced by examining gas-sensitive plants is approximately one part per million. Hydrogen fluoride causes injury if there is one part of the gas to a hundred million parts of air. Plant injuries offer such an efficient method of measuring toxic gases that it is surprising that the method has not been used more.

The effects of air pollution are quite obvious in many cases. It is well known that sulfur dioxide can cause extensive damage to vegetation. All smelters reducing sulfide ores to the metals produce enormous amounts of this gas. Before men realized the evils of air pollution, the smelters would release quantities of sulfur dioxide into the atmosphere. The surrounding vegetation would be killed; the extent of the damage would depend upon the size and location of the smelter. In some cases an area of a hundred square miles or so around the smelter would be blighted. Not a plant would grow there. Dead tree trunks would protrude from the bare soil — or rather bare rock, because there would be no plants to hold the soil. Cheerless settlements would house the factory workers, who could stand sulfur dioxide concentrations that no plant could tolerate. Appropriate control measures have now been adopted. The result is that the vegetation is returning to formerly blighted areas.

Los Angeles County Air Pollution Control District

Univ. of California, Riverside, Cal.

**Silver leaf is a silvery discoloration brought about in plants by smog. Above we see the effects of silver leaf on a petunia (left) and on spinach leaves (right).**

**Below: the endive plant at the left was subjected to smog; the other, just as old, was grown in filtered air.**

Los Angeles County Air Pollution Control District

Smelters or other factories producing sulfur dioxide used to produce great damage to growing crops in agricultural areas. This led to a profusion of damage suits, and farmers were often awarded compensation. Faced with the prospect of mounting damage suits, certain companies carried out research programs to determine how high a concentration of toxic gases a plant could tolerate. Once the limit had been established, settlements were often made out of court. The companies decreased their output of toxic gases by using control devices, which we shall describe later.

A famous case of air pollution by a big smelter — the Trail Smelter in British Columbia, Canada — once led to international complications. This huge smelting plant is just across the border from the United States; its fumes caused damage to crops and forests on the American side of the boundary. The matter was finally settled by agreement between the United States and Canada.

At present the release of hydrogen fluoride by aluminum reducing plants and other factories represents a serious problem. This gas produces even more damaging effects upon both animals and plants than does sulfur dioxide. Progress has been made in reducing the amount of fluorides released by factories.

The effects of air pollution are particularly noticeable in or near industrial cities. In a number of cases human deaths have been traced directly to the effects of smog in such localities. Death struck the old and the weak in each instance. In the winter of 1930, a low fog hugged the

Smog obscuring the main business district of Donora, Pennsylvania. This photograph was taken on October 31, 1948, during the time when smog was taking a heavy toll of the inhabitants. It is estimated that in the period from October 27 to 31, 20 persons died from the effects of the air pollution and 5,000 others became very ill.

World Wide photo

ground for days in the industrialized Meuse valley near Liége, in Belgium, causing throat irritations, but no fatalities. About ten miles south of the city, where there was less industrial activity, the fog became particularly heavy one day. It caused the death of cows and dogs who were out of doors. Many old people succumbed the next day.

These deaths were generally attributed to sulfur dioxide vapors. If this were so, why did no deaths occur in the heavily industrialized area closer to Liége, where the sulfur dioxide concentration was undoubtedly higher? In 1950 I noted that plants had been damaged by hydrogen fluoride near the place where the deaths had occurred. I am convinced that the 1930 fatalities south of Liége were caused by hydrogen fluoride.

In the period from October 27 to 31, 1948, there was a heavy fog in Donora, Pennsylvania. It was made up, in part at least, of wastes discharged to the atmosphere from factories in the area. The fog caused widespread distress and took a heavy toll. Twenty persons died as a result of it and more than 5,000 became seriously ill. The deaths were attributed to sulfur dioxide, though it was agreed that other toxic gases might also have been present.

During a week of very heavy fog in December 1952, some 4,000 Londoners lost their lives. Sulfur dioxide was blamed, I believe wrongly, for the tragic loss of life. I happened to be in London in September and November of 1952. At that time I did not find any plant damage due to sulfur dioxide. However, both inside the city limits and as far as forty miles to the north of the city there was extensive damage to smog-sensitive plants due to concentrations of hydrocarbons (compounds containing hydrogen and carbon). On the foggy days in December there was a very low temperature inversion in the atmosphere above London. This caused the polluting elements in the air to be concentrated in a thin air blanket only a few hundred feet high. Naturally they reached an abnormally high concentration.

In November, then, only hydrocarbons had caused damage to plants, while the sulfur dioxide concentration had not harmed plant life. I assume that the same conditions were present in the following month. These hydrocarbons were undoubtedly contributed by gasoline vapors — the substances responsible for the smog that has plagued the residents of Los Angeles.

In the 1940's the inhabitants of this California city and others near it began complaining about a peculiar kind of smog that made their eyes smart. The "Los Angeles smog" became a favorite topic of conversation. As time went on, conditions became so bad that the city authorities decided to intervene. They ordered proper control devices to be installed in factory smokestacks and barred the use of soft coal for heating and cooking purposes.

Experts predicted that this approach to the problem would not work in Los Angeles. They did not see how the smog could be caused by soft coal, since this fuel was not much used in Southern California. They did not believe, either, that sulfur dioxide had anything to do with the smog. The only sulfur dioxide damage that had been reported was in the immediate vicinity of a sulfuric acid factory.

The experts proved to be right. Through control measures, the total emission of sulfur dioxide fumes in the Los Angeles air was reduced by one half. Yet the smog was as bad as ever, though the only apparent sign of it was a light blue haze. Evidently, the Los Angeles smog was quite different from the known types of air pollution. A special research program was started to study this unusual kind of smog.

It was noted that plants in the Los Angeles area showed a completely new type of damage to leaves. It was usually referred to as silver leaf. It consisted of a silvery discoloration of the lower sides of leaves, particularly those of spinach, beets and tobacco. Silver-leaf symptoms were found all over the Los Angeles metropolitan area after days when the smog had been particularly heavy.

At last it was discovered that silver leaf in plants is due to hydrocarbons in the air, particularly to partially oxidized gasoline vapors. The vapors themselves do not cause the damage. However, when they are exposed to strong light in the presence of even small amounts of nitrogen oxide, they give rise to chemical substances called ozonides and peroxides. It is these substances that are so toxic to plants. If they are present even in the proportion of one part to ten million parts of air, plants exposed to them for only a few hours will develop the silver-leaf symptoms.

Fortunately the ozonides and peroxides soon react with one another and lose their harmful qualities. They form various condensation products that become visible as a light blue haze. Therefore, the haze that is typical of Los Angeles smogs is actually their end product. It is harmless as far as plants are concerned.

The type of smog we have just described is not peculiar to the Los Angeles area. It occurs in many other parts of the world. I have found symptoms of it in San Francisco, Chicago, New York, Philadelphia, Baltimore, Washington, London, Manchester, Copenhagen, Cologne, Paris, São Paulo, Bogotá, Sydney and Melbourne. Apparently the increased use of gasoline has caused a highly toxic material to be added to the atmosphere of cities. It has even been suggested that the offending hydrocarbons may be responsible for some of the cases of lung cancer for which tobacco smoking has been blamed.

Various measures have been taken to combat air pollution. Settling chambers have been provided for smokestacks or chimneys in order to trap smoke particles. The chambers are generally built into the base of smokestacks or chimneys; the particles are allowed to settle, through the force of gravity, from the gases with which they are mixed. To eliminate smaller particles, flue gases are sometimes led

Both lower photos, Los Angeles County Air Pollution Control District

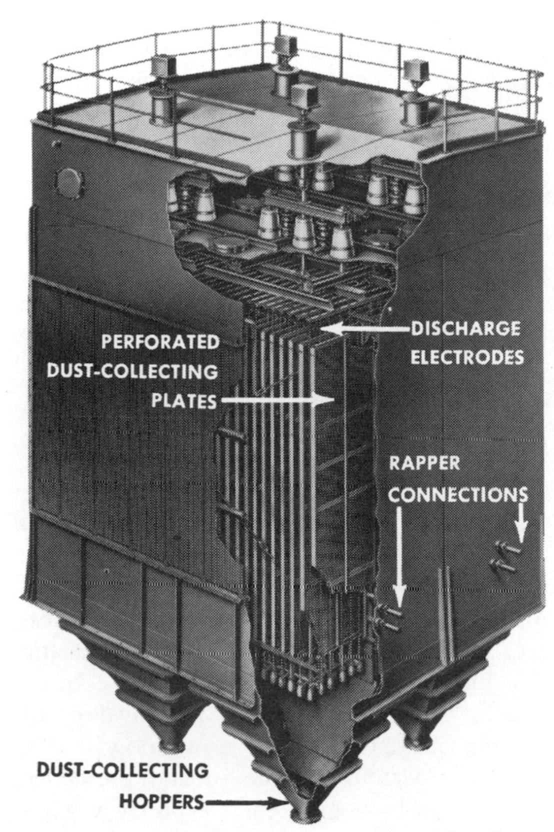


Research-Cottrell, Inc.

**The electrostatic precipitator, shown above, traps the particles contained in smoke. There are a number of dust-collecting plates, arranged within an outer shell; wires called discharge electrodes are set between the plates. The smoke that is to be treated is led into the precipitator. An electric charge is passed through the wires and causes an electric field to be set up; the particles in the smoke become charged. The plates have the opposite charge; the particles collect upon them. The charged particles are later dislodged from the plates by rapping; they drop into hoppers and are removed. Electrostatic precipitators frequently recover valuable materials, such as metallic oxides, which otherwise would have passed, together with the smoke, to the outer air.**

through a series of baffle plates; these provide more time for the smaller particles to settle. Cyclone chambers have also proved effective. In these the flue gases enter a circular chamber in such a way as to form whirling eddies. The particles that are present are caught up in these eddies. Filtering devices have been employed to trap solid particles. Automatic stokers for furnaces burning soft coal help prevent air pollution; so does the use of smokeless fuel such as anthracite.

The device called the electrical precipitator, or electrostatic precipitator, is particularly effective in removing particles of less than one micron in diameter. This device was developed early in the present century by Frederick Gardner Cottrell, an American chemist. On this page we give a diagram of one type of electrostatic precipitator, together with a brief explanation of how it works.

It is possible to remove certain harmful gases from smoke by causing them to pass through a water spray in a device called a scrubber. They are trapped in this device by becoming dissolved in the liquid. It is much more difficult to remove the gases if they are not readily soluble in water. For example, sulfur dioxide ($SO_2$) must first be combined chemically with another oxygen atom to form sulfur trioxide ($SO_3$) before it can be trapped in a scrubber.

One way of counteracting the dangers of high concentrations of toxic smokestack gases is to build lofty smokestacks and to discharge the gases high in the air. If this

**In the photo on the preceding page we see what was once a familiar sight at Los Angeles petroleum refineries. An accumulation of oil-bearing gases had to be burned in a flare to prevent an explosive build-up; plumes of heavy smoke were produced. Today, as shown in the lower photo on this page, all emergency flares are so designed that almost no smoke is passed to the outer air.**

is done, the gases will become so diluted that they will not cause trouble unless a temperature inversion causes them to be trapped in smog.

Researchers are trying to find out how to prevent the escape of unburned gasoline from the exhaust pipes of cars, especially when the motor is idling or when the driver is accelerating. One method would be to improve combustion in the motor. Another would be to install an afterburner, such as we find in jet engines. This device would burn the partially oxidized vapor when it emerged from the exhaust pipe of the car.

If a city's air pollution is due for the most part to sulfur dioxide and to carbon particles derived from soft coal, antismoke legislation can bring about a great improvement. The first ordinance giving an American city government the authority to prevent smoke was adopted by Chicago in 1881. Since that time many cities have passed smoke abatement ordinances.

The example of Pittsburgh shows what can be accomplished in this way. For a long time this city was notorious for the dense smoke released into the atmosphere from steelmaking furnaces and from the smokestacks of other industrial plants. Pittsburgh was called, and with good reason, the Smoky City. In 1947, a drastic anti-smoke ordinance went into effect.* Factory owners were compelled to adopt adequate control devices, such as electrical precipitators and scrubbers; the use of soft coal for heating and cooking was barred. Conditions improved amazingly in short order. Today Pittsburgh is no longer the Smoky City.

There has been a gratifying reduction in the number of cities that are covered with a pall of black smoke coming from unprotected smokestacks. But the bluish haze due to hydrocarbons still floats over many large communities. This haze varies in density accordingly to the severity of the smog. The Dutch-American biologist A. J. Haagen-Smit of the California Institute of Technology was the first to prove that the haze is caused by gasoline vapors. When he combined the vapors with ozone, he produced a typical Los Angeles smog, complete with its characteristic haze.

How can we account for the blue summer or heat haze that hangs over the North American continent, especially during the summer? It looks very much like a dilute Los Angeles smog. Yet it cannot be due to unburned gasoline vapor, because it was present long before the development of automobiles or gasoline. It attracted the attention of early white settlers, who were struck by the bluish color of distant mountains. They named many mountain ranges after this color—the Blue Ridge Mountains in Virginia and Georgia, the Great Smoky Mountains in North Carolina and Tennessee, the Blue Mountains in Oregon and Washington.

Scientists have solved the mystery of the blue heat haze. We now know that it is due to certain volatile substances, called terpenes, which give each type of vegetation its peculiar smell. These substances account for the odor of pine forests, of sagebrush, of decaying leaves in autumn. Like gasoline, terpenes consist largely of hydrogen and carbon. When we drop slightly crushed pine needles, or eucalyptus leaves or sage leaves into a bottle containing some ozone, a blue haze emanates from the bottle. The source of the blue summer haze has to be sought, therefore, in the vegetation that flourishes at this time. There is much less of this haze in winter. There is none at all in areas, such as the oceans or deserts, that have no plants.

We know very little at the present time about the effect of the natural blue haze on living things. It cuts down the total amount of radiation emanating from the sun, particularly the ultraviolet rays. In this way, of course, it affects both animals and plants. There are certain indications that the blue haze may injure the leaves of shade-grown tobacco in Connecticut, Ontario and Virginia. At the present time, however, it does not present a particularly serious problem.

*See also Vol. 10, p. 277:* "Air Pollution."

* It was passed in 1941. In that year the United States entered World War II. The ordinance became a dead letter for a time; people felt that national defense was more urgent than smoke abatement.

# ANTS, BEES AND WASPS

## Social Organization in the Insect World

AMONG the most remarkable of all insects are those that form elaborate social organizations — the ants, and some of the bees and wasps. These insects are known, appropriately enough, as social insects. Their co-operative enterprises are rivaled only by those of man himself. Certainly there is no counterpart, among other living creatures, to the military, food-gathering, cattle-keeping and slave-making activities of the ants, or to the perfectly ordered system of the beehive.

The social insects belong to the order Hymenoptera, which is perhaps the most highly developed group of insects. The order also includes gallflies, sawflies, horntails and ichneumons (parasitic wasps that lay their eggs within the bodies of other insects' larvae). More than 100,000 species are known, and probably as many more still remain to be named. These insects have mouth parts specialized for biting, chewing, scraping, lapping or sucking. Most members of the Hymenoptera have two pairs of stiff and quite narrow wings; most ants, however, are wingless. Of all the Hymenoptera, only the ants and certain bees and wasps are social insects. The rest are solitary insects — that is, they do not live in communities.

The keynote of the social organization of ants is specialization. In the case of solitary insects, each one must perform all the tasks necessary for its survival — collecting food, seeking shelter, fighting off enemies. Ants live in large nests, or ant cities, and divide their labor. Each has its own duties; the rule of one for all and all for one is successfully applied. The members of the nest are divided into three distinct classes, on the basis of the work they do; there are various subdivisions (some twenty-five in all) among these three classes.

The largest class comprises the wingless, normally infertile females, which function as workers or soldiers. These insects are usually the smallest members of the community. They enlarge, maintain and defend the nest, gather food and feed and nurse the queen and young. Occasionally a worker, if carefully fed and attended, may lay fertile eggs. The soldiers are workers with large heads and strong jaws, which are employed in fighting and in crushing hard foods, such as seeds.

The second class is composed of the queen ants, which are large, winged individuals. They are the mainstay of the nest, producing eggs almost constantly, except during the cold season. The winged males are the third class, with small heads and small jaws. Some species have both winged and wingless queens or males.

Ants live almost everywhere — in the deserts and fields, in the forests and mountains, along the seashore and in the villages and dwellings of men. Their homes vary from galleries and chambers excavated in the soil to nests built in hollow stumps. Some nests, constructed of paperlike material, are suspended from trees.

An ant colony may be formed either by a solitary queen or by several queens together. Before the nuptial, or mating, flight, each queen has stored immense energy reserves within her body. On the appointed day, which varies with the different ant species, the winged males and the queens swarm together; mating occurs during or at the end of the flight. It is the queen's only flight. When she alights, she is ready to begin her egg-laying duties. The males swarm, as we noted, on the same day as the queens. After the nuptial flight is over, they seek shelter under stones or sticks or fallen leaves. The males make no effort to return to the nest. They may emerge from their places of refuge and fly around a little; they may find some pollen on which to feed. In most cases, however, they are unable to fend for themselves. Even if they survive the onslaught of their natural enemies, they succumb in a matter of days.

When the queen alights after the nuptial flight, she may be surrounded by worker ants of her own species; these workers will escort her to their home nest and she will become a member of a full-fledged colony. Often, however, the young queen must found her own colony. First she casts off her wings and seeks an appropriate shelter — the underside of a stone or a stick or simply a hole in the ground. She digs until she has hollowed out a burrow, with a chamber at one end of it. The insect then blocks the entrance to the burrow and for a considerable period rests quietly, shut off from the outer world. The wing muscles are gradually broken down and converted into fat bodies, for energy.

Finally the queen begins to lay eggs, and in a comparatively short time the first larvae hatch. The queen is a devoted mother to the first generation of her offspring. She forces saliva, containing fat bodies, into the mouths of the young larvae and they grow rapidly. Should her food supply fail, the queen may even eat some eggs and larvae so as to maintain herself and the nest. If mating occurred in late summer, cold weather comes while the brood is quite young. The queen and her larvae then hibernate until spring.

USDA

Queen ant of a species in the genus *Formica*. The genus includes mound-building wood ants and some allied forms.

While the larvae of certain species are quite small, they spin cocoons and, enveloped in these wrappings, become pupae. (See Index, under Pupae.) Incidentally, the so-called "ant eggs" that pet shops sell as food for baby turtles are not really eggs, but the cocoon-clad pupae of ants. As each pupa becomes mature, it is freed from the cocoon by the mother; it is now a small but perfectly formed worker. In some ant groups, the larva does not spin a cocoon and the pupa is naked.

The workers of the first generation dig their way to the surface of the ground and go in search of food. They bring it to the queen, which is in a rather sad state because she has fed her brood with her own bodily substance. She soon recovers her strength and continues her egg-laying activities. Gradually the workers take over the task of caring for the young. Eventually the queen becomes a sort of egg-laying machine; workers feed her by cramming food regurgitated from their stomachs into her mouth.

As times goes on, the workers become more numerous and larger. Soldiers first make their appearance some time after the queen has established her colony; new queens appear upon the scene last of all. It is the workers that carry on the essential work of the colony. They may move it to a more suitable site again and again; they carry their egg-laying queen, immature workers, larvae and pupae to the new location. Although most workers are sterile, an occasional worker may lay eggs. which will give rise to male or worker offspring.

Male ants are usually smaller than fertile females. Like the drones of bees, they do not play an active part in the work

Philip Gendreau

**Carpenter ants at home in a log, which has been split open to reveal the nest. Using their powerful jaws, the workers have excavated an intricate system of chambers and tunnels in the wood. The cocoons enclose pupae.**

of the community. The one important event of their lives is the nuptial flight which, as we have seen, they do not survive.

It has been shown that much of a young worker ant's behavior results from imitating its elders. The novice must learn the lay of the land around its nest so that it may return to the nest after a foraging expedition. Perhaps it will lose its way and have to be carried home by the more experienced workers. Certain individuals of a colony seem to learn better than others. These key workers initiate the various activities in the colony and thus set an example to the other ants. If these choice workers are removed from the nest, the activity of the colony is retarded; fewer tasks are accomplished and there is less exploration for food. The colony fails to prosper; the health of the queen and of other workers is bound to suffer.

Differences of opinion exist as to whether ants respond only by innate impulses or whether they also possess intelligence and resort to reason. Some students of ant behavior think that an ant acts solely by automatic responses or by responses based on past experiences in obtaining food. Others believe that, at times, an ant will show a very complicated type of behavior that may change as different situations arise. Such modification of behavior is called intelligence. The opinion that ants behave intelligently was boldly set forth by the English author of science works Sir John Lubbock: "When we see an anthill, tenanted by thousands of industrious inhabitants, excavating chambers, forming tunnels, making roads, guarding their home, gathering food, feeding the young, tending their domestic animals, each one fulfilling its duties without con-

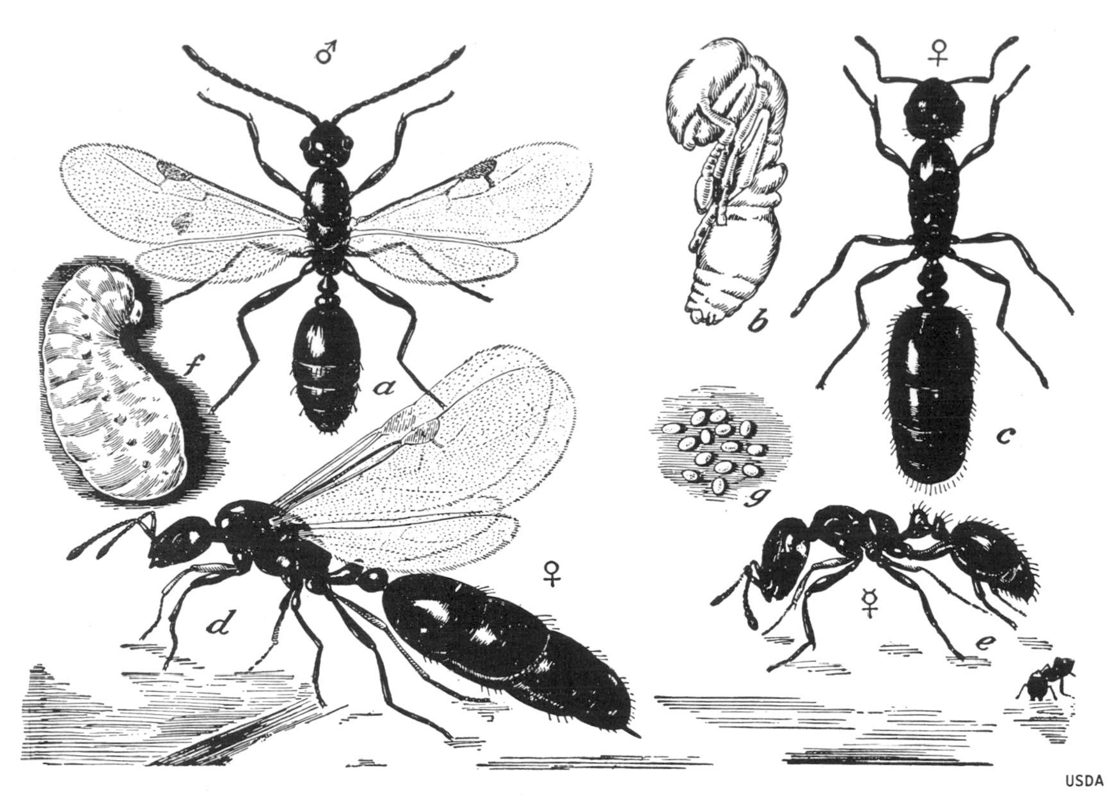


Various stages in the development of ants. *a*, adult male; *b*, pupa; *c*, adult female; *d*, female with wings; *e*, worker; *f*, larva; *g*, eggs. A circle with an arrow indicates a male; a circle with a plus sign, a female.

fusion, it is difficult . . . to deny them the gift of reason; and the preceding observations tend to confirm the opinion that their mental powers differ from those of men not so much in kind as in degree." Modern authorities are not so convinced as was Sir John that his conclusions were correct.

We still have much to learn about the senses of ants. There is general agreement, however, that they have a strong sense of smell. The olfactory organs, or organs of smell, are little sensory pits in the antennae. It is through the sense of smell that ants recognize the members of their own nest and those of other species that they treat as enemies.

### Noises made by various ants

It is thought that at least certain species of ants may be able to hear, because they can make certain sounds that may be a mode of communication with their kind. The ant known as *Myrmica rubra* produces sounds with a strigil, or file, on the seventh abdominal segment. Ants belonging to another genus (*Polyrhachis*) tap on the surface of a leaf with their heads. An Assamese species produces a noise by scraping the end of its abdomen on the dry leaves of its nest.

### Food exchange among the ants

Much of the apparently friendly behavior of ants and other social insects stems from the mutual exchange of pleasing food secretions. The larvae of ants, for example, exude substances that are greatly appreciated by the workers who feed the young and receive these as a "reward." The adults of social insects also engage in this practice among themselves. The material may consist of regurgitated food or glandular secretions. The entire habit in question is called *trophallaxis* (Greek for "food exchange"). Furthermore, foreign insects that live as "guests" in ant and termite nests are tolerated because of their "payment" of food matter desired by their hosts. Certain ants "herd" other insects for the secretions the latter provide.

### Various species of ants keep and milk "cows"

One of the most amazing things about the ant is that it keeps and milks "cows." This practice is carried on by more than one species of ant; the "cow" in question is really an insect, the aphid, or plant louse, which deposits a secretion called honeydew upon the foliage and stems of vegetation. The ants milk their little "cows" by gently stroking them with their antennae, thus coaxing them to give droplets of honeydew. The ants carry their insect livestock to different parts of the plant or to different plants in the garden to make sure that they have enough to eat. Some ants dig tunnels in the soil for the convenience of root-sucking plant lice. Certain ants take the plant lice into their own nests for the winter.

Aphids are not the only forms of animal life that find hospitality in the nests of ants. A great number of beetles, cockroaches, flies and arachnids take up their abode there. They are allowed by their willing or unwilling hosts to feed on the excretions of the ants or on their food. The fostering instincts of ants seem to be extended in various degrees to their guests. More than 1,500 species of Arthropods alone are known to live in more or less cordial relations with ants.

### Ants are sometimes troublesome pests

When ants make their way into our homes, they become unmitigated pests. As they swarm in our kitchens and pantries in almost unbelievable numbers, the

The black-and-yellow mud-dauber wasp (*Sceliphron*) and its nest of clay.

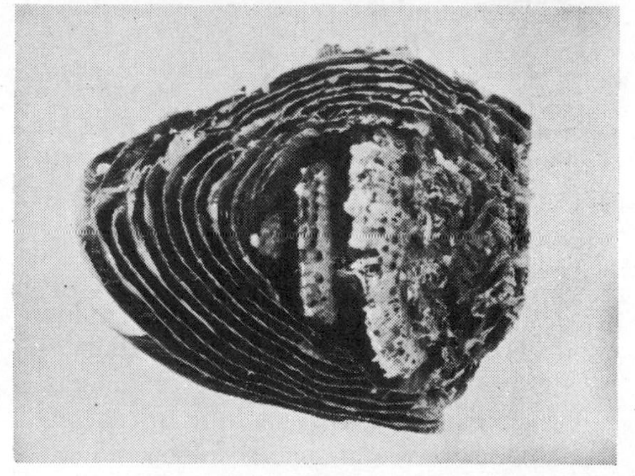

The bald-faced hornet (*Vespula maculata*) and its nest, made of paper.

insects work havoc with our food. The Pharaoh's ant (*Monomorium pharaonis*) is a notorious culprit, nesting in the foundations and often completely overrunning a house from the cellar to the attic. These tiny yellowish or reddish ants quickly sense the presence of food and rush in hordes to consume any sweets, pastry, butter, bread, fat or meat left accessible to them. The

Bernard L. Gluck from Nat. Audubon Soc.

**A female ichneumon wasp. This wasp is not social but parasitic, laying eggs in the young of other insects.**

North American carpenter ant (*Camponotus*), too, enters our homes to live in ceiling beams, porch columns and window sills. Here it may do considerable damage by excavating its galleries and chambers. Unlike termites, carpenter ants do not eat wood but feed on plant juices, animal remains and the honeydew of aphids.

The termites, or white ants, are not ants at all but deserve mention as social insects. They are of the order Isoptera, which refers to the equal size and shape of the wings of those forms having wings. Termites, unlike ants, are not slender-waisted and their antennae are not elbowed.

Termite society is based on a caste system. A queen and king form the royal couple. The queen has an enormously enlarged, cylinder-shaped abdomen. During a year, she lays at least three million eggs and probably more. The small king is devoted to his queen throughout her life, fertilizing her eggs at intervals. Aside from this royal pair, there are individuals capable of reproducing who are kept in readiness for the queen's eventual decline. Workers are soft-bodicd, pale creatures that labor constantly at constructing the nest, collecting food, caring for the eggs and feeding the queen, soldiers and young. With some species, nymphs instead of workers do these tasks. The soldiers defend the nest against invading ants. Some termite soldiers have enlarged, armored heads with formidable jaws; others, with small jaws, have a snoutlike projection which ejects an acrid or sticky fluid on an enemy. Each termite caste includes individuals of all ages and of both sexes.

Termites nest either below ground or above the surface in mounds, trees and stumps. They build enclosed mud runways to their sources of food, which consists of dead tree stumps and limbs and the wood of our houses. This wood is broken down in their digestive tract by protozoa that convert the cellulose into a usable food. Termites also eat fungi and grasses.

Among the true ants, we find gardeners, living receptacles, slaveholders and other interesting varieties. The members of the genus *Atta* — dwellers in the New World — cultivate various species of the fungi known as Basidiomycetes, and eat the fruits of these "gardens." The ants provide a garden bed made of cut sections of leaves. A large ant expedition sets out to obtain this material, generally in the late part of the afternoon. The insects climb up nearby trees and cut off more or less circular pieces

By "Life" photographer Andreas Feininger, © "Time Inc."

Paper wasps swarming over their nest, feeding the larvae chewed-up caterpillars. The larvae are in the open cells. Those cells that are topped with silken caps contain pupae, which will soon mature.

House wasps (genus *Polistes*) building a paper-walled nest. They chew wood pulp and manfacture a strong paper from it.

Photos, Hugh Spencer

of leaves. As the ants return to the colony with their booty, each one carries its piece of leaf overhead like a parasol; for this reason the insects are sometimes called parasol ants. When the leaf sections have been brought to the nest they are thoroughly chewed and then deposited on the floor of a large chamber. As the leaf layers are built up on the floor, they acquire a sponge-like structure. Soon they are covered with the desired fungus growths.

The body of the fungus, called the mycelium, is made up of a great many slender branched threads, known as hyphae. Small spherical swellings develop on the hyphae, and these swellings provide food for the ants. The smaller workers in the colony rarely leave the nest, but spend most of their time "weeding" the fungus garden — that is, removing unwanted growths.

The black honey ant (*Camponotus inflatus*) has a curious but effective way of storing honey. Certain workers, known as repletes, gorge themselves until their abdomens are greatly distended. They then serve as animated honey jars, ready to serve the needs of the other ants. When food becomes scarce, the members of the colony stroke the abdomens of the repletes and devour the droplets of honey that are regurgitated. Repletes are also used to store honey by various species of the genus *Myrmecocystus,* found in the southwestern United States, Central America and South America. After the abdomen of the repletes has become swollen, they cling to the roofs of underground chambers; they are relieved of their loads of honey as occasion requires. The Indians in the areas where *Myrmecocystus* honey ants occur seek out the nests of these ants in order to lay in a supply of repletes, which form a treasured dessert. The Indians bite off and swallow the distended abdomens of the insects.

The tropical driver (army) ants are a carnivorous group whose natural prey consists of insects and other small invertebrates, though they will attack any living thing in their path. Generally, animals such as mammals (including man) and birds can easily avoid the driver ants. They fall victim to the little carnivores only if they are injured and cannot keep out of the way of the ants.

The two genera — *Eciton* and *Dorylus* — that make up the driver ants are found in widely separated areas. *Eciton* dwells in the American tropics; *Dorylus* in the African tropics. Though these ants are completely blind, they advance in long columns with remarkable precision; their foragers fan out on all sides of the main columns. The ants form temporary bivouacs, consisting of clusters of insects hanging from the branches of shrubs and bushes.

The activities of driver ants are often beneficial to man. The natives in the areas where the insects occur simply abandon their huts when the ants appear on the scene and let the insects swarm at will through their homes. When the driver ants finally depart, the premises have been cleared of insect pests, scorpions and various other undesirable tenants.

Parasitism plays an extremely important part in the life of ants. Different species or groups may live off others temporarily or even permanently. For example, various colonies of the genus *Formica,* in the temperate zone, may establish themselves in the nests of other *Formica* communities. A single queen seduces the workers of a foreign colony (of a similar species) to care for her and for the offspring from the eggs she lays in their nest. The original queen of the colony may remain, may be killed or may be driven off, even by her own workers. As the invading queen's larvae mature, they eventually replace all the original workers, who naturally have been dying off. At this point, there is little evidence that the thriving colony had begun by parasitism.

Ant slavery, or dulosis (from the Greek word for "slave"), is carried out by militant colonies raiding and capturing members of other colonies, often of a different ant species. Workers of *Formica sanguinea* carry off the larvae and pupae of the horse ant. These are tended by ants that had been reared from previously captured young. As the new captives mature, they care for the larvae, forage for food and so on in the slaveholding community. However, *Formica sanguinea* workers carry on the main duties of the colony.

In other ant species, such as *Polyergus rufescens,* slavery is an absolute necessity if the colony is to survive. *Polyergus rufescens* slave makers, or amazons as they are called, invade a nest of ants belonging to the species *Formica fusca,* kill the queen and introduce their own queen into the nest. Thereafter the *Formica* workers care for the amazon brood. The *Polyergus* workers do nothing but raid other nests in order to make off with larvae and pupae and thus keep the slave population at the appropriate level. The slaves carry on all the work of the colony.

Most species of wasps do not form colonies; they are known as solitary. The social wasps all belong to the family Vespidae; the most familiar representatives of the order are the common wasps and hornets, of the genus *Vespa.* The young *Vespa* queens mate in the autumn. They alone of all the members of wasp colonies survive the winter season, hibernating until the coming of spring. Roused by the first warm spring days, each young queen

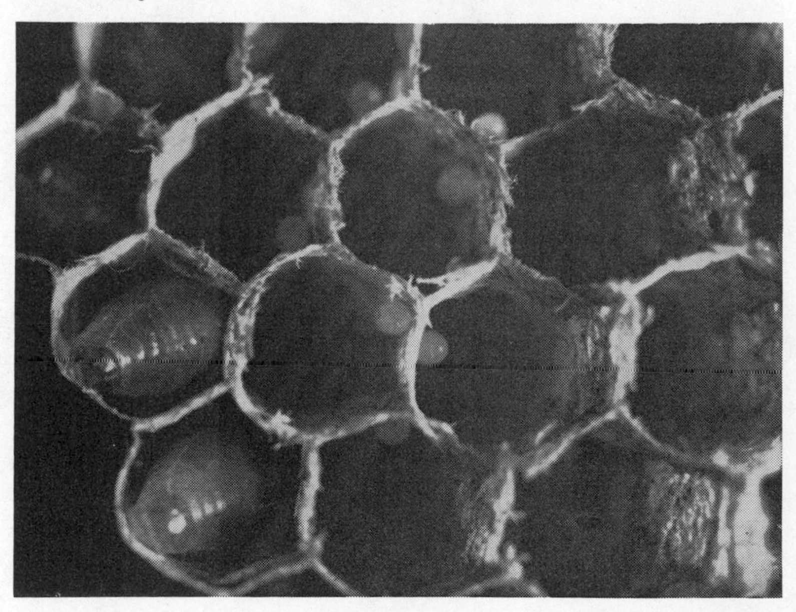

Eggs and larvae in the comb of a house wasp. An egg is deposited in each cell; then larvae (grubs) hatch from these eggs.

seeks a place for a nest — perhaps under an eave, or in a hollow tree or upon a bush. There she rapidly constructs a series of cells of paperlike structure, formed from woody fibers and other vegetable matter. She deposits an egg in each cell. When the larvae hatch, the queen feeds them with chewed caterpillars or flies or fruit juices. When the time comes for the larvae to assume the pupa form, they close their cells, undergo their metamorphosis and emerge as undeveloped females, or workers. The workers assist the founder of the colony in making more cells and in feeding the larvae. Tier after tier is made, until the nest becomes a truly imposing structure.

Toward the end of the summer young queens, and after that the males, appear. The earlier generations have all been undeveloped females, which, as we saw, are also called workers. Some of these are capable of laying eggs that may develop into males. The young queens and the young males finally leave the nest, never to return. The workers that remain destroy the larvae that are left in the nest; then they themselves await death.

The social wasp, whose life cycle we have just described, eats great numbers of insects that are harmful to man. On the other hand, it is quite destructive of fruit and sometimes attacks human beings, particularly when it is molested or fancies that it is molested. The hornet is our largest wasp, and its sting is particularly formidable to man.

Not all wasps form their nests of paperlike material. Some of the solitary wasps make nests of sand, clay or mud — quite efficient dwellings, though less imposing than the beautiful structure fashioned by the social paper wasps. Some of the solitary wasps have a rather singular method of supplying their young with food. Since

A worker bee cooling the air of the hive with its wings.

Both photos, Edwin Way Teale

Queen honeybee surrounded by a retinue of worker bees.

there are no workers among these wasps, the mother wasps have to provide food to which the larvae, upon hatching, may help themselves. This is brought about by the capture of various insects, particularly caterpillars, and spiders. A mother wasp stings the prey in a vital spot and paralyzes it; then she drags it off to her nest. The wretched victims remain alive, but motionless and helpless, to await the attack of the larvae, as soon as hatching takes place. Of course, in this way the young of the solitary wasps are assured of a bountiful supply of fresh meat!

Other remarkable members of the hymenopterous order include wasps or wasplike forms such as the sphex, which preys in the main upon members of the grasshopper tribe; the wasps belonging to the genus *Pompilus,* which do not hesitate to attack even such large spiders as the wolf spiders; the gall wasp, or gallfly, which lays its eggs in plants and causes galls to form. Finally, there is the ichneumon fly, which deposits its eggs on or beneath the skin of various insect larvae. When these eggs hatch, the young feed on the tissues or juices of the victims. Because the ichneumon fly in this way kills great numbers of larvae of insects that are harmful to man, it ranks as one of man's most valuable allies in the insect world.

### Bees — solitary, parasitic and social

We now come to the Hymenoptera called bees, which store pollen and honey in their nests to provide food for their young. Pollen is collected by the hairs on a bee's legs and body and also by specialized structures, called pollen brushes, which are found on the hindlegs and sometimes on the abdomen of female bees. After pollen is collected, it is brushed off by the insect's head and feet, dampened with dew or some other form of moisture, mixed with honey from the bee's mouth and formed into tiny pellets. The pellets are then pushed into the so-called pollen baskets. These baskets, consisting of long and stiff hairs on the hindlegs, hold a considerable number of pollen pellets.

Honey is a product of nectar, a sweet liquid secreted by the glands found in certain flower petals. The bee sips nectar and swallows it. The substance is transformed into honey in the insect's crop, or honey sac, and is later disgorged. Honey is made up chiefly of the sugars levulose and dextrose and of water; it also contains dextrins, gums, vitamins, enzymes, pollen grains and various minerals. Bees also collect a sticky substance called propolis, derived from the resinous secretions of various trees. Propolis is used as a cement in the building of nests.

### The bee's ovipositor is a most useful organ

The abdomen of bees is joined to the thorax by a rather slender "waist." Females have an organ called an ovipositor at the end of the abdomen. This serves as an egg-layer and also as a weapon; it can inflict a painful sting. In the case of honeybee workers, the ovipositor, or sting, has little barbs that turn inward; if the worker stings a foe, the sting is generally left in the body of the victim and the bee dies. The ovipositor of the honeybee queen is smooth; this bee can sting its enemies repeatedly without harm to itself.

The bees make up a superfamily, the Apoidea, comprising some 20,000 species. Contrary to popular belief, most species do not form communities but are solitary in habit. The female of the solitary bee builds her nest cell by cell. She stocks each cell with pollen mixed with nectar and then lays an egg on this food supply. She seals each nest before she goes on to the next one; after the entire nest has been completed, she closes it up and flies away, never to return. The larvae that are hatched from the eggs go through a series of molts, become pupae and finally emerge from the nest as adult bees.

There are some interesting varieties of solitary bees. The small carpenter bees (Ceratinidae) and the large carpenter bees (Xylocopidae) dig through wood and make nests in the resulting burrows. The mason bee (*Chalicodoma*) of southern Europe constructs its nest of soil and tiny pebbles, mixed with saliva. The leaf-cutter bees, belonging to the family Megachilidae, form their nests of pieces of leaves and petals cut from roses and other plants.

Certain bees lay their eggs in the nests of other bees; their young become unwelcome lodgers. These parasitic bees are known as inquilines, from the Latin word *inquilinus,* meaning lodger. For example, the parasitic bee *Stelis* lays its eggs in the cells of *Osmia.* Larvae of both species are hatched; ultimately the *Stelis* larvae devour those of *Osmia* and then they take over the nest as undisputed proprietors.

Canadian Agr. Sc. Serv.

**Swarm of bees entering a man-made hive. Led by their queen, bees leave an old hive when it is overpopulated.**

The social bees — bees that form communities — have developed distinct castes, which correspond more or less to those found among ants. In each hive there is a mature queen bee, a number of males, which are known as drones, and a great many workers, or undeveloped females.

Perhaps the best-known of the social bees is the honeybee (*Apis mellifica*). It builds an elaborate hive with wax secreted from eight wax pockets on the underside of the abdomen. The honeycombs in this hive are set vertically, side by side; each comb consists of thousands of hexagonal cells. Some of the cells contain eggs, or larvae or pupae; others are used for the storage of honey and pollen.

At the beginning of the spring, the honeybee hive contains a queen and a comparatively small number of workers. The queen begins to lay eggs. Some of these will develop into workers; others into drones; a few into queens. The latter are reared in special enlarged cells, which are called royal cells. When the eggs hatch (after about three days), the larvae, or grubs, are all fed for the first two or three days with royal jelly, a secretion from certain glands of the workers. After this period is over, the prospective workers and drones are gorged with honey and pollen. The larvae that are to be reared as queens, or rather princesses, continue to be fed with royal jelly.

#### Cocoons spun by bee larvae

The larvae grow rapidly. After about six days, the cell is sealed by the attendant workers, and the larva becomes a pupa. The larvae that are to become workers or drones spin a complete cocoon about themselves before entering the pupal stage. The cocoons spun by the larvae that are to develop into queens enclose only the head, thorax and a small part of the abdomen.

After twelve days or so, the workers cast off the pupal skin and chew their way out of the cell. It takes the drones about two days longer to pass through the pupal stage. Princesses develop from pupae to adults in about seven days. They do not emerge from their cells; attendant bees make a small hole in each royal cell and continue to feed the occupant.

In time the honeybee colony becomes overpopulated and a form of emigration, called swarming, takes place. The old queen, accompanied by many of the workers, leaves the hive and seeks a new nest; this may be the hollow of a tree or a man-made hive. One of the young princesses now emerges from her cell in the old hive. She makes her way to all the other royal cells and slays their occupants. If two or more princesses emerge at the same time,

## THE BEEKEEPER AT WORK

Beekeeper blowing smoke into a beehive containing a comb that he wishes to examine. The smoke has a calming effect on bees.

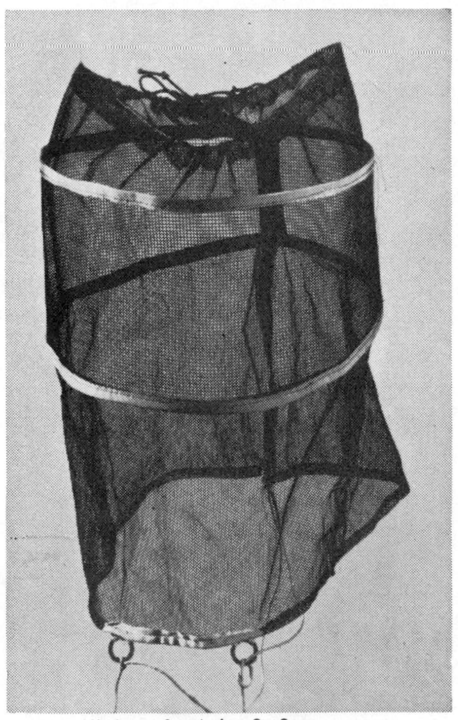

All photos, Canada Agr. Sc. Serv.

A bee veil (above), which a keeper may wear while he handles bees. Made of wire cloth, it protects the head.

The bee smoker (right) is a cylindrical "stove" provided with a small bellows and a nozzle. It blows out smoke.

they fight until only one remains alive. The newly established queen then flies from the nest, mates with a drone, returns to the nest again and begins to lay eggs. The drone dies immediately after mating. The rest of the drones are tolerated in the hive for a time, though they perform no communal tasks. If the food supply dwindles, however, they are driven out of the nest and are not permitted to re-enter. Since they are dependent on the food brought in by the workers, they are doomed to die once they are excluded from the nest.

Workers have different tasks. Some of them collect pollen and nectar; others care for the young; still others attend the queen. A certain number fan the hive by means of rapid wing movements. This is a most important task; if the premises become too hot, the wax cells would melt.

As a result of the researches of the German zoologist Karl von Frisch, we now know that a worker bee conveys information about a promising food source to the other bees by means of a series of dances. When the food supply is near at hand, the returning bee performs a round dance on one of the vertical combs of the hive. It circles to the right and then to the left, over and over again.

If the supply of food is a hundred yards or more distant, the returning bee performs a tail-wagging dance. First it makes a short straight run up or down the comb, wagging its abdomen from side to side; it then circles to the left. Again it makes a straight run with a tail-wagging motion and this time circles to the right. The bee tells how far away the food is by the speed of the dance; the slower the dance, the farther away the food. It indicates the direction of the food by the direction of its straight run. If the dancer heads directly upward during the straight part of the dance, it means that the feeding place is in the same direction as the sun. If the insect heads directly downward in its straight run, it means that the workers must fly away from the sun to reach the food. Suppose the bee goes 45° to the right of vertical in the straight part of its run. This would indicate that the feeding place is located at 45° to the right of the sun.

The honey produced by the honeybee has been used as a food since prehistoric times; the culture of bees, or beekeeping, probably goes back to the days of the early Egyptians. For many centuries the techniques of beekeeping did not differ greatly from those described by Vergil in the fourth book of the GEORGICS. A revolution in beekeeping practice was brought about in 1851 when the American L. L. Langstroth invented a movable frame for beehives.

Today the Langstroth type of hive is widely used, particularly in the United States. It consists of two or more boxes without top or bottom, set on top of one another. The undermost box, to which a bottom board is fitted, is used as a brood chamber and storehouse. A number of movable wooden frames are set vertically in the box. A thin sheet of beeswax is fitted in each frame and provides a base upon which the bees construct their cells. The beekeeper never removes honey from the box serving as the brood chamber.

Honey is collected from a similar box, called the superbox; it is set upon the brood chamber and is provided with a top. Often the two boxes are separated by a perforated zinc sheet. The holes are just large enough to permit the workers to make their way through the holes. The queen is too large to pass through them; therefore she must remain in the brood chamber, where she will continue to lay eggs. To collect honey, the superbox is removed and an empty one is put in its place. A series of superboxes may be set in position, one on top of the other.

Honeybees are quite gentle when handled by an experienced beekeeper; however, certain precautions must be taken. To make bees tractable, smoke is blown into the entrance and over the frames by the bee smoker shown on the previous page. The beekeeper generally uses a bee veil made of wire cloth to protect his face, but he handles the bees with his bare hands. Even the most experienced beekeepers are stung now and then by the insects; they soon become immune to the poison injected by

Douglas Grundy from Three Lions

Above, a beekeeper scrapes an empty comb from a frame. Beeswax, secreted by bees to make the comb in which honey is stored, is a valuable by-product used in making candles.

Left, frames are placed in the centrifuge used to extract honey from the comb. The combs have earlier been uncapped of the sealing agent deposited over the surface by the honeybees.

Below, honey is drawn off through an opening at the bottom of the centrifuge in which it has just been extracted. It must be strained at least twice to remove all foreign particles.

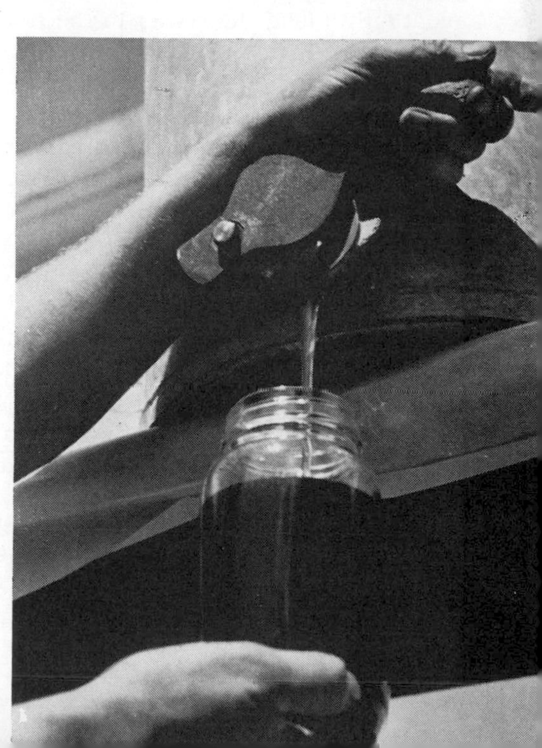

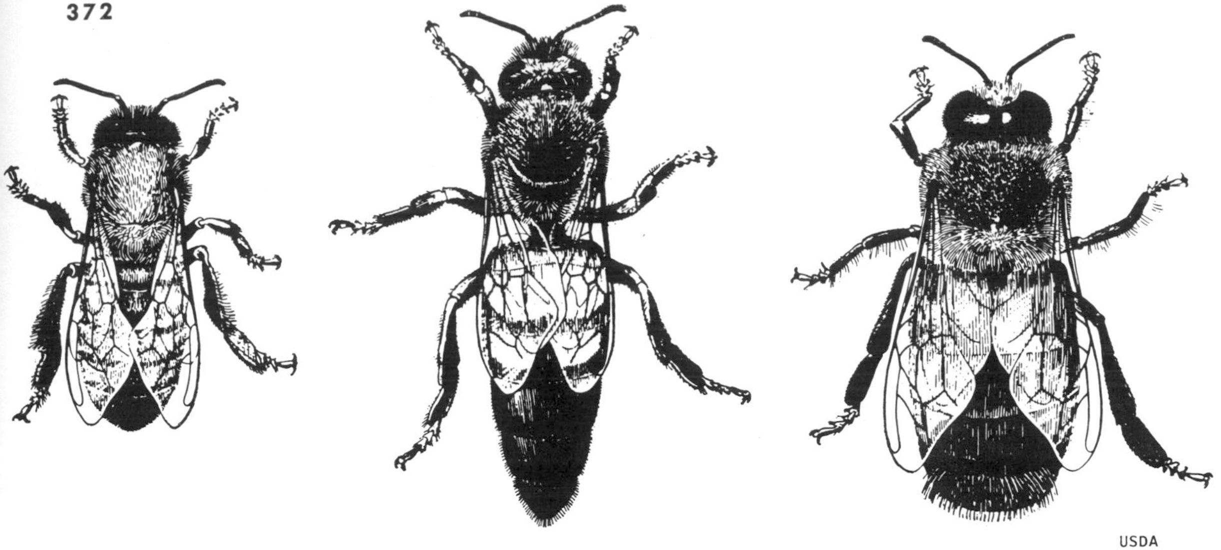


The three castes of the honeybee. Left: a worker; center: a queen (female); right: a drone (male).

the insect's ovipositor, so that no swelling takes place following the sting.

Swarming takes place from man-made hives as it does in nature. Sometimes the beekeeper waits for the bees to cluster on the limb of a tree or bush; then he saws off the limb and carries it to a new hive. Artificial swarming is often practiced nowadays. An empty hive is set near the old one. The queen and one or more brood combs are removed from the old hive and put in the new one; the old hive is then set in another location. Bees on the wing as well as the older bees in the old hive will eventually make their way to the new one.

The marketable products of beehives are honey for human consumption and beeswax. Honey in solid form is called comb honey. Liquid honey is prepared from combs in a device called a honey extractor. The comb is put in the extractor and is then set spinning at a high rate of speed. The honey flies out of the comb by centrifugal force against the sides of the extractor and is drawn from the machine in the form of a liquid. Beeswax, secreted by bees and used in building the comb, serves for waterproofing, modeling wax, cosmetics, ointments, lithographic inks, sacramental candles and various other purposes.

The social bees known as bumblebees (Bombidae) range in length from less than three eighths of an inch to an inch or more. Bumblebee colonies, like those of wasps, must be started anew each year. Each colony is established in the spring by a young fertilized queen, which has spent the winter under brush or debris or in a hole in a log. The queen seeks a site in or on the ground; this is often the vacated nest of a field mouse or chipmunk or other small mammal. She may have to fight other bumblebee queens for possession of the site. Once established in her new home, the queen prepares balls of honey and pollen on which she lays her eggs. With the coming of cold weather the old queen, her retinue of workers and the drone hangers-on all perish; only the young queens survive to continue the race.

The stingless bees (Meliponidae) are a third important group of social bees; their ovipositors are atrophied. These bees are much smaller than honeybees; some species are only a twelfth of an inch long. The Meliponidae build combs in horizontal sections. Some make their nests on the ground; others in the hollows of trees; still others establish themselves in the nests of termites or ants. In constructing their abodes the bees use earth and leaf particles, dung and other substances in addition to the wax that they secrete. They store honey in fairly large receptacles. Human beings occasionally use the honey as food — a dangerous practice, as this substance is sometimes highly poisonous.

*See also Vol. 10, p. 275:* "Insects."

# SCIENCE GROWS UP (1600-1765) III

BY JUSTUS SCHIFFERES

### *THE PLANT AND ANIMAL KINGDOMS*

THE invention of the microscope was only one of several factors that opened men's eyes to the infinite variety and complexity of living things on the face of the globe. A new era of scientific exploration re-emphasized this fact. French, English, Dutch, Portuguese and Spanish ships sailed the seven seas; they touched at hitherto unknown islands in the East and West Indies and they also cruised along the coasts, often uncharted, of North America, South America, Africa and Australia.

The explorers returned to their home ports with vast amounts of new information and heaps of animal and plant specimens. Of outstanding importance, for example, was quinine bark sent back to Europe from the jungles of Peru for the treatment of malarial fevers. Red Indians from North America were also brought back and exhibited in Europe as "noble savages." The native soil of Europe, too,

Vessels like this French frigate brought back new information and heaps of plant and animal specimens from the New World.

was re-examined for species of plants and animals that had never before been closely observed and described; many notable finds were made.

All this was new and exciting; but it was also extremely confusing. By the middle of the eighteenth century, the steady accumulation of facts and specimens was bringing about a state of chaos in the study of natural history, forerunner of the science of biology. In order to bring order out of this chaos, some way had to be found to name and classify living things. This method would have to be simple, practical and useful; furthermore it would have to win the acceptance of botanists and natural historians the world over. Without agreement on names that completely identify objects under discussion, scientists cannot communicate effectively with one another.

The man who did more than any other to provide living things with useful scientific names and to classify them according to a logical system was the "prince of botanists" — the Swedish naturalist Carl von Linné, better known as Carolus Linnaeus. (This was the Latinized version of his name.) The son of a poor clergyman, who took his family name from the linden tree growing in his yard, he was born plain Carl Linné, in 1707, at Rashult, in the Swedish province of Smaland. He was an enthusiastic student of natural history from an early age; after studying at the University of Upsala, he was appointed assistant lecturer in botany and was given the entire direction of the University's botanical garden. Two years afterward he explored Lapland in search of botanical specimens. Upon his return, he described his finds in his PLANTS OF LAPLAND, the first of a series of treatises that were to bring him fame. Later he made a similar tour of the Swedish region of Dalecarlia.

**A doctor of medicine who turned to botany**

In 1737 Linnaeus received a medical degree from the University of Harderwijk, in Holland. There he spent much time in botanical studies. He worked as curator of a private botanical garden owned by a rich Dutch merchant. After considerable travel in Holland, France and England, Linnaeus settled in Stockholm as a physician. In 1741 he was named professor of medicine at Upsala; the next year he exchanged this professorship for one in botany. His fame as a naturalist now spread far and wide. Travelers from distant lands kept him supplied with an incessant flow of new species of plants and animals; students flocked from all parts of Europe to hear his lectures and to accompany him on field trips. In 1761 he was ennobled — the first man of science to be so honored in Sweden — and from that time he was known in his native land as Carl von Linné ("von" indicating nobility). He died in Upsala in 1778.

Throughout his scientific career, Linnaeus was absorbed by the problem of naming and classifying the numberless

CAROLUS LINNAEUS

members of the plant and animal kingdoms. He was born with an instinct for sorting and arranging things. Indeed, in jest, he once went so far as to classify all the living botanists of his day by assigning them military rank, according to their merit; his name, with the title of general, led all the rest!

Other biologists before Linnaeus, including the Swiss Caspar Bauhin (1550-1624) and the Englishman John Ray (1628–1705), had interested themselves in the problem of classifying living things. Linnaeus was the first, however, to provide a practical and widely adopted classification system. It was presented in four treatises: THE SYSTEM OF NATURE (1735), THE CLASSES OF PLANTS (1738), BOTANICAL PHILOSOPHY (1751) and THE SPECIES OF PLANTS (1753).

Linnaeus' classification of plants was based on the observation of their visible

flowers, more particularly on the number and arrangement of their reproductive organs (stamens and carpels). He divided all plants into classes according to the number of stamens (male elements) in each flower; each class was divided into orders according to the number of carpels (female elements) in each flower. Orders were further subdivided into genera (plural of genus); genera into species. Linnaeus arranged animals in six classes as follows: (1) mammals (give birth to living young); (2) birds (produce eggs); (3) amphibians (breathe by lungs); (4) fishes (breathe by gills); (5) insects (have antennae); (6) worms (have tentacles). Linnaeus' "sexual system" of plant classification was simple and easily applicable and was very popular for years to come; but his animal classification did not win much favor. Both systems have long since been discarded. (For the modern system of plant and animal classification, see the article The Roll Call of Living Things, in Volume 1.)

Linnaeus' most lasting contribution to systematic classification was his use of a genus and species name for each living

COUNT DE BUFFON

The plate at the left, showing a squirrel, appeared in Volume 7 of the Count de Buffon's lavishly illustrated *Natural History*.

thing — a system called binomial nomenclature (two-term naming). In this system, which endures to the present day, a lion is *Felis leo;* a dromedary, *Camelus dromedarius;* a white oak, *Quercus alba;* and man himself, *Homo sapiens.* These terms are in international use; thus, a lion is *Felis leo* in France, Sweden and Australia as well as in the United States. Linnaeus gave simple and concise descriptions of the living objects for which he selected names; he used obvious and visible characteristics that made identification easy. Thus, of *Camelus dromedarius* he wrote: "A second chambered stomach for pure water, providing for a long time in the thirsty desert. Carries burdens, makes haste slowly, when weary lies down on its breast."

Modern botanists have accepted as the basis for naming plants the binomial nomenclature in the first edition of Linnaeus' SPECIES OF PLANTS (1753); the tenth edition of his SYSTEM OF NATURE (1758) is the basis for naming animals.

Linnaeus held that the various species of animals and plants are fixed and immutable; that is, once and for all created and not subject to change. When the Darwinian evolutionary theory of the origin of species came into vogue, Linnaeus' reputation suffered considerably. Yet he will always be honored for his contributions to plant and animal classification. His name lives on, too, in the famed Linnaean Society of London, which has furthered botanical studies. The society was founded in 1788 to purchase and bring from Upsala Linnaeus' famous collection of plants (herbarium) and his library.

#### An aristocratic French naturalist

Another great naturalist of the eighteenth century was the aristocratic Frenchman Georges Louis Leclerc, Count de Buffon (1707–88), a brilliant, handsome, witty man of the world. He was little interested in classification systems like those of Linnaeus; it was his aim, rather, to set forth in glowing prose the whole story of living things and their world.

In his youth, disregarding his father's wish that he follow the profession of law, he devoted himself to the study of physics, mathematics and agriculture. He considered himself first of all a physicist. Like Voltaire, his friend and admirer, he was tremendously interested in the works of Newton and translated his FLUXIONS into French. However, a tour of the European continent in the company of a young Englishman, Lord Kingston, awakened Buffon's enthusiasm for nature. In 1732 he came into a considerable property, inherited from his mother, and from that time on he devoted himself almost entirely to natural history. In 1739 he became Keeper of the Royal Gardens (Jardin du Roi) and of the Royal Museum. Thereafter he divided his time between the Royal Gardens in Paris, where he had a residence, and the little village of Montbard, in Burgundy, where he was born.

#### The *Natural History*—Buffon's masterpiece

In Montbard he started his monumental work on natural history, in which he hoped to sum up everything then known about the subject. The NATURAL HISTORY, GENERAL AND PARTICULAR in forty-four huge volumes, copiously and beautifully illustrated, was published over a period of fifty-five years (1749–1804); the last eight volumes, edited by B. G. E. Lacépède, appeared after Buffon's death. The first volume is devoted to general problems; the next fourteen deal mainly with mammals. Following are seven supplementary volumes, including his famous EPOCHS OF NATURE, which we take up in Science in Revolution III, in Volume 4. Then come nine volumes on birds and five on minerals. The last eight volumes take up reptiles, fishes and cetaceans (mammals that live in the water).

The NATURAL HISTORY was immensely successful from the appearance of the first three volumes in 1749. It was the first important popularization of natural history; the story of living things and of the earth is told clearly and vividly. As a describer of nature, Buffon is in many respects still unexcelled. As a literary stylist he ranks high; his NATURAL HISTORY is considered a masterpiece of French literature as well as of scientific writing.

To him style was as important as subject matter. In the DISCOURSE ON STYLE, an address that he delivered when he was received into the French Academy in 1753, he declared that "a man's style is the very man himself." He went so far as to claim that "well written books are the only ones that will pass on to posterity." This may seem exaggerated; yet it is indeed true that scientific reputations are often properly made by the ability to express difficult thoughts in simple words.

The NATURAL HISTORY is more than a delightful popularization of scientific knowledge. Buffon laid the foundation of anthropology (the study of man) in the modern sense by discussing the peculiarities of wild tribes, the abnormalities existing among civilized peoples, the development of speech in the child, the influence of emotions upon facial expression. He made notable contributions to geology, too; in his EPOCHS OF NATURE, he gave a plausible account of the origin of the earth — the best exposition given up to that time.

Buffon is recognized as one of the forerunners of evolutionary theory. In a celebrated passage of the NATURAL HISTORY he remarks that all animals might be regarded "as forming one and the same family [which contains] other small families, produced by time . . . If one admits this, it could be said likewise that all animals have come from one animal, which, in the course of time, by perfecting itself and by degenerating, has produced all the races of other animals." Again and again in the NATURAL HISTORY he points to the close similarities between kindred species, such as the wolf and the dog, the horse and the ass; he implies that one species may well have developed from the other. We even find in the NATURAL HISTORY a sort of first draft of the doctrine of the survival of the fittest. For Buffon called attention to the great fertility of living things and to the fact that only the checks provided by natural enemies and by the

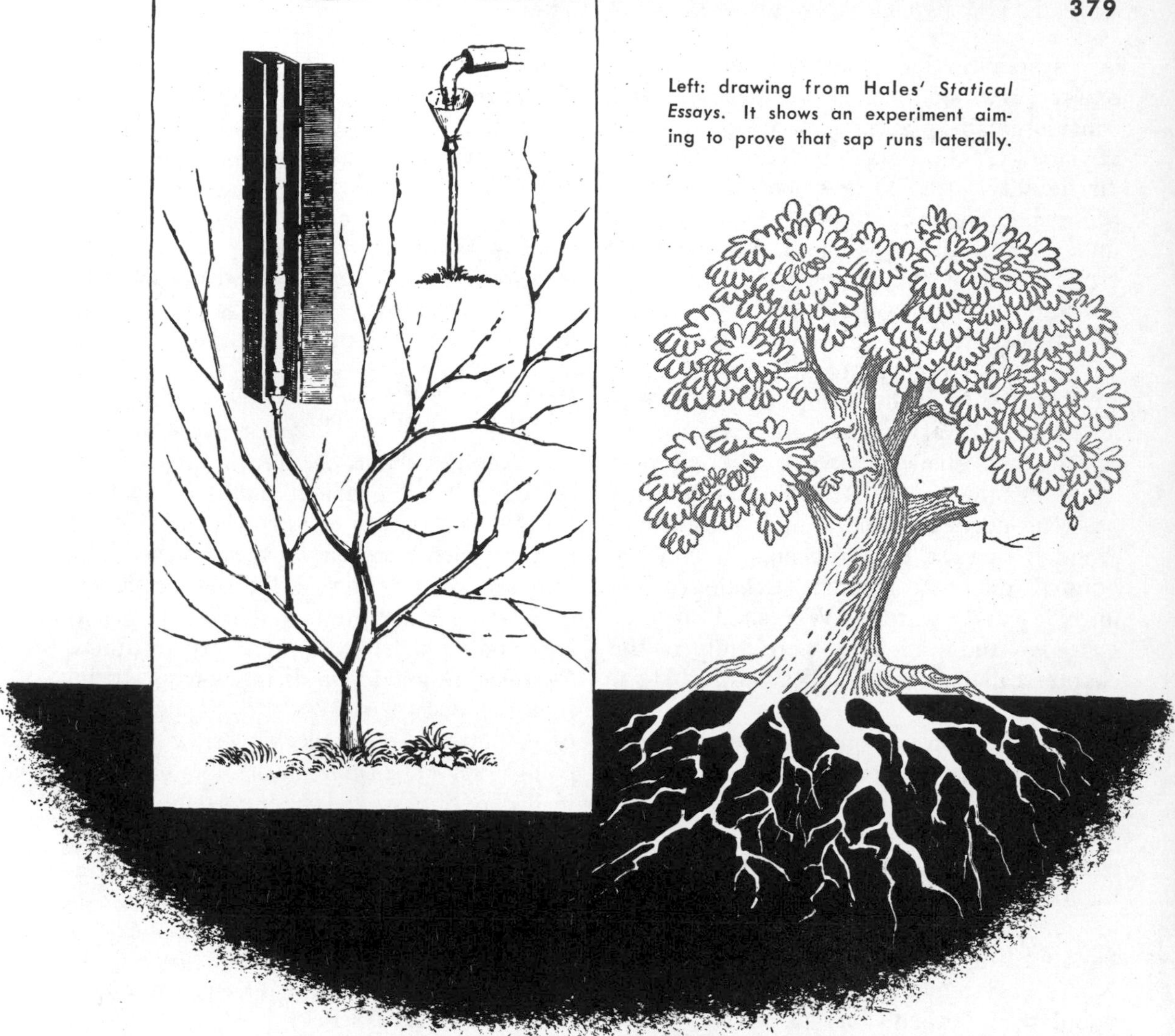

Left: drawing from Hales' *Statical Essays*. It shows an experiment aiming to prove that sap runs laterally.

environment prevent serious overcrowding.

Linnaeus and Buffon were the greatest names in the history of biology in the eighteenth century, but there were many others. Particularly fine work was done in the field of plant and animal physiology. The great Dutch physician Hermann Boerhaave (1668–1738), who taught medicine at the University of Leyden, set the pattern for physiological investigation with his MEDICAL INSTITUTIONS (1708). In this work, he sought to explain the functioning of the body in terms of physical and chemical laws. Boerhaave was also a distinguished chemist; his ELEMENTS OF CHEMISTRY (1732) was a highly regarded textbook, based upon the author's own experiments.

The English clergyman Stephen Hales (1677–1761) advanced some surprisingly modern ideas about plant nutrition and circulation. He anticipated the modern theory of photosynthesis — the formation of food elements in leaves and other plant tissues in the presence of light. (See Index, under Photosynthesis.) His account of the movement of sap in plants is quite modern. In one experiment he measured the quantities of water taken in by the roots and given off by the leaves. Hales had a vividly inquiring mind, directed to many subjects. For example, he wrote one of the first books on ventilation and he performed experiments to measure blood pressure in animals.

Albrecht von Haller (1708–77), a Swiss physician, analyzed the nature of living substance and the action of the nervous system in his ELEMENTS OF PHYSIOLOGY (1759–66). This work is generally considered to be a cornerstone of modern physiology. In it we first find the term "irritability" used to describe the property of responding to stimuli. (See Index, under Irritability.) We find, too, the modern concept of the nerves as channels of sensation leading to the brain.

In his EXPERIMENTS AND OBSERVATIONS OF DIFFERENT KINDS OF AIR (1774), the English chemist Joseph Priestley showed that when plants are put in water, they give off oxygen (he called it "dephlogisticated gas"); he noted that this gas was necessary for animal respiration. Another chemist, the Frenchman Lavoisier, pointed out that animals exhale carbon dioxide and water. We shall discuss Priestley and Lavoisier more fully in the chapter called Science in Revolution II, in Volume 4.

Apart from its contribution to biological science, the eighteenth century marked an advance in the utilization of plants by mankind. There was great progress in agriculture. Better horse-drawn agricultural implements — especially plows, seeders and hoers — were invented, many by a talented English farmer, Jethro Tull. The Norfolk system of rotating crops became popular; it called for successive yearly plantings of clover, wheat, turnips and barley. Plants were successfully introduced to new soils. The turnip became acclimated in England; the potato, in France. The rubber plant and the breadfruit were brought from South America to the East Indies.

Digitalis, derived from the plant called the foxglove, was introduced into medical practice in the century. An English physician, William Withering, analyzed a concoction of herbs brewed by an old Shropshire woman for the treatment of dropsy. He discovered that there was one element in the brew that had real medicinal value — the leaves of the foxglove. From these leaves he prepared the drug digitalis, which is still of sovereign importance in the treatment of heart disease.

**Scurvy, formerly the bane of mariners**

Another plant, the lime, a citrus fruit, helped solve a problem that had plagued ships' crews for centuries. The chief limiting factor on sea voyages had been the disease called scurvy, which is characterized by bleeding gums and internal bleeding and which sometimes proves fatal. Modern authorities in dietetics have shown that the disease results from a deficiency of Vitamin C, or ascorbic acid; but seagoing men of former days knew only that it broke out when fresh vegetables and fruits had not been available for a long time.

After arguing for nearly forty years, James Lind (1716–94), a Scottish physician, persuaded the British Admiralty, in 1794, to add lime juice to sea rations in order to combat scurvy. As a result the disease was almost entirely eliminated among ships' crews of the British Navy. From that time British sailors have been nicknamed "Limeys," by way of tribute to their lime-juice rations. Today we know that this juice is so effective in the prevention and treatment of scurvy because it is rich in Vitamin C.

*Continued on page 32, Volume 4.*

# THE CIRCULATORY SYSTEM

**How Blood Is Transported through the Body**

**BY CHALMERS L. GEMMILL**

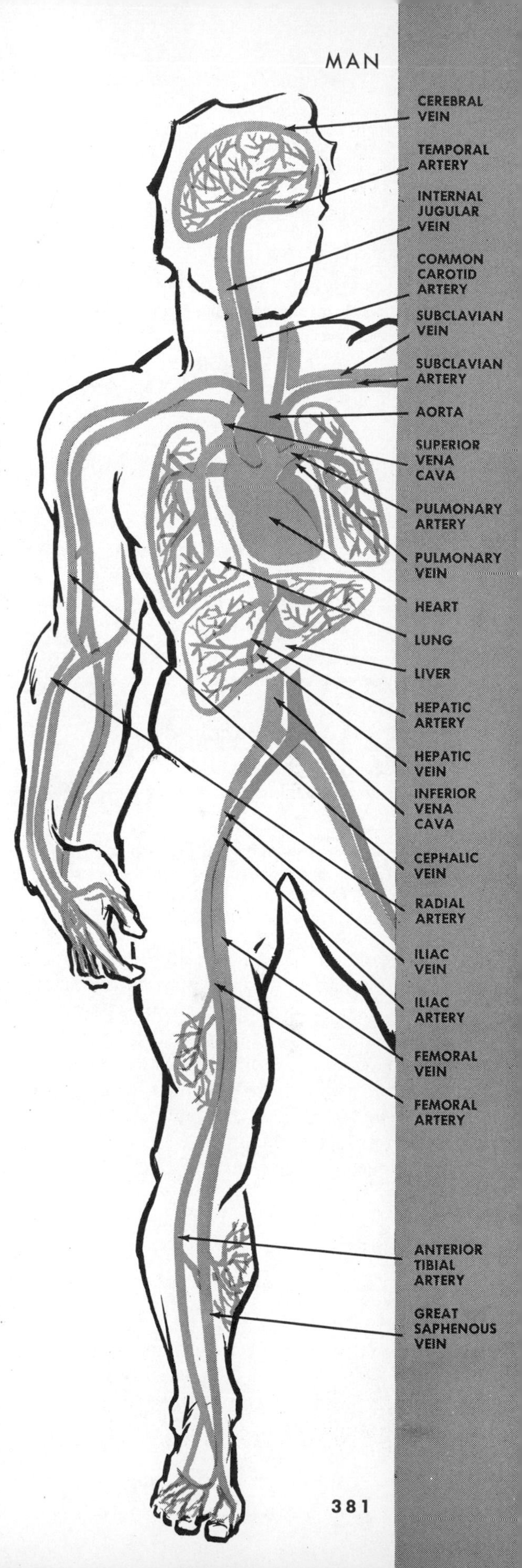


IN ANOTHER article in this volume (The Life Stream), we dealt with the composition and properties of the blood. It is carried to every part of the body in an intricate system of tubes, called the circulatory system. This consists of a pump — the heart — and a network of blood vessels — arteries, capillaries and veins. All arteries lead *from* the heart; all veins lead *to* the heart. Arteries and veins are connected by capillaries.

Briefly, this is what happens as blood circulates. Part of the blood in the heart has just received a fresh store of oxygen from the lungs. This blood is pumped into a large artery — the aorta — and from the aorta it is carried into a branched system of smaller arteries. From the arteries it passes into the capillaries. Here oxygen and food materials (which have been absorbed from the small intestine and liver) are given up to the tissues. Waste materials, including the gas carbon dioxide, are received. The blood then passes to the veins. It is returned to the heart by way of two large veins.

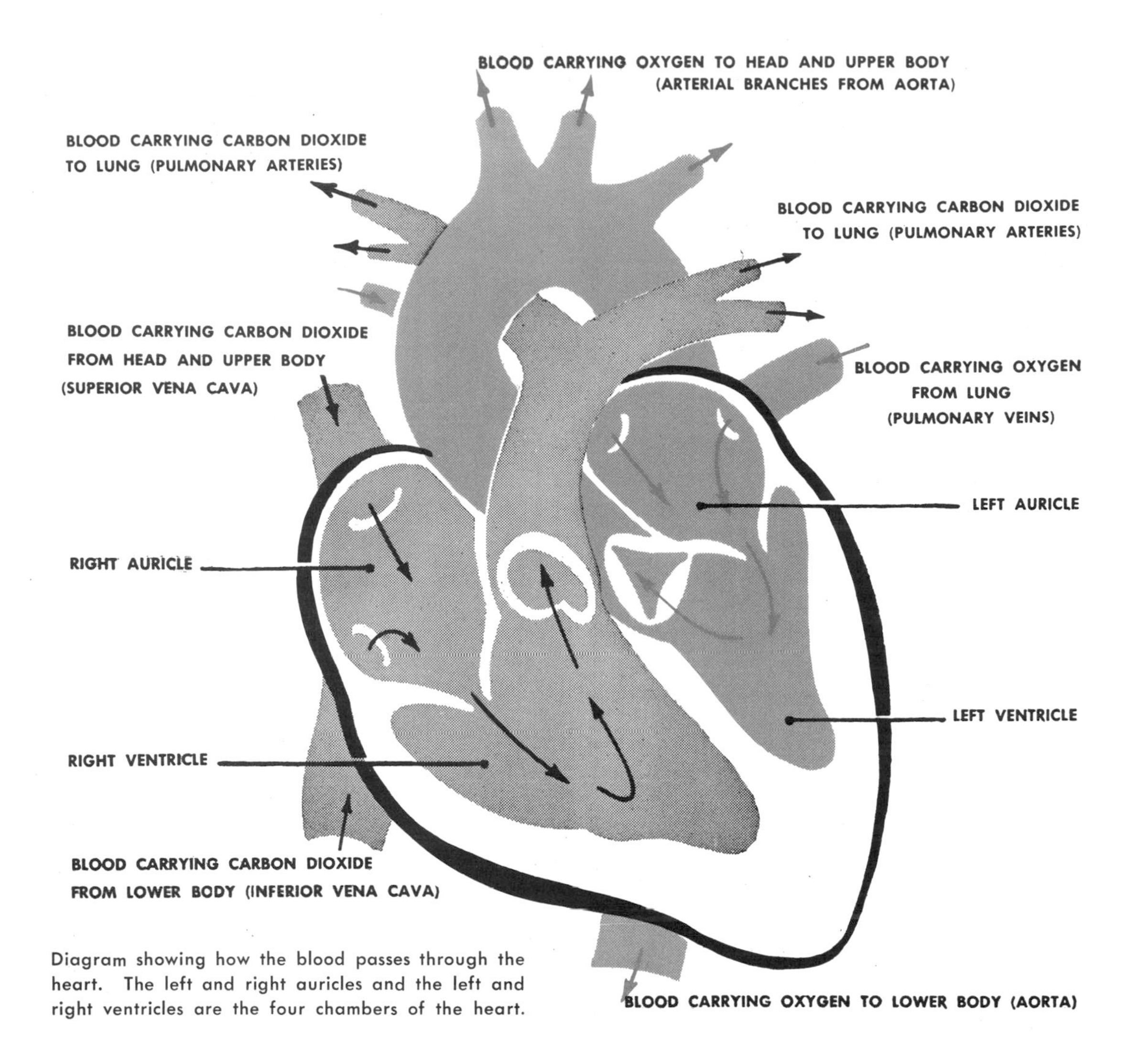


Diagram showing how the blood passes through the heart. The left and right auricles and the left and right ventricles are the four chambers of the heart.

Next, the blood is pumped from the heart through the large pulmonary artery to the lungs. (Pulmonary comes from the Latin *pulmo*: "lung.") In the lungs carbon dioxide is discharged and oxygen is received. The blood is then returned to the heart through the pulmonary veins, and another cycle begins.

There are two major circulation systems in the body. In the greater, or systemic circulation, blood is transported from the heart to every part of the body and back to the heart. In the lesser, or pulmonic circulation, blood travels from the heart to the lungs and from the lungs to the heart. There are various minor systems, such as the hepatic circulation, through the liver, and the cerebral circulation, through the brain.

The blood that has received fresh supplies of oxygen from the lungs is called arterial blood. The name venous blood is given to blood that has passed through the capillaries and has become charged with carbon dioxide. Generally speaking, arterial blood is transported in arteries; venous blood in veins. There are exceptions, however. The blood that passes from the heart through the pulmonary arteries on its way to the lungs is venous blood. It becomes arterial blood as it receives a fresh charge of oxygen in the lungs. It is then carried to the heart, as we have seen, through the pulmonary veins.

Let us now have a closer look at the works of the circulatory system. The heart, the pumping station of the system, is a muscular organ, about as large as the closed fist. Its average weight is about 12 ounces in men and 9 ounces in women. Normally it lies somewhat to the left in the chest. A wall, called the septum, divides the heart into a left half and a right half. Each half, in turn, is divided into an upper chamber — the auricle, or atrium — and a lower chamber — the ventricle. The auricles receive blood from the veins; the ventricles pump blood into the arteries. Completely enclosing the heart is a double sac — the pericardium.

There is an opening between the auricle and the corresponding ventricle on each side of the heart. The two openings are guarded by valves, made up of thin, membranous flaps. The valve between the right auricle and right ventricle has three flaps. It is called the tricuspid valve ("valve with three cusps, or projections"). There are only two flaps in the valve between the left auricle and left ventricle. This is called the bicuspid valve ("valve with two cusps"). It is also known as the mitral valve, because with its two flaps it looks something like a bishop's miter.

Two other valves, the aortic and the pulmonary, guard the exits from the ventricles. The aortic valve is set at the beginning of the aorta, the big artery from which blood flows to every part of the body. The pulmonary valve is located at the beginning of the pulmonary artery, which carries blood to the lungs.

Each pumping action of the heart consists of two phases — systole and diastole. The heart contracts in systole; it relaxes in diastole. During the relaxing phase, blood enters the right auricle from two big veins, the superior (upper) vena cava and the inferior (lower) vena cava. The upper one of these veins leads from the head, neck and arms; the lower one, from the abdomen, pelvis and lower limbs. At the same time that blood flows into the right auricle from the venae cavae, it enters the left auricle from the pulmonary veins, leading from the lungs (No. 1, in diagram).

The valves between the auricles and ventricles on both sides of the heart open (No. 2); blood rushes into the ventricles. When these are quite filled, or nearly filled, the auricles contract and empty their contents into the ventricles, giving an extra thrust to the blood (No. 3); this is the start of systole. The ventricles contract in their turn. As they do so, the valves between the ventricles and the auricles close (No. 4). For a time each ventricle, still contracting, forms a closed container for blood, since the valves on either side of it are shut.

Finally the aortic valve and the pulmonary valve, shutting off the ventricles from the arteries, are forced open (No. 5). Blood spurts from the right ventricle to the pulmonary artery and is sent on its way to the lungs. At the same time, blood is forced from the left ventricle into the aorta. From here it will be transported by way of the other arteries to the different parts of the body. The ventricles relax, and diastole sets in. The aortic and pulmonary valves then close.

The arteries, into which the ventricles pump blood, are tough and elastic vessels. Their walls are made up of three coats. The inner coat consists of a delicate and thin cellular lining. The middle one is much thicker; it is made up of smooth muscle fibers and elastic connective tissue. The connective tissue of the outer coat bears nerves and small blood vessels. Generally the arteries are overfilled with blood and therefore their walls are somewhat stretched. They are stretched still more when the ventricle ejects blood into them. During the relaxing phase, or diastole, the walls of the arteries contract. As they do so, they press upon the blood and force it along the arterial system.

Let us follow the course of the blood as it leaves the left ventricle and enters

The pumping action of the heart, as described on this page.

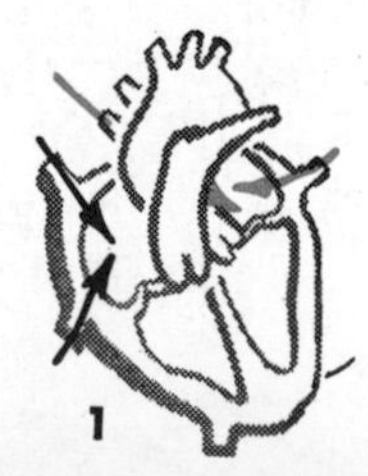


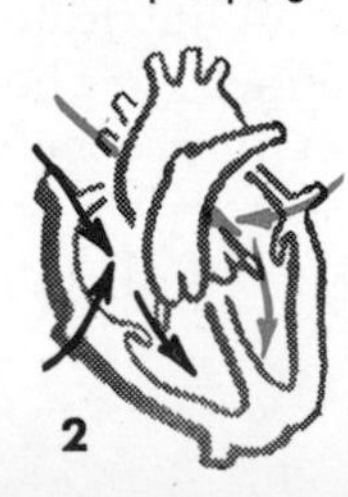

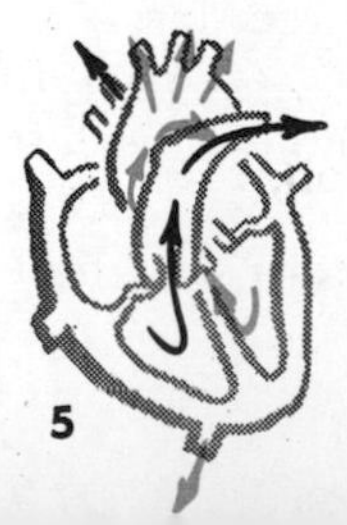

the aorta, from which it will circulate throughout the body. As we have seen, it is now rich in oxygen. This gas, absorbed in the lungs, has combined with the hemoglobin in the red blood cells, to form a compound called oxyhemoglobin. (See Index, under Hemoglobin.) Because of the renewed supply of oxygen, the red blood cells are now bright scarlet, and they impart this color to the blood.

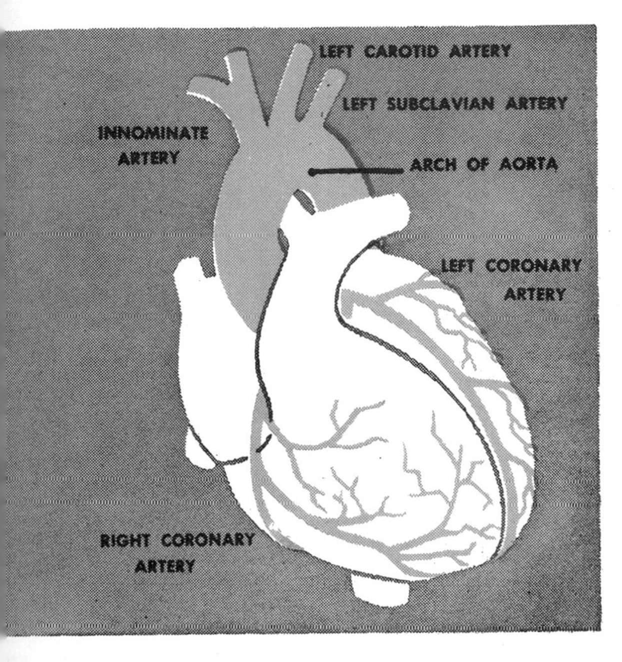


**Blood is pumped from the heart to the big artery called the aorta. The coronary arteries, leading from the aorta, near its base, supply the heart with blood. Beyond the coronary arteries, the aorta forms a distinct arch.**

The aorta, the first and largest of the arteries, is more than an inch in diameter at the place where it begins at the upper part of the left ventricle. Two arteries, leading from the aorta near its base, supply the heart muscle with blood. They are called the coronary arteries, because they encircle the heart like a crown. (*Corona* means "crown" in Latin.)

Beyond the coronary arteries the aorta forms an arch about two inches high. Three large arterial branches lead from the top part of the arch. They are the innominate, left carotid and left subclavian arteries. Blood flows through them to arteries in the head, neck, upper limbs and chest. The descending part of the aortal arch, passing near the vertebral column, leads from the chest cavity into the abdominal cavity. There are branches from this part of the aorta to the intestines, kidneys and other organs of the abdomen. The descending aorta finally divides into two terminal (end) branches. These are the right and left iliac arteries, which carry blood to the legs. Some of the more important arteries are shown in the diagram on the first page of this article.

As they branch out, the arteries grow smaller. The tiniest tubes in the arterial system are not much thicker than hairs; they are only about a fifth of a millimeter in diameter. These wee blood vessels are called arterioles, or little arteries. They are plentifully supplied with encircling smooth muscle. When the muscle fibers contract, they reduce the diameter of the arterioles, or close them off entirely. The diameter of the arterioles is increased when the muscle fibers relax.

The arterioles play an important part in distributing blood in appropriate quantities to the different parts of the body. When the skeletal muscles, for example, are employed in strenuous exercise, they use up food substances and oxygen more rapidly. They also produce more carbon dioxide and other wastes. More blood must then flow through the muscles to supply the necessary oxygen and food materials and to carry off wastes.

More blood is supplied as the arteriole muscles relax; the flow of blood is cut down as the muscles contract. The nerves that trigger this muscle action form part of the autonomic nervous system (see Index). They respond to chemical stimuli; they are also affected by external stimuli such as cold, heat and massage.

From the arterioles, blood flows into the capillaries. These are the finest of all blood vessels; their diameter is only about a tenth that of the arterioles. The capillaries form an intricate network of connecting channels. Though each of the blood vessels in the network is tiny, their combined capacity is greater than that of the arteries. To all intents and purposes, they offer a wider channel for the passage of blood than do the arteries. As a result, the flow of blood through the capillaries is comparatively slow.*

As we pointed out, the exchange of materials between the blood and the tissues of the body takes place only in the capillaries. Their walls are made up of a single layer of cells and are porous. They allow fluid to pass from the blood and to mix with the fluid present in the tissues. They also permit tissue fluid to enter the capillaries and to mix with the blood.

The exchange of materials is brought about by diffusion through the capillary walls. Suppose that a given material is more highly concentrated on one side of these walls than on the other. It will pass through the walls from the area of greater concentration. For example, there is more oxygen in the blood that flows into the capillaries than there is in the tissue fluid. Therefore the oxygen will pass through the capillary walls into this fluid. To do so, it breaks loose from the hemoglobin with which it has been combined in the red blood cells. Dissolved food materials transported by the blood will also pass into the tissue fluid.

Food materials and oxygen diffuse into the body cells from the tissue fluid. In the cells food substances are converted into a variety of new materials. Energy is released as food materials are oxidized — that is, as they combine with oxygen. The energy provided in this way enables the body to carry on its activities. The heat that accompanies the energy release helps maintain body temperature.

Carbon dioxide is one of the end products when food materials are oxidized. It passes from the cells to the tissue fluid and from the tissue fluid into the capillaries. Various wastes, such as urea and uric acid, also make their way from the cells to the capillaries. They are carried in the blood to the kidneys and are later passed from the body in the urine.

The diameter of the capillaries can be decreased or increased. It was once thought that these changes were due to arteriole contraction or dilation, which of course would govern the quantity of blood entering the capillaries. We now know that the capillaries nearest the arterioles (and perhaps other capillaries) are provided with smooth muscle. As this contracts or dilates, it helps control the quantity of blood passing through the capillaries.

Changes in the diameter of the arterioles and capillaries account for various changes in the color of the skin. For example, when these tiny blood vessels are dilated and are filled with blood, the face becomes hot and flushed. When they contract and hold less blood, the face becomes pale and cool.

* In a system of closed tubes, a liquid travels more slowly when it passes through a wide channel and more rapidly when the channel is narrow.

The blood flowing through the arteries ultimately makes its way into smaller arteries, or arterioles, and from these into capillaries. It then passes to the venules, or small veins, and from the venules into larger veins.

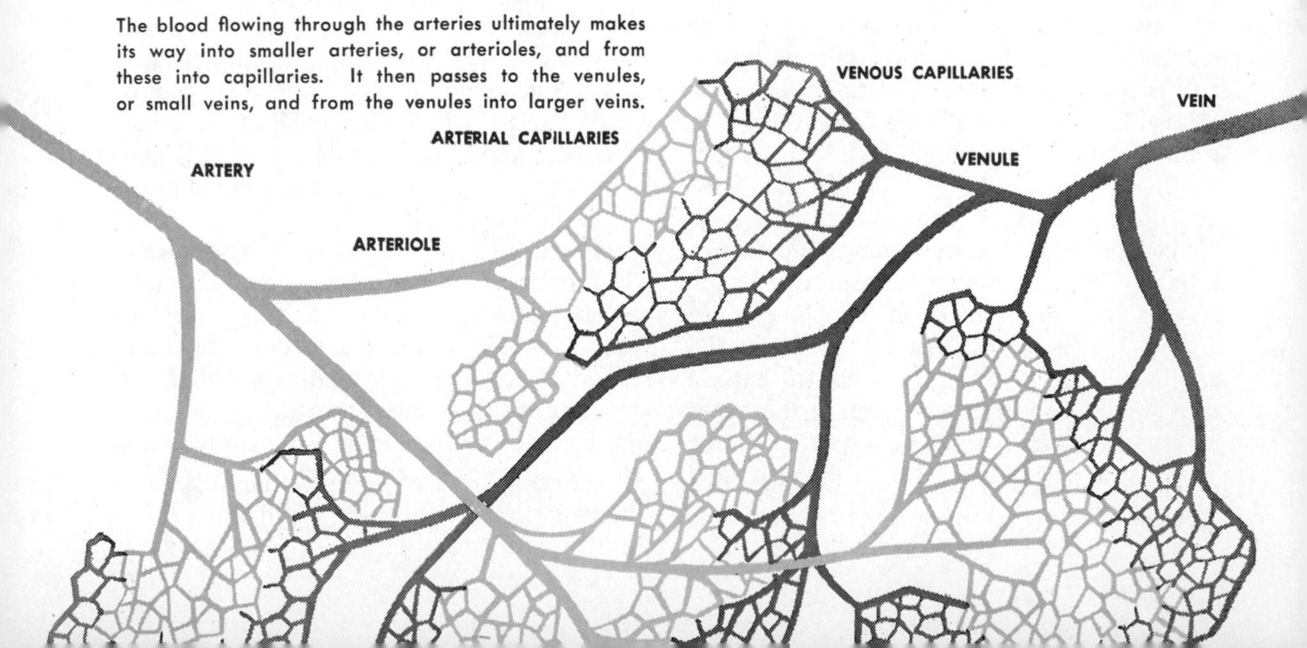

The blood passes from the capillaries to very small veins, called venules. By this time the red blood cells have given up a part of their oxygen load and as a result they have become dark red. This accounts for the dark red color of venous blood. There are thousands upon thousands of venules. They gradually merge and form larger veins; these combine in their turn. As the blood reaches the larger veins, its speed increases. The venous system ends with the two large veins — the superior and inferior venae cavae — that lead to the right auricle. We show the venae cavae and other important veins in the diagram on page 381.

The structure of the veins is much like that of the arteries. However, their walls are thinner and have less smooth muscle. Many of the larger veins contain valves that permit the blood to flow toward the heart but not backward toward the arteries. The action of the skeletal muscles helps maintain the flow of blood in the veins. When the muscles contract, they press upon the veins, forcing blood forward toward the heart and cutting off the flow from the arteries. When the muscles relax, blood flows into the veins again, from the capillaries. The movements made in breathing also help to pump blood through the veins. As the volume of the chest cavity is increased in inhalation, pressure is reduced in the large veins of the body. This causes blood to be drawn into them from the capillaries.

From the right auricle, as we have seen, the blood enters the right ventricle and is then forced into the large pulmonary artery. This divides into a right branch, leading to the right lung, and a left branch, leading to the left lung. Each branch subdivides into smaller arteries and arterioles. At last the blood enters a network of capillaries. These capillaries surround the alveoli, or air spaces, in the lung (see Index). As blood travels through the capillaries, it gives up a load of carbon dioxide; it also releases water to the outer air in the form of water vapor. At the same time the blood receives a fresh load of oxygen.

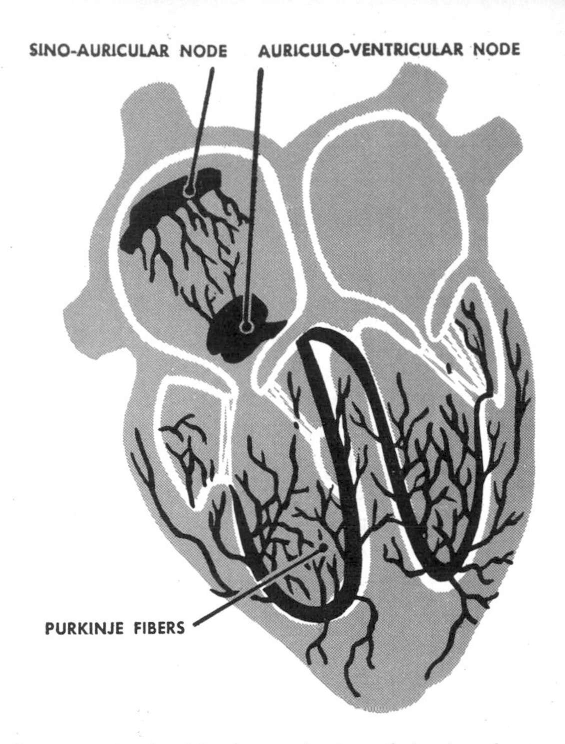


Structures involved in the production of the heartbeat. Their operation is described in the text on this page.

From the capillaries of the lung, blood flows into small veins, or venules, and from these into larger ones. It makes its way at last into the left auricle through four large pulmonary veins. (In some cases there are only three.) Finally blood flows into the left ventricle, from which it spurts out into the aorta. It takes the blood only about a minute to make a complete circuit of the body.

### A close-up of the heart at work

Medical science has not yet found out what causes the heart to beat. We do know, however, how impulses are transmitted within this organ. They arise in a small bundle of tissue, called the sino-auricular node, or sino-atrial node, and located in the upper part of the right auricle. The impulses sent out by this node form what is called a wave of excitation. This spreads through the right and left auricles, causing their muscle fibers to contract. The wave of excitation reaches a second node — the auriculo-ventricular, or atrio-ventricular, node — in the upper part of the wall between the two ventricles. The impulses are then relayed by way of a bundle of tissue to a fiber network, known as the Purkinje system. This network transmits the impulses to the muscle fibers of the ventricle and causes them to contract.

Certain electrical changes accompany the wave of excitation spreading through the heart. The blood and tissue fluids are good conductors of electric current. Therefore changes in electric potential arising in the heart are carried to all parts of the body. Their pattern is recorded in the form of waves in the device called the electrocardiograph. We describe this instrument elsewhere (see Index). By analyzing the waves recorded by the electrocardiograph, the physician can obtain valuable information about the condition of the heart.

The heartbeat is not initiated by the action of nerves, but it is affected by two sets of nerves arising in the medulla oblongata, a part of the brain. (See Index, under Medulla oblongata.) These are the vagus nerves and the accelerator nerves; they both belong to the autonomic nervous system. The vagus nerves act as brakes; they slow up the heart and reduce the strength of its beats. The accelerator nerves, on the contrary, speed up the heart and increase the strength of the heart beats.

With each heartbeat, a pulse wave spreads all over the arterial system. The succession of waves can be felt as a series of taps or thuds if the finger is pressed lightly over any large artery. To count the number of heartbeats per minute or fraction of a minute, the physician generally presses on the radial artery, which passes through the wrist.

The pulse rate varies greatly in individuals. The average rate for adults who are relaxed mentally and physically is from 65 to 70 in men and from 70 to 75 in women. It is much higher in babies. The count slows up somewhat in sleep. It soars when a person works hard or becomes greatly excited; the count may reach 200 per minute. Various bodily conditions, such as surgical shock, hemorrhage and fever also cause a marked increase in the pulse rate.

When a physician applies his stethoscope to a person's chest, he can generally hear two sounds as the heart beats. The first is softer and lower in pitch than the second, and it lasts longer. It suggests the syllable *lub.* The second sound — something like *dup* — is short and sharp. The two sounds occur in quick succession and are followed by a slight pause. The physician listening to the stethoscope hears something like *lub-dup — lub-dup — lub-dup.* We show you elsewhere how to make a simple stethoscope. (See Index, under Stethoscopes — Homemade.)

The *lub* sound is really made up of two sounds that merge. One of these represents the vibrations that arise as the muscle fibers of the heart contract. The other accompanies the closing of the valves between the auricles and the ventricles. The *dup* sound is produced by the vibrations caused by the closing of the aortic and pulmonary valves. Since both the *lub* and *dup* sounds result, in part at least, from the action of the heart valves, the sounds will be distorted if anything is wrong with the valves. The stethoscope, therefore, is valuable in detecting and diagnosing various types of heart diseases.

Sometimes a third sound is heard. It follows the *dup* sound and is very faint. According to one theory it is due to the rapid flow of blood from the auricle to the ventricle. Some authorities hold that it represents a set of "after vibrations" — a sequel to the vibrations heard as the *dup* sound is produced.

### Blood pressure in the circulatory system

As blood is ejected from the left ventricle into the aorta, there is an increase in blood pressure — that is, the pressure of the blood upon the walls of the blood vessels. After the ventricle has started to relax, the pressure falls. The highest point in the pressure range is called the systolic pressure; the lowest point, the diastolic pressure. There is a gradual decrease in pressure as blood circulates from the arteries to the capillaries and from the capillaries to the veins. At the venae cavae,

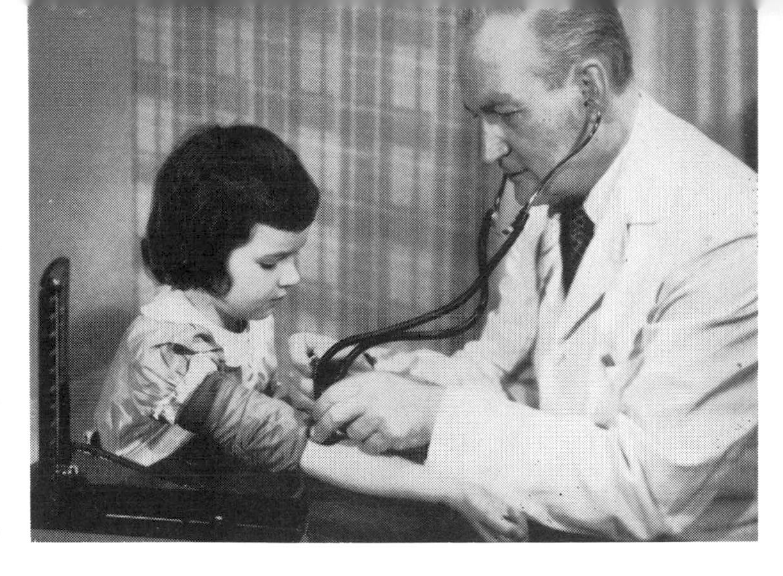

Measuring blood pressure with a sphygmomanometer. A cuff that contains an airtight bag is wrapped around the upper arm. Two rubber tubes lead from the bag. One is connected to a rubber bulb; the other leads to a manometer — a U-shaped glass bulb containing mercury. An upright of the U is attached to a millimeter scale, shown at the extreme left of the picture.

Ewing Galloway

the two large veins that empty into the right auricle, the pressure is relatively slight.

To take a person's blood pressure, the physician generally measures the pressure in a large artery of the arm — the brachial artery. An instrument called a sphygmomanometer is used for this purpose. (See photograph on this page.) A cuff containing an airtight bag is wrapped around the upper arm. Two rubber tubes lead from the bag. One of them is connected to a rubber bulb. The other leads to a manometer, a **U**-shaped glass bulb containing mercury. One of the uprights of the **U** is fixed to a millimeter scale.

The physician inflates the cuff by pressing the bulb. Increasing pressure within the cuff causes the mercury to rise in the part of the manometer that is attached to the millimeter scale. The physician continues to pump air into the cuff until the pulse at the wrist disappears. This means that no blood is passing in the brachial artery below the cuff. He then applies a stethoscope to the artery just below the cuff. The air pressure in the cuff is reduced as the physician opens a valve on the tubing near the bulb. As soon as blood begins to move again in the artery below the cuff, a tapping sound is heard.

A reading is now taken; the physician notes the height of the mercury column in the glass tube. Suppose that the top of the column is opposite the 120-millimeter mark on the scale. The reading will be "120 mm. Hg," or "120 millimeters of mercury." (Hg is the symbol for mercury.) This reading represents the systolic pressure.

The physician releases more air from the cuff, and the pressure is lowered still more. The sound of the blood flowing through the artery becomes louder; then it dwindles to a murmur and finally stops entirely. A reading is taken just before the sound disappears. This reading is the diastolic pressure.

In the young male adult, the average systolic pressure is 120 mm. Hg; the average diastolic pressure is 80 mm. Hg. The figures are somewhat lower in women until the time of menopause. At that time there is a sharp increase; thereafter, blood pressure is a little higher than in men. In both men and women, pressure gradually mounts with increasing age.

As in the case of pulse rate, there are pretty wide variations in blood pressure among perfectly healthy persons. Any systolic pressure over 150 or diastolic pressure over 100 is considered abnormal. This condition is called hypertension, or high blood pressure. In some persons, the blood pressure is lower than normal. Such persons are said to have low blood pressure (hypotension).

### The lymphatic system of the body

The so-called lymphatic system, found in all parts of the body, is intimately connected with the blood circulation. The lymphatic system is made up of a series of ducts — the lymph vessels — and it carries a fluid called lymph. This resembles blood plasma, the liquid part of the blood; its protein content, however, is much lower. Lymph is derived from the tissue fluid that occupies the space between cells in body tissues. Not all of this fluid enters the lymph system. Some of it passes into the capillary network of the blood circulation and is carried along in the veins until it reaches the heart. It is transported, therefore, in the regular venous system.

The lymphatic system begins in the lymphatic capillaries, through which tissue fluid passes. These ducts differ in several respects from ordinary capillaries. They are not connected, of course, to a system of arteries; they are much more porous than ordinary capillaries. The lymphatic capillaries lead to larger lymph vessels. These have valves that direct the flow of lymph away from the capillaries. The largest ducts of the system empty the lymph into certain large veins of the neck — the left subclavian and the left jugular veins. Here the lymph merges with the blood stream.

Most of the larger lymph vessels have enlarged structures, called lymph nodes, at intervals along their course. These nodes consist of a meshwork of connective tissue, packed tightly with lymphocytes and phagocytes. Lymphocytes are white blood cells; phagocytes are cells that attack and engulf invading organisms. Lymph must pass through the intricate meshes of the nodes as it proceeds to the subclavian veins.

The nodes act as effective filters. They entrap various particles of matter, including bacteria, and therefore remove them from the blood stream. Bacteria caught and held in the meshes are destroyed in great numbers by the phagocytic cells. The lymph nodes, therefore, represent an important line of defense against germs that have penetrated the blood stream. Lymph nodes manufacture great numbers of lymphocytes. They also produce antibodies, which help protect the body against disease.

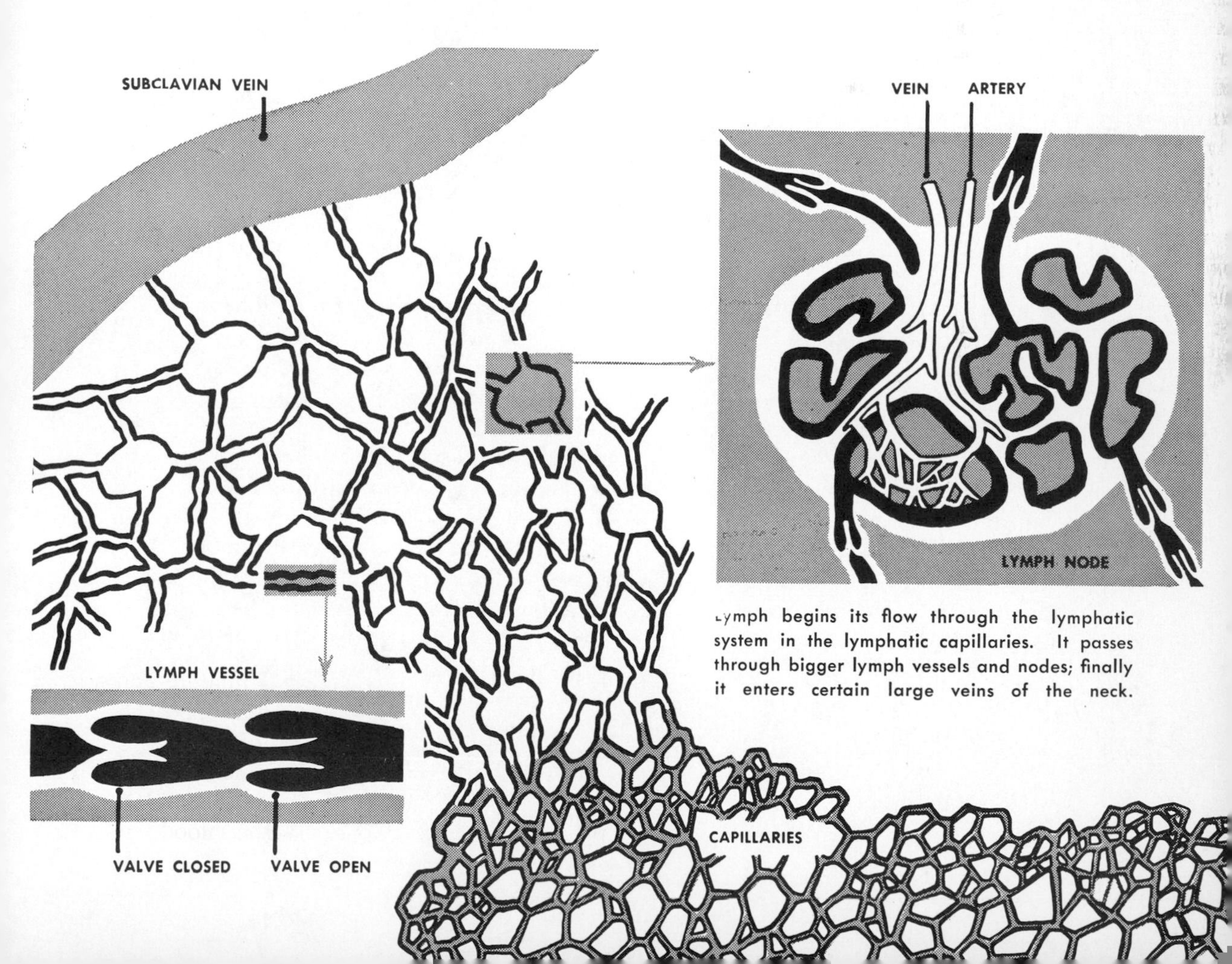


Lymph begins its flow through the lymphatic system in the lymphatic capillaries. It passes through bigger lymph vessels and nodes; finally it enters certain large veins of the neck.

Of course, since germs are enmeshed in the lymph nodes, they will collect there in great numbers in the event of a serious infection. The phagocytic cells may not be able to dispose of them effectively enough. The result may be a local infection at the nodes. For example, if the throat is infected, the lymph nodes of the neck may swell and become tender. The doctor, therefore, examines the nodes in such cases to find out how severe the infection is and how widely it has spread.

The lymph system carries away cancerous tissue from a malignant growth. Therefore it plays an important part in the spread of cancer throughout the body. If the doctor suspects that a growth in any area of the body is cancerous, he carefully examines the neighboring lymph nodes. In this way he can find out if the tumor tissue is being spread by the lymphatic system.

#### Disorders of the circulatory system

The heart, the hardest-working organ of the body, is subject to a great many disorders. Disease affecting the heart has been called the "great killer"; it claims more lives annually than any other type of disease. Many disorders of the heart are due to changes brought on by old age. Hence heart disease is bound to claim even more victims as the life expectancy of man increases.

When the heart cannot maintain effective circulation of the blood during rest or mild exercise, the condition called heart failure results. It is often due to defective valves. Valves are sometimes attacked by such diseases as rheumatic fever and scarlet fever. They may be partly destroyed or may become stiff and deformed.

If the valves do not function properly, the heart will have to perform more work to keep the blood circulating properly. The heart muscle will then become thicker, a condition called hypertrophy. The heart may be able to compensate in this way for the defective valve or valves. The sufferer may remain in fairly good health for years. In time, however, the heart will no longer supply enough oxygen for the tissues.

Heart failure may be due to various other causes. It may result, among other things, from infections of the respiratory tract, excessive exercise, lung disorders accompanied by violent coughing and too rapid heart action.

The first symptom of heart failure may be shortness of breath during mild exercise. Later, there is rapid pulse, swelling of the legs, especially around the ankles, coughing and general weakness. The venous pressure rises. Cyanosis may develop — that is, the surface of the body may become blue because of the lack of oxygen. The liver and spleen may become enlarged.

Various drugs are used to relieve sufferers. Among the most effective is digitalis, derived from the foxglove plant. This drug stimulates the failing heart and improves the circulation. This causes fluid to be removed from tissues swollen with water and relieves other symptoms of heart failure. We have told elsewhere how digitalis was introduced into medical practice. (See Index, under Withering, William.)

Coronary disease affects the coronary arteries, which supply the heart muscle with blood. These arteries may become clogged up, especially in the later years of life, because of faulty metabolism. In this condition, called arteriosclerosis, the middle layer of the artery degenerates. There is an overgrowth of fibers in the innermost layer, and calcium deposits form along the walls. The clogging up of the coronary arteries is a gradual process. Often, as one of these arteries is narrowed, new blood vessels develop in the tissues and help supply the heart with blood.

Sometimes the clogged-up arteries may close entirely as the result of a sudden spasm. This brings on the terrible agony of angina pectoris ("pain in the chest," in Latin). Angina pectoris occurs in the region of the heart. Often it spreads to the left shoulder and down the left arm. Sometimes angina pectoris is caused by an insufficient supply of oxygen to the tissues. In such cases, the coronary arteries may be quite normal. Angina pectoris is relieved

## NORMAL AND DEFECTIVE CIRCULATORY STRUCTURES

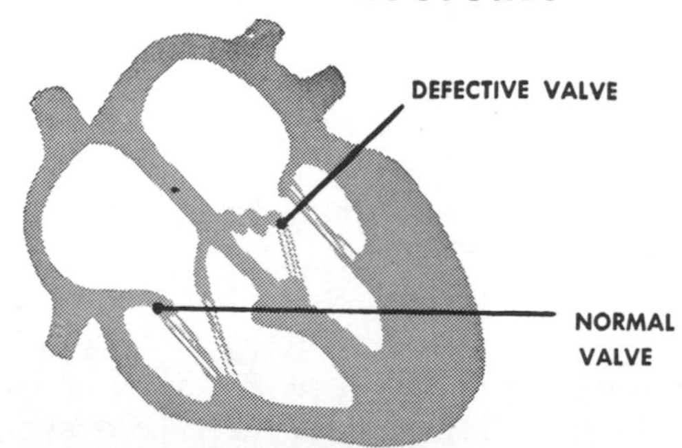


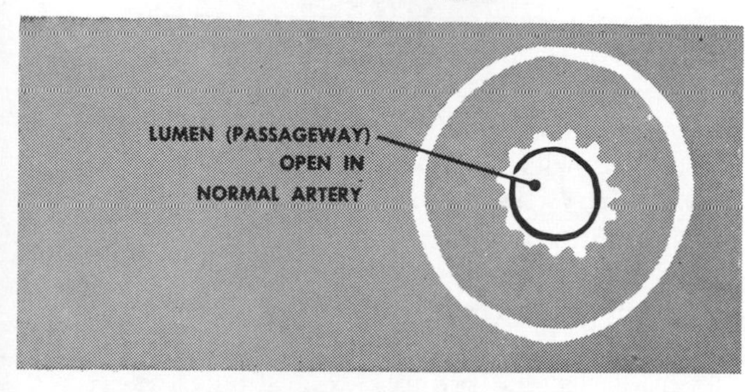


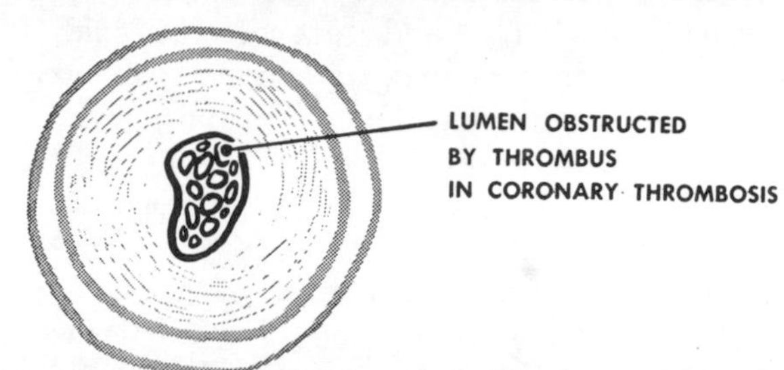


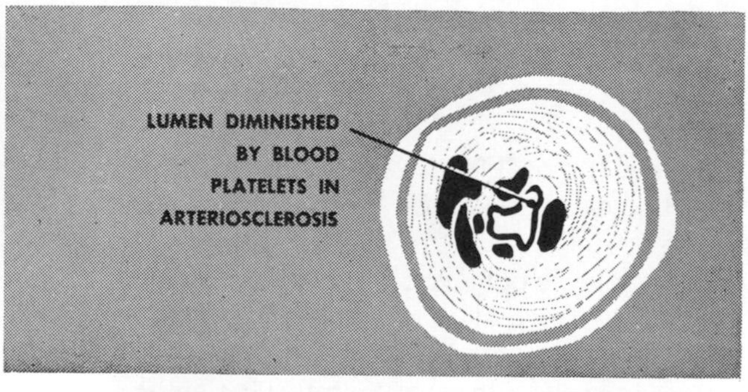


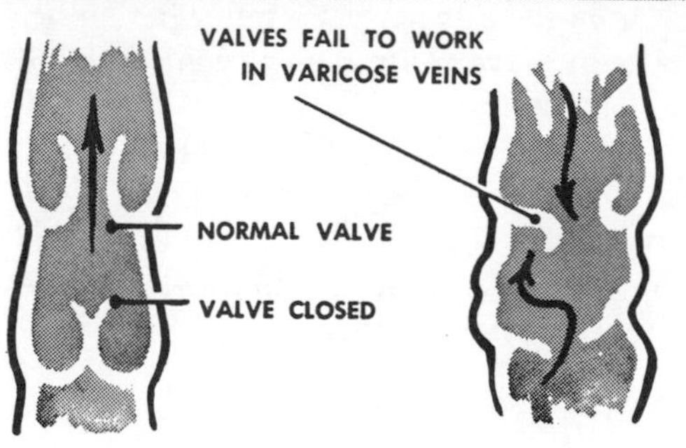


by various drugs, including amyl nitrite and nitroglycerine.

In the disease called coronary thrombosis, a blood clot forms in one part of the coronary arteries, shutting off part of the blood supply to the heart muscles. There is great pain, which lasts much longer than the temporary pain of angina pectoris. Coronary thrombosis is often fatal; however a sufferer may survive several attacks. In a coronary embolism, a blood vessel is suddenly plugged up by a blood clot that has broken loose from some other part of the body and that has been transported to the coronary arteries in the blood stream.

In the disorder called auricular fibrillation, the beat of the auricle is rapid and entirely irregular. Auricular fibrillation results from rheumatic fever and certain other diseases. It is relieved by the drug quinidine, which causes the heart muscle to lengthen the period between beats. Digitalis is also effective in treating this condition.

Some children are born with various defects of the heart. The pulmonary valve, between the right ventricle and the pulmonary artery, may be too small. There may be a hole between the wall separating the right and left auricles or the right and left ventricles. In the fetus * there is a duct between the pulmonary artery and the aorta; this duct normally atrophies and disappears by the time of birth. Sometimes, however, atrophy does not take place and the duct remains open. Such congenital conditions (conditions present at birth) can often be corrected by surgery.

Blood circulation will be affected if the arteries or veins are attacked by disease. The aorta may be infected as a result of syphilis. It may become permanently dilated, if a person has had high blood pressure for many years. If the walls of the aorta are weak, they may rupture. Arteriosclerosis is likely to develop in this area with advancing years.

The middle-sized arteries are particularly apt to be affected by arterioscle-

* The unborn young in the latter stages of development.

rosis. As we saw, this condition is particularly dangerous when it develops in the coronary arteries. Arteriosclerosis also attacks the smallest arteries, the arterioles. It cuts down the flow of blood and increases blood pressure.

Perhaps the most familiar disorder involving the veins is the condition known as varicose veins. It most often occurs in the lower limbs, though it is found elsewhere. It is brought about when some obstruction or other impedes the return of the venous blood to the heart. Varicose veins may be caused by tumors within the abdomen, by enlargement of the liver, by pregnancy or even by the use of tight garters. The veins become swollen; the valves of the veins fail to work properly. Ulcers may form; in extreme cases the veins may burst.

Varicose veins may sometimes be cured by removing the cause of the obstruction. Treatment often consists of rest in a horizontal position, bathing the affected limb in cold water. Sufferers may find relief by wearing a bandage, or a laced stocking or a snugly fitting elastic stocking. In advanced cases it may be necessary to remove the obstructed portion of a vein.

Phlebitis, or inflammation of the veins, is a serious disorder. It may result from a wound or abscess outside of the vein. In certain cases the inflammation arises within the vein, because a clot has formed on the inner wall. Parts of the inflamed tissue may break off. They may be carried to other parts of the body by the blood circulation, causing similar inflammation elsewhere. The inflammation may cause a vein to be entirely stopped up; death may result if a large vein is affected in this way. To treat this condition, medicines that counteract inflammation are prescribed. The patient is kept perfectly still, with the affected part raised. Abscesses within reach are opened; fresh dressings are frequently applied to wounds.

Some persons with high blood pressure, or hypertension, may lead quite normal lives. In other cases this condition has serious consequences. The high pressure of the blood against the walls of small arteries and capillaries may cause them to rupture. A break of this type in the brain is called a cerebral hemorrhage. If a person who has extremely high blood pressure has a weak heart, the added work required of the heart may lead to a serious case of heart failure.

Among the symptoms of hypertension are general weakness, headache, shortness of breath, palpitation of the heart and irritability. Hypertension may be a symptom of a disease, such as goiter or arteriosclerosis. In that case the disease must be treated to bring about relief. If high blood pressure is not connected with any other disorder, relaxation and freedom from worry are helpful. Sedatives, such as bromides and barbiturates, have provided relief. So have salt-free diets.

Persons with low blood pressure, or hypotension, generally tire quickly. In severe cases, the sufferer's digestion is affected; his hands and feet become cold. He is apt to feel dizzy when he gets up suddenly after lying down. In most cases persons with hypotension get along very well if they learn not to tax their strength. They should have enough sleep, eat nourishing meals and exercise in moderation.

Hemorrhage (bleeding) from arteries or veins, either within or outside of the body, is always potentially dangerous. If more than 30 per cent or so of the blood volume is quickly lost, death will result unless blood transfusion takes place. If the amount of blood lost is not excessive, the body can generally repair the damage pretty rapidly.

In the article A First-Aid Primer, in Volume 10, we show what can be done to stanch the flow of blood in external bleeding. It is usually easy to tell whether the blood comes from an artery, vein or capillary. If it is from an artery, it escapes in spurts and is bright scarlet in color. Blood from a vein is darker and flows more slowly; it wells up at the site of a wound. If the blood is from capillaries, it oozes slowly and quickly clots.

*See also Vol. 10, p. 276:*
"General Works"; "Anatomy"; "Physiology."

# KEEPING AN ANT COLONY

## How to Spy upon the Private Lives of Ants

by CHARLES D. MICHENER and MARY H. MICHENER

IT IS fascinating to watch those eternally active little insects, the ants, scurrying about out-of-doors. Yet by far the greater part of the ants' complex social activity goes on inside the nest out of our sight. The only way we can watch them care for their young, distribute food, dig and build and carry, meet intruders and perform other acts of their private lives is to provide an artificial observation nest for a colony or part of a colony.

No one type of artificial nest will allow us to observe all kinds of ant behavior to best advantage. In nests containing soil we can watch ants make tidy homes with a network of galleries. To study their feeding, growth, grooming, fighting and other intimate behavior, however, it is better to have a nest without soil.

In making any kind of artificial nest we must keep in mind a few basic needs of all ants. Fresh air and moisture are very important. If the soil and atmosphere become too dry, the ants will die; if too wet, molds will develop. Usually the feeding area of the nest can be well lighted, for most ants forage for food in bright daylight; but ants will not rear their young or behave naturally if the brood rooms and galleries within the nest are not dark enough. Fortunately for human observers, however, ants cannot see light in the longer wave lengths (the red end of the spectrum). A piece of red glass or cellophane laid over the clear glass of the observation nest will make it possible to watch the ants without disturbing them. With a little experimenting you can find a shade that you can see through easily but that will be dark enough so that the ants continue to work normally. If the workers pick up larvae and pupae and scurry about as though to seek shelter, it is too light for them. If you do not have a red glass or cellophane cover, you can keep the brood area covered with cardboard or cloth, removing the cover for short periods of observation.

As a beginning or emergency nest, you may simply place ants and earth in a shallow dish set in a pan of water, which will prevent the ants from escaping. Add small amounts of food and water directly to the soil. A nest with a separate food

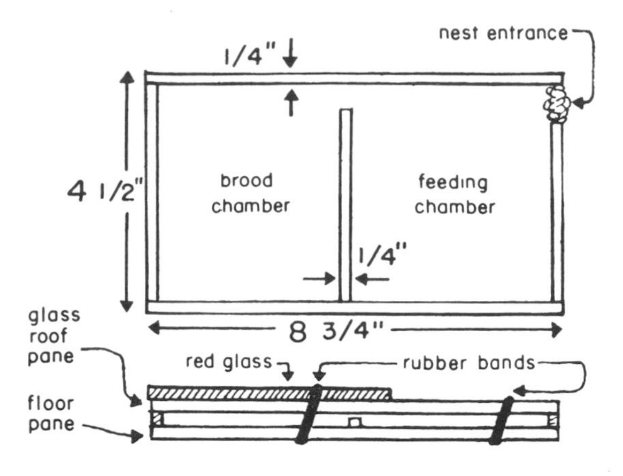

Top and side views of all-glass nest.

chamber will be much more satisfactory than a dish. You can watch the individual foragers more readily, you can easily remove discarded or excess food before it spoils, and if you should want to clear the brood chamber of ants for any reason, you can expose it to light and darken the feeding area so that the ants will migrate into the latter for the time being.

The size of your nest is not particularly important, except that it should be large enough for the size and number of ants you want to keep. It should also be convenient to carry. A nest with two chambers, each 4 inches square (inside measurement), will do for a small colony. For this you will need two panes of glass 4½ by 8¾ inches for the roof and floor. Glue strips of glass about ¼ inch wide and ¼ inch high along the edges of the lower pane to form walls. Glue another strip across the middle to separate the brood chamber from the feeding chamber. Be sure to leave a gap at one end of this partition so that ants may go from one chamber to the other. Also leave part or all of one end open as an entrance. The roof pane should be fitted over the walls and held securely in place by a couple of stout, wide rubber bands. You can remove it easily to add food and water or to clean the nest.

The depth of the nest is important. For one containing soil, ¼ inch is probably the maximum; too great a depth will permit the ants to tunnel in the soil out of sight. Soilless nests may be as deep as ½ inch, but the shallower ¼-inch nest is preferable because it is easier to focus a hand lens on the ants when you want to observe them closely.

The entrance serves as a doorway for the ants. If you are not using soil in your nest, you must induce the ants to move in by themselves in a way to be described later. More important, however, the entrance opening lets in fresh air and moisture. When you want to keep the ants in, you close the entrance with cotton or with

Nest made with plaster of Paris walls.

blotting paper that does not have a paper backing. This plug can be wetted from the outside from time to time so that it will conduct moisture into the nest.

You can use plaster of Paris for the walls of the ant nest, particularly if it is to be very small or very shallow. First oil the upper pane of glass so that it will not stick to the plaster. On the lower pane draw out ridges of a fairly liquid plaster-and-water mixture along the edges and across the center. Leave a gap in the partition and in one end wall as in the case of the all-glass nest. Now press the oiled surface of the roof pane firmly onto the plaster walls until the two panes are as close together as you desire. Be sure the plaster touches the cover at every point along the edge, leaving no cracks through which ants might escape. Remember that plaster of Paris sets very rapidly. Have your nest planned and everything laid out before you mix the dry plaster with water; then work quickly. Mix ½ cup powdered plaster for a 4-by-8-inch shallow nest.

The plug of cotton or blotting paper at the entrance of the nest may not provide enough moisture. Fortunately there are various ways to make good the deficiency. You can glue narrow strips of toweling along the tops of the walls and partition and dampen these strips from time to time. The toweling will also help to fill in any unevenness if your cover pane does not lie smoothly on the walls. You can place little pieces of wet sponge in the nest, both to humidify the atmosphere and for the ants to lick.

Partial colonies of very tiny ants, such as Pharaoh's ants, can be kept between microscope slides and may be observed through a binocular microscope. To watch mound-building or crater-making ants at their engineering, you will need an entirely different sort of nest. Put several inches of soil in a box or dish and cover it with a glass lid or bowl or a bell jar. You can also set an uncovered dish colony in a pan of water or devise some other moat arrangement, such as a water-filled edging molded of plasticine (modeling clay) or warmed paraffin, to keep the ants at home. If you make the moat accessible to the ants, they can bring water from it to moisten the soil when necessary. Otherwise, give them a dish or jar lid of water.

All ants should be supplied with water in one way or another unless the food given them is quite liquid. A bit of bread soaked in syrup or honey can be placed in the nest for the sweets-loving ants; a similar method of feeding may be used for grease ants. Harvester ants will take small seeds and crumbled breakfast cereals. All ants will welcome and should have an occasional meal of protein — scraped meat, raw egg yolk, dead insects.

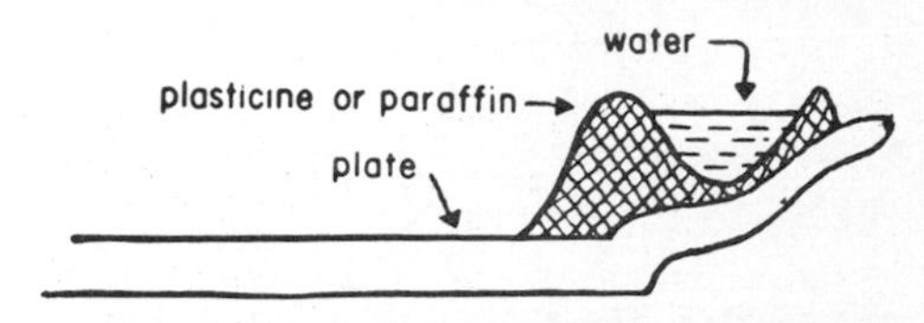

Moat arrangement for an uncovered nest.

A trowel for digging and a cloth or paper sack or wide-mouthed jar for carrying are all you will need to collect an ant colony. Forceps may be useful in some cases. Put a few twigs or some other springy material in the sack or jar to keep the soil from shifting too much and possibly injuring the ants. Get the heart of the nest, ignoring fleeing workers, for you must have some brood or a fertile queen if you want to watch the colony develop. Look for nests under broad, flat rocks or old boards lying on the ground. Often you will see small nests in the soil beside paths and walks. Sometimes you can locate a nest by following ants home.

If the colony is quite small and not mixed with too much soil, you can simply dump it into the feeding chamber of the artificial nest. Leave this area light and dry and the ants will retreat, carrying their brood into the comfortable darkness of the brood chamber. After a few hours you can clean the soil and remove any dead or injured ants from the feeding area.

Often, however, it is necessary to get the ants to move into the nest by themselves. Any space large enough to contain the artificial nest and the captured colony with its soil can be used if you make a moat or some other arrangement to keep the ants from getting away. One method is to use the Forel arena. This is a ring

of dry, powdered plaster of Paris, the inner side graded steep and smooth with a knife so that the ants cannot climb it. Place the artificial nest and the colony with its soil in the enclosure formed by the moat or wall of powder. The nest should be covered so that it is quite dark inside. Spread out the soil to encourage the ants to seek the darkness and moisture that the artificial nest offers them. Watch the ants carry their young — eggs, larvae and pupae — into the new home. Incidentally, the objects that most people call ant eggs are really the pupae. Ant eggs are very small and not easily seen without a lens. Watch, too, for so-called ant guests. These may be beetles or other small insects that share the ants' quarters.

Once your nest is established, you can find the answers to dozens of questions: Is there a queen? Is she laying eggs? Do the workers carry the eggs about? Lick them? Do they keep the eggs, the larvae of different sizes and the pupae in separate parts of the nest? Are the pupae enclosed in cocoons? Do workers assist the new adults to emerge from their cocoons? Are the workers of distinctly different sizes?

The answers to these questions are not the same for all species of ants and not always identical for different colonies of the same species. Much depends on the age of the colony, the conditions under which it must live and other factors.

It is possible to mark individual ants with tiny spots of bright enamel, fingernail polish or model-airplane dope. Then you can see whether it is the same few individuals that do all the foraging, or digging or other chores. Probably you will find that, just as in human society, some individuals loll about, doing nothing.

If you put strangers of the same species or of other species into your ant colony, you will see some savage battles. Try introducing an ant taken from the original outdoor nest of your colony, if any survive. Sometimes ants recognize their nestmates after a long separation and accept them peaceably. Try to form a compound colony of two species of ants by adding pupae of another species. Do the introduced workers, when they mature, behave more like their hosts or like their blood relatives? In naturally formed compound nests, it sometimes happens that a timid species, if matured in a nest of aggressive ants, will be as bold as the workers around it.

Watch for signs of communication among the ants. When a forager finds the food you have put out, does this ant bring other foragers back with it for more? Do others come out and find the food successfully if you remove the original finder? If so, do they seem to come upon the food accidentally, or at least independently, or do they seem to be following a scent trail? If a trail seems important, can you throw the ants into confusion by wiping your fingers across the trail? You can make little mazes for your ants to solve in order to find food. How soon does a forager learn the route so that it can go directly to the food without taking wrong turns? How long can it remember what it has learned?

Have your ants identified by an expert, if possible, so that your notes on their behavior can be compared with the observations of others on the same species or other species. There is no end to the questions you can ask your ants.

Here are the titles of several books on ant behavior:

John Lubbock Avebury, 1st Baron, *Ants, Bees and Wasps;* E. P. Dutton, New York, 1929.

Charles D. Michener and Mary H. Michener, *American Social Insects;* D. Van Nostrand, New York, 1951.

William Morton Wheeler, *Ants;* Columbia University Press, New York, 1926.

# THE SEED PLANTS

## Dominant Forms of the Earth's Vegetation

BY HARRY J. FULLER

IN ALMOST every type of natural landscape most of the plants you see belong to the group known as seed plants. These plants (which are classified in the phylum Tracheophyta, or vascular plants) number more than 250,000 known species — more than those of all other plant groups, such as algae, bacteria, fungi, mosses and ferns, put together. Since many regions of the earth have not yet been thoroughly explored, it is probable that when all living species of seed plants have been discovered, the total number will exceed 350,000. Botanists have been particularly interested in the seed plants not only because of their numbers but also because of their variety and outstanding economic importance. As a result, they have been more thoroughly investigated than any other plant group.

The seed plants show truly amazing variety in structure, growth and reproductive processes; yet they all have certain features in common.

(1) Their characteristic reproductive structures are seeds, produced by flowers or cones. Each seed contains a tiny plant, the embryo, which results from a process of sexual reproduction: that is, the fusion of a male sex cell (sperm) with a female sex cell (egg). Upon sprouting, the embryo grows into a mature plant.

(2) The sperm makes its way to the egg by way of a structure called a pollen tube, which is found only in seed plants. In most other plant groups — algae, ferns, mosses, liverworts and many fungi — sperms can reach eggs only by swimming through water. Water is not necessary for fertilization in seed plants because of the pollen tube. This fact accounts at least in part for the dominance of seed plants in the earth's vegetation; they can complete their reproduction under a much greater variety of environmental conditions than lower plants.

(3) Seed plants possess complex vascular, or conducting, tissues, which serve to transport water, minerals, foods and other substances within the plant. Many of the lower plants also have vascular tissues; but these are much less complex and less varied than the corresponding tissues of seed plants.

(4) Virtually all seed plants possess the green pigment chlorophyll, which is essential in photosynthesis, the basic process of food manufacture in plants. Only a few species of seed plants lack chlorophyll and therefore cannot carry on photosynthesis. Such plants are either parasites, obtaining their food from the tissues of other living plants, or saprophytes, deriving food from decomposing organic materials in soils, the dead stumps of trees and various other places.

### The structure and physiology of seed plants

The bodies of practically all seed plants consist of stems, roots, leaves and cones or flowers. One or more of these parts may be absent in a few species. For example, the asparagus plant and certain parasitic flowering plants lack leaves; in asparagus, the functions of leaves have been taken over by small, green stem branches.

Roots, stems and leaves are called vegetative organs, because they are concerned with maintaining the life of the

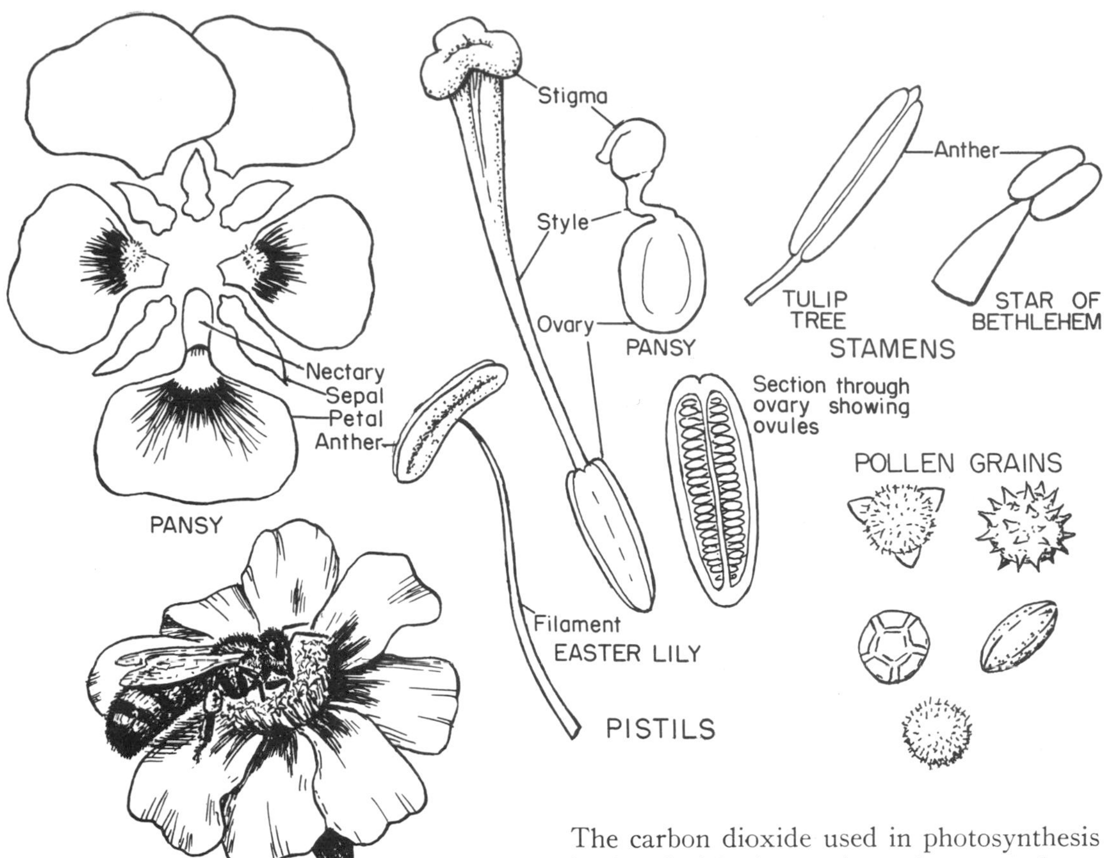


The chief parts of flowers are shown here. Notice the distinctive shapes they take in the different blossoms.

individual plant, rather than with reproduction. Roots anchor the plant body firmly in the soil, absorb water and mineral nutrients from the soil, conduct these absorbed materials upward into stems and transport food downward from stems. Stems produce and support leaves and flowers or cones, conduct water and nutrients upward and foods downward; they also store food. Leaves manufacture food in the process of photosynthesis.

In photosynthesis, carbon dioxide reacts with water, utilizing the energy of light absorbed by chlorophyll; as a result of this reaction, a simple sugar — glucose — is synthesized and oxygen is released. The process is represented by the simplified equation at the bottom of the page. The carbon dioxide used in photosynthesis is absorbed by leaves from the atmosphere, the gas passing into leaves through numerous minute pores in the leaf surface. The water required for the process is absorbed by roots from the soil; it travels upward through the conducting tissue of roots and stems into leaf veins, which are continuations of the conducting tissues of stems.

Since most of the chlorophyll of plants occurs in leaves, most photosynthesis obviously takes place in these plant structures. In some plants with green stems (such as tomatoes and petunias), the process also occurs in green stem tissues. Photosynthesis is the only major process for converting inorganic substances into food; it is also the only major source of oxygen in the earth's atmosphere. Without photosynthesis, all animal life, as well as almost all plant life, would come to an end. Animals depend upon this process for their food, since they derive various essential nutrients from plants or by devouring ani-

$$6CO_2 + 6H_2O + \text{light energy} \rightarrow C_6H_{12}O_6 + 6O_2$$

Six molecules of carbon dioxide + six molecules of water + light energy yield one molecule of glucose + six molecules of oxygen

$$C_6H_{12}O_6 + 6O_2 \rightarrow 6CO_2 + 6H_2O + \text{energy}$$

| One molecule of glucose | + | six molecules of oxygen | yield | six molecules of carbon dioxide | + | six molecules of water | + | energy |
|---|---|---|---|---|---|---|---|---|

mals that eat plants; they also depend upon photosynthesis to maintain the supply of oxygen, essential for respiration.

Like all other living organisms, seed plants carry on respiration. In this process foods combine chemically with oxygen and release their stored energy, chiefly as chemical energy. (A small quantity of heat energy is also released.) The energy derived from respiration is utilized in growth, reproduction and various other processes. Respiration is essentially the reverse of photosynthesis, as the simplified equation at the top of the page shows.

Photosynthesis converts light energy (chiefly from the sun) into the stored chemical energy of foods; respiration releases this stored energy for active use. Respiration differs from photosynthesis in other respects. It takes place in all living cells, while photosynthesis occurs only in chlorophyll-containing cells. Respiration occurs at all times, while photosynthesis occurs only in the presence of light.

During the day, the rate of photosynthesis in green leaves exceeds the rate of respiration. Hence the release of oxygen and the intake of carbon dioxide in photosynthesis overbalance the use of oxygen and the release of carbon dioxide in respiration. At night, when photosynthesis ceases, respiration continues; the result is that green plants absorb oxygen and release carbon dioxide. This characteristic gas exchange of green plants is responsible for the superstition that green plants "purify" air during the day and "poison" it during the night. The practice of removing plants from hospital rooms at night is based on this belief. While theoretically sound, the practice is really quite silly, since the amount of oxygen absorbed by a few plants at night is insignificant and carbon dioxide is not a poisonous gas.

All seed plants obviously grow. The process of growth, which results in the formation of new tissues, occurs chiefly at the tips of roots and in the buds of stems. In many seed plants a growth tissue called cambium is located in stems and roots between the two conducting tissues: xylem (which conducts water and soil nutrients upward) and phloem (which conducts foods manufactured in leaves downward). The growth activity of cambium leads to the production of new tissues transversely (crosswise) in stems and roots, and, therefore, causes these organs to grow in diameter. Growth processes are influenced by many factors, including the hereditary nature of the individual plant, growth hormones, water supply, nutrient supply, temperature and light.

As we have already pointed out, the reproductive structures of seed plants are seeds, produced in flowers or cones. We describe these structures later.

### The classification of seed plants

Living seed plants are classified in two main groups: Gymnospermae, or gymnosperms, and Angiospermae, or angiosperms. Gymnospermae ("naked seeds," in Greek) are so called because their seeds are produced on the surface of cone scales. In the Angiospermae ("seeds in a vessel," in Greek) seeds develop within structures called fruits. The gymnosperms are represented by comparatively few species — 750, as against the approximately 250,000 known species of angiosperms.

There are obvious differences between these two plant groups, in addition to the manner of seed production. The reproductive processes of gymnosperms occur in cones; of angiosperms, in flowers. Embryos of gymnosperms have more than two seed leaves (cotyledons); those of angiosperms have either one or two seed leaves. The ovules (structures that develop into seeds) of gymnosperms bear their eggs in tiny female sex organs celled archegonia; archegonia are lacking in angiosperm ovules. The xylem tissue of most gymnosperms is relatively simple; it consists usu-

U. S. Forest Service

The majestic ponderosa pine, one of the larger conifers.

ally of one kind of cell, called a tracheid, which conducts sap and provides support for the stem. Xylem of angiosperms is more complex in structure. It is generally made up of several types of cells and cell groups: tracheids, vessels (long, continuous conducting tubes), fibers and parenchyma (storage) cells.

**The gymnosperms, or cone-bearing plants**

The gymnosperms are woody plants — that is, trees or shrubs; they possess roots, stems and leaves, as well as the reproductive structures called cones. In most gymnosperms there is a dominant main stem, with conspicuously smaller branches; in some species there are no branches at all. The main stems of certain gymnosperms, such as the redwoods of California, may reach a height of more than three hundred feet. Trunk diameters are sometimes enormous.

The leaves of most gymnosperms are needlelike or scalelike; they contain chlorophyll and therefore carry on photosynthesis. In certain gymnosperms, such as cycads and ginkgoes, the leaves are broad and thin, rather than needlelike. Most gymnosperms are evergreen: that is, they retain their leaves throughout the year and often for several years (as in pines and firs). A few species, including larches, bald cypresses and ginkgoes, are deciduous; they lose their leaves in autumn, remain leafless throughout the winter and produce a new crop of leaves in the spring.

Many species of gymnosperms produce essential oils; these are volatile and highly scented. The essential oils, which appear to be waste products, are responsible for the characteristic odors of the needles and wood of many gymnosperms, including pines and cedars. When essential oils oxidize (combine with oxygen), they form extremely viscous liquids or brittle solids called resins. Resins are important raw materials of industry. They are obtained by cutting gashes into the sapwood of tree trunks; the resins ooze out of the cuts and are caught in containers. Among the most valuable gymnosperm

Brooklyn Botanic Garden

Cycas circinalis, at the left, is a typical cycad.

resins are the turpentines, derived chiefly from several species of pines and used principally in the preparation of paints and varnishes.

Botanists divide the living species of gymnosperms into several groups, called orders: the Cycadales, or cycads, the Ginkgoales, or maidenhair trees, and the Coniferales, or conifers. (A fourth order, that of the Gnetales, is not so well known to nonprofessional readers; it is made up chiefly of tropical shrubs, vines and small trees.) It is believed that the gymnosperms evolved from some group or groups of extinct ferns, probably by way of an order of curious plants called seed ferns. These plants had fernlike leaves and stems, but reproduced by means of seeds; the seeds were produced, not in cones, but directly on the leaves. The study of rocks of great geological age reveals the fossils of many gymnosperm species that have become extinct.

The order Cycadales consists of about a hundred species. The plants belonging to this group have unbranched, somewhat woody stems and compound, fernlike leaves, growing in a tuft from the tip of the stem. The leaves are subdivided into flat, thin structures, called leaflets. Many cycads resemble huge pineapples; they have a squat, thick stem, surmounted by a crown of leaves. Cycads are chiefly tropical and subtropical; they thrive in southern Florida, Mexico, the West Indies, tropical Asia, Australia and South Africa.

The leaves and fruit of the tree *Ginkgo biloba.*

U. S. Forest Service

The cycads are often cultivated as ornamental plants in display greenhouses and in tropical gardens. One Asiatic cycad produces a valuable gum. Another species, celled the, sago palm, which is extensively cultivated in Malaya and Indonesia, is the source of sago starch, or pearl sago, widely used as a food.

The reproductive processes of cycads occur in cones: large and usually woody female cones and smaller, more delicate male cones. The male cone bears scales that produce pollen grains; these are carried by wind to the ovules borne on scales of the female cone. Pollen tubes, produced by the grains of pollen, grow into the ovules; sperms from the pollen tubes then fertilize the eggs contained in the ovules. A fertilized egg grows into the embryo of the seed. When it is mature, the seed falls to the ground, where it sprouts.

UN

In the composite family, of which the sunflower (left) is a familiar example, many small flowers are crowded on a disk-shaped receptacle. Burpee's yellow cosmos (below) belongs to this group.

The Ginkgoales, or maidenhair trees, numbered many species in past geological ages; today there is only one living species — *Ginkgo biloba,* the Chinese maidenhair tree. This tree sometimes attains a height of ninety feet or more and its trunk may have a diameter of four feet. The leaves are crowded on short spur branches; they are from two to three inches long and are broad and fan-shaped, with prominent rib-like veins. Some trees produce small male cones; others, female cones of simple structure. A female cone bears two tiny cup-like scales, each containing one ovule. Pollination is by wind; after fertilization, each ovule develops into a seed with a single embryo. The surface tissues of the seed become fleshy at maturity and give off a very unpleasant odor. Ginkgo trees have no economic value. They have been widely planted, however, in many parts of the world in parks and gardens and along boulevards as ornamental trees. The maidenhair trees of Washington, D.C., are particularly well known.

The nearly 600 living species of the Coniferales, or conifers, are much more familiar than the members of the other orders of gymnosperms. Among the better known conifers are pines, spruces, Douglas firs, true firs, yews, incense cedars, true cedars, junipers, cypresses, sequoias (redwoods and California big trees) and larches, or tamaracks. Pines, which number about eighty species, make up the largest subgroup of the order.

W. Atlee Burpee Co.

Most conifers are trees; a few species are shrubs. They usually have straight, dominant trunks, often reaching heights of more than a hundred feet. The leaves of most conifers are needles or small scales. Often they are borne singly on the stems; sometimes they are produced in clusters of from two to five needles. The majority of conifers are evergreen; a few are deciduous.

The reproduction processes of most coniferous trees are quite similar. A pine tree, for example, produces two kinds of cones. There are small, rather soft male cones, each bearing a number of diminutive pollen-producing scales; there are also large, quite woody female cones, which have numerous cone scales. Each female cone scale bears two ovules on its upper surface. Pollen grains, shed by the mature scales of male cones, are carried by wind to female cones. The grains of pollen land on an ovule and form pollen tubes; these penetrate the ovule through a tiny pore (the micropyle) in the ovule coat. The pollen tubes carry sperms into the ovule; the eggs in the archegonia of the ovule are fertilized by the sperms. Since several archegonia and eggs are present in an ovule, multiple fertilization may occur; usually, however, only one fertilized egg develops into an embryo in the ovule. The embryo and the surrounding parts of the ovule grow, and ultimately a mature seed is formed; this falls from the female cone scale to the soil.

The conifers are the most valuable economically of all the gymnosperms. They furnish large quantities of softwood lumber (pine, redwood, fir, spruce and so on); they produce resins used in paints and varnishes; they are widely used as ornamentals. A few species, including the piñon pine of the southwestern United States, produce edible seeds. The bark of some species yields tannins, used in tanning hides. Conifers growing in extensive forests break the force of falling rain and bind the soil, thus preventing soil erosion.

Coniferous trees are very widely distributed on the surface of the earth, particularly in the cooler parts of the temperate zones and in elevated areas of the tropics. In the Americas, coniferous forests occur principally in the Rocky Mountains, in western and southeastern Canada, in Alaska, in the northeastern United States, in various highland areas of Mexico, Central America and the West Indies and in the Andes Mountains of South America. Some species thrive in warmer climates; coniferous trees are found in large numbers in the Gulf States of the United States.

### The angiosperms, or flowering plants

As we have already pointed out, the 250,000 known species of angiosperms produce their seeds within fruits. The fruits may become dry when they are mature, forming structures such as pea pods, bean pods and lily capsules; or they may become soft and fleshy, as in the case of tomatoes, grapes, watermelons and peaches.

There are two subclasses of angiosperms: the Monocotyledonae (monocotyledons, monocots) and the Dicotyledonae (dicotyledons, dicots). There are various differences between the two groups. Monocots have only one seed leaf (cotyledon) in the embryo; dicots have two such seed leaves. The flower parts of monocots are in threes or in multiples of three; those of dicots are most often in fives, or multiples of five, less frequently in fours, rarely in twos and threes. The leaves of most monocots, including grasses and irises, are much longer than they are wide; the veins usually run in the same direction lengthwise in the leaf. Leaves of most dicots

W. Atlee Burpee

**The beautiful flower of *Lilium henryi*, a monocotyledon.**

(elms, lettuce, maples and others) are about as long as they are broad or usually not more than two or three times longer than they are broad; the veins form an extensively branched network in the leaf. The stems of monocots generally lack cambium and consequently they increase but little in diameter. (The palms form a notable exception.) Most species of dicots have cambium in their stems and roots; as a result their girth, especially in the case of trees, is likely to be very impressive.

How did the angiosperms originate? The prevailing view, based largely on speculation, is that the earliest angiosperms to appear on earth had their origin in the seed ferns or some other group of primitive and now extinct gymnosperms. It is generally agreed that the most primitive living angiosperms are probably the magnolias, tulip trees and their relatives, and that other types of angiosperms may have evolved from primitive magnolialike ancestors. Most botanists regard the monocots as derivatives of primitive dicots of the magnolia type. The older view that monocots are more primitive than dicots is no longer accepted.

U. S. Forest Service

Burr of chestnut *(Castanea dentata)*, showing the seed.

There are about fifty orders of angiosperms in all; these orders are subdivided into about three hundred families, which cannot be described in a brief account of this type. The monocot orders include cattails, grasses, sedges, lilies, tulips, narcissuses, irises, amaryllids, palms, bananas and orchids. Among the dicot orders are almost all broad-leaf trees (willows, elms, oaks, maples, apples, walnuts, mahoganies, locusts, plum trees and the like), most shrubs (lilacs, privets, roses, forsythias, snowballs), most woody vines (grapes, wisterias) and thousands of species of plants that generally have herbaceous stems (sweet potatoes, bindweed, morning-glories, petunias, marigolds, zinnias, celery, soybeans, sweet peas, larkspurs). The monocots number about 50,000 species; the dicots, nearly 200,000.

All angiosperms have well differentiated conducting, food-making, storage and strengthening tissues. The typical angiosperm body consists of four major parts: roots, stems, leaves and flowers.

### Angiosperms have varied leaf forms

Leaf forms are infinitely varied in angiosperms. The leaves of most species are broad and thin; in some flowering plants they are modified into climbing organs called tendrils; in still others they form protective scales over buds. Leaves of certain species, such as the Venus's-flytrap, are specialized for trapping and digesting the bodies of insects. Century-plant leaves function primarily in food or water storage. There are many variations among stems too. Certain stems are modified into tendrils or into organs of food storage (bulbs of onions, Irish-potato tubers) or of vegetative reproduction (runners of strawberry plants, rhizomes of quack grass). Roots are just as varied. For example, grass roots are slender, fibrous structures; those of carrots, sugar beets and turnips become fleshy and enlarged as food and water are stored in them.

The reproductive processes that lead to seed formation in angiosperms occur in flowers. The most familiar type of angiosperm flower, found in plants such as irises, snapdragons, roses and petunias, bears four kinds of parts: sepals, petals, stamens and pistil. A flower having all four parts is known as complete. The sepals, known collectively as the calyx, are the outermost parts of the flower. They are usually small, inconspicuous and green in color and seem in most cases to serve no particular purpose. In a few angiosperms, including the tulips, the sepals are large, conspicuous and distinctively colored and probably function in the same way as petals.

Inside the sepals are the petals (known collectively as the corolla); they are the largest and most conspicuous parts of most complete flowers. They play an important part in the reproductive function of flowers. Because of their showy appearance, their frequently bright colors, their odor and their nectar (a viscous, sweet liquid) they attract insects, which bring about pollination.

The stamens, or male parts, are found inside the petals. Each stamen usually consists of a slender stalk (filament), surmounted by an enlarged pollen-producing structure, the anther. When anthers are mature, they open, releasing the pollen grains that have developed within them.

In the very center of the complete flower is the female part, or pistil. (Some flowers, such as those of buttercups and raspberries, have more than one pistil.) The typical pistil consists of an enlarged base, the ovary, within which ovules are produced, a slender stalk (style) rising from the top of the ovary and a somewhat enlarged tip (stigma) at the apex of the style.

Flowers that do not have a complete complement of four parts (sepals, petals, stamens and pistil) are known as incomplete. They are found in a minority of angiosperm species, including such well-known forms as wheat, elm, cottonwood and oak. In a perfect flower both stamens and pistil are present; an imperfect flower has either stamens or pistil, but not both. Plants with imperfect flowers (corn, cottonwood, walnut, among others) have two kinds of flowers. One has stamens, but no pistil; the other has a pistil, but lacks stamens. The presence of both pistil and stamens in the same flower, or the absence of either pistil or stamens makes it possible to distinguish between different species of angiosperms. There are other distinguishing features — the shape of petals, the number of parts, the size of parts, color patterns, odors and so on.

The mode of pollination — the transfer of pollen grains from stamens to the stigmas of pistils — depends upon the type of flowers found on a plant. If flowers are complete, if they are conspicuous because of the size or color of their petals, if they are fragrant and if they produce nectar, insect pollination takes place. Most incomplete flowers are small, greenish or

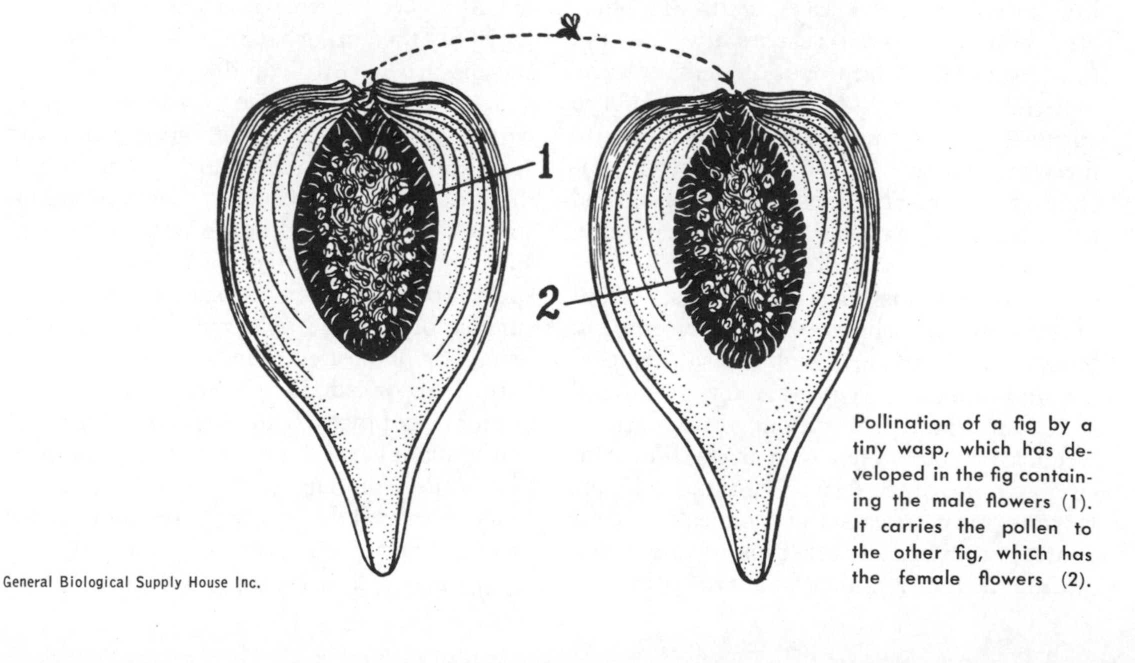


General Biological Supply House Inc.

Pollination of a fig by a tiny wasp, which has developed in the fig containing the male flowers (1). It carries the pollen to the other fig, which has the female flowers (2).

greenish yellow in color and have no odor or nectar. They do not attract insects; they are wind-pollinated.

Pollination is the first stage in the reproductive process. When a pollen grain, carried by insects or other animals (such as hummingbirds) or by the wind, lands on a stigma, it begins to grow, forming a long, slender pollen tube. This grows downward through the style into the ovary of the pistil. Within the ovary, the ovules await fertilization. Each ovule contains a central structure — the embryo sac — within which a single female sex structure, the egg, is produced. A pollen tube enters a pore (micropyle) in the surface tissues of the ovule, and discharges sperms into the embryo sac. A sperm fertilizes the egg, and the fertilized egg, or zygote, begins its development; it will ultimately become the embryo of the seed.

#### Later stages of the reproductive process

Foods are transported into the ovule; it grows and its surface cell layers develop into the seed coat. In the meantime foods and often water move into the tissues of the surrounding ovary, which ultimately forms the mature fruit. If this fruit is dry, it may split open at last along definite seams or it may split irregularly. If the fruit is fleshy, its tissues ultimately disintegrate. In either case, seeds are liberated. When a seed reaches the soil and finds suitable conditions of temperature, moisture and oxygen supply, its embryo sprouts and in time grows into a mature flowering plant. In some angiosperms, including navel oranges, pineapples and bananas, fertilization does not occur and, consequently, the fruits are seedless.

Angiosperms thrive in just about every type of climate and environment. They grow in tropical, subtropical, temperate and subarctic regions; they are found at sea level and at elevations of many thousands of feet. They constitute the principal plants of deserts, grasslands, alpine regions, most forests and the bleak tundras near the permanent ice caps of the poles. Although the angiosperms are primarily land plants, they include some species of water plants, which grow principally in the fresh water of lakes, ponds and streams. Among aquatic or semiaquatic fresh-water angiosperms are water lilies, cattails, duckweed, pondweed, tape grass and frogbit. Very few angiosperms grow in salt water.

Angiosperms play a more important part in our lives than any other plant group. Generally they are beneficial to mankind. We derive some of our most useful plant products from angiosperms. They are the source of such foods as fruits, grains, vegetables and nuts. They help provide us with meat, since meat-producing animals build their flesh at the expense of the plants upon which they graze. Angiosperms also supply us with hardwoods (oak, hickory, maple, cherry, mahoganies), nonalcoholic beverages (coffee, tea, cocoa), essential oils used in perfumes, colognes, soaps and cosmetics, drugs (quinine, digitalis, cocaine, belladonna), latex products (rubber, chicle, gutta-percha), fibers (cotton, flax, hemp), tannins and dyes. They are responsible, too, for the flavors of spices.

Angiosperms are beneficial to man in less direct ways. Their roots check soil erosion and thus aid in preventing floods. They provide shelter and food for some of the wild animals that are valuable to man. They lend beauty to natural landscapes as well as to parks and gardens.

On the other hand, certain angiosperms are harmful or disadvantageous to man. Some species are weeds, competing with crop plants for space, light, water and soil nutrients; if they grow unchecked, they reduce crop yields. Certain angiosperms liberate wind-borne pollens responsible for various kinds of allergies. Other species produce toxic substances that cause surface or internal poisoning, when human beings or domestic animals come in contact with them or eat them. Still other species (opium poppy, *Erythroxylon coca,* the hemp plant) produce narcotic substances that cause human misery and economic loss. Fortunately for man, the angiosperm species that are harmful to him are few.

*See also Vol. 10, p. 272:* "General Works."

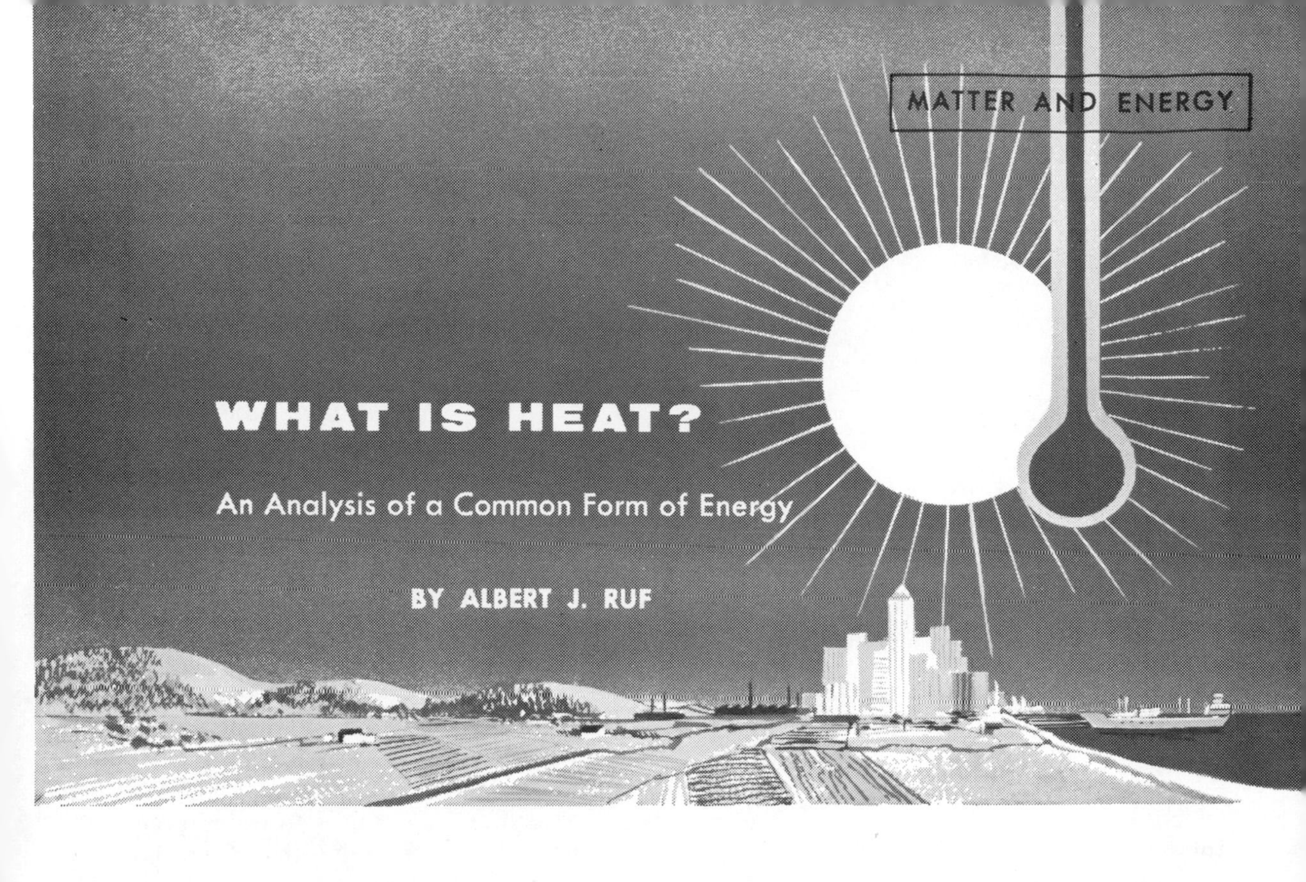


# WHAT IS HEAT?

## An Analysis of a Common Form of Energy

BY ALBERT J. RUF

HEAT plays an all-important part in our daily lives. Heating systems keep us comfortably warm in winter; in the summer, we seek to keep heat in its proper place by means of air-conditioning systems. Heat engines supply power for automobiles, trucks, diesel-electric locomotives, tugboats, ocean liners, airplanes and rockets. Certain heat engines run dynamos, which generate electricity; the electricity operates factory machines, television sets, telephone systems and a host of other devices. We owe life itself to heat. Our bodies, in a sense, are heat engines; the food we eat is fuel that keeps us warm and supplies us with energy for our various activities.

The sun is the source of *most* of the heat known to us. (The heat derived from man-made devices for producing atomic fission or fusion has nothing to do with the sun.) As a direct source of heat, the sun maintains life upon this planet, 93,000,000 miles distant. It is also an indirect source of heat. Since the earth is derived from the sun, it is to the latter that we must trace earth's internal heat, revealing itself through the action of volcanoes, geysers and hot springs.

The sun is also the source of the heat stored in fuels such as coal and petroleum. Coal is derived from vegetation that flourished on the earth during the Carboniferous period, more than 200,000,000 years ago. At that time, huge trees and various other plants grew in the swamps that covered much of the earth's land surface. Through the process of photosynthesis (see Index), they manufactured food in the presence of sunlight and thus stored the sun's radiant energy. They died and their remains accumulated, only partly decayed, in the swamps. In later ages, this vegetation became part of the earth's crust and under the action of heat and pressure was transformed into coal. Like coal, petroleum is derived from plants (and also probably from animals) that lived in the remote past. The solar energy stored up in coal and petroleum is released when we burn these fuels.

Even the heat produced by electricity is ultimately derived from the sun. Much of it is generated in steam plants through the use of coal and of liquid fuels derived from petroleum. The electricity produced by the force of running water may also

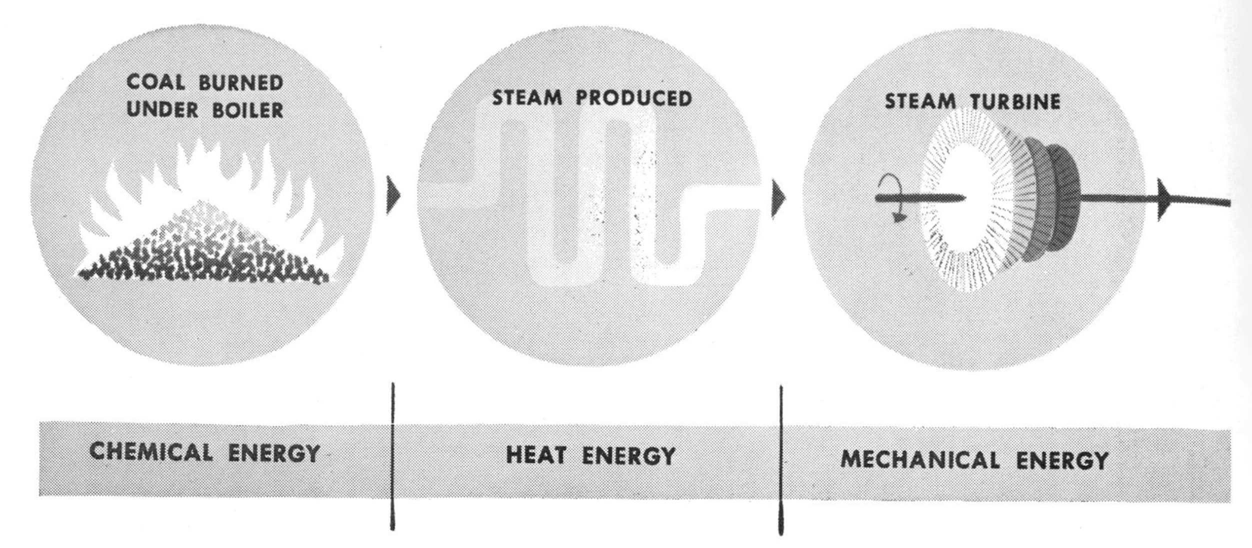


be traced to the sun. The reason is that our star plays the most important part in the water cycle, or hydrological cycle, of which running water is a phase. (See Index, under Water cycle.)

The scientists of earlier days had some rather curious ideas concerning the nature of heat. They regarded it as a fluid, which they called "caloric." All substances were supposed to contain caloric in various degrees, depending on how hot they were. It was maintained that when one body was heated by another, it was because caloric was transferred from the hotter body. Apparently this idea was confirmed by the fact that as fuels burn, they decrease in weight. But when scientists sought to determine the change in weight in heated substances, they found that the results were confused and often contradictory. This did not cause them to abandon the caloric theory; they simply revised it somewhat, by making caloric a weightless fluid.

It remained for the physicist Count Rumford (1753-1814), born Benjamin Thompson, to advance what is currently accepted as the correct theory concerning the nature of heat. While Rumford was supervising the boring of a cannon in the late 1790's, he noted that the gun became very hot after being bored for even a short time. He put an insulated and watertight box containing about 2½ gallons of water in contact with the gun barrel. As the barrel was bored, the water in the box became hotter and hotter and in two and a half hours it began to boil. It continued to do so even when the tool became so dull that it was no longer cutting. Apparently the rotation of the drill in the bore of the cannon in some way transferred heat to the water.

It occurred to Rumford that perhaps the mechanical energy released by the rotation of the drill was transformed into heat. He sought to determine the relationship between the amount of work done and the quantity of water that boiled away. His measurements were so crude, however, that no definite relationship could be established. Sir James Prescott Joule (1818-89) later proved definitely by refined experiments that a given amount of mechanical energy always produced the same amount of heat. This gave rise to a new concept of heat as a form of energy and to the development of the kinetic-molecular theory.

According to this theory, the heat that a body possesses is directly related to the kinetic energy, or energy of motion, of the molecules composing the body. ("Kinetic" comes from the Greek *kinein:* "to move.") The greater the kinetic energy involved, the hotter the body is.

In solids, the molecules attract one another strongly and their positions are fixed. Their motion, therefore, consists

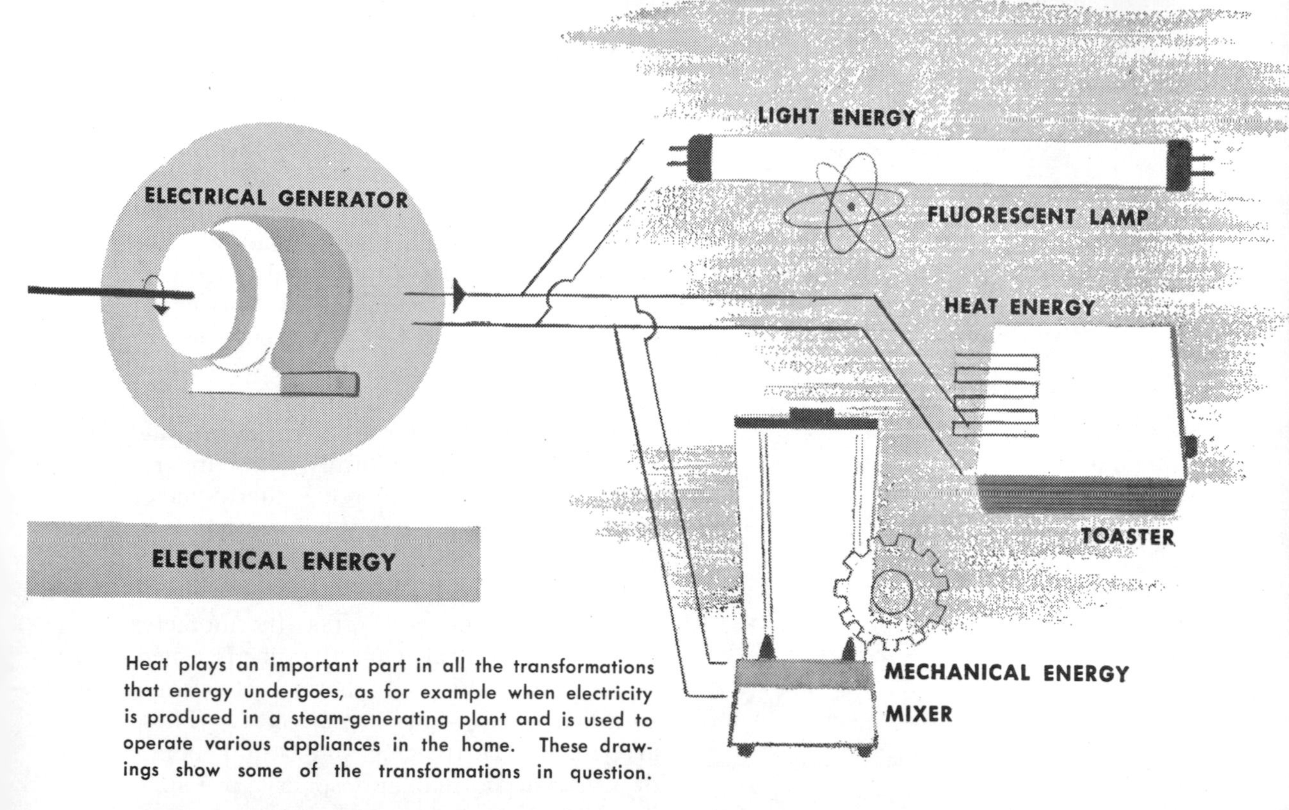


Heat plays an important part in all the transformations that energy undergoes, as for example when electricity is produced in a steam-generating plant and is used to operate various appliances in the home. These drawings show some of the transformations in question.

of vibration about a fixed point. The molecules of liquids can move about quite freely from one place to another, since they are not in fixed positions. They attract one another, but not quite so strongly as the molecules of solids. The molecules of liquids, because they can move, are able to assume the shape of the containing vessel, though the force of gravity keeps them from rising above a certain level in the container. In gases, the molecules are so far apart that the force of attraction between them is negligibly small. They move about even more freely than the molecules of liquids; they occupy the entire volume of the container.

Heat, then, is energy of motion. How shall we define cold? Actually, there is no such thing from the scientific point of view. When we say that object A is colder than object B, we mean that it has less heat. The molecules of A are vibrating, if A is a solid, or moving about, if it is a liquid or gas; but they are vibrating or moving about less energetically than the molecules of B. Molecular motion stops altogether only at the very low temperature called absolute zero. We shall refer to this temperature later in the article. It has never been attained as far as we know, though scientists have come very close to it in the laboratory.

Strictly speaking, then, all objects are hot, in the sense that they contain at least a certain amount of heat. To avoid confusion, however, we use such words as "cool," "cooler" and "cold" to indicate certain degrees of heat.

### How heat is measured

We must measure heat in order to acquire exact knowledge about it. Our senses cannot give us an accurate idea of how hot an object is, for they often mislead us. This can be illustrated by a simple trick often used at initiations to fraternities. The subject is blindfolded and is told that he is to be burned with a hot iron. Actually, the "hot iron" is only a piece of ice; yet when it is applied to the subject's skin, he is generally convinced that he is being branded.

As solids, or liquids or gases become hotter, they generally expand; as they become cooler (less hot), they generally contract.* These effects provide an effective

* There are some exceptions, as we shall see.

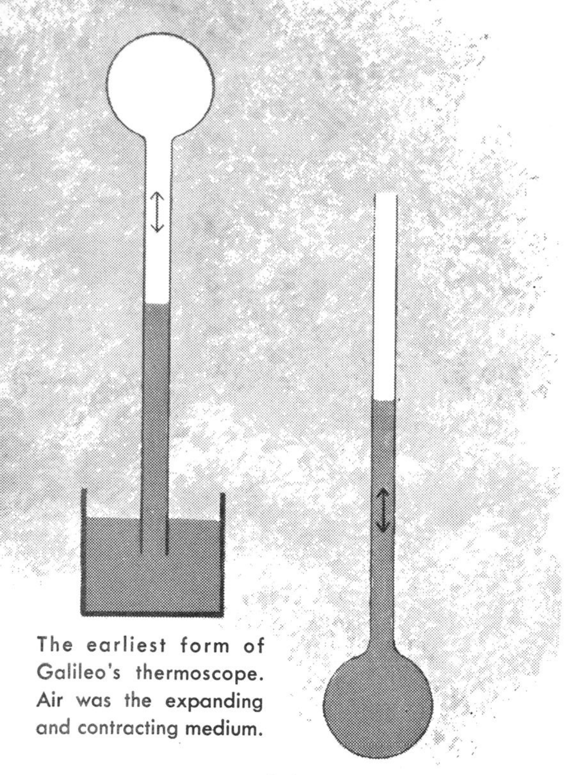

The earliest form of Galileo's thermoscope. Air was the expanding and contracting medium.

A later version of the thermoscope, with water serving as the expanding and contracting medium.

tool for measuring temperature — that is, the "degree of hotness" or "intensity of heat" of a body. The instruments designed to measure temperature are called thermometers, from the Greek *thermos,* meaning "heat," and *meter,* meaning "measure."

The great Italian scientist Galileo Galilei (1564-1642) was among the first to devise such an instrument — a crude affair called a thermoscope. It consisted of a glass bulb provided with a long glass stem, as shown in the diagram. The bulb was heated and the stem was dipped in water contained in another vessel. When the air in the bulb cooled, it contracted, and the water would rise up into the stem. The instrument could be used, then, to show changes in temperature. Galileo's thermoscope had two major disadvantages. It had no scale, and therefore could only indicate whether temperatures were rising or falling. Also it was sensitive to fluctuations in atmospheric pressure, which undoubtedly had something to do with the rise or fall of water in the stem. Somewhat later, water was used instead of air for the expansive medium, and the thermoscope was inverted. The water filled the bulb and a part of the stem.

Still later, alcohol was substituted for water, as it did not freeze so readily. The end of the stem was also closed to prevent the alcohol from evaporating. Another refinement was the use of mercury as the heat-measuring medium. Mercury is ideal for this purpose. It is opaque, has a low freezing point and high boiling point, and its volume changes uniformly with temperature. The ordinary mercury thermometer is made by fusing a glass bulb to the lower end of a narrow-bore glass tube. Mercury is then introduced into the tube through a funnel at the top end. The thermometer is boiled and annealed and the tube is sealed.

In determining any thermometric scale, so-called fixed points are selected. These are usually the freezing and boiling points of water at normal atmospheric pressure. The Swedish astronomer Anders Celsius (1701-44) described a thermometer in which the value 100 was assigned to the boiling point of water and 0 to the freezing point. The range from 0 to 100 was divided into 100 parts, called degrees and indicated by the symbol °. It is not certain that Celsius developed this thermometer himself; some authorities attribute it to the biologist Carolus Linnaeus (1707-1778). It is called the centigrade ("hundred-degree") or Celsius thermometer, and is used extensively, particularly for scientific work.

In the scale developed by the French naturalist René-Antoine Ferchault de Réaumur (1683-1757), the freezing point of water is set at 0° and the boiling point at 80°. The Réaumur (sometimes spelled Reaumur) scale is in common use in Germany and other Teutonic countries.

The German physicist Gabriel Daniel Fahrenheit (1686-1736) chose other fixed points. The lowest temperature he was able to attain with a mixture of ice and salt he called 0°. He gave the value 96° to the normal temperature of the human body.*

* We know now that the normal temperature of the body is about 98.4° in the Fahrenheit scale.

After his death the values of the freezing and boiling points were set at 32° and 212°, respectively. The Fahrenheit thermometer is employed widely in English-speaking countries. (The simpler and more convenient centigrade thermometer is commonly used in continental European nations.)

The range of the common mercury thermometer is limited to the interval between the freezing point of mercury, −38.87° C. (centigrade), and its boiling point, 356.9° C. Higher temperatures may be read if the space above the mercury is filled with an inert gas such as nitrogen or helium. As the column of mercury rises, it compresses the gas and raises the boiling point of the mercury.* The range of a thermometer may be extended downward by substituting for mercury a liquid with a lower freezing point.

There are different kinds of thermometers. A widely known variety, the clinical thermometer, is used to determine body temperature. Its outstanding feature is that the mercury column remains at the highest temperature reached. This is obviously most desirable, since otherwise the recorded temperature would change appreciably in the interval between removal from the body and reading. In the clinical thermometer, there is a very fine constriction (narrow part) in the tube near the bulb. Although the mercury can rise freely in the tube, surface tension prevents the mercury above the constriction from returning to the bulb. (See Index, under Surface tensions.) After the reading has been taken, the mercury can be shaken down again by swinging it quickly in a small arc. The centrifugal force causes the mercury to pass through the constriction. (See Index, under Centrifugal force.)

The Beckman differential thermometer is used for very precise temperature measurement. This instrument may be adjusted by means of a small mercury reservoir so as to cover a given portion — say 5 degrees — of its total range, to an accuracy of 1/100 of a degree.

* We shall point out later in this article that the higher the pressure upon a liquid, the higher is its boiling point.

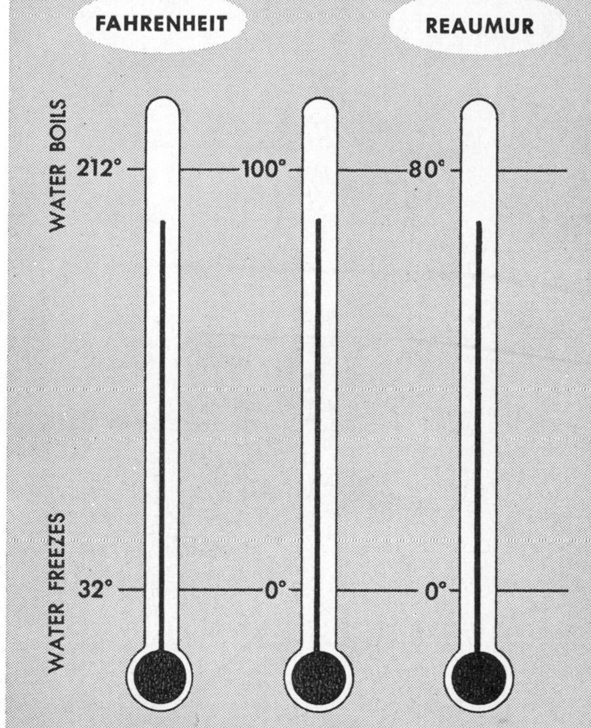


A comparison of the Fahrenheit, centigrade and Reaumur thermometers.

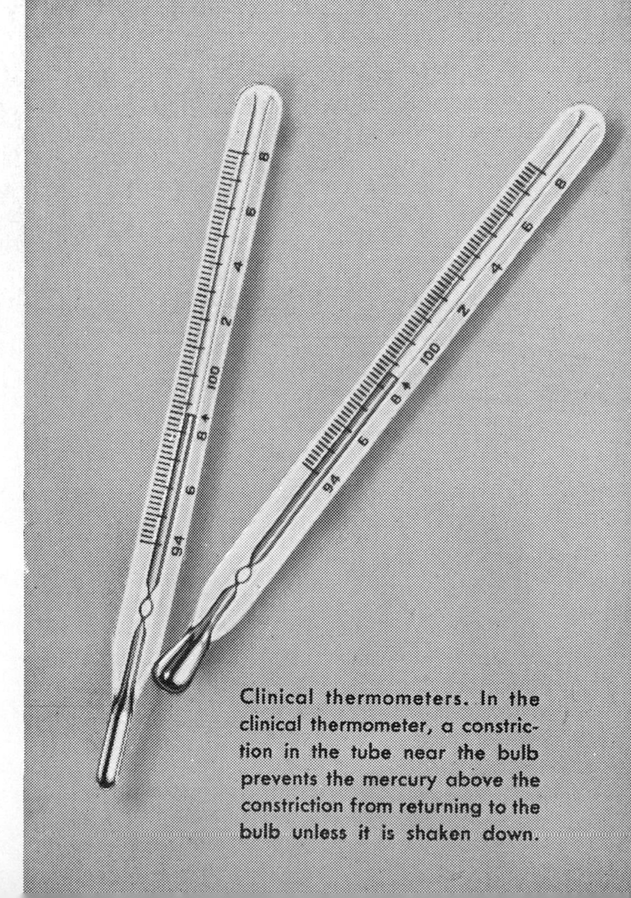

Clinical thermometers. In the clinical thermometer, a constriction in the tube near the bulb prevents the mercury above the constriction from returning to the bulb unless it is shaken down.

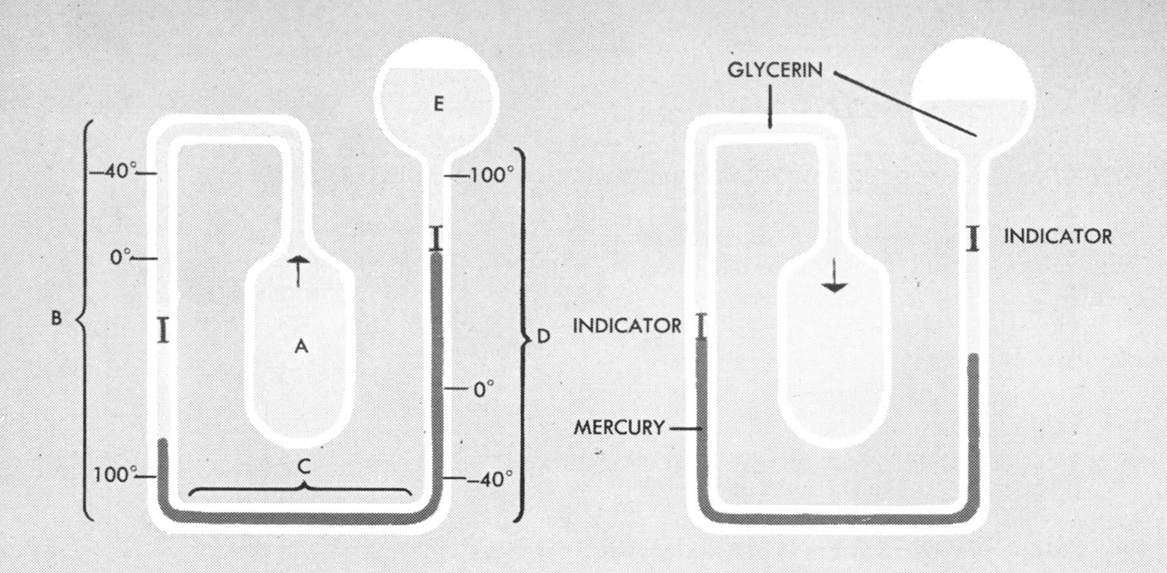

Diagram of maximum-minimum thermometer, described on this page.

Another interesting thermometer is the maximum-minimum type, shown in the diagram on this page. Chamber A and part of section B of the tube are filled with glycerin (or alcohol). The rest of section B, all of section C and part of section D contain mercury. Above the mercury column in section D there is some more glycerin, which does not quite fill the top chamber, E. As the temperature rises, the glycerin in chamber A expands, pushing the mercury up section D of the tube. The rising mercury column pushes ahead of it a small steel indicator. Because it is provided with small springs, the indicator remains in place in the tube, when the temperature becomes lower and the mercury column falls. With decreasing temperatures, the glycerin in chamber A contracts, and the mercury in section B of the tube rises. It pushes up a steel indicator set in section B, and thus indicates the minimum temperature. The indicators can be reset by means of a magnet.

To measure slight differences in temperature, a thermocouple, or thermopile, may be used. In this device, two different metals, such as iron and copper, are joined together in a circuit, forming two junction points. If one junction point is hotter than the other, an electromotive force is produced; this sends a current through the circuit. The electromotive force increases as the difference in temperature increases.

All of these devices measure the temperature, or the intensity of heat, but they are not a measure of the *quantity* of heat. Perhaps a simple example will illustrate the difference involved. Suppose you draw a ladleful from a vessel containing a quantity of hot water. The water in the vessel and in the ladle are at the same temperature but they contain different amounts of heat. If we immersed two identical specimens of iron, one in the ladle and the other in the vessel, the temperature of the iron in the vessel would be raised more than that of the iron in the ladle, because it would be subjected to more heat.

To measure the quantity of heat in the metric system, we use the unit called the gram-calorie, or calorie. This represents the amount of heat required to raise the temperature of a gram of water 1° C. The unit of heat in the English system is known as the British Thermal Unit, usually abbreviated as B.T.U. or Btu. It is defined as the amount of heat required to raise the temperature of a pound of water 1° F. (Fahrenheit).

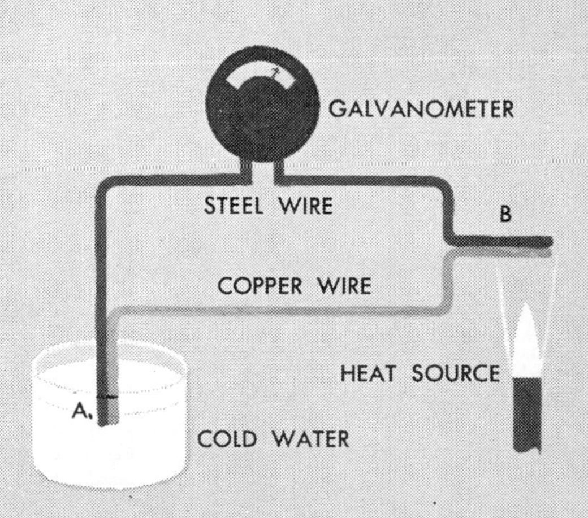


How a simple thermocouple works. Two different metals are joined together at junction points A and B. Since B is hotter than A, a current passes through the circuit, as is shown by the galvanometer. The greater the difference in temperature between junction points A and B, the stronger the current will be.

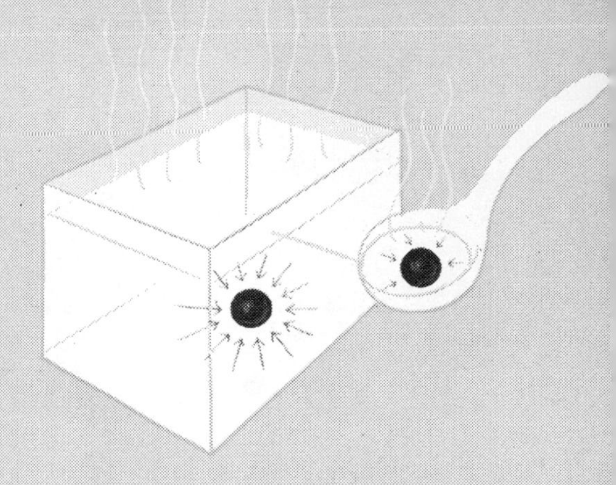

A ladleful of hot water is drawn from the container at the left. Two identical specimens of iron are immersed, one in the container and the other in the ladle. The temperature of the iron in the container will be raised more than that of the iron in the ladle, since it is subjected to a greater quantity of heat.

These definitions are accurate enough for most engineering work. It should be noted, however, that the amount of heat required to raise a unit mass of water through a given temperature change — say from 50° C. to 52° C. — will differ according to the initial temperature (50° C. in the above example). If a high degree of accuracy is required, the mean calorie is used. This is one hundredth of the heat required to raise the temperature of a gram of water from 0° to 100° C. The value in question is approximately equal to the amount of heat required to raise one gram from 15° C. to 16° C.

Another heat unit — the kilogram calorie — is widely used in the field of dietetics. It represents the heat required to raise the temperature of a kilogram of water 1° C. The calorie and the kilogram-calorie are sometimes referred to as the small and large calorie, respectively.

Not all materials absorb heat at the same rate when they are subjected to equal changes in temperature. For example, if equal weights of water, iron, mercury and gold at the same initial temperature undergo an equal temperature rise, each will absorb a different quantity of heat. The unit called specific heat is used to indicate this heat-absorbing capacity. The specific heat of a substance is the number of calories needed to raise a gram of it through 1° C.* In the case of water, exactly 1 calorie is required; the specific heat of water, therefore, is 1. The specific heat of any substance other than water is less than 1. For pure iron it is 0.11; for mercury, 0.033; for gold, 0.0316.

To show how we use units such as these, let us see how much heat would be required to raise the temperature of 10 grams of copper through 50° C. The specific heat of copper is 0.09: that is, 0.09 calories are required to raise the temperature of a gram of copper 1° C. The amount of heat required to raise the temperature of 10 grams of copper through 50° C., then, would be $10 \times .09 \times 50$, or 45 calories.

The specific heat of a material is generally determined by heating a sample of the material to a certain temperature and then putting it in a quantity of water whose temperature is also known. We carefully note the maximum temperature rise of the water. Since the heat gained by the water is equivalent to the heat lost by the sample

* It may also be defined as the number of British thermal units needed to raise a pound of a substance through 1° F.

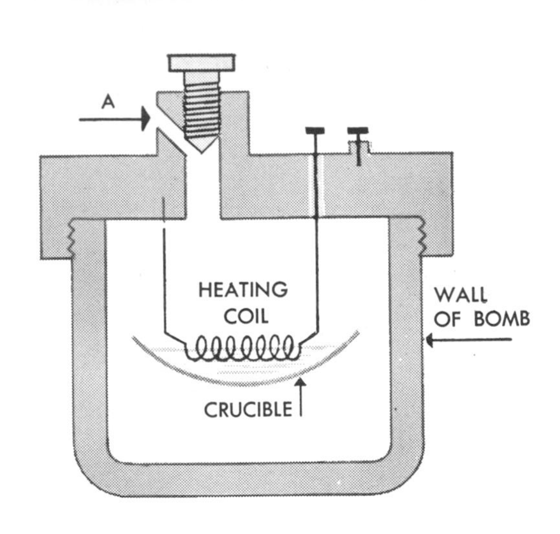


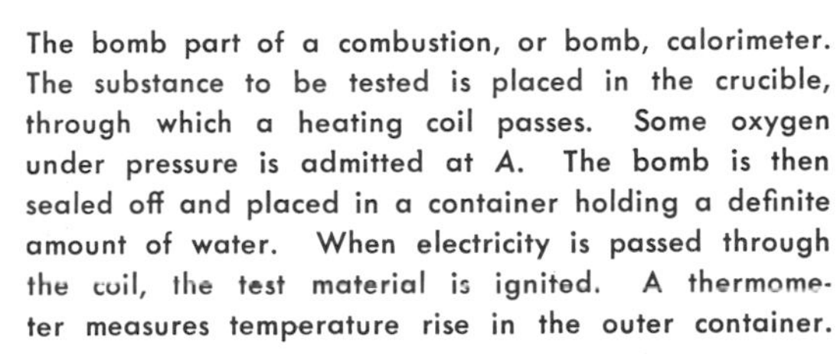
The bomb part of a combustion, or bomb, calorimeter. The substance to be tested is placed in the crucible, through which a heating coil passes. Some oxygen under pressure is admitted at A. The bomb is then sealed off and placed in a container holding a definite amount of water. When electricity is passed through the coil, the test material is ignited. A thermometer measures temperature rise in the outer container.

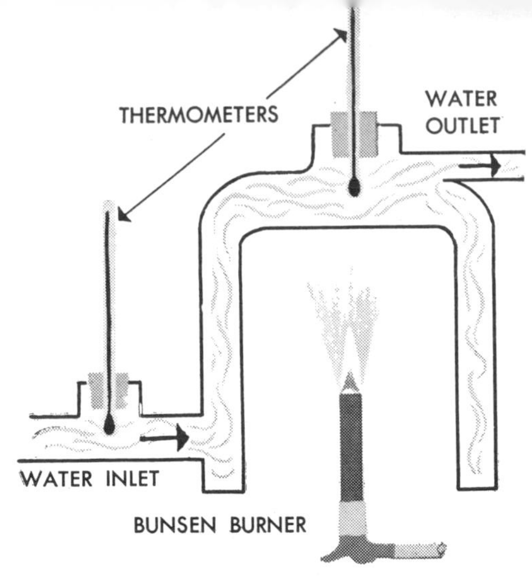


A continuous-flow calorimeter, which is used to measure the quantity of heat that is produced by a gas flame. Gas is burned at a constant rate in a special chamber which is almost completely surrounded by water. This water is made to circulate at a uniform rate. The heat of combustion can be calculated by recording the temperature of the water as it enters the calorimeter and the temperature of the water as it leaves the device.

of material, we can calculate the specific heat of the latter.

It is important to know the heat content of a fuel — that is, the heat made available when combustion, or burning, takes place. This quantity is called the heat of combustion; it is measured in heat units per unit weight of the material — that is, in so many calories per gram.

The heat of combustion of most solid and liquid fuels is measured by means of a combustion calorimeter, or bomb calorimeter. The substance that is to be tested is put in a crucible mounted in a heavy steel container called a bomb; this is lined with a corrosion-resistant material such as nickel or platinum. A small heating coil is connected to the substance that is being tested. Oxygen under pressure is then admitted to the bomb, which is then sealed and placed in a container containing a known amount of water. A current is passed through the heat coil, igniting the sample. A thermometer measures the temperature rise of the water in which the bomb is immersed.

The heat of combustion of gaseous fuels is measured by means of a continuous-flow calorimeter. In this apparatus, gas is burned at a constant rate in a chamber surrounded by water which is circulating uniformly. We can determine the heat of combustion by noting the temperature of the water as it enters and leaves the chamber.

**Expansion and contraction of solids, liquids and gases**

We have already pointed out that most solids expand when they are heated and contract when they become cooler. When railroad tracks are laid, a short space must be left between adjoining rails to allow for expansion of the steel in summer. The rivets used to join structural steel members are a good example of the effects of cooling upon solids. The rivet is first heated and is then driven into place while still hot. When it becomes cooler, it contracts, drawing the two members together firmly.

The expansion of a solid may be expressed in terms of one, two or all three of its dimensions. Thus we speak of linear (one-dimension), areal (two-dimension) and volume (three-dimension) expansion of solids. It is particularly important to know the linear expansion; once we have this value, we can easily find out the areal and volume expansions.

We measure linear expansion in terms of the coefficient of linear expansion —

the change in length of a unit length of a material for every unit change in temperature. This coefficient has been measured quite accurately for most materials. For example, if the temperature of a steel rod one foot long is raised or lowered 1° F., it will suffer a change in length of six-millionths of a foot. Hence a steel rod 40 feet long, in a location where the temperature ranges from 32° F. (in winter) to 90° F. (in summer) will change in length by .0139 feet (.000006 × [90 − 32] × 40) between the extreme temperatures. It is important to note that the amount of expansion depends on the length as well as the temperature change. A two-foot length of rod will expand twice as much as a one-foot rod subjected to the same variation in temperature.

Because different metals have different coefficients of expansion, one metal will expand more than another when it is subjected to the same degree of heat. An interesting illustration of this unequal expansion is the compound bar. It consists of two strips of different metals, such as copper and iron, riveted or spot-welded together so as to form a single bar. When such a bar is subjected to heat, one metal will expand more than the other, and the bar will bend or curl.

The compound bar finds numerous applications. The bimetallic thermometer commonly seen on kitchen ranges is made up of a compound bar wound into a spiral. The bar has copper on the inside and steel on the outside of the spiral.* When the bar is heated, it tends to straighten out, and this causes an attached pointer to move over a scale. Compound bars are widely used as thermostats in temperature control. We describe the workings of these devices elsewhere. (See Index, under Thermostats.)

Sometimes the expansion of solids as they are heated produces startling effects. It is well known that ordinary glassware will break when it is subjected rapidly to extreme temperature changes, as for example when one pours hot fat into a cold jar. This is because glass is a poor conductor of heat — another way of saying that heat is transmitted slowly through glass. The sides of the glass become unequally heated and consequently expand at a different rate. A strain is set up, and since ordinary glass is a brittle material, it will fracture. The glassware called Pyrex is subjected to a long cooling process called annealing, which results in a lower coefficient of expansion. This type of glass, therefore, is less liable to break when there is a rapid drop or rise in temperature. For this reason Pyrex was used for the 200-inch mirror of the great Palomar Mountain telescope.

Liquids generally expand much more than solids when heated. They exert terrific pressures, therefore, if they are prevented from expanding by being confined, say, in a container. In the case of liquids, we measure cubical rather than linear expansion. The coefficient of cubical expansion is the expansion per unit volume at 0° C. for a 1-degree rise in the temperature of a given liquid.

Water expands when heated and contracts when cooled like other liquids at temperatures above 4° C. From 0° C. to 4° C., however, it expands when it is cooled and contracts when it is heated. This means that water at 4° C. will expand and become lighter whether it is cooled or warmed.

When gases are subjected to heat, they show greater and more regular expansion

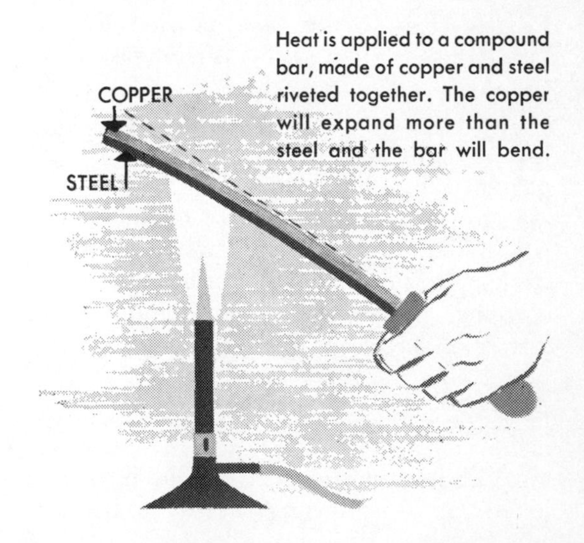

Heat is applied to a compound bar, made of copper and steel riveted together. The copper will expand more than the steel and the bar will bend.

* Copper has a much higher coefficient of expansion than steel.

than solids and liquids. The French physicist Jacques Charles (1746-1823) discovered in 1787 that all gases expand by the same amounts as they rise in temperature. He then worked out what is now called Charles's law. It states that a given volume of gas at 0° C. will expand approximately 1/273 of its volume when heated 1° C.; it will contract an equal amount when cooled 1° C. If we were to cool a gas to about −273° C., it would theoretically have no volume at all. −273° C. (more exactly, −273.16° C.) is called absolute zero (see Index). Actually, a gas would turn into a liquid before this temperature would be reached. Charles's law holds only if the pressure of a gas remains constant during the temperature change.

What happens if the pressure varies while the temperature remains constant? This relationship was investigated by Robert Boyle (1627-91) and independently by Edmé Mariotte (1620-84). The result of their researches was the formulation of the law that now bears Boyle's name. According to this law, the volume of a gas varies inversely as the pressure on it, provided the temperature remains constant.

If the *volume* of a gas remains constant, the pressure will increase 1/273 of its original value for each degree rise in temperature. This relationship is sometimes called Gay-Lussac's law, after the French chemist Joseph-Louis Gay-Lussac (1778-1850), who first formulated it.

The three laws described above apply exactly only in the case of what we call "ideal gases." An ideal gas is one in which the space occupied by the molecules and the attractive forces between them are negligible. For most gases at comparatively low pressures and high temperatures, the three laws are accurate enough for ordinary purposes. There are cases, however, in which we have to take into account both the space occupied by molecules and the forces of attraction between them.

### Heat and change of state

An important effect of heat is change of state. This means that as a body is heated, it passes from one to another of the three states, or phases, in which matter can exist — solid, liquid and gas.

What happens when a body passes from the solid to the liquid phase, as when ice is converted into water? We must recall that the molecules of a solid are closely packed together and exert considerable attractive force upon one another; that is why the original shape of the solid is maintained. Now, as heat is applied to the solid, the vibratory energy of the molecules is increased, and the individual molecules break loose from the bonds that formerly held them. What was formerly a solid has now become a liquid.

The cooling effect of ice is a consequence of the change of state from solid to liquid. The heat required for the melting of ice is absorbed from the surrounding objects, thus lowering their temperature. When ice cubes are added to a beverage, the cubes gradually melt as they absorb heat from the liquid; this causes the liquid to become cooler.

The temperature at which a solid changes to a liquid is called the melting point of the solid. The name "freezing point" is applied to the temperature at which a liquid solidifies. For a given crystalline material (one which forms crystals in the solid state), these two temperatures are the same. They are not identical in the case of certain substances, such as fats and glasses. (Glass is not a true solid, but a supercooled liquid.)

The melting points of different materials vary widely. Mercury, for example, melts at − 38.87° C.; iron, at 1535° C.; tungsten, at 3370° C. The low melting point of some alloys, such as Wood's metal, makes them valuable in fire-control systems. The sprinkler valves in such systems are held shut by plugs of the alloy. If the temperature in the room rises above 70° C. (158° F.), the plug melts, causing the sprinklers to go into action. Wood's metal is often used in making trick spoons. A spoon of this sort will melt when it is used to stir a hot liquid.

The amount of heat required to melt a unit mass of a substance when it reaches the melting point is called the heat of fusion. This varies with different substances. Ice, for example, absorbs 79.71 calories for each gram of ice melted. Aluminum absorbs 94 calories; copper, 49 calories; lead, only 5.47. The heat energy absorbed at this stage does not show itself in a rise in temperature; hence the heat of fusion is sometimes referred to as latent (hidden) heat.

If the temperature of a liquid is lowered, it will become a solid. Here too, the kinetic molecular theory furnishes an adequate explanation of what takes place. As the temperature of the liquid drops, the molecules of which it consists possess less energy. They move more sluggishly and they undergo, to a greater extent than before, the attraction of adjacent molecules. Finally, they are so strongly attracted by these molecules that they acquire fixed positions.

Most liquids contract upon freezing. A notable exception is water, which expands considerably when freezing takes place. A cubic foot of liquid water will become 1.09 cubic feet of ice; of course the ice will be less dense than liquid water. This phenomenon has certain fortunate effects from our viewpoint. Consider what happens when a body of water, such as a river or lake, is subjected to freezing temperatures. When the surface freezes, the ice, being less dense than water, remains on top of the remaining liquid water. The body of water, then, becomes frozen from the top downward. Since the surface ice serves to insulate the water below, total freezing rarely takes place, except in the case of very shallow streams or pools.

If the ice were denser than water, it would sink to the bottom, leaving the top layer of water exposed to freezing temperature. Ultimately the whole body of water would be frozen solid. Most aquatic life in it would be destroyed. A long period of thaw would be required before the river or lake would be navigable. Some bodies of water would probably remain permanently frozen.

Sometimes the expansive force produced by water as it freezes brings about certain undesirable effects. If the water in the pipes of our water-supply systems freezes, the expansion of the ice will cause the pipes to crack. This condition will first be revealed when thawing takes place. Water will then again flow through the pipes, which will leak badly. The freezing of water in the cooling system of cars may produce breaks in radiator tubes and may even crack the engine block. It is for this reason that anti-freeze, with a lower freezing point than water, is used.

When a liquid is heated, a different sort of change takes place. At any specific temperature, a given liquid contains molecules possessing different amounts of energy. Some of the molecules will be energetic enough to pass through the bound-

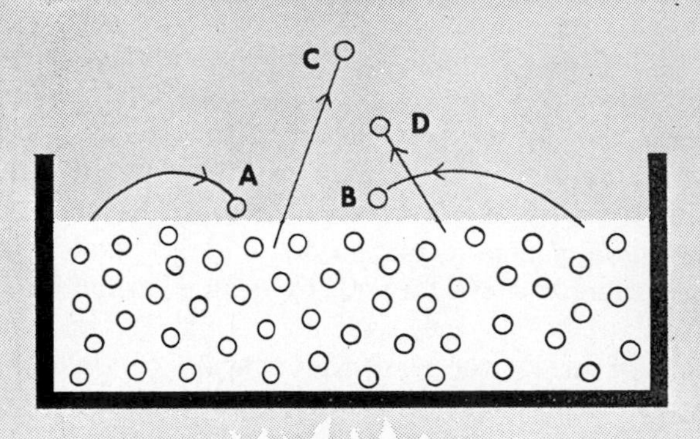


When a liquid is heated, certain molecules (A, B) pass out of the liquid but are drawn back to it again through the attraction of the molecules at the surface. Other high-energy molecules (C, D) escape from the attraction of the surface molecules; they form a gas.

ary surface of the liquid; but the attraction of the surface molecules will draw the escaping molecules back again. Certain molecules, possessing even more energy, will pass to the outer air and will travel beyond the attractive influence of the molecules at the surface. The escaping molecules will constitute a gas, or vapor. The process of passing from the liquid to the gaseous state is called evaporation.

Evaporation goes on at all temperatures. It continues until the liquid disappears or until the space above the liquid becomes saturated. This condition occurs when as many molecules leave the liquid as return to it from the space above, thus bringing about a condition of equilibrium.

Since evaporation consists of the loss of high-energy molecules from the liquid, it is obvious that the average energy of the remaining molecules will decrease and the temperature will drop. This effect of evaporation has been known since ancient times and is still utilized in various tropical countries in order to cool drinking water. The water is put in porous earthenware jugs or animal skins. As it evaporates from the surface of such containers, the water inside becomes colder.

Water evaporating from the skin helps maintain body temperature. When the weather is warm, we perspire; as the perspiration evaporates, the skin is cooled. Evaporation is slowed up in damp weather, since the concentration of water vapor in the air approaches the saturation point. That is why we feel cooler on dry days than on damp days, even though the temperature may be the same.

As the temperature of a liquid is increased, it will reach a point at which the liquid will begin to boil. Boiling may be considered as evaporation taking place throughout the body of the liquid rather than just at the surface. Bubbles of vapor are formed in the interior of the liquid; they rise and then break through the surface.

Boiling can be brought about more readily either by increasing the temperature or lowering the pressure. An increase in pressure will raise the boiling point. On the other hand, when the pressure is lessened, the boiling point is lowered. At high altitudes, the decrease in atmospheric pressure is such that certain cooking processes are slowed up and others become practically impossible. Such difficulties can be eliminated through the use of a pressure cooker. This is a closed vessel in which pressure is built up by the steam that forms as water in the vessel is heated. The pressure is regulated by means of a relief valve so that the contents can be made to boil at any desired temperature. The pressure cooker is also useful in speeding up many cooking operations and particularly in canning.

The reduced boiling point at low pressures finds considerable practical application in the field of vacuum evaporation (evaporation under low pressure). This process is of primary importance in the sugar industry. Boiling off the water from the syrup at normal atmospheric pressure would char the sugar. However, the pressure is kept so low in vacuum evaporation that the water may be removed at comparatively low temperatures.

The amount of heat required to change a unit mass of liquid to its vapor state at its boiling point is called the heat of vaporization of the substance. Each material has its own heat of vaporization. For example, it takes nearly 540 calories to vaporize a gram of water at 100° C.; only 204 calories are required in the case of ethyl (grain) alcohol.

In going from the solid to the gaseous state, most substances pass through the intermediate liquid state. Certain substances, however, go directly from solid to vapor form — a process known as sublimation. Naphthalene moth balls, iodine crystals and the insecticide paradichlorobenzene are good examples of this phenomenon. Solid carbon dioxide also sublimes under ordinary conditions. It is often called "dry ice" because it is used to keep objects cold. It is excellent for this purpose; since it sublimes, instead of melting, it does not wet the substances (such as ice cream and frozen confections) that it refrigerates.

*See also Vol. 10, p. 281:* "Heat."

# THE CRUST OF OUR PLANET

## The Basic Structure of Earth's Overlying Rocks

BY TERENCE T. QUIRKE

WHEN we see rocky ridges and sheer cliffs breaking harshly into a pleasing pattern of green forest and undulating fields, we are apt to think of the rock formations as interlopers. Yet such formations make up the real crust of the earth, in land areas and sea areas alike. The soil of the land, upon which plant life and animal life flourish, is an exceedingly thin film composed chiefly of bedrock broken into fragments by weathering and erosion. In the ocean areas, deposits of mineral matter and of plant and animal remains also form a superficial covering upon the underlying bedrock of the sea bottom.

The rocky crust of the earth is made up of three kinds of rock formations: igneous, sedimentary and metamorphic rocks. The igneous rocks were formed by the solidifying of molten, rock-producing matter. The sedimentary rocks consist of sediments, or fragments, that have been transformed into rocks in the course of the centuries. The metamorphic rocks were originally igneous or sedimentary rocks, transformed by changes in temperature and pressure and other factors operating within the crust of the earth.

### *IGNEOUS ROCKS*

The molten, seething mass from which the igneous rocks have been formed is called magma. It contains various gases in solution, but for the most part it is made up of rock-forming minerals, dissolved in haphazard fashion.

According to most theories of the earth's formation, our planet was once a mass of swirling gases cast off from the sun. As these gases gradually cooled, they were converted into liquid form. This liquid — the magma — began to cool; the minerals it contained began to crystallize. Heavy minerals tended to sink in the still-liquid mass of the magma; lighter minerals floated on top of the heavier ones. As the cooling continued, the rocks began to solidify; in time they formed a solid crust.

This represented only one stage in the formation of igneous rock. Vast quantities of magma at extremely high temperatures were imprisoned under the crust. The gases contained in the magma exerted enormous pressures not only upon the igneous rocks that had already been formed but also upon an entirely different kind of formation — the sedimentary rocks. The surface layers of igneous rock had been constantly acted upon by erosive forces, such as changes in temperature, running water and the blasting effects of fragments hurled by the wind. As a result, some of the igneous rock at the surface had been broken down into fragments. These fragments, swept along by winds, glaciers, streams and shore currents, had been deposited in lakes or in shallow parts of the sea. They had been pressed together under the weight of later accumulations and had been gradually transformed into sedimentary rock. Sedimentary rock had also been formed from plant and animal remains and through the evaporation of sea water.

The original igneous rocks of the earth's crust and the sedimentary rocks that were formed later were sometimes unable to resist the pressure of the magma under the crust. The liquid rock would then make its way up to higher levels.

In some cases the advance of the magma was stopped by the rock layers in its path; only its gases escaped. The magma gradually cooled and solidified under the rock layers that made up the upper part of the earth's crust. Igneous rocks that have been formed in this way are called intrusive masses. In other cases the magma was not checked in its upward progress by the surrounding rocks; it reached the surface of the earth and was either discharged through a simple opening, or volcanic vent, or through a crack in the rocks. Igneous rocks formed from magma extruded, or thrust out, from the earth are known as extrusive masses.

Intrusive masses of igneous rock form a considerable portion of the earth's crust. On the basis of their shape and their relation to the rocks that surround them, they may be divided into several classes — dikes, sills, laccoliths, volcanic necks, batholiths and stocks.

Dikes have been formed by magma that has cut its way through rock layers and that has been solidified in the fissures produced in this way. In certain instances dikes are exposed to view, either because the original fissures extended to the surface or because the rocks that once covered them have eroded away in the course of the ages. In certain areas dikes are very numerous and close together. They may be parallel to one another or may radiate from a common center, like the spokes of a wheel.

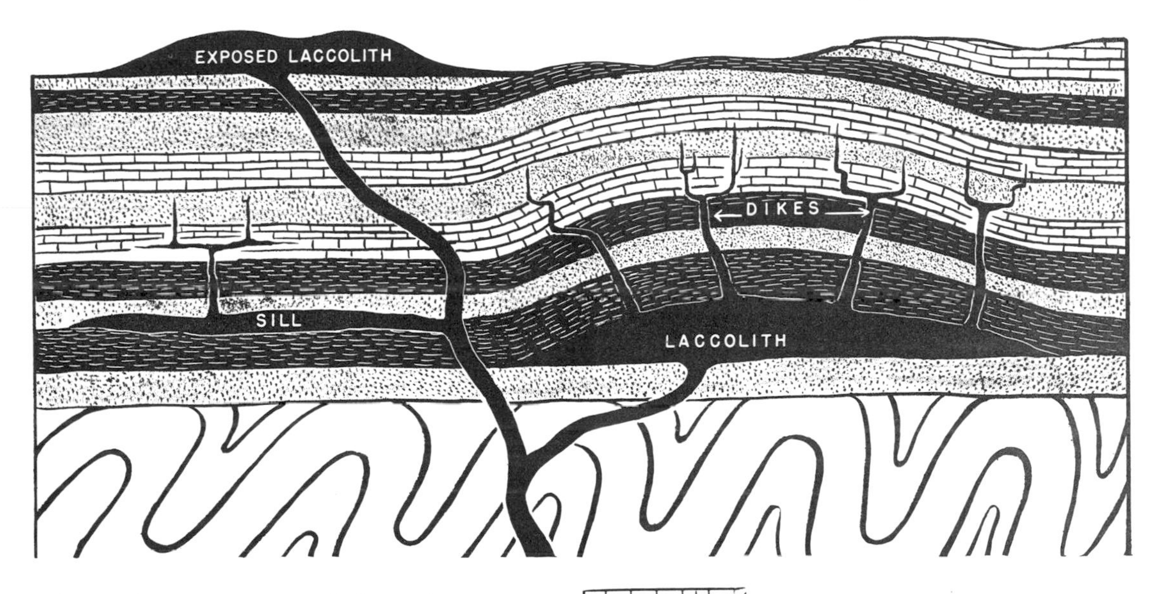


Molten rock moving across rock layers forms dikes. When it passes between the layers it produces sills or, if the overlying rock bed is arched upward, the result is a laccolith.

## KEY

The chart at the right shows diagramatic representations of various rocks. It is a key to the identification of rocks in the figure above.

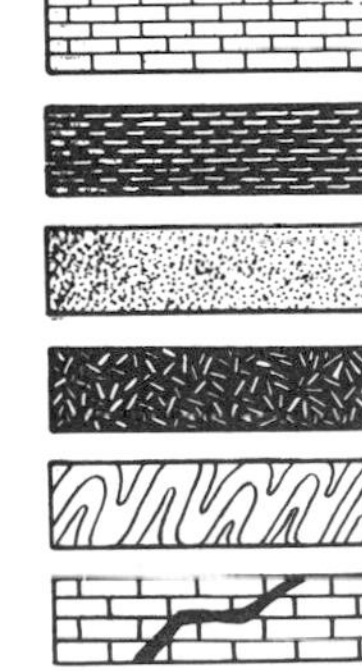

U. S. Geological Survey

Great dike running north from West Spanish Peak, Colorado. This dike has been exposed because of erosion.

The mass of igneous rock called a sill is formed in much the same way as a dike, except that the magma of a sill thrusts itself into the spaces between the layers of the surrounding rock; it is parallel to these layers. The magma has displaced the two adjacent beds between which it has been inserted; its thickness measures the extent to which the surrounding rock has been shifted. The resulting sill may be thousands of feet thick.

A sill may be horizontal or it may lie tilted along with the rock layers to which it is parallel. In the latter case, it may mean that the sill was either intruded into already tilted, or dipping, rocks or into flat-lying beds that, together with the sill, were later shifted about by forces in the earth's crust. The bluffs known as the Palisades, along the western shore of the Hudson River in eastern New Jersey, are the edge of an igneous sill hundreds of feet thick, tipped downward to the west some time after its formation.

Sometimes, during the formation of a sill, the magma does not spread easily enough and causes the overlying layer of host rock to arch upward. The resulting igneous mass, having the shape of a lens in cross section, is called a laccolith (Greek for "reservoir of stone"). It has a flat floor, connected by a "stem" to the magma source below, so that as a whole it resembles a toadstool in side view.

The vent of a dead volcano may be filled with a cylindrical mass of solid lava known as a volcanic plug, or neck, which may be up to a mile thick. It may connect with a large underground mass of igneous rock called a stock, which may have supplied lava to the erupting volcano in the magmatic stage. Dikes and sills often emanate from the plug. If the surrounding rocks have been eroded away, the plug stands out as a neck, or tower.

Huge intrusive masses of igneous rock enlarging and extending downward to unknown depths are called batholiths ("deep rocks," in Greek). They differ from laccoliths in that they do not show signs of a

Batholiths are gigantic masses of igneous rock usually lying well below the surface of the earth. They may extend downward for miles. Sometimes they are exposed to view by erosion and form impressive mountain ranges.

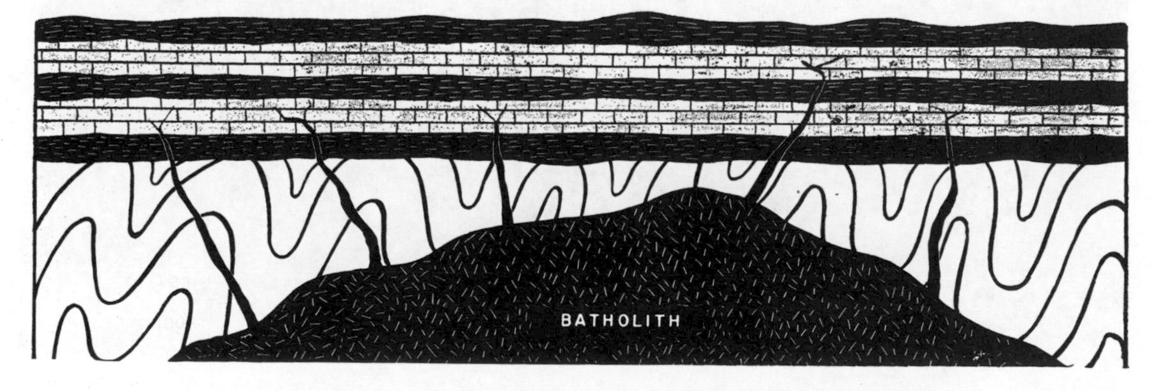

definite floor, as do the laccoliths; besides, the magma from which they were derived made its way through rock layers and not between them. Of course, as in the case of other intrusive masses, the batholiths are exposed to view only when the overlying rocks are eroded away. One of the world's largest batholiths forms the Sierra Nevada range, California; it has been uncovered over wide areas by erosion. Batholiths form the foundations of the continental masses of the earth, as well as the cores of many mountain ranges.

Like batholiths, stocks are intrusive masses that have no known floor. The two formations differ only in size; a stock is a batholith that covers an area of less than 100 square kilometers — that is, about 38.6 square miles.

There are several kinds of extrusive masses. In some cases magma has been extruded from the depths of the earth with explosive force, forming a spray of atomized rock stuff shooting high into the air; the fragments have later been converted into rock. In other cases the magma has not been blown from a volcanic vent, but has welled up from it and has poured out over the countryside; it has been transformed into igneous rock upon cooling.

The minerals that have contributed to the formation of igneous rocks are rather limited in number and in variety. Commonest of all is quartz, a very hard mineral which is a compound of silicon and oxygen. The feldspars are also abundant. These are light to dark in color and contain potassium, sodium or calcium, as well as aluminum, silicon and oxygen. Pyroxene and hornblende, containing different metals plus silicon and oxygen, are darker and heavier. The micas are sheetlike minerals composed of aluminum, other metals, oxygen and silicon. Magnetite, an iron-oxygen compound, is heavy and magnetic. Olivine, consisting of iron, magnesium, oxygen and silicon, is a green mineral limited to darker igneous rocks.

Igneous rocks differ from one another in texture: that is, in the size, shape and arrangement of the particles of which they consist. The size of the individual grains depends upon the rate at which the magma cooled. If the process was slow, the mineral crystals had a longer time to form and so they grew comparatively large; the result is a coarse-grained rock. If, however,

The Devil Postpile, a national monument in California. It consists of symmetrical columns of blue-gray basalt.

Chicago Nat. Hist. Mus.

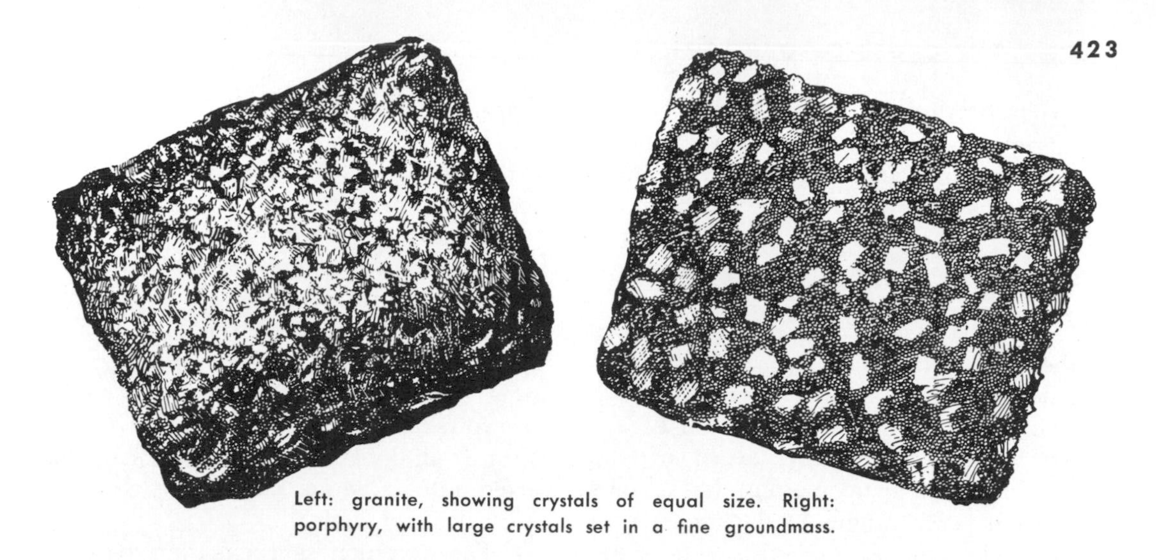

Left: granite, showing crystals of equal size. Right: porphyry, with large crystals set in a fine groundmass.

the magma cooled quite rapidly, the crystals did not have much time to grow; the resulting rock formation is fine-grained. Generally there is a relationship between the depth at which the rock was formed and the coarseness of its grain; the greater the depth, the coarser the grain.

In some igneous rocks we find a number of comparatively large crystals set in a mass of finer-grained ones, known as a matrix, or ground mass. A rock like this, in which coarse and fine grains are mixed together, is called a porphyry.

There are many varieties of igneous rocks; several hundred are recognized in some classifications. However, the crust of the earth is made up principally of only a few kinds. One of the most common is the plutonic, or deep-seated, rock called granite, which occurs in the form of huge, irregular masses. Granite consists chiefly of quartz and feldspar; it also contains some hornblende and mica (the black variety called biotite). This type of rock generally has a comparatively even texture. The variety of granite called pegmatite is often found in dikes; it is very coarsely granular. Pegmatite is the source of much of the white mica used in commerce.

Like granite, diorite is an even-textured rock. However, unlike granite, it contains no quartz; it is made up of feldspar and of one or more dark minerals, chiefly hornblende, pyroxene and biotite. In the rock called gabbro, these dark minerals predominate; feldspar is also present but not in large quantities. Syenite is another igneous rock formation that contains no quartz. The name is derived from Syene (now called Aswan), on the Nile, where there is a striking formation of this rock.

The dark and heavy rock called peridotite is composed entirely of ferromagnesian minerals (compounds of iron and magnesium). Peridotite is to be found chiefly in intrusive masses such as dikes and sills. Considerable deposits of nickel, chromium, platinum and iron are to be found in this distinctive rock formation.

The basalts are the commonest of all the extrusive igneous rocks; they are black, brown, dark gray or dark green in color. The basalt formation in the Columbia Plateau, in the northwestern part of the United States, covers an area of over 200,000 square miles and ranges up to 4,000 feet in thickness. There are also vast basalt formations in the Deccan (in western India) and the northern British Isles.

Rhyolite, like granite in composition, is an extrusive rock with a very fine (*felsitic*) grain; or it may have a porphyritic texture if there are some large crystals of feldspar, quartz or mica. Andesite is the extrusive form of diorite; darker than rhyolite, it too may be porphyritic. It receives its name from the Andes Mountains, where it is abundant.

Among the most striking of the glassy rocks found among the extrusive masses are obsidian and pitchstone. Obsidian is a lustrous rock, black, gray, yellow or brown in color. It is found in many different parts of the world; one of the largest formations is Obsidian Cliff, in Yellowstone National Park. Pitchstone is much less lustrous than obsidian.

## SEDIMENTARY ROCKS

As we have seen, the sedimentary rocks of the earth have been formed from rock fragments and plant and animal remains, as well as through the evaporation of sea water. The sedimentary layers of the earth's crust make up most of its surface area; each layer ranges in thickness from a few inches to several yards. In some places there are only a few layers; in others there are vast accumulations of beds totaling several miles in thickness.

Generally speaking, the older the bed, the more thoroughly the sedimentary rocks have been cemented. The fragments of young rocks are so loosely held together that they may sometimes be quarried by digging with a spade. The older rocks, however, are generally so solid that they cannot be quarried unless they are drilled, blasted with dynamite and broken into still smaller fragments with pick or sledge hammer.

The sedimentary rocks that are made up entirely of particles of other rocks are known as fragmental, or clastic, rocks. ("Clastic" comes from the Greek *klastos,* meaning "broken.") The fragments from which the fragmental rocks are derived are generally classified, on the basis of size, in four groups: gravel, sand, silt and mud.

Gravel is the coarsest sediment of all; it consists of fragments that are at least 2 millimeters in diameter. Gravel fragments range from small pebbles to big boulders; in between there are the particles called cobbles. Sand sediments consist of rounded grains ranging from 2 millimeters in diameter to 1/16 millimeter. Particles smaller than 1/16 millimeter and at least 1/256 millimeter in diameter are called silt. The finest sediments, muds and clays, consist of fragments under 1/256 millimeter in diameter.

When more or less rounded gravel particles — pebbles, cobbles and boulders — are cemented together, they form the type of rock called conglomerate. In the variety known as breccia, the gravel fragments are not rounded, but angular.

Cemented sand grains give rise to the porous formation known as sandstone. The pores of this rock may make up as much as 30 per cent of the total volume. Liquids move quite freely through the pores; for this reason sandstones are often reservoirs for petroleum deposits, as well as for ground water. Silt particles are converted into siltstone; mud particles, into mudstone and shale.

The commonest rocks derived from plants are the various varieties of coal. All coal started as leafy and woody material — the remains of plants that thrived aeons ago in swampy areas. The remains were buried under later sediments; the various strata were folded, broken up, sunk under ancient seas and raised up again by movements of the earth's crust. As a result of tremendous pressure, heat and the exclusion of air, the organic materials were transformed in the course of time into coal. Peat, lignite, bituminous coal and anthracite represent successive stages in this transformation.

Animal remains have contributed to the formation of various familiar kinds of sedimentary rocks. Coral reefs and islands represent the skeletons of countless coral

Coquina, a sedimentary rock composed of cemented fragments of sea shells. They consist mostly of calcite.

Amer. Mus. of Nat. Hist.

polyps and other organisms; one generation after another has contributed to these huge deposits, which have been piled up hundreds of feet deep.

The rock called limestone is composed of calcite, or calcium carbonate, a mineral compound of calcium, carbon and oxygen. Calcite is commonly found in the shells of animals such as snails, clams, brachiopods and other shellfish. These shells may go into the making of limestone; much of the latter, however, is made up of calcium carbonate that has been precipitated directly from solution in water. There are marine and fresh-water limestones. As the waters retreat, limestone deposits become part of the rocks on land.

Sedimentary deposits formed from salts precipitated out of solution in evaporating water are called evaporites. Evaporites may develop from sea water or from the water of streams and lakes. Many of these salts are found in both oceanic and inland waters, but their concentrations may differ, and the mineral deposits they form may differ accordingly. Evaporites indicate dry, hot conditions. An arm of the sea may be cut off and become a lake of concentrated brine as the water evaporates. In desert regions, temporary streams and ponds may be saturated salt solutions, leaving behind deposits when the sun dries them up.

Once evaporation starts, the least soluble compounds precipitate (separate out of solution) first. In deposits derived from a body of sea water, calcium carbonate and iron oxide form the first, or lowest, layer. Then comes gypsum, a sulfur-oxygen compound (sulfate) of calcium. Common salt, or rock salt, is the next to precipitate; afterwards, various other compounds, such as sulfates of magnesium and sodium. These four groups of dissolved minerals, if present together, would then form four or more successive layers of evaporites from the bottom up. If dry conditions last, these deposits remain; otherwise, returning waters or rain will redissolve them or cover them over with layers of mud and sand.

Sodium nitrate (a nitrogen-oxygen compound of sodium) forms vast evaporite deposits in the deserts of northern Chile. This nitrate was originally produced by certain bacteria in plants; then water flooded the region and dissolved the nitrate from the soil. Later, the waters dried up, leaving this compound behind in such large quantities that it can be profitably exploited.

As sediments accumulate, they may contain remains of dead organisms. These turn into stone or leave their imprints in the rocks formed from sediments, and so become fossils. Scientists study them to discover the evolution of life through the ages. All known fossil-containing sedimen-

Shale, made up of flaky clay particles compacted into thin layers. This rock may be brittle.

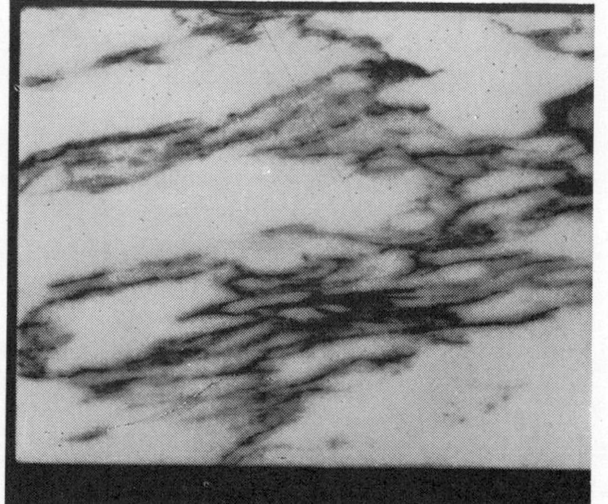

Marble, the metamorphosed form of limestone. The mottling is caused by colored impurities.

Amer. Mus. of Nat. Hist.

tary rocks, arranged layer on layer, make up what is known as the stratigraphic column. It is the geologist's historical yardstick, used to date geological events.

## *METAMORPHIC ROCKS*

The third great division of rocks — the metamorphic rocks — is made up, as we have pointed out, of igneous or sedimentary rocks that have been transformed within the earth's crust into rocks of a quite different sort. Generally some traces of the original structures have been preserved.

One of the most important factors in metamorphism — the formation of metamorphic rock — is pressure. It may be applied by overlying sedimentary beds; it may be caused by magma making its way into surrounding rock layers; it may be due to the mountain-building forces that deform the earth's crust. As pressure is applied, tremendous heat is generated; this quickens the chemical reactions taking place and heightens their effects. The presence of water is another factor in metamorphism. Even if the water is so scanty that it forms a mere film around the particles, it provides a medium in which rock substances can pass into solution and from which they can condense on the surface of new and growing crystals.

Under extreme heat and pressure the original rock particles are forced into new arrangements. In some cases the rock constituents recombine with those in the immediate vicinity and form new minerals, many of which grow with nearly perfect crystal form. One such mineral, which is frequently found in metamorphic rocks, is the common garnet.

Metamorphic rock may exhibit layers resembling those of sedimentary rock. But metamorphic layering, or foliation, is due to mechanical and chemical changes in the original rock. Grains, crystals and fossils are shifted or broken up and strung out into linear series; parallel rows of platelike minerals may form that are not at all related to the original bedding. Shale is changed by pressure into slate, which may later become a rock called phyllite and, with continued pressure, a crystalline foliated rock known as schist. Other kinds of rock, such as sandy shales and granites, for example, are transformed into a more coarsely foliated, crystalline rock called gneiss.

Slate is a dark rock with an invisibly fine grain; it splits readily into thin smooth slabs. Phyllite represents a stage intermediate between slate and schist. Its grain is very fine, consisting primarily of mica; but here and there a few larger crystals appear. The foliation is rougher and wavier than that of slate. Schist is visibly crystalline, with wavy foliation. It consists of such flakelike or tabular minerals as mica, chlorite, hornblende or talc, as well as distinct crystals of quartz and garnet. Gneiss looks irregular and streaky because it has alternating layers of different minerals. Gneisses originate from several different kinds of rock. Each type of gneiss has much the same composition as the mother rock. Granite gneiss, for example, is derived from granite.

Other kinds of metamorphic rock show no foliation at all. Pure sandstones are changed into a more compact mass called quartzite, where pore spaces have been compressed, making a very hard and durable rock. Limestone is converted by heat and pressure into marble, where the carbonate grains become visibly crystalline. Pure marble is white; but most limestones contain impurities that often react under metamorphism to produce new minerals. These often impart striking colors or mottling to the resulting marble, making it more attractive as a building stone. Marble is very plastic and flows easily under pressure, rupturing other rocks as it does so.

Metamorphism in reverse also occurs. Once the transforming forces cease, metamorphic rocks tend to return to their original unaltered condition. Rocks may be remelted into magmas, and then the whole cycle begins anew. For the rock-forming processes described above are still going on; and new rocks are replacing those destroyed by the effects of air and water in a continuing geologic cycle.

*See also Vol. 10, p. 270*: "Rock Structure."

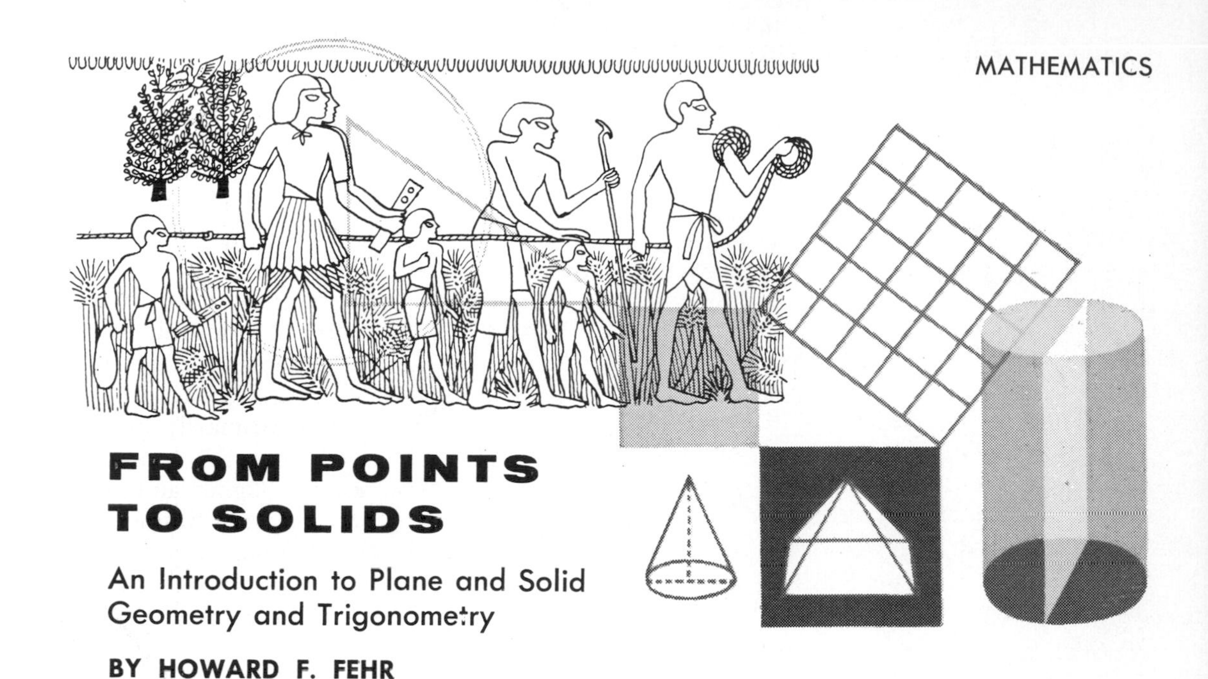

# FROM POINTS TO SOLIDS

## An Introduction to Plane and Solid Geometry and Trigonometry

BY HOWARD F. FEHR

WHEN we pass from arithmetic and algebra to geometry, we enter a world of shapes occurring in space — a world of points, lines, surfaces and solids. We study the properties of these shapes and the relations between them; we learn to measure them. At the outset, geometry was used to solve specific problems; but in the course of its development it became a thoroughly abstruse subject. However, this abstruse branch of mathematics can often be put to practical use, as we shall see.

The beginnings of geometry go back far into prehistory. As the inhabitants of a given region grew in number, the natural caves and other dwelling places available did not suffice. It became necessary to build shelters, big enough to house families and strong enough to withstand winds, rain and storms. To make his shelter the proper size, a man had to compare lengths. Thus the roof had to be higher above the ground than the top of the head of the tallest man, when he was standing. The door or entrance had to be not only tall enough but also wide enough to enable people to enter without discomfort. From these constructions men learned to measure. They also succeeded in finding shapes that would withstand storms. This was the beginning of geometry.

The ancient Babylonians were pioneers in this branch of mathematics. Writings on baked clay tablets dug from various ruins in Babylonia show that the early inhabitants had rules for finding the areas of rectangles, triangles and circles. The land between the Tigris and Euphrates rivers, where the Babylonians dwelt, was originally marshland. Canals were built to drain the marshes and to catch the overflow of the rivers. For the purposes of canal construction, it was necessary to survey the land. In so doing, the Babylonians developed rules for finding areas. These rules were not exact by modern standards,* but the results they gave sufficed for canal construction.

In Egypt, the people who had farms along the banks of the Nile River were taxed according to their holdings. In the rainy season, the river would overflow its banks and spread over the land, washing away all landmarks. It became necessary, therefore, to remeasure the land so that each owner would have his rightful share. After the floods had subsided, specially trained men, called rope-stretchers, would establish new landmarks. They would use ropes knotted at equal intervals so that they could measure out desired lengths and divide the land into triangles, rectangles and trapezoids.**

*One rule, for example, was that the circumference of a circle is 3 times the diameter. Actually, it is about 3 1/7 times the diameter.

** A trapezoid is a four-sided figure with one pair of parallel sides.

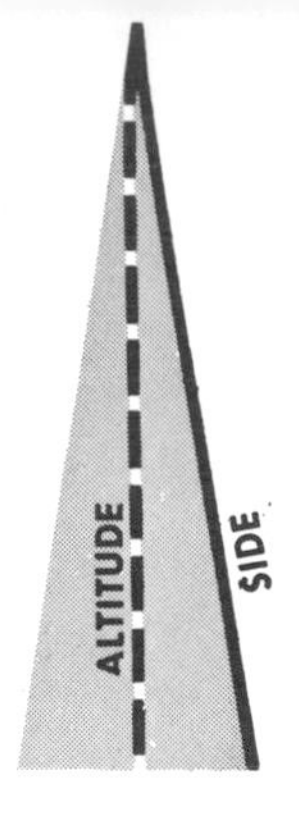


**1. In this long and narrow triangle, the side is not a great deal longer than the altitude. The Egyptians used such triangles in surveying.**

They devised practical rules for the areas of these figures. The rules were of the rough-and-ready variety and were often inexact. We know today, for example, that the area of any triangle is one-half the product of its altitude, or height, and its base. The Egyptians erroneously gave this area as one-half the product of the base and a side. However, most of the triangles used in their surveying work were long and narrow (Figure 1); and in such triangles there is not too much difference in length between the side and the altitude. Hence the results of the Egyptians' calculations served as a pretty fair basis for the allotting of land and the taxation of landowners.

The Greeks called the early Egyptian surveyors geometers, or earth-measurers (from the Greek *ge*: "earth" and *metria*: "measurement"). The geometers found out many facts about triangles, squares, rectangles and even circles. These facts became a body of knowledge that the Greeks called geometry, or "the study of the measurement of the earth." Geometry today involves much more than it did at that early stage; yet it is still concerned with the sizes, shapes and positions of things.

The Greeks made important advances in the field of geometry. They not only corrected many of the faulty rules of the Egyptians, but also studied the different geometrical figures in order to work out relationships. Thales (640?–546 B.C.), the first of the seven wise men of Greece, discovered that no matter what diameter one draws in a circle, it always bisects the circle — that is, cuts it into two halves (Figure 2). He also noticed that if two straight lines cross each other, the opposite angles are always equal, no matter at what angle the lines cross (see *a* and *b* in Figure 3). This was the beginning of the study of figures for the sake of discovering their properties rather than for practical use. The Greeks changed geometry from the study of land measurement to the study of the relations between different parts of the figures existing in space. This is what geometry means today.

After Thales, other Greek mathematicians discovered and proved facts about geometric figures; they set forth these facts in statements called theorems. They also devised various instruments for drawing figures. By custom, the only instruments allowed in the formal study of geometry were an unmarked straightedge (ruler) for drawing straight lines and a pair of compasses for drawing circles and transferring measurements (Figure 4).

The Greeks proposed various construction problems, to be solved with only the straightedge and compasses. Among these problems were the following: (1) Construct a square whose area exactly equals that of a given circle (called squaring the circle). (2) Construct the edge of a cube whose volume will be exactly twice the volume of a given cube (called duplicating the cube). (3) Construct an angle equal to exactly one-third of a given angle (called trisecting the angle). For over twenty-two centuries, mathematicians attempted to solve these problems, without success. Finally, in

**2. Each of the two diameters that we show in the drawing cuts the circle in halves.**

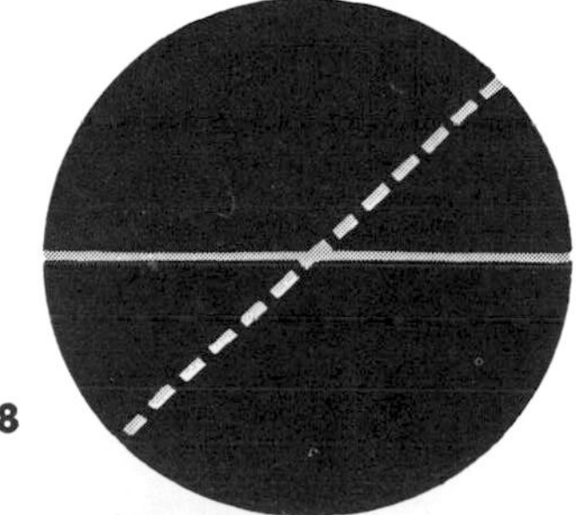

**3. If two lines cross, the opposite angles are always equal. In *a*, angle 1 = angle 2 and angle 3 = angle 4; in *b*, 5 = 6 and 7 = 8.**

4. By custom, the ancient Greeks used only an unmarked straightedge and compasses in the formal study of geometry.

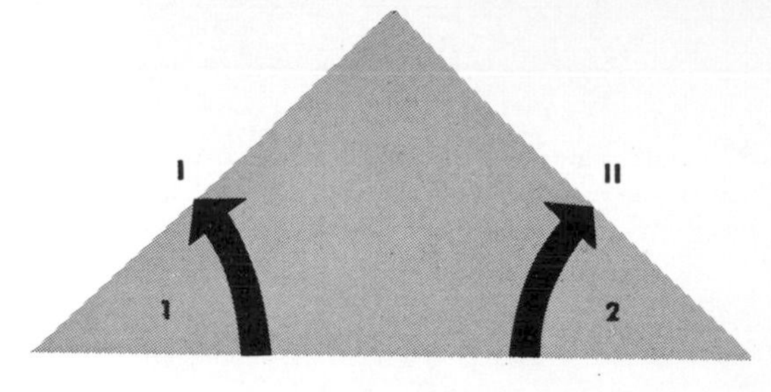


5. If two sides of a triangle are equal, the angles opposite these sides are also equal: that is, if I = II, 1 = 2.

the nineteenth century, it was proved that it is impossible to square the circle, duplicate the cube or trisect an angle if one uses only the straightedge and compasses.*

By the fourth century B.C., there had grown up a vast body of facts concerning geometric figures, but for the most part these facts were unrelated. There were many theorems about triangles and circles; some about similar figures and areas; but no orderly arrangement. The learned Greek mathematician Euclid, who taught at the Museum of Alexandria, in Egypt, about 300 B.C., was the first man to apply a logical development to the mathematical knowledge of his time. He presented this development in his ELEMENTS OF GEOMETRY.

Euclid realized that it is not possible to prove every single thing we say and that we must take certain things for granted. He assumed that everybody knows and uses properly such words as "between," "on," "point" and "line"; hence it is not necessary to define them. He used these *undefined terms* to give *definitions* of various figures; thus he defined a circle as "a closed curved line every point of which is the same distance from a fixed point called the center." Again, Euclid noted that one cannot prove certain statements of relations between geometric figures. An example would be: "Only one straight line can be drawn between two points." Euclid called such statements common notions; today we call them *postulates.*

Euclid used undefined terms, definitions and postulates to prove theorems about geometric figures. A theorem is a statement that gives certain facts about a figure and that concludes from these facts that a certain other fact must be true. A typical theorem is "If two sides of a triangle are equal, the angles opposite these sides must be equal" (Figure 5). The theorem states the facts that (1) there is a triangle and that (2) two sides of the triangle are equal. It then draws the conclusion that two of the angles of the triangles are equal. Once a theorem is proved, it can be used to prove other theorems.

Euclid built up a logical chain of theorems and it introduced order in what had been a chaos of more or less unrelated facts. Besides organizing a vast body of knowledge about geometric figures, he introduced a method of treatment that became a model for the development of other branches of mathematics and pure science. This method is as valid today as ever.

* Certain instruments that will make the three constructions in question have been invented; they fall in the domain of higher geometry.

## PLANE GEOMETRY

The first branch of geometry we shall consider is plane geometry — the study of points, lines and figures occurring in planes. Just what do we mean by these terms?

A point is the simplest element in geometry. It has neither length nor width nor thickness, which is another way of saying that it has no dimensions at all. We can represent a point by a dot, made with a lead pencil or a piece of chalk. Such a dot is not a geometric point but a physical point, since it has length, width and thickness, however small these dimensions may be. In geometric constructions, we have to use physical points, such as pencil dots, to represent geometric points, because it would be impossible for us to set down on paper a point with no dimensions.

A B C D

6. Points *A* and *B* and the length *CD* are parts of the line shown above and must always stay within the line.

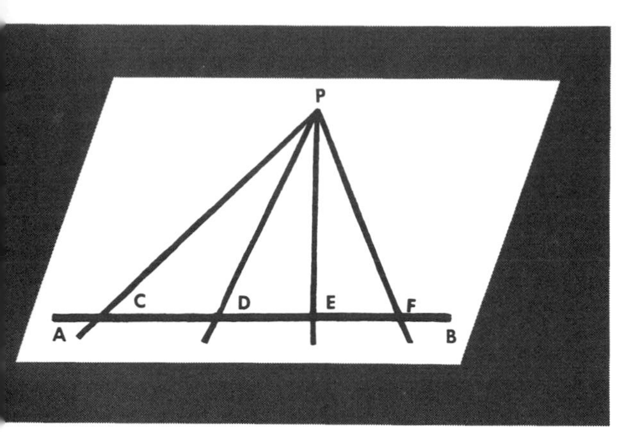


7. Lines drawn from *P* to the line *AB* create a number of figures (such as *PCD* and *PEF*) occurring in a plane.

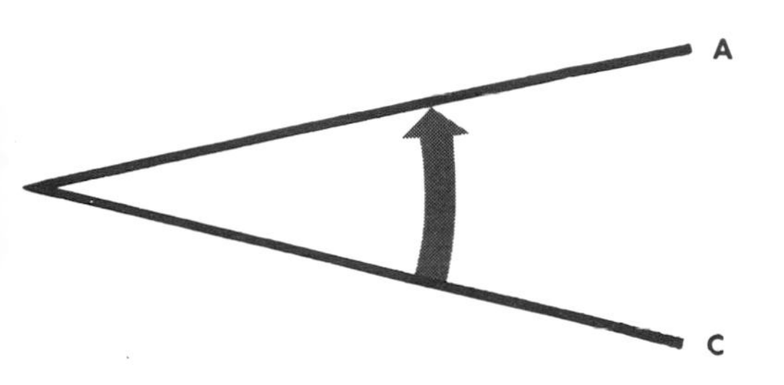


8. In the above drawing, *AB* and *BC* are two rays having the same starting point, *B*. They form the angle *ABC*.

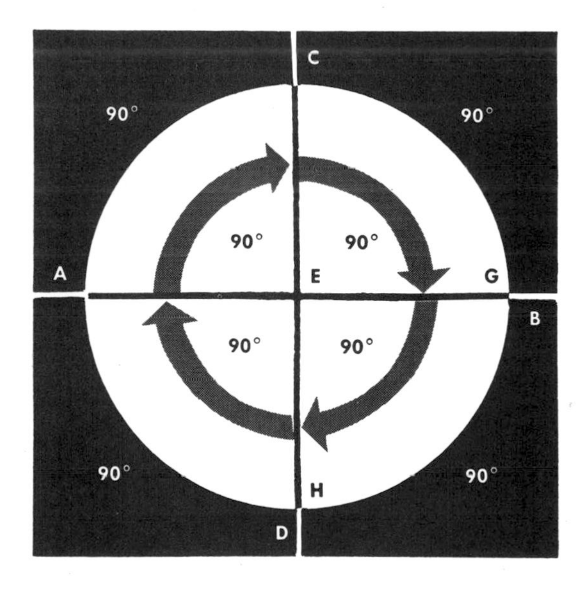


9. Lines *AB* and *CD* are perpendicular to each other; the angles at *E* are all right angles, each having 90°. The arcs that they cut off on the circle also have 90° each.

If there are two different points, the shortest distance between them is a straight line. This line segment has only one dimension, called length; it does not have width or thickness. A straight line that we draw on paper with a pencil *does* have width and thickness. Hence, when we draw a line in constructing a geometric figure, we are again giving a physical representation of a geometric element.

If we were confined to a world having only one dimension, such as length, we would have a rather dull time of it. We would be points on a line, being able to move only forward and backward and always bumping into points ahead of us or behind us. In Figure 6, the points *A* and *B* and the segment *CD* are all parts of the line shown in the figure and must always stay within the line.

Suppose now that we selected a point *P* outside the line. The lines drawn through the point *P* and meeting the original line create a series of figures existing in a plane — a surface having the two dimensions of length and width (Figure 7). The surface of a table top is a plane; a continuation of the surface would represent part of the same plane. If we were points in a two-dimensional world, we could move freely in any direction, except out of the plane. Our world would have other points like ourselves, and also lines. There would also be a great variety of figures — made up of combinations of points and lines — figures such as triangles, squares, circles and so on.

### Angles in plane geometry

The name *ray* is given to the part of a line that starts at a given point. A plane figure formed by two rays having the same starting point is called an angle. In Figure 8, *AB* and *BC* are two rays with the same starting point, *B*. The angle formed by the two rays is *ABC*. You will note that letter *B*, standing for the starting point, is inserted between letter *A*, on one of the two rays and letter *C*, on the other ray. That is how angles are always indicated.

If two lines meet so that all the angles formed are equal, the lines are said to be perpendicular and the angles are called right angles. In Figure 9, $AB$ is perpendicular to $CD$ and the four angles — $AEC$, $BEC$, $AED$ and $BED$ — are all equal. If we draw a circle about point $E$, its length, called its circumference, can be divided into 360 units, called degrees and written with the symbol °. The parts of the circle labeled $IG$, $GH$, $HF$ and $FI$ are called arcs; each arc has 90°. The angle at the center of the circle has the same number of degrees as the arc it cuts off on the circle; hence each of the four angles we mentioned above has 90°. These angles are known as *right angles.*

If an angle is less than a right angle — that is, if it has less than 90° — it is called *acute.* It is *obtuse* if it is greater than a right angle — that is, if it has more than 90°. When the obtuse angle becomes so large that its sides form a straight line, it is a *straight angle* and has 180°. An angle larger than a straight angle is called a *reflex angle*: of course it must have more than 180°. Figure 10 shows these different kinds of angles. Angles can be measured by the instrument called the protractor; it consists of a semicircle divided into 180 parts, each part representing a degree of angle at the center O (Figure 11). As we shall see, angles play an all-important part in the study of geometry.

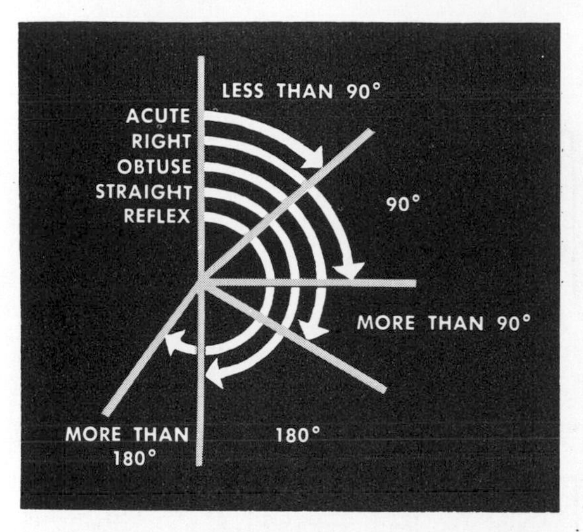


10. The different types of angles, from acute to reflex.

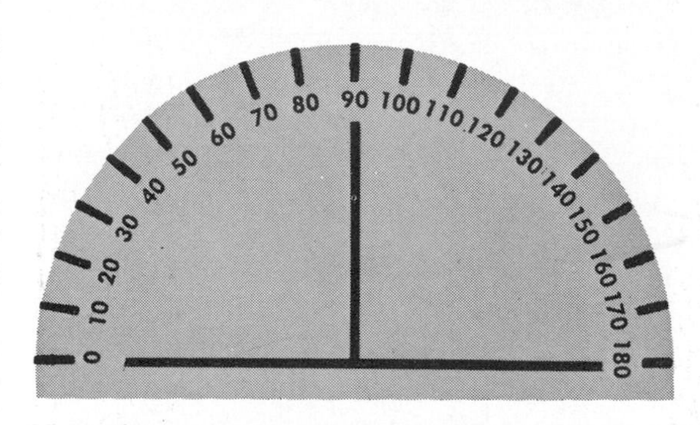


11. Angles are measured by the device called the protractor.

### The study of triangles

A distinct part of a line — the distance between two particular points — is called a segment. When three line segments connect three points in a plane, they form a triangle. Figure 12 shows the different kinds of triangles. There are literally thousands of theorems about the sides, angles and lines in triangles.

One of the first theorems proved in plane geometry is "If three definite lengths are given, such that the sum of any two lengths is greater than the third length, it is possible to use the lengths in making a triangle that will have a definite size and shape" (Figure 13). Since the shape never varies, a construction built in the form of a triangle will be rigid and will not "give."

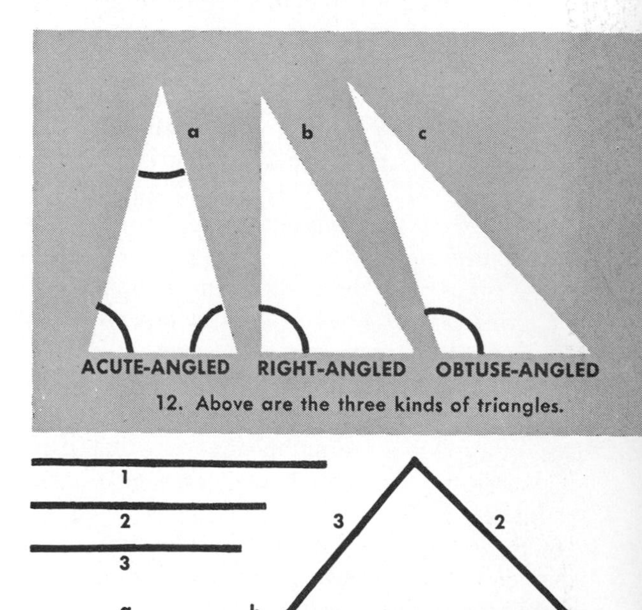


12. Above are the three kinds of triangles.

13. *a* shows three line segments (1, 2 and 3). In *b*, these segments are joined together to form a triangle.

Because of this property, interlocking triangles are used in all bridge and building design in order to prevent a structure from collapsing. A figure of four sides, each of a definite length, could have many different shapes, as shown in Figure 14; a construction built in this shape would be collapsible. Hence it cannot be used in rigid construction unless it is braced by a diagonal. Each diagonal makes two triangles of a four-sided figure (Figure 15) and each of these triangles is rigid.

The most famous and perhaps the most important theorem in plane geometry is one dealing with a right triangle (a triangle having a right angle). It is called the Pythagorean theorem, after its discoverer, the Greek philosopher Pythagoras, who lived in the sixth century B.C. This theorem states that "The sum of the squares on two sides of a right triangle is equal to the square on the hypotenuse (the side opposite the right angle)." In the right angle in Figure 16, the sides are 3, 4 and 5 units in length, the side with 5 units being the hypotenuse. We draw the three large squares as shown: the first with a side ($AB$) consisting of 3 units; the second with a side ($BC$) consisting of 4 units; the third with a side ($AC$) consisting of 5 units. According to the Pythagorean theorem, $(AB)^2 + (BC)^2 = (AC)^2$. In this case $3^2 + 4^2 = 5^2$, or $9 + 16 = 25$. We can verify this by counting the small squares, each with a side a unit long, in the three large squares. There are 9 small squares in square *ABED,* 16 in square *BCGF* and 25 in square *ACIH.*

Figure 17 shows how the two smaller squares on the sides of a right triangle can be cut up so as to form the square on the hypotenuse. This is another confirmation of the Pythagorean theorem.

It follows from this theorem that if a triangle has sides such that the sum of the squares of the two smaller sides is equal to the square of the largest side, the angle opposite the largest side is a right angle. This theorem has various practical applications. The carpenter uses it to see whether a wall is perpendicular to the floor. If boards of 6, 8 and 10 feet, for example, are

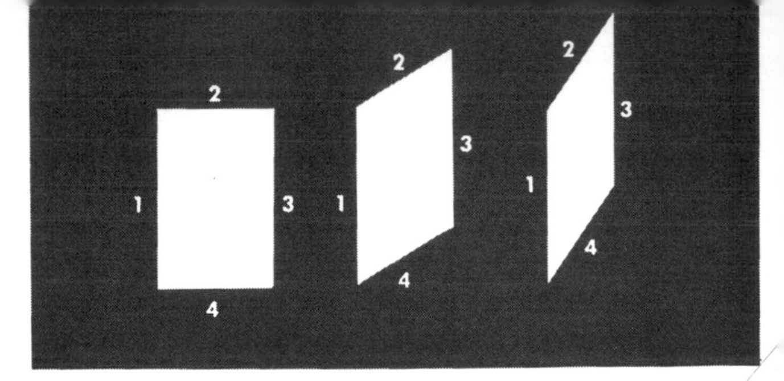


14. In the four-sided figures we give above, the sides marked 1 are equal; so are the sides marked 2, 3 and 4.

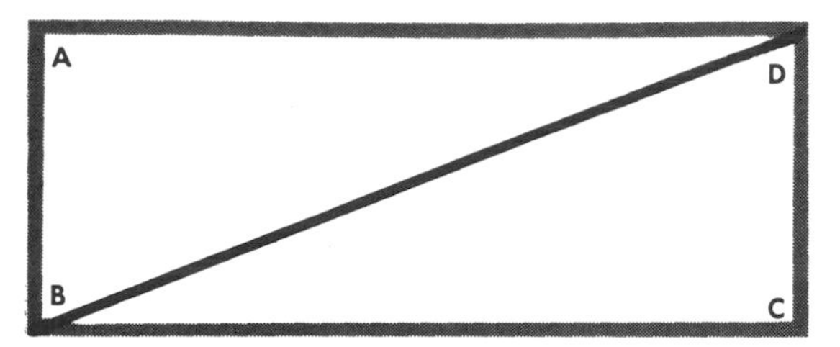


15. To brace the four-sided construction *ABCD*, the diagonal piece *BD* is inserted, making two rigid triangles.

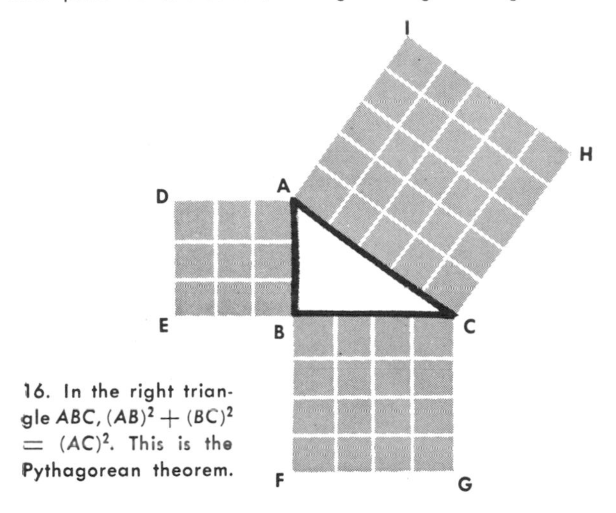


16. In the right triangle *ABC*, $(AB)^2 + (BC)^2 = (AC)^2$. This is the Pythagorean theorem.

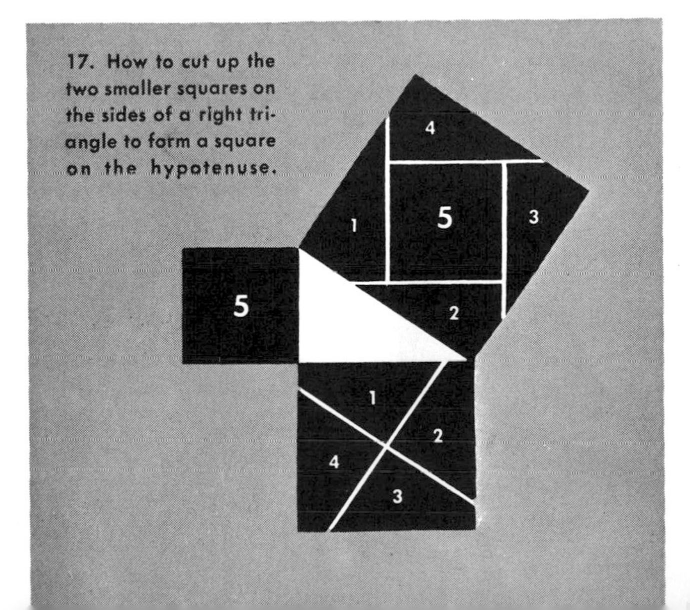


17. How to cut up the two smaller squares on the sides of a right triangle to form a square on the hypotenuse.

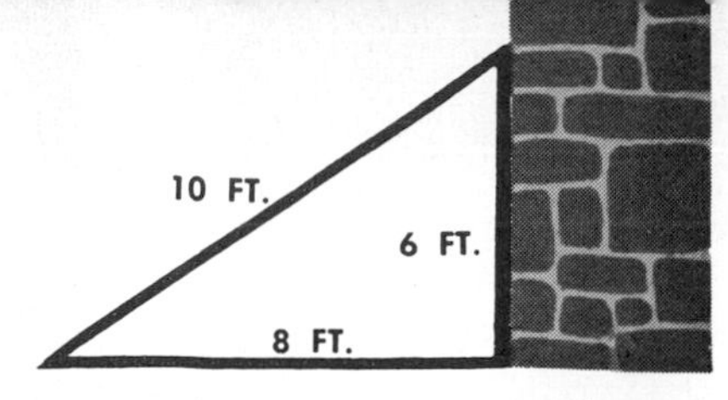

18. These boards form a right triangle. If the wall is perpendicular to the floor, the triangle will fit snugly.

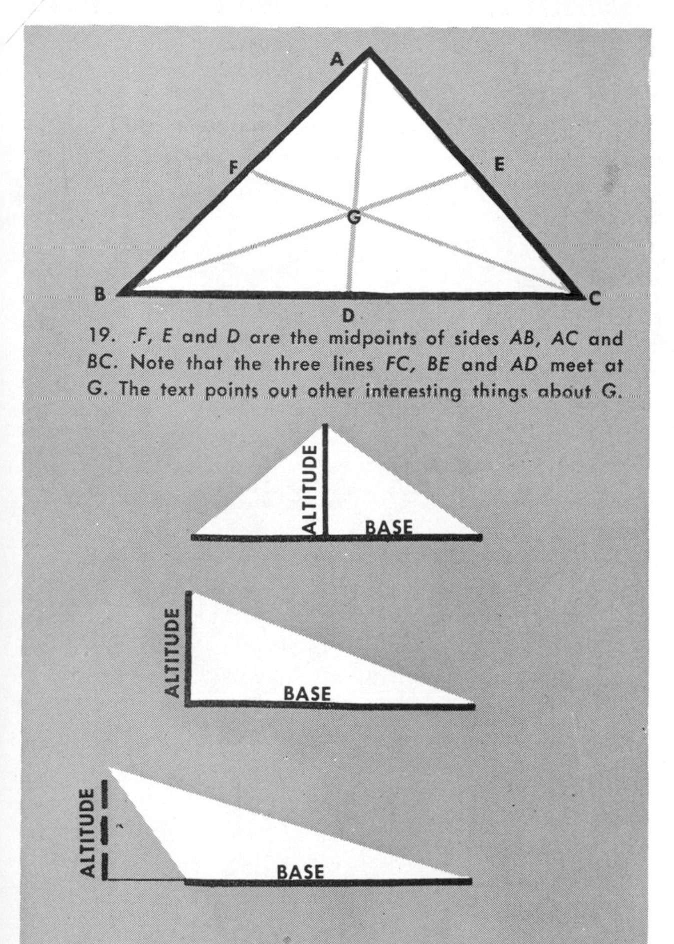

19. *F, E* and *D* are the midpoints of sides *AB, AC* and *BC*. Note that the three lines *FC, BE* and *AD* meet at *G*. The text points out other interesting things about *G*.

20. The three triangles above have equal bases and altitudes. Hence the areas of the triangles are also equal.

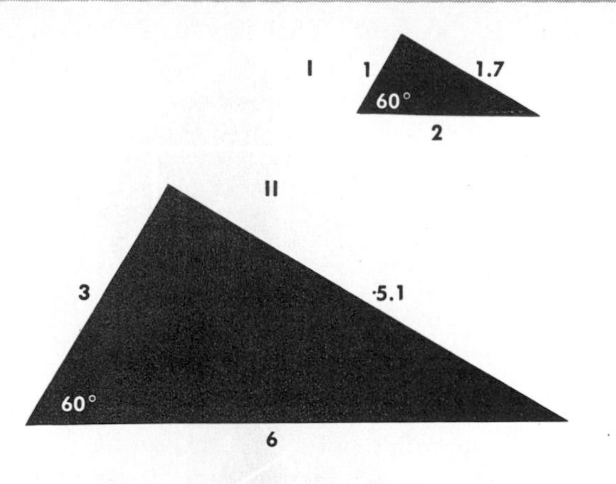

21. Triangles with equal corresponding angles are similar.

joined together as shown in Figure 18, the angle between the 6- and 8-foot lengths must be a right angle, since $6^2 + 8^2 = 10^2$, or $36 + 64 = 100$. If the wall is truly perpendicular to the floor, the triangle will fit snugly.

The Pythagorean theorem is one of many that reveal an unexpected and important relationship. Here is another instance of such a theorem. In the triangle $ABC$, in Figure 19, $D$, $E$ and $F$ are the midpoints of the sides $BC$, $AC$ and $AB$. If we connect the vertex* $A$ to the midpoint, $D$, of the opposite side, $BC$, the line $AD$ is called a median. Let us draw the two other medians of the triangle, $CF$ and $BE$. We learn what we had not suspected — that these three medians all pass through the same point $G$, inside the triangle; also that $G$ on any of the medians is two-thirds the distance from the vertex to the opposite side. Plane geometry offers proofs of these statements. Here is another interesting fact about $G$. If we cut out a triangle of cardboard and draw the three medians, as in Figure 19, we can balance the triangle on the blunt end of a lead pencil if we put this end directly under $G$. This point is called the center of gravity. We discuss it in detail elsewhere (see Index, under Gravity — Center of).

### Equal triangles and similar triangles

When triangles have the same size, or area, they are called equal triangles. All triangles with equal bases and altitudes are equal although they may have many different shapes (Figure 20).

Some triangles have the same shape, but are different in size; they are known as similar triangles. The corresponding angles of similar triangles are equal and their corresponding sides are always in the same ratio. Thus the two triangles I and II shown in Figure 21 are similar because the corresponding angles are equal. Each side of triangle II is 3 times as great as the corresponding side of triangle I.

* The vertex of a triangle is a point where two sides meet — a point opposite the base. Since each side in turn may be taken as the base, there are three vertexes in a triangle.

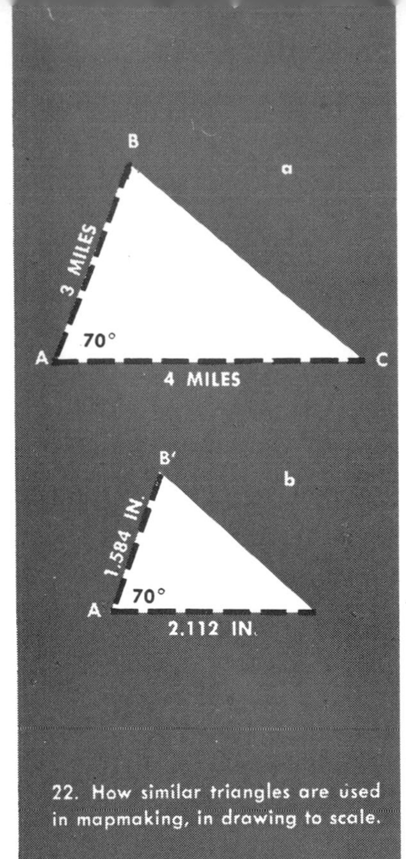

22. How similar triangles are used in mapmaking, in drawing to scale.

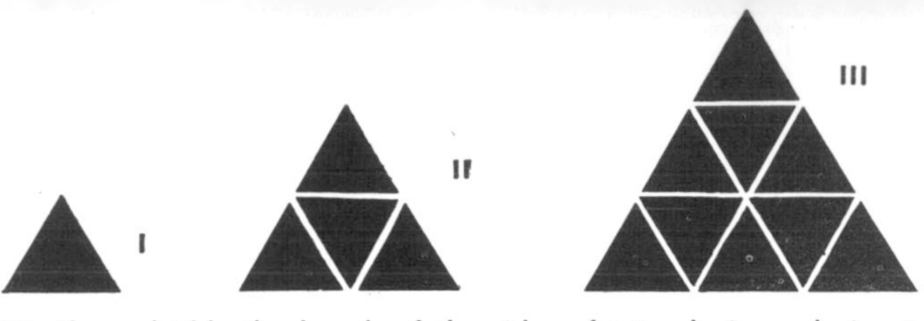

23. If we double the length of the sides of triangle I, producing triangle II, the area of II will be four times as great as the area of I. The area of triangle III, above, is nine times as great as that of I.

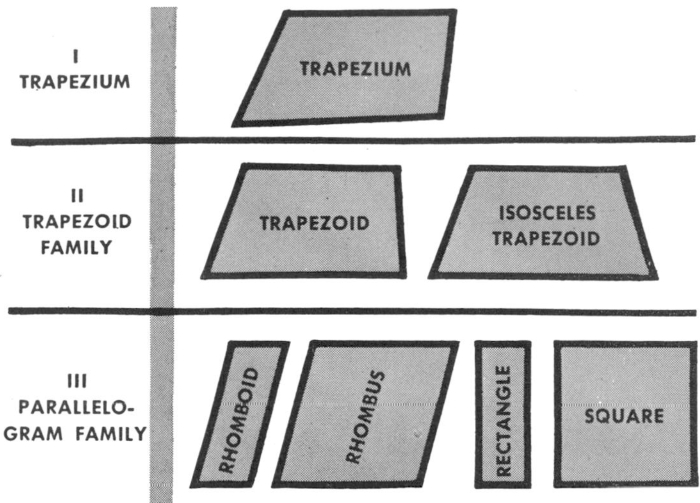

24. The three basic types of quadrilaterals, or four-sided figures.

Similar triangles are used in drawing to scale. In making a map, for example, we represent a large area of land on a small piece of paper. Suppose site *A*, in Figure 22*a*, is 3 miles from site *B* and 4 miles from site *C* and that the angle *CAB* is 70°. We are to show these sites on a map, where the scale is to be 1 inch = 10,000 feet. First, using a protractor, we draw an angle of 70°. Three miles is equal to 15,840 feet, since there are 5,280 feet in a mile. Since 1 inch represents 10,000 feet, to represent 15,840 feet we divide 15,840 by 10,000, giving 1.584. Hence we measure 1.584 inches (about 1⁹⁄₁₆ inch) on one side of the angle to represent 3 miles. Four miles is 5,280 × 4 feet = 21,120 feet; 21,120 ÷ 10,000 = 2.112. We measure off 2.112 inches (about 2⅛ inches) on the other side of the 70° angle. Joining the ends of these segments, we have a map (Figure 22*b*) representing the triangular area formed by the three sites.

All maps, models and photographs are similar to the original objects they represent. Hence the angles in these representations are exactly the same size as in the originals, and all lines are changed in the same ratio. Their areas will also have a definite ratio; they will vary as the squares of the corresponding sides. If we double each of the sides of a triangle, the area will be 4 times as great; if we triple each of the sides, the area will be 9 times as great. Figure 23 shows that this is so.

If a film 1 inch square is projected on a screen so that the picture on the screen is 40 inches square, the projection is $40^2$ or 1,600 times as large as the original. The light used in projecting the film must cover 1,600 times as much area as the film; hence its intensity on the screen is only 1/1600 as great as at the film. This is a striking illustration of the manner in which the geometry of similar figures can be applied to the study of photographic phenomena.

#### Four-sided figures — quadrilaterals

A figure with four sides is called a quadrilateral. The quadrilateral family is shown in Figure 24. Examining these figures, we see that there are really three basic types of quadrilaterals: the trapezium, which has no parallel sides; the trapezoid, which has one pair of parallel sides; and

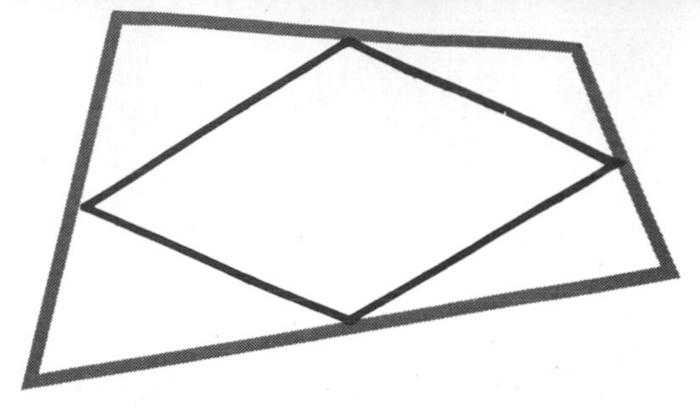

25. In a quadrilateral, the inner figure formed by joining together the midpoints, of the sides is a parallelogram.

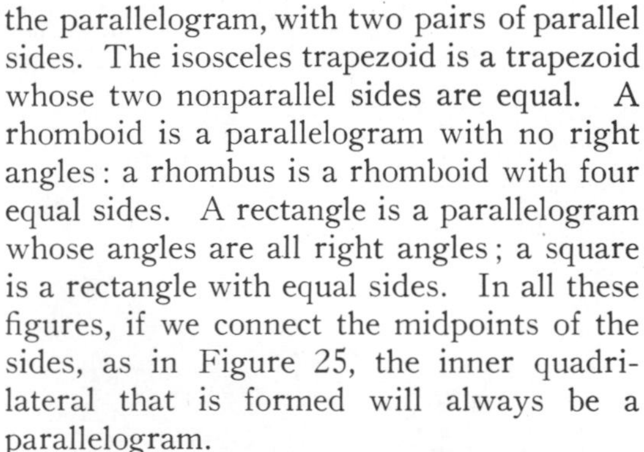

26. The corresponding sides of these parallelograms are equal. The diagonals of both figures bisect each other.

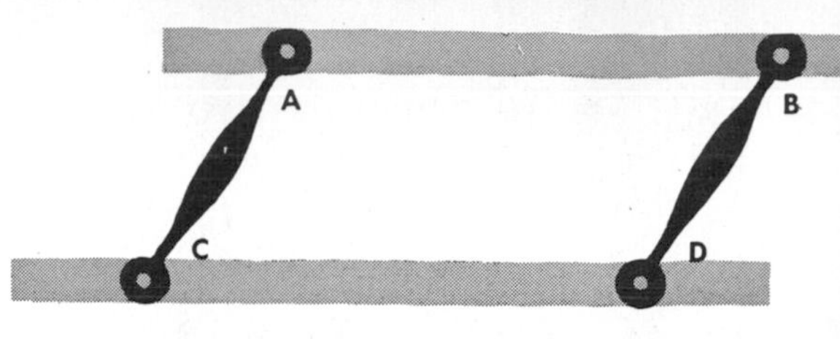


27. The device called parallel rulers.

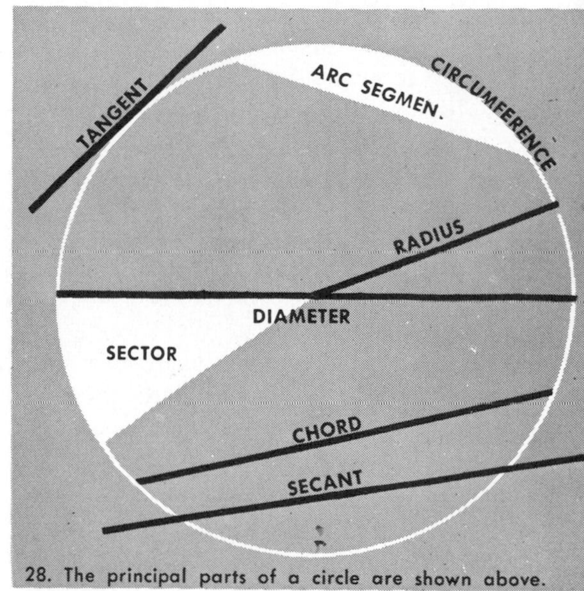


28. The principal parts of a circle are shown above.

the parallelogram, with two pairs of parallel sides. The isosceles trapezoid is a trapezoid whose two nonparallel sides are equal. A rhomboid is a parallelogram with no right angles: a rhombus is a rhomboid with four equal sides. A rectangle is a parallelogram whose angles are all right angles; a square is a rectangle with equal sides. In all these figures, if we connect the midpoints of the sides, as in Figure 25, the inner quadrilateral that is formed will always be a parallelogram.

In a parallelogram, the diagonals bisect each other, no matter how we distort the figure (Figure 26). This is a good example of an *invariant* — a property of a figure that remains true under all distortions. Another invariant is "The opposite sides of a parallelogram are equal."

The draughtsman makes use of this invariant in the instrument called the parallel rulers (Figure 27). It consists of two straightedges, joined together by two equal rods $AB$ and $CD$ in such a way that $AC = BD$. This device is flexible; hence $AC$ can be at varying distances from $BD$. However, no matter how $AC$ is moved it always remains parallel to $BD$. Hence, using this device, we can draw a parallel to a given line at any accessible point in a plane.

By drawing diagonals from the vertices of a polygon* containing more than four sides, the polygon can be divided into quadrilaterals and triangles. Hence the study of the two latter figures forms much of the subject matter of plane geometry.

### The study of the circle

The study of the circle is also important. In Figure 28, we show its important parts. You will note that the closer a chord gets to the center of the circle, the larger it becomes. The diameter is really a chord that passes through the center; it consists of two radii joined together so as to form a straight line. To measure the length of a circle — its circumference — we find the number of diameters in it. This number is about $3\frac{1}{7}$ or, more accurately, 3.1416; its exact value is indicated by the Greek letter $\pi$ ("pi"). The circumference of the circle, then, is $\pi d$ or $\pi$ times the diameter; its area is $\pi r^2$, or $\pi$ times the radius squared.

* A polygon is a closed figure having three or more angles and therefore sides. The name "polygon" is used especially to indicate a figure with more than three sides.

A half circle is called a semicircle. A simple and quite surprising theorem involving a semicircle is this: "If any point on a semicircle is joined to the ends of the diameter, an angle of 90° is formed at the point." In Figure 29, $AB$ is the diameter and $P$ is a point anywhere on the semicircle. It is easy to prove that $APB$ is a 90° angle, or right angle. First connect $P$ to the center of the circle ($O$). $AO$, $PO$ and $OB$ are all radii of the circle and are equal. In the triangle $APO$, since side $AO = PO$, the two angles marked x° are equal, because if two sides of a triangle are equal, the opposite angles must also be equal. Likewise in the triangle $POB$, sides $OP$ and $OB$ are equal, and so the angles marked y° must be equal. Ignoring the line $OP$, we have the triangle $APB$, whose angles must total 180°, since the sum of the angles of a triangle is 180°. There are two x° angles and two y° angles in the triangle $APB$; hence one x° angle and one y° angle must give half of 180°, or 90°. Since the angle $APB$ is composed of an x° angle and a y° angle, it must be equal to 90°; it must be a right angle.

There are various applications of this theorem. For example, a pattern maker can determine if the core box shown in Figure 30 is a true semicircle. He places a square in the box. If it makes firm contact at three points, as shown, he knows that the pattern will give a true semicircle.

An angle at the center of a circle has as many degrees as the arc that it intercepts, or cuts off, on the circle; or, as a mathematician would say, "A central angle is measured by its intercepted arc." In Figure 31, the obtuse angle at $O$ is equal to 110°, and the arc it intercepts is also equal to 110°.

An angle whose vertex is on the circumference of a circle and which cuts off or intercepts an arc is called an inscribed angle (angle $ABC$ in Figure 32). No matter where we place the vertex $B$ in this arc, the angle will remain the same size. In other words, all inscribed angles intercepting the same arc are equal.

This invariant is used by navigators in order to steer clear of obstacles. In Fig-

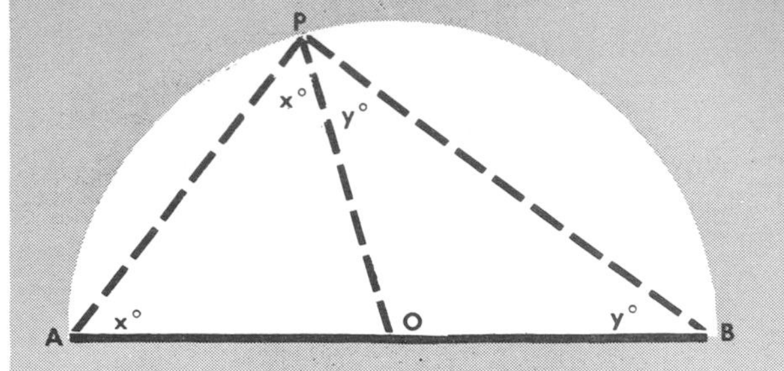


29. If point P on the semicircle is joined to the diameter at A and B, angle APB will be a 90° angle.

30. How one determines by means of a carpenter's square whether a core box gives a true semicircle.

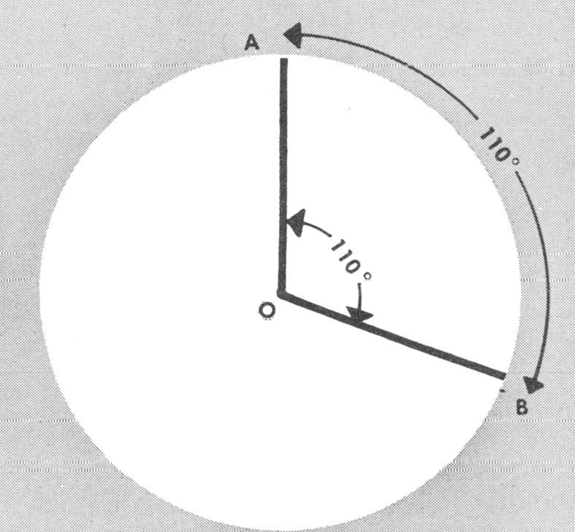


31. Both the angle at O and the arc that it intercepts (AB) on the above circle are equal to 110°.

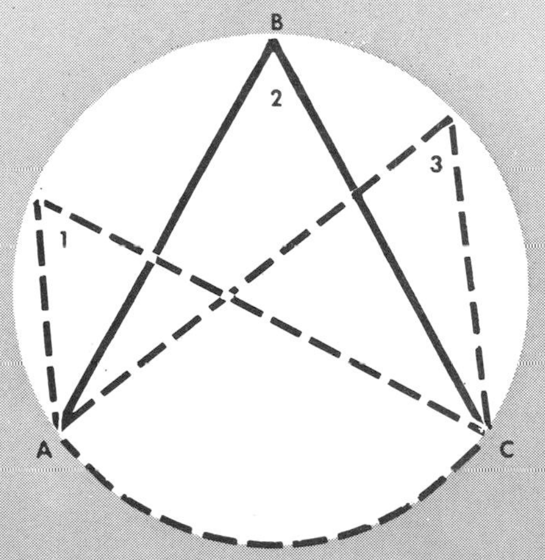


32. The angles marked 1, 2 and 3 intercept the same arc (AC) in this circle; therefore they are equal.

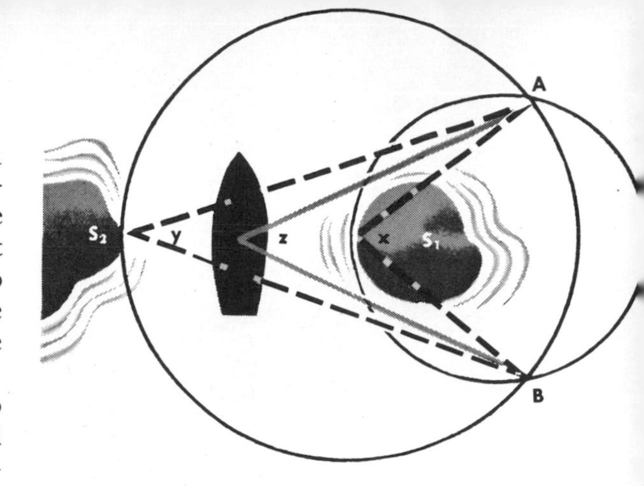


33. The ship can be steered so as to avoid the shoals $S_1$ and $S_2$. The explanation is provided on this page.

ure 33, $A$ and $B$ are two landmarks and $S_1$ and $S_2$ are two shoals. We draw two circles, each having points $A$ and $B$ on the circumference. The smaller circle will not only pass through $A$ and $B$ but will also have within it the shoal called $S_1$; the other shoal, $S_2$, will be outside of the large circle.

To stay between these shoals, a ship must keep outside the smaller circle and inside the larger one. The angle $x$ is always the same size, wherever it may be in the smaller circle, since it always intercepts the same arc; the angle $y$ is always the same size for the same reason. The navigator, by the use of a sextant, sees to it that as the ship sails between $S_1$ and $S_2$, the angle $z$ is smaller than the angle $x$ (which keeps him outside the smaller circle) but greater than the angle $y$ (which keeps him inside the larger circle). Thus the ship avoids both shoals.

**The locus, the path of a moving point**

It is often necessary in plane geometry to determine the path that a point describes in a plane when it moves according to a fixed rule. If, for example, a point must always remain 3 inches from a fixed point, it travels in a circle around the fixed point. The mathematician gives the name "locus" to the path described by a point. (*Locus* means "position" in Latin.) By studying the paths of moving points, we determine how machine parts move and how heavenly bodies appear to move.

We can illustrate the use of the locus by a very simple treasure-hunt problem. A treasure is reported to be buried 20 feet from an oak tree and also equally distant from two intersecting roads (Figure 34). Since it is 20 feet from the oak tree, it is somewhere on the circumference of a circle with a radius of 20 feet and with the tree as the center. Now we also know that the treasure is equidistant between the two intersecting roads. We may think of it as a moving point remaining at the same distance from each road. It can be proved in geometry that a point moving so as to be equally distant from the sides of an angle, traces a line that bisects the angle. Hence the treasure must be on a line bisecting the angle made by the two roads. This line cuts the circle, as shown in Figure 34, at two places, $T$ and $G$. The treasure, therefore, must be at one or the other of these two points.

Machines that are designed to trace moving points are called linkages because they consist of linked bars. A pair of compasses is the simplest linkage; the path it traces is a circle. Another linkage, called Peaucellier's Cell, changes circular motion into straight-line motion (Figure 35). As $A$ in the figure moves around the circle, the point $B$ moves up and down the straight line $CD$. There are other types of linkages that transform circular motion into linear motion. A study of linkage was necessary to help solve the problem of providing smooth motion in a locomotive, where the straight-line motion of the drive shaft had to be converted into the circular motion of the wheels.

34. How to find a treasure 20 feet from a tree and equally distant from two roads. For details, see the text.

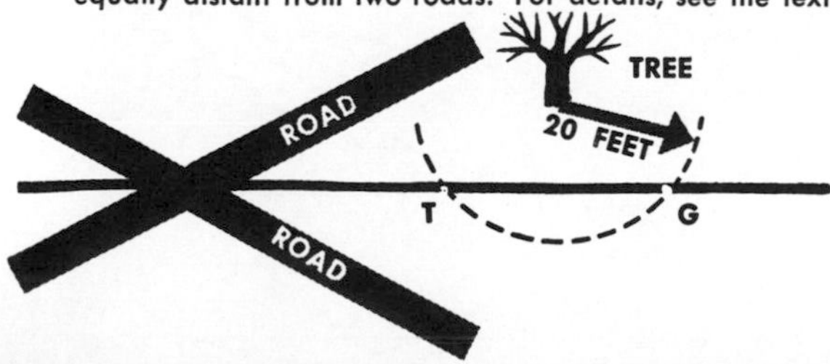


35. Peaucellier's Cell, which changes circular motion into straight-line motion. The device is called a linkage.

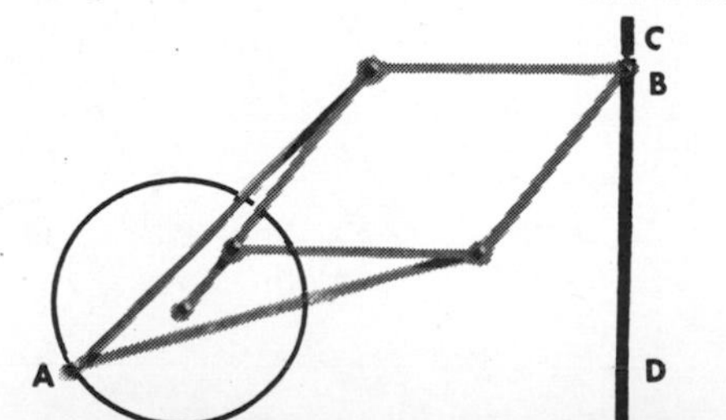

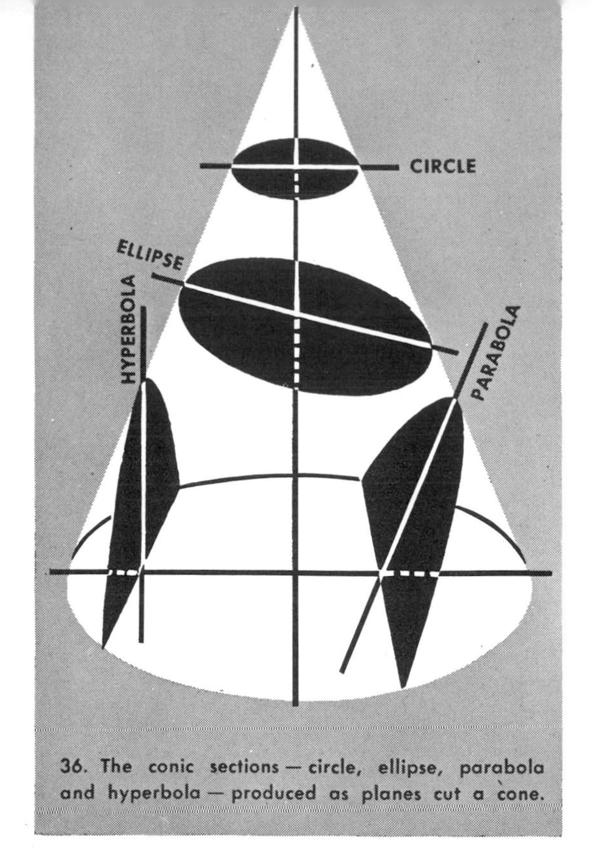

36. The conic sections — circle, ellipse, parabola and hyperbola — produced as planes cut a cone.

### The conic sections — ellipses, parabolas and hyperbolas

The circle is the most common type of curve, but there are other kinds. The Greek geometers noticed very early that when a cone was cut by planes at different angles, the intersections gave different kinds of curves: circles, ellipses, parabolas and hyperbolas (Figure 36). Because the three latter kinds of curves were first described in connection with cones, they were called *conic sections*. Apollonius of Perga, who lived in the third century B.C., wrote a treatise on the properties of these curves. In more recent times, it was discovered that they could also be defined as paths made in a plane by points moving according to certain rules. Such definitions are particularly meaningful when we put the curves to practical use.

*The ellipse.* An ellipse is the path traced by a point which moves so that the sum of its distances from two fixed points is always the same. The two fixed points are called focuses, or foci. It is easy to draw an ellipse, using the method illustrated in Figure 37. First we insert thumbtacks at two fixed points, $F_1$ and $F_2$. We then take a piece of string that is larger than the distance between $F_1$ and $F_2$; we attach one end of the string to the thumbtack at $F_1$ and the other to the thumbtack at $F_2$ (Figure 37*a*). We draw the string taut and insert a pencil as shown in Figure 37*b*. As we move the pencil, its point will trace an ellipse (Figure 37*c*). The sum of the distances from the moving pencil point to the fixed points will remain constant. In figure 37*c*, $P_1F_1 + P_1F_2 = P_2F_1 + P_2F_2$.

If a billiard table were elliptical in shape, any ball hit from one focus would rebound through the other focus. In an elliptical room, any sound issuing from one focus will be reflected by the walls to the other focus. This is the principle of the "whispering gallery." The elliptically shaped Mormon tabernacle at Salt Lake City is a good example. The focuses in the tabernacle are clearly marked. A person standing at one focus can distinctly hear a whisper coming from a person at the other focus; those standing nearby hear nothing.

The ellipse has found many practical applications. Power punching machines use elliptical gears. At the narrow ends of the ellipse, the gears move faster, giving a quick return. At the flat parts, the gears move slower, exerting a greater force. Storage tanks and transportation tanks are made elliptical in cross section so as to lower the center of gravity and to lessen the danger of overturning.

The ellipse also serves to explain the movements of various heavenly bodies. All

37. It is easy to draw an ellipse, using two thumbtacks and a piece of string as shown below.

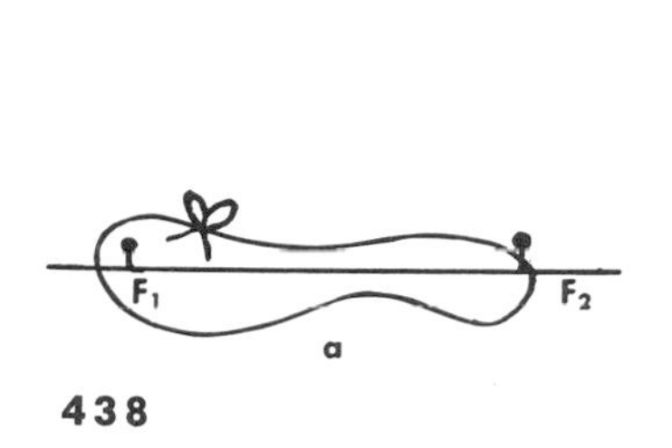

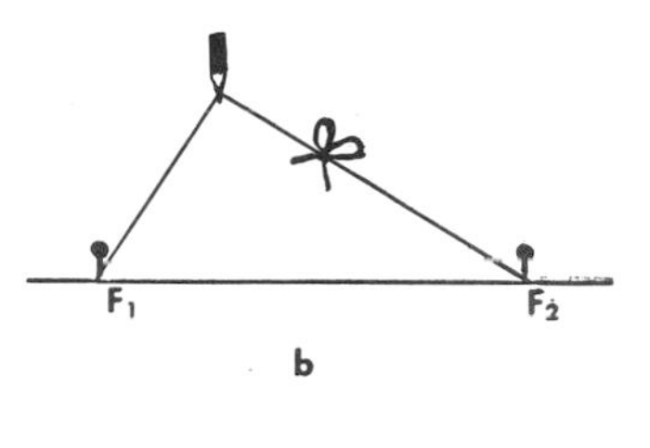

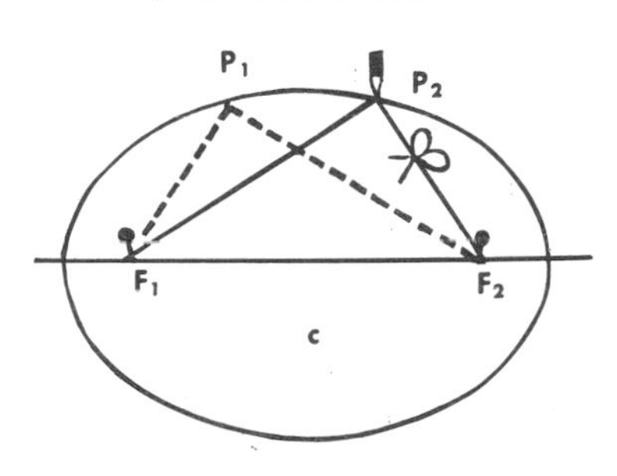

the planets move in elliptical orbits with the sun at one focus (Figure 38). A planet moves along the orbit so that the radius from the sun to the planet sweeps through equal areas in the same time. Knowing the elliptical orbit of any planet, astronomers can accurately predict, to the nearest second, the position of the planet in its orbit at any time.

*The parabola.* The parabola is the path of a point which moves so that its distance from a fixed line, called the directrix, always equals its distance from a fixed point, called the focus. Thus in Figure 39, as a point moves along the parabola, occupying positions $P_1$, $P_2$ and $P_3$ in turn, $AP_1 = P_1O$; $BP_2 = P_2O$; $CP_3 = P_3O$. A reflecting searchlight has a parabolic surface with the light source at the focus. All light beams emanating from the focus are reflected from the parabola in parallel rays (Figure 40). Sound detectors have parabolic surfaces; sound waves are reflected upon striking the surface and are concentrated at the focus. The mirror of a reflecting telescope is in the form of a parabola.* Parallel rays of light from a distant heavenly body strike the parabolic surface and, reflected from it, meet at the focus within the telescope tube.

* The surface of the mirror is really a paraboloid of revolution — that is, the surface generated by a parabola as it is rotated about its axis.

*The hyperbola.* The hyperbola is the path of a point moving so that the distance to one fixed point minus the distance to another fixed point is always the same. The diagram in Figure 41 shows a hyperbola in which $F_1$ and $F_2$, called the foci, are the fixed points. $P_1$ is a point on the hyperbola. $P_1F_1$ minus $P_1F_2$ equals $A_1A_2$. $P_2$ is another point on the hyperbola. $P_2F_1$ minus $P_2F_2$ is also equal to $A_1A_2$.

The hyperbola is applied to Loran, or long-distance navigation by the use of radar. We can give a general explanation of Loran by referring again to Figure 41. There are two radar stations, one called the master station ($F_2$ in the diagram) and the other the slave station ($F_1$); they are located on land about 300 miles apart. Electric pulsations are sent out from each station. It takes longer for such pulsations to travel from $F_1$ to $P_1$ than from $F_2$ to $P_1$ and from $F_1$ to $P_2$ than from $F_2$ to $P_2$. The difference in time is a fixed constant for all points on the hyperbola. If the difference in time is greater or less, we have a different hyperbola.

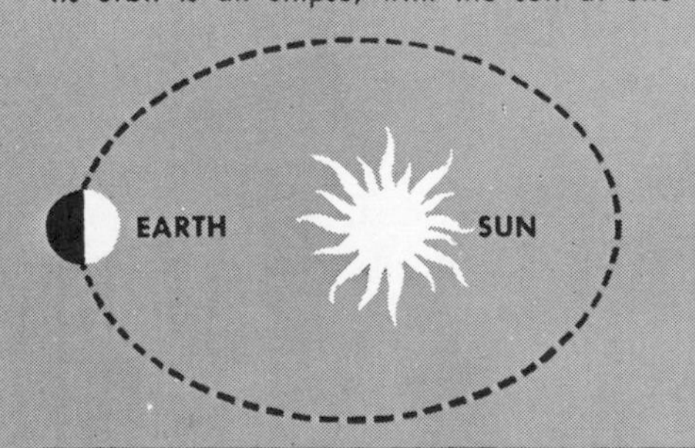

38. As the planet earth travels around the sun, its orbit is an ellipse, with the sun at one focus.

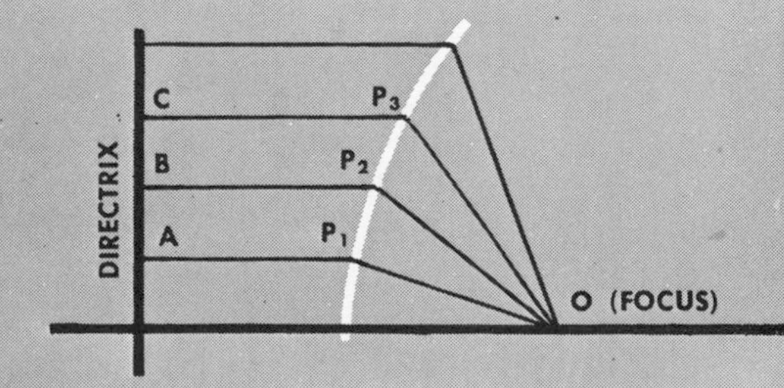

39. As a point moves along a parabola, its distance from the directrix and focus (both fixed) is equal.

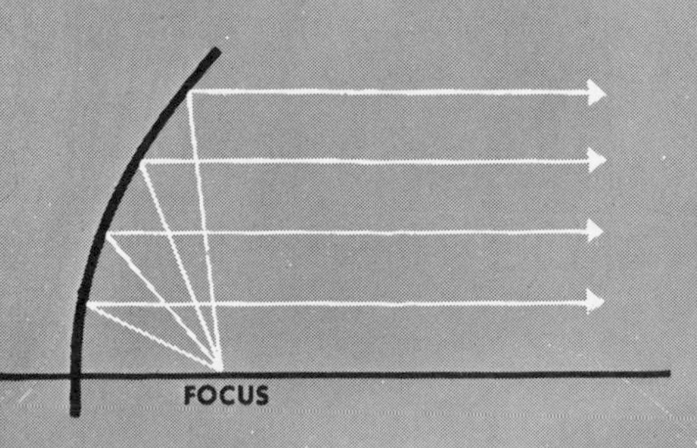

40. The light beams emanating from the focus of the searchlight that we show below are reflected from the parabolic surface in a series of parallel rays.

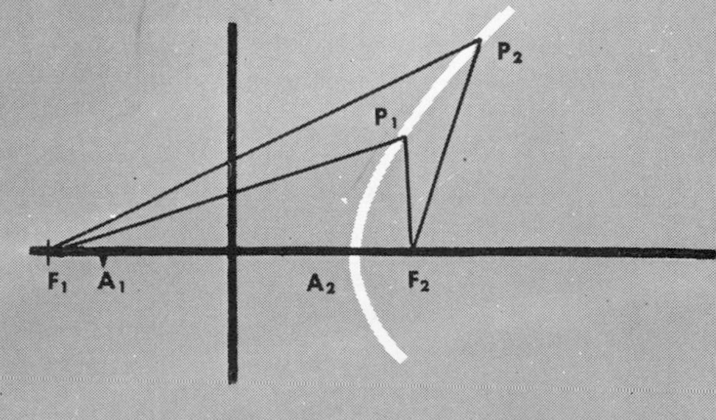

41. $P_1$ and $P_2$ are two positions of a point moving along a hyperbola. The distance to $F_1$ minus the distance to $F_2$ always equals the distance $A_1A_2$.

An airplane or surface vessel has a radar reception instrument which picks up this difference in time between the pulsations from the two stations $F_1$ and $F_2$. The navigator consults a map upon which are drawn the various hyperbolas corresponding to the various differences in time. He consults the map and locates his own craft on a hyperbola. The hyperbola passing through the "home port" is then picked out, and the difference in the time of the pulsations is noted. The course is changed, until the radar receiver indicates that the craft is on the "home-port" hyperbola. The difference in pulsations is kept constant and the craft sails home along the hyperbola.

#### The use of plane geometry in design

The figures and shapes that we study in plane geometry lend themselves well to design (Figure 42). The ancient Egyptians used designs based on geometrical principles on their pottery; we find such designs also in the vases and temples of the Greeks and in the blankets of the American Indians. Many artists today use geometrical principles in creating their works. Geometrical figures are to be found, too, in various industrial products, such as wallpaper and gift-wrapping paper, in which the design repeats itself endlessly. The study of geometrical formations as applied to art and as observed in living organisms is usually called dynamic geometry.

42. Geometric figures and shapes, such as the ones that we show above, lend themselves excellently to design.

## SOLID GEOMETRY

The two-dimensional world of plane geometry would not suffice to explain the world in which we live. It is a world of three dimensions. In it, there are many planes, which are boundless and extend in every conceivable direction; there are also many kinds of curved surfaces. We must consider not only north, south, east and west but also up and down. To explain this three-dimensional world, the branch of mathematics called solid geometry has been developed. We use this kind of geometry in building machines, skyscrapers, airplanes, steamships, bridges and automobiles and also in explaining the phenomena of the heavens.

In solid geometry, there are many more possible relationships between geometric elements than in plane geometry. In a single plane, two lines are either always parallel or else they intersect. In solid geometry, too, two lines may be parallel or else they may intersect; but they may also be *skew lines*, which are not in the same plane, are never parallel and never intersect. Only one line, in a plane, can be drawn perpendicular to another line at a given point. In solid geometry, any number of such perpendicular lines can be drawn. For example, the spokes of a wheel are lines every one of which is perpendicular to the axle at the same point. In a plane, all points at a fixed distance from a fixed point, are on a circle; in three-dimensional space, however, they are on a sphere, containing an infinite number of circles passing through the center.

#### Angles in solid geometry

The simplest angle in solid geometry is called a dihedral ("two-faced") angle, and is formed by two intersecting planes. The size of this angle is measured by the plane angle. This is formed by two lines, one in each face, meeting the edge (the intersection of the two planes) at right angles (Figure 43). When an airplane banks its wings, the angle of bank is a dihedral angle between the horizontal and dip position of the wings (Figure 44). The dihedral angle is measured by an instrument in the plane, and the size of this angle determines in part the speed with which the airplane will change its direction of travel.

When three planes meet at a point, they form a trihedral angle (Figure 45). Each of the angles making up a trihedral angle is called a face angle; in Figure 45, *ADC*, *CDB* and *ADB* are all face angles. If more than three planes meet in a point, the angle is called a polyhedral (many-faced) angle. The sum of the face angles of a polyhedral angle must be less than 360°. As the sum of the angles gets closer to 360°, the angle becomes less pointed until at 360°, it becomes a plane (Figure 46). The crystals of minerals show many kinds of polyhedral angles. An analysis of these angles makes it possible to identify the various minerals. (See the article on crystals — Symmetry Unlimited — in Volume 5.)

**Prisms, cylinders, pyramids, cones and spheres**

The major part of the study of solid geometry is based on five common solids: the prism, the cylinder, the pyramid, the cone and the sphere.

*The prism.* In a prism, all of the sides, or side faces, are parallelograms; the bases are parallel and equal polygons. Several kinds of prisms are shown in Figure 47. The most common of all is the cube, in which all the faces are square (Figure 47*b*). In a triangular prism (Figure 47*a*), the bases are triangles. Triangular prisms made of glass are used in studying the refraction of light. (See the article An Introduction to Optics, in Volume 7.)

43. The intersecting planes below form a dihedral angle. a is the plane angle — the angle between the planes.

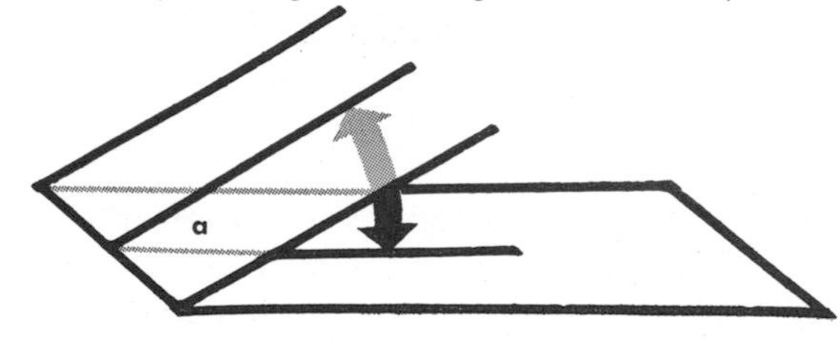


44. The dihedral angle between the horizontal and dip positions of an airplane's wing is the angle of bank.

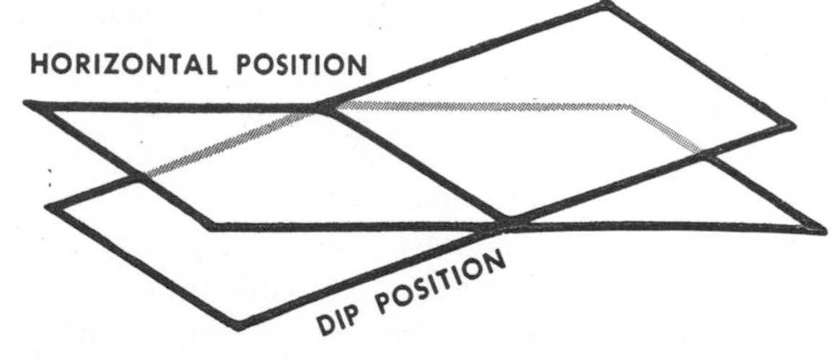


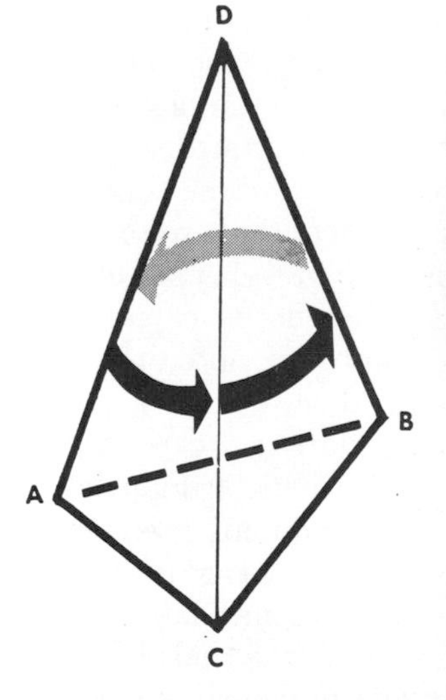


45. When three planes meet at a point, they form a trihedral angle, like this one.

46. A series of polyhedral angles. As the sum of the face angles becomes greater, the polyhedral becomes less and less pointed. At 360° it is a plane.

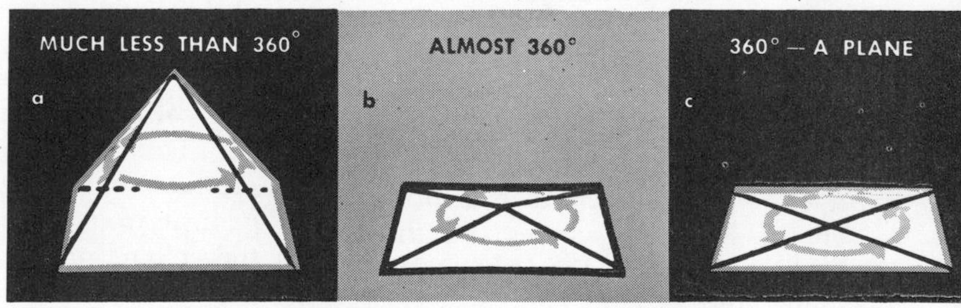


47. The triangular prism, cube and hexagonal prisms, shown below, are known as regular prisms. In all of them, the bases are at right angles to the sides.

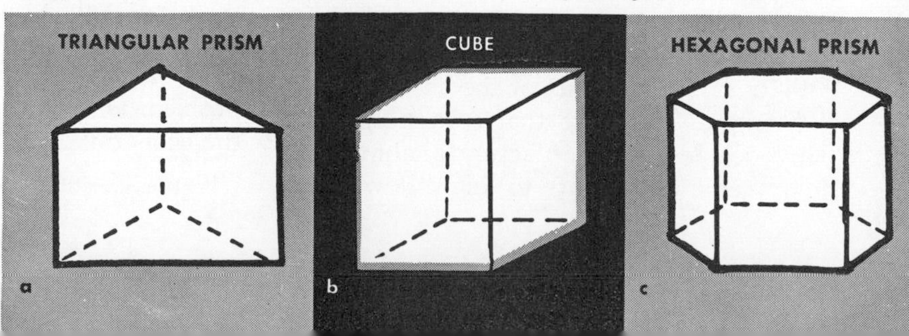

The most important properties of a prism are its area, or surface measure, and its volume, or space-filling measure. The lateral area (area of the sides) is equal to the perimeter of the base times the height. Of course, to find the total area of a prism, one adds the areas of the two bases to the lateral areas. The interior of a room is often a prism. If we want to decorate the walls, for example, we can find their total area by first finding the perimeter of the floor and then multiplying it by the height of the room. If we want to paint both the walls and ceiling of a room, we find the lateral area of the room and then the area of the floor (which, in the ordinary room, is the same as the area of the ceiling) ; then we add the two. It is by making such calculations that a painter or decorator estimates the cost of the materials needed in the work.

The volume of a prism is found by multiplying the area of the base by the altitude. This calculation is very important in the building of a house. Most builders estimate the construction cost as so much per cubic foot. To estimate how much it will cost to build a home, you must first find the total volume of the prisms of which the house will consist. If the volume in question is 28,000 cubic feet and the builder gives an estimate of 70 cents per cubic foot, the cost will be approximately 28,000 × $.70, or $19,600.

*The cylinder.* If one rotates a rectangle completely about one of its sides, as in Figure 48, it will define (mark the boundaries of) the solid called a cylinder. An ordinary tin can is a good example of a cylinder — a right circular cylinder, in which the bases are circles and the sides are perpendicular to the bases. It is not the only kind of cylinder, however. The bases of cylinders can have elliptical shapes and the sides are not always at right angles at the bases.

The lateral area of a right circular cylinder is $2\pi rh$, in which $\pi$ is approximately 3.1416, $r$ is the radius of the base and $h$ is the height. Suppose that a canning company decides to manufacture a number of quart cans, and wants to know how much

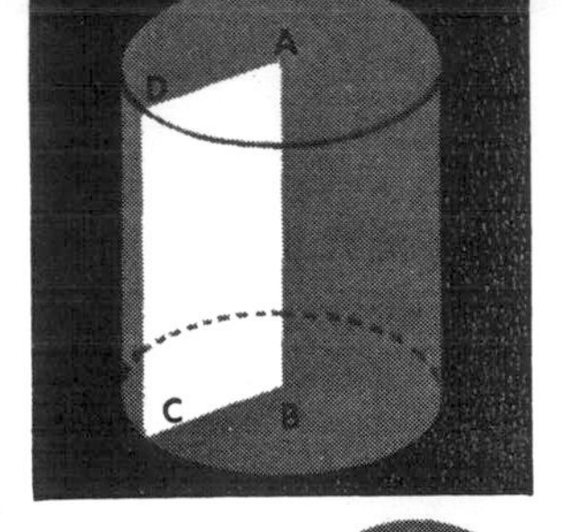


48. If rectangle *ABCD* is rotated about side *AB*, it will mark the boundaries of a cylinder, which will have *AB* as the axis.

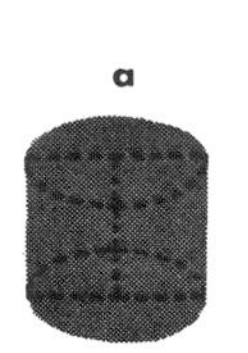

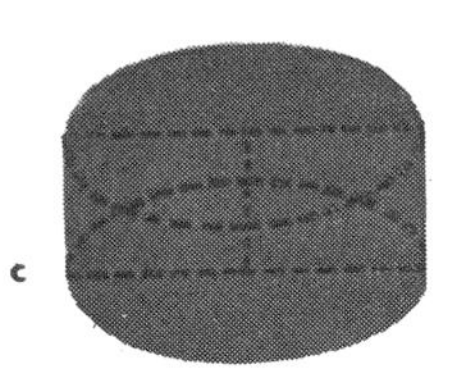


49. If the height of cylinder *a* is doubled and the base remains the same, as in *b*, the volume is doubled. If the height remains the same and the diameter is doubled, as in c, the volume of the cylinder is increased fourfold.

metal will be needed for the cans. The required quantity is found by adding the areas of the two circles that make up the bases * to the area of the side.

The volume of a cylinder is found by multiplying the area of the base by the height; the formula is $\pi r^2 h$, in which $r$ is the radius of the base and $h$ is the height. It is important to find the volume of a cylinder in computing the content of tin cans, gas-storage tanks, reservoirs and so on, and also in determining the rate of flow and pressure in pipes containing liquids.

If the height of a cylinder is doubled, and the diameter remains the same, its volume will also be doubled. You can see that this is so by placing one can on top of another just like it. If, however, the height of a cylinder remains the same and the diameter of the base is doubled, the new cylinder will hold four times as much as the original one (Figure 49).

To see whether the economy size of a product sold in cans provides a real bargain, it would be very helpful to calculate the volume of the regular size and that of the economy size and then compare the two. Suppose the regular-size can is 5 inches tall and has a base with a diameter of 6 inches; suppose the economy-size can is also 5 inches tall and has a base with a diameter of 8 inches. The regular size costs $1.00 and the economy size $1.50. The problem is: will we save money if we buy the economy size?

* Remember that the area of a circle is $\pi r^2$.

We know that the volume of a cylinder is $\pi r^2h$. To find the volume of the regular can we substitute 3 for $r$ (the radius is half the diameter) and 5 for $h$. To simplify our calculations we give $\pi$ the value $3\frac{1}{7}$. We would then have $3\frac{1}{7} \times 9 \times 5 = 141\frac{3}{7}$ cubic inches; this is the volume of the regular can. The volume of the economy-size can would be $3\frac{1}{7} \times 16 \times 5 = 251\frac{3}{7}$ cubic inches. The economy-size can would hold about $1\frac{3}{4}$ times as much as the regular can ($251\frac{3}{7} \div 141\frac{3}{7}$) and would cost $1\frac{1}{2}$ times as much. It would therefore represent quite a good bargain.

*The pyramid and the frustum of a pyramid.* In the solid called the pyramid, the sides are triangles whose vertices meet at a common point and whose bases form a plane (Figure 50). If the base is a square we have a square pyramid. The great pyramids of Egypt are square pyramids; so are the two Egyptian obelisks called "Cleopatra's needles." *

When the top of a pyramid is cut off by a plane parallel to the base, the lower part is called a frustum (Figure 51). Army squad tents and coal hoppers, among other things, have the shape of the frustum of a pyramid. To calculate the volume of a frustum, a rather complex formula is required. Yet there is evidence that the ancient Egyptians had an exact formula for making such a calculation; they used it in determining the amount of granite required to build different sections of their pyramids.

*The cone and the frustum of a cone.* A cone is formed by holding one end of a line fixed and rotating the line, following a circular path (Figure 52). If the fixed end of the line is held directly over the center of the circle, the cone is called a right circular cone (Figure 53*a*). If the vertex (topmost part) of a cone is not directly over the center of the circle, the cone is said to be oblique (Figure 53*b*). The height of a cone is the perpendicular distance from the vertex to the base (Figure 53). Its volume is equal to one-third the product of the area of the base and the altitude (height).

If we divide a cone in two parts by passing through it a plane parallel to the base, the lower part is called a frustum (Figure 54). Many machine parts are in the form of cones or frustums of cones.

* One of them is now in Central Park, New York; the other is on the Thames Embankment, in London.

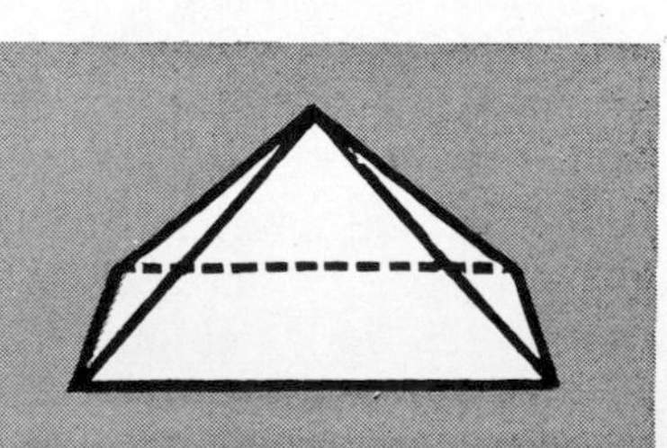

50. Square pyramid. It has a square base.

51. The frustum of a square pyramid.

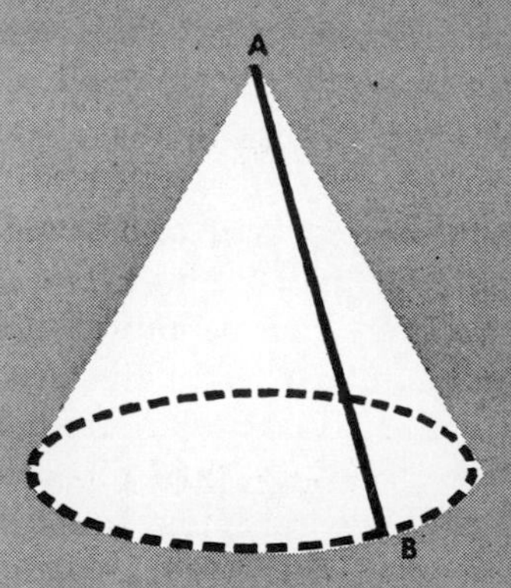


52. A is held fixed while point B follows a circular path. A cone is formed.

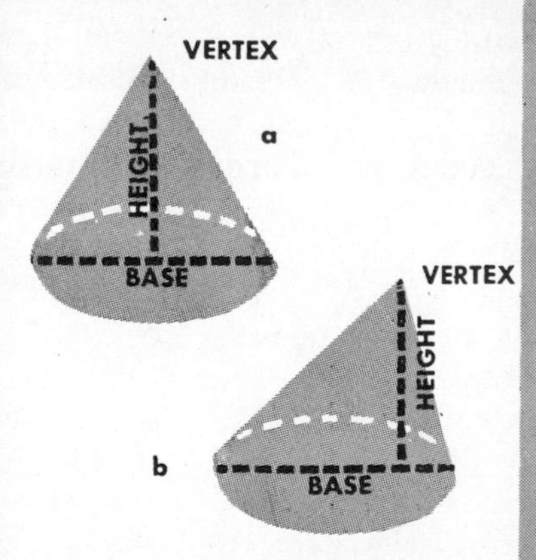


53. Two kinds of cones. *a*: right circular cone; *b*, oblique cone.

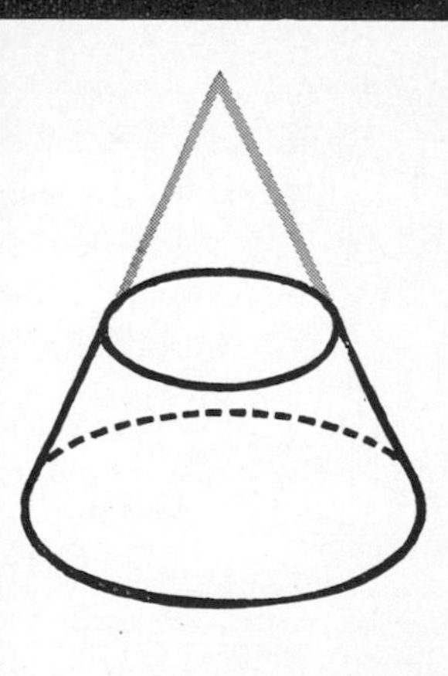

54. Above is the frustum of a right circular cone.

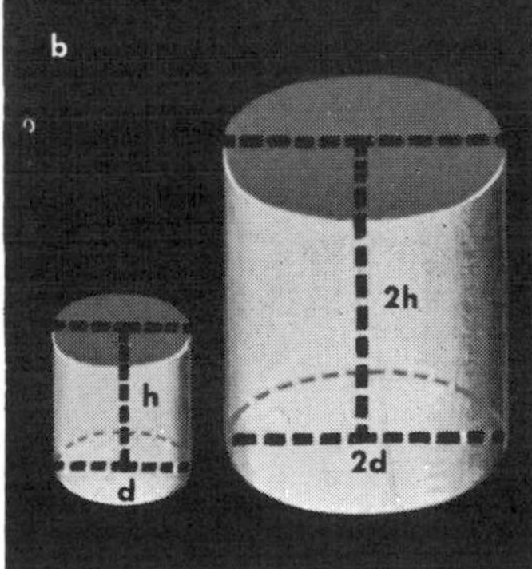


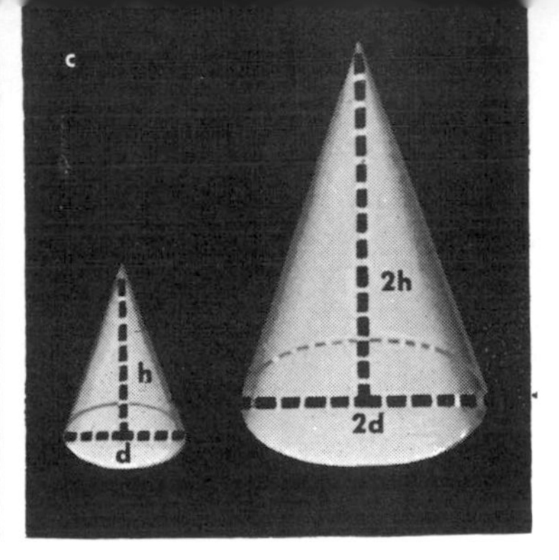


**55. Similar solids. Similar squares are shown in *a*; similar cylinders, in *b*; similar cones, in *c*.**

**Characteristics of similar solids**

Solids which are of the same shape but of different sizes are said to be similar. The corresponding polyhedral angles of similar solids are equal; the corresponding lines are in proportion.

The areas of similar solids have the same ratio as the squares of the corresponding linear parts (see Figure 55). In each of the sets of similar figures in Figure 55, the surface area of the larger figure is four times the surface area of the smaller, because the linear parts are twice as large. If the linear parts were enlarged three times, the area would be nine times as great.

The volume of similar solids have the same ratio as the cubes of the linear parts. In the sets of similar solids shown in Figure 55, the larger figure has eight times the contents of the smaller figure. In the case of the cubes in Figure 55*a*, you can count eight small cubes in the larger cube.

The ratio of areas and volumes of similar solids have many practical applications. All spheres are similar. If oranges 3 inches in diameter sell for 30 cents a dozen, while oranges of the same kind and 4 inches in diameter sell for 50 cents, which would be the better buy? The volume of a large orange is $\left(\frac{4}{3}\right)^3$, or $\frac{64}{27}$ or more than $2\frac{1}{3}$ times the volume of a small orange. Obviously in this case the large oranges would be a much better buy, since they cost less than twice as much as the smaller oranges but have more than twice the volume.

*The sphere.* If a semicircle is rotated about a diameter (Figure 56), the solid defined in this way is called a sphere. When the sphere is cut by a plane, the intersection is a circle; Figure 57 shows various circles formed in this way. If the plane passes through the center of the circle, the circle of intersection has the same radius as the radius of the sphere. A circle such as this is called a great circle; all the other circles are small circles.

The earth may be considered as a sphere * in which the north and south poles are the ends of a diameter called the axis (Figure 58). The circles passing through both the north and south poles are great circles; they are known as circles of longitude. All the planes (except one) that cut the earth at right angles to the axis form small circles, called circles of latitude. There is just one plane that passes through the center of the earth and at right angles to the axis; it forms a great circle called the equator.

Between any two points on the earth (not including the poles) only one great circle can be drawn (Figure 59). All other circles passing through the two points will be small circles. The shortest of all the arcs between the two points is the arc of the great circle (*AB*, in Figure 59). Pilots of planes, as far as possible, steer a course determined by the arc of a great circle between the starting point of their flight and the destination.

If we think of the earth as a rubber ball and cut this ball along one half of a circle of longitude, we can stretch the ball to form a flat rectangular sheet. The circles of longitude will then become parallel vertical lines and the circles of latitude parallel and equal horizontal lines (Figure 60). This sheet now represents a rectangular map of

* Actually the earth is somewhat flattened at the poles.

the world. A map of this type is called a Mercator projection, after the Flemish geographer Gerardus (or Gerhardus) Mercator* (1512–94), who developed it. The great longitudinal circle passing through Greenwich, England, is given the value 0° longitude; the equator is given the value 0° latitude. On such a map any place on earth can be located by determining the longitude and latitude. Also, on such a map, the farther we go from the equator, the more we find the original area stretched, so that land areas near the poles seem much larger on the map than they really are on the earth. Users of the map must take such distortions into account.

The area of a sphere is four times the area of a great circle; the formula is $4 \pi r^2$. The earth's radius is approximately 3,963 miles; hence the total surface of the earth is $4 \times \pi \times 3{,}963^2$ square miles or about 197,000,000 square miles.

The volume of a sphere can be expressed by the formula $\frac{4}{3} \pi r^3$. Since the radius of the earth is 3,963 miles, the volume of our planet is $\frac{4}{3} \times \pi \times 3{,}963^3$, which gives about 260,000,000,000 cubic miles.

* The Latinized form of Gerhard Kremer.

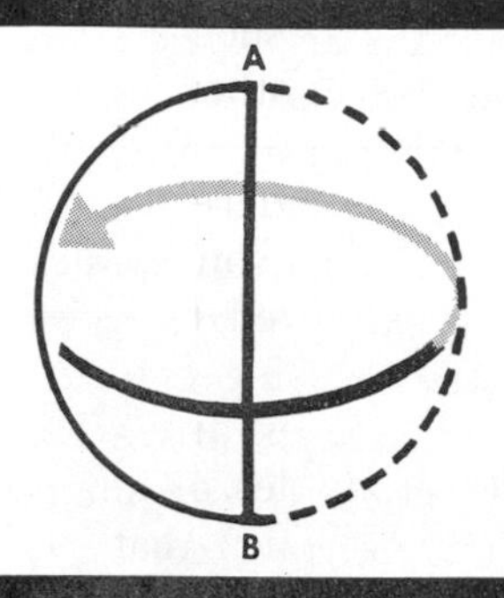


56. A spherical surface is formed as the semicircle is rotated around AB, which is the diameter.

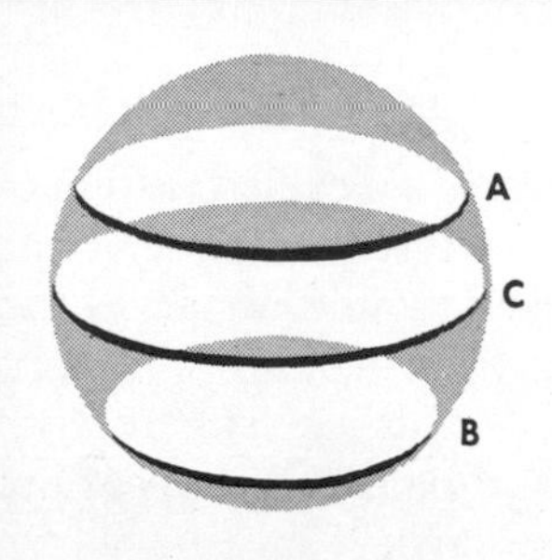


57. Plane C, passing through the center of the circle, is a great circle. A and B are small circles.

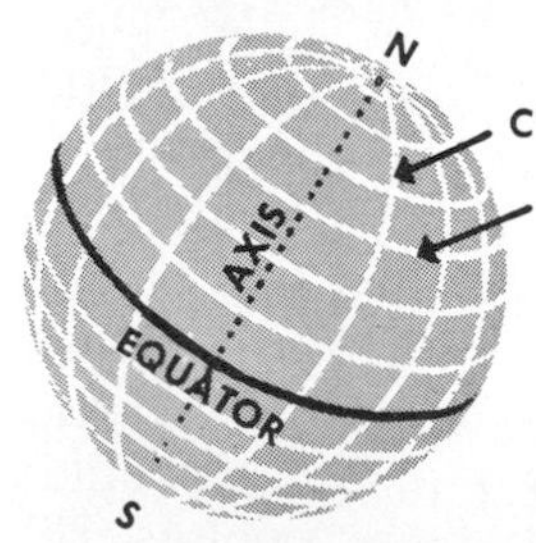


58. The earth as a sphere. The north and south poles are the ends of the axis, a diameter.

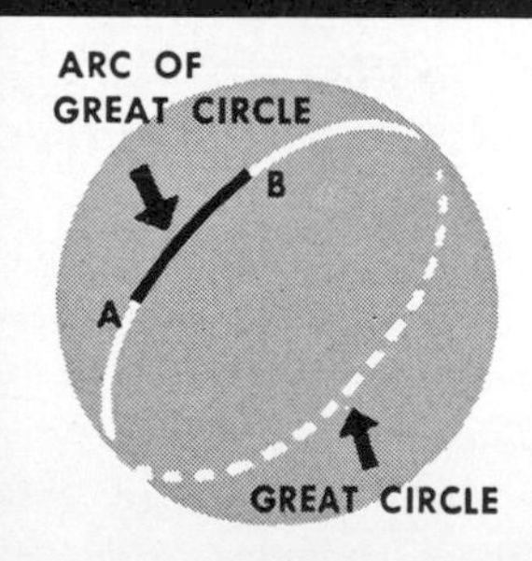


59. The shortest distance between A and B on the sphere is arc AB, forming part of a great circle.

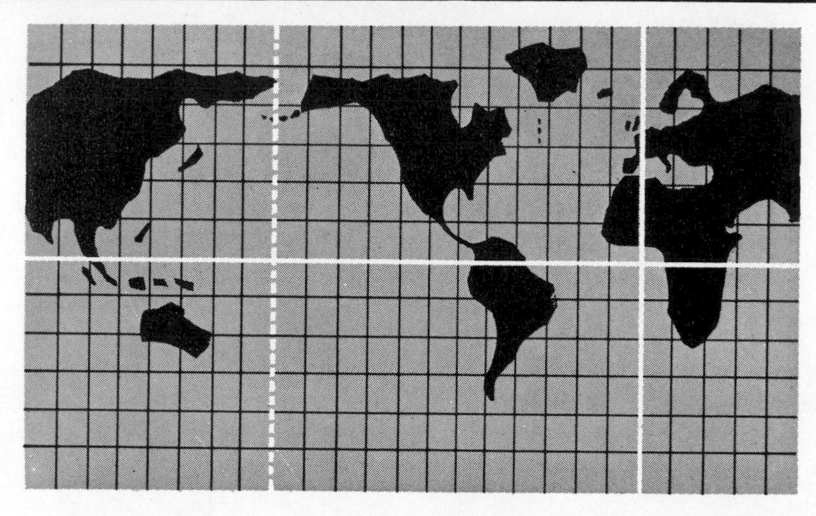

60. The Mercator projection. On such a map, any place on earth can be located by determining the latitude and longitude.

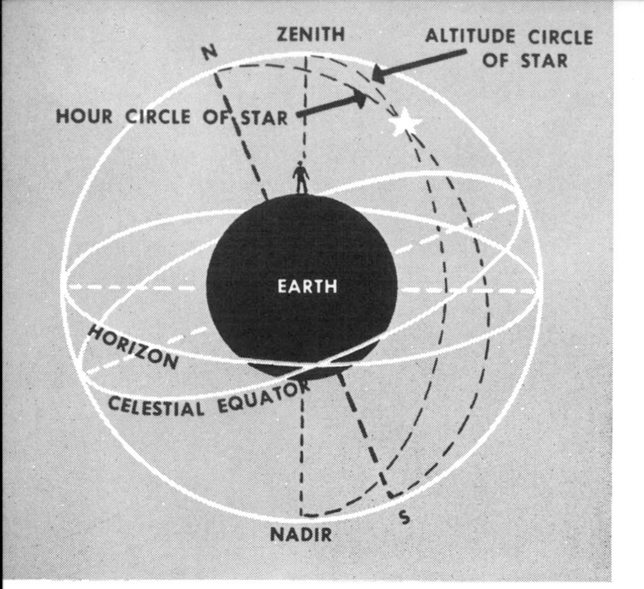


61. Simplified diagram showing the universe as a sphere.

**Interpreting the universe by means of solid geometry**

Solid geometry has enabled astronomers to give a useful interpretation of the heavens and to calculate the distance and position of the heavenly bodies. The universe is conceived of as a huge sphere with an infinitely great radius, which appears to revolve around the earth. In Figure 61, we give a greatly simplified presentation of such a sphere as seen from the vantage point of a man who is stationed at latitude 50°. This man stands on a much smaller sphere, which of course is the earth. Directly overhead is the zenith; directly below him is the nadir. The line where the sky seems to meet the earth is called the horizon; it divides the universe into two hemispheres. If the axis of the earth, which passes through the north and south poles, is extended, it will meet the outer bounds of our imaginary celestial sphere at the celestial poles — north and south. The line connecting the two celestial poles is the celestial axis. The plane of the earth's equator will cut the outer limits of the celestial sphere in a great circle called the celestial equator. A great circle passing through the poles and a star is the hour circle of the star. The altitude circle of the same star is a great circle passing through the zenith and the star.

These are but some of the features of the celestial sphere.* They provide a frame of reference that enables the astronomer to determine the positions and to trace the motions of celestial objects. This is one of the outstanding contributions of solid geometry to science.

* For a further discussion of the celestial sphere, see the article The Face of the Sky, in Volume 1.

## TRIGONOMETRY

An important offshoot of geometry is trigonometry, or triangle measurement. In trigonometry, when certain parts of triangles are known, one can determine the remaining parts and thus solve a great variety of problems.

Boy scouts have occasion to use trigonometry in their field work. A common problem that is put to them is to find the height of a tree. The scout first measures a distance, say fifty feet, from the base of the tree, as shown in Figure 62. This will be his "base line." At $A$, he measures the angle from the ground to the tree by means of a protractor. Let us suppose that this angle is 35°. The scout now knows two facts about the large right triangle formed when he connects points $B$ (the base of the tree), $A$ (the end of the line drawn from the tree) and $C$ (the top of the tree). He knows that $AB$ is 50 feet and that angle $BAC$ is 35°.

62. The problem is to find the height of the tree when the angle at A (35°) and the distance AB (50 feet) are known. We show on this page how to solve the problem in question by means of similar triangles ABC and A'B'C'.

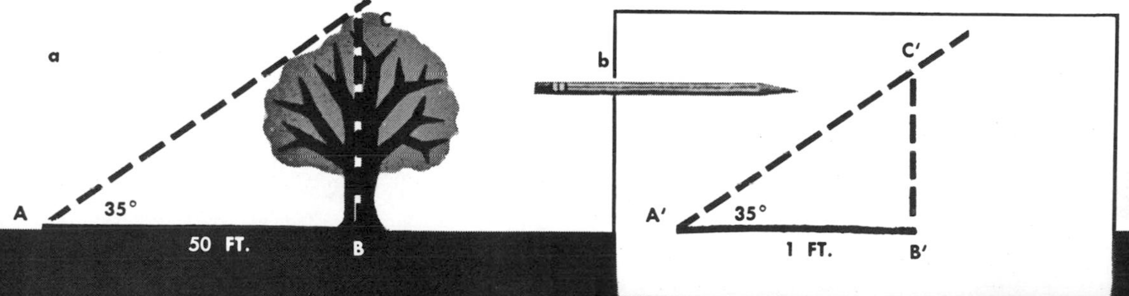

On paper our scout now makes a triangle similar to the large one in the field. First he draws a line $A'B'$ a foot long, and at $A'$ he draws an angle of 35° — angle $B'A'D'$ — with a protractor. Next he erects a perpendicular to line $A'B'$ at $B'$. This line will intersect $A'D'$ at $C'$; and the angle $A'B'C'$ will be a right angle. The corresponding angles of the large and small triangles are equal: angle $CAB$ = angle $C'A'B'$; angle $ABC$ = $A'B'C'$; angle $ACB$ = $A'C'B'$. Hence we have two similar triangles, and the corresponding sides will be proportionate. The scout now measures line $B'C'$ and finds that it is 0.7 foot. Since the sides of the similar triangles are in the same ratio, $AB$ is to $A'B'$ as $BC$ is to $B'C'$. We know all these quantities except $BC$, which we can call $x$. We now have the proportion 50 is to 1 as $x$ is to 0.7, which we can write as $50 : 1 :: x : 0.7$. In any proportion, the product of the extremes (the two outer terms) is equal to the product of the means (the two inner terms); hence $x = 35$. The height of the tree, then, is 35 feet. The scout solved this problem knowing only one side and an acute angle of a right triangle. He used the basic methods of trigonometry, though a mathematician, as we shall see, would not go at the problem in that particular way.

Trigonometry is based on the use of the right triangle. It can be applied to any triangle because by drawing an altitude (a perpendicular from the vertex to the base) we can always convert it into right triangles. In Figure 63*a,* for example, the altitude $AD$ divides the triangle $ABC$ into the right triangles $ADB$ and $ADC$; in Figure 63*b,* the altitude $GH$ converts the triangle $GEF$ into right triangles $GHF$ and $GHE$.

**63. How to convert any triangles into right triangles. One draws a perpendicular from the vertex to the base.**

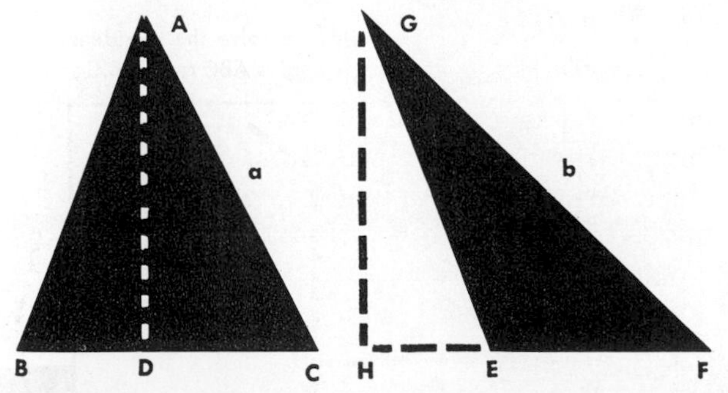


Certain basic ratios or relationships between the sides of a right triangle are the very heart of the study of trigonometry. Among these ratios are the sine, cosine, tangent and cotangent. To understand what these terms mean, let us draw a typical right triangle with angles 1, 2 and 3 and sides $a$, $b$, and $c$ (Figure 64). Angle 3 is a right

**64. This is a typical right triangle, with c as the hypotenuse.**

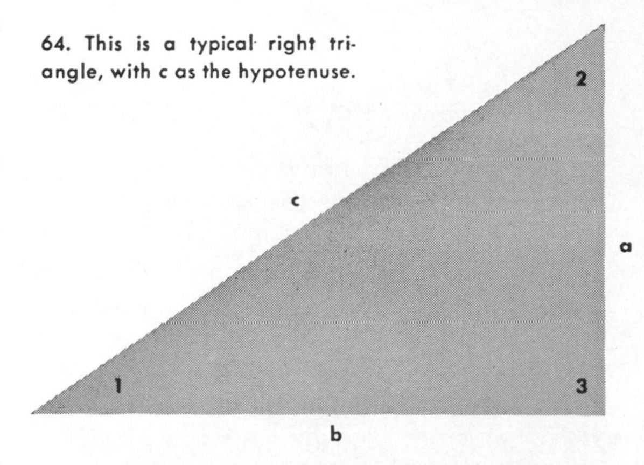


angle; the other two angles are acute angles (that is angles of less than 90°). Side $c$, which is opposite the right angle, is the hypotenuse; the other two sides, $a$ and $b$, are called legs. We can now define sine, cosine, tangent and cotangent as follows:

The *sine* of either of the acute angles is the ratio of the opposite leg to the hypotenuse. The sine of angle 1 is $a/c$; the sine of angle 2 is $b/c$.

The *cosine* of either of the acute angles is the ratio of the adjacent leg to the hypotenuse. The cosine of angle 1 is $b/c$; the cosine of angle 2 is $a/c$.

The *tangent* of either of the acute angles is the ratio of the opposite leg to the adjacent leg. The tangent of angle 1 is $a/b$; the tangent of angle 2 is $b/a$.

The *cotangent* of either of the acute angles is the ratio of the adjacent leg to the opposite leg. The cotangent of angle 1 is $b/a$; the cotangent of angle 2 is $a/b$.

A sine of an angle is said to be a trigonometric function of that angle, because its value depends upon the size of the angle.* The cosine, tangent and cotangent are also trigonometric functions.

* In the language of mathematics, a function is a magnitude so related to another magnitude that to values of the latter there correspond values of the former.

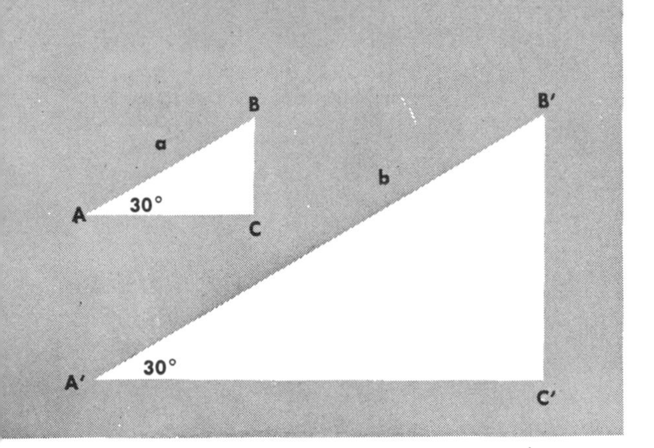


65. The sine of angle BAC (30°) is equal to the sine of the similar triangle A'B'C'. The sine of 30° is .50000.

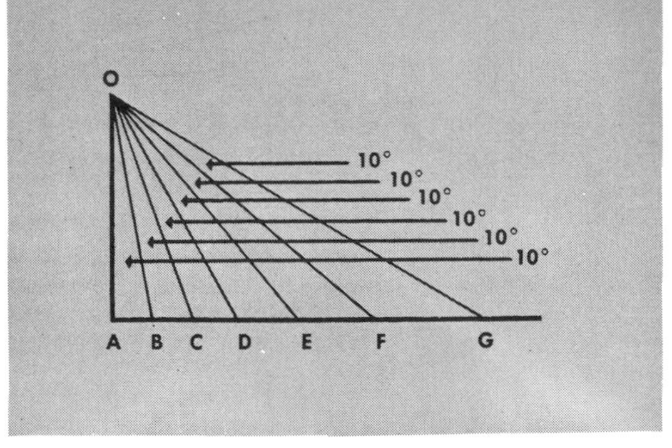


66. This diagram shows how mathematicians have found the tangents of angles. The explanation is given in the text.

A given trigonometric function, such as the sine, is always the same for a given acute angle in a right-angle triangle. In Figure 65*a*, for example, angle $BAC$ of the right triangle $ABC$ is 30°. This means that angle $ABC$ must be 60° since angle $ACB$ is 90° and the sum of the interior angles of a triangle is 180°. Angle $B'A'C'$ of the right triangle $A'B'C'$ is 30° and angle $A'B'C'$ must be 60°. Hence the corresponding angles of the two triangles are equal, and the corresponding sides must be in the same ratio. Since this is so, $\frac{BC}{AB}$ (the sine of the 30° angle $BAC$) and $\frac{B'C'}{A'B'}$ (the sine of the 30° angle $B'A'C'$) must be equal. Hence the sine of 30° is always the same no matter how large or how small the right triangle in which it occurs.

Mathematicians have worked out the values of the trigonometric functions. By way of example, the sine of 40°, to five decimal places, is .64279; its cosine, .76604; its tangent, .83910; its cotangent, 1.1918. To give some idea of how such figures are derived, let us examine the procedure for finding out the tangents of different acute angles. You will recall that the tangent of an angle is the ratio of the opposite leg to the adjacent leg.

In Figure 66, side $OA$ is equal to exactly one inch. Each of the small angles at $O$ is equal to exactly 10°. Angle $BOA$, therefore, is 10°; angle $COA$, 20°; angle $DOA$, 30°; angle $EOA$, 40°; and so on. We now measure $AB$ and find that it is about .18 inches.* Since the tangent of an angle $BOA$ is $\frac{AB}{OA}$ and since $OA = 1$, the tangent of the angle is $\frac{.18}{1}$, or .18. This is the tangent of the angle 10°, whether the side $OA$ is an inch, or a mile or 1,000,000 miles. Measuring $AC$, $AD$, $AE$ and so on in turn, we can find the tangents of 20°, 30°, 40° and the rest.

* Measured more accurately it is .17633 inches.

The values of the trigonometric functions are to be found in special tables. Armed with these tables, it is possible for one to work out a great variety of measurements with great ease. Let us return, for a moment, to the boy-scout problem as presented in Figure 62*a*. We know that $AB$ is 50 feet and that angle $CAB$ is 35°. We are to find out $BC$ (the height of the tree). The tangent of the 35° angle $CAB$ is $\frac{BC}{AB}$. We know that $AB$ is 50 feet; since $BC$ is the unknown quantity, we call it $x$. Hence the tangent of angle $CAB$ is $\frac{x}{50}$. Looking up the table of trigonometric functions, we find that the tangent of 35° is .7. We then have the equation $\frac{x}{50} = .7$. Multiplying both sides of the equation by 50, we have $\frac{50x}{50} = 35$. $x = 35$.

In the foregoing problem, the unknown quantity was a part of the tangent ratio. In other trigonometric problems, the cotangent, or the sine or the cosine might be involved. In still other cases, the unknown quantity might be an angle, as in the following problem.

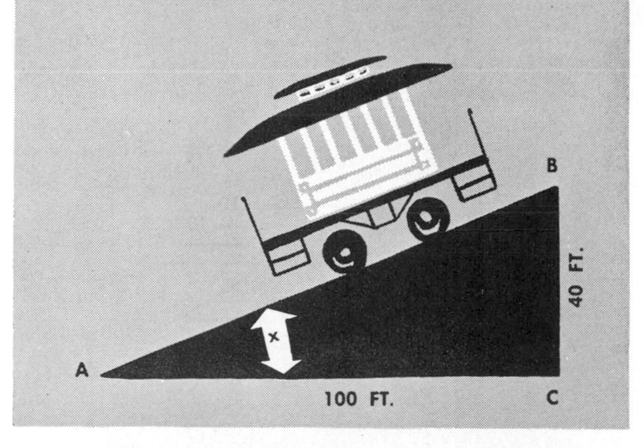


67. The cable car rises 40 feet in a horizontal distance of 100 feet. The problem here is to find the angle at A.

A cable car going up a hill in a uniform slope rises 40 feet in a horizontal distance of 100 feet. What is the angle of the slope to the nearest degree? We diagram the problem as in Figure 67. We want to find angle $x$. We know that the tangent of $x$ is $\frac{BC}{AC} = \frac{40}{100} = .4$. Consulting the table of trigonometric functions, we note that .4 is the tangent of the angle 22° (to the nearest degree). Therefore the angle of the slope is 22°.

The surveyor makes extensive use of trigonometry in his measurements. He tries to get a fixed line that has no obstruction so that he can measure it fairly accurately. He calls this the base line and uses it in his calculations. His other measurements, as far as possible, are measurements of angles. Trigonometry is also of vital importance in engineering, navigation, artillery fire, mapping and astronomy.

The founder of trigonometry was the Greek astronomer Hipparchus of Nicaea, who lived in the second century B.C. Hipparchus attempted to measure the size of the sun and moon and their distances from the earth. He felt the need for a type of mathematics that, by applying measurements made on the earth, would enable him to measure objects far out in space. He was led to the invention of trigonometry.

Trigonometry is used nowadays in ways that Hipparchus never thought of. For one thing it serves in the study of various periodic phenomena. Any phenomenon that repeats itself in regular intervals of time is called periodic. The tides, for example, are periodic, since they rise and fall in regular sequence; the motion of a pendulum bob is also periodic as it swings back and forth. Let us show how we describe all periodic phenomena in terms of the sine of an angle.

We all know that the spoke of a moving wheel sweeps through 360° as it makes a complete turn. It repeats the same sweep in the second complete turn and in the third complete turn and so on. Obviously such rotation is periodic. The spoke of a wheel is really the radius of a circle. We can analyze the motion of the radius around the center of the circle by examining the diagrams in Figure 68. We are to suppose that a rod, $CD$, is kept in a vertical position at the end of the radius as the latter moves around the circle. You will note that a series of right angles is formed as the rod maintains its vertical position. The line $AP$, joining the points where $CD$ meets the circumference of the circle and the diameter $EF$, grows longer and then shorter. Angle $POA$, which is called the angle of rotation, also changes as the radius goes around the center of the circle. The sine of $POA$, the angle of rotation, is $\frac{AP}{OP}$. $OP$, the hypotenuse of the right triangle $APO$, is the radius of the circle and of course never changes. If we give it the value unity (that is, 1), $AP$ will represent the sine of the angle of rotation, since the sine is equal to $\frac{AP}{OP}$.

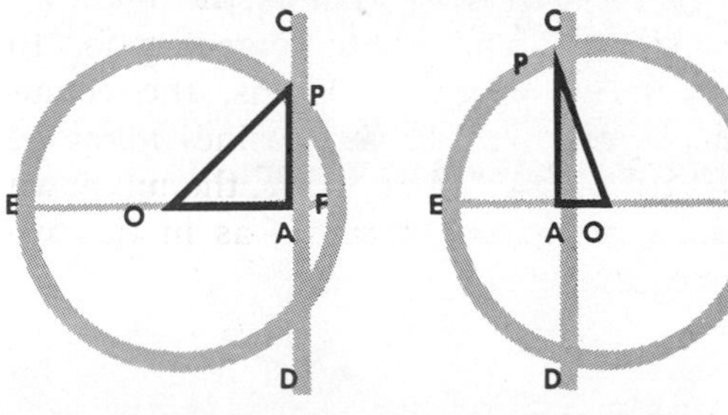


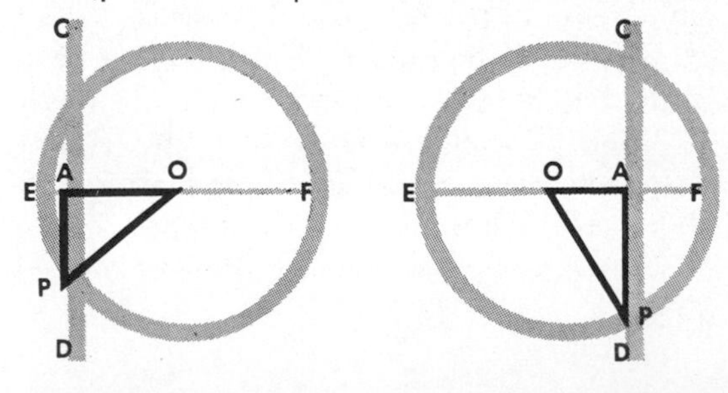


68. As radius $OP$ goes around the center of the circle in a counterclockwise direction, it makes a series of triangles with the diameter of the circle and with the rod $CD$, kept in a vertical position at the end of the radius.

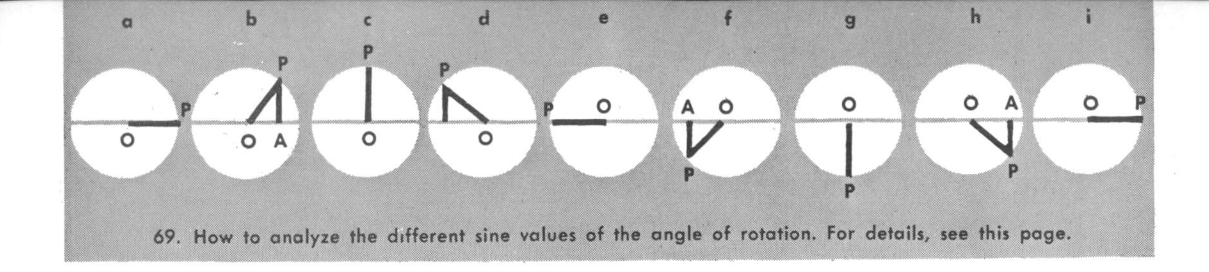


69. How to analyze the different sine values of the angle of rotation. For details, see this page.

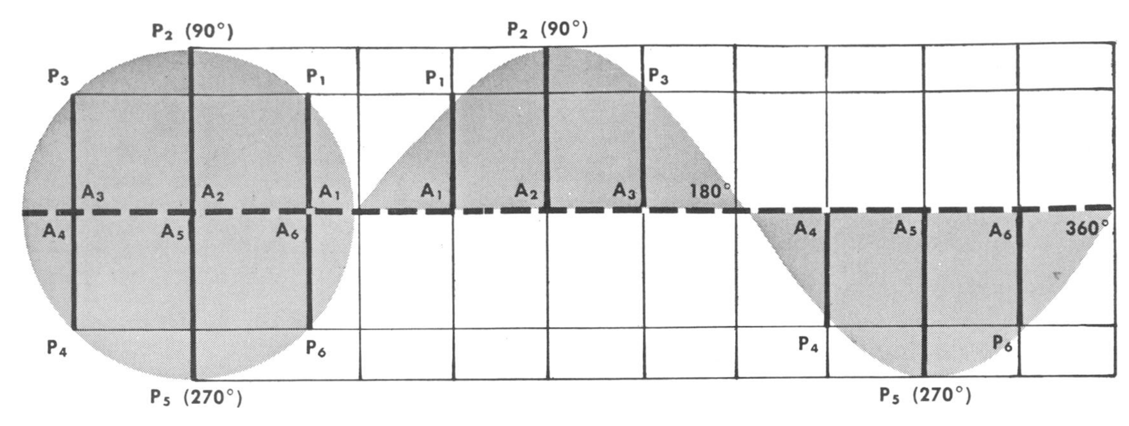


70. Variations in sine value are shown here by means of a graph. The different values of the sine *PA* are given here as $P_1A_1$, $P_2A_2$, $P_3A_3$ and so on. The curved line at the right of the circle is known as the sine curve.

Let us now analyze the different sine values of the angle of rotation as the radius sweeps around the circle. Sine values above the diameter are expressed as positive values; values below the diameter are negative values. In Figure 69*a*, the sine, *PA*, is zero, corresponding to a zero angle of rotation. The sine continues to grow as the radius revolves around the center (*b*) until it reaches the value 1 in *c*. It becomes smaller (*d*) until it reaches zero again in *e*. Then it goes below the diameter and is assigned negative values. It increases in size (*f*) until it reaches the value −1 in *g*. It becomes smaller thereafter (*h*) until, after a complete rotation has taken place, it becomes zero again (*i*).

We can show the variation in the sine by the line graph in Figure 70. We give the different values of the sine, *PA*, as $P_1A_1$, $P_2A_2$, $P_3A_3$ and so on. The line at the right of the circle is called the sine curve; it repeats itself every 360°.

The sine curve can be applied, among other things, to the periodic phenomena of sound. For example, when a tuning fork is struck, it vibrates and the vibration results in sound waves being sent out. If, immediately after being struck, a tuning fork which gives a tone of middle *C* is drawn very rapidly over a sheet of paper covered with soot, the vibration of the fork will describe a series of sine curves (Figure 71). Two hundred sixty-four complete oscillations are produced in a second; hence middle *C* corresponds to 264 vibrations per second; we say that it has a frequency of 264 cycles per second. Suppose we strike a tuning fork that gives a tone one octave higher than middle *C* and draw it over the paper as before. In this case 528 sine curves will be produced in a second. That means that the frequency is twice as great as before.

Sine curves of various amplitudes and frequencies are used to explain phenomena of electricity, light (both polarized and plane) and force, as well as those of sound.*

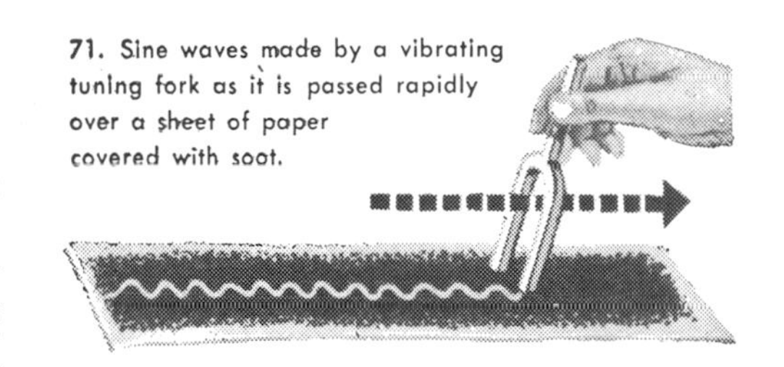

71. Sine waves made by a vibrating tuning fork as it is passed rapidly over a sheet of paper covered with soot.

Thus trigonometry helps to explain and control our physical environment.

*See also Vol. 10, p. 279*: "Mathematics."

* Cosine curves may also be used.

# THE VIRUSES

## Tiny Disease Agents in Life's Twilight Zone

BY C. A. KNIGHT

VIRUS diseases have undoubtedly occurred from time immemorial among the peoples and animals of the world. We know that smallpox, which still kills thousands of people annually in some regions, afflicted the inhabitants of ancient India; it is believed to have been the cause of a serious epidemic in China in 1122 B.C. Early Greek records indicate that rabies, another virus disease, was causing wild seizures in dogs and hydrophobia in man several centuries before the birth of Christ.

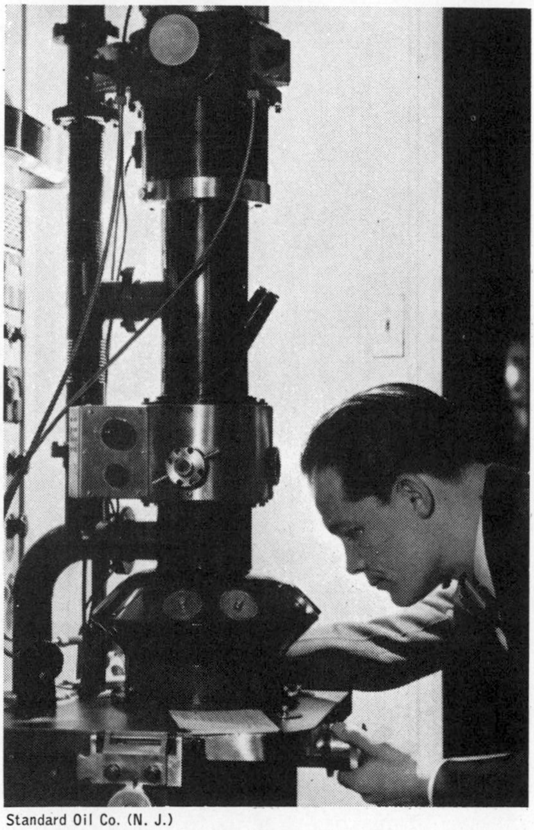

Standard Oil Co. (N. J.)

Electron microscope for viewing viruses.

Plant virus diseases also probably go back to ancient times. But though the diseases caused by viruses have been familiar to man for so many centuries, it has only been within the last generation or two that scientists have come to understand the real nature of the viruses themselves.

During the latter half of the nineteenth century the brilliant researches of Pasteur, Koch and others established that microscopic plants and animals (bacteria, fungi and protozoans) caused infection. Were all contagious diseases due to organisms of this kind? The answer was to be forthcoming before long.

Standing before the Russian Academy of Sciences on February 12, 1892, Dimitri Iwanowski presented a short paper entitled On the Mosaic Disease of the Tobacco Plant. Toward the end of the paper Iwanowski made the following comment on an experiment he had carried out: "I have found that the sap of leaves attacked by the mosaic disease retains its infectious qualities even after filtration through Chamberland filter candles." The filter to which Iwanowski referred was a newly developed piece of bacteriological equipment. It was a sort of porcelain sieve through which various liquids could move, but whose pores were too fine to permit bacteria or other known microorganisms to pass. Since the agent of mosaic disease had gone through the filter, the obvious conclusion was that it was considerably smaller than any known microorganism.

Curiously enough, Iwanowski apparently did not accept this conclusion until some years later. The Dutchman M. W. Beijerinck (or Beigerinck), who was unaware of Iwanowski's work and who had carried on filtration experiments of his own, was the first to advance the theory that tobacco mosaic disease was caused by a unique type of infectious agent — a "contagious living fluid," as he called it.

Beijerinck's observations were reported near the turn of the century. At the same time, F. A. J. Loeffler and P. Frosch showed that foot-and-mouth disease of cattle is caused by a filterable agent (that is, one that can pass through a bacterial filter), and Giuseppe Sanarelli established that a disease of rabbits called infectious myxomatosis was due to a similar cause. Walter Reed and his co-workers demonstrated in 1901 that yellow fever is also due to a filterable agent. Later investigations showed that such agents cause a number of other diseases.

How did these disease agents come to be known as viruses? Before the middle of the nineteenth century, the word "virus" was commonly applied to all toxic, or poisonous, substances, including snake venom. With the development of the germ theory of disease, the term was expanded to include infectious agents of all kinds. When it was discovered that certain infectious agents could pass through bacterial filters, they became known as filterable viruses, as opposed to other kinds of viruses. In the twentieth century "viruses" has come to have a more restricted meaning; the term now applies *only* to filterable viruses.

Viruses are distinguished from poisons and venoms because of their infectious quality. They differ from bacteria in several important respects. As we have noted, bacteria are held back on filters that allow viruses to pass through. Though bacteria are very small, they may be observed with an ordinary microscope. With few exceptions, viruses are submicroscopic — that is, too small to be seen with the ordinary microscope; they have become visible only with the development of a special instrument, the electron microscope. Bacteria may prey on living cells, but they are usually not dependent upon them and can often be grown in the laboratory in broth solutions or on nutrient jellies. Viruses must get into living cells to survive and to reproduce their kind. For this reason they are known as obligate parasites. ("Obligate" means "limited to a single condition.")

In addition to smallpox and yellow fever, viruses cause such human diseases as mumps, measles, poliomyelitis (infantile paralysis), chicken pox, Japanese B encephalitis, infectious hepatitis, herpes, dengue, influenza, certain warts and probably the common cold. One of these diseases, influenza, was responsible for one of the worst epidemics in the history of mankind; in 1918–19 it killed an estimated twenty million people and affected hundreds of millions of others.

Domestic animals are subject to many virus diseases. Among them are cowpox, pseudorabies (mad itch), Newcastle disease of chickens, dog distemper, horse encephalomyelitis, louping ill of sheep and hog cholera.

The names applied to plant virus diseases generally describe the disease symptoms and are rather picturesque. Some examples are sugar-beet curly top, tomato bushy stunt, spinach blight, tomato spotted wilt, tulip break, camellia mosaic, buckskin disease of cherry, potato yellow dwarf, little peach and quick decline of citrus. Virus diseases have also been observed in lower forms of life, such as insects, worms and even bacteria.

Almost all living things seem to be susceptible to virus infection. No virus diseases have yet been reported for algae, fungi, mollusks, beetles and a few other groups. Closer investigation, however, may reveal that even these groups are not immune to diseases caused by viruses.

**Various symptoms are the result of virus infection**

Virus infections cause an amazing variety of disease symptoms. Some of these are not noticeable. For example, older varieties of American potatoes are almost universally infected with a virus called potato virus X; yet in many cases the symptoms of infection are so negligible that the virus has been referred to as the "healthy potato virus." Only when it is transmitted to a different host, such as tobacco, can the virus be shown to exist in infected but symptomless potato plants. It appears, likewise, that poliomyelitis virus causes little damage in infected cells of the digestive tract. While a mild illness results from such infection, the virus is actually helpful under these conditions; it stimulates the body to develop a defense mechanism against a more serious attack by this disease agent.

On the other hand, some virus diseases cause the complete destruction of the infected cells. The dreadful paralysis caused by the polio virus is due to the widespread disintegration of nerve cells in the central nervous system. There is a similar destructive effect when bacteria are infected with certain viruses known as bacteriophages. In fact, the existence of these viruses was first recognized because of the spectacular disruption (lysis) of infected cells.

Among the most striking symptoms of virus infection is the uncontrolled development of tissues. In plants there is sometimes a wild growth of secondary shoots, producing a formation called witches'-broom. In some plants the overgrowths caused by viruses take the form of galls or tumors. Such symptoms are found in Fiji disease of sugar cane, wallaby-ear disease of corn and wound-tumor of clover.

Animals also suffer from virus tumors These include the malignant kidney tumor of the leopard frog, several fowl sarcomas, a couple of rabbit tumors, a mouse leukemia and a mouse mammary sarcoma (cancer of the breast). Viruses are responsible for benign tumors of the human skin — warts and molluscum contagiosum. It is not known to what extent viruses are involved in human cancers; but there is reason to suspect them.

What is the precise nature of these insidious disease agents that cause such a wide variety of afflictions? Thirty years or more after it was recognized that virus diseases represent a distinct class of infections, researchers could only guess at the physical nature of viruses. One reason was their submicroscopic size. Then too, viruses, unlike other infectious agents, frequently resemble in size and shape some of the constituents of the cells in which they multiply. This makes it difficult even for modern researchers to recognize them.

#### Viruses can be isolated in the form of crystals

An entirely new approach to the study of viruses was presented when, in 1935, the American biochemist Wendell M. Stanley reported that he had isolated tobacco mosaic virus in the form of needlelike crystals. This startling announcement upset the comfortable notion that all infectious diseases were caused by small living organisms. Here was an agent that reproduced its kind like a living thing in the host. Yet it had been obtained in crystalline form, a condition that no bacterium or any other living thing had ever assumed.

Several other viruses have been isolated in the form of crystals, but not all by any means. Each of the crystals represents thousands of individual virus particles. Authorities are still debating the question: "Are these particles living or nonliving?" Many of them are inclined to accept Stanley's view that viruses exist in "the twilight zone of life." According to this viewpoint, they possess properties of both living matter and lifeless matter, and form a link between them.

By isolating viruses in the highly purified form represented by the crystals, researchers have been able to determine the chemical composition of these infectious agents. It has been shown by chemical analysis that tobacco mosaic virus is a nucleoprotein. This means that it is a combination of two organic materials — nucleic acid and protein. Nucleic acid is an organic substance present in the nuclei of all cells; it is particularly abundant in germ cells, such as the spermatozoa. Protein is a major constituent of living matter; such things as lean meat and the whites of egg are largely protein.

All viruses that have been chemically analyzed thus far have contained nucleic acid and protein; apparently these are the

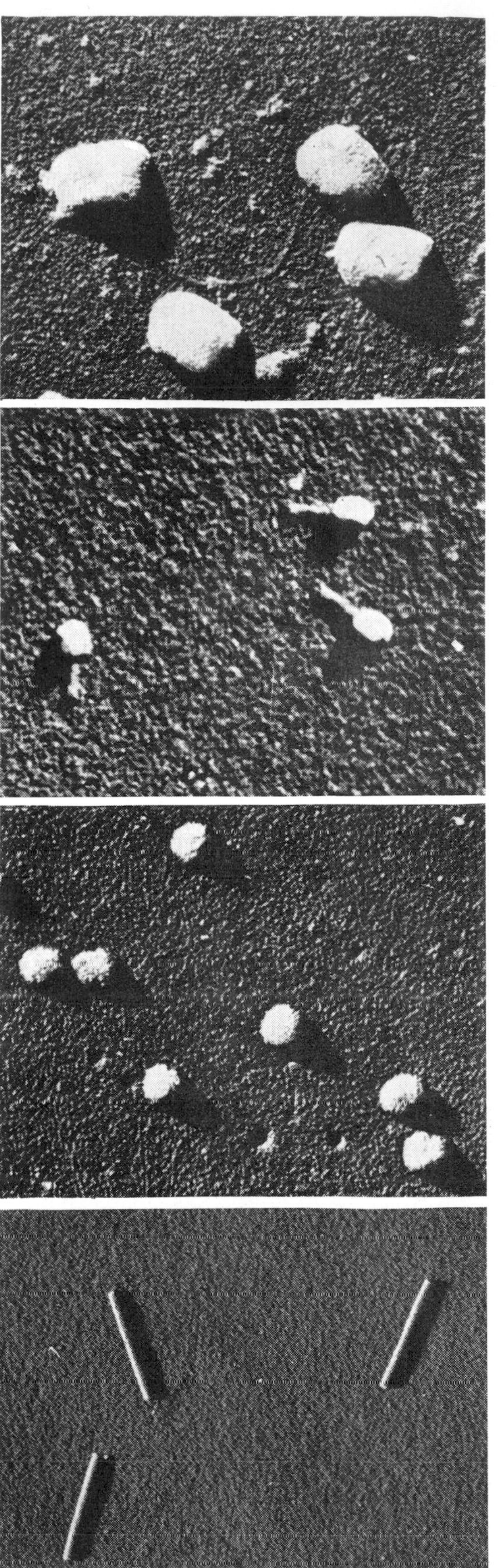

R. C. Williams

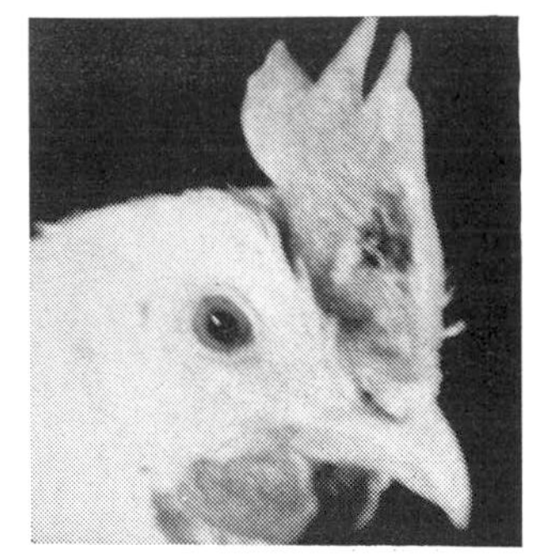

C. A. Knight

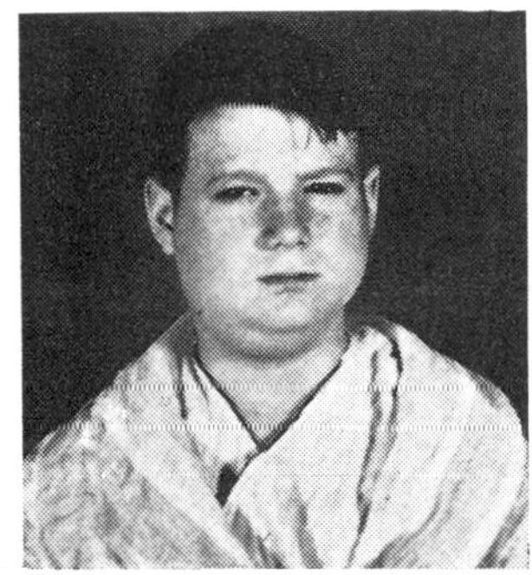

Clay Adams Inc.

Viruses are the cause of fowl pox of domestic fowl, mumps in man and mosaic disease of tobacco. At the right we show photomicrographs of (top to bottom) vaccinia virus, T2 bacteriophage, influenza virus and tobacco mosaic virus.

C. A. Knight

basic constituents. The proportions of nucleic acid and protein vary in the different groups of viruses; so does the chemical structure of these substances.

Even the most complex of the viruses are chemically much simpler than the bacteria. The plant viruses are the simplest of all; they seem to consist only of nucleoprotein. Most animal viruses contain fatty substances in addition to nucleoprotein. At least one virus, that of influenza, seems to include a sugar as well. Vaccinia virus, which is used in a vaccine against smallpox, appears to have in it a small amount of a vitamin and traces of other substances. These differences in chemical complexity, together with variations in structural features, probably account for the wide variety of diseases caused by the various groups of viruses.

All living things have the ability to mutate — that is, to undergo a sudden variation in some well-marked character. Viruses, which may or may not be living things, also have this property. Usually a virus is reproduced true to type for thousands of generations. Occasionally, however — perhaps once in a hundred thousand duplications — a sudden, spontaneous change occurs. This results in a new virus, whose properties are similar to those of the parent virus in some respects and different in others. If such a change is heritable, it is a mutation; the product of mutation is a new virus strain.

Mutation is a random process and no one can closely predict the direction it will take. Some new virus strains are milder than the original virus in a given host; others are more severe. Isolation of milder strains, occurring in nature, has provided material for successful vaccines against smallpox and yellow fever. On the other hand, mutation of a virus to a more deadly strain probably accounts for the great world-wide scourge of influenza during the years 1918–19.

In the course of extensive chemical studies on strains of tobacco mosaic virus, the writer showed that the protein parts of different virus strains sometimes, but not always, contain different amounts of amino acids. (These acids are the building blocks of proteins.) By way of contrast, the nucleic acid portions of such strains appear to be perfectly uniform in composition. Apparently the vital chemical differences among virus strains are due to the ways in which the building blocks are arranged to form the individual virus particles.

Because of the relative chemical simplicity of viruses, it should be possible to determine exactly what chemical changes occur in a virus when it mutates. Knowing this, it will be possible some day, perhaps, to create desired strains in the laboratory — let us say, for the purpose of vaccine production.

Before the development of the electron microscope, about the only virus that could be seen by researchers was the vaccinia virus; this could be detected with a particularly powerful light microscope (one employing beams of ordinary light). The individual particles of other viruses were too small to be made out. Hence the first estimates of the size and shape of highly purified tobacco mosaic virus were obtained by ingenious indirect methods. These were based on such properties of the virus as osmotic pressure, X-ray scattering, viscosity, diffusion rate and behavior when subjected to great centrifugal forces. (A centrifugal force tends to propel things outward from a center of rotation.)

About 1940, virus researchers began to use the electron microscope, which had been developed in the preceding decade in Germany and the United States. This instrument, using a beam of electrons in place of a beam of ordinary light, extended the limits of visibility to a size range that, fortunately, includes all the viruses known at the present time. With the best light microscopes it is possible to magnify the sizes of objects up to about 2,000 times. With the electron microscope, a direct magnification of 10,000 times can be obtained. When clear photographic enlargements are made, the original virus particles appear magnified 200,000 times.

Examination of highly purified viruses in the electron microscope has revealed that the particles of each type of virus possess

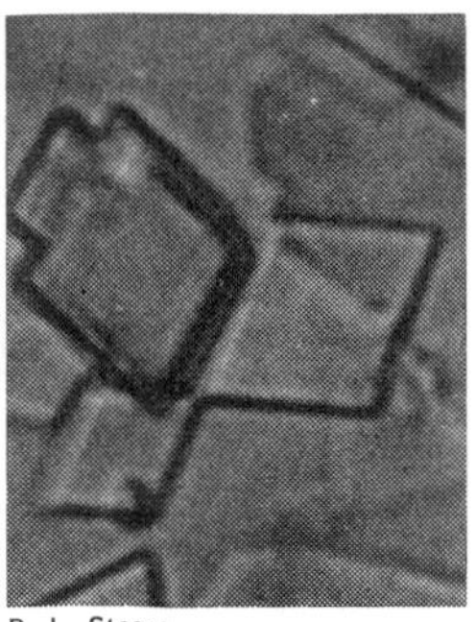

R. L. Steere

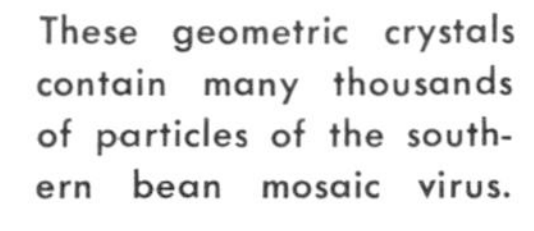

These geometric crystals contain many thousands of particles of the southern bean mosaic virus.

To the right is shown the crystallized form of the virus responsible for tomato bushy stunt disease.

a characteristic shape and size, but that different kinds of viruses show striking varieties of shapes and sizes. The particles of tobacco mosaic virus are rodlike. Micrographs of potato virus X and of an orchid mosaic virus show filamentous, or threadlike, forms. The relatively big elementary bodies of vaccinia virus are shaped like loaves of bread. Influenza virus particles are globular; so are a number of the smaller viruses, such as those of tomato bushy stunt and poliomyelitis. The particles of tobacco ring-spot virus have hexagonal faces.

Perhaps the most bizarre shapes are found among the bacterial viruses; some of them appear to be spermlike in form, with prism-shaped heads and knobby tails. There are also curious forms among the insect viruses, particularly those responsible for the so-called polyhedral and capsule diseases of insect larvae. For example, a dying silkworm suffering from silkworm jaundice becomes swollen with a milky fluid containing polyhedral, or many-sided, bodies. These are masses of protein in which stubby rods of nucleoprotein are embedded. The rods are the individual virus particles.

Ever since the earliest filtration experiments, it has been recognized that all viruses are exceedingly small. It is now possible to be more specific. Known viruses range in size from a few millionths of an inch to about a hundred thousandth of an inch. The influenza virus is of about average size; it would be possible to line up 500 particles of this virus on the point of a pin.

It is not possible, using the electron microscope, to see viruses attacking a host and spawning new virus particles. The reason is that the electron microscope operates with dried specimens and in a high vacuum; these conditions immobilize viruses and kill cells. However, using other means of investigation, researchers have pieced together odds and ends of evidence and have succeeded in clearing up somewhat the relationship between virus and host.

Viruses are generally transmitted from one victim to another by direct or indirect contact. Among men and animals, the cough, the sneeze and physical contact are familiar means of transmission. A few instances have been reported of the spread of viruses by means of contaminated food. The rabies virus is usually transmitted by the bite of a rabid animal. Human hands carry all sorts of germs, among which are viruses picked up from nasal and oral discharges and from feces. It is now well known that viruses are passed in the stools of persons suffering from poliomyelitis.

Contact or mechanical transmission also occurs with a certain number of plant viruses. In experimental work, such viruses as tobacco mosaic and tomato bushy stunt are introduced into normal plants simply by rubbing a leaf with a gauze pad, wet with sap from an infected plant. In the tobacco fields, mosaic virus is frequently spread when cultivating machinery brushes against an infected plant and later comes in direct contact with other plants. Mosaic virus is often unwittingly introduced into tomato plants and others in the home garden by gardeners addicted to smoking. As the gardener puffs at a cigarette, the virus comes out of moistened bits of tobacco onto the smoker's fingers and from there goes into the plants he touches.

Bacterial viruses are also spread by contact. Infected bacterial cells burst when infection has progressed to a certain stage; the released virus particles are then free to infect other susceptible bacteria with which they happen to collide.

Insect vectors, or carriers, transmit many kinds of viruses. Flies, ticks and mosquitoes help spread both human and animal virus diseases. Mosquitoes are the most deadly of these vectors; they are responsible for the transmission of yellow fever, dengue, St. Louis encephalitis, equine encephalomyelitis, Japanese B encephalitis and Rift Valley fever. Most plant virus diseases are spread by insects. The most prominent vectors are aphids and leaf hoppers; others include thrips, earwigs, beetles, white flies, mealy bugs and grasshoppers.

The mechanism of infection by viruses and the precise details of multiplication are not yet entirely clear. In certain cases, insects introduce viruses into host cells as they feed. In other instances, particularly in man and animals, viruses seem able to penerate many of the cells with which they come into contact. With plant viruses, infection by mechanical means is generally effective only when the plant cells are slightly injured, so that the virus can readily enter them.

The infection of *Escherichia coli* bacteria by certain bacterial viruses, or phages, as they are called, represents a fascinating sequence of events. The virus particles in question are sperm-shaped, with heads and tails; a single one of them is capable of infecting a cell. First the phage attaches itself to the bacterium by its tail; in technical language, it is adsorbed by its tail to the surface of the bacterium. The nucleic acid contained in the phage's head then makes its way into the bacterium, presumably by being injected through the tail. The invading material probably comes in contact with the nucleus of the bacterial cell. The end product is a crop of new virus particles. The multiplication process begins immediately after the virus nucleic acid enters the cell. In about a half-hour, the cell dissolves and hundreds of new virus particles emerge.

It is not clear what happens in the period between the adsorption of the infecting phage particle and the emergence of the new viruses. If the infected cells are artificially broken open before they dissolve, particularly in the earliest stage of infection (the "dark period"), nothing recognizable as virus is found. In the later stages of infection, incomplete viruses can be observed. When viewed in the electron microscope, they are about as large as the heads of the mature, infectious phages; they are composed of protein. Apparently the virus particles become mature after a tail has been formed and the head has been filled with nucleic acid. The viruses that spill out when the bacterial cell bursts open are mature; they are able to attack and infect other bacteria.

We do not know if other kinds of viruses multiply in just this way. Certain features of the infecting process seem to be common to several virus groups. For example, the "dark period" immediately following infection, when there are not yet any recognizable virus particles, has also been observed in influenza virus infections, as studied in the chick embryo. More positive evidence is needed, however, before the mechanism of reproduction of any virus will be fully understood.

### Progress is being made in the control of virus diseases

The control of virus diseases is inadequate at present, though progress has been made, particularly in supplying protection against certain human and animal viruses. In general, these disease agents are not affected by antibiotic drugs, such as penicillin, aureomycin and streptomycin. As a matter of fact, a convenient way of transferring certain human viruses to an experimental host in the presence of bacteria is to treat the infectious fluids with antibiotics. For example, throat washings, taken from an influenza patient and treated with penicillin and streptomycin, can be inoculated directly into chick embryos in order to make a culture of the virus, for diagnostic or other purposes. The antibiotics do not affect the influenza virus but prevent the bacteria always found in throat washings from growing. Without such treatment the bacteria would grow wild on the chick embryo and in its fluids and the virus would be lost.

Chemical substances, either synthetic or derived from microorganisms, prevent the growth of viruses in some cases. However, certain authorities doubt the effectiveness of antivirus drugs. They maintain that many viruses have already done extensive damage by the time a diagnosis of the infection has been made; drugs, they say, can do little in such cases.

In a constant effort to eliminate virus diseases in man and animals, vaccines are employed to aid the body defenses. There are two major types of virus vaccine; one contains "live" and the other "killed" virus. In the fight against polio, both types have been used.

The successful vaccines used to prevent smallpox and yellow fever contain mild "live" strains of virus. When these are scratched into the skin or injected under it, minor infections set in; the body responds to these infections by producing antibodies. (See Index, under Antibodies.) The antibodies, which sometimes persist for years, are capable of combining with and neutralizing the more severe virus strains to which the vaccinated person may later be exposed.

The vaccines used in America against poliomyelitis, influenza and some other diseases contain "killed" virus. This has been sterilized by treatment with a chemical such as formaldehyde, but it can still stimulate the production of antibodies. Usually, more virus must be used in the "killed" vaccines than in "live" vaccines, in which the virus can multiply after it has been injected, providing a fresh source of antibody-stimulating material. In "killed" virus vaccines, the amount of virus injected is the entire source of stimulus.

Protection is never absolute with either type of vaccine, because individuals vary greatly in their response to vaccination. The possible mutation of viruses is a disturbing factor. If the strains of virus to which people are subjected are too different from those in the vaccine, the vaccine becomes worthless.

Plants do not give an antibody response; consequently vaccination cannot be used to combat plant virus diseases. However, it is often possible to breed varieties of plants that are more resistant than others to a particular virus. The sugarcane crop was threatened by mosaic disease at one time in certain areas of the world; it was saved by the development of mosaic-resistant varieties. Since several plant and animal viruses are spread by insects, a considerable measure of control is achieved in such cases by destroying the insects, usually by timely spraying.

Two groups of infectious microorganisms — the rickettsiae and the psittacosislike organisms — deserve mention in passing because they resemble viruses in some respects. Typhus, Rocky Mountain spotted fever and Q fever are familiar rickettsial diseases of man. The diseases caused by psittacosislike organisms include psittacosis (parrot fever), a cat pneumonia and infectious abortion of sheep.

Like viruses, rickettsiae and psittacosislike organisms reproduce only in living cells. They are smaller than most bacteria and have fewer independent activities; they depend, like viruses, on the cells in which they are parasites. In most respects, however, they resemble bacteria rather than viruses. They are held back by bacterial filters, are large enough to be seen with the light microscope, stain with common dyes, produce toxins, have some independent metabolic activities, show growth forms similar to those of bacteria and respond to some of the antibiotic drugs. It has been suggested that the rickettsiae and psittacosislike organisms represent microbial forms intermediate between viruses and bacteria.

The study of viruses is of the utmost importance because of the alarming toll taken by these disease agents. It is important, too, for quite a different reason. In the scheme of nature the viruses are the simplest substances possessing the ability to reproduce and to mutate. Because of their very simplicity they offer a unique opportunity to study these vital processes.

*See also Vol. 10, p. 278:*
"Medicine, Progress of."

Amer. Mus. of Nat. Hist.